MATHEMATICAL METHODS *for* MOLECULAR SCIENCE

Theory and Applications, Visualizations and Narrative

JOHN E. STRAUB

Boston University

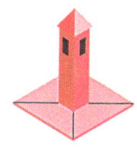

University Science Books · New York

University Science Books
An Imprint of AIP Publishing
uscibooks.aip.org

Printer & Binder: Integrated Books International
Copyright © 2022 by University Science Books
First Printing, First Edition

Print ISBN 978-1-940380-13-1
eBook ISBN 978-1-940380-12-4

Front Cover: The physical form of a vertically polarized electromagnetic wave may be described by a vector-valued function. The oscillating electric field (red) forms a right angle with the oscillating magnetic field (blue), and both are perpendicular to the direction of propagation.

Library of Congress Cataloging-in-Publication Data

Straub, John Edward, author.
Mathematical methods for molecular science / John E. Straub.
Description: First edition. | [New York] : University Science Books, [2022] Includes bibliographical references.
Identifiers: LCCN 2022016219 (print) | LCCN 2022016220 (ebook) |
           ISBN 9781940380131 (paperback) | ISBN 9781940380124 (ebook)
Classification: LCC QH506 .S8265 2022 (print) | LCC QH506 (ebook) | DDC 572.8–dc23/eng/20220701

Printed in the United States of America

10 9 8 7 6 5 4 3 2 1

To my parents, Jack and Maurine, and my family, Mary, Philip, and Joshua.

CONTENTS

Introduction

It is widely acknowledged that the traditional calculus sequence required of most molecular science majors, consisting of a year of differential and integral calculus and possibly a semester of multivariate calculus, does not provide the mathematical background needed for success in the quantum mechanics and statistical thermodynamics courses that follow.[1]

Mastery of key ideas in quantum mechanics requires knowledge of operators, differential equations, multidimensional integration, vector algebra, and functions of complex numbers. Statistical thermodynamics and kinetics require facility with partial differentiation, extremizing functions subject to constraints, and knowledge of differential equations. However, a treatment of these topics through specialized courses offered by mathematics departments is prohibitively time-consuming for many molecular science majors in the fields of chemistry, biochemistry, biophysics, and materials science.

Mathematical Methods for Molecular Science is designed to support a one-semester course that builds on the introductory calculus sequence and covers critical topics in multivariate calculus, probability and statistics, ordinary and partial differential equations, and linear algebra. In addition, advanced topics including partial differential equations, Fourier analysis, and group theory are introduced.[2]

Philosophy

Mathematicians are justly fond of *abstract* mathematical problems involving rather *simple* functions.[3] In the physical sciences our challenges are typically quite different. We face *concrete* mathematical problems involving *complicated* functions with multiple variables and many physical constants.[4] Solutions are found using standard algebraic operations that must be carefully applied. And without careful calculation, we are doomed! These observations inform the approach taken by this text:

> The narrative and exercises are focused solely on the mathematical functions and methods most relevant to molecular science.

> Examples demonstrate how complicated functions can be made manageable through the proper substitution of variables, simplifying expressions and facilitating problem solving.

> The only way to develop mastery of mathematical methods is through devoted practice. Many exercises are provided so that we can practice, practice, and practice some more.

The text offers a concise narrative, examples worked in full detail, ample visualizations, and plenty of homework problems for practice-to-mastery of course concepts.[5]

How to use this book

This text is written to serve a third-semester course in calculus. While it assumes the student has had a thorough introduction to differential and integral calculus of functions of one variable, it also provides a summary of

[1] It is common to have specialized courses in introductory calculus for the Social Sciences or Business. It is also common for Physics departments to offer a course in *Mathematical Methods for Physicists* taught by physicists. However, until recently it was unusual to see a course in *Mathematical Methods for Molecular Science* taught by molecular scientists. The tide is turning.

[2] Student A: "Great for chemists - they get a taste of physical chemistry before the course. Prepares them better than if they'd just taken the regular math courses."

[3] Prove that there are no natural numbers x, y, and z such that $x^n + y^n = z^n$ when n is a natural number greater than 2.

[4] Consider the function

$$C_V(T) = 3R \left(\frac{h\nu_E}{k_B T} \right)^2 \frac{e^{-\frac{h\nu_E}{k_B T}}}{\left(1 - e^{-\frac{h\nu_E}{k_B T}} \right)^2}$$

Show that for $T \gg \frac{h\nu_E}{k_B}$, $C_V \approx 3R$.

[5] Student B: "This class was definitely more difficult than MA225 [*Multivariate Calculus*] would have been, but I will be better prepared for physical chemistry."

key topics with examples and exercises to test and consolidate the student's knowledge.[6]

CHAPTER NARRATIVES include *qualitative* conceptual discussions and *quantitative* exercises that are solved in full detail. The chapters are concise and should be read before lecture. Completely.[7] The student should review each example and then repeat the exercise to verify that the example is well understood before moving on to the end-of-chapter exercises.

VISUALIZATIONS are used throughout the text. More than 400 original figures explore mathematical concepts that can be challenging to learn through equations alone. A variety of graphics are offered, from line drawings

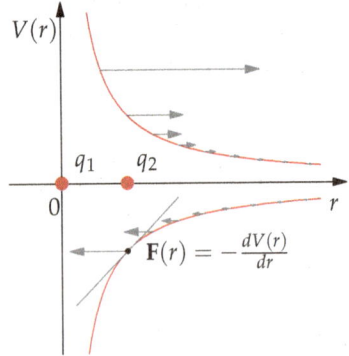

to surfaces and volumes, vectors and vector fields, and even atoms and molecules (click this interactive 3D figure using Adobe Acrobat Reader).

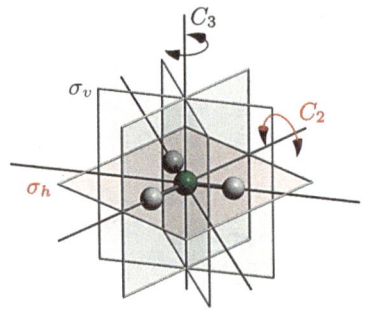

The visualizations offer examples of how quantitative information can be effectively displayed and analyzed.[8]

CHAPTER COMPLEMENTS are used to introduce advanced topics. The main body of each chapter presents the foundation of information on a particular topic. At the end of each chapter, a set of *complements* exploring advanced topics is included.[9] Each complement is self-contained and understandable given the information provided in the main body of the chapter. Complements may be assigned at the discretion of the instructor.

QUALITATIVE THEORY is emphasized in problem solving. The narrative often challenges the reader to anticipate the general form of a result before

[6] Student C: "At first I wasn't the biggest fan of the text, but it is extremely helpful if you work out the examples (as suggested)!"

[7] Student D: "I will say the lecture is only helpful to me if I thoroughly read the chapter beforehand."

[8] Student E: "Explains concepts very well and draws out diagrams which helped me as a visual learner."

[9] Student F: "Thank you for creating a text that I can trust to be succinct and teach me lessons beyond the scope of the course."

beginning work on the exact solution. When a problem is associated with an application in the molecular sciences, physical intuition and principles can also serve to define the general form and behavior of the solution. This approach is particularly valuable in the solution of ordinary and partial differential equations. Extensive visualizations support the qualitative exploration of mathematical problems and their solutions.

END-OF-CHAPTER EXERCISES are the heart of the text. They are divided into three classes.

WARM-UPS explore the key ideas presented in each chapter at a basic level. All warm-ups should be completed by all students.

HOMEWORK EXERCISES can be assigned in part or in whole depending on the topics emphasized by the instructor.

BONUS PROBLEMS explore advanced concepts introduced in the complements. They are marked by an asterisk (*).

Sitting in front of a problem you cannot solve is psychologically challenging. Even painful. However, finding a solution to the problem brings both relief and reward. Learn to love the process and it will serve you well.[10]

NARRATIVE ENRICHMENTS are included through two principal means. The main narrative is to be read and understood by all students.[11] When there is brief additional detail that can enrich a discussion, it is included as a *margin note* that may be read or not depending on the interest of the student. In addition, *historical notes* are included to highlight the chronological development of the topic as well as the diverse backgrounds and interests of the individuals who made seminal contributions to the subject.[12]

SUPPLEMENTS at the end of the text include a compilation of mathematical symbols and constants, collections of geometric and trigonometric formulas, and tables of power series, definite integrals, indefinite integrals, error functions, and Fourier transform pairs. Students will benefit from printing this material and using it as a valuable study resource.

Coauthors and contributors

I have benefitted from the advice provided by a number of individuals who are effectively coauthors of this text. Joshua Straub contributed to the content of every topic. His advice helped establish the tone of the narrative and the emphasis on providing simple and complete explanations and fully worked examples. Wei-Lun "Amy" Tsai [蔡瑋倫] reviewed the text and provided valuable advice from the perspective of a former student of the course. Mi Kyung Lee [이미경] assisted with typesetting and typography. Estelle Qi provided detailed advice on the chapter concerning Fourier analysis.

It is with deep appreciation that I acknowledge the many fine students at Boston University with whom I have worked and from whom I have learned so much. I am indebted to Jane Ellis, of University Science Books, for her encouragement, constructive criticism, and many good ideas that were essential to the production of this text. I thank Katherine Hastie, for extensive

[10] Student G: "I have really struggled through this course, but I have learned a ton."

[11] Student H: "The text is also VERY helpful and provides excellent supplemental material for homework and general curiosity."

[12] Student I: "This class has helped me actually enjoy math again!"

copy editing that improved the clarity and completeness of the narrative, and Felicity Henson, of AIP Publishing, for her assistance and technical support.

I am grateful to Geoffrey Hutchison and Dave Ewing of the University of Pittsburgh, Tamar Schlick and Alexej Jerschow of New York University, Ksenia Bravaya of Boston University, and Amy Mullin and Pratyush Tiwary of the University of Maryland, who kindly adopted a developmental version of this text. I have also benefitted from advice received from Marcia Levitus of Arizona State University, Arthur G. Palmer III of Columbia University, Ken Jordan of University of Pittsburgh, Arun Yethiraj of University of Wisconsin, Eitan Geva and Roseanne Sension of University of Michigan, and F. Lee Eiseman of Charlestown, Massachusetts.

Warm regards and gratitude to Nicole Zizzi and Lisa Giancola who lead Evolve Dynamicz. I had the pleasure of collaborating with and learning from them in the development of their Mathematical Methods dance project, which explores fundamental concepts in mathematics through movement. Finally, I thank the National Science Foundation for support provided in the development of this text.

Sources and suggestions

The selection of topics was informed by my experience with *Applied Mathematics for Physical Chemistry* by James Barrante[13] as well as the advice and feedback of students and teaching fellows of CH225 *Mathematical Methods for Chemists* at Boston University. Certain end-of-chapter exercises were adapted from problems appearing in Barrante's text, as well as exercises in *Mathematics for Physical Chemistry*[14] and *Quantum Chemistry*[15] by Donald McQuarrie, and *Molecular Driving Forces* by Dill and Bromberg.[16]

There are advanced texts that I recommend for further reading. The excellent text *Mathematics for Physical Chemistry* by Donald McQuarrie offers a more advanced presentation of topics covered in this text. In addition, many fine exercises related to applications to problems in physical chemistry are provided. McQuarrie's writing sets the standard for the clear and complete presentation of complex information in the physical sciences.

The fine text *Mathematical Methods in the Physical Sciences* by Mary L. Boas offers a rigorous and complete treatment of many topics covered in this text.[17] In addition, the text discusses advanced topics in vector calculus, calculus of variations, contour integration, tensor analysis, and special functions. The encyclopedic text *Mathematical Methods for Physicists: A Comprehensive Guide* by George B. Arfken, Hans J. Weber, and Frank E. Harris is a great book to have on your shelf when your internet access to Wolfram MathWorld is compromised.[18]

I am grateful to the teachers and students who use this text. Suggestions for improvements are welcome. In addition, a reward of $3.14 will be paid to the first person to note any error, whether mathematical, historical, or typographical.

[13] James R. Barrante. *Applied Mathematics for Physical Chemistry [1974]*. Pearson Prentice Hall, third edition, 2004. ISBN 978-0131008458

[14] Donald A. McQuarrie. *Mathematics for Physical Chemistry [2008]*. University Science Books, first edition, 2008. ISBN 978-1891389153

[15] Donald A. McQuarrie. *Quantum Chemistry [2007]*. University Science Books, second edition, 2007. ISBN 978-1891389504

[16] Ken A. Dill and Sarina Bromberg. *Molecular Driving Forces [2010]*. Garland Science, second edition, 2010. ISBN 978-0815344308

[17] Mary L. Boas. *Mathematical Methods in the Physical Sciences [2005]*. Wiley, third edition, 2005. ISBN 978-0471198260

[18] George B. Arfken, Hans J. Weber, and Frank E. Harris. *Mathematical Methods for Physicists: A Comprehensive Guide [2012]*. Academic Press, seventh edition, 2012. ISBN 978-0123846549

June 2022
Cambridge, Massachusetts

J.E.S.
straub@bu.edu

1 FUNCTIONS AND COORDINATE SYSTEMS

1.1 Survey of common functions of continuous variables

THERE IS A SMALL SET OF FUNCTIONS that are commonly used to describe physical systems and their properties. These include polynomial functions (such as a line or parabola), exponential and logarithmic functions, trigonometric functions (such as sine and cosine), and the gaussian (a special case of the exponential function). In this section, we explore the general properties of the most commonly used functions of a single variable.

1.1.1 Linear functions and slope

A great deal of information is summarized in even the simplest graph. Consider the example of a *linear equation*

$$y(x) = mx + b \tag{1.1}$$

where the function $y(x)$ depends on the single variable x and two constants m and b. We can plot the function $y(x)$ as a function of x using the *cartesian coordinate system*.[1] The value of the variable x is measured by the x-axis (the *abscissa*) and the value of the function $y(x)$ is measured by the y-axis (the *ordinate*) as shown in Figure 1.1.

[1] Named for the French philosopher and mathematician *René Descartes* (1596-1650) who developed analytical geometry as a bridge between algebra and geometry.

What can we learn from this graph? Each point on the xy-plane is defined by an *ordered pair* (x, y). At the point $(x, y) = (0, -2)$ the red line crosses the y-axis. We call that point the *y-intercept*. It defines the parameter b in Equation 1.1 since $y(0) = b = -2$. The point at which the red line crosses the x-axis is called the *x-intercept*. At that point $y(x) = 0 = mx + b$ so that $-b = mx$ and $x = -b/m = 1$. Since $b = -2$ we find the slope of the line $m = -(-2)/1 = 2$. Combining these results, the equation for the red line is

$$y(x) = 2x - 2$$

We can also determine the slope of the red line by calculating the *rise* of the function $\Delta y = y(x_2) - y(x_1)$ divided by the *run* of the function $\Delta x = x_2 - x_1$. For the red line in Figure 1.1 we can use the two points $(2, 2)$ and $(3, 4)$ to find the slope as

$$\text{slope} = \frac{\Delta y}{\Delta x} = \frac{y(x_2) - y(x_1)}{x_2 - x_1} = \frac{4 - 2}{3 - 2} = 2$$

This definition of the slope is called a *finite difference* equation. It defines the slope in terms of a finite difference in the function Δy divided by a finite difference in the variable Δx.

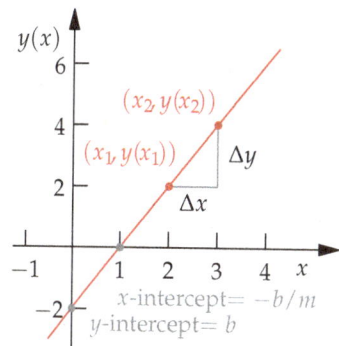

Figure 1.1: The red line defined by the linear equation $y(x) = 2x - 2$ has slope $m = 2$ and y-intercept $b = -2$. All figures in this text were drawn using the vector graphics language *Asymptote*.

1.1.2 Quadratic functions and the quadratic formula

Now consider a *quadratic equation*

$$y(x) = ax^2 + bx + c \tag{1.2}$$

where a, b and c are constants. Several examples are plotted in Figure 1.2. In each case, the curve takes the form of a *parabola*, which may point upward or downward and may intercept the x-axis twice (red curve), once (black curve), or not at all (blue curve).

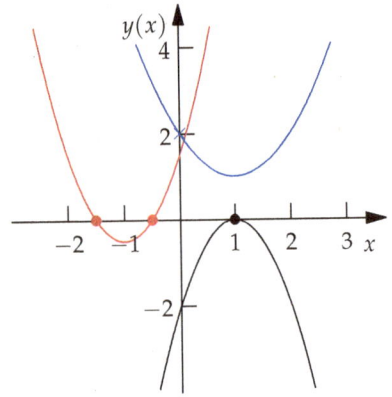

Figure 1.2: Quadratic functions having a single real zero at $x = 1$ (black curve and dot), two real zeros at $x_\pm = -1 \pm 1/2$ (red curve and dots), and no real zeros but two complex roots at $x_\pm = 1 \pm i$ (blue curve).

By setting the equation equal to zero we have *the* quadratic equation

$$y(x) = ax^2 + bx + c = 0 \tag{1.3}$$

The values of x that satisfy the equation are defined by the *quadratic formula*[2]

$$x_\pm = \frac{-b \pm \sqrt{b^2 - 4ac}}{2a} \tag{1.4}$$

At the points $x = x_+$ and $x = x_-$ the function $y(x_\pm) = 0$. As such, we call $x_\pm$ the *roots* of the quadratic equation.

The argument of the square root $b^2 - 4ac$ is known as the *discriminant*. Depending on the value of the discriminant the equation can yield two distinct real roots ($b^2 - 4ac > 0$, red curve), one real root ($b^2 - 4ac = 0$, black curve), or two complex roots ($b^2 - 4ac < 0$, blue curve).[3] We will return to the quadratic formula again and again. Take it to heart.

1.1.3 Application of the quadratic formula

Solutions to the quadratic equation are important in the study of chemical equilibrium. Consider the unimolecular decomposition reaction

$$\text{PCl}_5(g) \longrightarrow \text{PCl}_3(g) + \text{Cl}_2(g)$$

with the equilibrium constant

$$K_p = \frac{p_{\text{PCl}_3} p_{\text{Cl}_2}}{p_{\text{PCl}_5} p_{\text{ref}}}$$

[2] The first systematic solution of the quadratic equation was presented by the Persian mathematician, astronomer, and geographer *Muhammad ibn Mūsā al-Khwārizmī* (780-850). The general solution was first obtained by the Flemish mathematician, physicist, and engineer *Simon Stevin* (1548-1620).

[3] There are algebraic solutions for the roots of quadratic, cubic, and quartic equations. However, no general algebraic solutions for the roots of quintic or higher degree polynomial equations exist. A necessary and sufficient condition for the algebraic solution of a polynomial equation was first provided by *Évariste Galois* (1811-1832) who died as the result of a duel at the age of 20 years.

If there is an initial pressure of PCl_5 of 1.0 atm and no PCl_3 or Cl_2, we can assign $x = p_{Cl_2} = p_{PCl_3}$ and $p_{PCl_5} = 1.0 - x$ so that

$$K_P = \frac{(x)(x)}{1.0 - x}$$

This is a quadratic equation that can also be expressed in the form

$$x^2 + K_P x - K_P = 0$$

The roots of this equation provide the pressures of the gases at equilibrium:

$$x = p_{Cl_2} = p_{PCl_3} = \frac{1}{2} K_P \left[1 + \sqrt{1 + 4/K_P} \right]$$

and the partial pressure $p_{PCl_5} = 1.0 - x$.

Note that while the quadratic equation has two solutions, only one is physically meaningful. We ignore the second root of the quadratic equation, as the predicted pressures would be negative and therefore would not represent a physically meaningful solution.

1.1.4 Polynomial functions of varying degrees

We have examined linear and quadratic functions. They are special cases of a more general function called a *polynomial* and written

$$y(x) = a_n x^n + a_{n-1} x^{n-1} + \ldots + a_2 x^2 + a_1 x + a_0 \tag{1.5}$$

where the $a_n, a_{n-1}, \ldots, a_1, a_0$ are constant coefficients. The highest power n is called the *degree* of the polynomial. So a linear function is a 1st degree polynomial, a quadratic function is a 2nd degree polynomial and so on. A few examples are shown in Figure 1.3.

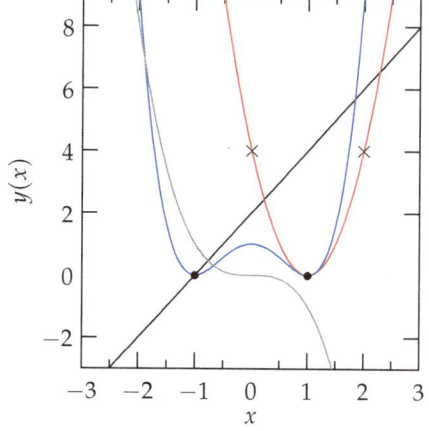

Figure 1.3: Polynomial functions of varying degrees, including linear or 1st degree (black), quadratic or 2nd degree (red), cubic or 3rd degree (gray), and quartic double well or 4th degree (blue) functions. Function zeros are marked with dots.

Let's derive an equation for the linear function plotted in Figure 1.3 in the form of Equation 1.1. The y-intercept is found at $y(0) = 2$. The slope can be determined from the points $(3, 8)$ and the x-intercept $(-1, 0)$ using the *finite difference* equation

$$\text{slope} = \frac{\Delta y}{\Delta x} = \frac{y(x + \Delta x) - y(x)}{\Delta x} = \frac{8 - 0}{3 - (-1)} = 2$$

From that information, we know that the equation describing the line is given by

$$y(x) = 2x + 2$$

The parabola in Figure 1.3 is described by the quadratic function having a minimum at $x = 1$. As such, we can express the quadratic function as

$$y(x) = a_2 \left(x^2 - 2x + 1\right) = a_2(x - 1)^2$$

where $x = 1$ is the minimum.[4] We can determine the constant coefficient a_2 by noting that the parabola passes though the points $(0, 4)$ and $(2, 4)$, where $y(0) = y(2) = a_2 = 4$. Our final equation for the parabola is

$$y(x) = 4(x - 1)^2$$

with its minimum at $x = 1$ where the function $y(1) = 0$.

[4] This compact form of the quadratic equation is formed by *completing the square*.

1.1.5 Odd and even symmetry of polynomial and sinusoidal functions

In addition to the degree of a polynomial function, it is important to consider the symmetry of the function. Functions for which $f(-x) = f(x)$ are said to have *even* symmetry while functions for which $f(-x) = -f(x)$ are said to have *odd* symmetry.[5]

Consider the cubic polynomial function

$$y(x) = -x^3$$

[5] The polynomial function $f(x) = x^n$ is an even function for which $f(-x) = f(x)$ when n is even power and an odd function for which $f(-x) = -f(x)$ when n is an odd power.

shown in Figure 1.3 (gray line). This cubic equation is an *odd function* of x since $y(-x) = -y(x)$. Another odd function of x is $\sin(-x) = -\sin(x)$. Also note that the cubic equation is *unbounded*.

The final polynomial shown in Figure 1.3 (in blue) is a quartic double well function that is very useful in modeling various physical processes. The form of the function is

$$y(x) = x^4 - 2x^2 + 1 = \left(x^2 - 1\right)^2 \tag{1.6}$$

The curvature of the function at its minimum located at $x = 1$ is 8. That fact explains the close agreement between the quartic double well function (blue line) and inverted parabolic function (red line) near the zero located at $x = 1$ (black dot). Note that the quartic double well is an *even function* of x since $y(-x) = y(x)$. Another even function of x is $\cos(-x) = \cos(x)$.

Note that every function $f(x)$ can be expressed as the sum of an even function $f_e(x)$ and an odd function $f_o(x)$ where

$$f_e(x) = \frac{f(x) + f(-x)}{2} \qquad f_o(x) = \frac{f(x) - f(-x)}{2} \tag{1.7}$$

so that $f(x) = f_e(x) + f_o(x)$. Figure 1.4 shows a function $f(x)$ (black) decomposed into functions $f_e(x)$ (red) and $f_o(x)$ (blue) possessing even and odd symmetry, respectively. This fundamental property will prove to be useful in the analysis of functions of importance in the physical sciences.

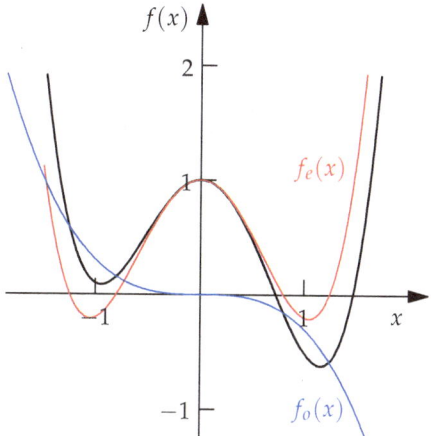

Figure 1.4: A function $f(x)$ (black line) decomposed into functions $f_e(x)$ (red line) and $f_o(x)$ (blue line) of even and odd symmetry, respectively, such that $f(x) = f_e(x) + f_o(x)$.

1.1.6 Tips for drawing functions by hand... and its value

If you want to become familiar with the form and properties of a function, draw it by hand. To plot a function $y(x)$ you must consciously consider the *domain* of the variable x and the associated *range* of the function $y(x)$. Is the function defined over a limited domain of values of x or over the whole real line? Is the range of the function *bounded* from above or below by a finite constant or it is *unbounded* and diverging to infinity? Is the function *continuous* or does it have discontinuities?

Even for a rough plot, it is helpful to know the exact value of the function for at least one point. You also need to know how the function behaves near that point, if it is increasing or decreasing, and if its curvature is positive or negative. It is also helpful to know how the function behaves asymptotically as it may converge to a constant, oscillate, or diverge to infinity.

With relatively little information about the function, you can develop a graphical representation that captures essential properties of the function's behavior that can be of real value in problem solving. Make it a habit to sketch your functions when you are solving a problem. Ask if your drawing captures the most essential properties qualitatively, such as the number of zeros, slope and curvature, in terms of the function's general behavior, and quantitatively, in terms of specific values of the function. We will find that as problems become increasingly complicated, being able to define the qualitative features of the solution can be an essential step in finding the quantitative solution to the problem.

1.1.7 Exponential functions describing growth and decay

A function widely used in describing physical processes is the *exponential function*

$$y(x) = e^x = \exp(x) \tag{1.8}$$

where the constant e is approximately $e = 2.718$.[6] In modeling physical processes, the exponential function can represent *exponential growth*. Since $e > 1$, the function $y(x) = e^x$ increases with increasing x. The exponential function can also be used to represent *exponential decay* as

$$y(x) = e^{-x} = \frac{1}{e^x} = \exp(-x)$$

Since e^x is an increasing function of x, e^{-x} decreases as x increases (see Figure 1.5).

Consider what we need to know to draw $y(x) = e^x$. When $x = 0$, the function $y(0) = e^0 = 1$. That gives us one point to draw at $(x, y) = (0, 1)$.[7] As x increases the function grows and as $x \to \infty$ the function approaches ∞. As x becomes more negative the function diminishes and as $x \to -\infty$ the function decreases to 0. That is enough information to draw a rough graph of $y(x) = e^x$. In addition, once we can draw $y(x) = e^x$ we can draw $y(x) = e^{-x}$ by symmetry (see Figure 1.5).

1.1.8 *Logarithm as the anti-exponential*

Another function commonly used to model physical systems is the *logarithmic function*. Let's explore the natural logarithm $y(x) = \ln(x)$, defined as the inverse of the exponential function

$$\ln(\exp(x)) = \exp(\ln x) = x \tag{1.9}$$

This property of the logarithm as the *anti-exponential* is shown in Figure 1.6. The natural logarithm $\ln(x)$ is the reflection of the exponential e^x across the diagonal defined by $y(x) = x$.

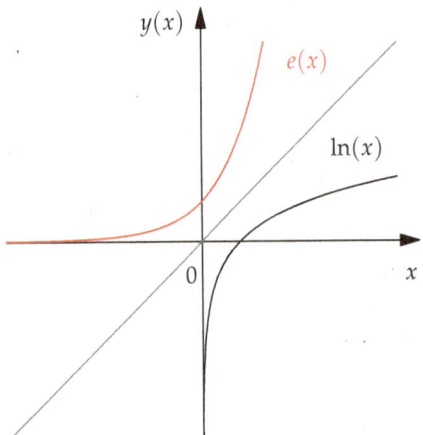

If a function $y(x) = \exp(-1/x)$ then the natural logarithm of the function is

$$\ln y(x) = \ln\left(e^{-1/x}\right) = -\frac{1}{x}$$

Using the logarithm, we can convert an exponential function into an algebraic function. We will explore many useful properties of logarithms in Chapter 2.

[6] The mathematical constant e was first estimated to be $2 < e < 3$ by Swiss mathematician *Jacob Bernoulli* (1655-1705). Today e is commonly known as *Euler's constant*.

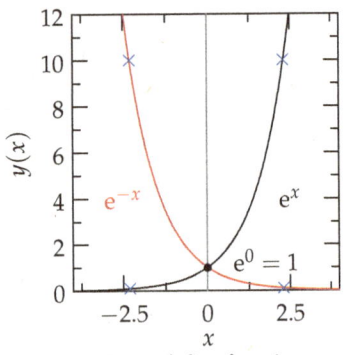

Figure 1.5: Lines define functions $y(x) = \exp(x)$ (black) and $y(x) = \exp(-x)$ (red) over a range of x. Note that e^x is the image of e^{-x} reflected across the vertical line $x = 0$ (gray line).

[7] If we want a slightly more accurate drawing, we might remember that $e^{2.3} \approx 10$ and $e^{-2.3} \approx 1/10$ (see blue crosses in Figure 1.5.

Figure 1.6: The natural logarithm (black) is the reflection of the exponential function (red) across the diagonal line $y(x) = x$ (gray).

1.1.9 Application of logarithms in visualizing exponential functions

Consider the exponential function used to model dependence of vapor pressure p above a liquid on the temperature T

$$p(T) = A \exp\left(-\frac{\Delta H}{RT}\right)$$

where the argument of the exponential function, $\Delta H/RT$, must be a dimensionless number. A plot of $p(T)$ as a function of T is shown as the black line in Figure 1.7. Note that $p(T)$ remains very small for small values of T and then rapidly increases.

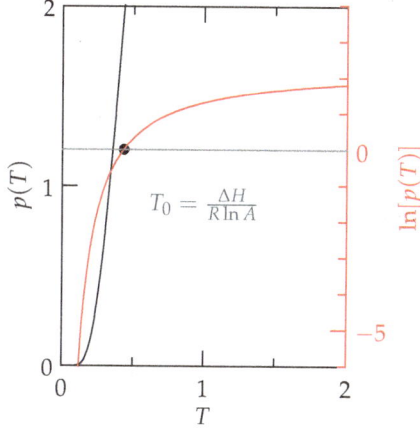

Figure 1.7: The dependence of $p(T)$ for small values of T (black line). As T increases, so does $p(T)$. The logarithm $\ln[p(T)]$ (red line) over the same range of T with x-intercept defined as T_0.

As $p(T)$ is an exponential function of $1/T$, the *logarithm* of $p(T)$ will be an *algebraic function* of T where

$$\ln[p(T)] = \ln A - \frac{\Delta H}{RT} \qquad (1.10)$$

Note that we have used the property of the natural logarithm that $\ln(xy) = \ln x + \ln y$. The result is also shown in Figure 1.7 (red line). Note that we can observe the behavior of $\ln[p(T)]$ (red line) over a much broader range of T.

The x-intercept of the function is located at the point $(T_0, 0)$ where

$$\ln[p(T_0)] = \ln A - \frac{\Delta H}{RT_0} = 0$$

Solving for T_0 we find

$$T_0 = \frac{\Delta H}{R \ln A}$$

as shown in Figure 1.8 (black dot).

Let's explore one more way to plot $p(T)$. Suppose we plot $\ln[p(T)]$ as a function of $1/T$ rather than T. We find that $\ln[p(T)]$ is a *linear function* of $1/T$. The result is shown in Figure 1.8.

While $p(T)$ increases with increasing T, it decreases with increasing values of $1/T$. The same is true for the logarithm $\ln[p(T)]$ that decreases linearly as $1/T$ increases. We say that the function $p(T)$ is a *log-linear* function of the

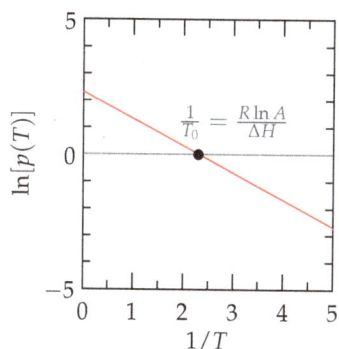

Figure 1.8: On a log-lin plot the function $\ln[p(T)]$ (red line) decreases linearly as a function of $1/T$. The x-intercept occurs at $(1/T_0, 0)$ where $1/T_0 = R \ln A/\Delta H$.

variable $1/T$, as it is a linear function on a *semi-log plot* (see Figure 1.8).[8] Let's now explore plotting the trigonometric sine and cosine functions.

[8] There are two types of semi-log plots. A *log-lin plot* uses a logarithmic scale on the ordinate and a linear scale on the abscissa. A *lin-log plot* uses a linear scale on the ordinate and logarithmic scale on the abscissa.

1.1.10 *Sinusoidal functions have maxima, minima, and zeros*

Consider the functions $\sin(2\pi x)$ and $\cos(\theta)$ that are defined over the whole real line, oscillate with a fixed period, and are bounded between 1 and -1. We begin by plotting $\sin(2\pi x)$ as x varies from 0 to 1. The sine function is periodic, repeating every 2π units of its argument. The argument of the function is $2\pi x$ so the period of oscillation is 1 as shown in Figure 1.9.

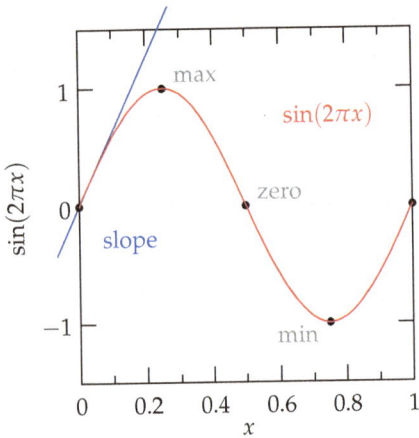

Figure 1.9: The function $\sin(2\pi x)$ as a function of x over $[0,1]$. The initial slope, maxima, minima, and zeros are noted.

The plot identifies a local maximum and local minimum, shows the function's initial slope, and marks and the function's zeros that appear at values of x satisfying

$$\sin(2\pi x) = 0$$

These zeros occur at $x = 0, \frac{1}{2}, 1$. These general properties can be easily read off of a plot of the function. Alternatively, those properties can be helpful when drawing the function by hand.

Let's use what we learned about the sine function to draw the cosine function over one period of oscillation. In the last example, we plotted $\sin(2\pi x)$ over one period, which was the range $x \in [0,1]$.[9] In this case we will plot $\cos(\theta)$, where one period of oscillation will occur over the range $\theta \in [0, 2\pi]$ as shown in Figure 1.10.

To get a rough idea of how the function $\cos(\theta)$ will vary as a function of θ, we mark five points that include the initial and final values that are local maxima at $(\theta, y(\theta)) = (0,1)$ and $(2\pi, 1)$, the local minimum $(\pi, -1)$, and two zeros $(\pi/2, 0)$ and $(3\pi/2, 0)$. We can make a rough sketch of the function by connecting the five points by straight lines (red line).

For some approximations, the red line would be good enough. It exactly captures the function's maxima, minima, zeros, and period of oscillation. For a more accurate representation, we can further improve this plot by adding information about the slope of the function. For example, at the minima and maxima the slope is zero. That means we should round off the sharp corners

[9] The symbol $\in$ should be read *in*.

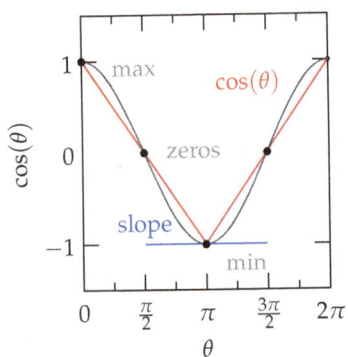

Figure 1.10: The function $\cos(\theta)$ as a function of θ over $[0, 2\pi]$. The initial sketch with straight lines (red) is smoothed to create a more continuous curve (gray).

where the line passes through the minima and maxima. The result is the gray line in Figure 1.10 representing a good approximation of the function $\cos(\theta)$.

We have reviewed exponential functions and sinusoidal functions. Let's explore the properties of products of these two fundamental functions.

1.1.11 Compound functions and the exponentially damped sinusoid

Many physical processes are described mathematically in terms of a product of common functions referred to as a *compound function*. One example is a cosine function multiplied by a decaying exponential:

$$f(t) = \exp(-t/\tau)\cos(2\pi\nu t) \qquad (1.11)$$

This compound function is called a *damped cosine*. Note that the argument of the sinusoidal function, $2\pi\nu t$, must be a dimensionless number. The behavior of this compound function $f(t)$ is depicted in Figure 1.11.

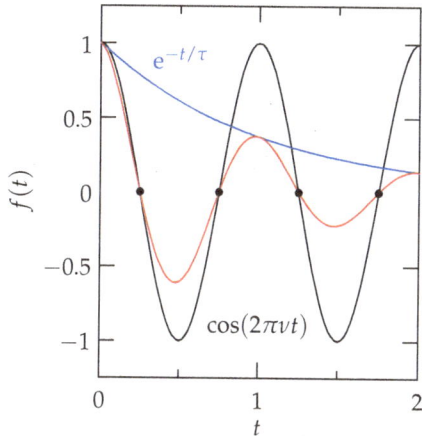

Figure 1.11: Compound function (red) formed by the product of an oscillating cosine (black), with a period $T = 1$, and decaying exponential (blue), with a decay time $\tau = 1$. The decaying exponential reduces the bounds of oscillation while the zeros of the function (black dots) are unchanged.

To interpret the behavior of this function, it is helpful to think of the variable t as time. The cosine function oscillates with a frequency ν and corresponding period

$$T = \frac{1}{\nu}$$

In a time t, the cosine undergoes $t/T = \nu t$ oscillations. The higher the frequency ν the shorter the period T. The oscillations of the cosine are bounded by a maximum of 1 and a minimum of -1. The exponential has a decay time τ. After a time τ the exponential is $1/e$ of its initial value. Since the compound function $f(t)$ is a product of the cosine and the exponential, it oscillates with a frequency ν and an amplitude that decays exponentially in time.

When $T/\tau \ll 1$, the exponential function controlling the amplitude of oscillation will change very little during a period of oscillation. We say the oscillation is *underdamped*. When $T/\tau \gg 1$, the amplitude of the function will diminish rapidly and the function oscillates little if at all. We say the oscillation is *overdamped*. Examples of underdamped and overdamped behavior are shown in Figure 1.12.

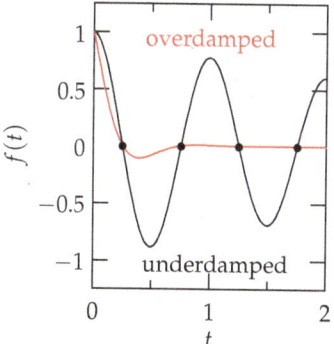

Figure 1.12: Damped cosine function displaying underdamped ($T/\tau = 1/5$, black) and overdamped ($T/\tau = 4$, red) behavior. Since T is constant while τ is varied, the zeros of the functions (black dots) are unchanged.

1.1.12 Gaussian functions and the bell curve

Another useful function in the physical sciences is the *gaussian function*,[10] an exponential function with a quadratic argument

$$p(x) = \exp\left(-\frac{1}{2\sigma^2}x^2\right) \tag{1.12}$$

[10] Named for the German mathematician *Carl Friedrich Gauss* (1777-1855).

The gaussian function has the shape of a *bell curve* and is used to form the *normal distribution* that appears throughout the physical and social sciences and can be used to describe everything from a distribution of course grades to the distribution of speeds of molecules in a gas.

Consider the gaussian function

$$p(x) = \exp\left(-\frac{1}{4}(x-3)^2\right) = \exp\left(-\frac{1}{2\sigma^2}(x-x_0)^2\right)$$

where the parameters $x_0 = 3$ and $\sigma^2 = 2$ (see Figure 1.13).

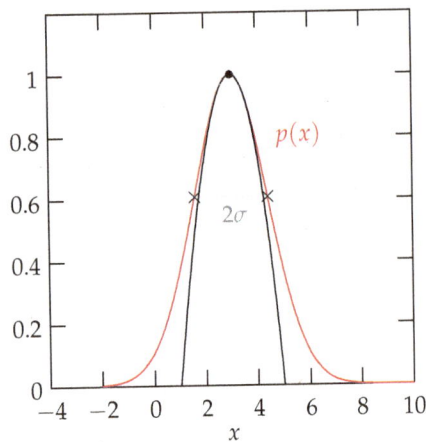

Figure 1.13: The gaussian function $p(x) = \exp(-(x-3)^2/2\sigma^2)$ where $\sigma = \sqrt{2}$ (red) is compared with the inverted parabola $y(x) = 1 - (x-3)^2/4$ (black).

This function has a single global maximum at $x = x_0 = 3$. The width is associated with the parameter $\sigma = \sqrt{2} = 1.414$. Asymptotically, the function approaches zero for large and small values of x:

$$\lim_{x \to \pm\infty} p(x) = 0$$

We will return to the gaussian function again and again in modeling a variety of physical processes.

We have explored a variety of functions including linear, quadratic, cubic and quartic polynomials, exponential, logarithmic and sinusoidal functions, and the gaussian function. We have seen how functions are defined by fundamental properties including slope, curvature, maxima, minima, placement of zeros, odd or even symmetry, continuity, periodicity, and bounds. Explore these functions. Vary the parameters to see how they change. Repeated visualization will deepen your understanding of a function's fundamental properties and serve as a valuable tool in problem solving.

1.2 Exploring coordinate systems and their utility

A VARIETY OF COORDINATE SYSTEMS may be used to represent and visualize the mathematical functions. A function that may be complicated in one set of coordinates appears simple in another. In this section, we explore the basic properties of commonly used coordinate systems including cartesian coordinates, plane polar coordinates, spherical polar coordinates, and cylindrical coordinates.

1.2.1 *Two-dimensional cartesian coordinates*

A point on the xy-plane can be represented as an ordered pair of cartesian coordinates (x, y), as shown in Figure 1.14. We use the two-dimensional cartesian coordinate system to represent a function $y(x)$ where the *abscissa* defines the variable x and the *ordinate* the function $y(x)$.

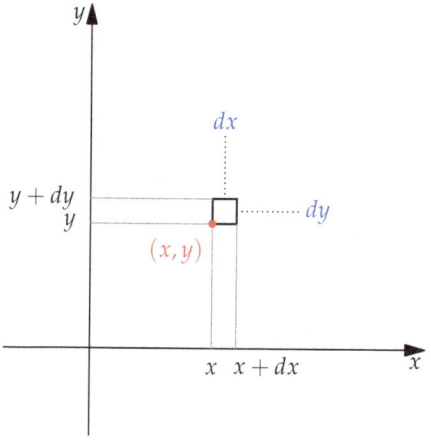

Figure 1.14: The planar cartesian co-ordinate system in two dimensions. The area element $dA = dxdy$ (bold) is the product of the two lengths dx and dy.

For functions of one variable, we represent an increment of length on the one-dimensional x-axis as dx. In the same way, we can represent a two-dimensional increment of area corresponding to a small increment dx along the x-axis and a small increment dy along the y-axis. That area increment is represented as the small square $dA = dxdy$ in Figure 1.14, reflecting the square symmetry of cartesian coordinates.

To get a better feeling for the square symmetry of the cartesian coordinate system, let's draw a round circle on the two-dimensional cartesian plane. The equation for a circle can be written

$$x^2 + y^2 = a^2 \tag{1.13}$$

where a is a constant defining the radius of the circle. To plot this function, we can solve for y as a function of x to find

$$y(x) = \pm\sqrt{a^2 - x^2} \ \text{ for } \ x \in [-a, a]$$

Note that we need to restrict the range of x, as it cannot be greater than the radius of our circle in the positive or negative direction. In addition, $y(x)$

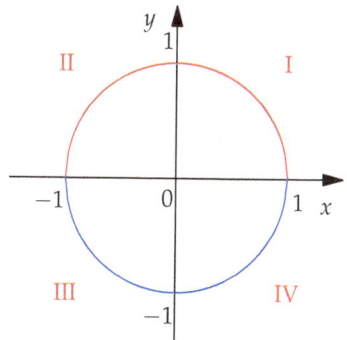

Figure 1.15: A unit circle drawn on the cartesian xy-plane. The four *quadrants* forming the xy-plane are labeled I, II, III, and IV.

is not a single-valued function of x. For every value of $x \in (-a, a)$, there are two corresponding values of y. For example, when $x = 0$ the equation can be satisfied by $y = \pm a$. For that reason, there is a $\pm$ in front of the radical.

The example of the *unit circle* where $a = 1$ is shown in Figure 1.15. In cartesian coordinates, the circle is defined by $y(x) = \sqrt{a^2 - x^2}$ in quadrants I and II and $y(x) = -\sqrt{a^2 - x^2}$ in quadrants III and IV. There is no single equation that describes the full unit circle in cartesian coordinates. The take home message is that a circle with round symmetry is not easy to describe using square cartesian coordinates.

1.2.2 *Two-dimensional plane polar coordinates*

Fortunately, there is a coordinate system in which *round functions* are easily expressed, known as *plane polar coordinates*. In plane polar coordinates a point in the two-dimensional plane is defined in terms of its radial distance from the origin, r, and the angle, θ, measured counter-clockwise from the x-axis to the line connecting the point to the origin (see Figure 1.16).

We can translate between the two representations using the equations

$$x = r \cos \theta$$
$$y = r \sin \theta \tag{1.14}$$

or conversely

$$r = \sqrt{x^2 + y^2}$$
$$\theta = \arctan\left(\frac{y}{x}\right) = \tan^{-1}\left(\frac{y}{x}\right) \tag{1.15}$$

The point $x = r \cos \theta$ is the projection of r on the x-axis and $y = r \sin \theta$ is the projection of r on the y-axis. A right triangle is formed by the triples $(0,0) - (r, \theta) - (x, 0)$ or $(0,0) - (r, \theta) - (0, y)$. Using *Pythagoras's theorem*,[11] we find

$$x^2 + y^2 = r^2 \cos^2 \theta + r^2 \sin^2 \theta = r^2 \tag{1.16}$$

By varying the radial distance $r \in [0, \infty)$ and the angle $\theta \in [0, 2\pi)$, we can reach any point on the infinite two-dimensional plane. Just as we could identify a unique address for any point on the two-dimensional plane in cartesian coordinates as (x, y), we can represent any point uniquely in plane polar coordinates as the ordered pair (r, θ).

Note that some care needs to be taken in determining the angle θ from the cartesian coordinates x and y, since $\tan^{-1}(y/x)$ is a mutlivalued function. Recall that $\theta \in [0, 2\pi)$. For the point $(x, y) = (1, 1)$, we find $\theta = \tan^{-1}(1) = \pi/4$, as we expect. However, for $(x, y) = (-1, 1)$, we expect $\theta = 3\pi/4$, but find that $\tan^{-1}(-1) = -\pi/4$. What gives?

It turns out it is not enough to determine the ratio y/x and its inverse tangent, we also need to consider the *quadrant* of the point (x, y) on the xy-plane (see Figure 1.15). Take the point $(-1, 1)$ in quadrant II for which $\tan^{-1}(y/x) = \tan^{-1}(-1) = -\pi/4$. We say $\theta = \pi + \tan^{-1}(-1) = 3\pi/4$. For the point $(-1, -1)$ in quadrant III, we say $\theta = \pi + \tan^{-1}(y/x) = 5\pi/4$. And for the point $(1, -1)$ in quadrant IV, we say $\theta = 2\pi + \tan^{-1}(y/x) = 7\pi/4$.

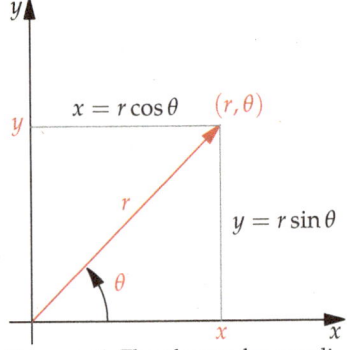

Figure 1.16: The plane polar coordinate system in two dimensions.

[11] This fundamental geometrical formula is named for the Greek philosopher *Pythagoras* (570-495 BCE).

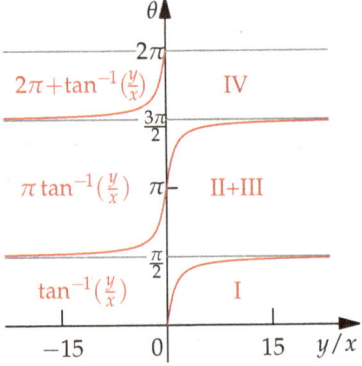

Figure 1.17: The value of θ as a function of the ratio y/x (red line). The identity relating the coordinates x and y to the variable $\theta \in [0, 2\pi)$ depends on the multivalued inverse tangent $\tan^{-1}(y/x)$.

The identities used to map the y/x to θ for the four quadrants forming the xy-plain are summarized in Figure 1.17.

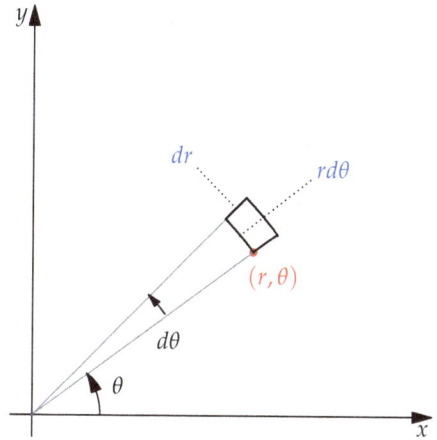

Figure 1.18: The plane polar coordinate system in two dimensions. The area element $dA = rdrd\theta$ (bold) is the product of the two lengths r and $rd\theta$.

What is the area element created by taking a small radial increment dr and small angular increment $d\theta$? The circumference of a circle of radius r is the length of an arc created by a change in angle of 2π equal to

$$C = 2\pi r$$

The length of an arc created by a small change in angle $d\theta$ is $rd\theta$. It follows that the area element is the length of the arc $rd\theta$ times the increment in the radius dr or

$$dA = r\,dr\,d\theta \tag{1.17}$$

That volume is shown as the small box in Figure 1.18.

You might say, "Wait a minute! You can't just multiply the arc length $rd\theta$ by dr. The arc is not a straight line!" It turns out that we can as long as our increments dr and $d\theta$ are *infinitesimal* increments in r and θ. An infinitesimal arc is essentially a straight line. As such, our estimate of the area element $dA = rdrd\theta$ will be accurate for infinitesimal dr and $d\theta$.

Now let's repeat the exercise of drawing a unit circle on the two-dimensional plane, but this time we will use plane polar coordinates. The equation for the circle of radius a is simply[12]

$$r = a \tag{1.18}$$

Since θ does not appear in this equation, the relation holds for all values of $\theta \in [0, 2\pi)$. The result is shown in Figure 1.19 (red circle). Shown for comparison is the unit circle with origin displaced along the x-axis (black circle). In cartesian coordinates this can be written

$$y(x) = \pm\sqrt{1 - (x-1)^2} \text{ for } x \in [0, 2]$$

but in plane polar coordinates it is simply written $r = 2\cos\theta$.[13]

Note that we can also have a function of two variables $f(r, \theta)$ in two-dimensional plane polar coordinates. In general, plotting this function

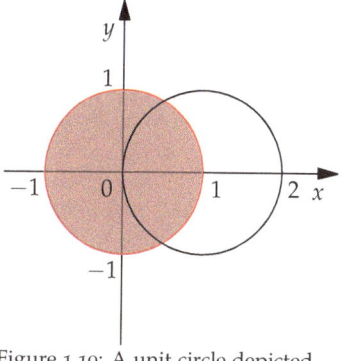

Figure 1.19: A unit circle depicted in plane polar coordinates as $r = 1$ (red) and $r = 2\cos\theta$ (black).

[12] Consider the definition used by the Greek mathematician *Euclid* (*circa* 300 BCE), "A circle is a plane figure bounded by one line, and such that all right lines drawn from a certain point within it to the bounding line, are equal. The bounding line is called its circumference and the point, its center." An equation is worth a thousand words.

[13] This can be shown by substituting $x = r\cos\theta$ and $y = r\sin\theta$ in $(1-x)^2 + y^2 = 1$. Expanding the expression and solving for r leads to $r = 2\cos\theta$ valid for $r \geq 0$.

requires two dimensions to define the point (r, θ), and a third dimension to represent the value of the function $f(r, \theta)$ at that point.

1.2.3 *Three-dimensional cartesian coordinates*

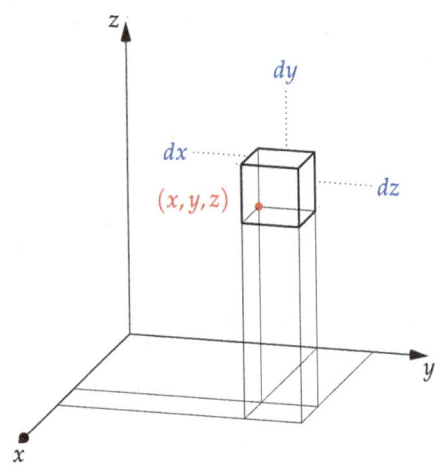

Figure 1.20: The cartesian coordinate system in three dimensions. The volume element $dV = dx\,dy\,dz$ (bold) is the product of the three lengths dx, dy, and dz.

Moving from two to three dimensions is a simple exercise in *cartesian coordinates*. Points in three-dimensional space are represented as an ordered triplet (x, y, z). The corresponding volume element is a cube of volume

$$dV = dx\,dy\,dz \tag{1.19}$$

These definitions are represented in Figure 1.20.

A function of one variable $y(x)$ is conveniently plotted in two-dimensional cartesian coordinates. At each point x the value of the function $y(x)$ is represented by the height along the y-axis. Similarly, a function of two variables $z(x, y)$ is conveniently plotted in three-dimensional cartesian coordinates. At each point (x, y) the value of the function $z(x, y)$ is represented by the height along the z-axis.

Let's use three-dimensional cartesian coordinates to represent a two-dimensional gaussian function

$$z(x, y) = e^{-\frac{1}{2}(x^2 + y^2)} = e^{-\frac{1}{2}x^2} e^{-\frac{1}{2}y^2}$$

where z is the magnitude of the function as shown in Figure 1.21. Just as a function $y(x)$ is represented as a one-dimensional line, a function $z(x, y)$ is represented as a two-dimensional surface.

Note that we can also have a function of three variables $f(x, y, z)$ in three-dimensional cartesian coordinates. In general, plotting this function requires three dimensions to define the point (x, y, z) and a fourth dimension to represent the value of the function $f(x, y, z)$ at that point. As such, in graphically representing functions of three variables, we often constrain one variable to a constant or represent the surface defined by $f(x, y, z) = $ constant.

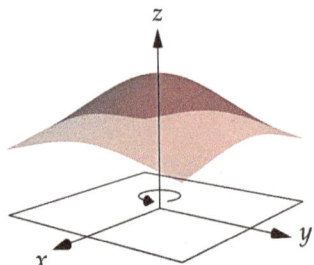

Figure 1.21: A gaussian function in the xy-plane represented in three-dimensional cartesian coordinates.

1.2.4 Three-dimensional spherical polar coordinates

Some functions in three dimensions are most easily represented using *spherical polar coordinates*, an extension of plane polar coordinates to three dimensions. Any point in the infinite three-dimensional space can be uniquely defined in terms of the ordered triplet (r, θ, φ) where $r \in [0, \infty]$ is the radial distance from the origin, $\theta \in [0, \pi]$ is the angle between a line connecting the point to the origin and the z-axis, and $\varphi \in [0, 2\pi]$ is the angle between the x-axis and a line connecting the origin to the projection of the point onto the xy-plane (see Figure 1.22).

We can translate between points in cartesian coordinates (x, y, z) and spherical polar coordinates (r, θ, φ) using

$$x = r \sin \theta \cos \varphi$$
$$y = r \sin \theta \sin \varphi$$
$$z = r \cos \theta \tag{1.20}$$

or conversely

$$r = \sqrt{x^2 + y^2 + z^2}$$
$$\theta = \arccos\left(\frac{z}{r}\right) = \cos^{-1}\left(\frac{z}{r}\right)$$
$$\varphi = \tan^{-1}\left(\frac{y}{x}\right) \tag{1.21}$$

Considering the right triangle formed by points $(0,0,0) - (r, \theta, \varphi) - (0,0,z)$, we see that $z = r \cos \theta$. The opposite side of that right triangle has length $r \sin \theta$. On the xy-plane we find two right triangles formed by the origin, the projection of the point (r, θ, φ) on the xy-plane, and the point $(x, 0, 0)$ or $(0, y, 0)$. From these two triangles, we recognize that $x = r \sin \theta \cos \varphi$ and $y = r \sin \theta \sin \varphi$ (see Figure 1.22).

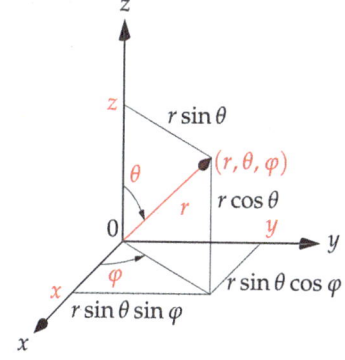

Figure 1.22: The spherical polar coordinate system in three dimensions.

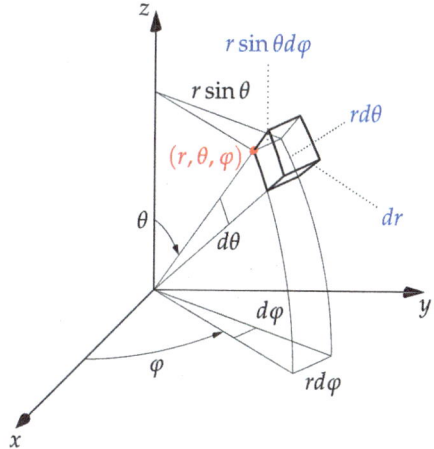

Figure 1.23: The spherical polar coordinate system in three dimensions. The volume element $dV = r^2 \sin \theta \, dr d\theta d\varphi$ (bold) is the product of the three lengths dr, $r d\theta$, and $r \sin \theta d\varphi$.

What is the volume element created by taking a small radial increment

dr and small angular increments $d\theta$ and $d\varphi$? We note that as we increase r, the length of an arc created by a change in θ increases linearly with growing r as $r d\theta$. Similarly, the length of an arc created by a change in φ is the arc's radius $r \sin \theta$ times the increment $d\varphi$. The volume element is then given by the length of the two angular arcs times the radial increment dr or

$$dV = r^2 \sin \theta \, dr \, d\theta \, d\varphi \qquad (1.22)$$

That volume is shown as the small box in Figure 1.23.

Note that there is more than one convention for defining the angles θ and φ. What we call θ, mathematicians call φ. And what we call φ, mathematicians reasonably call θ, as an extension of the definition of θ in two-dimensional plane polar coordinates (see Figure 1.16). For some reason, physical scientists use the opposite convention. We will use the physical scientist's convention for θ and φ, defined in Figure 1.22.

Just as a circle was simply represented in plane polar coordinates, a spherical surface is simply represented in spherical polar coordinates. A spherical shell of radius a defined in cartesian coordinates is

$$x^2 + y^2 + z^2 = a^2 \qquad (1.23)$$

In spherical polar coordinates, the relation is simply $r = a$, with the implication that the set of points includes $r = a$ for all values of $\theta \in [0, \pi]$ and $\varphi \in [0, 2\pi)$. Figure 1.24 provides an example of the two-dimensional spherical surface in three-dimensional space defined by

$$f(x, y, z) = a^2 = \text{constant}$$

This is an example of a three-dimensional function graphically represented as the surface defined by the function set to a constant.

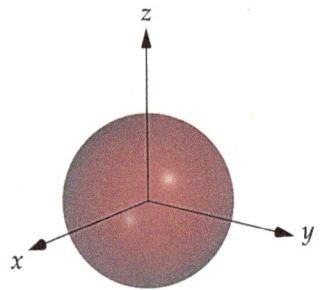

Figure 1.24: A spherical shell of fixed radius is simply represented in spherical polar coordinates.

1.2.5 *Three-dimensional cylindrical coordinates*

Another three-dimensional coordinate system that is useful in describing physical systems is *cylindrical coordinates*, in which the plane polar coordinates covering the xy-plane are extended linearly along the z-axis.

In cylindrical coordinates a point in space is uniquely defined in terms of the ordered triplet (r, θ, z) where $r \in [0, \infty)$, $\theta \in [0, 2\pi]$, and $z \in (-\infty, \infty)$ (see Figure 1.25). A point expressed in cartesian coordinates (x, y, z) can be transformed to cylindrical coordinates (r, θ, z) as

$$x = r \cos \theta$$
$$y = r \sin \theta$$
$$z = z \qquad (1.24)$$

Or conversely, expressing (r, θ, z) in terms of (x, y, z), we find

$$r = \sqrt{x^2 + y^2}$$
$$\theta = \arctan\left(\frac{y}{x}\right) = \tan^{-1}\left(\frac{y}{x}\right)$$
$$z = z \qquad (1.25)$$

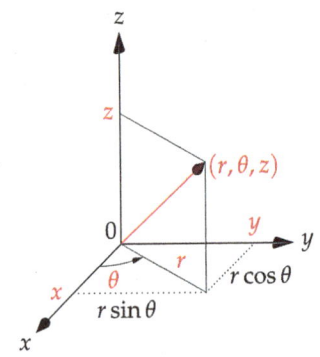

Figure 1.25: The cylindrical coordinate system in three dimensions.

As in plane polar coordinates, right triangles are formed by the triples $(0,0,0) - (r,\theta,0) - (x,0,0)$ and $(0,0,0) - (r,\theta,0) - (0,y,0)$. The point $x = r\sin\theta$ is the projection of r on the x-axis and $y = r\cos\theta$ is the projection of r on the y-axis (see Figure 1.25).

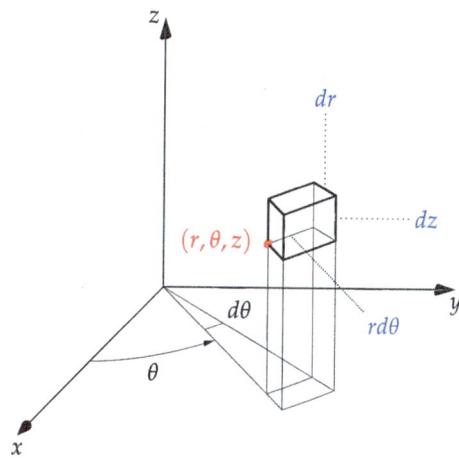

Figure 1.26: The cylindrical coordinate system in three dimensions. The volume element $dV = rdrd\theta dz$ (bold) is the product of the three lengths dr, $rd\theta$, and dz.

The volume element defined by a small radial increment dr, small angular increment $d\theta$, and small increment in height dz is simply the area increment $rdrd\theta$ in the xy-plane times the height dz, or

$$dV = r\,dr\,d\theta\,dz \qquad (1.26)$$

That volume element is shown as the small box in Figure 1.26.

Let's consider a helical curve represented in cylindrical coordinates as shown in Figure 1.27. For a helical curve of radius a and pitch 1.0 (so that one helical turn occurs every increment of 1.0 in z), we write

$$r = a \qquad \theta = 2\pi z \qquad (1.27)$$

Note that a parametric function can be used to define the *unit circle*. Defining a parameter $t \in [0,1)$, the equations

$$x(t) = \cos(2\pi t)$$
$$y(t) = \sin(2\pi t)$$

define a unit circle as shown in Figure 1.15. The value of the function is repeated periodically with a period of 1 as t increases.

Similarly, we can express the helical curve shown in Figure 1.27 as

$$x = x(z) = a\cos(2\pi z)$$
$$y = y(z) = a\sin(2\pi z) \qquad (1.28)$$

In these equations, z is treated as a parameter. As z is varied, the values of x and y will also change. We say that the cartesian coordinates x and y depend *parametrically* on the parameter z. As $z \in (-\infty, \infty)$ the equations describe an infinite helix traced on the surface of an infinitely tall cylinder of radius a.

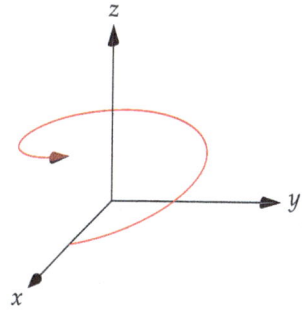

Figure 1.27: A helical curve is simply represented in cylindrical coordinates as $r = a$ and $\theta = 2\pi z$.

A₁ End-of-chapter problems

*If I were again beginning my studies,
I would follow the advice of Plato and
start with mathematics.*

Galileo Galilei

Warm-ups

1.1 Determine the values of the cartesian coordinates (x, y) for the following points expressed in terms of plane polar coordinates (r, θ).

(a) $r = 4.65, \theta = 225°$ (b) $r = 4.00, \theta = 130°$ (c) $r = 3.00, \theta = 0°$

1.2 Determine the values of the plane polar coordinates (r, θ) for the following points expressed as (x, y).

(a) $(0, 2)$ (b) $(-2, -2)$ (c) $(3, 0)$

1.3 Determine the values of the spherical polar coordinates (r, θ, φ) for the following points expressed in three-dimensional cartesian coordinates (x, y, z).

(a) $(0, 0, 2)$ (b) $(-2, 0, 1)$ (c) $(1, -2, -3)$

1.4 We constructed the cylindrical coordinate system by extending a z-axis from the origin of a plane polar coordinate system perpendicular to the xy-plane. A point in this system is expressed as (r, θ, z). Derive the differential volume element dV for the cylindrical coordinate system.

1.5 Identify the roots of the following equations.

(a) $2(y - 2) = -6x$ (b) $y = x^2 + x - 12$ (c) $y = \cos x$

1.6 Plot the following functions in plane polar coordinates for $\theta \in [0, 2\pi]$, noting that $r \geq 0$.

(a) $r = 2$ (b) $r = \theta/36°$ (c) $r = 2\cos\theta$

1.7 Plot the following functions using the two-dimensional cartesian coordinate system. Select a range of the variable such that your plot displays the essential features of the function.

(a) $y(x) = x^2 - x + 3$ (b) $p(V) = 8/V$ (c) $C(t) = 3e^{-t}$

1.8 Draw by hand the following functions over the specified range of x.

(a) $y(x) = -1 + 3x, \ [-2, 2]$ (b) $y(x) = 2x^2 - 4x + 2, \ [-1, 3]$ (c) $y(x) = x^3, \ [-2, 2]$
(d) $y(x) = 3e^{-\frac{x}{2}}, \ [0, 6]$ (e) $y(x) = \ln(2x + 1), \ [0, 10]$ (f) $y(x) = 2e^{-\frac{1}{x}}, \ [0, 10]$
(g) $y(x) = \sin(x), \ [-4\pi, 4\pi]$ (h) $y(x) = e^{-x}\cos(2\pi x), \ [0, 3]$ (i) $y(x) = e^{-(x-1)^2}, \ [-2, 4]$

Homework exercises

1.9 Plot the following functions using the two-dimensional cartesian coordinate system. Select a range of the variable such that your plot displays the essential features of the function.

(a) $V(r) = -2/r, r > 0$

(b) $E(v) = \frac{1}{2}v^2, v \geq 0$

(c) $F(r) = 1/r^2, r > 0$

(d) $k_r(T) = e^{-1/T}, T \geq 0$

(e) $A(t) = e^{-t}, t \geq 0$

(f) $A(t) = \dfrac{1}{1+t}, t \geq 0$

1.10 Plot the functions in the previous problem, choosing the coordinates so that the resulting plot is a straight line. For example, plotting the function $y = 1 - x^2$ as a function of x^2 for $x > 0$ results in a straight line with an x-intercept at $x = 1$.

1.11 Plot the function $f(x) = |x| - x$ over a range $-4 \leq x \leq 4$. Determine if $f(x)$ is a continuous function of x.

1.12 Prove that the hyperbolic sine function

$$\sinh(x) = \frac{e^x - e^{-x}}{2}$$

is an odd function of x by demonstrating that $\sinh(-x) = -\sinh(x)$. Further show that the hyperbolic cosine function

$$\cosh(x) = \frac{e^x + e^{-x}}{2}$$

is an even function of x by demonstrating that $\cosh(-x) = \cosh(x)$. Those functions are depicted below.

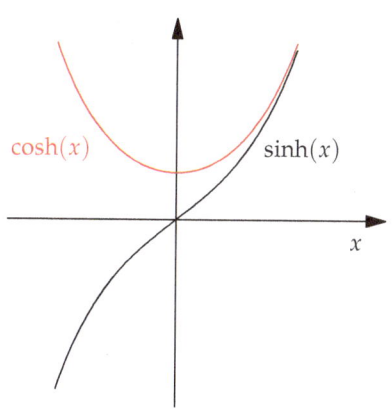

1.13 Characterize each function as being an odd function of x, an even function of x, or neither even nor odd.

(a) $\tanh(x)$

(b) $e^x \sin(x)$

(c) $\dfrac{e^x}{(e^x + 1)^2}$

(d) $\cos(x) + \sin(x)$

1.14* Consider the function

$$f(x) = \frac{1}{(1 - e^{-\frac{1}{x}})}$$

Evaluate $f(x)$ in the following one-sided limits (a) $\lim\limits_{x \to 0^+} f(x)$, in which x approaches 0 from above, and (b) $\lim\limits_{x \to 0^-} f(x)$, in which x approaches 0 from below.

1.15[*] The Heaviside step function is defined

$$\theta(x) = \begin{cases} 0 & x < 0 \\ 1 & x \geq 0 \end{cases}$$

and shown in the figure below.

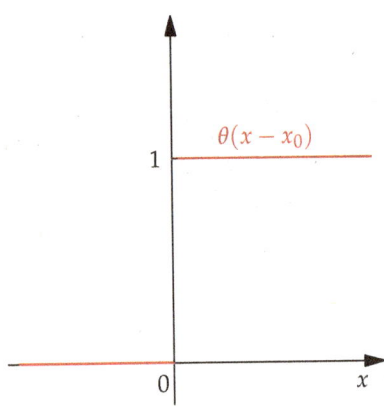

The function is characterized by the one-sided limits $\lim\limits_{x \to 0^+} \theta(x) = 1$, in which x approaches 0 from above, and $\lim\limits_{x \to 0^-} \theta(x) = 0$, in which x approaches 0 from below. Plot $f(x) = \theta(x) - \theta(x - 1)$ over $-1 < x < 2$.

2 COMPLEX NUMBERS AND LOGARITHMS

2.1 Complex numbers and the complex plane

ALTHOUGH COMPLEX NUMBERS HAVE IMAGINARY COMPONENTS, their occurrence and usefulness in modeling properties of physical systems are quite real. We encounter complex numbers in the classical mechanical theory of an oscillating mass on a spring and throughout the quantum mechanical theory of atoms and molecules. This section explores the properties of complex numbers, including the algebra of complex numbers and how complex numbers can be graphically represented on the complex plane.

2.1.1 Imaginary numbers, complex numbers, and the complex plane

Consider the quadratic equation

$$x^2 + 1 = 0$$

with two roots at $x_\pm = \pm\sqrt{-1}$. This equation confused mathematicians for hundreds of years, because no one could think of a number that could be squared to equal the negative number -1. Solving this problem required the discovery of *imaginary numbers*.[1]

Imaginary numbers are defined using the definition

$$i = \sqrt{-1} \tag{2.1}$$

It follows that $i^2 = -1$ and that $(-i)(i) = (i)(-i) = 1$. If we multiply two real numbers, the result is real. If we multiply two imaginary numbers, the result is also real, as $i^2 = -1$ is a real number. However, if we multiply a real number by an imaginary number, such as $3 \times 2i = 6i$, the result is imaginary.

Imaginary numbers can be combined with real numbers to create *complex numbers*, a number with a real part and an imaginary part.[2] Any complex number can be written as

$$z = x + iy \tag{2.2}$$

where the real part of the complex number z is $\text{Re}(z) = x$ and the imaginary part of the complex number z is $\text{Im}(z) = y$. In other words

$$z = \text{Re}(z) + i\,\text{Im}(z) \tag{2.3}$$

[1] We say *discovery* rather than *invention*. Just as no one invented gravity, mathematical truths *exist* in the space of possible ideas, waiting to be discovered.

[2] Complex numbers were discovered by Italian mathematician, physicist, chemist, biologist, physician, philosopher, and gambler *Gerolamo Cardano* (1501-1576) in his attempts to solve the cubic equation.

We need two dimensions to represent a complex number graphically. One dimension is used to represent the real component and another dimension to represent the imaginary component.

This can be done on the *complex plane* (see Figure 2.1). The ordinate y-axis is used to represent the imaginary part of the complex number. The abscissa x-axis is used to represent the real part of the complex number.

2.1.2 *The magnitude of a complex number and the complex conjugate*

What is the magnitude of a complex number? For a real number, the magnitude is the *absolute value* of the number, computed as $|x| = \sqrt{x^2}$. The magnitude $|x|$ is the distance of the point x from the origin $x = 0$ on the real axis. That definition works well if the number is positive, negative, or zero.

Let's apply that definition to a complex number $z = x + iy$. We find

$$\sqrt{z^2} = \sqrt{(x + iy)^2} = \sqrt{x^2 - y^2 + 2ixy}$$

but that is not the magnitude of z. The magnitude $|z|$ must be a real number, as it is the distance of the point $z = x + iy$ from the origin on the complex plane, which can be defined as $|z| = \sqrt{x^2 + y^2}$.

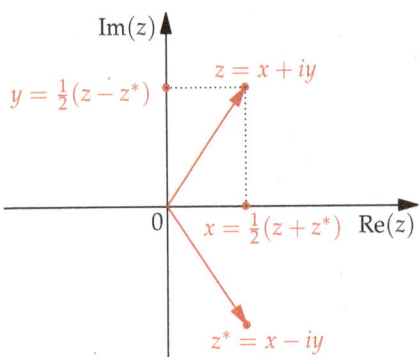

With this insight, we can define a general formula to compute the magnitude $|z|$. To do that, we introduce the *complex conjugate* of $z = x + iy$, written as $z^* = x - iy$. To create the complex conjugate of z we simply change the sign of the imaginary component. Why is this useful? Consider the product $(z^*)(z) = z^*z$, computed as

$$z^*z = (x - iy)(x + iy) = x^2 + y^2$$

We see that z^*z is a real number equal to the square of the distance of the point z from the origin on the complex plane (the length of the red arrow in Figure 2.1). We call this product the absolute value of z. It is written

$$|z| = \sqrt{z^*z} = \sqrt{x^2 + y^2} \tag{2.4}$$

Note that the complex conjugate has other interesting properties. For example, $z + z^* = 2x = 2\operatorname{Re}(z)$ and $z - z^* = 2iy = 2i\operatorname{Im}(z)$. These relationships are represented graphically in Figure 2.2.

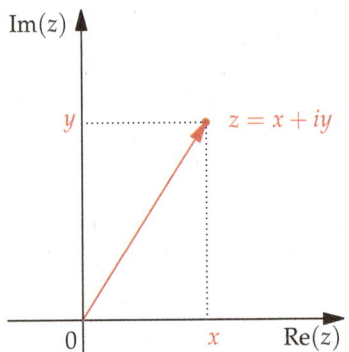

Figure 2.1: On the complex plane the x-axis is the real number line and the y-axis is the imaginary number line. A point on the two-dimensional complex plane $z = x + iy$ has real (x) and imaginary (y) components.

Figure 2.2: Visualizing the relationship between a complex number z and its complex conjugate z^*.

2.1.3 *An alternate view of the complex plane and Euler's formula*

We can use our definition of the magnitude of z and knowledge of coordinate systems explored in Chapter 1 to find an alternative way to represent complex numbers. Suppose we think of the complex plane in terms of plane polar coordinates. The radial distance of the point z from the origin is $r = |z| = \sqrt{x^2 + y^2}$, and the angle between the x-axis and the arrow connecting z to the origin is θ. Any complex number $z = x + iy$ can be represented as a point on the complex plane (r, θ), where $r \in [0, \infty)$ and $\theta \in [0, 2\pi)$ (see Figure 2.3).

Using our knowledge of the plane polar coordinate system and Equation 1.14, we can translate between the two representations using the equations

$$x = \text{Re}(z) = r \cos \theta$$
$$y = \text{Im}(z) = r \sin \theta \tag{2.5}$$

or conversely[3]

$$r = \sqrt{x^2 + y^2}$$
$$\theta = \tan^{-1}\left(\frac{y}{x}\right) \tag{2.6}$$

With these definitions, we can express any complex number as

$$z = x + iy = r \cos \theta + ir \sin \theta \tag{2.7}$$

Power series representations of functions are explored in Chapter 8 and summarized in Supplement S_4. For now, recall that the power series representations of $\cos \theta$ and $\sin \theta$:

$$\cos \theta = 1 - \frac{\theta^2}{2!} + \frac{\theta^4}{4!} - + \ldots$$

$$\sin \theta = \frac{\theta}{1!} - \frac{\theta^3}{3!} + \frac{\theta^5}{5!} - + \ldots$$

where *n factorial* is defined $n! = 1 \times 2 \times 3 \times \ldots \times (n-1) \times n$.[4] Using the properties of i, we can rewrite these equations in terms of a real part

$$\cos \theta = 1 + i^2 \frac{\theta^2}{2!} + i^4 \frac{\theta^4}{4!} + \ldots$$

and an imaginary part

$$i \sin \theta = i\theta + i^3 \frac{\theta^3}{3!} + i^5 \frac{\theta^5}{5!} + \ldots$$

Note that we have used the fact that $i^2 = -1$, $i^3 = -i$, $i^4 = 1$ and so on. If we sum the two series, we find

$$\cos \theta + i \sin \theta = 1 + i\theta + \frac{(i\theta)^2}{2!} + \frac{(i\theta)^3}{3!} + \frac{(i\theta)^4}{4!} + \frac{(i\theta)^5}{5!} + \ldots$$

which is the power series of the exponential function $\exp(i\theta)$.[5] The result is

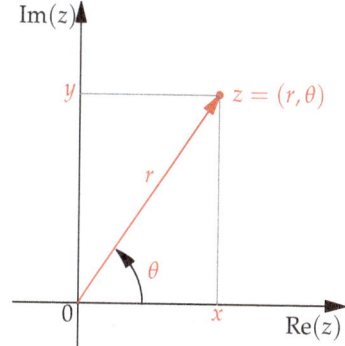

Figure 2.3: The complex number $z = x + iy$ represented as the ordered pair (r, θ) in the plane polar coordinate system.

[3] The exact relation used to map the ratio y/x to θ depends on the quadrant containing the point (x, y). See the discussion surrounding Figure 1.17.

[4] These power series representations of cosine and sine may be more readily recognized when written as $\cos x = 1 - x^2/2 + x^4/4! - + \ldots$ and $\sin x = x - x^3/3! + x^5/5! - + \ldots$.

[5] The power series of $\exp(x) = 1 + x + x^2/2! + x^3/3! + \ldots$.

the remarkable relation

$$e^{i\theta} = \cos\theta + i\sin\theta \qquad (2.8)$$

known as *Euler's formula.*[6]

What does this have to do with complex numbers? Starting from

$$z = x + iy$$

we can use Equation 2.7 to substitute $x = r\cos\theta$ and $y = r\sin\theta$. We find

$$z = r\cos\theta + ir\sin\theta = r(\cos\theta + i\sin\theta) = re^{i\theta}$$

This result has great utility in the algebra of complex numbers. As we will see, it is especially useful in determining products, quotients, and roots of complex numbers.[7]

2.1.4 Properties of Euler's formula and Euler's identity

The complex conjugate of z can be written using Euler's formula as

$$z^* = r\cos\theta - ir\sin\theta = re^{-i\theta}$$

Here we have used the fact that $\cos\theta = \cos(-\theta)$ is an even function of θ and $\sin\theta = -\sin(-\theta)$ is an odd function of θ. Combining the last two relations[8] we find the satisfying result:

$$|z| = \sqrt{z^*z} = \sqrt{(re^{-i\theta})(re^{i\theta})} = \sqrt{r^2} = r$$

The magnitude of z equals the radial distance r.

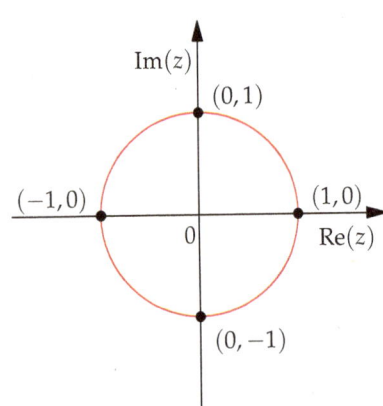

Figure 2.4: The unit circle defined by $z^*z = x^2 + y^2 = r^2 = 1$ shown on the complex plane. Four points on the circle are identified as intersections of the unit circle with the x-axis or y-axis.

Let's explore other properties of Euler's formula (see Figure 2.4). For the point $(x, y) = (1, 0)$, we find $r = 1$ and $\theta = 0$, so that

$$z = 1 = \cos(0) + i\sin(0) = e^0$$

For the point $(x, y) = (0, 1)$, we find $r = 1$ and $\theta = \pi/2$, so that

$$z = i = \cos\left(\frac{\pi}{2}\right) + i\sin\left(\frac{\pi}{2}\right) = e^{i\frac{\pi}{2}}$$

For the point $(x, y) = (-1, 0)$, we find $r = 1$ and $\theta = \pi$, so that

$$z = -1 = \cos(\pi) + i \sin(\pi) = e^{i\pi}$$

And for the point $(x, y) = (0, -1)$, we find $r = 1$ and $\theta = 3\pi/2$, so that

$$z = -i = \cos\left(\frac{3\pi}{2}\right) + i \sin\left(\frac{3\pi}{2}\right) = e^{i\frac{3\pi}{2}}$$

The result is the set of identities $1 = e^0$, $i = e^{i\pi/2}$, $-1 = e^{i\pi}$, $-i = e^{i3\pi/2}$, and back to $1 = e^{i2\pi}$.

Suppose we sum two complex numbers z_1 and z_2 using the identity $z = x + iy$. We find

$$z_1 + z_2 = (x_1 + iy_1) + (x_2 + iy_2) = (x_1 + x_2) + i(y_1 + y_2)$$

The magnitude of the sum is

$$|z_1 + z_2| = [(x_1 + x_2)^2 + (y_1 + y_2)^2]^{1/2}$$

Now let's repeat that sum using Euler's formula, $z = re^{i\theta}$. We find

$$z_1 + z_2 = r_1 e^{i\theta_1} + r_2 e^{i\theta_2}$$

which is difficult to reduce to the simple form $z_1 + z_2 = re^{i\theta}$. We find Euler's formula is not helpful when adding or subtracting complex numbers.

On the other hand, multiplying two complex numbers is easily performed using Euler's formula where

$$z_1 z_2 = r_1 e^{i\theta_1} \times r_2 e^{i\theta_2} = (r_1 r_2)\, e^{i(\theta_1 + \theta_2)} = re^{i\theta}$$

The point on the complex plane representing the product will be found a distance $r = \sqrt{r_1 r_2}$ from the origin, rotated through an angle $\theta = \theta_1 + \theta_2$ counterclockwise from the x-axis.

Powers of complex numbers are readily represented using Euler's formula. The square root of the complex number $z = x + iy = re^{i\theta}$ can be written

$$z^{1/2} = \left(re^{i\theta}\right)^{1/2} = r^{1/2} e^{i\theta/2}$$

This idea can be extended to any power n using the identity

$$z^n = r^n e^{in\theta} \tag{2.9}$$

Finally, let's further consider the point $z = -1$, which can be represented by $r = 1$ and $\theta = \pi$ as

$$e^{i\pi} = -1 \tag{2.10}$$

This remarkable relation can also be written

$$e^{i\pi} + 1 = 0 \tag{2.11}$$

which is known as *Euler's identity*. Euler's identity is considered to be one of the most beautiful mathematical equations, as it almost magically relates the fundamental numbers e, π, i, 1, and 0.

2.1.5 The origin of complex roots of polynomial equations

In exploring polynomial equations and their zeros, we discussed how the *quadratic formula*, Equation 1.4, can yield roots that are real, imaginary, or complex numbers depending on the particular values of the coefficients. Real zeros of a polynomial can be identified graphically as those points where $y(x) = 0$ and the quadratic function intersects the x-axis (see Figure 1.2). However, when the roots are complex numbers, it is possible to satisfy the equation $y(x) = 0$, even though the function never intersects the x-axis on a plot of $y(x)$ versus x.

How can we visualize complex roots of polynomial equations? Let's return to the quadratic equation

$$y(x) = x^2 - 2x + 2 = 0 \tag{2.12}$$

represented in Figure 1.2 (in blue). This parabola never intersects the x-axis, but the quadratic equation has two roots $x_\pm = 1 \pm i$. These roots are *complex numbers* with real and imaginary parts (see Figure 2.5).

While we think of the variable x in this quadratic equation as a real number, the fact that we accept complex numbers as solutions to the equation $y(x) = 0$ reveals that x is actually complex. Let's make this explicit by writing the same quadratic equation as $f(z) = 0$, where we substitute $z = x + iy$ for x. We find

$$f(z) = z^2 - 2z + 2 = 0 \tag{2.13}$$

If we expand this equation we see that

$$f(z) = (x + iy)^2 - 2(x + iy) + 2$$
$$= x^2 - y^2 - 2x + 2 + i(2xy - 2y) = 0$$

Just as the complex variable z must be plotted in two-dimensions to portray its real and imaginary parts, the *complex function* $f(z)$ also has a real part

$$\text{Re}[f(z)] = x^2 - y^2 - 2x + 2$$

and an imaginary part

$$\text{Im}[f(z)] = 2xy - 2y$$

If we evaluate $f(z)$ at the complex roots $z_\pm = 1 \pm i$, we find $f(1 + i) = f(1 - i) = 0$. This demonstrates that we can search for solutions to $f(z) = 0$ on the two-dimensional complex plane. Those solutions will be equal to the roots provided by the solutions to two quadratic formulas. Solving

$$\text{Im}[f] = 2xy - 2y = 0$$

yields $x = 1$. Solving

$$\text{Re}[f] = x^2 - y^2 - 2x + 2 = 0$$

for $x = 1$, which results in $1 - y^2 = 0$, yields $y = \pm 1$. As such, our solutions to $f(z) = 0$ are $z_\pm = 1 \pm i$. The challenge of graphically representing the complex roots of a complex function is explored in the complements.

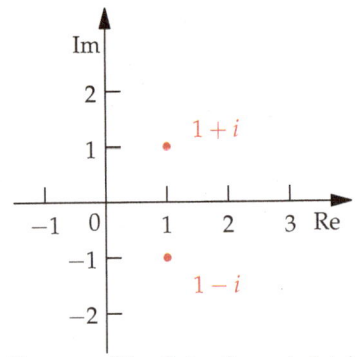

Figure 2.5: Visualizing the roots $1 + i$ and $1 - i$ on the two-dimensional complex plane.

2.2 Special properties of logarithms

THE SCALES OF PHYSICAL PROPERTIES INCLUDING ENERGY, LENGTH, AND TIME vary over many orders of magnitude. The *electromagnetic spectrum* includes wavelengths ranging from x-rays (10^{-10} meters) to visible light (10^{-7} meters) to microwaves (10^{-2} meters) to radio waves ($10^0 = 1$ meter). That variation of wavelength over 10 *orders of magnitude* is challenging to visualize on a linear scale. The *logarithmic function* has special properties that are useful in analyzing and visualizing functions that vary over many orders of magnitude. The properties of logarithms are explored in this section.

2.2.1 *The concept of base in logarithms*

We have explored how the *natural logarithm* can be used to transform a rapidly varying exponential function to a slowly varying linear function (see Figure 1.6).[9] That example demonstrated that the logarithm acts as the inverse of an exponential function.

[9] Logarithms were discovered by Scottish mathematician, physicist, and astronomer *John Napier* (1550-1617).

Consider the exponential function

$$y(x) = a^x$$

This function can be inverted using a logarithmic function as

$$\log_a[y(x)] = \log_a[a^x] = x$$

where a is the *base* of the exponential and logarithm. The base should be greater than zero and not equal to one.

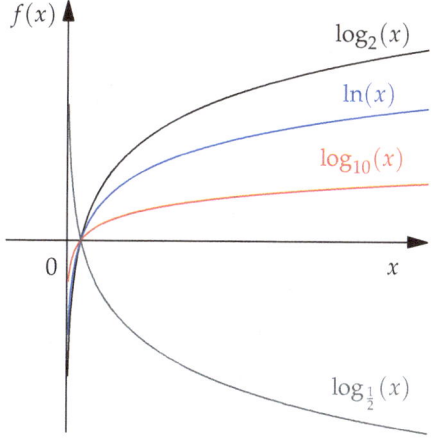

Figure 2.6: Comparison of binary (black), natural (blue), and common (red) logarithmic functions. Shown for comparison is the exotic base $\frac{1}{2}$ logarithm (heavy gray) where $\log_{1/2}(x) = -\log_2(x)$.

[10] Various notations are used. For natural logarithms, you might see $\log_e(x) = \ln(x)$. For common logarithms, you might see $\log_{10}(x) = \log(x)$.

In the physical sciences, the most common bases are $a = e$ (base e of the natural logarithm), $a = 10$ (base 10 of the *common logarithm*), and $a = 2$ (base 2 of the *binary logarithm*).[10] Physical processes are often modeled using the exponential function, making it common to use the natural logarithm (see Equation 1.10). In capturing the range of a function varying over many orders of magnitude, we often use the base 10 logarithm.[11]

[11] As it is common to have 10 fingers, humans find it natural to count in base 10.

The binary, natural, and common logarithms are plotted in Figure 2.6. We can make a number of observations based on this plot. All logarithmic functions equal zero when $x = 1$, as $a^0 = 1$ for any base a. For bases $a > 1$ and $x > 1$, the logarithm increases faster when the base is smaller. However, for bases $a > 1$ and $x < 1$, the opposite is true. Finally, there is a reciprocal relationship between a logarithm of base a and a logarithm of base $1/a$ as $\log_a(x) = -\log_{1/a}(x)$.[12] This is demonstrated in Figure 2.6 for the case $a = 2$.

[12] We note that $\log_a(x) = \log_{\frac{1}{a}}(x)\log_a(\frac{1}{a}) = -\log_{\frac{1}{a}}(x)$. We will further explore the algebra of logarithmic functions later in this section.

2.2.2 The algebra of logarithmic functions

The logarithm transforms the exponential function into a linear function as

$$\ln(e^x) = x$$

Suppose we have a function that is a product of exponentials

$$10^m 10^n = 10^{m+n}$$

If we take the logarithm of this product using the common logarithm, we find

$$\log_{10}(10^m 10^n) = \log_{10}(10^{m+n}) = m + n$$
$$= \log_{10}(10^m) + \log_{10}(10^n)$$

We write this fundamental property of logarithms as

$$\log(xy) = \log(x) + \log(y) \tag{2.14}$$

We define a similar relation for the logarithm of quotients. If we take the logarithm of a quotient using the natural logarithm, we find

$$\log_e\left(\frac{e^m}{e^n}\right) = \log_e(e^{m-n}) = m - n = \log_e(e^m) - \log_e(e^n)$$

We write this fundamental property of logarithms as

$$\log(x/y) = \log(x) - \log(y) \tag{2.15}$$

Finally, if we take the logarithm of a number such as 2 raised to a power, we find

$$\log_2\left[(2^m)^k\right] = \log_2\left(2^{mk}\right) = mk = k\log_2(2^m)$$

We write this fundamental property of logarithms as

$$\log(x^k) = k\log(x) \tag{2.16}$$

Through these examples, we see that logarithms can turn complicated multiplication and division into simple addition and subtraction.

2.2.3 How logarithms can be used to slow functions down

Let's explore the numerics of the common logarithm by examining a *logarithmic table* (see Table 2.1). This table clearly demonstrates the power of a logarithm to slow a function down. The first two columns show that a million-fold variation in a function, from 0.001 to 1000, is captured by a change in the logarithm from -3 to 3.

Base 10 logarithms of numbers and exponentials			
$10^{-3} = 0.001$	$\log(0.001) = -3$	$10^{0.3010} = 2$	$\log(2) = 0.3010$
$10^{-2} = 0.01$	$\log(0.01) = -2$	$10^{0.4343} = e$	$\log(e) = 0.4343$
$10^{-1} = 0.1$	$\log(0.1) = -1$	$10^{0.4771} = 3$	$\log(3) = 0.4771$
$10^{0} = 1$	$\log(1) = 0$	$10^{0.6990} = 5$	$\log(5) = 0.6990$
$10^{1} = 10$	$\log(10) = 1$	$10^{0.8451} = 7$	$\log(7) = 0.8451$
$10^{2} = 100$	$\log(100) = 2$	$10^{0.9542} = 9$	$\log(9) = 0.95421$
$10^{3} = 1000$	$\log(1000) = 3$	$10^{1.4771} = 30$	$\log(30) = 1.4771$
$10^{4} = 10000$	$\log(10000) = 4$	$10^{2.4969} = 314$	$\log(314) = 2.497$

Table 2.1: Brief table of base 10 logarithms for numbers over six orders of magnitude. Note $\log(x) = \log_{10}(x)$.

For base 10 logarithms, we can express any number in terms of the *mantissa*, a number between 1 and $9.\overline{99}$, and a *characteristic*, a power of 10.[13] For example, using our table of logarithms we write

$$\log_{10}(314) = \log_{10}\left(3.14 \times 10^2\right) = \log_{10}(3.14)\log_{10}\left(10^2\right)$$
$$= 0.497 + 2 = 2.497$$

[13] The notation $9.\overline{99}$ implies an infinity of repeated 9s following the decimal place.

How many *significant figures* does the result 2.497 have? The characteristic 2 is simply the power of 10. The power is an exact integer and does not contribute to the number of significant figures. The mantissa 3.14 carries all the significant figures, which in this case is three. Those significant figures appear as $\log_{10}(3.14) = 0.497$.

2.2.4 The very special properties of the natural logarithm

Let's explore the origin of the special properties of the natural logarithm. We can define the logarithm in terms of the equation

$$\frac{\Delta \ln(x)}{\Delta x} = \frac{1}{x} \tag{2.17}$$

This expression can be recognized as the slope of the logarithm, since

$$\text{slope} = \frac{\text{rise}}{\text{run}} = \frac{\Delta \ln(x)}{\Delta x} = \frac{1}{x}$$

Figure 2.7 shows the natural logarithm with the slope identified at three points $x = \frac{1}{3}$, 1, and 3, where $\frac{1}{x} = 3$, 1, and $\frac{1}{3}$. We can rewrite Equation 2.17 as

$$\Delta \ln(x) = \ln(x + \Delta x) - \ln(x) = \frac{\Delta x}{x} \tag{2.18}$$

This is known as the *finite difference* estimate of the change in the function, $\Delta \ln(x)$, resulting from a finite change in the variable, Δx.

This estimator can be useful in plotting the function $\ln(x)$. For example, if we know the value of $\ln(x)$ at some point x, we can estimate $\ln(x + \Delta x)$ using Equation 2.18 rewritten here as the finite difference equation

$$\ln(x + \Delta x) = \ln(x) + \frac{\Delta x}{x} \tag{2.19}$$

As x grows larger, the incremental growth of $\ln(x)$ diminishes in proportion

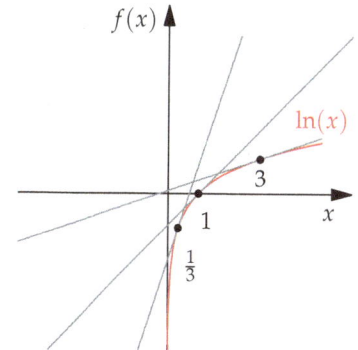

Figure 2.7: The natural logarithm (red) with slope $\frac{1}{x}$ shown as gray lines at $x = \frac{1}{3}$ (slope=3), $x = 1$ (slope=1), and $x = 3$ (slope=$\frac{1}{3}$).

Estimating values of ln(x) using finite differences				
	ln(x)			
x	$\Delta x = 3$	$\Delta x = 1$	$\Delta x = 0.5$	exact
0.0		-1.000	-1.500	-∞
0.5			-0.500	-0.693
1.0	0.000	0.000	0.000	0.000
1.5			0.500	0.405
2.0		1.000	0.833	0.693
2.5			1.083	0.916
3.0		1.500	1.283	1.099
3.5			1.450	1.253
4.0	3.000	1.833	1.593	1.386
4.5			1.718	1.504
5.0		2.083	1.829	1.609
5.5			1.929	1.705
6.0		2.283	2.020	1.792
6.5			2.103	1.872
7.0	3.750	2.450	2.180	1.946
7.5			2.252	2.015
8.0		2.593	2.318	2.079
8.5			2.381	2.140
9.0		2.718	2.440	2.197
9.5			2.495	2.251
10.0	4.179	2.829	2.548	2.303

Table 2.2: Estimates of ln(x) generated by the finite difference estimator Equation 2.19 for varying values of Δx.

to $1/x$. This is the fundamental definition of the natural logarithm. Table 2.2 shows estimates of $\ln(x)$ generated from an initial value of $x = 1$ where $\ln(1) = 0$ for three values of Δx.

As Δx decreases, the estimator improves in accuracy. However, as x approaches 0 the function $\ln(x)$ approaches $-\infty$. Our estimator does a better job of estimating $\ln(x)$ for $x > 1$ than for $1 > x > 0$. Figure 2.8 shows how the points generated by Equation 2.18 compare with the exact value of $\ln(x)$.

2.2.5 Converting between logarithms of different bases

What is the relationship between $\log_e(x)$ and $\log_{10}(x)$? For the function $y(x) = e^x$, we know that $\log_e y(x) = \log_e(e^x) = x$. Now consider

$$\log_{10} y(x) = \log_{10} e^x$$

We use the algebra of logarithms and Table 2.1 to find

$$\log_{10} e^x = x \log_{10} e = \ln y(x) \log_{10} e = 0.4343 \ln y(x)$$

which we can also write

$$\ln y(x) = 2.303 \log_{10} y(x)$$

This can also be written more generally as

$$\log_a(x) = \log_a(b) \log_b(x) \tag{2.20}$$

These relations demonstrate how we can convert between one base and another using logarithms.[14]

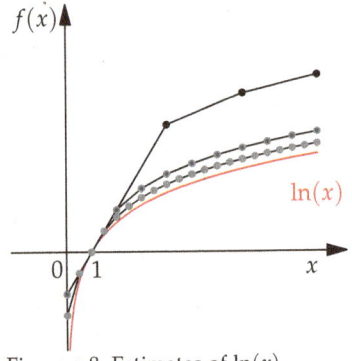

Figure 2.8: Estimates of $\ln(x)$ generated by the finite difference formula Equation 2.18 for $\Delta x = 3$ (black), 1 (heavy gray), and 0.5 (gray).

[14] In this example, $a = e$ and $b = 10$ so that $\log_e(x) = \log_e(10) \log_{10}(x)$ where $\log_e(10) = 2.303$.

A₂ Visualizing complex functions of complex variables

How can we graphically represent the behavior of $f(z)$ as a function of $z = x + iy$? This is challenging because the complex function $f(z)$ is two dimensional, having independent real and imaginary components. So is the complex variable z. To plot the dependence of $x = \text{Re}(z)$, $y = \text{Im}(z)$, $\text{Re}[f(z)]$, and $\text{Im}[f(z)]$, we need a four-dimensional coordinate system, something beyond our ability to visualize.

While we can't visualize $f(z)$ as function of z, we can visualize the square of its magnitude:

$$|f(z)|^2 = f^*f = f(z^*)f(z)$$

The second equality in this equation is true for a polynomial function where $f(0) = f^*(0)$. This function has the nice property that it is everywhere positive, as $|f(z)|^2 \geq 0$. Therefore, we can identify the roots of the equation $f(z) = 0$ as local minima of the function $|f(z)|^2$.

Let's return to the problem of visualizing the roots of Equation 2.13

$$f(z) = z^2 - 2z + 2 = 0$$

by plotting $|f(z)|^2$ and identifying the zeros on the complex plane. The result is shown in Figure 2.9. The roots appear as two points on the complex plane $z_\pm = 1 \pm i$ at which $f(z) = 0$ and therefore $|f(z)|^2 = 0$. Note that the original quadratic equation $x^2 - 2x + 2$ does not intersect the complex plane.

Our graphical approach to identifying roots of quadratic functions is general and also applies to quadratic functions with real roots. Consider the quadratic equation

$$f(x) = x^2 - 1 = (x - 1)(x + 1) = 0$$

This function has two real roots at $x_\pm = \pm 1$ where $f(x)$ intersects the x-axis and $f(\pm 1) = 0$. Suppose we interpret this quadratic equation as a complex function

$$f(z) = (z - 1)(z + 1)$$

where we have substituted $z = x + iy$ for x. The square of the magnitude of this function $|f|^2 = f(z^*)f(z)$ is shown in Figure 2.10. The two roots of the quadratic equation appear as local minima of $|f|^2$ located at $(1, 0)$ and $(-1, 0)$ on the complex plane.

Interestingly, along the imaginary y-axis, $|f|^2$ is a monotonically increasing function of y. For example, if we set $x = 0$, we find $|f(0, y)|^2 = 1 + 2y^2 + y^4$, which increases steeply as a function of y. Along the real x-axis, $|f|^2$ is a quartic double well function of x. For example, if we set $y = 0$, we find $|f(x, 0)|^2 = 1 - 2x^2 + x^4 = (1 - x^2)^2$. This is the same quartic double well function we encountered earlier in our exploration of polynomial functions with minima located at $x = \pm 1$ (see Equation 1.6) .

Note that the original quadratic equation intersects the two local minima of $|f(x, y)|^2 = 0$ at the roots $x = \pm 1$. On the complex plane, those points

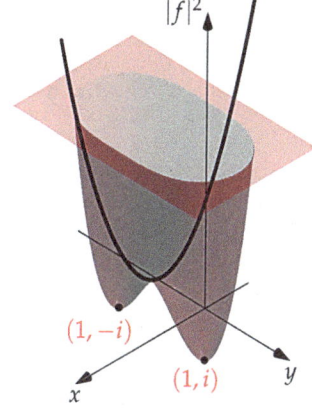

Figure 2.9: The square of the magnitude of $f(z) = (z - (1 + i))(z - (1 - i))$. This quadratic equation has roots at $z_\pm = 1 \pm i$ (dots). The quadratic equation $x^2 - 2x + 2$ is shown for comparison (black line).

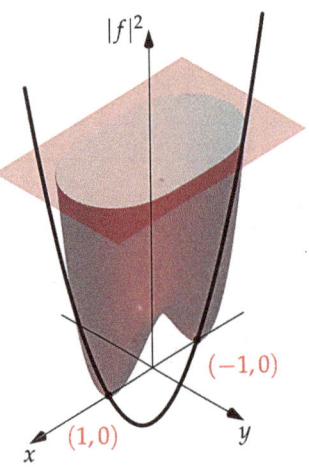

Figure 2.10: The square of the magnitude of $f(z) = (z-1)(z+1)$. This quadratic equation has roots at $z_{\pm} = \pm 1$ (dots). The quadratic equation $x^2 - 1$ is shown for comparison (black line).

correspond to $(-1,0)$ and $(1,0)$, which are tangentially intersected by the surface $|f(x,y)|^2$. So you see how that works.

B₂ Application of logarithms and the logarithmic scale

For certain physical variables, the range of values varies over many orders of magnitude. Examples include the acidity of a solution, the strength of an earthquake, and the wavelength or frequency of light. In those cases, it is convenient to work with the logarithm of the variable and a *logarithmic scale*.

Consider the wavelength of light, which varies over many orders of magnitude (see Figure 2.11). Light of wavelength 10^{-3} nm is an x-ray, light of wavelength 10^3 nm is in the visible spectrum, and light of wavelength 10^{11} nm is a radio wave. This total variation in wavelength spans 14 orders of magnitude. As such, it is convenient to represent the spectrum using a logarithmic scale.

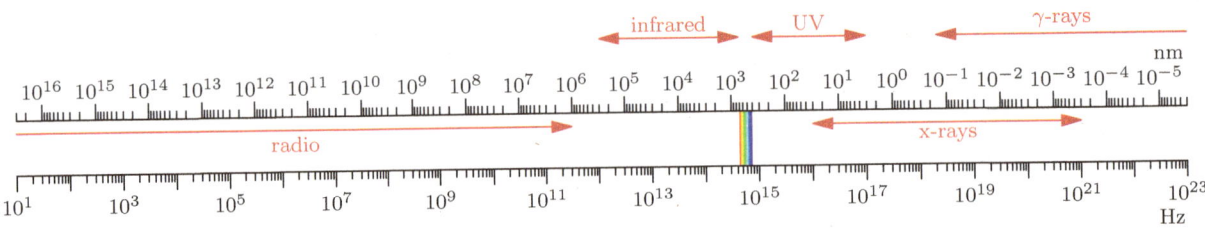

Figure 2.11: The electromagnetic spectrum displayed as a function of wavelength, in nanometers (nm), and frequency, in Hertz (Hz), on a logarithmic scale. Over 20 orders of magnitude are displayed.

Now consider the acidity of an aqueous solution measured in terms of the hydrogen ion concentration $[H^+]$ and its dimensionless activity $a_{H^+} = [H^+]/1$ M. An acidic solution may have $[H^+] = 10^{-2}$ M while a basic solution may have $[H^+] = 10^{-12}$ M, a difference of ten-billion-fold. To simplify the discussion of acidity, it is convenient to introduce the logarithmic *pH scale* as

$$\text{pH} = -\log_{10}(a_{H^+})$$

The acidic solution with $[H^+]=10^{-2}$ M has pH=2 while the basic solution with $[H^+]=10^{-12}$ M has pH=12. The exponential variation over many orders of magnitude in $[H^+]$ is reduced to a linear scale over a factor of 10 in pH. The pH scale is shown graphically in Figure 2.12.

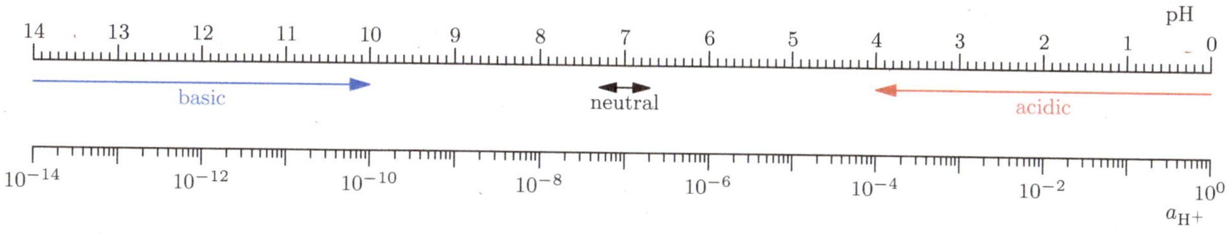

Figure 2.12: The relationship between the logarithmic scale of hydrogen ion activity, a_{H^+}, and the linear scale of pH, where pH $= -\log(a_{H^+})$. Note that the a_{H^+} varies over 14 orders of magnitude.

Similarly, the optical *transmittance*, T, of a solution is found to vary over many orders of magnitude. As such, it is convenient to define the optical *absorbance*, A, in terms of the logarithm of the transmittance using

$$A = -\log_{10} T$$

So you see how it works.

Length scales of objects are also found to vary over many orders of magnitude. Consider the variation in diameter of an atomic nucleus of 10^{-15} m, an atom of 10^{-10} m, a cell of 10^{-6} m, an apple of 10^{-1} m, the Earth of 10^{7} m, the Milky Way galaxy of 10^{21} m, and the known Universe of 10^{26} m. There are 41 orders-of-magnitude variation in the size of these objects defining the shape and composition of our World.

C2 Logarithms and Stirling's approximation

Consider the *Boltzmann entropy formula*

$$S = k_B \ln W$$

where k_B is the Boltzmann constant and W is the number of states available to a system.[15] Think of the number of states of the system as the number of ways to order the system. For example, suppose we have a deck of 52 unique cards. There are $W = 52! = 52 \times 51 \times 50 \times \ldots \times 2 \times 1$ ways of arranging the deck of cards.[16] In another example, suppose we lay N coins on a table. Each coin can show a head or a tail. There are $W = 2^N$ ways of ordering the N coins. These simple examples demonstrate that the number of ways of ordering N objects in a system grows exponentially as N increases.

It can be very challenging to represent numbers as large as $W = 52! \approx 8 \times 10^{67}$. When we consider the number of ways we can order N atoms and

[15] Named for the Austrian physicist and philosopher *Ludwig Boltzmann* (1844-1906).

[16] Here $n!$ is defined $n! = n \times (n-1) \times (n-2) \times \ldots \times 1$ and pronounced *n factorial*.

molecules with $N = 6 \times 10^{23}$ we can forget about determining $N!$ since it is much too large! Instead, we work directly with the logarithm $\ln(N!)$.

Let's consider $\ln(N!)$ where N is very large. We write

$$\ln N! = \ln[1 \times 2 \times \ldots \times (N-1) \times N]$$
$$= \ln 1 + \ln 2 + \ldots + \ln(N-1) + \ln N$$

which we can express as the sum

$$\ln N! = \sum_{n=1}^{N} \ln(n) \Delta n$$

where $\Delta n = 1$. For very large N, we can approximate this sum in terms of an integral[17]

$$\ln N! \approx \int_1^N \ln(n) dn = [n \ln n - n]_1^N = N \ln N - N + 1 \qquad (2.21)$$

This result is known as *Stirling's approximation*.[18] It provides a useful way to estimate logarithms of very large numbers, such as those that appear in the physical sciences.

Let's compute $\ln(N!)$ where $N = 6.02 \times 10^{23}$. If we first try to determine $N!$, we will fail. It's just too big! But if we directly estimate $\ln(N!)$, we find

$$\ln(6.02 \times 10^{23}) \approx (6.02 \times 10^{23}) \ln(6.02 \times 10^{23}) - 6.02 \times 10^{23} + 1$$
$$= 3.30 \times 10^{25} - 6.02 \times 10^{23} + 1$$
$$= 3.24 \times 10^{25}$$

The dominant term is $N \ln N$, with a small correction from $-N$, and a negligible contribution from 1.

Note that we can alternatively write Stirling's approximation as

$$N! = e^{\ln N!} \approx e^{(N \ln N - N)} = e^{N \ln N} e^{-N} \qquad (2.22)$$

Using the properties of the exponential function,[19] we find

$$e^{N \ln N} = \left(e^{\ln N} \right)^N = N^N$$

Returning to Equation 2.22, we find

$$N! = e^{\ln N!} \approx N^N e^{-N} = \left(\frac{N}{e} \right)^N$$

This is another useful form of Stirling's approximation.

Finally, note that if we perform this derivation more carefully, we find $N! \approx \sqrt{2\pi N} \, (N/e)^N$ so that

$$\ln N! \approx N \ln N - N + \frac{1}{2} \ln(2\pi N)$$

for very large N. For $N = 6.02 \times 10^{23}$ this improved estimate is $3.30 \times 10^{25} - 6.02 \times 10^{23} + 28.3 = 3.24 \times 10^{25}$ leads to no change in our result.

[17] You can prove this result by noting that

$$\frac{d}{dn}(n \ln n - n) = \ln n + 1 - 1 = \ln n$$

where we use the identity $d \ln(n)/dn = 1/n$.

[18] Named for the Scottish mathematician *James Stirling* (1692-1770).

[19] We use the identity $(e^x)^n = e^{nx}$ and the fact that $e^{\ln x} = x$.

D$_2$ Connecting complex numbers and logarithms

Recall Euler's formula for expressing a complex number

$$z(r,\theta) = re^{i\theta}$$

It follows that the logarithm of a complex number can be written

$$\ln(z) = \ln(re^{i\theta}) = \ln(r) + i\theta$$

Constant values of $\ln(r)$ are circles on the xy-plane, while constant values of $i\theta$ are radial lines (see Figure 2.13). This result demonstrates the intimate connection between the Euler relation, logarithms, and complex numbers.

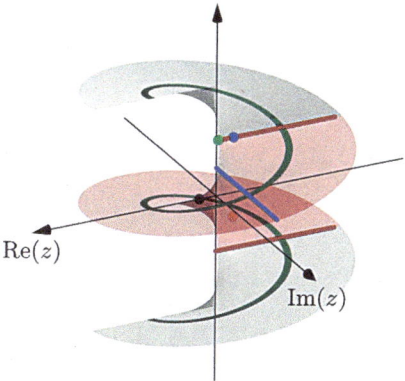

Figure 2.13: The Riemann surface for the complex logarithm $w = \ln(z) = \ln(r) + i\theta = \ln(r) + i\theta + i2\pi n$. Highlighted are the region of $-\pi < \theta \le \pi$ (red surface), $r = 2$ and $-\infty < \theta < \infty$ (green line), and $\theta = \frac{\pi}{2}$ (blue line).

Due to the fact that $e^{i\theta} = e^{i(\theta + 2n\pi)}$ where n is an integer, $\ln(z)$ is a periodic function of θ. As such, the logarithm of a complex number

$$w = \ln(r) + i\theta = \ln(r) + i\theta + i2\pi n$$

where n is any integer. The complex logarithm is a multivalued function of r and θ. This property is shown graphically through the *Riemann surface* in Figure 2.13.[20] To remove this ambiguity, in practice we restrict the value of θ to be $-\pi < \theta \le \pi$. An alternative representation of the Riemann surface is provided in Figure 2.14, where the axes are taken to be $\ln(r)$ and $i\theta$.

To appreciate the differences between the two mappings, consider the following examples of z and $\ln(z)$ pairings, where each point is also marked in Figures 2.13 and 2.14. The logarithm of $z = 1$ where $r = 1$ and $\theta = 0$ is $\ln(1) = 0$ (black dot). For $z = -1$ where $r = 1$ and $\theta = \pi$ we find $\ln(-1) = i\pi$, just as $e^{i\pi} = -1$ (blue dot), while for $z = -0.1$ where $r = 0.1$ and $\theta = \pi$ we find $\ln(-0.1) = -2.30 + i\pi$ (green dot). Finally, for $z = -1 - i$ where $r = \sqrt{2}$ and $\theta = \frac{5\pi}{4}$ we find $\ln(-1 - i) = \ln(\sqrt{2}) + i\frac{5\pi}{4} = 0.346 - i\frac{3\pi}{4}$ as $-\pi < \theta \le \pi$ (red dot). These examples demonstrate the fundamental properties of the complex logarithm.

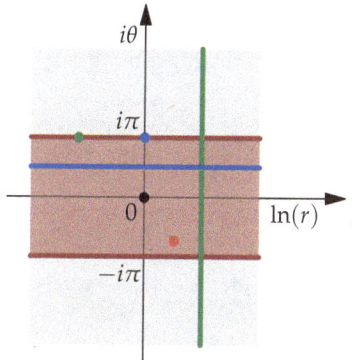

Figure 2.14: Alternative visualization of the Riemann surface for the complex logarithm $w = \ln(z) = \ln(r) + i\theta = \ln(r) + i\theta + i2\pi n$. Highlighted are the region of $-\pi < \theta \le \pi$ (red strip), $\ln(r) = 0.693$ and $-\infty < \theta < \infty$ (green line), and $\theta = \frac{\pi}{2}$ (blue line).

[20] *Bernard Riemann* (1826-1866) was a German mathematician whose contributions to our understanding of the geometry of surfaces profoundly impacted the development of physical science.

E$_2$ End-of-chapter problems

> To those who do not know mathematics
> it is difficult to get across a real feeling
> as to the beauty, the deepest beauty, of
> nature.
>
> Richard Feynman

Warm-ups

2.1 Consider the following complex numbers $z = x + iy$. By Euler's formula we can also write z in polar form as $z = re^{i\theta}$. Determine the modulus, r, and phase angle, θ, for each complex number.

(a) 1   (b) $4i$ (c) $1 - i$
(d) $3 - 4i$ (e) $-2 + 3i$ (f) $-3 - 3i$

2.2 Demonstrate that $e^{-i\theta} = \cos\theta - i\sin\theta$.

2.3 Show that $\cos\theta = \frac{1}{2}\left(e^{i\theta} + e^{-i\theta}\right)$ and $\sin\theta = \frac{1}{2i}\left(e^{i\theta} - e^{-i\theta}\right)$.

2.4 Show for $z = x + iy$ that

$$\frac{1}{z} = \frac{x}{x^2 + y^2} - \frac{iy}{x^2 + y^2}$$

2.5 Plot the function $|z - 1| = 3$ on the complex plane.

2.6 The pH of an aqueous solution is defined $pH = -\log_{10} a_{H^+}$ where the activity of the hydrogen ion $a_{H^+} = [H^+]/1$ M. Find the pH of the following solutions defined in terms of a_{H^+}.

(a) $a_{H^+} = 1.00 \times 10^{-7}$ (b) $a_{H^+} = 8.54 \times 10^{-10}$ (c) $a_{H^+} = 0.13$ (d) $a_{H^+} = 1.15$

2.7 Given the following values for the pH, find $[H^+]$ for each solution.

(a) pH = 0   (b) pH = 2.156   (c) pH = 0.234   (d) pH = 7.876

Homework exercises

2.8 Evaluate zz^* and z^2 for each of the following complex numbers z.

(a) 1   (b) $1 + i$ (c) $2i$
(d) $-2 + 2i$ (e) -4 (f) $-4 - 4i$

2.9 Express the following complex numbers in polar form using Euler's formula (a) $z = 1 + i$ and (b) $z = -1 - i$.

2.10 Using Euler's formula, derive *de Moivre's identity*

$$\cos(n\theta) + i\sin(n\theta) = (\cos(\theta) + i\sin(\theta))^n$$

2.11 Prove the following identities by reexpressing the functions using Euler's formula.

(a) $\cos(ix) = \cosh(x)$ (b) $\sinh(ix) = i\sin(x)$

(c) $\sin(ix) = i\sinh(x)$ (d) $\cosh(ix) = \cos(x)$

2.12 Use Euler's formula to evaluate $(i)^i$ as a real number.

2.13 Use Euler's formula to derive the identities (a) $\cos(\alpha)\cos(\beta) = \frac{1}{2}\cos(\alpha + \beta) + \frac{1}{2}\cos(\alpha - \beta)$ and (b) $\sin(\alpha)\sin(\beta) = \frac{1}{2}\cos(\alpha - \beta) - \frac{1}{2}\cos(\alpha + \beta)$.

2.14 For the isothermal, reversible expansion or compression of an ideal gas, the pressure-volume work done on the gas is given by

$$w = -nRT\ln\frac{V_2}{V_1}$$

where V_1 and V_2 are the initial and final volumes of the gas. Here n is the number of moles of gas, R is the gas constant $= 8.314\,\text{J/mol K}$, and T is the absolute temperature. Find the work done on the gas in the isothermal, reversible expansion of 1.00 mole of ideal gas at 298.15K from a volume of 3.00 liters to a volume of 7.00 liters.

2.15 Radioactive decay is observed to follow the integrated rate law

$$\ln\frac{[A](t)}{[A]_0} = -kt$$

where $[A](t)$ is the concentration of A at time t, $[A]_0$ is the initial concentration $[A](t = 0) = [A]_0$, and k is the rate constant (see figure below). The predominant isotope of carbon, $^{12}_{6}C$, is stable, while $^{14}_{6}C$ is radioactive. The fraction of $^{14}_{6}C$ in a sample of wood ash from an archaeological dig was found to be 0.249. How old is the wood ash, given that $k = 1.21 \times 10^{-4}\,\text{yr}^{-1}$ for the radioisotope $^{14}_{6}C$?

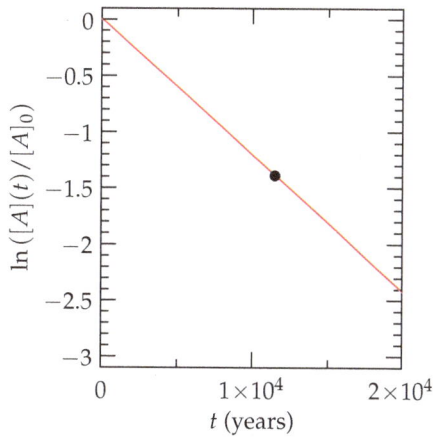

2.16 The change in entropy associated with the expansion or compression of an ideal gas is given by

$$\Delta S = nC_V\ln\frac{T_2}{T_1} + nR\ln\frac{V_2}{V_1}$$

where n is the number of moles of gas, C_V is the molar heat capacity of the gas at constant volume, V is the volume of the gas, and T is the absolute temperature. The subscripts indicate the initial (1) and final (2) states. In the expansion of 1.00 mole of an ideal gas from 1.00 liter to 3.00 liters, the temperature falls from 300K to 284K. Determine the change in entropy, ΔS, for the ideal gas in this process. Take $C_V = \frac{3}{2}R$ and $R = 8.314\,J/mol\,K$.

2.17* Evaluate the logarithm $\ln(z)$ for each of the following complex numbers z.

(a) 1

(b) $1 + i$

(c) $2i$

(d) $-2 + 2i$

(e) -4

(f) $-4 - 4i$

Recall that the logarithm of a complex number is a multivalued function. Restrict the value of the imaginary component to be between $-i\pi$ and $i\pi$.

2.18 The Soviet mathematical physicist *Lev Landau* (1908-1968) proposed a base-10 *logarithmic scale* to measure ability in theoretical physics. He placed Isaac Newton at 0, Albert Einstein at 0.5, and Paul Dirac at 1. Landau eventually ranked himself at 2. By Landau's own estimate, how many times better a theoretical physicist was Einstein than Landau?

3 DIFFERENTIATION IN ONE AND MANY DIMENSIONS

3.1 Differentiating functions of one variable

A FOUNDATION IDEA IN CALCULUS is the concept of the derivative. The derivative provides a measure of how much a function changes as a result of a change in its variable. In modeling physical processes, we are often interested in understanding how one property (a function) varies when another property (the variable) is changed. This might involve the change in concentration (the function) after a change in time (the variable). Or the change in pressure (the function) due to a change in volume (the variable). In this section, we explore the definition of the derivative and review the principal methods of differentiation of a function of one variable.

3.1.1 The concept of the limit and the definition of the derivative

In Chapter 1, we defined the slope of a function $f(x)$ in terms of the change in the value of the function, Δf, resulting from an incremental change in the variable, Δx, as

$$\text{slope} = \frac{\text{rise}}{\text{run}} = \frac{\Delta f}{\Delta x} = \frac{f(x + \Delta x) - f(x)}{\Delta x} \tag{3.1}$$

This is a *finite difference* estimate of the slope defined in terms of a difference in the function, Δf, divided by a finite change in the variable, Δx.

Figure 3.1 graphically depicts the measure of the slope (black line) resulting from the calculation of the rise of the function $\Delta f = f(x_2) - f(x_1)$ divided by the run of the variable $\Delta x = (x_2 - x_1)$ between two points $(x_1, f(x_1))$ and $(x_2, f(x_2))$. The line connecting the two points is known as the *secant line*. For a linear function $f(x) = mx + b$, the slope is defined by the constant m and is independent of the position x. However, when a function is non-linear the slope will vary as x varies. For example, in Figure 3.1 the function is steeper at x_1 than at x_2. As such, our estimate of the slope will depend on the value of Δx.

To get a precise estimate of the slope at a single point using Equation 3.1, we need to know the rise of the function, Δf, for a run in the variable, Δx, that is as small as we like. Even zero! This is the concept of the *derivative*.

Figure 3.2 shows a series of secant lines defined over an interval $x = a$ to $x = a + \Delta x$ for varying values of Δx. As Δx decreases the slope of the secant line increases. Taking the *limit* as Δx approaches 0[1] leads to the definition of the derivative

$$\lim_{\Delta x \to 0} \frac{f(x + \Delta x) - f(x)}{\Delta x} \equiv \frac{df}{dx} \tag{3.2}$$

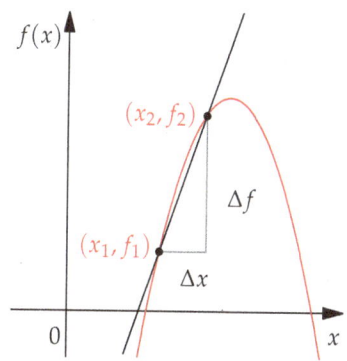

Figure 3.1: The rise in the function, $\Delta f = f_2 - f_1 = f(x_2) - f(x_1)$, for the run in the variable, $\Delta x = x_2 - x_1$. The secant line provides a measure of the slope of the function over the region $[x_1, x_2]$.

[1] This concept of Δx becoming so small that it cannot be measured is related to the theory of *infinitesimals* explored by Greek mathematician, physicist, engineer and philosopher *Archimedes* (*circa* 287-212 BCE).

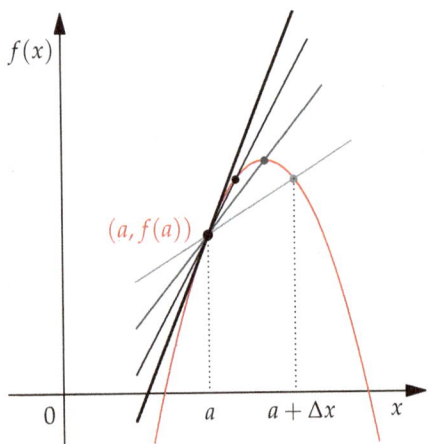

Figure 3.2: A series of secant lines providing a measure of the slope of $f(x)$ at $x = a$ for decreasing values of Δx (light gray to heavy gray). The thick black line shows the tangent line representing the exact slope at $x = a$.

The derivative of $f(x)$ with respect to x defines the slope of the function at a single point x represented by the heavy black *tangent line*[2] in Figure 3.2. The definition of the *derivative* is an essential foundation of calculus.[3]

The derivative of $f(x)$ can be written using a variety of notations, a popular form being[4]

$$\frac{df}{dx} = f'(x)$$

We call this the *first derivative* of $f(x)$. Taking the derivative of the first derivative leads to the *second derivative*:

$$\frac{d^2 f}{dx^2} = \frac{d}{dx} f'(x) = f''(x)$$

While the first derivative defines the rate of change in $f(x)$ as x varies, the second derivative defines the rate of change in $f'(x)$. This idea can be generalized to form the third derivative

$$\frac{d^3 f}{dx^3} = f'''(x)$$

and derivatives of arbitrary order.

There is a small set of rules for differentiating functions of one variable that can be used to determine the derivatives we need in modeling physical processes.[5] The next section provides a concise summary of those rules.

3.1.2 *Rules for differentiating functions of one variable*

Let's review the rules for differentiating functions of one variable. It is important to practice and master these rules as they form the foundation for applying calculus in the physical sciences.

1. $f(x) = c$. As the constant c does not vary with x, the slope of $f(x)$ and its derivative are zero:

$$\frac{df}{dx} = \frac{dc}{dx} = 0 \tag{3.3}$$

[2] While the tangent line is an ancient concept considered by *Archimedes* and *Euclid*, its modern development owes much to the French lawyer and mathematician *Pierre de Fermat* (1607-1665).

[3] The discovery of calculus is typically attributed to German philosopher and mathematician *Gottfried Leibniz* (1646-1716) and to British physicist, mathematician, astronomer, and theologian *Isaac Newton* (1642-1727).

[4] The derivative may be written using a variety of equivalent *derivative notations*, including $df(x)/dx$ (Leibniz's notation), $\dot{f}$ (Newton's notation), f' (Lagrange's notation), and $D_x f(x)$ (Euler's notation).

[5] Most functions of interest in the physical sciences are *continuous* and *smooth* with derivatives of arbitrary order. A function $f(x)$ is *continuous* if arbitrarily small changes in x result in arbitrarily small changes in $f(x)$. A function $f(x)$ is *smooth* if its derivatives of all orders are continuous.

2. $f(x) = cx$. The first derivative of a linear function has a constant slope:

$$\frac{df}{dx} = \frac{d}{dx}(cx) = c\frac{dx}{dx} = c \tag{3.4}$$

3. $f(x) = x^n$. The *power rule* can be applied to differentiate functions of arbitrary powers of x (see Figure 3.3). It is written

$$\frac{df}{dx} = \frac{d}{dx}x^n = nx^{n-1} \tag{3.5}$$

The power rule can also be applied to negative powers of x:

$$\frac{df}{dx} = \frac{d}{dx}x^{-n} = -nx^{-n-1}$$

As an example, consider the derivative of $p(V) = nRT/V$ as a function of V:

$$\frac{d}{dV}p(V) = \frac{d}{dV}\left(\frac{nRT}{V}\right) = nRT\frac{d}{dV}\frac{1}{V} = -nRT\frac{1}{V^2}$$

4. $f(x) + g(x)$. The derivative of a sum of two functions is the sum of the derivatives:

$$\frac{d}{dx}[f(x) + g(x)] = \frac{df}{dx} + \frac{dg}{dx} \tag{3.6}$$

Consider the derivative of the function $f(T) = A - \frac{\Delta H}{RT}$ as a function of T where A and ΔH are constants:

$$\frac{df}{dT} = \frac{dA}{dT} - \frac{d}{dT}\left(\frac{\Delta H}{RT}\right) = \frac{\Delta H}{RT^2}$$

5. $\sin(ax)$. The derivative of $\sin(ax)$ is constant proportional to $\cos(ax)$ (see Figure 3.4). It is written

$$\frac{d}{dx}\sin(ax) = a\cos(ax) \tag{3.7}$$

Consider the derivative of

$$\varphi(x) = \left(\frac{2}{L}\right)^{\frac{1}{2}}\sin\left(\frac{n\pi x}{L}\right)$$

where L and n are constants, written

$$\frac{d}{dx}\varphi(x) = \left(\frac{2}{L}\right)^{\frac{1}{2}}\left(\frac{n\pi}{L}\right)\cos\left(\frac{n\pi x}{L}\right)$$

6. $\cos(ax)$. The derivative of $\cos(ax)$ is proportional to $-\sin(ax)$.[6] It is written

$$\frac{d}{dx}\cos(ax) = -a\sin(ax) \tag{3.8}$$

Consider the derivative of $\varphi(t) = A\cos(2\pi\nu t)$ where A and ν are constant:

$$\frac{d}{dt}\varphi(t) = -A(2\pi\nu)\sin(2\pi\nu t)$$

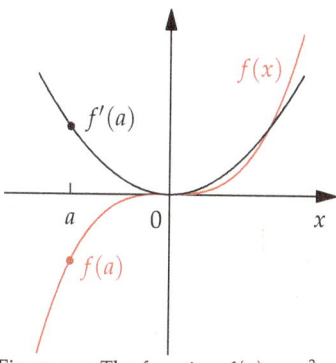

Figure 3.3: The function $f(x) = x^3$ (red) and its first derivative $f'(x) = 3x^2$ (black). At $x = a$ the function $f(a)$ is negative (red dot) while the corresponding derivative $f'(a)$ is positive (black dot).

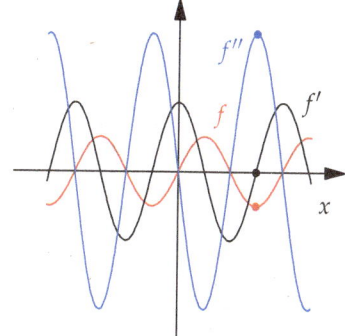

Figure 3.4: The function $f(x) = \sin(2x)$ (red), its first derivative $f'(x) = 2\cos(2x)$ (black) and second derivative $f''(x) = -4\sin(2x)$ (blue). At the minimum of the function (red dot) the first derivative is zero (black dot) and the second derivative is positive (blue dot).

[6] You can remember that $d\sin(x)/dx = +\cos(x)$ and $d\cos(x)/dx = -\sin(x)$ by thinking graphically of the sign of the slope of the function that is being differentiated. Near $x = 0$ the slope of $\sin(x) > 0$ so $d\sin(x)/dx = +\cos(x)$. Near $x = \pi/2$ the slope of $\cos(x) < 0$ so $d\cos(x)/dx = -\sin(x)$ (see Figure 3.4).

Returning to the prior example, we can combine the rules for differentiating sine and cosine to write the second derivative of

$$\varphi(x) = \left(\frac{2}{L}\right)^{\frac{1}{2}} \sin\left(\frac{n\pi x}{L}\right)$$

in the form

$$\frac{d^2}{dx^2}\varphi(x) = -\left(\frac{2}{L}\right)^{\frac{1}{2}} \left(\frac{n\pi}{L}\right)^2 \sin\left(\frac{n\pi x}{L}\right)$$

The property that sine and cosine functions are proportional to their own second derivatives is profound. The sine and cosine functions appear in the solution of many fundamental problems in the physical sciences, particularly those involving waves.[7]

7. exp(ax). The derivative of a real or complex exponential function is proportional to the exponential function itself:

$$\frac{d}{dx}e^{ax} = ae^{ax} \tag{3.9}$$

An example is shown in Figure 3.5.

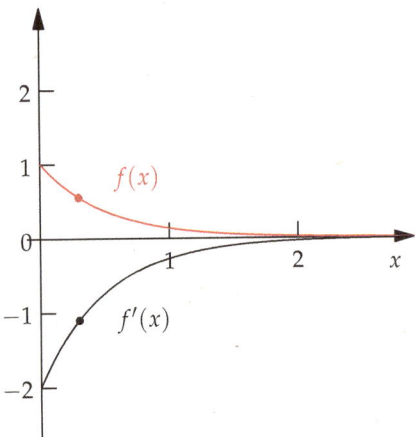

Consider the complex exponential function $\Phi(\varphi) = Ae^{im\varphi}$ for which

$$\frac{d}{d\varphi}\Phi(\varphi) = imAe^{im\varphi}$$

and

$$\frac{d^2}{d\varphi^2}\Phi(\varphi) = (im)^2 Ae^{im\varphi} = -m^2 Ae^{im\varphi}$$

We often find that the rate of change in a property is proportional to the property itself. For example, suppose the rate of birth in a population is proportional to the population, or that the rate of change in a chemical species is proportional to its concentration. In that case, the population will be an exponential function of time. This explains why the exponential function is often used to model processes in the physical sciences.

[7] One example is the *particle in a box* model in quantum mechanics where we encounter the equation

$$-\frac{\hbar^2}{2m}\frac{d^2}{dx^2}\varphi(x) = E\varphi(x)$$

where $\hbar$, m, and E are constants. Since $\varphi''(x)$ is proportional to $\varphi(x)$, $\varphi(x)$ must be a sinusoidal function of x.

Figure 3.5: The function $f(x) = \exp(-2x)$ (red) and its first derivative $f'(x) = -2\exp(-2x)$ (black). As $f(x)$ is a decreasing function of x (red dot) the first derivative is negative (black dot).

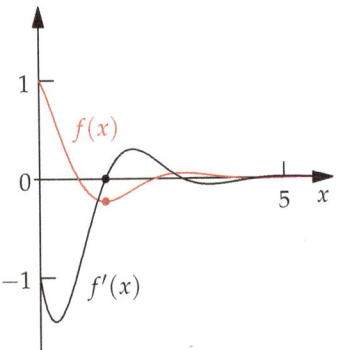

Figure 3.6: The function $f(x) = \exp(-x)\cos(2x)$ (red) and its first derivative $f'(x) = -\exp(-x)\left[2\sin(2x) + \cos(2x)\right]$ (black). At the minimum (red dot) the first derivative is zero (black dot).

8. $f(x)g(x)$. The derivative of the product of two functions $f(x)g(x)$ is defined by the *product rule* to be

$$\frac{d}{dx}[f(x)g(x)] = f(x)\frac{dg}{dx} + g(x)\frac{df}{dx} \qquad (3.10)$$

Consider the compound function

$$\varphi(x) = e^{-ax}\cos(bx)$$

where a and b are constants. We find

$$\begin{aligned}
\frac{d\varphi}{dx} &= e^{-ax}\frac{d}{dx}\cos(bx) + \cos(bx)\frac{d}{dx}e^{-ax}\\
&= -e^{-ax}b\sin(bx) - a\cos(bx)e^{-ax}\\
&= -e^{-ax}[b\sin(bx) + a\cos(bx)]
\end{aligned}$$

This result is shown for the case $a = 1$ and $b = 2$ in Figure 3.6.

One common mistake is to use the false relation

$$\frac{d}{dx}[f(x)g(x)] = f'g'$$

as if it were the true product rule. If you are unsure whether you have remembered the product rule correctly, take a simple example such as $f(x) = x$ and $g(x) = x$ for which

$$\frac{d}{dx}[f(x)g(x)] = \frac{d}{dx}x^2 = 2x$$

You will find that the true product rule gives the correct result

$$\frac{d}{dx}[f(x)g(x)] = fg' + gf' = x + x = 2x$$

while the false product rule

$$\frac{d}{dx}[f(x)g(x)] = f'g' = 1$$

does not.

9. $\ln(x)$. The derivative of the natural logarithm is

$$\frac{d}{dx}\ln(x) = \frac{1}{x} \qquad (3.11)$$

This property of the natural logarithm, introduced through Figure 2.7 and the surrounding discussion, is intimately related to the fundamental definition of the natural logarithm

$$d\ln(x) = \frac{dx}{x}$$

The fractional change in the function, $\ln(x)$, resulting from a change in the variable, dx, is proportional to $\frac{1}{x}$. This result is shown in Figure 3.7. Note that as x increases, the rate of change in $\ln(x)$ decreases. This makes the logarithm $\ln(x)$ a slowly converging function of x.

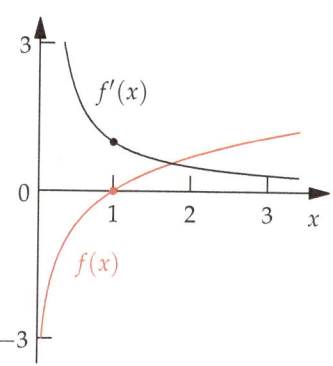

Figure 3.7: The function $f(x) = \ln(x)$ (red) and its first derivative $f'(x) = \frac{1}{x}$ (black). At $x = 1$ the function $\ln(1) = 0$ (red dot) and the derivative is positive (black dot).

10. $f(g(x))$. The derivative of a *composite function* is defined by the *chain rule*:

$$\frac{d}{dx}f(g(x)) = \frac{df}{dg}\frac{dg}{dx} \qquad (3.12)$$

Take the function $\Phi(\theta) = 3\cos^2(\theta) - 1$. We define $g(\theta) = \cos(\theta)$ so that $\Phi(g(\theta)) = 3(g(\theta))^2 - 1$. By the chain rule, the derivative of $\Phi(g(\theta))$ with respect to θ is

$$\frac{d}{d\theta}\Phi(g(\theta)) = \frac{d\Phi}{dg} \times \frac{dg}{d\theta} = 6g(\theta) \times (-\sin(\theta)) = -6\cos(\theta)\sin(\theta)$$

So we see how this works.

Now let's consider the challenging problem of differentiating the function

$$U(T) = \frac{\varepsilon}{2} + \frac{\varepsilon}{\exp(\varepsilon/T) - 1} \qquad (3.13)$$

with respect to T, where ε is a constant. The function $U(T)$ and its derivative dU/dT are shown in Figure 3.8.

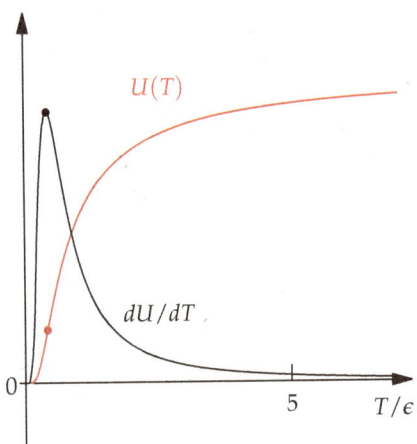

Figure 3.8: The function $U(T)$ (red) and its first derivative dU/dT (black). At the point of inflection (red dot) the first derivative is a maximum (black dot) and the second derivative is zero.

Differentiation of this function is complicated by the fact that the argument of the exponential varies as $1/T$ rather than T. To simplify our work, we introduce a variable $\beta = 1/T$. Applying the chain rule for differentiation of composite functions, we can write[8]

$$\frac{dU}{dT} = \frac{dU}{d\beta}\frac{d\beta}{dT} \qquad (3.14)$$

[8] These functions appear in Einstein's theory of the heat capacity of solids. The thermodynamic energy is $U(T)$ and the heat capacity is $C_V(T) = dU/dT$.

We have converted the daunting problem of evaluating dU/dT into the easier problem of finding $dU/d\beta$, and the even easier problem of finding $d\beta/dT$.

Let's evaluate Equation 3.14, starting with

$$\frac{dU}{d\beta} = \frac{d}{d\beta}\left(\frac{\varepsilon}{2} + \frac{\varepsilon}{e^{\beta\varepsilon} - 1}\right) = \frac{-\varepsilon}{(e^{\beta\varepsilon} - 1)^2}\frac{d}{d\beta}(e^{\beta\varepsilon} - 1)$$

Note that we have used the power rule: $f'(x) = nx^{n-1}$ for $f(x) = x^n$ with

$n = -1$. Further evaluation leads to

$$\frac{dU}{d\beta} = \frac{-\varepsilon}{(e^{\beta\varepsilon} - 1)^2} \frac{d}{d\beta}\left(e^{\beta\varepsilon} - 1\right) = \frac{-\varepsilon^2 e^{\beta\varepsilon}}{(e^{\beta\varepsilon} - 1)^2}$$

The second part of Equation 3.14 can be evaluated as

$$\frac{d\beta}{dT} = \frac{d}{dT}\left(\frac{1}{T}\right) = -\frac{1}{T^2} = -\beta^2$$

Combining these results we find

$$\frac{dU}{dT} = \frac{dU}{d\beta}\frac{d\beta}{dT} = (\varepsilon\beta)^2 \frac{e^{\beta\varepsilon}}{(e^{\beta\varepsilon} - 1)^2}$$

As discussed in Chapter 1, in the physical sciences we commonly encounter exponential functions of the form $e^{E/T}$, where E is an energy and T is temperature. When we explore the temperature dependence of those functions, using differentiation with respect to T, the approach used above can be very useful.

11. a^x. The derivative of a constant a raised to the power of x is

$$\frac{d}{dx}a^x = a^x \ln a \tag{3.15}$$

The earlier result for the derivative of e^x is a special case of this formula[9] where $a = e$ and $\ln e = 1$. Consider a function θ^t where θ is a constant. The first derivative is

$$\frac{d}{dt}\theta^t = \theta^t \ln\theta$$

The second derivative is

$$\frac{d^2}{dt^2}\theta^t = \frac{d}{dt}\left(\frac{d}{dt}\theta^t\right) = \ln\theta\,\frac{d}{dt}\theta^t$$

Since $\ln\theta$ is a constant we can pull it in front of the derivative and

$$\frac{d}{dt}\theta^t \ln\theta = \ln\theta\left(\frac{d}{dt}\theta^t\right) = \ln\theta\left(\theta^t \ln\theta\right) = \theta^t \left(\ln\theta\right)^2$$

This process can be generalized to the nth order derivative to find

$$\frac{d^n}{dt^n}\theta^t = \theta^t \left(\ln\theta\right)^n$$

12. $f(x)/g(x)$. The derivative of a *compound function* $f(x)/g(x)$ involving a quotient can be evaluated using the *product rule* and the *chain rule*:

$$\frac{d}{dx}\left[\frac{f(x)}{g(x)}\right] = \frac{1}{g(x)}\frac{df}{dx} + f(x)\frac{d}{dx}\left[\frac{1}{g(x)}\right]$$

$$= \frac{1}{g^2(x)}g(x)\frac{df}{dx} + f(x)\left[-\frac{1}{g^2(x)}\frac{dg}{dx}\right]$$

$$= \frac{1}{g^2(x)}\left[g(x)\frac{df}{dx} - f(x)\frac{dg}{dx}\right] \tag{3.16}$$

[9] The form of the result $\frac{d}{dx}a^x = a^x \ln a$ provides a hint of its proof. Since

$$\ln a^x = x \ln a$$

we find

$$\frac{d}{dx}\ln a^x = \frac{d}{dx}x \ln a = \ln a$$

Applying the chain rule results in

$$\frac{d}{dx}\ln a^x = \frac{da^x}{dx}\frac{d}{da^x}\ln a^x = \frac{da^x}{dx}\frac{1}{a^x}$$

Combining these results leads to $\frac{d}{dx}a^x = a^x \ln a$.

There is no need for a special formula or *quotient rule*. We can simply remember the product rule and power rule and apply them.

Consider the function

$$p(\beta) = \frac{e^{-\beta\varepsilon}}{1+e^{-\beta\varepsilon}}$$

shown in Figure 3.9, which we recognize to be $f(x)/g(x)$ where $f(x) = e^{-\beta\varepsilon}$ and $g(x) = 1 + e^{-\beta\varepsilon}$. Applying our rules for differentiating compound functions we have

$$\frac{dp}{d\beta} = -\varepsilon e^{-\beta\varepsilon}\frac{1}{1+e^{-\beta\varepsilon}} - e^{-\beta\varepsilon}\frac{-\varepsilon e^{-\beta\varepsilon}}{(1+e^{-\beta\varepsilon})^2}$$

$$= \frac{1}{(1+e^{-\beta\varepsilon})^2}\left[-\varepsilon e^{-\beta\varepsilon}\left(1-\varepsilon e^{-\beta\varepsilon}\right)+\varepsilon e^{-2\beta\varepsilon}\right]$$

$$= \frac{-\varepsilon e^{-\beta\varepsilon}}{(1+e^{-\beta\varepsilon})^2}$$

The function and its derivative are shown in Figure 3.9.

Further consider the compound function $\tan(x) = \sin(x)/\cos(x)$. To execute the derivative of $\tan(x)$ with respect to x, we first apply the product rule

$$\frac{d}{dx}\tan(x) = \frac{d}{dx}\left[\frac{\sin(x)}{\cos(x)}\right]$$

$$= \frac{1}{\cos(x)}\frac{d}{dx}\sin(x) + \sin(x)\frac{d}{dx}\frac{1}{\cos(x)}$$

The derivative of $\cos^{-1}(x)$ can be evaluated using the chain rule as

$$\frac{d}{dx}\frac{1}{\cos(x)} = \frac{d}{d\cos(x)}\frac{1}{\cos(x)} \times \frac{d}{dx}\cos(x)$$

$$= -\frac{1}{\cos^2(x)} \times (-\sin(x))$$

Combining these results we find

$$\frac{d}{dx}\tan(x) = \frac{\cos(x)}{\cos(x)} + \sin(x)\left[\frac{\sin(x)}{\cos^2(x)}\right]$$

$$= 1 + \frac{\sin^2(x)}{\cos^2(x)}$$

$$= 1 + \tan^2(x) = \sec^2(x)$$

The result is shown graphically in Figure 3.10.

The rules above provide the foundation of knowledge needed to carry out differential calculus on functions of one variable. In the next section, we will find that these rules provide the tools needed to compute derivatives of functions of many variables common to the physical sciences. These rules should be practiced on simple functions and complicated functions until mastery of each rule is achieved.

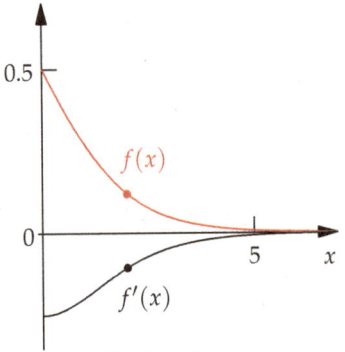

Figure 3.9: The function $f(x) = \exp(x)/(\exp(x)+1)$ (red) and its first derivative $f'(x) = -\exp(x)/(\exp(x)+1)^2$ (black). As the function decreases monotonically (red dot) the first derivative is negative (black dot).

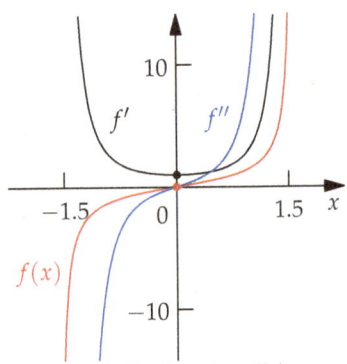

Figure 3.10: The function $f(x) = \tan(x)$ (red) and its first and second derivatives. At the point of inflection (red dot), f' (black) is a minimum (black dot), and f'' (blue) changes sign passing through zero.

3.2 Partial derivatives of functions of many variables

FUNCTIONS OF MORE THAN ONE VARIABLE are common in modeling properties of physical systems. For an ideal gas, the pressure can be written as a function of three variables (the volume, temperature, and number of particles). If we want to understand how the pressure changes with volume, we hold the temperature and number of particles fixed and vary only the volume. That is the concept of a partial derivative. For a function of many variables, we hold all but one variable fixed and measure the change in the function resulting from the change in that one variable. In this section, we explore the definition of the partial derivative and survey common methods for taking partial derivatives of functions of many variables.

3.2.1 The concept of the partial derivative

For a function of one variable, $f(x)$, we defined the first derivative at a point x in terms of the limit:

$$f'(x) = \frac{df}{dx} = \lim_{\Delta x \to 0} \frac{f(x + \Delta x) - f(x)}{\Delta x}$$

We then interpreted $f'(x)$ as the slope of the function at the point x. For a function of two variables, $f(x, y)$, we have slopes in the x- and y-directions that must be defined (see Figure 3.11).

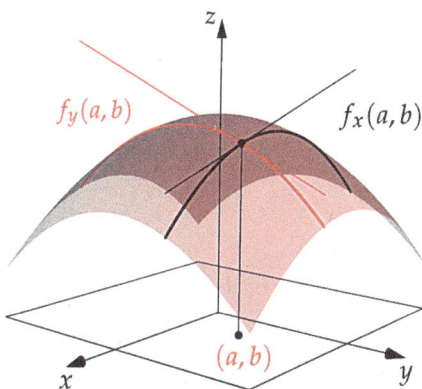

Figure 3.11: The inverted parabolic function $f(x, y)$ shown as a surface over the xy-plane. The black point on the surface marks $f(a, b)$ and the curves (thick lines) are the functions of one variable $f(a, y)$ where $x = a$ (red line) and $f(x, b)$ where $y = b$ (black line). The slopes of the tangent lines at the point (a, b) are

$$f_x = \left(\frac{\partial f}{\partial x}\right)_y \qquad f_y = \left(\frac{\partial f}{\partial y}\right)_x$$

where we have used a convenient shorthand notation for first order partial derivatives.

To assess the rate of change in the function $f(x, y)$ due to a change in x, we hold y constant, as indicated by the y-subscript, and define the derivative of $f(x, y)$ with respect to x as

$$\left(\frac{\partial f}{\partial x}\right)_y = \lim_{\Delta x \to 0} \frac{f(x + \Delta x, y) - f(x, y)}{\Delta x} \qquad (3.17)$$

This is known as a *partial derivative*,[10] as it provides partial information on the rate of change in the function with respect to the change in one of many variables. Similarly, to know the rate of change in $f(x, y)$ due to a change in y, we hold x constant and define the derivative of the function $f(x, y)$ with respect to y as[11]

[10] The partial derivative is distinguished from the total derivative using a stylized letter d written ∂. This notation was introduced by French philosopher, mathematician, and political scientist *Nicolas de Condorcet* (1743-1794).

[11] The concept of the partial derivative was developed by French mathematician *Adrien-Marie Legendre* (1752-1833).

$$\left(\frac{\partial f}{\partial y}\right)_x = \lim_{\Delta y \to 0} \frac{f(x, y + \Delta y) - f(x,y)}{\Delta y} \tag{3.18}$$

Figure 3.11 depicts a function of two variables $f(x,y)$ and its first order partial derivatives $f_x(a,b)$ and $f_y(a,b)$. The heavy lines drawn over the surface represent the one-dimensional functions $f(x,b)$ and $f(a,y)$. The thin straight lines are *tangent lines* representing the partial derivatives $f_x(a,b)$ and $f_y(a,b)$ taken at the point (a,b). The one-dimensional functions $f(x,b)$ and $f(a,y)$ and associated tangent lines defined by $f_x(a,b)$ and $f_y(a,b)$ at the point (a,b) are compared in Figure 3.12.

We can extend this concept of the first order partial derivative to create higher-order partial derivatives of a multivariate function. For a function of $n = 2$ variables $f(x,y)$ there will be 2×1 first order partial derivatives, f_x and f_y, and 2×2 second order partial derivatives:

$$f_{xx} = \frac{\partial^2 f}{\partial x^2} \quad f_{yy} = \frac{\partial^2 f}{\partial y^2} \quad f_{xy} = \frac{\partial^2 f}{\partial x \partial y} = \frac{\partial}{\partial x} f_y \quad f_{yx} = \frac{\partial^2 f}{\partial y \partial x} = \frac{\partial}{\partial y} f_x$$

In general, for mth-order differentiation of a function of n variables, there will be $n \times m$ partial derivatives.

3.2.2 *Techniques for partial differentiation*

In the expression

$$\left(\frac{\partial f}{\partial x}\right)_y \tag{3.19}$$

the subscript is used to explicitly note that when the partial derivative of $f(x,y)$ is taken with respect to x, the variable y is held constant. For example, suppose $f(x,y) = 3xy^2 + x + 2y^2$. In computing

$$\left(\frac{\partial f}{\partial x}\right)_y = \frac{\partial}{\partial x}\left(3xy^2 + x + 2y^2\right) = \frac{\partial}{\partial x}3xy^2 + \frac{\partial}{\partial x}x + \frac{\partial}{\partial x}2y^2$$

we treat the factors $3y^2$ and $2y^2$ as constants that can be pulled in front of the derivative. This leads to

$$\left(\frac{\partial f}{\partial x}\right)_y = 3y^2\frac{\partial}{\partial x}x + \frac{\partial}{\partial x}x + 2y^2\frac{\partial}{\partial x}1 = 3y^2 + 1$$

Similarly, for the partial derivative with respect to y

$$\left(\frac{\partial f}{\partial y}\right)_x = \frac{\partial}{\partial y}\left(3xy^2 + x + 2y^2\right) = \frac{\partial}{\partial y}3xy^2 + \frac{\partial}{\partial y}x + \frac{\partial}{\partial y}2y^2$$

$$= 3x\frac{\partial}{\partial y}y^2 + x\frac{\partial}{\partial y}1 + 2\frac{\partial}{\partial y}y^2 = 6xy + 4y$$

We record our final results as

$$\left(\frac{\partial f}{\partial x}\right)_y = \frac{\partial f}{\partial x} = 3y^2 + 1 \qquad \left(\frac{\partial f}{\partial y}\right)_x = \frac{\partial f}{\partial y} = 6xy + 4y$$

The presence of the partial derivative notation already indicates that the derivative is taken with respect to only one variable, implying that all

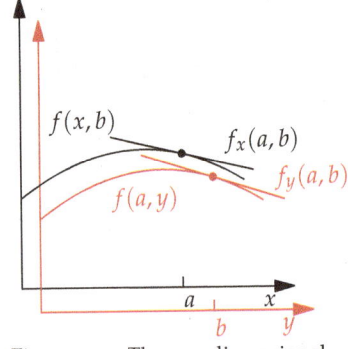

Figure 3.12: The one-dimensional functions $f(x,b)$ (black), a function of x for fixed $y = b$, and $f(a,y)$, a function of y for fixed $x = a$, for the two-dimensional function $f(x,y)$ in Figure 3.11. The slope lines represent the partial derivatives f_x (black) and f_y (red) taken at the point (a,b).

other variables are held constant. As such, we will typically omit the explicit subscript. In practice, we will reserve the subscript notation for cases in which variables are held constant to specific values or in cases where we want to emphasize the fact that specific variables are held constant. For example, for the function above we write

$$\left(\frac{\partial f}{\partial x}\right)_{y=1} = 4 \qquad \left(\frac{\partial f}{\partial y}\right)_{x=2} = 16y \qquad \left(\frac{\partial f}{\partial y}\right)_{x=2, y=1} = 16$$

where the subscripts are used to assign specific values to one or both of the variables x and y.

Now consider the function of two variables $p(V, T)$ where

$$p(V, T) = \frac{RT}{V}$$

for which

$$\frac{\partial p}{\partial V} = -\frac{RT}{V^2} \qquad \frac{\partial p}{\partial T} = \frac{R}{V}$$

The function $p(V, T)$ and the first derivatives $p_V(V, T)$ and $p_T(V, T)$ are shown graphically in Figure 3.13 for positive values of V and T. Note the positive upward slope of $p_T(V, T)$ and negative downward slope of $p_V(V, T)$ that increase in magnitude as V decreases toward zero.

Note that the partial derivatives $p_V(V, T)$ and $p_T(V, T)$ are functions of two variables that form surfaces over the xy-plane, as is the function $p(V, T)$. Evaluating the partial derivative $p_T(V, T)$ at a particular value of T results in a function of one variable V (gray line), while evaluating $p_V(V, T)$ at a particular value of V results in a function of one variable T (black line). Evaluating the partial derivatives at a specific point on the VT-plane results in numbers (gray and black points).

The second partial derivatives of $p(V, T)$ are evaluated as

$$\frac{\partial^2 p}{\partial V^2} = \frac{2RT}{V^3} \qquad \frac{\partial^2 p}{\partial T^2} = 0$$

and

$$\frac{\partial^2 p}{\partial T \partial V} = \frac{\partial^2 p}{\partial V \partial T} = -\frac{R}{V^2}$$

If a function $f(x, y)$ has *continuous* mixed partial derivatives f_{xy} and f_{yx} then

$$f_{xy} = f_{yx}$$

This is referred to as *Clairaut's theorem*.

This equality of mixed partial derivatives is true for most functions encountered in the physical sciences, including $p(V, T)$, which has continuous second partial derivatives so that

$$p_{TV} = p_{VT}$$

This concept can be extended to mixed partial derivatives of any order. For example, if f_{xxy}, f_{xyx}, and f_{yxx} are continuous then $f_{xxy} = f_{xyx} = f_{yxx}$.

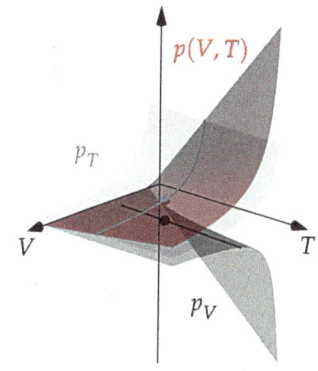

Figure 3.13: The function $p(V, T) = \frac{RT}{V}$ (red) shown with the first derivatives $p_V(V, T) = -\frac{RT}{V^2}$ (black) and $p_T(V, T) = \frac{R}{V}$ (gray).

3.3 Infinitesimal change and the total differential

THE DERIVATIVE PROVIDES a measure of the rate of change in a function resulting from a small change in a variable. In one dimension, if we multiply the first derivative of a function, df/dx, by a small change in the variable, dx, the result is an estimate of the change in the function, df. The extension of this idea to functions of many variables leads to the concept of the total differential. The concept of the total differential is essential in the physical sciences and plays a central role in thermodynamics.

3.3.1 The concept of the total differential

The first derivative of a function $f(x)$ with respect to x written

$$\frac{df}{dx}$$

can be thought of as a fraction, formed by a change in the function, df, divided by a change in the variable, dx. Now consider the equality

$$\frac{df}{dx} = \frac{df}{dx}$$

Multiplying each side by dx and canceling terms leads to

$$\left(\frac{df}{dx}\right) dx = \left(\frac{df}{dx}\right) dx = df \qquad (3.20)$$

and the identity

$$df \equiv \left(\frac{df}{dx}\right) dx$$

The term df is known as the *differential* of f. The first derivative is the proportionality constant relating the infinitesimal change in the function, df, resulting from an infinitesimal change in the variable, dx.

To extend this concept to a multivariate function, $f(x,y)$, we write

$$df = \left(\frac{\partial f}{\partial x}\right) dx + \left(\frac{\partial f}{\partial y}\right) dy$$

Here df is known as the *total differential* of f.[12] The total change in the function df is the sum of the change in f resulting from a change in x, proportional to f_x, and the change in f resulting from a change in y, proportional to f_y.

In an arbitrary number of dimensions, N, the total differential of a function $f(x_1, x_2, \ldots x_N)$ can be written

$$df = \sum_{k=1}^{N} \left(\frac{\partial f}{\partial x_k}\right) dx_k \qquad (3.21)$$

The total differential defines the infinitesimal change in the function, df, resulting from the infinitesimal changes in the variables $dx_1, dx_2, \ldots$ and dx_N.

As an example, let's determine the total differential, dV, for the volume of a cylinder $V(r,h) = \pi r^2 h$ as a function of changes in its radius r and height h.

[12] The total differential was developed by German mathematician and physicist *Rudolf Clausius* (1822-1888) in the context of thermodynamics.

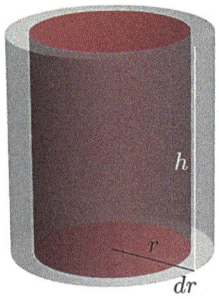

Figure 3.14: A cylinder of radius r and height h (red). The incremental volume $dV = 2\pi r h dr$, resulting from an increase in radius dr, is shown in gray.

Using Equation 3.21 we write the incremental change in volume, dV, due to an incremental change in the cylinder's radius, dr, and height, dh, as

$$dV = \left(\frac{\partial V}{\partial r}\right) dr + \left(\frac{\partial V}{\partial h}\right) dh = 2\pi rh \, dr + \pi r^2 \, dh$$

The concept of the total differential of the volume of a cylinder is explored graphically for a change in radius, dr, in Figure 3.14, and a change in radius, dh, in Figure 3.15. Extending the height by dh increases the volume by $\pi r^2 \, dh$. Expanding the radius by dr increases the volume by $2\pi rh \, dr$.

3.3.2 Extending the chain rule for multivariate composite functions

Let's consider a function of three variables $f(x, y, z)$ where the variables $x(u)$, $y(u)$, and $z(u)$ each depend on a parameter u. We say that f is a *composite function* of u. The function

$$f(x(u), y(u), z(u))$$

has indirect dependence on u through $x(u)$, $y(u)$ and $z(u)$, but no direct dependence on u.

The total differential of $f(x(u), y(u), z(u))$ can be written

$$df = \left(\frac{\partial f}{\partial x}\right) dx + \left(\frac{\partial f}{\partial y}\right) dy + \left(\frac{\partial f}{\partial z}\right) dz$$

Using our rule for differentiating composite functions we find

$$df = \left[\left(\frac{\partial f}{\partial x}\right)\frac{dx}{du} + \left(\frac{\partial f}{\partial y}\right)\frac{dy}{du} + \left(\frac{\partial f}{\partial z}\right)\frac{dz}{du}\right] du \qquad (3.22)$$

which is the total differential of the composite function.[13]

Let's determine the differential of the composite function $U(z(x)) = mgz$ in which m and g are constants and $z = x^2$ depends parametrically on the variable x. We write the differential

$$dU = \frac{dU}{dz}\frac{dz}{dx}dx = mg \times 2x \times dx = 2mgx \, dx$$

defining the change in U resulting from a small change in x.

Now consider the composite function

$$U(x(t), y(t)) = x^2(t) + y^2(t)$$

where the variables $x = \sin(t)$ and $y = \cos(t)$ depend parametrically on the variable t. We write the total differential

$$dU = \left(\frac{\partial U}{\partial x}\frac{dx}{dt} + \frac{\partial U}{\partial y}\frac{dy}{dt}\right) dt = \left(2x\frac{dx}{dt} + 2y\frac{dy}{dt}\right) dt$$
$$= [2x\cos(t) + 2y(-\sin(t))] \, dt$$

where we must keep in mind that $x(t)$ and $y(t)$ are functions of time. This result can be written in a way that explicitly expresses that time dependence:

$$dU = [2\sin(t)\cos(t) - 2\cos(t)\sin(t)] \, dt = 0$$

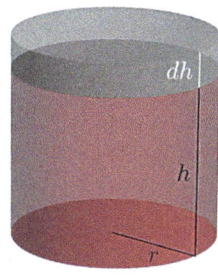

Figure 3.15: A cylinder of radius r and height h (red). The incremental volume $dV = \pi r^2 dh$, resulting from an increase in height dh, is shown in gray.

[13] As the variables $x(u)$, $y(u)$, and $z(u)$ depend on the single variable u, we write their variation with respect to u as total derivatives $\frac{dx}{du}$, $\frac{dy}{du}$, and $\frac{dz}{du}$.

We could anticipate this result from the beginning by noting that

$$U = \sin^2(t) + \cos^2(t) = 1$$

This tells us that U is a constant, so that $dU = 0$. This result is shown graphically in Figure 3.16.

3.3.3 The total derivative and the partial derivative

Consider a composite function $f(x(t), y(t), t)$ that has *direct dependence* on the parameter t as well as *indirect dependence* on t through the variables $x(t)$ and $y(t)$. The total differential df can be written

$$df = \left(\frac{\partial f}{\partial t}\right) dt + \left(\frac{\partial f}{\partial x}\right) dx + \left(\frac{\partial f}{\partial y}\right) dy$$

If we divide each term by dt we find an expression for the *total derivative* of the function with respect to t:

$$\frac{df}{dt} = \left(\frac{\partial f}{\partial t}\right)\frac{dt}{dt} + \left(\frac{\partial f}{\partial x}\right)\frac{dx}{dt} + \left(\frac{\partial f}{\partial y}\right)\frac{dy}{dt}$$
$$= \frac{\partial f}{\partial t} + \left(\frac{\partial f}{\partial x}\right)\frac{dx}{dt} + \left(\frac{\partial f}{\partial y}\right)\frac{dy}{dt} \qquad (3.23)$$

This total derivative of the function $f(x, y, t)$ provides a measure of the rate of change in the function resulting from a change in t, including direct dependence on t and indirect dependence on t through $x(t)$ and $y(t)$.[14]

For example, consider the exponentially damped sinusoidal function

$$f(x(t), y(t), t) = e^{-t}(x + y) \qquad (3.24)$$

where $x(t) = \sin(at)$ and $y(t) = \cos(bt)$ and a and b are constants. The total derivative of this composite function is

$$\frac{df}{dt} = \frac{\partial f}{\partial t} + \left(\frac{\partial f}{\partial x}\right)\frac{dx}{dt} + \left(\frac{\partial f}{\partial y}\right)\frac{dy}{dt}$$
$$= -e^{-t}(x + y) + e^{-t}\frac{dx}{dt} + e^{-t}\frac{dy}{dt}$$
$$= -e^{-t}[\sin(at) + \cos(bt)] + e^{-t}[a\cos(at) - b\sin(bt)]$$

The resulting function and its total derivative are shown in Figure 3.17 for the special case of $a = 1$ and $b = 2$.

3.3.4 Identifying exact differentials using Euler's test

We established that the total differential of a function $f(x, y)$ can be written

$$df(x, y) = \left(\frac{\partial f}{\partial x}\right) dx + \left(\frac{\partial f}{\partial y}\right) dy$$

Suppose someone proposes a differential

$$s(x, y)\, dx + t(x, y)\, dy$$

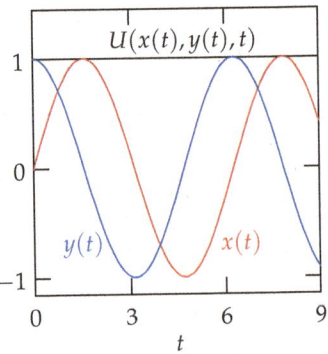

Figure 3.16: Compound composite function $U(x(t), y(t), t) = x^2(t) + y^2(t)$ (black) with $x(t)$ (red) and $y(t)$ (blue).

[14] The total derivative appears in fundamental equations of classical mechanics and quantum mechanics. It is surprisingly common to see the total derivative misrepresented as a partial derivative and *vice versa*. Don't let this happen to you!

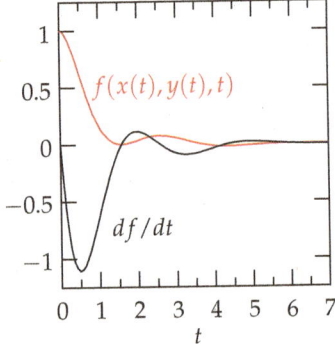

Figure 3.17: Compound composite function $f(x(t), y(t), t)$ (red) defined by Equation 3.24 and its total derivative $\frac{df}{dt}$ (black).

and claims that there exists a function $f(x,y)$ for which the differential above is the *total differential*. That is, they claim that

$$df(x,y) = f_x \, dx + f_y \, dy = s(x,y) \, dx + t(x,y) \, dy$$

How can we know if their claim is valid? Sometimes the answer is obvious. And sometimes it's not.

We can answer this question using a simple procedure known as *Euler's test*. If the equation

$$\frac{\partial}{\partial y} s(x,y) = \frac{\partial}{\partial x} t(x,y) \qquad (3.25)$$

is satisfied, we say that $s(x,y) \, dx + t(x,y) \, dy$ is an *exact differential*. In that case, there exists a function $f(x,y)$ for which

$$df = s(x,y) \, dx + t(x,y) \, dy$$

Otherwise, we say $s(x,y) \, dx + t(x,y) \, dy$ is an *inexact differential*. In that case, there exists no function with that corresponding total differential.

Let's consider a few examples starting with the differential

$$2x \, dx + 2y \, dy$$

where $s(x,y) = 2x$ and $t(x,y) = 2y$. Applying Euler's test, we find

$$\frac{\partial}{\partial y} s(x,y) = \frac{\partial}{\partial y} 2x = 0 \qquad \frac{\partial}{\partial x} t(x,y) = \frac{\partial}{\partial x} 2y = 0$$

demonstrating that $2x \, dx + 2y \, dy$ is an exact differential.[15] In this case, the corresponding function is

$$f(x,y) = x^2 + y^2$$

To verify this result, we evaluate the total differential[16]

$$df(x,y) = \left(\frac{\partial f}{\partial x} \right) dx + \left(\frac{\partial f}{\partial y} \right) dy = 2x \, dx + 2y \, dy$$

Now consider the differential

$$x^2 y \, dx + y \, dy \qquad (3.26)$$

Is this an exact differential? Applying Euler's test where $s(x,y) = x^2 y$ and $t(x,y) = y$ we find

$$\frac{\partial}{\partial y} s(x,y) = \frac{\partial}{\partial y} x^2 y = x^2$$

$$\frac{\partial}{\partial x} t(x,y) = \frac{\partial}{\partial x} y = 0$$

which fails the test. Equation 3.26 is an inexact differential. There exists no function $f(x,y)$ for which the total differential corresponds to Equation 3.26.

Exact differentials play an important role in thermodynamics. We expect the variation of state functions, such as the energy and entropy, resulting from changes in state variables, such as pressure and temperature, to be described by exact differentials.

[15] Any total differential of the form $a(x)dx + b(y)dy$ satisfies Euler's test since

$$\frac{da(x)}{dy} = \frac{db(y)}{dx} = 0$$

As such, there will necessarily be a function $f(x,y)$ with total differential $df(x,y) = a(x)dx + b(y)dy$.

[16] Any function of the form $f(x,y) = g(x)h(y)$ will have a total differential of the form $df(x,y) = g_x(x)h(y)dx + g(x)h_y(y)dy$ that satisfies Euler's test. If you recognize the form $g_x(x)h(y)dx + g(x)h_y(y)dy$ in a differential you can conclude that $f(x,y) = g(x)h(y)$.

3.3.5 Application of Euler's test to thermodynamics

For a function $f(x,y)$, Euler's test is equivalent to the condition that

$$\frac{\partial^2 f}{\partial x \partial y} = \frac{\partial^2 f}{\partial y \partial x} \tag{3.27}$$

This is a property of all functions $f(x,y)$ that have continuous mixed second partial derivatives. All thermodynamic equations of state or *state functions* have this property. The change in a state function depends only on the value of the function at the initial and final points, not on the path taken between the two.

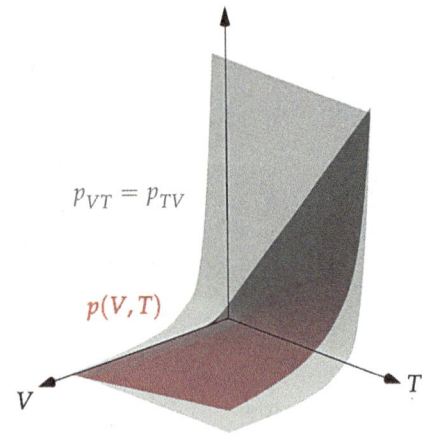

Figure 3.18: The function $p(V,T) = \frac{RT}{V}$ (red) plotted with the second derivative $p_{VT} = p_{TV} = -\frac{R}{V^2}$ (gray).

To explore this idea, let's return to the thermodynamic state function

$$p(V,T) = \frac{RT}{V}$$

with total differential

$$dp = \frac{\partial p}{\partial V}\, dV + \frac{\partial p}{\partial T}\, dT$$
$$= -\frac{RT}{V^2}\, dV + \frac{R}{V}\, dT$$

We find that

$$\frac{\partial^2 p}{\partial V \partial T} = \frac{\partial^2 p}{\partial T \partial V} = -\frac{R}{V^2}$$

which satisfies Euler's test as we expect for a thermodynamic state function. The result is shown graphically in Figure 3.18.

In thermodynamics, we encounter the state functions energy U and entropy S. Changes in these functions are represented by *exact differentials*. We also encounter functions such as work and heat that are not state functions. Changes in these functions are represented by *inexact differentials*. Knowledge of the properties of total differentials, as well as exact and inexact differentials, is essential to the study of thermodynamic properties of physical systems.

A₃ Euler's theorem for homogeneous functions

We say that the function $f(x)$ is an nth order *homogeneous function* if

$$f(\lambda x) = \lambda^n f(x)$$

where λ is a constant. Here n is known as the degree of homogeneity. For such a function, we find

$$nf(x) = x\left(\frac{df}{dx}\right)$$

This is known as *Euler's theorem* for homogeneous functions.

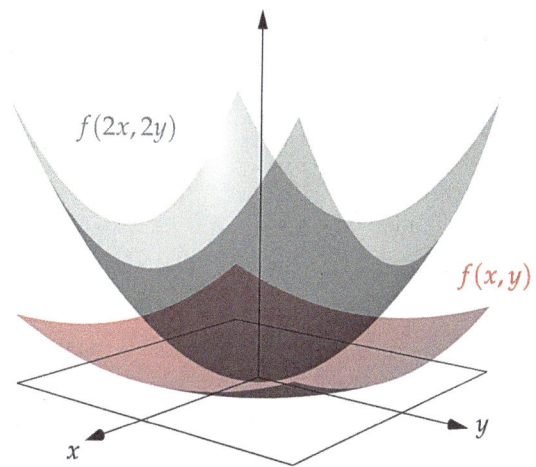

$f(2x, 2y)$

$f(x, y)$

Figure 3.19: A second order homogeneous function $f(x, y) = x^2 + y^2$ (red) shown as a surface over the xy-plane along with $f(\lambda x, \lambda y)$ (gray) where $\lambda = 2$.

As an example, consider the function $f(x) = x^n$ where $f(\lambda x) = (\lambda x)^n = \lambda^n x^n = \lambda^n f(x)$. For this nth order homogeneous function, we find

$$x\left(\frac{df}{dx}\right) = x\left(\frac{d}{dx}x^n\right) = x(nx^{n-1}) = nx^n = nf(x)$$

More generally, consider that if $f(\lambda x) = \lambda^n f(x)$ then

$$\frac{df}{d\lambda} = n\lambda^{n-1}f(x)$$

and

$$\frac{df}{d\lambda} = \frac{df}{d(\lambda x)}\frac{d(\lambda x)}{d\lambda} = x\frac{df}{d(\lambda x)}$$

Setting $\lambda = 1$ we find

$$nf(x) = x\frac{df}{dx}$$

We can extend this result to functions of more than one variable. If $f(x, y)$ is an nth order homogeneous function of x and y for which

$$f(\lambda x, \lambda y) = \lambda^n f(x, y)$$

then

$$nf(x, y) = x \left(\frac{\partial f}{\partial x} \right) + y \left(\frac{\partial f}{\partial y} \right) \tag{3.28}$$

Consider the following examples. For the function $f(x, y) = x^2 + y^2$ we find

$$f(\lambda x, \lambda y) = (\lambda x)^2 + (\lambda y)^2 = \lambda^2 (x^2 + y^2) = \lambda^2 f(x, y)$$

The function $f(x, y)$, shown in Figure 3.19, is a homogeneous function of order $n = 2$. Now consider the function $g(x, y) = \frac{x}{x^2 + y^2}$ for which

$$g(\lambda x, \lambda y) = \frac{\lambda x}{(\lambda x)^2 + (\lambda y)^2} = \frac{\lambda}{\lambda^2} \frac{x}{x^2 + y^2} = \lambda^{-1} g(x, y)$$

The function $g(x, y)$ is a homogeneous function of order $n = -1$. Finally, the function $h(x, y) = x^2 + y^3$, where $h(\lambda x, \lambda y) \neq \lambda^n h(x, y)$ is not a homogeneous function.

First order homogeneous functions are common to thermodynamics. Consider the thermodynamic energy $U(S, V)$ that is known to be an *extensive property* of the system and is itself a function of the extensive variables entropy, S, and volume, V. If you double S and V, you also double $U(S, V)$. We can capture this property mathematically as

$$U(\lambda S, \lambda V) = \lambda U(S, V)$$

From Euler's theorem and Equation 3.28 where $n = 1$, we find

$$U(S, V) = S \left(\frac{\partial U}{\partial S} \right) + V \left(\frac{\partial U}{\partial V} \right)$$

This equation relates the thermodynamic energy $U(S, V)$ to its various first partial derivatives.

Zeroth order homogeneous functions also appear in thermodynamics. Consider the temperature $T(S, V)$, which is an *intensive property* of the system and a function of entropy, S, and volume, V. We find

$$T(\lambda S, \lambda V) = T(S, V) = \lambda^0 T(S, V)$$

This means that $T(S, V)$ is a zeroth order homogeneous function of S and V. In thermodynamics, equations expressing intensive variables, such as temperature or pressure, in terms of extensive variables, such as entropy and volume, are known as *equations of state*.

B₃ Geometric interpretation of the total differential

For a function of one variable, $f(x)$, following Equation 3.20 we define the differential

$$df = df \frac{dx}{dx} = \frac{df}{dx} dx$$

This identity provides a geometric interpretation of the differential, df, in terms of the first derivative of f and the incremental change in the variable, dx, defining the *tangent line* shown in Figure 3.2. Let's develop a similar geometric interpretation of the total differential of a function of two variables.

Consider the parabolic function $f(x,y)$ (red surface) shown in Figure 3.20.

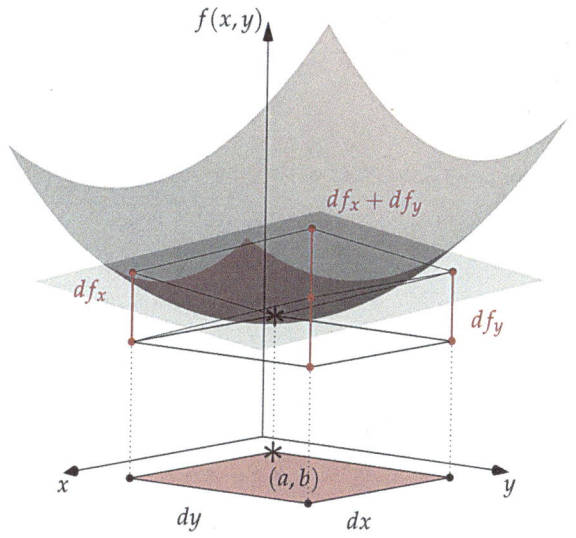

Figure 3.20: The parabolic function $f(x,y)$ shown as a red surface over the xy-plane. Asterisks mark the point $f(a,b)$ on the surface directly above the point (a,b) on the plane. The tangent plane (gray surface) is defined by the first order partial derivatives of the function at the point (a,b).

Following the work of Tall,[17] we write the total differential of this function as

$$df = df_x + df_y$$

Moving a distance dx in the x-direction on the *tangent plane* (gray surface) leads to a displacement df_x. Moving a distance dy in the y-direction on the tangent plane leads to a displacement df_y. The sum $df = df_x + df_y$ is the total displacement defining the total differential (see Figure 3.20).

This equality can be reformed as

$$df = \frac{df_x}{dx}\,dx + \frac{df_y}{dy}\,dy$$

Combining this result with the definition of the total differential

$$df = \left(\frac{\partial f}{\partial x}\right)dx + \left(\frac{\partial f}{\partial y}\right)dy$$

leads to the identities

$$df_x = \frac{\partial f}{\partial x}\,dx \qquad df_y = \frac{\partial f}{\partial y}\,dy$$

This is a geometrically intuitive definition of the total differential, df. Geometric interpretations of finite difference formulations of derivatives can be useful in developing numerical methods and approximations.

[17] David O. Tall. Visualizing differentials in two and three dimensions. *Teaching Mathematics and its Applications*, 11:1–7, 1992

C_3 End-of-chapter problems

> Mathematics knows no races or geographic boundaries; for mathematics, the cultural world is one country.
>
> David Hilbert

Warm-ups

3.1 Differentiate the following functions, assuming that any parameter not noted as a variable of the function is constant.

(a) $y(x) = x^3 + 5x^2 - 2x + 3$

(b) $f(T) = \dfrac{-\Delta H}{RT} + K$

(c) $d(v) = \dfrac{M}{v}$

(d) $r(\theta) = 5\tan 2\theta$

(e) $p(T) = A\,e^{-B/RT}$

(f) $f(x) = x^2\tan(2x)$

(g) $y(x) = x^2\,e^{-x}\cos(x)$

(h) $p(x) = \exp[-(x^2 + a^2)^{\frac{1}{2}}]$

(i) $u(r) = \dfrac{A}{r^{12}} - \dfrac{B}{r^6}$

(j) $y(x) = x^5\sqrt{1 - e^{2x}}$

(k) $y(x) = x^{-2}(1 - e^x)$

(l) $j(x) = e^{-\sin(x)}$

(m) $\dfrac{1}{x}\sin(x)$

(n) $W(n) = N\ln N - n\ln n$

(o) $s(t) = e^{-3t}\,\ln t$

(p) $g(p) = \dfrac{A}{p} + p\ln p$

(q) $e(z) = \dfrac{E^2}{A}\left(z^2 - \tfrac{27}{8}z\right)$

(r) $\varphi(x) = 2A\cos\left(\tfrac{n\pi x}{L}\right)$

3.2 Determine the slope of each of the following functions for the value of the variable indicated.

(a) $y(x) = 3\ln x$ at $x = 3$

(b) $y(x) = x^3$ at $x = 3$

(c) $y(x) = x^3 - 5x^2 + 2x - 3$ at $x = -1$

(d) $y(x) = (x^2 - 6)^{1/2}$ at $x = 4$

(e) $r(\theta) = 5\cos\theta$ at $\theta = \pi$

(f) $r(\theta) = 3\sin\theta\cos\theta$ at $\theta = \pi/2$

3.3 Twice differentiate the following functions with respect to x.

(a) $x^2\cos(x)$

(b) $e^{-x}\sin(x)$

(c) $x^2\ln(x)$

(d) $y(x) = \ln(1 - e^x)$

(e) $y(x) = \sqrt{1 - x^2}$

(f) $(2 + x)e^{-x^2}$

3.4 Evaluate the following partial derivatives.

(a) $p(V, T) = nRT/V$; $\partial p/\partial V$

(b) $\left(p + \dfrac{n^2 a}{V^2}\right)(V - nb) = nRT$; $\partial p/\partial V$

(c) $\rho(p, T) = \dfrac{pM}{RT}$; $\partial \rho/\partial T$

(d) $H = a + bT + cT^2 + \dfrac{d}{T}$; $\partial H/\partial T$

(e) $r(x, y, z) = \sqrt{x^2 + y^2 + z^2}$; $\partial r/\partial y$

(f) $y(r, \theta, \varphi) = r\sin\theta\cos\varphi$; $\partial y/\partial\varphi$

(g) $q(E_1, E_2, T) = e^{-E_1/kT} + e^{-E_2/kT}$; $\partial q/\partial E_1$

(h) $q(E_1, E_2, T) = e^{-E_1/kT} + e^{-E_2/kT}$; $\partial q/\partial T$

3.5 Use Euler's test to demonstrate that the total differential of the volume of a cylinder of radius r and height h written

$$dV(r, h) = \pi r^2\,dh + 2\pi rh\,dr$$

is an exact differential.

Homework exercises

3.6 Determine the slope of each of the following functions for the value of the variable indicated.

(a) $x(t) = \frac{1}{2}at^2$ at $t = 30$ seconds with $a = 9.8$ m/s^2 is constant
(b) $C_P(T) = 25.9$ Jmol^{-1} K^{-1} $+ 33.0 \times 10^{-3}\, T$ Jmol^{-1} K^{-2} $- 30.4 \times 10^{-7}\, T^2$ Jmol^{-1} K^{-3} at $T = 298.$ K
(c) $\ln[p(T)/p_0] = -\Delta H/RT + B$ with $T = 298.$ K, $\Delta H = 30,820.$ J/mol, $R = 8.314$ J/mol·K, and $B = 2.83$
(d) $A(t) = A_0\, e^{-kt}$ at $t = 10.0$ minutes with $A_0 = 0.050$ M and $k = 0.021$ min^{-1}

3.7 The rate constant $k(T)$ for a chemical reaction is found to vary with temperature according to the Arrhenius equation

$$k(T) = A\, e^{-E^{\ddagger}/RT}$$

where A, $E^{\ddagger}$, and R are constants. Find an expression for the change in $k(T)$ with respect to T.

3.8 The density of an ideal gas is found to vary with respect to temperature according to the equation

$$\rho(T) = \frac{pM}{RT}$$

where p, M, and R are constants. Find an expression for the slope of ρ versus T.

3.9 Consider the van der Waals equation of state for a real gas

$$\left(p + \frac{n^2 a}{V^2}\right)(V - nb) = nRT$$

where a and b are constants. The figure below shows the compressibility factor $Z = pV/nRT$ as a function of V and $1/V$ for the real gas (red line) and ideal gas (black line).

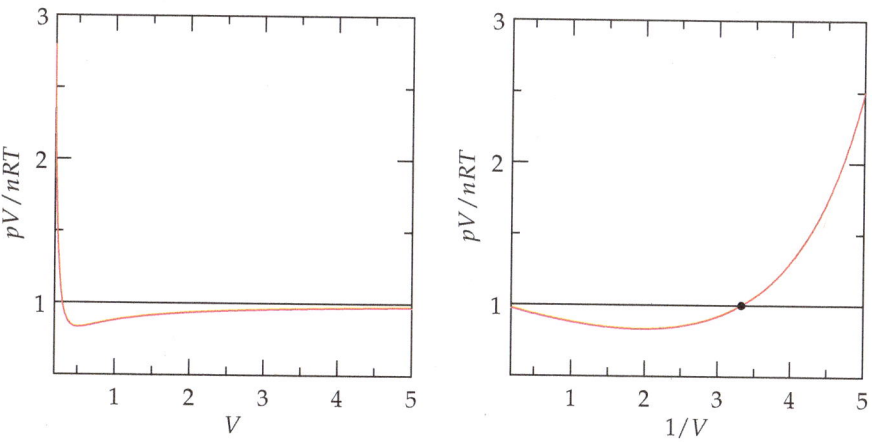

Find the partial derivative of Z with respect to V and with respect to $1/V$.

3.10 A certain gas obeys the equation of state

$$p(V - nb) = nRT$$

where n and R are constants. Determine the coefficient of thermal expansion

$$\alpha = \left(\frac{1}{V}\right)\left(\frac{\partial V}{\partial T}\right)_p$$

3.11 For the function $f(\theta) = \ln[\cos(\theta) + i\sin(\theta)]$ show that $\dfrac{df}{d\theta} = i$.

3.12* Determine the derivative of $h(x) = x^x$ with respect to x. HINT: Take the derivative of $d\ln[h(x)]/dx$.

3.13 Consider the van der Waals equation for one mole of gas

$$p = \frac{RT}{(V-b)} - \frac{a}{V^2}$$

Find $\left(\dfrac{\partial p}{\partial T}\right)_V$ and $\left(\dfrac{\partial p}{\partial V}\right)_T$.

3.14 Show that $V(x,y,z) = 1/\sqrt{x^2 + y^2 + z^2}$ satisfies

$$\frac{\partial^2 V}{\partial x^2} + \frac{\partial^2 V}{\partial y^2} + \frac{\partial^2 V}{\partial z^2} = 0$$

which is known as *Laplace's equation*.

3.15* Consider an ideal gas for which $pV = nRT$. Determine the change in pressure, p, for $n = 1.0$ mole of an ideal gas when the temperature, T, is changed from 273.15 K to 274.00 K, and the volume, V, is changed from 10.00 L to 9.90 L. The gas constant is $R = 8.314$ J/(mol K). Express your answer in atmospheres. HINT: Use an expression for the total differential $dp(T,V)$. Evaluate your partial derivatives at the point (V_1, T_1) with the increments in volume $dV = (V_2 - V_1)$ and $dT = (T_2 - T_1)$.

3.16 Suppose $u(x,y) = x^2 + y + xy^2$ with $x(t) = te^{-t}$ and $y(t) = e^{-t}$. Evaluate

$$\frac{du}{dt} = \frac{\partial u}{\partial x}\frac{dx}{dt} + \frac{\partial u}{\partial y}\frac{dy}{dt}$$

for the composite function $u(x(t), y(t))$.

3.17 Suppose $u(x,y) = ye^{-x} + xy$ with $x(s,t) = s^2 t$ and $y(s,t) = e^{-s} + t$. Find $\dfrac{\partial u}{\partial s}$ and $\dfrac{\partial u}{\partial t}$ where u is a function of the variables s and t. HINT: Note that

$$\frac{\partial u}{\partial s} = \frac{\partial u}{\partial x}\frac{\partial x}{\partial s} + \frac{\partial u}{\partial y}\frac{\partial y}{\partial s} \qquad \frac{\partial u}{\partial t} = \frac{\partial u}{\partial x}\frac{\partial x}{\partial t} + \frac{\partial u}{\partial y}\frac{\partial y}{\partial t}$$

3.18 Suppose $u(x,y) = e^{x+y}$ with $x(s,t) = te^s$ and $y(s,t) = \sin(s)$. Find $\dfrac{\partial u}{\partial s}$ and $\dfrac{\partial u}{\partial t}$ where u is a function of s and t.

3.19 Show that for an ideal gas

$$\left(\frac{\partial V}{\partial T}\right)_{n,p} = \left(\frac{\partial T}{\partial V}\right)_{n,p}^{-1}$$

3.20 Verify that the following relation, known as the *cyclic rule*, is true for an ideal gas.

$$\left(\frac{\partial T}{\partial V}\right)_p \left(\frac{\partial V}{\partial p}\right)_T \left(\frac{\partial p}{\partial T}\right)_V = -1$$

3.21 Use Euler's test to determine if the following differentials are exact differentials or inexact differentials

(a) $(x^3 + y)\,dx + (y^3 + x)\,dy$

(b) $e^{-s}\cosh(t)\,ds - e^{-s}\sinh(t)\,dt$

(c) $e^y\cos(x)\,dx - e^x\sin(y)\,dy$

(d) $\frac{1}{y}\,dx - \frac{x}{y^2}\,dy$

(e) $\frac{1}{V-b}\,dT + \frac{T}{(V-b)^2}\,dV$

(f) $-\frac{2T}{V-b}\,dT + \frac{T^2}{(V-b)^2}\,dV$

3.22 For the exact differential

$$df(x,y) = a(x,y)\,dx + b(x,y)\,dy$$

the following identity follows from Euler's test

$$\left(\frac{\partial a}{\partial y}\right)_x = \left(\frac{\partial b}{\partial x}\right)_y$$

Now consider the thermodynamic relation

$$dU(S,V) = T\,dS - p\,dV$$

For this exact differential, it follows from Euler's test that

$$\left(\frac{\partial T}{\partial V}\right)_S = -\left(\frac{\partial p}{\partial S}\right)_V$$

This identity is known as a *Maxwell relation*.

(a) Starting from $dH(S,p) = T\,dS + V\,dp$ complete the Maxwell relation

$$\left(\frac{\partial V}{\partial S}\right)_p =$$

(b) Starting from $dA(T,V) = -S\,dT - p\,dV$ complete the Maxwell relation

$$\left(\frac{\partial S}{\partial V}\right)_T =$$

(c) Starting from $dG(T,p) = -S\,dT + V\,dp$ complete the Maxwell relation

$$\left(\frac{\partial V}{\partial T}\right)_p =$$

3.23 For the exact differential

$$df(x,y,z) = a(x,y,z)\,dx + b(x,y,z)\,dy + c(x,y,z)\,dz$$

the following identites follow from Euler's test

$$\left(\frac{\partial a}{\partial y}\right)_{x,z} = \left(\frac{\partial b}{\partial x}\right)_{y,z} \qquad \left(\frac{\partial b}{\partial z}\right)_{x,y} = \left(\frac{\partial c}{\partial y}\right)_{x,z} \qquad \left(\frac{\partial c}{\partial x}\right)_{y,z} = \left(\frac{\partial a}{\partial z}\right)_{x,y}$$

Now consider the thermodynamic relation

$$dU(S,V,N) = T\,dS - p\,dV + \mu\,dN$$

For this exact differential, it follows from Euler's test that

$$\left(\frac{\partial T}{\partial V}\right)_{S,N} = -\left(\frac{\partial p}{\partial S}\right)_{V,N} \qquad \left(\frac{\partial p}{\partial N}\right)_{S,V} = -\left(\frac{\partial \mu}{\partial V}\right)_{S,N} \qquad \left(\frac{\partial \mu}{\partial S}\right)_{V,N} = \left(\frac{\partial T}{\partial N}\right)_{S,V}$$

These identities are known as *Maxwell relations*.

(a) Starting from $dH(S, p, N) = TdS + Vdp + \mu dN$ complete the Maxwell relation

$$\left(\frac{\partial V}{\partial S}\right)_{p,N} =$$

(b) Starting from $dA(T, V, N) = -SdT - pdV + \mu dN$ complete the Maxwell relation

$$\left(\frac{\partial S}{\partial N}\right)_{T,V} =$$

(c) Starting from $dG(T, p, N) = -SdT + Vdp + \mu dN$ complete the Maxwell relation

$$\left(\frac{\partial \mu}{\partial T}\right)_{p,N} =$$

3.24* Consider that the total differential

$$dU(S, V, N) = TdS - pdV + \mu dN$$

Further consider the function $G = \mu N = U - TS + pV$. Using your knowledge of exact differentials, derive an expression for dG and substitute the result for dU to prove that

$$-SdT + Vdp - Nd\mu = 0$$

This result is known as the *Gibbs-Duhem equation*.

4 SCALARS, VECTORS, AND VECTOR ALGEBRA

4.1 Fundamental properties of scalars and vectors

SOME PHYSICAL PROPERTIES ARE DESCRIBED BY A SCALAR NUMBER with the properties of magnitude and dimension (units). Examples include mass, charge, and energy. Other physical properties must be described by a vector with the properties of magnitude, dimension, and the additional property of direction. Examples include velocity, momentum, and force. In this chapter, we explore the basic properties of vectors and vector algebra.

4.1.1 Scalars and vectors in cartesian coordinates

Scalars are numbers that have magnitude and sign. For example, the mass of an electron can be described by the scalar number $m_e = 9.109 \times 10^{-31}$ kg, where the magnitude is 9.109×10^{-31} and the dimension is kg. Another example is the charge of an electron expressed by the scalar number $-e = -1.609 \times 10^{-19}$ C, where the *magnitude* of the charge is the absolute value $|-e| = 1.609 \times 10^{-19}$ C, which must be a positive real number.

Vectors are numbers that have magnitude but also direction. Consider the two-dimensional cartesian plane in Figure 4.1. The arrow with its tail at the origin $(0,0)$ and its head at the point (v_x, v_y) is a vector **v**. It has both magnitude and direction. The magnitude of the vector $|\mathbf{v}|$ is the Euclidean distance between the tail of the vector at the origin $(0,0)$ and the head of the vector at (v_x, v_y) given by the Pythagorean theorem where

$$|\mathbf{v}| = \sqrt{v_x^2 + v_y^2}$$

as $|\mathbf{v}|$ is the length of the hypotenuse of the right triangle with sides v_x, v_y, and $|\mathbf{v}|$. The orientation of **v** is defined in terms of its magnitude in the x-direction, v_x, and its magnitude in the y-direction, v_y. To express this mathematically, we define *unit vectors* in the x-direction, $\hat{\mathbf{x}}$, and y-direction, $\hat{\mathbf{y}}$ (see Figure 4.1). The unit vector $\hat{\mathbf{x}}$ has its tail at the origin and its head at $(1,0)$ and the unit vector $\hat{\mathbf{y}}$ has its tail at the origin and its head at $(0,1)$.[1] Vectors with unit magnitude are said to be *normalized*. Using the unit vectors, we write **v** in *vector notation* as[2]

$$\mathbf{v} = v_x\,\hat{\mathbf{x}} + v_y\,\hat{\mathbf{y}}$$

Note that we are literally adding the point $v_x\,\hat{\mathbf{x}} = (v_x, 0)$ to $v_y\,\hat{\mathbf{y}} = (0, v_y)$ to obtain the ordered pair (v_x, v_y).[3]

To represent a vector in a three-dimensional space, we define unit vectors $\hat{\mathbf{x}}$, $\hat{\mathbf{y}}$, and $\hat{\mathbf{z}}$ oriented along the positive x-axis, y-axis, and z-axis, respectively.

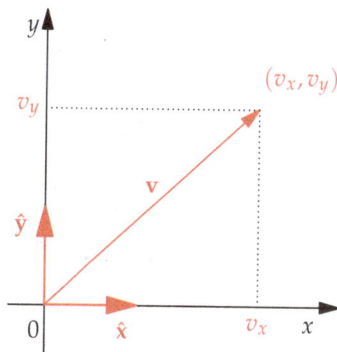

Figure 4.1: The vector $\mathbf{v} = v_x\,\hat{\mathbf{x}} + v_y\,\hat{\mathbf{y}}$ shown with its individual scalar components v_x and v_y in the two-dimensional cartesian coordinate system.

[1] For any vector **a** with magnitude $|\mathbf{a}|$ there is a unit vector

$$\hat{\mathbf{a}} = \frac{1}{|\mathbf{a}|}\mathbf{a}$$

oriented in the direction of **a** with unit magnitude.

[2] There is a variety of notations used for vectors, including **v** and $\vec{v}$. The latter notation is commonly used when working with vectors on pen and paper.

An arbitrary vector in three-dimensional space with its tail at the origin $(0,0,0)$ and its head at a point (w_x, w_y, w_z) can be defined in vector notation:

$$\mathbf{w} = w_x\,\hat{\mathbf{x}} + w_y\,\hat{\mathbf{y}} + w_z\,\hat{\mathbf{z}}$$

The magnitude of the vector is defined

$$|\mathbf{w}| = \sqrt{w_x^2 + w_y^2 + w_z^2}$$

An example of a vector $\mathbf{w}$ and its decomposition into its cartesian components is shown in Figure 4.2.

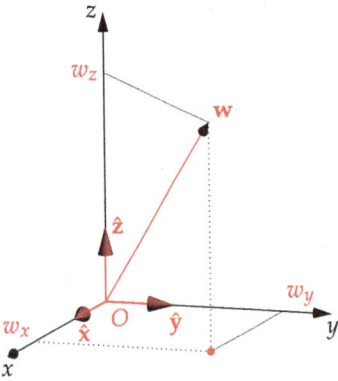

Another popular notation for unit vectors along the cartesian axes is $\hat{\mathbf{i}} = \hat{\mathbf{x}}$, $\hat{\mathbf{j}} = \hat{\mathbf{y}}$, and $\hat{\mathbf{k}} = \hat{\mathbf{z}}$. Using this notation, the vector $\mathbf{w}$ is written

$$\mathbf{w} = w_x\,\hat{\mathbf{i}} + w_y\,\hat{\mathbf{j}} + w_z\,\hat{\mathbf{k}}$$

Physical properties described using vectors include a particle's position in space, which can be represented as $\mathbf{r} = x\,\hat{\mathbf{i}} + y\,\hat{\mathbf{j}} + z\,\hat{\mathbf{k}}$. A particle's linear momentum can be represented as $\mathbf{p} = p_x\hat{\mathbf{i}} + p_y\hat{\mathbf{j}} + p_z\hat{\mathbf{k}}$. The scalar variables x, y, z and p_x, p_y, p_z define the components of position or momentum in the x-, y-, and z-directions, respectively.

Other physical properties described using vectors include the electric field, $\mathbf{E} = E_x\hat{\mathbf{i}} + E_y\hat{\mathbf{j}} + E_z\hat{\mathbf{k}}$, and magnetic field, $\mathbf{B} = B_x\hat{\mathbf{i}} + B_y\hat{\mathbf{j}} + B_z\hat{\mathbf{k}}$. Force is a physical property that has both magnitude and direction, and can be written in vector notation as $\mathbf{F} = F_x\hat{\mathbf{i}} + F_y\hat{\mathbf{j}} + F_z\hat{\mathbf{k}}$. For example, an electric field pointing in the z-direction with magnitude $|\mathbf{E}| = E_0$ can be written $\mathbf{E} = E_0\hat{\mathbf{k}}$, and a force pointing in the x-direction with magnitude $|\mathbf{F}| = F_0$ can be written $\mathbf{F} = F_0\hat{\mathbf{i}}$. In the next section, we explore the addition and subtraction of vectors.

4.1.2 Addition and subtraction of vectors

Consider two vectors $\mathbf{v}_1 = x_1\hat{\mathbf{i}} + y_1\hat{\mathbf{j}}$ and $\mathbf{v}_2 = x_2\hat{\mathbf{i}} + y_2\hat{\mathbf{j}}$ in two-dimensional cartesian space. We add the two vectors by separately adding the x-components and y-components, respectively, as

[3] This algebra should remind you of adding the complex numbers $z_1 = x$ and $z_2 = iy$, represented by the ordered pairs $z_1 = (x, 0)$ and $z_2 = (0, y)$. Just as $v_x\,\hat{\mathbf{x}} + v_y\,\hat{\mathbf{y}} = (v_x, v_y)$ we have $z_1 + z_2 = (x, y)$.

Figure 4.2: The vector $\mathbf{w}$ shown with its individual scalar components w_x, w_y, and w_z in the three-dimensional cartesian coordinate system. Also shown are the unit vectors $\hat{\mathbf{x}}$, $\hat{\mathbf{y}}$, and $\hat{\mathbf{z}}$.

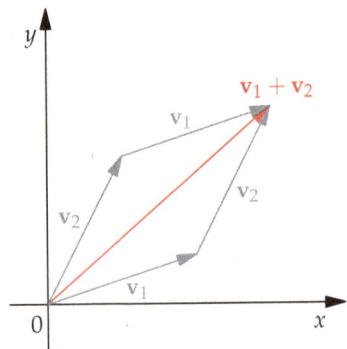

Figure 4.3: The addition of vectors $\mathbf{v}_1$ and $\mathbf{v}_2$ leading to vector $\mathbf{v}_1 + \mathbf{v}_2 = \mathbf{v}_2 + \mathbf{v}_1$ in the two-dimensional cartesian coordinate system.

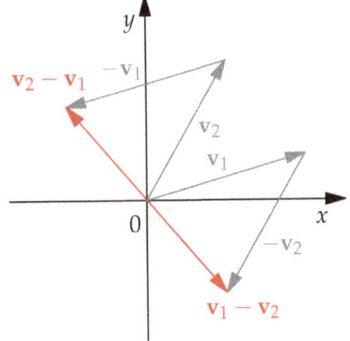

Figure 4.4: The subtraction of vectors $\mathbf{v}_1$ and $\mathbf{v}_2$ leading to vector $\mathbf{v}_1 - \mathbf{v}_2 = -(\mathbf{v}_2 - \mathbf{v}_1)$ in the two-dimensional cartesian coordinate system.

$$\mathbf{v}_1 + \mathbf{v}_2 = (x_1 + x_2)\,\hat{\mathbf{i}} + (y_1 + y_2)\,\hat{\mathbf{j}}$$

As shown in Figure 4.3, adding $\mathbf{v}_1$ and $\mathbf{v}_2$ is equivalent to placing the tail of $\mathbf{v}_2$ at the head of $\mathbf{v}_1$, or *vice versa*. The sum does not depend on the order of addition:

$$\mathbf{v}_1 + \mathbf{v}_2 = \mathbf{v}_2 + \mathbf{v}_1 = (x_1 + x_2)\,\hat{\mathbf{i}} + (y_1 + y_2)\,\hat{\mathbf{j}}$$

This is demonstrated graphically in Figure 4.3.

Considering the subtraction of the vectors $\mathbf{v}_1$ and $\mathbf{v}_2$, we find the results

$$\mathbf{v}_1 - \mathbf{v}_2 = (x_1 - x_2)\,\hat{\mathbf{i}} + (y_1 - y_2)\,\hat{\mathbf{j}} = -(\mathbf{v}_2 - \mathbf{v}_1)$$

This is represented in Figure 4.4. Unlike the vector sums, the difference of two vectors depends on the order of subtraction. Similar rules are followed for adding or subtracting vectors in any dimension. The addition of two vectors $\mathbf{w}_1 = x_1\hat{\mathbf{i}} + y_1\hat{\mathbf{j}} + z_1\hat{\mathbf{k}}$ and $\mathbf{w}_2 = x_2\hat{\mathbf{i}} + y_2\hat{\mathbf{j}} + z_2\hat{\mathbf{k}}$ in the three-dimensional cartesian space results in

$$\mathbf{w}_1 + \mathbf{w}_2 = \mathbf{w}_2 + \mathbf{w}_1 = (x_1 + x_2)\,\hat{\mathbf{i}} + (y_1 + y_2)\,\hat{\mathbf{j}} + (z_1 + z_2)\,\hat{\mathbf{k}}$$

The addition of the two vectors is represented graphically in Figure 4.5.

4.1.3 Scalar multiplication of vectors

When a vector is multiplied by a scalar number, each component of the vector is multiplied by the same scalar number. Consider the vector

$$\mathbf{b} = b_x\hat{\mathbf{x}} + b_y\hat{\mathbf{y}} + b_z\hat{\mathbf{z}}$$

If we multiply the vector by the scalar constant c, all of the components of the vector are scaled by the same constant as

$$c\mathbf{b} = cb_x\hat{\mathbf{x}} + cb_y\hat{\mathbf{y}} + cb_z\hat{\mathbf{z}}$$

The result of scalar multiplication of a vector is shown graphically in Figure 4.6.

Note that the magnitude of the vector $|\mathbf{b}| = (b_x^2 + b_y^2 + b_z^2)^{1/2}$ is scaled by the absolute value of the scalar constant as

$$|c\mathbf{b}| = \left[(cb_x)^2 + (cb_y)^2 + (cb_z)^2\right]^{1/2}$$
$$= |c|\left[(b_x)^2 + (b_y)^2 + (b_z)^2\right]^{1/2} = |c||\mathbf{b}|$$

When the scalar constant is negative, the orientation of the vector is reversed as shown in Figure 4.6.

4.1.4 Representing vectors in other coordinate systems

We found in Chapter 1 that certain functions are most naturally represented in a coordinate system other than cartesian coordinates. For example, the radially dependent coulomb potential energy is most conveniently represented in spherical polar coordinates (see Figure 1.21), while the helical curve is most conveniently represented in cylindrical coordinates (see Figure 1.24). How

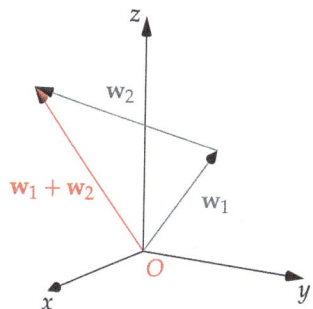

Figure 4.5: The addition of vectors $\mathbf{w}_1$ and $\mathbf{w}_2$ leading to vector $\mathbf{w}_1 + \mathbf{w}_2$ in the three-dimensional cartesian coordinate system.

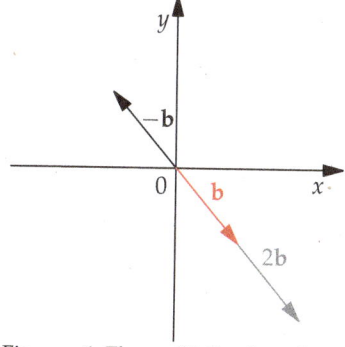

Figure 4.6: The multiplication of a vector $\mathbf{b}$ (red) by a scalar number c results in a scaled vector $c\mathbf{b}$, shown for scalar constants $c = 2$ (gray) and $c = -1$ (black).

can we best represent a vector in spherical polar coordinates or cylindrical coordinates?

We can represent any vector in three-dimensional space using the cartesian unit vectors $\hat{\imath}$, $\hat{\jmath}$, and $\hat{k}$. Suppose we have a vector beginning at the origin and ending at a point (r, θ, φ) expressed in spherical polar coordinates. We write

$$\mathbf{a} = a_x\,\hat{\imath} + a_y\,\hat{\jmath} + a_z\,\hat{k}$$

where

$$a_x = r\sin\theta\cos\varphi$$
$$a_y = r\sin\theta\sin\varphi$$
$$a_z = r\cos\theta$$

and conversely

$$r = \sqrt{a_x^2 + a_y^2 + a_z^2}$$
$$\theta = \arccos\left(\frac{a_z}{r}\right) = \cos^{-1}\left(\frac{a_z}{r}\right)$$
$$\varphi = \tan^{-1}\left(\frac{a_y}{a_x}\right)$$

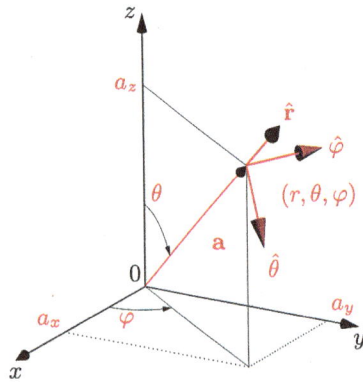

Figure 4.7: A vector $\mathbf{a}$ represented in the spherical polar coordinate system. The unit vectors $\hat{r}, \hat{\theta}$ and $\hat{\varphi}$ depend on the coordinates θ and φ.

For spherical polar coordinates, it is possible to define unit vectors centered at a point (r, θ, φ) defining displacements in the radial distance (dr), inclination $(d\theta)$, and azimuth $(d\varphi)$ between points (r, θ, φ) to $(r + dr, \theta + d\theta, \varphi + d\varphi)$ (see Figure 1.22). The three unit vectors are shown in Figure 4.7.

$$\hat{r} = \sin\theta\cos\varphi\,\hat{x} + \sin\theta\sin\varphi\,\hat{y} + \cos\theta\,\hat{z} \qquad (4.1)$$
$$\hat{\theta} = \cos\theta\cos\varphi\,\hat{x} + \cos\theta\sin\varphi\,\hat{y} - \sin\theta\,\hat{z} \qquad (4.2)$$
$$\hat{\varphi} = -\sin\varphi\,\hat{x} + \cos\varphi\,\hat{y} \qquad (4.3)$$

Unlike the cartesian unit vectors that are independent of the magnitude and orientation of the vector $\mathbf{a}$, the unit vectors $\hat{r}$, $\hat{\theta}$, and $\hat{\varphi}$ depend on the position (r, θ, φ). As the magnitude or orientation of the vector $\mathbf{a}$ changes, so will the orientation of the unit vectors. For this reason, when we describe a vector in three-dimensional space in spherical polar coordinates, we do so using the cartesian unit vectors as

$$\mathbf{a} = a_x\,\hat{\mathbf{i}} + a_y\,\hat{\mathbf{j}} + a_z\,\hat{\mathbf{k}} = r\sin\theta\cos\varphi\,\hat{\mathbf{i}} + r\sin\theta\sin\varphi\,\hat{\mathbf{j}} + r\cos\theta\,\hat{\mathbf{k}}$$

We draw the same conclusions for a vector $\mathbf{a}$ expressed in cylindrical coordinates (r,θ,z). Using the cartesian unit vectors $\hat{\mathbf{i}}$, $\hat{\mathbf{j}}$, and $\hat{\mathbf{k}}$, we write

$$\mathbf{a} = a_x\,\hat{\mathbf{i}} + a_y\,\hat{\mathbf{j}} + a_z\,\hat{\mathbf{k}}$$

where

$$a_x = r\cos\theta$$
$$a_y = r\sin\theta$$
$$a_z = z$$

or conversely

$$r = \sqrt{a_x^2 + a_y^2}$$
$$\theta = \arctan\left(\frac{a_y}{a_x}\right) = \tan^{-1}\left(\frac{a_y}{a_x}\right)$$
$$z = a_z$$

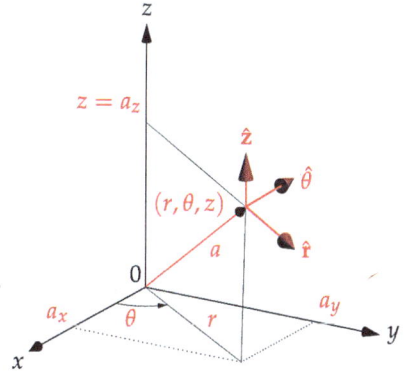

Figure 4.8: The cylindrical coordinate system in three dimensions. The unit vectors $\hat{\mathbf{r}}$ and $\hat{\boldsymbol{\theta}}$ depend on the coordinate θ.

We define the unit vectors describing the direction of displacements in the radial distance r, azimuth, θ, and elevation z as

$$\hat{\mathbf{r}} = \cos\theta\,\hat{\mathbf{x}} + \sin\theta\,\hat{\mathbf{y}}$$
$$\hat{\boldsymbol{\theta}} = -\sin\theta\,\hat{\mathbf{x}} + \cos\theta\,\hat{\mathbf{y}}$$
$$\hat{\mathbf{z}} = \hat{\mathbf{z}} \tag{4.4}$$

as shown in Figure 4.8. As in the case of the unit vectors $\hat{\mathbf{r}}$, $\hat{\boldsymbol{\theta}}$, and $\hat{\boldsymbol{\varphi}}$ in spherical polar coordinates, the unit vectors $\hat{\mathbf{r}}$ and $\hat{\boldsymbol{\theta}}$ depend on the coordinates (r,θ,z). However, the unit vector $\hat{\mathbf{z}}$ is independent of the position (r,θ,z).

4.2 Multiplication of vectors

THERE ARE TWO WAYS TO MULTIPLY TWO VECTORS, one resulting in a scalar number and the other resulting in a vector. Both methods of vector multiplication are widely used in the physical sciences. We convert the momentum (a vector) into kinetic energy (a scalar) using the scalar product. For a mass moving in a circular orbit, we convert its velocity (a vector) and radial position (a vector) into its angular momentum (a vector) using the vector product. The properties of the scalar and vector products are explored in this section.

4.2.1 Multiplying vectors and the scalar dot product

The multiplication of the two vectors $\mathbf{v} = v_x\,\hat{\mathbf{i}} + v_y\,\hat{\mathbf{j}} + v_z\,\hat{\mathbf{k}}$ and $\mathbf{w} = w_x\,\hat{\mathbf{i}} + w_y\,\hat{\mathbf{j}} + w_z\,\hat{\mathbf{k}}$ by the *scalar product* is defined as

$$\mathbf{v} \cdot \mathbf{w} = v_x w_x + v_y w_y + v_z w_z \qquad (4.5)$$

It is formed by a sum of products of each vector's scalar coefficients. This is referred to as a *dot product* and pronounced *v dot w*. The dot product is *commutative*, as $\mathbf{v} \cdot \mathbf{w} = \mathbf{w} \cdot \mathbf{v}$. Taking the dot product of a vector with itself leads to

$$\mathbf{v} \cdot \mathbf{v} = v_x^2 + v_y^2 + v_z^2 = |v|^2$$

which is the square of the magnitude of the vector. Furthermore, if we multiply the vector $\mathbf{v}$ by a scalar constant c and take the dot product of $c\mathbf{v}$ and $\mathbf{w}$, we find

$$(c\mathbf{v}) \cdot \mathbf{w} = (cv_x)w_x + (cv_y)w_y + (cv_z)w_z = c\,(\mathbf{v} \cdot \mathbf{w})$$

demonstrating that scalar constants can be pulled out of dot products.

The dot product can also be written

$$\mathbf{v} \cdot \mathbf{w} = |\mathbf{v}||\mathbf{w}| \cos\theta \qquad (4.6)$$

where θ is the angle between the two vectors (see Figure 4.9). Note that if the two vectors are perpendicular, the angle $\theta = \pi/2$ and $\mathbf{v} \cdot \mathbf{w} = 0$. In the later case, we say that the vectors are *orthogonal*. If the two vectors are parallel, the angle $\theta = 0$ and $\mathbf{v} \cdot \mathbf{w} = |\mathbf{v}||\mathbf{w}|$. That is the case for the dot product of a vector and itself, as $\mathbf{v} \cdot \mathbf{v} = |\mathbf{v}|^2$.

4.2.2 Geometric interpretation of the dot product

Let's explore graphically the dot product between two vectors $\mathbf{v}$ and $\mathbf{w}$ (see Figure 4.10). We can interpret $\mathbf{v} \cdot \mathbf{w} = |\mathbf{v}||\mathbf{w}| \cos\theta$ in two ways. We can write

$$\mathbf{v} \cdot \mathbf{w} = (|\mathbf{v}| \cos\theta)\,|\mathbf{w}| = v_w |\mathbf{w}|$$

where $v_w = |\mathbf{v}| \cos\theta$ is the magnitude of the vector $\mathbf{v}$ projected onto the unit vector $\hat{\mathbf{w}} = \mathbf{w}/|\mathbf{w}|$ as shown in the Figure 4.10. We can write this *vector projection* of $\mathbf{v}$ onto $\hat{\mathbf{w}}$ as

$$v_w = \mathbf{v} \cdot \hat{\mathbf{w}} = \mathbf{v} \cdot \left(\frac{1}{|\mathbf{w}|}\mathbf{w}\right) = \frac{1}{|\mathbf{w}|}(\mathbf{v} \cdot \mathbf{w}) = \frac{1}{|\mathbf{w}|}|\mathbf{v}||\mathbf{w}| \cos\theta = |\mathbf{v}| \cos\theta$$

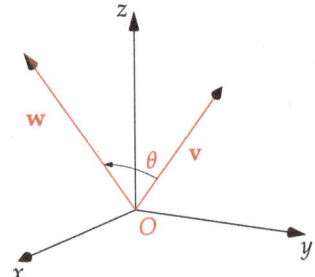

Figure 4.9: The scalar product of vectors $\mathbf{v}$ and $\mathbf{w}$ defined as $\mathbf{v} \cdot \mathbf{w} = |\mathbf{v}||\mathbf{w}| \cos\theta$.

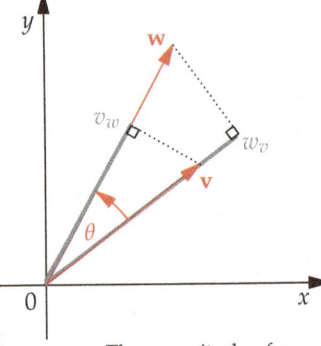

Figure 4.10: The magnitude of a scalar product $|\mathbf{v} \cdot \mathbf{w}| = |\mathbf{v}||\mathbf{w}| \cos\theta$. The magnitude can also be written $|\mathbf{v} \cdot \mathbf{w}| = v_w\,|\mathbf{w}| = |\mathbf{v}|w_v$ where $w_v = |\mathbf{w}| \cos\theta$ and $v_w = |\mathbf{v}| \cos\theta$ are shown as gray bars.

Equivalently, we can write the dot product as

$$\mathbf{v} \cdot \mathbf{w} = |\mathbf{v}| \left(|\mathbf{w}| \cos \theta \right) = |\mathbf{v}| w_v$$

where $w_v = |\mathbf{w}| \cos \theta$ is the magnitude of the vector $\mathbf{w}$ projected onto the unit vector $\hat{\mathbf{v}} = \mathbf{v}/|\mathbf{v}|$ as

$$w_v = \mathbf{w} \cdot \hat{\mathbf{v}} = \mathbf{w} \cdot \left(\frac{1}{|\mathbf{v}|} \mathbf{v} \right) = \frac{1}{|\mathbf{v}|} \left(\mathbf{w} \cdot \mathbf{v} \right) = \frac{1}{|\mathbf{v}|} |\mathbf{w}||\mathbf{v}| \cos \theta = |\mathbf{w}| \cos \theta$$

The dot product $\mathbf{v} \cdot \mathbf{w}$ can be considered the projection of $\mathbf{v}$ onto $\mathbf{w}$ or equivalently the projection of $\mathbf{w}$ onto $\mathbf{v}$. The dot product $\mathbf{v} \cdot \mathbf{w} = |\mathbf{v}| w_v = v_w |\mathbf{w}|$ varies from a maximum of $|\mathbf{v}||\mathbf{w}|$, when the vectors are parallel ($v_w = |\mathbf{v}|$ and $w_v = |\mathbf{w}|$), to a minimum of $-|\mathbf{v}||\mathbf{w}|$, when the vectors are antiparallel ($v_w = -|\mathbf{v}|$ and $w_v = -|\mathbf{w}|$). When the vectors are perpendicular $|\mathbf{v} \cdot \mathbf{w}| = 0$, as $v_w = w_v = 0$. This interpretation of the dot product as a vector projection is valuable when applying the dot project to model physical processes.[4]

4.2.3 *Vector projection in general*

The vector projection is valuable in expressing arbitrary vectors in terms of the x, y, and z-components. Consider the unit vectors $\hat{\mathbf{x}}$, $\hat{\mathbf{y}}$, and $\hat{\mathbf{z}}$. Each unit vector is parallel to itself so that $\hat{\mathbf{x}} \cdot \hat{\mathbf{x}} = \hat{\mathbf{y}} \cdot \hat{\mathbf{y}} = \hat{\mathbf{z}} \cdot \hat{\mathbf{z}} = 1$. Each unit vector is orthogonal to the other two unit vectors so that $\hat{\mathbf{x}} \cdot \hat{\mathbf{y}} = \hat{\mathbf{y}} \cdot \hat{\mathbf{z}} = \hat{\mathbf{z}} \cdot \hat{\mathbf{x}} = 0$.

We can define the x-component of an arbitrary vector v as $v_x = \mathbf{v} \cdot \hat{\mathbf{x}}$ with similar relations for the y-component, $v_y = \mathbf{v} \cdot \hat{\mathbf{y}}$, and z-component, $v_z = \mathbf{v} \cdot \hat{\mathbf{z}}$. As such, we can express any vector v as

$$\mathbf{v} = (\mathbf{v} \cdot \hat{\mathbf{x}}) \, \hat{\mathbf{x}} + (\mathbf{v} \cdot \hat{\mathbf{y}}) \, \hat{\mathbf{y}} + (\mathbf{v} \cdot \hat{\mathbf{z}}) \, \hat{\mathbf{z}}$$
$$= v_x \, \hat{\mathbf{x}} + v_y \, \hat{\mathbf{y}} + v_z \, \hat{\mathbf{z}}$$

This identity is valuable when we are given an arbitrary vector that we wish to express in the cartesian coordinate system. An example of the decomposition of a vector into its cartesian components is shown in Figure 4.11.

Similar relations can be used to decompose the vector $\mathbf{v}$ in spherical polar coordinates

$$\mathbf{v} = (\mathbf{v} \cdot \hat{\mathbf{r}}) \, \hat{\mathbf{r}} + (\mathbf{v} \cdot \hat{\boldsymbol{\theta}}) \, \hat{\boldsymbol{\theta}} + (\mathbf{v} \cdot \hat{\boldsymbol{\varphi}}) \, \hat{\boldsymbol{\varphi}}$$
$$= v_r \, \hat{\mathbf{r}} + v_\theta \, \hat{\boldsymbol{\theta}} + v_\varphi \, \hat{\boldsymbol{\varphi}}$$

and in cylindrical coordinates

$$\mathbf{v} = (\mathbf{v} \cdot \hat{\mathbf{r}}) \, \hat{\mathbf{r}} + (\mathbf{v} \cdot \hat{\boldsymbol{\theta}}) \, \hat{\boldsymbol{\theta}} + (\mathbf{v} \cdot \hat{\mathbf{z}}) \, \hat{\mathbf{z}}$$
$$= v_r \, \hat{\mathbf{r}} + v_\theta \, \hat{\boldsymbol{\theta}} + v_z \, \hat{\mathbf{z}}$$

We must keep in mind, however, that the unit vectors $\hat{\mathbf{r}}$, $\hat{\boldsymbol{\theta}}$, and $\hat{\boldsymbol{\varphi}}$ are coordinate dependent.

4.2.4 *Application of the dot product*

As an example of the utility of the dot product in the physical sciences, consider the mechanical *work* done when a force acts over a distance. The

[4] The full scalar product $\mathbf{v} \cdot \mathbf{w} = \left(v_x \hat{\mathbf{i}} + v_y \hat{\mathbf{j}} + v_z \hat{\mathbf{k}} \right) \left(w_x \hat{\mathbf{i}} + w_y \hat{\mathbf{j}} + w_z \hat{\mathbf{k}} \right) = v_x w_x \hat{\mathbf{i}} \cdot \hat{\mathbf{i}} + v_x w_y \hat{\mathbf{i}} \cdot \hat{\mathbf{j}} + v_x w_z \hat{\mathbf{i}} \cdot \hat{\mathbf{k}} + v_y w_x \hat{\mathbf{j}} \cdot \hat{\mathbf{i}} + v_y w_y \hat{\mathbf{j}} \cdot \hat{\mathbf{j}} + v_y w_z \hat{\mathbf{j}} \cdot \hat{\mathbf{k}} + v_z w_x \hat{\mathbf{k}} \cdot \hat{\mathbf{i}} + v_z w_y \hat{\mathbf{k}} \cdot \hat{\mathbf{j}} + v_z w_z \hat{\mathbf{k}} \cdot \hat{\mathbf{k}}$ Since $\hat{\mathbf{i}}$, $\hat{\mathbf{j}}$, and $\hat{\mathbf{k}}$ are mutually orthogonal this reduces to $\mathbf{v} \cdot \mathbf{w} = v_x w_x + v_y w_y + v_z w_z$.

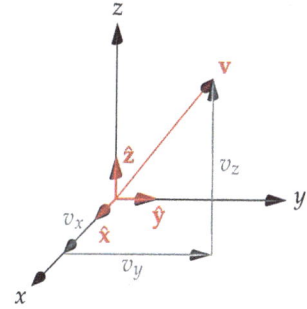

Figure 4.11: The decomposition of a vector $\mathbf{v} = v_x \, \hat{\mathbf{x}} + v_y \, \hat{\mathbf{y}} + v_z \, \hat{\mathbf{z}}$ in terms of its cartesian components.

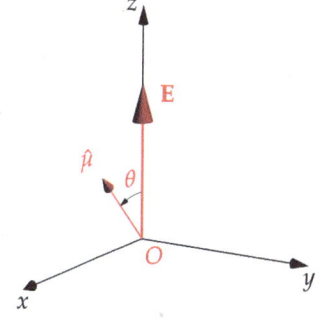

Figure 4.12: The scalar product of vectors $\boldsymbol{\mu}$ and $\mathbf{E}$ defined $\boldsymbol{\mu} \cdot \mathbf{E} = |\boldsymbol{\mu}||\mathbf{E}| \cos \theta$ in terms of the angle θ separating the two vectors and the magnitude of each vector.

work is a scalar w. The force has magnitude and direction and is expressed as a vector $\mathbf{f} = f_x\,\hat{\mathbf{x}} + f_y\,\hat{\mathbf{y}} + f_z\,\hat{\mathbf{z}}$ as is the distance $\mathbf{d} = d_x\,\hat{\mathbf{x}} + d_y\,\hat{\mathbf{y}} + d_z\hat{\mathbf{z}}$. The work is defined as the dot product of the force and distance:

$$w = \mathbf{f} \cdot \mathbf{d} = f_x d_x + f_y d_y + f_z d_z = |\mathbf{f}||\mathbf{d}|\cos\theta$$

where θ is the angle between the force $\mathbf{f}$ and displacement $\mathbf{d}$. When the displacement is perfectly in line with the force, $\theta = 0$, $w = |f||d|$, and the maximum work is done. When the displacement is perpendicular to the force, $\theta = \pi/2$, $w = 0$, and no work is done.[5]

Now consider the energy of an electric dipole $\boldsymbol{\mu}$ interacting with an electric field $\mathbf{E}$, each characterized by a magnitude and orientation as shown in Figure 4.12. The energy is a scalar $V = -\boldsymbol{\mu} \cdot \mathbf{E}$. Suppose that the electric field is oriented in the z-direction with magnitude E_0. The electric field vector is written $\mathbf{E} = E_0\,\hat{\mathbf{z}}$. If the electric dipole is $\boldsymbol{\mu} = \mu_y\,\hat{\mathbf{y}} + \mu_z\,\hat{\mathbf{z}}$, the energy of interaction is $V = -\mathbf{E} \cdot \boldsymbol{\mu} = -E_0\mu_z$. In general for an electric field $\mathbf{E} = E_x\,\hat{\mathbf{x}} + E_y\,\hat{\mathbf{y}} + E_z\hat{\mathbf{z}}$ and electric dipole $\boldsymbol{\mu} = \mu_x\,\hat{\mathbf{x}} + \mu_y\,\hat{\mathbf{y}} + \mu_z\,\hat{\mathbf{z}}$, the interaction energy is

$$V(\theta) = -\mathbf{E} \cdot \boldsymbol{\mu} = -|\mathbf{E}||\boldsymbol{\mu}|\cos\theta$$

where θ is the angle between the electric field and the electric dipole as shown in Figure 4.12.

The variation in the potential energy described by the dot product over a range of θ is shown in Figure 4.13. When the vectors $\boldsymbol{\mu}$ and $\mathbf{E}$ are parallel ($\theta = 0, 2\pi$), $V = -|\mathbf{E}||\boldsymbol{\mu}|$ is a minimum. When the vectors $\boldsymbol{\mu}$ and $\mathbf{E}$ are antiparallel ($\theta = \pi$), $V = |\mathbf{E}||\boldsymbol{\mu}|$ is a maximum. When the vectors are perpendicular ($\theta = \pi/2, 3\pi/2$), we find $V = 0$.

4.2.5 Multiplying vectors and the vector cross product

[5] For example, in circular motion the inward pointing *centripetal force* is perpendicular to the direction of motion. The centripetal force changes the direction of motion but does no work.

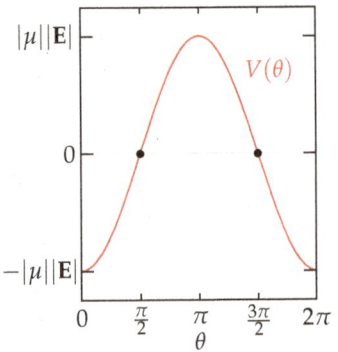

Figure 4.13: The dot product $V(\theta) = -\boldsymbol{\mu} \cdot \mathbf{E} = |\boldsymbol{\mu}||\mathbf{E}|\cos\theta$ is shown as a function of θ.

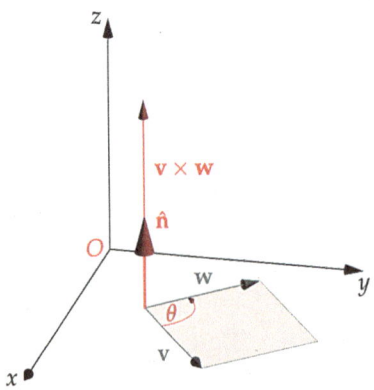

Figure 4.14: The vector product of $\mathbf{v}$ and $\mathbf{w}$ is defined $\mathbf{v} \times \mathbf{w} = |\mathbf{v}||\mathbf{w}|\hat{\mathbf{n}}\sin\theta$ in terms of the angle θ separating the two vectors and the magnitude of each vector. The unit vector $\hat{\mathbf{n}}$ is normal to the plane containing $\mathbf{v}$ and $\mathbf{w}$ in the direction determined by the right hand rule.

Consider the vectors $\mathbf{v} = v_x\,\hat{\mathbf{i}} + v_y\,\hat{\mathbf{j}} + v_z\,\hat{\mathbf{k}}$ and $\mathbf{w} = w_x\,\hat{\mathbf{i}} + w_y\,\hat{\mathbf{j}} + w_z\,\hat{\mathbf{k}}$. We can multiply $\mathbf{v}$ and $\mathbf{w}$ by taking the product of the magnitude of each vector as $|\mathbf{v}||\mathbf{w}|$. We can also multiply $\mathbf{v}$ and $\mathbf{w}$ using the *dot product* $\mathbf{v} \cdot \mathbf{w} = |\mathbf{v}||\mathbf{w}|\cos\theta$, resulting in a scalar. A third way to multiply $\mathbf{v}$ and $\mathbf{w}$ is the *vector product* written $\mathbf{v} \times \mathbf{w}$ and defined as

$$\mathbf{v} \times \mathbf{w} = (v_y w_z - v_z w_y)\,\hat{\mathbf{i}} + (v_z w_x - v_x w_z)\,\hat{\mathbf{j}} + (v_x w_y - v_y w_x)\,\hat{\mathbf{k}} \qquad (4.7)$$

The vector product is always itself a vector. It is also referred to as a *cross product* and pronounced *v cross w*.[6] An example of a cross product between two vectors $\mathbf{v}$ and $\mathbf{w}$ is shown in Figure 4.14. In this example, the vectors $\mathbf{v}$ and $\mathbf{w}$ are found in the xy-plane and the cross product $\mathbf{v} \times \mathbf{w}$ is in the direction of the positive z-axis.

A bit of algebra can be used to demonstrate that the magnitude of the vector cross product is related to the alternative scalar products $|\mathbf{v}||\mathbf{w}|$ and $\mathbf{v} \cdot \mathbf{w}$ by

$$|\mathbf{v} \times \mathbf{w}|^2 = |\mathbf{v}|^2 |\mathbf{w}|^2 - (\mathbf{v} \cdot \mathbf{w})^2$$

Since $|\mathbf{v} \cdot \mathbf{w}| = |\mathbf{v}||\mathbf{w}| \cos\theta$ it follows from the identity $1 - \cos^2\theta = \sin^2\theta$ that

$$|\mathbf{v} \times \mathbf{w}| = |\mathbf{v}||\mathbf{w}||\sin\theta| \qquad (4.8)$$

which is the magnitude of the shaded area in Figure 4.14. The magnitude of the cross product is equal to the area of the parallelogram defined by the two vectors in the plane containing those vectors.

Note that if the two vectors are parallel, the angle $\theta = 0$ and $|\mathbf{v} \times \mathbf{w}| = 0$. That is the case for the cross product of a vector and itself as $|\mathbf{v} \times \mathbf{v}| = 0$. If the two vectors are orthogonal, the angle $\theta = \pi/2$, and $|\mathbf{v} \times \mathbf{w}| = |\mathbf{v}||\mathbf{w}|$, which is the largest possible magnitude of the cross product. In practice, the dot product and Equation 4.6 provide the most convenient way to determine the angle between two vectors.

4.2.6 Getting the direction of the cross product right

The cross product defined in Equation 4.7 can also be written as

$$\mathbf{v} \times \mathbf{w} = |\mathbf{v}||\mathbf{w}|\hat{\mathbf{n}} \sin\theta$$

where θ is the angle between the two vectors and $\hat{\mathbf{n}}$ is a unit vector that is *normal* to the plane containing the vectors $\mathbf{v}$ and $\mathbf{w}$ (see Figure 4.14).

It is standard convention to restrict $\theta \in [0, \pi]$ so that $0 \leq \sin\theta \leq 1$. The direction of the unit vector $\hat{\mathbf{n}}$ can then be determined by the *right hand rule*:

#1: point the index finger of your right hand along vector $\mathbf{v}$

#2: point the middle finger of your right hand along vector $\mathbf{w}$

#3: your thumb points in the direction of the cross product $\mathbf{v} \times \mathbf{w}$

Apply the right hand rule to determine the direction of the cross product taken between the vectors $\mathbf{v}$ and $\mathbf{w}$ in Figure 4.14. The vectors are found in the xy-plane, so the cross product $\mathbf{v} \times \mathbf{w}$ will be orthogonal to the xy-plane, directed along the positive or negative z-axis. You should find that the right hand rule dictates that the unit vector $\hat{\mathbf{n}} = \hat{\mathbf{z}}$ is in the direction of the positive z-axis.[7]

After you succeed, try your hand at determining the direction of the cross product $\mathbf{a} \times \mathbf{b}$ shown in Figure 4.15. In this case, the relative orientation of

[6] Vector multiplication was discovered independently by American physicist, chemist, and mathematician *Josiah Willard Gibbs* (1839-1903) and English physicist and engineer *Oliver Heaviside* (1850-1925). Gibbs developed the modern field of statistical mechanics and is often considered, along with *Benjamin Franklin*(1706-1790), to be one of the two greatest American scientists.

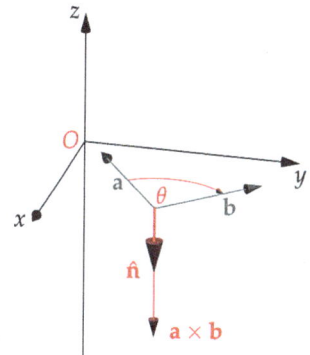

Figure 4.15: The vector product of $\mathbf{a}$ and $\mathbf{b}$ is defined as $\mathbf{a} \times \mathbf{b} = |\mathbf{a}||\mathbf{b}|\hat{\mathbf{n}} \sin\theta$. Here $\mathbf{a}$ and $\mathbf{b}$ are in the xy-plane and the direction of $\hat{\mathbf{n}}$ determined by the right hand rule is $-\hat{\mathbf{z}}$.

[7] We have followed the convention of determining the direction of

$$\mathbf{v} \times \mathbf{w} = |\mathbf{v}||\mathbf{w}|\hat{\mathbf{n}} \sin\theta$$

by restricting the angle θ to the range $\theta \in [0, \pi]$ so that $0 \leq \sin\theta \leq 1$ and using the right hand rule to determine the direction of $\hat{\mathbf{n}}$. However, one can also arbitrarily assign the direction of $\hat{\mathbf{n}}$ to be positive and compute $\sin\theta$ over the range $\theta \in [0, 2\pi]$. In this way

$$\hat{\mathbf{n}} \sin\theta$$

determines the direction of the cross product and there is no need for the right hand rule.

the two vectors leads to a direction of the cross product defined by $\hat{n} = -\hat{z}$. Finally, apply the right hand rule to the vector product shown in Figure 4.16. These three examples should prepare you well for the application of the right hand rule in the calculation of vector products.

4.2.7 Properties of the cross product and its efficient calculation

The cross product is *anticommutative* as

$$\begin{aligned}
\mathbf{v} \times \mathbf{w} &= (v_y w_z - v_z w_y)\,\hat{\mathbf{i}} + (v_z w_x - v_x w_z)\,\hat{\mathbf{j}} + (v_x w_y - v_y w_x)\,\hat{\mathbf{k}} \\
&= -(w_y v_z - v_y w_z)\,\hat{\mathbf{i}} - (v_x w_z - v_z w_y)\,\hat{\mathbf{j}} - (v_y w_x - v_x w_y)\,\hat{\mathbf{k}} \\
&= -\mathbf{w} \times \mathbf{v}
\end{aligned}$$

Furthermore, if we multiply the vector $\mathbf{v}$ by a scalar constant c and take the cross product of $c\mathbf{v}$ and $\mathbf{w}$ we find

$$\begin{aligned}
(c\mathbf{v}) \times \mathbf{w} &= ((cv_y)w_z - (cv_z)w_y)\,\hat{\mathbf{i}} + ((cv_z)w_x - (cv_x)w_z)\,\hat{\mathbf{j}} \\
&\quad + ((cv_x)w_y - (cv_y)w_x)\,\hat{\mathbf{k}} \\
&= c(v_y w_z - v_z w_y)\,\hat{\mathbf{i}} + c(v_z w_x - v_x w_z)\,\hat{\mathbf{j}} \\
&\quad + c(v_x w_y - v_y w_x)\,\hat{\mathbf{k}} = c\,\mathbf{v} \times \mathbf{w}
\end{aligned}$$

This demonstrates that scalar constants can be pulled out of cross products.

The expression we first used to define the cross product in Equation 4.7 is rather long. In calculating a cross product, it is easy to get vector indices confused or the order of operations wrong. To avoid these difficulties, it is convenient to represent the cross product as a *determinant*. First consider the definition of the 2×2 determinant

$$\begin{vmatrix} a_1 & a_2 \\ b_1 & b_2 \end{vmatrix} = a_1 b_2 - a_2 b_1$$

A 3×3 determinant can be evaluated in terms of three 2×2 determinants as[8]

$$\begin{vmatrix} c_1 & c_2 & c_3 \\ a_1 & a_2 & a_3 \\ b_1 & b_2 & b_3 \end{vmatrix} = c_1 \begin{vmatrix} a_2 & a_3 \\ b_2 & b_3 \end{vmatrix} - c_2 \begin{vmatrix} a_1 & a_3 \\ b_1 & b_3 \end{vmatrix} + c_3 \begin{vmatrix} a_1 & a_2 \\ b_1 & b_2 \end{vmatrix}$$

With this result, we can express our cross product conveniently using the determinant

$$\mathbf{v} \times \mathbf{w} = \begin{vmatrix} \hat{\mathbf{i}} & \hat{\mathbf{j}} & \hat{\mathbf{k}} \\ v_x & v_y & v_z \\ w_x & w_y & w_z \end{vmatrix}$$

$$= \hat{\mathbf{i}} \begin{vmatrix} v_y & v_z \\ w_y & w_z \end{vmatrix} - \hat{\mathbf{j}} \begin{vmatrix} v_x & v_z \\ w_x & w_z \end{vmatrix} + \hat{\mathbf{k}} \begin{vmatrix} v_x & v_y \\ w_x & w_y \end{vmatrix} \tag{4.9}$$

Evaluating the three 2×2 determinants leads to our original result in Equation 4.7:

$$\mathbf{v} \times \mathbf{w} = (v_y w_z - v_z w_y)\,\hat{\mathbf{i}} + (v_z w_x - v_x w_z)\,\hat{\mathbf{j}} + (v_x w_y - v_y w_x)\,\hat{\mathbf{k}}$$

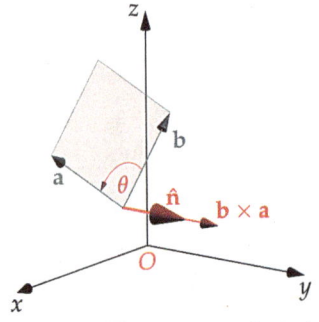

Figure 4.16: The vector product of $\mathbf{a}$ and $\mathbf{b}$ defined as $\mathbf{b} \times \mathbf{a} = |\mathbf{a}||\mathbf{b}|\hat{n}\sin\theta$. Note the order of multiplication and recall that $\mathbf{b} \times \mathbf{a} = -\mathbf{a} \times \mathbf{b}$. Reversing the order of the vectors in the cross product reverses the direction of the resulting vector.

[8] This convention for evaluating the determinant introduces the negative sign (red).

We will return to the use of determinants in the future. It is good to gain practice now evaluating determinants in the calculation of vector products.

4.2.8 Geometric interpretation of the cross product

Let's explore the magnitude of the vector product $\mathbf{v} \times \mathbf{w}$ and its dependence on the relative orientation of the vectors $\mathbf{v}$ and $\mathbf{w}$ graphically (see Figure 4.17). In the case of the scalar product we found

$$\mathbf{v} \cdot \mathbf{w} = v_w |\mathbf{w}| = |\mathbf{v}| w_v$$

where v_w is the component of $\mathbf{v}$ that is parallel to $\mathbf{w}$, and w_v is the component of $\mathbf{w}$ that is parallel to $\mathbf{v}$. In the case of the cross product we find the complementary result

$$|\mathbf{v} \times \mathbf{w}| = v_\perp |\mathbf{w}| = |\mathbf{v}|\, w_\perp$$

where $v_\perp$ is the component of $\mathbf{v}$ that is perpendicular to $\mathbf{w}$, and $w_\perp$ is the component of $\mathbf{w}$ that is perpendicular to $\mathbf{v}$.

From these relations, we see that the magnitude of the vector product

$$|\mathbf{v} \times \mathbf{w}| = |\mathbf{v}||\mathbf{w}| \sin \theta$$

is the area of the parallelogram defined by the vectors $\mathbf{v}$ and $\mathbf{w}$. This is shown as the shaded region in Figure 4.17. This geometric interpretation of the magnitude of the cross product is useful in considering the magnitude of cross products representing physical processes. The area of the parallelogram is greatest when $\mathbf{v}$ is orthogonal to $\mathbf{w}$ (the vectors form a rectangle of area $|\mathbf{v}||\mathbf{w}|$) and a minimum when the two vectors are parallel (the area is zero).

Let's end our discussion of vector projection by considering the unit vectors $\hat{\mathbf{x}}$, $\hat{\mathbf{y}}$, and $\hat{\mathbf{z}}$. Each unit vector is parallel to itself, so that

$$\hat{\mathbf{x}} \times \hat{\mathbf{x}} = \hat{\mathbf{y}} \times \hat{\mathbf{y}} = \hat{\mathbf{z}} \times \hat{\mathbf{z}} = 0 \qquad (4.10)$$

Each unit vector is orthogonal to the other two unit vectors, leading to the identities

$$\hat{\mathbf{x}} \times \hat{\mathbf{y}} = \hat{\mathbf{z}} \qquad \hat{\mathbf{y}} \times \hat{\mathbf{z}} = \hat{\mathbf{x}} \qquad \hat{\mathbf{z}} \times \hat{\mathbf{x}} = \hat{\mathbf{y}} \qquad (4.11)$$

By the anticommutation relation

$$\mathbf{v} \times \mathbf{w} = -\mathbf{w} \times \mathbf{v}$$

we find the relations $\hat{\mathbf{y}} \times \hat{\mathbf{x}} = -\hat{\mathbf{z}}$, $\hat{\mathbf{z}} \times \hat{\mathbf{y}} = -\hat{\mathbf{x}}$, and $\hat{\mathbf{x}} \times \hat{\mathbf{z}} = -\hat{\mathbf{y}}$.

4.2.9 Application of the cross product

The cross product is used to express a number of fundamental physical properties. Consider the *angular momentum*, defined in terms of the radial distance from the center of motion $\mathbf{r}$ and the linear momentum $\mathbf{p} = m\mathbf{v}$ as

$$\mathbf{l} = \mathbf{r} \times \mathbf{p}$$

where m is the particle mass and $\mathbf{v}$ is the velocity. Writing the radial position as $\mathbf{r} = x\,\hat{\mathbf{i}} + y\,\hat{\mathbf{j}} + z\,\hat{\mathbf{k}}$ and the linear momentum $\mathbf{p} = p_x\,\hat{\mathbf{i}} + p_y\,\hat{\mathbf{j}} + p_z\,\hat{\mathbf{k}}$, the

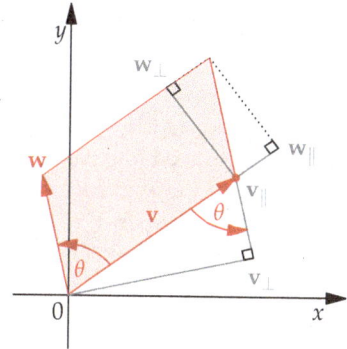

Figure 4.17: The magnitude of a vector product $|\mathbf{v} \times \mathbf{w}| = |\mathbf{v}||\mathbf{w}| \sin \theta$. The magnitude can also be written $|\mathbf{v} \times \mathbf{w}| = v_\perp |\mathbf{w}| = |\mathbf{v}|\, w_\perp$.

resulting angular momentum is

$$\mathbf{l} = \mathbf{r} \times \mathbf{p} = \begin{vmatrix} \hat{\mathbf{i}} & \hat{\mathbf{j}} & \hat{\mathbf{k}} \\ x & y & z \\ p_x & p_y & p_z \end{vmatrix}$$

$$= (y p_z - z p_y)\,\hat{\mathbf{i}} + (z p_x - x p_z)\,\hat{\mathbf{j}} + (x p_y - y p_x)\,\hat{\mathbf{k}}$$

As an example, consider clockwise circular motion of a mass, m, in the xy-plane. The velocity $\mathbf{v}$ of the mass is tangent to the circular trajectory and orthogonal to the radial vector $\mathbf{r}$ with $\theta = \pi/2$. This creates a maximum magnitude of the angular momentum oriented in the negative z-direction $\mathbf{l} = m\,(\mathbf{r} \times \mathbf{v}) = -m|\mathbf{r}||\mathbf{v}|\hat{\mathbf{z}}$. This result is depicted in Figure 4.18 where the area of the shaded region is the magnitude of the vector product.

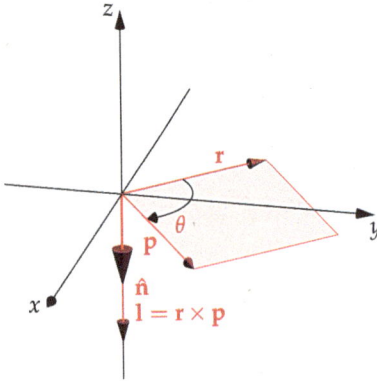

Figure 4.18: The vector product $\mathbf{r} \times \mathbf{p} = |\mathbf{r}||\mathbf{p}|\hat{\mathbf{n}} \sin\theta$ in terms of the angle θ separating the two vectors and unit vector $\hat{\mathbf{n}}$ normal to the plane containing $\mathbf{r}$ and $\mathbf{p}$. The shaded area is proportional to the magnitude of the cross product $|\mathbf{r}||\mathbf{p}| \sin\theta$. The resulting vector $\mathbf{l} = \mathbf{r} \times \mathbf{p}$ is oriented along the negative z-axis by the right hand rule.

Now consider a particle of mass m with charge q moving with a velocity $\mathbf{v} = v_x\,\hat{\mathbf{i}} + v_y\,\hat{\mathbf{j}} + v_z\,\hat{\mathbf{k}}$ in an external magnetic field $\mathbf{B} = B_x\,\hat{\mathbf{i}} + B_y\,\hat{\mathbf{j}} + B_z\,\hat{\mathbf{k}}$. A total force $\mathbf{F}$ acts on the particle with magnitude and orientation defined by

$$\mathbf{F} = q\,(\mathbf{v} \times \mathbf{B})$$

This is known as the *Lorentz force*.[9] This force is oriented perpendicular to both the direction of motion and the magnetic field. Consider a charge q with velocity $\mathbf{v} = v_x\,\hat{\mathbf{x}}$ in the x-direction in the presence of a magnetic field $\mathbf{B} = B_0\,\hat{\mathbf{z}}$ oriented along the z-axis. The force acting on the charge will be

$$\mathbf{F} = q\,(\mathbf{v} \times \mathbf{B}) = q\,[(v_x\,\hat{\mathbf{x}}) \times (B_0\,\hat{\mathbf{z}})] = q \begin{vmatrix} \hat{\mathbf{x}} & \hat{\mathbf{y}} & \hat{\mathbf{z}} \\ v_x & 0 & 0 \\ 0 & 0 & B_0 \end{vmatrix} = -q v_x B_0\,\hat{\mathbf{y}}$$

where the direction is determined by $\hat{\mathbf{x}} \times \hat{\mathbf{z}} = -\hat{\mathbf{y}}$.

In general, when the motion of the charge is parallel to the magnetic field, the force is zero. The greatest force occurs when the motion of the charge is perpendicular to the magnetic field. The variation in the cross product over a range of θ is shown in Figure 4.19.

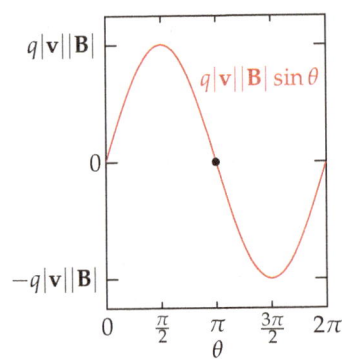

Figure 4.19: The cross product $\mathbf{F} = q\mathbf{v} \times \mathbf{B} = q|\mathbf{v}||\mathbf{B}|\,\hat{\mathbf{n}} \sin\theta$. The factor $q|\mathbf{v}||\mathbf{B}| \sin\theta$ is shown as a function of θ. When $\sin\theta$ changes sign the cross product changes direction from $\hat{\mathbf{n}}$ to $-\hat{\mathbf{n}}$.

[9] Named for the Dutch mathematical physicist *Hendrik Lorentz* (1853-1928) who derived the Lorentz transformation, underpinning the Special Theory of Relativity, and the Lorentz Force, used in electromagnetism. He received the 1902 Nobel Prize in Physics.

A$_4$ Building orthogonal vectors using Gram-Schmidt orthogonalization

In this complement, we explore a general procedure for forming a set of orthonormal basis vectors in a space of arbitrary dimension. We discussed how an arbitrary vector $\mathbf{v}$ in two dimensions can be expressed in terms of a linear combination of the orthonormal basis vectors $\hat{\mathbf{x}}$ and $\hat{\mathbf{y}}$ as

$$\mathbf{v} = v_x\,\hat{\mathbf{x}} + v_y\,\hat{\mathbf{y}}$$

The weighting coefficients v_x and v_y are defined in terms of the projections of the vector $\mathbf{v}$ onto the unit vectors $\hat{\mathbf{x}}$ and $\hat{\mathbf{y}}$:

$$v_x = \frac{1}{\hat{\mathbf{x}} \cdot \hat{\mathbf{x}}}(\hat{\mathbf{x}} \cdot \mathbf{v}) = \hat{\mathbf{x}} \cdot \mathbf{v} \qquad v_y = \frac{1}{\hat{\mathbf{y}} \cdot \hat{\mathbf{y}}}(\hat{\mathbf{y}} \cdot \mathbf{v}) = \hat{\mathbf{y}} \cdot \mathbf{v}$$

Suppose we are given two vectors $\mathbf{v}_1$ and $\mathbf{v}_2$ that are not orthogonal. That is, $\mathbf{v}_1 \cdot \mathbf{v}_2 \neq 0$. We can transform the vector $\mathbf{v}_2$ into a vector that is orthogonal to $\mathbf{v}_1$ using the Gram-Schmidt orthogonalization procedure.

Let's call the orthogonal vectors that we plan to form $\mathbf{u}_1$ and $\mathbf{u}_2$. To start, we define[10]

$$\mathbf{u}_1 = \mathbf{v}_1$$

[10] This choice was arbitrary. We could just as well have taken $\mathbf{u}_1 = \mathbf{v}_2$.

We can form the vector $\mathbf{u}_2$, that is orthogonal to $\mathbf{u}_1$, by subtracting from $\mathbf{v}_2$ its overlap with $\mathbf{u}_1$:

$$\mathbf{u}_2 = \mathbf{v}_2 - \left(\frac{\mathbf{u}_1 \cdot \mathbf{v}_2}{\mathbf{u}_1 \cdot \mathbf{u}_1}\right)\mathbf{u}_1$$

We can demonstrate that our new vector $\mathbf{u}_2$ is orthogonal to $\mathbf{u}_1$ by evaluating the dot product

$$\mathbf{u}_1 \cdot \mathbf{u}_2 = \mathbf{u}_1 \cdot \mathbf{v}_2 - \left(\frac{\mathbf{u}_1 \cdot \mathbf{v}_2}{\mathbf{u}_1 \cdot \mathbf{u}_1}\right)\mathbf{u}_1 \cdot \mathbf{u}_1 = \mathbf{u}_1 \cdot \mathbf{v}_2 - \left(\frac{\mathbf{u}_1 \cdot \mathbf{v}_2}{\mathbf{u}_1 \cdot \mathbf{u}_1}\right)\mathbf{u}_1 \cdot \mathbf{u}_1 = 0$$

Finally, the resulting orthogonal vectors $\mathbf{u}_1$ and $\mathbf{u}_2$ can be normalized:

$$\mathbf{e}_1 = \frac{1}{\sqrt{\mathbf{u}_1 \cdot \mathbf{u}_1}}\mathbf{u}_1 \qquad \mathbf{e}_2 = \frac{1}{\sqrt{\mathbf{u}_2 \cdot \mathbf{u}_2}}\mathbf{u}_2$$

So you see how this works.

Consider the example depicted in Figure 4.20. We begin with two vectors

$$\mathbf{v}_1 = 3\hat{\mathbf{x}} + \hat{\mathbf{y}} \qquad \mathbf{v}_2 = 2\hat{\mathbf{x}} + 2\hat{\mathbf{y}}$$

We define

$$\mathbf{u}_1 = \mathbf{v}_1 = 3\hat{\mathbf{x}} + \hat{\mathbf{y}}$$

We then form a vector $\mathbf{u}_2$ that is orthogonal to $\mathbf{u}_1$:

$$\mathbf{u}_2 = \mathbf{v}_2 - \left(\frac{\mathbf{u}_1 \cdot \mathbf{v}_2}{\mathbf{u}_1 \cdot \mathbf{u}_1}\right)\mathbf{u}_1 = (2\hat{\mathbf{x}} + 2\hat{\mathbf{y}}) - \frac{8}{10}(3\hat{\mathbf{x}} + \hat{\mathbf{y}}) = -\frac{2}{5}\hat{\mathbf{x}} + \frac{6}{5}\hat{\mathbf{y}}$$

We demonstrate that $\mathbf{u}_1$ and $\mathbf{u}_2$ are orthogonal by evaluating the dot product $\mathbf{u}_1 \cdot \mathbf{u}_2 = -\frac{6}{5} + \frac{6}{5} = 0$. Finally, we normalize the vectors to arrive at our final

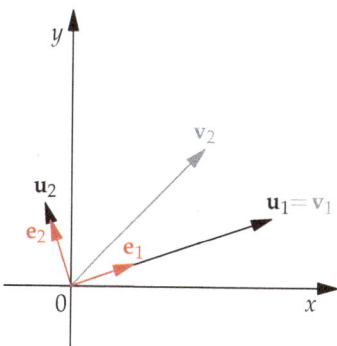

Figure 4.20: The initial non-orthogonal vectors $\mathbf{v}_1$ and $\mathbf{v}_2$ are transformed into orthogonal vectors $\mathbf{u}_1$ and $\mathbf{u}_2$. $\mathbf{e}_1$ and $\mathbf{e}_2$ are the normalized forms of $\mathbf{u}_1$ and $\mathbf{u}_2$.

result

$$\mathbf{e}_1 = \frac{1}{\sqrt{\mathbf{u}_1 \cdot \mathbf{u}_1}} \mathbf{u}_1 = \frac{1}{\sqrt{10}} (3\hat{\mathbf{x}} + \hat{\mathbf{y}}) \qquad \mathbf{e}_2 = \frac{1}{\sqrt{\mathbf{u}_2 \cdot \mathbf{u}_2}} \mathbf{u}_2 = \frac{1}{\sqrt{10}} (-\hat{\mathbf{x}} + 3\hat{\mathbf{y}})$$

Now let's extend this procedure to three dimensions. Suppose we start from three vectors $\mathbf{v}_1$, $\mathbf{v}_2$, and $\mathbf{v}_3$. We first define $\mathbf{u}_1 = \mathbf{v}_1$. As before, we then form a vector $\mathbf{u}_2$ that is orthogonal to $\mathbf{u}_1$:

$$\mathbf{u}_2 = \mathbf{v}_2 - \left(\frac{\mathbf{u}_1 \cdot \mathbf{v}_2}{\mathbf{u}_1 \cdot \mathbf{u}_1} \right) \mathbf{u}_1$$

In a similar way, we generate a vector $\mathbf{u}_3$, that is orthogonal to both $\mathbf{u}_1$ and $\mathbf{u}_2$, by subtracting from $\mathbf{v}_3$ its overlap with $\mathbf{u}_1$ and its overlap with $\mathbf{u}_2$:

$$\mathbf{u}_3 = \mathbf{v}_3 - \left(\frac{\mathbf{u}_1 \cdot \mathbf{v}_3}{\mathbf{u}_1 \cdot \mathbf{u}_1} \right) \mathbf{u}_1 - \left(\frac{\mathbf{u}_2 \cdot \mathbf{v}_3}{\mathbf{u}_2 \cdot \mathbf{u}_2} \right) \mathbf{u}_2$$

We can demonstrate that $\mathbf{u}_3$ is orthogonal to $\mathbf{u}_1$ since

$$\mathbf{u}_1 \cdot \mathbf{u}_3 = \mathbf{u}_1 \cdot \mathbf{v}_3 - \left(\frac{\mathbf{u}_1 \cdot \mathbf{v}_3}{\cancel{\mathbf{u}_1 \cdot \mathbf{u}_1}} \right) \cancel{\mathbf{u}_1 \cdot \mathbf{u}_1} - \left(\frac{\mathbf{u}_2 \cdot \mathbf{v}_3}{\mathbf{u}_2 \cdot \mathbf{u}_2} \right) \cancel{\mathbf{u}_1 \cdot \mathbf{u}_2} = 0$$

and that $\mathbf{u}_3$ is orthogonal to $\mathbf{u}_2$ since

$$\mathbf{u}_2 \cdot \mathbf{u}_3 = \mathbf{u}_2 \cdot \mathbf{v}_3 - \left(\frac{\mathbf{u}_1 \cdot \mathbf{v}_3}{\mathbf{u}_1 \cdot \mathbf{u}_1} \right) \cancel{\mathbf{u}_2 \cdot \mathbf{u}_1} - \left(\frac{\mathbf{u}_2 \cdot \mathbf{v}_3}{\cancel{\mathbf{u}_2 \cdot \mathbf{u}_2}} \right) \cancel{\mathbf{u}_2 \cdot \mathbf{u}_2} = 0$$

Here we have used the fact that $\mathbf{u}_1 \cdot \mathbf{u}_2 = 0$.

Consider the example depicted in Figure 4.21. We begin with three vectors

$$\mathbf{v}_1 = \hat{\mathbf{x}} + \hat{\mathbf{y}} + \hat{\mathbf{z}} \qquad \mathbf{v}_2 = \hat{\mathbf{x}} + \hat{\mathbf{z}} \qquad \mathbf{v}_3 = \hat{\mathbf{x}} + \hat{\mathbf{y}}$$

We define $\mathbf{u}_1 = \mathbf{v}_1 = \hat{\mathbf{x}} + \hat{\mathbf{y}} + \hat{\mathbf{z}}$ and then form a vector $\mathbf{u}_2$ that is orthogonal to $\mathbf{u}_1$:

$$\mathbf{u}_2 = \mathbf{v}_2 - \left(\frac{\mathbf{u}_1 \cdot \mathbf{v}_2}{\mathbf{u}_1 \cdot \mathbf{u}_1} \right) \mathbf{u}_1 = (\hat{\mathbf{x}} + \hat{\mathbf{z}}) - \frac{2}{3} (\hat{\mathbf{x}} + \hat{\mathbf{y}} + \hat{\mathbf{z}}) = \frac{1}{3} (\hat{\mathbf{x}} - 2\hat{\mathbf{y}} + \hat{\mathbf{z}})$$

Similarly, we form a vector $\mathbf{u}_3$ that is orthogonal to both $\mathbf{u}_1$ and $\mathbf{u}_2$:

$$\mathbf{u}_3 = \mathbf{v}_3 - \left(\frac{\mathbf{u}_1 \cdot \mathbf{v}_3}{\mathbf{u}_1 \cdot \mathbf{u}_1} \right) \mathbf{u}_1 - \left(\frac{\mathbf{u}_2 \cdot \mathbf{v}_3}{\mathbf{u}_2 \cdot \mathbf{u}_2} \right) \mathbf{u}_2$$

$$= (\hat{\mathbf{x}} + \hat{\mathbf{y}}) - \frac{2}{3}(\hat{\mathbf{x}} + \hat{\mathbf{y}} + \hat{\mathbf{z}}) - \left(-\frac{1}{2} \right) \frac{1}{3} (\hat{\mathbf{x}} - 2\hat{\mathbf{y}} + \hat{\mathbf{z}}) = \frac{1}{2}(\hat{\mathbf{x}} - \hat{\mathbf{z}})$$

We can demonstrate that the three vectors $\mathbf{u}_1$, $\mathbf{u}_2$, and $\mathbf{u}_3$ are mutually orthogonal, since $\mathbf{u}_1 \cdot \mathbf{u}_2 = \mathbf{u}_2 \cdot \mathbf{u}_3 = \mathbf{u}_3 \cdot \mathbf{u}_1 = 0$. Finally, the resulting orthogonal vectors can be normalized to create three orthonormal vectors

$$\mathbf{e}_1 = \frac{1}{\sqrt{3}}(\hat{\mathbf{x}} + \hat{\mathbf{y}} + \hat{\mathbf{z}}) \quad \mathbf{e}_2 = \frac{1}{\sqrt{6}}(\hat{\mathbf{x}} - 2\hat{\mathbf{y}} + \hat{\mathbf{z}}) \quad \mathbf{e}_3 = \frac{1}{\sqrt{2}}(\hat{\mathbf{x}} - \hat{\mathbf{z}})$$

This procedure can be generalized to the orthogonalization of vectors in arbitrary dimension. The Gram-Schmidt procedure is revisited in the complements to Chapter 13 in the context of creating sets of orthogonal polynomial functions.

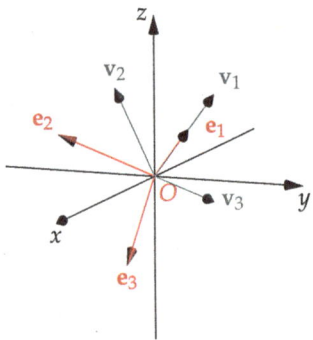

Figure 4.21: The initial non-orthogonal vectors $\mathbf{v}_1$, $\mathbf{v}_2$, and $\mathbf{v}_3$ are transformed into orthonormal vectors $\mathbf{e}_1$, $\mathbf{e}_2$, and $\mathbf{e}_3$.

B_4 End-of-chapter problems

A mathematician may say anything he pleases, but a physicist must be at least partially sane.

J. Willard Gibbs

Warm-ups

4.1 Determine the magnitude, $|\mathbf{a}|$, and direction, in terms of θ in two dimensions and θ and φ in three dimensions, of the following vectors.

(a) $\mathbf{a} = \hat{\mathbf{x}} + 3\hat{\mathbf{y}}$ (b) $\mathbf{a} = 2\hat{\mathbf{x}} - 2\hat{\mathbf{y}}$ (c) $\mathbf{a} = -3\hat{\mathbf{x}} + 2\hat{\mathbf{y}}$

(d) $\mathbf{a} = -2\hat{\mathbf{x}} + \hat{\mathbf{z}}$ (e) $\mathbf{a} = -\hat{\mathbf{x}} - 5\hat{\mathbf{y}} + 3\hat{\mathbf{z}}$ (f) $\mathbf{a} = \hat{\mathbf{x}} + \hat{\mathbf{y}} + 3\hat{\mathbf{z}}$

4.2 For each of the following sums $\mathbf{c} = \mathbf{a} + \mathbf{b}$, determine the magnitude, $|\mathbf{c}|$, and direction, in terms of θ in two dimensions and θ and φ in three dimensions, of the resulting vector.

(a) $\mathbf{a} = \hat{\mathbf{x}} + 3\hat{\mathbf{y}}$ $\mathbf{b} = 3\hat{\mathbf{x}} + \hat{\mathbf{y}}$ (b) $\mathbf{a} = -\hat{\mathbf{x}} + 2\hat{\mathbf{y}}$ $\mathbf{b} = 2\hat{\mathbf{x}} + 2\hat{\mathbf{y}}$

(c) $\mathbf{a} = -2\hat{\mathbf{x}} + 3\hat{\mathbf{y}}$ $\mathbf{b} = -\hat{\mathbf{x}} - 4\hat{\mathbf{y}} - 6\hat{\mathbf{z}}$ (d) $\mathbf{a} = 2\hat{\mathbf{x}} + 3\hat{\mathbf{z}}$ $\mathbf{b} = -3\hat{\mathbf{x}} + 6\hat{\mathbf{y}} - 9\hat{\mathbf{z}}$

4.3 Determine the scalar number resulting from the following dot products $\mathbf{a} \cdot \mathbf{b}$.

(a) $\mathbf{a} = \hat{\mathbf{x}} + 4\hat{\mathbf{y}}$ $\mathbf{b} = 4\hat{\mathbf{x}} + \hat{\mathbf{y}}$ (b) $\mathbf{a} = -\hat{\mathbf{x}} + 2\hat{\mathbf{y}}$ $\mathbf{b} = 2\hat{\mathbf{x}} + 2\hat{\mathbf{y}}$

(c) $\mathbf{a} = 3\hat{\mathbf{y}} + 4\hat{\mathbf{z}}$ $\mathbf{b} = -\hat{\mathbf{x}} - 4\hat{\mathbf{y}} - 6\hat{\mathbf{z}}$ (d) $\mathbf{a} = -2\hat{\mathbf{x}} + 3\hat{\mathbf{z}}$ $\mathbf{b} = -3\hat{\mathbf{x}} + 6\hat{\mathbf{y}} - 9\hat{\mathbf{z}}$

4.4 For each of the following vector products $\mathbf{c} = \mathbf{a} \times \mathbf{b}$, determine the magnitude, $|\mathbf{c}|$, and direction, in terms of θ in two dimensions and θ and φ in three dimensions, of the resulting vector.

(a) $\mathbf{a} = 2\hat{\mathbf{x}} + 3\hat{\mathbf{y}}$ $\mathbf{b} = 3\hat{\mathbf{x}} + 2\hat{\mathbf{y}}$ (b) $\mathbf{a} = -\hat{\mathbf{x}} + 2\hat{\mathbf{y}}$ $\mathbf{b} = 2\hat{\mathbf{x}} - 2\hat{\mathbf{y}}$

(c) $\mathbf{a} = -2\hat{\mathbf{x}} + 4\hat{\mathbf{z}}$ $\mathbf{b} = \hat{\mathbf{x}} - 4\hat{\mathbf{y}} - 5\hat{\mathbf{z}}$ (d) $\mathbf{a} = 2\hat{\mathbf{x}} + 3\hat{\mathbf{z}}$ $\mathbf{b} = -3\hat{\mathbf{x}} + 6\hat{\mathbf{y}}$

Homework exercises

4.5 Use the dot product relation $\mathbf{a} \cdot \mathbf{b} = |\mathbf{a}||\mathbf{b}| \cos \theta$ to determine the angle θ between the following vectors.

(a) $\mathbf{a} = 3\hat{\mathbf{x}} - \hat{\mathbf{y}}$ $\mathbf{b} = 4\hat{\mathbf{y}}$ (b) $\mathbf{a} = \hat{\mathbf{x}} + \hat{\mathbf{y}} + \hat{\mathbf{z}}$ $\mathbf{b} = 2\hat{\mathbf{x}} + 3\hat{\mathbf{y}} + 4\hat{\mathbf{z}}$

4.6 Use the cross product relation $|\mathbf{a} \times \mathbf{b}| = |\mathbf{a}||\mathbf{b}| \sin \theta$ to determine the angle θ between the following vectors.

(a) $\mathbf{a} = 3\hat{\mathbf{x}} - \hat{\mathbf{y}}$ $\mathbf{b} = 4\hat{\mathbf{y}}$ (b) $\mathbf{a} = \hat{\mathbf{x}} + \hat{\mathbf{y}} + \hat{\mathbf{z}}$ $\mathbf{b} = 2\hat{\mathbf{x}} + 3\hat{\mathbf{y}} + 4\hat{\mathbf{z}}$

4.7 Show that scalar multiplication is commutative and vector multiplication is not. That is, show for $\mathbf{a} = a_x\hat{\mathbf{x}} + a_y\hat{\mathbf{y}} + a_z\hat{\mathbf{z}}$ and $\mathbf{b} = b_x\hat{\mathbf{x}} + b_y\hat{\mathbf{y}} + b_z\hat{\mathbf{z}}$ that

$$\mathbf{a} \cdot \mathbf{b} = \mathbf{b} \cdot \mathbf{a}$$

but

$$\mathbf{a} \times \mathbf{b} \neq \mathbf{b} \times \mathbf{a}$$

4.8 Show that $\mathbf{a} \cdot \mathbf{a} = |a|^2$

4.9 Show that

$$\mathbf{a} + (\mathbf{b} + \mathbf{c}) = (\mathbf{a} + \mathbf{b}) + \mathbf{c}$$

4.10* Show that

$$\mathbf{a} \cdot (\mathbf{b} \times \mathbf{c}) = \mathbf{b} \cdot (\mathbf{c} \times \mathbf{a}) = \mathbf{c} \cdot (\mathbf{a} \times \mathbf{b})$$

4.11* Show that

$$\mathbf{a} \times (\mathbf{b} \times \mathbf{c}) = (\mathbf{a} \cdot \mathbf{c})\mathbf{b} - (\mathbf{a} \cdot \mathbf{b})\mathbf{c}$$

4.12* Show that

$$(\mathbf{a} \times \mathbf{b}) \cdot (\mathbf{c} \times \mathbf{d}) = (\mathbf{a} \cdot \mathbf{c})(\mathbf{b} \cdot \mathbf{d}) - (\mathbf{a} \cdot \mathbf{d})(\mathbf{b} \cdot \mathbf{c})$$

4.13 Angular momentum is given by the equation $\mathbf{l} = \mathbf{r} \times \mathbf{p}$, where $\mathbf{r} = x\,\hat{\imath} + y\,\hat{\jmath} + z\,\hat{k}$ is the radius of curvature of the motion and $\mathbf{p} = p_x\,\hat{\imath} + p_y\,\hat{\jmath} + p_z\,\hat{k}$ is the linear momentum of the body in motion. Assuming that

$$\mathbf{l} = l_x\,\hat{\imath} + l_y\,\hat{\jmath} + l_z\,\hat{k}$$

find the components of the angular momentum in the x-, y-, and z-directions.

4.14 Consider the three vectors $\mathbf{a} = 2\hat{x} + \hat{y}$, $\mathbf{b} = -\hat{x} + \hat{y}$, and $\mathbf{c} = \hat{x} + \hat{y} + 3\hat{z}$ shown in the figure below.

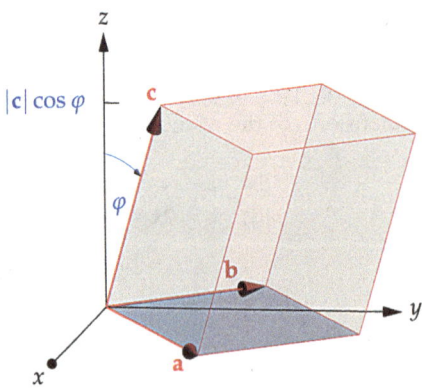

The volume of the shaded parallelepiped is given by the *scalar triple product*

$$V = \mathbf{a} \times \mathbf{b} \cdot \mathbf{c}$$

where $|\mathbf{a} \times \mathbf{b}|$ is the area of the base of the parallelepiped (shaded blue) in the xy-plane and $|\mathbf{c}| \cos \varphi$ is the parallelepiped's altitude. Determine the volume V.

4.15 Show that the vectors $\mathbf{a} = \frac{1}{2}\mathbf{q}_1 + \frac{1}{2}\mathbf{q}_2 + \frac{1}{2}\mathbf{q}_3 + \frac{1}{2}\mathbf{q}_4$ and $\mathbf{b} = \frac{1}{2}\mathbf{q}_1 - \frac{1}{2}\mathbf{q}_2 + \frac{1}{2}\mathbf{q}_3 - \frac{1}{2}\mathbf{q}_4$, where $\mathbf{q}_1$, $\mathbf{q}_2$, $\mathbf{q}_3$, and $\mathbf{q}_4$ are orthogonal unit vectors, are orthogonal and normalized.

4.16 For a particle of charge, q, moving with velocity, $\mathbf{v}$, in an external electric field, $\mathbf{E}$, and magnetic field, $\mathbf{B}$, the force acting on the particle is the Lorentz force defined as

$$\mathbf{F} = q\left(\mathbf{E} + \mathbf{v} \times \mathbf{B}\right)$$

Suppose a particle of mass, m, and charge, q, is moving with velocity $\mathbf{v} = v_z\,\hat{\mathbf{k}}$ in the presence of an electric field $\mathbf{E} = E_x\,\hat{\mathbf{i}} + E_y\,\hat{\mathbf{j}}$ and magnetic field $\mathbf{B} = B_x\,\hat{\mathbf{i}}$. Determine the total Lorentz force, $\mathbf{F}$, acting on the particle.

4.17 Consider a particle of mass, m, moving on a circle of radius, r_0, with velocity

$$\mathbf{v} = v_x\,\hat{\mathbf{i}} + v_y\,\hat{\mathbf{j}}$$

at a position $\mathbf{r} = r_x\,\hat{\mathbf{i}} + r_y\,\hat{\mathbf{j}}$ where $r_0 = \sqrt{r_x^2 + r_y^2}$ (see figure below). Determine the angular momentum of the particle

$$\mathbf{l} = \mathbf{r} \times \mathbf{p}$$

where the momentum, $\mathbf{p} = m\mathbf{v}$.

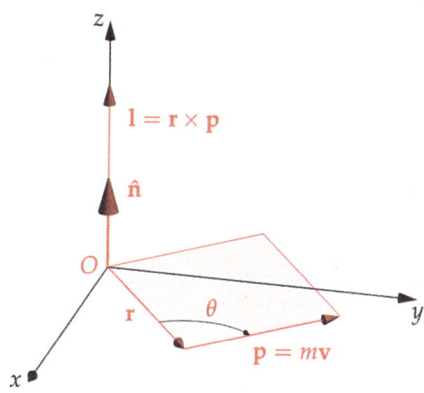

4.18 For an electric dipole, $\boldsymbol{\mu} = \mu_E\,\hat{\mathbf{j}}$, in an external electric field, $\mathbf{E} = E_x\,\hat{\mathbf{i}} + E_y\,\hat{\mathbf{j}}$, determine the energy defined

$$V = -\boldsymbol{\mu} \cdot \mathbf{E}$$

4.19 For a magnetic dipole, $\boldsymbol{\mu} = \mu_B\,\hat{\mathbf{i}}$, in an external magnetic field, $\mathbf{B} = B_z\,\hat{\mathbf{k}}$, determine the torque on the magnetic dipole defined

$$\mathbf{T} = -\boldsymbol{\mu} \times \mathbf{B}$$

4.20* Suppose you are given two non-orthogonal vectors

$$\mathbf{v}_1 = \hat{\mathbf{x}} - 2\hat{\mathbf{y}} \qquad \mathbf{v}_2 = 3\hat{\mathbf{x}} + 2\hat{\mathbf{y}}$$

(a) Use the Gram-Schmidt orthogonalization procedure to construct the orthogonal vectors $\mathbf{u}_1$ and $\mathbf{u}_2$. Assume $\mathbf{u}_1 = \mathbf{v}_1$.

(b) Determine the orthonormal vectors

$$\mathbf{e}_1 = \frac{1}{\sqrt{\mathbf{u}_1 \cdot \mathbf{u}_1}}\mathbf{u}_1 \qquad \mathbf{e}_2 = \frac{1}{\sqrt{\mathbf{u}_2 \cdot \mathbf{u}_2}}\mathbf{u}_2$$

4.21* Suppose you are given three non-orthogonal vectors

$$\mathbf{v}_1 = \hat{\mathbf{x}} - 2\hat{\mathbf{y}} + \hat{\mathbf{z}} \qquad \mathbf{v}_2 = \hat{\mathbf{x}} - \hat{\mathbf{y}} \qquad \mathbf{v}_3 = 3\hat{\mathbf{y}} - 2\hat{\mathbf{z}}$$

(a) Use the Gram-Schmidt orthogonalization procedure to construct the orthogonal vectors $\mathbf{u}_1$, $\mathbf{u}_2$, and $\mathbf{u}_3$. Assume $\mathbf{u}_1 = \mathbf{v}_1$.

(b) Determine the orthonormal vectors

$$\mathbf{e}_1 = \frac{1}{\sqrt{\mathbf{u}_1 \cdot \mathbf{u}_1}} \mathbf{u}_1 \qquad \mathbf{e}_2 = \frac{1}{\sqrt{\mathbf{u}_2 \cdot \mathbf{u}_2}} \mathbf{u}_2 \qquad \mathbf{e}_3 = \frac{1}{\sqrt{\mathbf{u}_3 \cdot \mathbf{u}_3}} \mathbf{u}_3$$

5   Scalar and vector operators

5.1 Scalar operators

Mathematical operations transform one function into another. Examples of mathematical operations include multiplying a function by a constant or taking the derivative of a function. The mathematical symbol or function that carries out the operation is known as an operator. In this section, we explore a variety of types of operators in action so that we understand the concept and its utility.

5.1.1 Operators transform one function into another

An *operator* transforms one mathematical function into another. Consider the *scalar operator* $\hat{A}$ acting on a function $f(x)$ as

$$\hat{A}f(x) = 2f(x)$$

The operator $\hat{A}$ acts on the function $f(x)$ by multiplying the function by the scalar number 2.[1] Alternatively, we might define an operator $\hat{B}$ that acts to square the function it operates on as

$$\hat{B}f(x,y) = (f(x,y))^2$$

[1] The hat notation used to designate an operator, such as $\hat{x}$, is distinct from the hat notation used to indicate a unit vector, such as $\hat{\mathbf{x}}$.

The operator $\hat{A}$ is called a *linear operator* for two reasons. First, its action is *distributed* over the sum of two functions as

$$\hat{A}\left(f(x) + g(x)\right) = \hat{A}f(x) + \hat{A}g(x)$$

Second, it acts on a scalar c times a function as

$$\hat{A}\left(cf(x)\right)) = c\hat{A}f(x)$$

The operator $\hat{B}$ is not a linear operator since

$$\hat{B}\left(f(x) + g(x)\right) = f^2(x) + 2f(x)g(x) + g^2(x)$$
$$\neq \hat{B}f(x) + \hat{B}g(x) = f^2(x) + g^2(x)$$

and

$$\hat{B}\left(cf(x)\right)) = c^2f^2(x) \neq c\hat{B}f(x) = cf^2(x)$$

The majority of operators of importance in modeling physical processes are linear operators.[2]

The idea of the operator can be generalized to include many mathematical operations, including taking a sum, taking a difference, raising a function to a power, taking a product, integrating a function, translating the coordinates of a function, or rotating the coordinates of a function to change the function's orientation in space. In applications to physical systems, the most commonly used operators are differential operators and rotational operators. Differential and rotational operators are fundamental to quantum theory and the evaluation of symmetry.

5.1.2 Algebra of differential operators

Let's explore how the derivative can be expressed using the idea of the operator. Consider the partial derivative of a function

$$\frac{\partial}{\partial x} f(x,y) = f_x(x,y)$$

We can identify

$$\hat{d}_x \equiv \frac{\partial}{\partial x}$$

as a *differential operator*[3] acting on the function $f(x,y)$ as

$$\hat{d}_x f(x,y) = \frac{\partial}{\partial x} f(x,y) = f_x(x,y)$$

We can similarly define the operator $\hat{d}_y$ as

$$\hat{d}_y f(x,y) = \frac{\partial}{\partial y} f(x,y) = f_y(x,y)$$

This differential operator is a linear operator since

$$\hat{d}_x \left(f(x,y) + g(x,y) \right) = \frac{\partial}{\partial x} \left(f(x,y) + g(x,y) \right) = \frac{\partial}{\partial x} f(x,y) + \frac{\partial}{\partial x} g(x,y)$$
$$= \hat{d}_x f(x,y) + \hat{d}_x g(x,y)$$

and

$$\hat{d}_x \left(cf(x,y) \right) = \frac{\partial}{\partial x} \left(cf(x,y) \right) = c \frac{\partial}{\partial x} f(x,y) = c\hat{d}_x f(x,y)$$

Operators can act sequentially on a function as

$$\hat{d}_y \hat{d}_x f(x,y) = \frac{\partial}{\partial y} \left(\frac{\partial}{\partial x} f(x,y) \right) = \frac{\partial^2}{\partial y \partial x} f(x,y) = f_{yx}(x,y)$$

or

$$\hat{d}_x \hat{d}_y f(x,y) = \frac{\partial}{\partial x} \left(\frac{\partial}{\partial y} f(x,y) \right) = \frac{\partial^2}{\partial x \partial y} f(x,y) = f_{xy}(x,y)$$

Note that the action of the operators can depend on the order of operation as $\hat{d}_x \hat{d}_y f(x,y)$ is not necessarily equal to $\hat{d}_y \hat{d}_x f(x,y)$. Taking the difference between the two results above results in a compact representation of Euler's

[2] The concept of the operator may seem abstract when we consider the operator on its own. However, the operator only has value when acting on a function. As such, it is often best to view the operator in action, acting on a function, so that its nature and practical significance are clear.

[3] The *differential operator* $\hat{d}_x$ has no value on its own. It only takes on value when operating on a function as in $\hat{d}_x f(x,y)$.

test:

$$\left(\hat{d}_x\hat{d}_y - \hat{d}_y\hat{d}_x\right) f(x,y) = \frac{\partial^2 f}{\partial x \partial y} - \frac{\partial^2 f}{\partial y \partial x} = f_{xy}(x,y) - f_{yx}(x,y) = 0$$

When this relation is satisfied, we say that the operators $\hat{d}_x$ and $\hat{d}_y$ *commute*.

It is common to use the compact notation

$$\left[\hat{d}_x, \hat{d}_y\right] = \left(\hat{d}_x\hat{d}_y - \hat{d}_y\hat{d}_x\right) \tag{5.1}$$

where $\left[\hat{d}_x, \hat{d}_y\right]$ is known as the *commutator*. Consider the commutation relation

$$[\hat{A}, \hat{B}] \, f(x) = \hat{A}\left(\hat{B}f(x)\right) - \hat{B}\left(\hat{A}f(x)\right) = 0$$

When this is true

$$\hat{A}\left(\hat{B}f(x)\right) = \hat{B}\left(\hat{A}f(x)\right)$$

and the order of operation is not important.

Let's explore commutator algebra. Consider the commutator

$$[\hat{A} + \hat{B}, \hat{C} + \hat{D}] = (\hat{A} + \hat{B})(\hat{C} + \hat{D}) - (\hat{C} + \hat{D})(\hat{A} + \hat{B})$$

Expanding the product we find

$$[\hat{A} + \hat{B}, \hat{C} + \hat{D}] = \hat{A}\hat{C} + \hat{A}\hat{D} + \hat{B}\hat{C} + \hat{B}\hat{D} - \hat{C}\hat{A} - \hat{C}\hat{B} - \hat{D}\hat{A} - \hat{D}\hat{B}$$

Regrouping pairs to form commutation relations, we arrive at the result

$$[\hat{A} + \hat{B}, \hat{C} + \hat{D}] = \hat{A}\hat{C} - \hat{C}\hat{A} + \hat{A}\hat{D} - \hat{D}\hat{A} + \hat{B}\hat{C} - \hat{C}\hat{B} + \hat{B}\hat{D} - \hat{D}\hat{B}$$
$$= [\hat{A}, \hat{C}] + [\hat{A}, \hat{D}] + [\hat{B}, \hat{C}] + [\hat{B}, \hat{D}]$$

So you see how that works.

5.1.3 *More complicated operators*

Let's apply the rules of operator algebra to more complicated differential operators. Extending the idea of a product of operators, we can take powers of operators. Consider the total differential operator

$$\hat{D}_x \equiv \frac{d}{dx}$$

raised to the third power as

$$(\hat{D}_x)^3 f(x) = \left(\frac{d}{dx}\right)^3 f(x) = \frac{d^3}{dx^3} f(x)$$

We can also consider more complicated linear differential operators:

$$\hat{C}f(x) = \left(\hat{D}_x^2 + 2\hat{D}_x + 3\right) f(x)$$
$$= \left(\frac{d^2}{dx^2} + 2\frac{d}{dx} + 3\right) f(x) = \frac{d^2 f}{dx^2} + 2\frac{df}{dx} + 3f$$

Linear differential operators play an important role in quantum theory, where they are used to measure physical properties including the linear momentum and energy. Examples include the *linear momentum operator*[4]

$$\hat{p}\psi(x) = -i\hbar\frac{d\psi(x)}{dx}$$

where the operator is defined as

$$\hat{p} \equiv -i\hbar\frac{d}{dx}$$

and the *energy operator*

$$\hat{H}\psi(x) = -\frac{\hbar^2}{2m}\frac{d^2\psi(x)}{dx^2} + V(x)\psi(x)$$

The energy operator can be written as a sum of the *kinetic energy operator* and *potential energy operator*

$$\hat{H} \equiv \hat{T} + \hat{V} = \frac{1}{2m}\hat{p}^2 + V(x)$$

where $\hat{V} = V(x)$ is a function of x.[5]

Consider the commutation of the operators $\hat{p}$ and $\hat{p}^2$ where

$$\left(\hat{p}\hat{p}^2 - \hat{p}^2\hat{p}\right)\psi(x) = \hat{p}\left(\hat{p}^2\psi(x)\right) - \hat{p}^2\left(\hat{p}\psi(x)\right)$$
$$= \hat{p}^3\psi(x) - \hat{p}^3\psi(x) = 0$$

The operators $\hat{p}$ and $\hat{p}^2$ commute. More generally, a linear operator will always commute with itself, multiples of itself, or arbitrary powers of itself.

Finally, consider the commutation of the operators $\hat{p}$ and $\hat{H}$ where

$$\left(\hat{p}\hat{H} - \hat{H}\hat{p}\right)\psi(x)$$

Expanding the operator $\hat{H}$, we find

$$\left(\hat{p}\hat{H} - \hat{H}\hat{p}\right)\psi(x) = \left[\hat{p}\left(\frac{1}{2m}\hat{p}^2 + V(x)\right) - \left(\frac{1}{2m}\hat{p}^2 + V(x)\right)\hat{p}\right]\psi(x)$$
$$= \left[\frac{1}{2m}\hat{p}^3 + \hat{p}V(x) - \frac{1}{2m}\hat{p}^3 - V(x)\hat{p}\right]\psi(x)$$

Canceling the two terms containing $\hat{p}^3$ leaves

$$\left(\hat{p}\hat{H} - \hat{H}\hat{p}\right)\psi(x) = \left(\hat{p}V(x) - V(x)\hat{p}\right)\psi(x)$$

Noting that the differential operator $\hat{p}$ acts on $V(x)$ as well as $\psi(x)$, we expand the expression above and find

$$\left[\hat{p}V(x) - V(x)\hat{p}\right]\psi(x) = -i\hbar\psi(x)\frac{dV}{dx} - i\hbar V(x)\frac{d\psi}{dx} + i\hbar V(x)\frac{d\psi}{dx}$$

Noting that the second and third terms cancel, we arrive at the final result

$$\left(\hat{p}\hat{H} - \hat{H}\hat{p}\right)\psi(x) = -i\hbar\psi(x)\frac{d}{dx}V(x)$$

If $dV(x)/dx=0$, the operators $\hat{p}$ and $\hat{H}$ commute. Otherwise, they do not.

[4] In quantum theory where $\psi(x)$ is the wave function, $\hat{p}\psi(x)$ is used to measure the linear momentum and $\hat{H}\psi(x)$ is used to measure the total energy.

[5] In quantum theory where $\psi(x)$ is the wave function, $\hat{T}\psi(x)$ is used to measure the kinetic energy and $\hat{V}\psi(x)$ is used to measure the potential energy.

5.2 Vector operators and the gradient and divergence

PHYSICAL PROPERTIES INCLUDING POSITION, MOMENTUM, AND FORCE have magnitude and orientation and are described as vectors. The concept of differentiation of scalar functions can be generalized when combined with the concept of vectors. We can describe the slope of a scalar function of two variables $f(x, y)$ at a point in space in terms of a vector defining the slopes in the x- and y-directions. Alternatively, we can define the scalar magnitude of the rate of change of a vector function by summing the rate of change of the function in each of its dimensions. In this section, we explore the differentiation of vector-valued functions and vector fields commonly used to describe physical processes.

5.2.1 Vector-valued functions

A scalar function $f(x, y)$ associates a scalar number with every point (x, y) on the xy-plane. That number has a magnitude $|f(x, y)|$ and sign. In contrast, a *vector-valued function* $\mathbf{f}(t)$ is a *vector function* that associates every point (x, y) with a vector with magnitude and orientation that may depend on the parameter t.

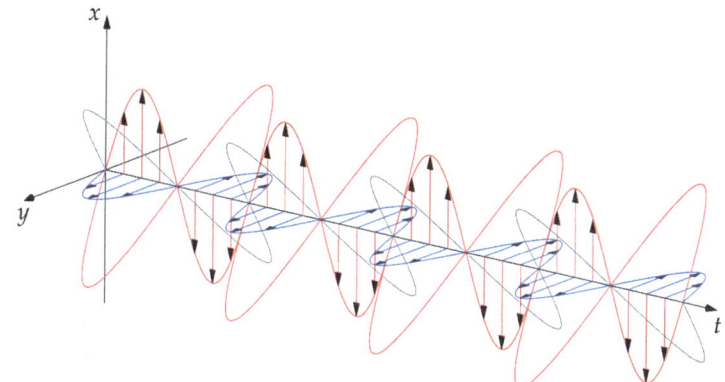

Figure 5.1: The vector-valued functions $\mathbf{x}(t) = \sin(2\pi t)\,\hat{\mathbf{x}}$ (blue) and $\mathbf{y}(t) = \sin(2\pi t)\,\hat{\mathbf{y}}$ (red). Also shown are $\mathbf{x}(t) + \mathbf{y}(t)$ (light gray) and $\mathbf{x}(t) - \mathbf{y}(t)$ (light red).

Consider the equation for the *plane waves* oscillating in the x- or y-direction as a function of the parameter t:

$$\mathbf{x}(t) = \sin(2\pi t)\,\hat{\mathbf{x}} \qquad \mathbf{y}(t) = \sin(2\pi t)\,\hat{\mathbf{y}}$$

Using vector notation, we can present both functions in a single three-dimensional plot as depicted in Figure 5.1. Oscillations in the x-direction are shown in red and those in the y-direction are shown in blue. The linear combinations $\mathbf{x}(t) + \mathbf{y}(t)$ (light gray) and $\mathbf{x}(t) - \mathbf{y}(t)$ (light red) appear on the $x = y$ plane and $x = -y$ plane, respectively.

Recall the helical curve written in cartesian coordinates in Equation 1.28 as

$$x = x(z) = a\cos(2\pi z)$$
$$y = y(z) = a\sin(2\pi z)$$

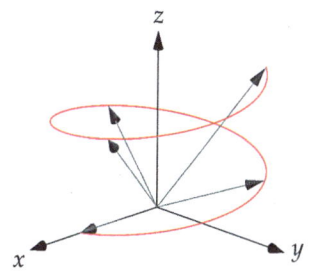

Figure 5.2: A helical curve $\mathbf{r}(t)$ (red) expressed in vector notation as the vector-valued function of the parameter t. The gray arrows mark the point $\mathbf{r}(t)$ for five values of t.

That same function can be written as a vector-valued function

$$\mathbf{r}(t) = a\cos(2\pi t)\,\hat{\imath} + a\sin(2\pi t)\,\hat{\jmath} + t\,\hat{\mathbf{k}} \qquad (5.2)$$

where we have introduced the parameter t and $z = t$. The result is shown in Figure 5.2.

5.2.2 *Vector functions and vector fields*

Consider the *vector function*

$$\mathbf{f}(x, y) = 2x\hat{\mathbf{x}} + 2y\hat{\mathbf{y}}$$

shown in Figure 5.3. This vector function assigns a vector to each point in space (x, y). This assignment of a vector to each point in space forms a *vector field*. All vector fields are vector-valued functions, but not all vector-valued functions are vector fields. For example, the helical curve $\mathbf{r}(t)$ defined by Equation 5.2 is a vector-valued function that is not a vector field as it assigns vectors given the value of a parameter t rather than to each point in space (x, y).

At the point $(1, 1)$, the value of the function is

$$\mathbf{f}(1, 1) = 2\hat{\mathbf{x}} + 2\hat{\mathbf{y}}$$

which is a vector of magnitude $2\sqrt{2}$ oriented outward from the origin along the line $x = y$. At the point $(0, -1)$, the value of the function is

$$\mathbf{f}(0, -1) = -2\hat{\mathbf{y}}$$

which is a vector of magnitude 2 oriented outward along the negative y-axis. At the point $(-2, 2)$, the value of the function is

$$\mathbf{f}(-2, 2) = -4\hat{\mathbf{x}} + 4\hat{\mathbf{y}}$$

which is a vector of magnitude $4\sqrt{2}$ oriented outward along the line $x = -y$. The value of $\mathbf{f}(x, y)$ at these three points is displayed in Figure 5.3.

If we add to the number of points presented, we can depict the variation of the vector field $\mathbf{f}(x, y)$ over a region of the xy-plane (see Figure 5.4). At each point, the vector function is defined in terms of a magnitude and orientation depicted by small arrows. The collection of arrows provides a sense of the changing magnitude and direction of the vector function over the xy-plane.

A vector field is used to model the force acting on a charge in an electric field. At each point in space, the force acting on the charge will have a specific magnitude and a direction. Vector fields are also used to model fluids for which each element of the fluid at some point in space has a given speed and direction of flow. Points from which vectors flow outward are *sources* and points to which vectors flow inward are *sinks*. Points with no net inward or outward flow are *saddle points*.

Now consider the periodic two-dimensional vector function

$$\mathbf{f}(x, y) = \cos x\,\hat{\mathbf{x}} + \cos y\,\hat{\mathbf{y}}$$

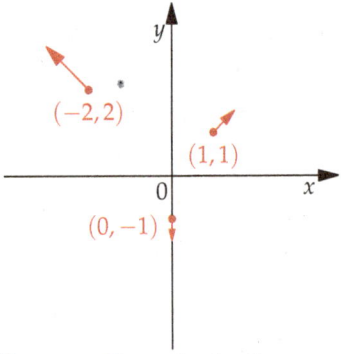

Figure 5.3: The vector function $\mathbf{f}(x, y) = 2x\hat{\mathbf{x}} + 2y\hat{\mathbf{y}}$ evaluated at three points on the xy-plane. Arrow length and orientation indicate magnitude and direction of the vector function.

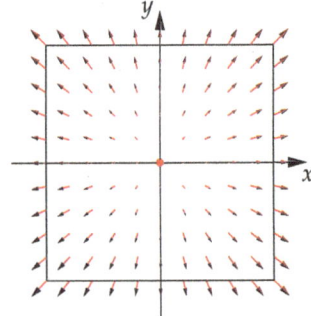

Figure 5.4: The vector field $\mathbf{f}(x, y) = 2x\,\hat{\mathbf{x}} + 2y\,\hat{\mathbf{y}}$ shown as an array of arrows on the xy-plane. The center of the graph is a source as vectors are directed outward from that point.

shown in Figure 5.5 below, with arrows indicating the continuous flow.

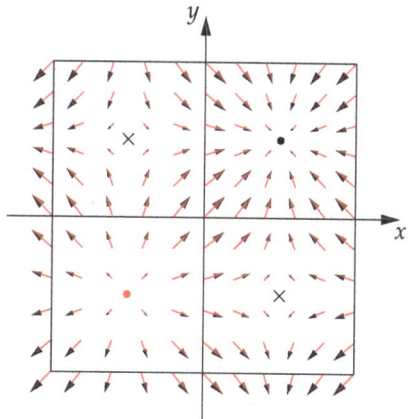

Figure 5.5: The vector field $\mathbf{f}(x,y) = \cos x\,\hat{\mathbf{x}} + \cos y\,\hat{\mathbf{y}}$ shown as an array of arrows on the xy-plane. The black dot represents a sink (upper right) and the red dot represents a source (lower left). The crosses represent saddle points.

In the upper right quadrant the vector field points inward toward a sink (black dot), while in the lower left quadrant the vectors point outward from a source (red dot). In the remaining quadrants, there are saddle points (crosses) with no net inward or outward flow. Overall, the collection of arrows represent the vector field $\mathbf{f}(x,y)$ over this region of the xy-plane.

5.2.3 Vector differentiation of a scalar function and the gradient

Consider the *vector function*

$$\mathbf{f}(x,y) = \frac{\partial V}{\partial x}\,\hat{\mathbf{x}} + \frac{\partial V}{\partial y}\,\hat{\mathbf{y}} = V_x\,\hat{\mathbf{x}} + V_y\,\hat{\mathbf{y}}$$

where $V(x,y)$ is a differentiable scalar function of x and y. The vector $\mathbf{f}(x,y)$ describes the slope of the scalar function $V(x,y)$ at the point (x,y) in the x- and y-directions. We say that $\mathbf{f}$ is the *gradient* of $V(x,y)$.

This expression can be written in the compact notation

$$\mathbf{f}(x,y) = \left(\hat{\mathbf{x}}\,\frac{\partial}{\partial x} + \hat{\mathbf{y}}\,\frac{\partial}{\partial y}\right) V(x,y) = \boldsymbol{\nabla} V(x,y) \qquad (5.3)$$

where

$$\boldsymbol{\nabla} \equiv \hat{\mathbf{x}}\,\frac{\partial}{\partial x} + \hat{\mathbf{y}}\,\frac{\partial}{\partial y} \qquad (5.4)$$

is the *gradient operator* in two dimensions. The gradient operator is a *vector operator* that acts on a scalar function and returns a vector function.

An example is shown in Figure 5.6 for the gradient $\boldsymbol{\nabla} V(x,y)$ of the function $V(x,y) = -\left(x^2 + y^2\right)$. The resulting vector field

$$\boldsymbol{\nabla} V(x,y) = V_x\,\hat{\mathbf{x}} + V_y\,\hat{\mathbf{y}} = \mathbf{f}(x,y) = -2x\,\hat{\mathbf{x}} - 2y\,\hat{\mathbf{y}}$$

is shown as an array of arrows on the xy-plane. The partial derivatives V_x and V_y determine the magnitude of the x- and y-components of the gradient as shown by the vectors at two points on the surface defined by $V(x,y)$. The magnitude of the gradient, $|\nabla V(x,y)| = (V_x^2 + V_y^2)^{\frac{1}{2}} = (4x^2 + 4y^2)^{\frac{1}{2}}$. In plane polar coordinates, $V = -r^2$, so that $\nabla V = -2r\,\hat{\mathbf{r}}$ and $|\nabla V| = 2r$.

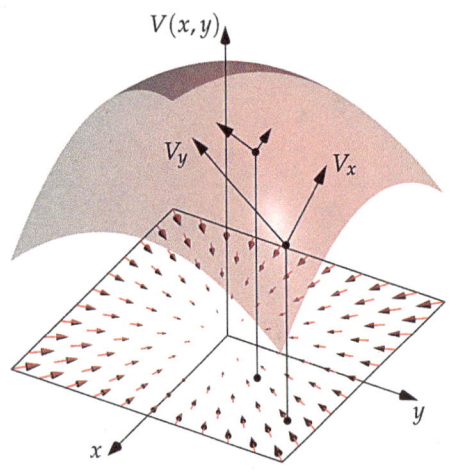

Figure 5.6: The inverted parabolic function $V(x,y) = -(x^2 + y^2)$ shown as a red surface. The corresponding vector field $\mathbf{f}(x,y) = V_x(x,y)\hat{\mathbf{x}} + V_y(x,y)\hat{\mathbf{y}}$ is depicted as an array of arrows on the xy-plane where $V_x(x,y) = -2x$ and $V_y(x,y) = -2y$.

Note that the gradient and its cartesian components point inward and upward toward the origin, which is the location of the maximum of the function. The direction of the gradient is the direction of fastest increase of the function at a given point in the xy-plane. The magnitude of the gradient provides a measure of the rate of increase of the function.

Another example is shown in Figure 5.7 for the gradient of the function

$$V(r,\theta) = \ln(r)$$

The gradient can be determined in cartesian coordinates as

$$\nabla V(x,y) = V_x\,\hat{\mathbf{x}} + V_y\,\hat{\mathbf{y}} = \mathbf{f}(x,y) = \frac{x}{r^2}\,\hat{\mathbf{x}} + \frac{y}{r^2}\,\hat{\mathbf{y}}$$

However, it is most naturally determined using the gradient operator in plane polar coordinates provided in the complements:

$$\nabla V(r,\theta) = \hat{\mathbf{r}}\frac{\partial V}{\partial r} + \hat{\boldsymbol{\theta}}\frac{1}{r}\frac{\partial V}{\partial \theta} = \mathbf{f}(r,\theta) = \frac{1}{r}\,\hat{\mathbf{r}}$$

The resulting vector field is shown as an array of arrows on the xy-plane. The partial derivatives V_x and V_y determine the magnitude of the x- and y-components of the gradient as shown by the vectors at a point on the surface.

The gradient operator in three dimensions is the natural extension of the gradient operator in two dimensions:

$$\nabla \equiv \hat{\mathbf{x}}\frac{\partial}{\partial x} + \hat{\mathbf{y}}\frac{\partial}{\partial y} + \hat{\mathbf{z}}\frac{\partial}{\partial z} \tag{5.5}$$

The forms of the gradient operator in two-dimensional plane polar co-

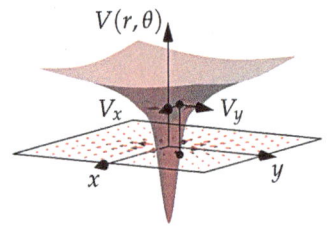

Figure 5.7: The logarithmic function $V(r,\theta) = \ln(r)$ shown as a red surface. The corresponding vector field $\mathbf{f}(r,\theta) = \frac{1}{r}\hat{\mathbf{r}} = V_x\hat{\mathbf{x}} + V_y\hat{\mathbf{y}}$ is depicted as an array of arrows on the xy-plane where $V_x(x,y) = \frac{x}{r^2}$ and $V_y(x,y) = \frac{y}{r^2}$.

ordinates and three-dimensional spherical polar coordinates, as well as three-dimensional cylindrical coordinates, are provided in the complements.[6]

As an example, let's see how the gradient operator acts on the three-dimensional function

$$V(x,y,z) = x^2 + y^2 + z^2 + 2xyz$$

The gradient of $V(x,y,z)$ is

$$\nabla V(x,y,z) = \left(\hat{\mathbf{x}} \frac{\partial}{\partial x} + \hat{\mathbf{y}} \frac{\partial}{\partial y} + \hat{\mathbf{z}} \frac{\partial}{\partial z} \right) V(x,y,z)$$

$$= \hat{\mathbf{x}} \frac{\partial V}{\partial x} + \hat{\mathbf{y}} \frac{\partial V}{\partial y} + \hat{\mathbf{z}} \frac{\partial V}{\partial z}$$

$$= (2x + 2yz)\hat{\mathbf{x}} + (2y + 2xz)\hat{\mathbf{y}} + (2z + 2xy)\hat{\mathbf{z}}$$

This is a vector function $\mathbf{f}(x,y,z) = \nabla V(x,y,z)$. These examples demonstrate the mechanics of calculating gradients and provide a sense of how to interpret the gradient as the direction and rate of fastest increase of a function.

5.2.4 *Vector differentiation of a vector function and the divergence*

Earlier we explored the dot product between two vectors $\mathbf{v}$ and $\mathbf{w}$ defined by

$$\mathbf{v} \cdot \mathbf{w} = v_x w_x + v_y w_y + v_z w_z \tag{5.6}$$

In the same way, we can take a dot product between two vector functions $\mathbf{f}(x,y) = a(x,y)\hat{\mathbf{x}} + b(x,y)\hat{\mathbf{y}}$ and $\mathbf{g}(x,y) = c(x,y)\hat{\mathbf{x}} + d(x,y)\hat{\mathbf{y}}$, leading to a scalar function

$$\mathbf{f}(x,y) \cdot \mathbf{g}(x,y) = a(x,y)c(x,y) + b(x,y)d(x,y) \tag{5.7}$$

For example, consider the vector field

$$\mathbf{r}(x,y) = \frac{x}{r}\hat{\mathbf{x}} + \frac{y}{r}\hat{\mathbf{y}} = \cos\theta\,\hat{\mathbf{x}} + \sin\theta\,\hat{\mathbf{y}}$$

that is directed radially outward in the xy-plane with a magnitude of unity at every point in space. Suppose we take the dot product $\mathbf{r}(x,y)$ with the unit vector $\hat{\mathbf{y}}$ as

$$\mathbf{r} \cdot \hat{\mathbf{y}} = \sin\theta$$

The resulting scalar is the projection of $\mathbf{r}$ on the unit vector $\hat{\mathbf{y}}$ that provides the magnitude of $\mathbf{r}$ in the y-direction. These examples show how we can use the dot product to multiply vectors and vector functions.

Suppose we take the dot product of the gradient operator

$$\nabla = \frac{\partial}{\partial x}\hat{\mathbf{x}} + \frac{\partial}{\partial y}\hat{\mathbf{y}}$$

and a general vector field in two dimensions

$$\mathbf{f}(x,y) = a(x,y)\hat{\mathbf{x}} + b(x,y)\hat{\mathbf{y}}$$

The result of this operation[7] is a scalar function

[6] Remember that the gradient operator ∇ is an action or operation that has no value on its own. It only takes on value when operating on a function as in $\nabla V(x,y)$.

[7] An alternative shorthand notation is $\nabla \cdot \mathbf{f}(x,y) = \text{div }\mathbf{f}(x,y)$.

$$\nabla \cdot \mathbf{f}(x,y) = \left(\frac{\partial}{\partial x} \,\hat{\mathbf{x}} + \frac{\partial}{\partial y} \,\hat{\mathbf{y}} \right) \cdot (a(x,y)\hat{\mathbf{x}} + b(x,y)\hat{\mathbf{y}}) = \frac{\partial a}{\partial x} + \frac{\partial b}{\partial y} \qquad (5.8)$$

that is the sum of the rate of change in the function in each direction. We call $\nabla \cdot \mathbf{f}(x,y)$ the *divergence* of $\mathbf{f}(x,y)$ and $\nabla \cdot$ is the divergence operator.[8] The divergence operator acts on a vector function and returns a scalar function.

For example, consider the vector function

$$\mathbf{f}(x,y) = \cos x \,\hat{\mathbf{x}} + \cos y \,\hat{\mathbf{y}}$$

examined previously and shown in Figure 5.6. The divergence of $\mathbf{f}(x,y)$ is the scalar function

$$\nabla \cdot \mathbf{f}(x,y) = \left(\hat{\mathbf{x}} \,\frac{\partial}{\partial x} + \hat{\mathbf{y}} \,\frac{\partial}{\partial y} \right) \cdot (\cos x \,\hat{\mathbf{x}} + \cos y \,\hat{\mathbf{y}})$$

$$= \frac{\partial}{\partial x} \cos x + \frac{\partial}{\partial y} \cos y = -(\sin x + \sin y)$$

Figure 5.8 depicts the vector field $\mathbf{f}(x,y)$ shown previously in Figure 5.6 as arrows in the xy-plane. The divergence of $\mathbf{f}(x,y)$ is represented as the continuously varying red surface. The local minimum in the divergence corresponds to the *sink* (black dot) where $\nabla \cdot \mathbf{f}(x,y) < 0$. The local maximum in the divergence corresponds to a *source* (red dot) where $\nabla \cdot \mathbf{f}(x,y) > 0$. The saddles on the surface (crosses) correspond to *saddle points*.

[8] The divergence operator $\nabla \cdot$ has no value on its own. It only takes on value when operating on a function as in $\nabla \cdot \mathbf{f}(x,y)$.

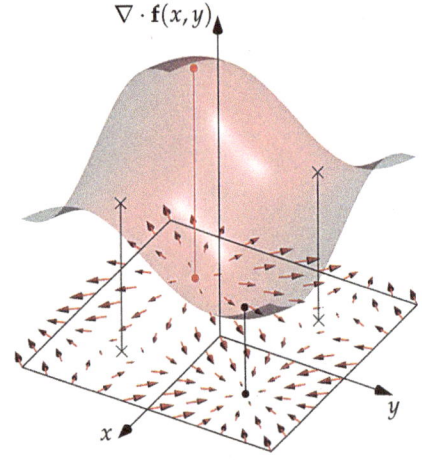

$$\nabla \cdot \mathbf{f}(x,y)$$

Figure 5.8: The divergence of the vector function $\mathbf{f}(x,y) = \cos x \,\hat{\mathbf{x}} + \cos y \,\hat{\mathbf{y}}$. The divergence is a scalar function $\nabla \cdot \mathbf{f}(x,y) = -(\sin x + \sin y)$ shown as a red surface. The vector field of $\mathbf{f}(x,y)$ is depicted as an array of arrows on the xy-plane. The surface includes a source (red dot) representing a maximum in the divergence, a sink (black dot) representing a minimum in the divergence, and two saddle points (crosses).

Think of the vector field in terms of the field lines. When the divergence at some point is positive, the overall flux of the field lines is outward and positive. When the divergence at some point is negative, the overall flux of the field lines is inward and negative. When the outward flux equals the inward flux, the net flux is zero and the divergence is zero.

Let's consider one last example of the divergence of the vector function

$$\mathbf{f}(x,y) = \frac{x}{r^2} \,\hat{\mathbf{x}} + \frac{y}{r^2} \,\hat{\mathbf{y}}$$

where $r^2 = x^2 + y^2$. The divergence is computed

$$\boldsymbol{\nabla} \cdot \mathbf{f}(x,y) = \left(\hat{\mathbf{x}} \frac{\partial}{\partial x} + \hat{\mathbf{y}} \frac{\partial}{\partial y} \right) \cdot \left(\frac{x}{r^2} \hat{\mathbf{x}} + \frac{y}{r^2} \hat{\mathbf{y}} \right) = \frac{\partial}{\partial x} \left(\frac{x}{r^2} \right) + \frac{\partial}{\partial y} \left(\frac{y}{r^2} \right)$$

We can compute the derivative

$$\frac{\partial}{\partial x} \left(\frac{x}{r^2} \right) = \frac{1}{r^2} \frac{\partial}{\partial x} x + x \frac{\partial}{\partial x} \frac{1}{r^2} = \frac{1}{r^2} - 2 \frac{x^2}{r^4}$$

A similar result can be found for the derivative

$$\frac{\partial}{\partial y} \left(\frac{x}{r^2} \right) = \frac{1}{r^2} - 2 \frac{y^2}{r^4}$$

Combining these terms leads to the final result for the divergence

$$\boldsymbol{\nabla} \cdot \mathbf{f}(x,y) = \left(\frac{1}{r^2} - 2 \frac{x^2}{r^4} \right) + \left(\frac{1}{r^2} - 2 \frac{y^2}{r^4} \right) = \frac{2}{r^2} \left(1 - \frac{x^2}{r^2} - \frac{y^2}{r^2} \right) = 0 \quad (5.9)$$

Let's consider the geometric meaning of the divergence being zero at all points in space.

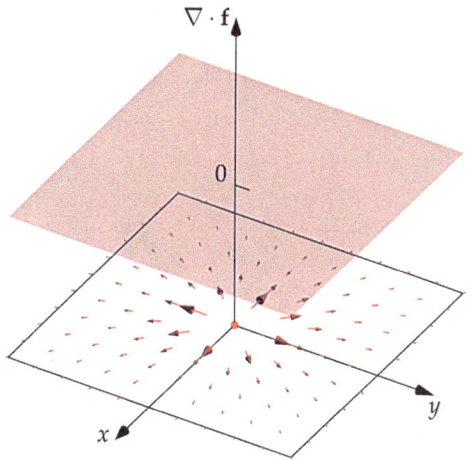

Figure 5.9: The vector field

$$\mathbf{f}(x,y) = \frac{x}{r^2} \hat{\mathbf{x}} + \frac{y}{r^2} \hat{\mathbf{y}}$$

depicted as an array of arrows on the xy-plane where $r^2 = x^2 + y^2$. The divergence of the vector field is a scalar constant $\boldsymbol{\nabla} \cdot \mathbf{f}(x,y) = 0$.

The vector field $\mathbf{f}(x,y)$ and its divergence $\boldsymbol{\nabla} \cdot \mathbf{f}(x,y)$ are shown in Figure 5.9. For the vector field shown in Figure 5.8, points at which the divergence was zero involved a balance of inward and outward pointing field lines. In this case, the field lines flow radially outward such that in any infinitesimal area the inward flux equals the outward flux and the divergence is zero. The forms of the divergence operator in plane polar coordinates, spherical polar coordinates, and cylindrical coordinates are provided in the complements.

5.2.5 Combining the gradient and the divergence to form the Laplacian

Let's explore a special combination of the gradient operator and the divergence operator that plays a central role in mathematical models in the physical sciences. Consider again the vector function

$$\mathbf{f}(x,y) = \frac{x}{r^2}\,\hat{\mathbf{x}} + \frac{y}{r^2}\,\hat{\mathbf{y}}$$

where $r = \sqrt{x^2 + y^2}$. We showed that the divergence is

$$\nabla \cdot \mathbf{f}(x,y) = 0$$

The vector field $\mathbf{f}(x,y)$ can be generated from the gradient of the scalar function $V(x,y) = \ln r$ as[9]

$$\nabla V(x,y) = \hat{\mathbf{x}}\,\frac{\partial V}{\partial x} + \hat{\mathbf{y}}\,\frac{\partial V}{\partial y}$$

$$= \hat{\mathbf{x}}\,\frac{\partial}{\partial x}\ln r + \hat{\mathbf{y}}\,\frac{\partial}{\partial y}\ln r = \hat{\mathbf{x}}\,\frac{x}{r^2} + \hat{\mathbf{y}}\,\frac{y}{r^2}$$

Combining these two results, we find

$$\nabla \cdot \mathbf{f}(x,y) = \nabla \cdot (\nabla V(x,y)) = \nabla^2 V(x,y) = 0$$

where

$$\nabla^2 \equiv \frac{\partial^2}{\partial x^2} + \frac{\partial^2}{\partial y^2} \qquad (5.10)$$

is called the *Laplace operator* and

$$\nabla^2 V(x,y) \equiv \frac{\partial^2 V}{\partial x^2} + \frac{\partial^2 V}{\partial y^2} = 0 \qquad (5.11)$$

is known as *Laplace's equation*.[10] The Laplace operator acts on a scalar function and returns a scalar function.[11] Our work above demonstrates that $V(x,y) = \ln r$ is a solution to Laplace's equation in two dimensions.

The three-dimensional version of Laplace's equation in cartesian coordinates is written

$$\nabla^2 V(x,y,z) \equiv \frac{\partial^2 V}{\partial x^2} + \frac{\partial^2 V}{\partial y^2} + \frac{\partial^2 V}{\partial z^2} = 0 \qquad (5.12)$$

For the scalar function $V(x,y,z) = \frac{1}{r}$ where $r = \sqrt{x^2 + y^2 + z^2}$ we find

$$\nabla^2 V = \frac{\partial^2 V}{\partial x^2} + \frac{\partial^2 V}{\partial y^2} + \frac{\partial^2 V}{\partial z^2}$$

$$= \frac{1}{r^5}\left(3x^2 - r^2\right) + \frac{1}{r^5}\left(3y^2 - r^2\right) + \frac{1}{r^5}\left(3z^2 - r^2\right) = 0$$

demonstrating that $V(x,y,z) = \frac{1}{r}$ is a solution to Laplace's equation in three dimensions.

The Laplace operator appears in many contexts in the physical sciences. The forms of Laplace's operator in two-dimensional plane polar coordinates and three-dimensional spherical polar coordinates, as well as three-dimensional cylindrical coordinates, are provided in the complements.

[9] Note that we equivalently use

$$V(x,y) = \ln r = \frac{1}{2}\ln r^2$$

where $r = \sqrt{x^2 + y^2}$. The latter form makes differentiation a bit easier. No square roots!

[10] The equation and associated operator are named for the French mathematician, physicist, and astronomer *Pierre-Simon Laplace* (1749-1827). Laplace made fundamental contributions to many areas of mathematics and physical science including mechanics, probability and statistics, and the theory of tides. He was first to express the philosophy of *scientific determinism* proposing that an all knowing being aware of the positions of all atoms and the forces acting on them, and able to submit that information to analysis, would have a certain knowledge of the past and the future.

[11] The Laplace operator ∇^2 has no value on its own. It only takes on value when operating on a function as in $\nabla^2 V(x,y)$.

A₅ The force and the potential energy

Force is something we experience directly. When Newton considered the response of a particle of mass m to a force $\mathbf{F}$, he reasoned that the force is equal to the mass times the acceleration:

$$\mathbf{F} = m\frac{d^2\mathbf{x}}{dt^2} = m\frac{d\mathbf{v}}{dt} = m\mathbf{a}$$

where $\mathbf{x}$ is the position, and the acceleration, $\mathbf{a}$, is the rate of change in the velocity, $\mathbf{v}$.

Newton discovered that the gravitational force between an object of mass m and another of mass M separated by a distance r follows the *inverse square law*[12]

$$\mathbf{F}(r) = -\frac{\mathcal{G}mM}{r^2}\,\hat{\mathbf{r}}$$

where $\mathcal{G}$ is the gravitational constant. In doing so, he provided a fundamental explanation of *Kepler's laws* for planetary motion around the Sun.[13] However, he never imagined the gravitational *potential energy*

$$V(r) = -\frac{\mathcal{G}mM}{r}$$

which is related to the force through the equation

$$\mathbf{F}(r) = -\frac{dV(r)}{dr}\,\hat{\mathbf{r}}$$

The gravitational potential energy and force are shown in Figure 5.10. The force is a vector that points in the direction of lower potential energy. If the slope of the potential energy is positive, the force is negative (pointing inward).

Suppose we have a force that depends only on the coordinates of the system. We can express the force in three dimensions using cartesian coordinates as

$$\mathbf{F} = \hat{\mathbf{x}}\,f_x + \hat{\mathbf{y}}\,f_y + \hat{\mathbf{z}}\,f_z = -\left(\hat{\mathbf{x}}\frac{\partial V}{\partial x} + \hat{\mathbf{y}}\frac{\partial V}{\partial y} + \hat{\mathbf{z}}\frac{\partial V}{\partial z}\right)$$

The cartesian components of the force are defined in terms of the respective partial derivatives of the potential energy

$$f_x = -\frac{\partial V}{\partial x} \qquad f_y = -\frac{\partial V}{\partial y} \qquad f_z = -\frac{\partial V}{\partial z}$$

We can also express the force as

$$\mathbf{F} = -\left(\hat{\mathbf{x}}\frac{\partial}{\partial x} + \hat{\mathbf{y}}\frac{\partial}{\partial y} + \hat{\mathbf{z}}\frac{\partial}{\partial z}\right)V(x,y,z) = -\boldsymbol{\nabla}V(x,y,z)$$

which is the negative gradient of the potential energy. The later result

$$\mathbf{F} = -\boldsymbol{\nabla}V$$

[12] The inverse square law is a foundation for the universal theory of gravitation presented by *Isaac Newton* (1642-1727) in *Philosophiae Naturalis Principia Mathematica* (1686) often known as *The Principia*.

[13] Discovered by German mathematician, astronomer, and astrologer *Johannes Kepler* (1571-1630). In 1611, Kepler published *On the Six-Cornered Snowflake* contemplating the hexagonal symmetry of snowflakes and anticipating modern ideas of molecular crystal structure.

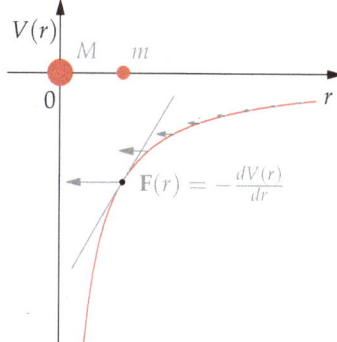

Figure 5.10: The potential energy function for gravitational interaction $V(r)$ between masses M and m and the corresponding force $\mathbf{F}(r)$. The length of each gray arrow is proportional to the magnitude of the force at the arrow's point of origin.

is completely general. Recognizing the natural symmetries of the system, one simply employs the appropriate form of the gradient operator for the coordinate system reflecting those symmetries. The forms of the gradient operator in plane polar coordinates, spherical polar coordinates, and cylindrical coordinates are provided in the complements.

B5 A survey of potential energy landscapes

Potential energy is an abstract concept. We do not experience potential energy in the same way we experience force. The most direct experience we have with potential energy is derived from moving in a hilly landscape. We know that we must do work to climb to the top of a hill. Doing this work, we are storing *potential energy* that has the potential to do work in the future. From the top of a hill, we coast downward without doing work. We convert our stored potential energy into the *kinetic energy* of motion.

When we consider potential energy in physical science, it can be useful to think in terms of a *potential energy landscape* that describes the potential energy as a function of position in space. We can then use our intuition developed by moving over hills, valleys, and planes to inform how we interpret the ups and downs of energy functions.

Consider a mass connected by a spring (see Figure 5.11). The resting point for the mass is $x = x_0$. For $x < x_0$ the spring is compressed and a force pushes outward to lengthen the spring. For $x > x_0$ the spring is stretched and a force pushes inward to shorten the spring. The force acting on the mass is described mathematically by *Hooke's law*[14]

$$\mathbf{F}(x) = -\kappa(x - x_0)$$

where κ is the spring's force constant. Hooke's law says the restoring force varies linearly with the displacement $(x - x_0)$ of the spring. For the same displacement $(x - x_0)$, a stiffer spring with a larger value of κ will result in a greater restoring force than a softer spring with a smaller value of κ. The corresponding potential energy

$$V(x) = \frac{1}{2}\kappa(x - x_0)^2$$

is quadratic in the displacement of the spring and related to the force as

$$\mathbf{F}(x) = -\frac{dV(x)}{dx} = -\kappa(x - x_0)$$

The potential energy function and corresponding force vectors, which vary in magnitude and direction, are shown in Figure 5.11. This quadratic potential energy function is used to model vibrations of bonds in molecules and small displacements of atoms in solids.

The interaction of two neutral atoms can be modeled by the *Lennard-Jones potential* energy function

$$V(r) = 4\varepsilon\left[\left(\frac{\sigma}{r}\right)^{12} - \left(\frac{\sigma}{r}\right)^{6}\right]$$

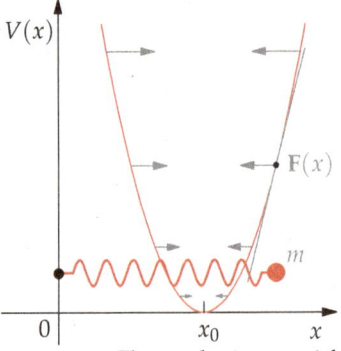

Figure 5.11: The quadratic potential energy function $V(x)$ and the corresponding force $\mathbf{F}(x)$.

[14] Named for English natural philosopher *Robert Hooke* (1635-1703). He published a popular book *Micrographia*, describing his seminal contributions to microscopy, and coined the word *cell*.

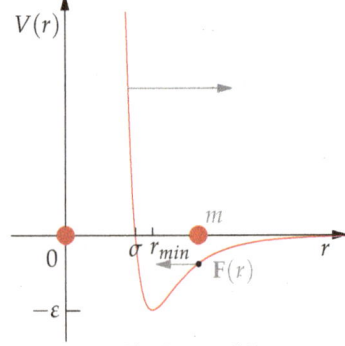

Figure 5.12: The Lennard-Jones potential energy function for pairwise atom-atom interaction $V(r)$ and the corresponding force $\mathbf{F}(r)$.

which depends on the radial distance r between the two atoms. The parameter σ represents the atomic diameter for which $V(\sigma) = 0$. The potential energy function has a minimum value at $r_{min} = \sqrt[6]{2}\sigma$ of $V(r_{min}) = -\varepsilon$ as shown in Figure 5.12. The force acting between the atoms is

$$\mathbf{F}(r) = -\frac{dV(r)}{dr}\,\hat{\mathbf{r}} = 4\varepsilon\left[12\left(\frac{\sigma}{r}\right)^{12} - 6\left(\frac{\sigma}{r}\right)^6\right]\frac{\hat{\mathbf{r}}}{r}$$

When $r < r_{min}$, the repulsive force proportional to $1/r^{12}$ dominates the total force, while at distances $r > r_{min}$, the attractive force proportional to $1/r^6$ is dominant.

Another important force in physical science is the coulomb force acting between charges q_1 and q_2 separated by a distance r. *Coulomb's law*[15] states that the force is proportional to the product of the charges divided by the square of the radial distance r between the charges

[15] Named for French military engineer and physicist *Charles-Augustin de Coulomb* (1736-1806).

$$\mathbf{F}(r) = \frac{\mathcal{C}q_1q_2}{r^2}\,\hat{\mathbf{r}}$$

where $\mathcal{C} = 1/4\pi\varepsilon_0$ is a constant and ε_0 is the *permittivity of a vacuum*. This force can be related to the *coulombic potential* energy function

$$V(r) = \frac{\mathcal{C}q_1q_2}{r}$$

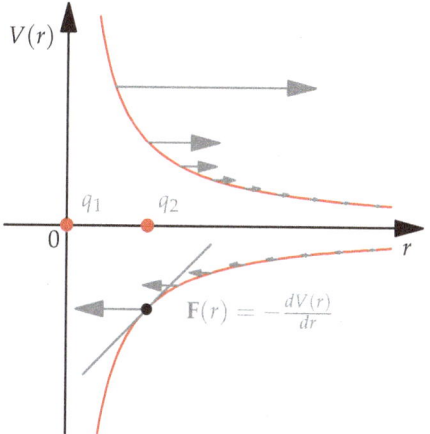

Figure 5.13: The coulomb potential energy function for charge-charge interaction $V(r)$ between charge q_1 and q_2 and the corresponding force $\mathbf{F}(r)$. When the charges have opposite signs (lower curve), the force is attractive. When the charges have like signs (upper curve), the force is repulsive.

For two opposite charges $q_1 = e$ and $q_2 = -e$, the force is attractive and pointing inward:

$$\mathbf{F}(r) = -\frac{\mathcal{C}e^2}{r^2}\,\hat{\mathbf{r}}$$

Conversely, for two like charges $q_1 = q_2 = -e$ the force is repulsive and pointing outward

$$\mathbf{F}(r) = \frac{\mathcal{C}e^2}{r^2}\,\hat{\mathbf{r}}$$

The attractive and repulsive coulomb potentials are shown in Figure 5.13 along with vectors indicating the varying magnitude and direction of the force acting between the charges. Just as in the case of the gravitational potential energy, the $1/r$ dependence of the potential energy leads to a $1/r^2$ dependence of the force.

The coulomb interaction is the essential force determining the properties of atoms and molecules. Negatively charged electrons are repelled by other negatively charged electrons but attracted to positively charged nuclei. This balance of forces gives rise to the rich variety of atoms and molecules that compose our world.

C$_5$ Explicit forms of vector operations

In this chapter, we explored the form of the gradient and divergence in two dimensions and three dimensions in cartesian coordinates. It is also possible to express the gradient and divergence in two dimensions in terms of plane polar coordinates and the gradient, divergence, and curl in three dimensions in spherical polar coordinates and cylindrical coordinates. Useful expressions for the gradient, divergence, curl, and Laplace operator are provided below.

Included for completeness are the differential displacement vector $d\mathbf{l}$, associated with an infinitesimal displacement, the differential area element dA, the change in surface area associated with an infinitesimal displacement in two dimensions, and the differential volume element dV, the change in volume associated with an infinitesimal displacement in three dimensions. In the next complement, a concise derivation of each vector operator is presented.

Cartesian coordinates (two dimensions)

Consider the two-dimensional scalar function $V(x, y)$ and vector function $\mathbf{f}(x, y) = \hat{\mathbf{x}}\, f_x + \hat{\mathbf{y}}\, f_y$ expressed in cartesian coordinates (x, y). The displacement vector and area element are written

$$d\mathbf{l} = \hat{\mathbf{x}}\, dx + \hat{\mathbf{y}}\, dy = \hat{\mathbf{x}}\, dl_x + \hat{\mathbf{y}}\, dl_y$$

$$dA = dl_x dl_y = dx\, dy$$

The operator relations for the gradient, divergence, and Laplace operator can be written

$$\nabla V = \hat{\mathbf{x}}\, \frac{\partial V}{\partial x} + \hat{\mathbf{y}}\, \frac{\partial V}{\partial y}$$

$$\nabla \cdot \mathbf{f} = \frac{\partial f_x}{\partial x} + \frac{\partial f_y}{\partial y}$$

$$\nabla^2 V = \frac{\partial^2 V}{\partial x^2} + \frac{\partial^2 V}{\partial y^2}$$

in two-dimensional cartesian coordinates. Let's generalize these formulas to the case of two-dimensional plane polar coordinates.

Plane polar coordinates (two dimensions)

Consider the two-dimensional scalar function $V(r, \theta)$ and vector function $\mathbf{f}(r, \theta) = \hat{\mathbf{r}} f_r + \hat{\boldsymbol{\theta}} f_\theta$ expressed in plane polar coordinates (r, θ) (see Figure 5.14). The displacement vector and area element are writen

$$d\mathbf{l} = \hat{\mathbf{r}} \, dr + \hat{\boldsymbol{\theta}} \, r d\theta = \hat{\mathbf{r}} \, dl_r + \hat{\boldsymbol{\theta}} \, dl_\theta$$

$$dA = dl_r dl_\theta = r \, dr \, d\theta$$

The operator relations for gradient, divergence, and Laplace operator are

$$\boldsymbol{\nabla} V = \hat{\mathbf{r}} \frac{\partial V}{\partial r} + \hat{\boldsymbol{\theta}} \frac{1}{r} \frac{\partial V}{\partial \theta}$$

$$\boldsymbol{\nabla} \cdot \mathbf{f} = \frac{1}{r} \frac{\partial}{\partial r} (r f_r) + \frac{1}{r} \frac{\partial f_\theta}{\partial \theta}$$

$$\nabla^2 V = \frac{1}{r} \frac{\partial}{\partial r} \left(r \frac{\partial V}{\partial r} \right) + \frac{1}{r^2} \frac{\partial^2 V}{\partial \theta^2}$$

in two-dimensional plane polar coordinates.

Cartesian coordinates (three dimensions)

Consider the three-dimensional scalar function $V(x, y, z)$ and vector function $\mathbf{f}(x, y, z) = \hat{\mathbf{x}} f_x + \hat{\mathbf{y}} f_y + \hat{\mathbf{z}} f_z$ expressed in cartesian coordinates (x, y, z) (see Figure 5.15). The displacement vector and volume element are written

$$d\mathbf{l} = \hat{\mathbf{x}} \, dx + \hat{\mathbf{y}} \, dy + \hat{\mathbf{z}} \, dz = \hat{\mathbf{x}} \, dl_x + \hat{\mathbf{y}} \, dl_y + \hat{\mathbf{z}} \, dl_z$$

$$dV = dl_x dl_y dl_z = dx \, dy \, dz$$

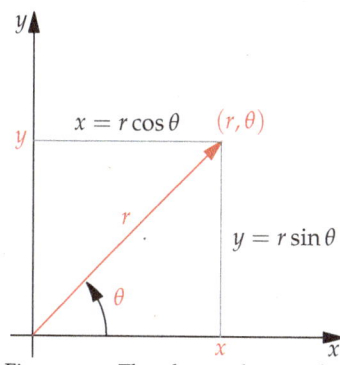

Figure 5.14: The plane polar coordinate system in two dimensions.

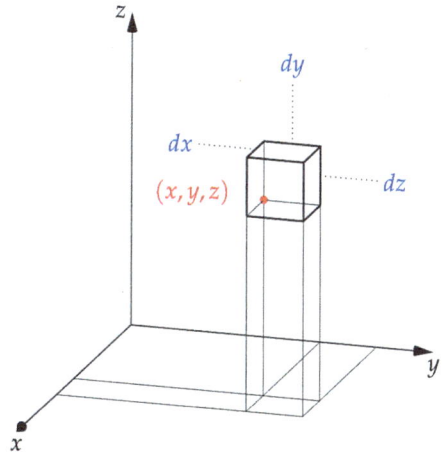

Figure 5.15: The cartesian coordinate system in three dimensions. The volume element $dV = dxdydz$ (bold) is the product of the three lengths dx, dy, and dz.

The operator relations for gradient, divergence, curl, and Laplace operator are

$$\boldsymbol{\nabla} V = \hat{\mathbf{x}} \frac{\partial V}{\partial x} + \hat{\mathbf{y}} \frac{\partial V}{\partial y} + \hat{\mathbf{z}} \frac{\partial V}{\partial z}$$

$$\boldsymbol{\nabla} \cdot \mathbf{f} = \frac{\partial f_x}{\partial x} + \frac{\partial f_y}{\partial y} + \frac{\partial f_z}{\partial z}$$

$$\boldsymbol{\nabla} \times \mathbf{f} = \hat{\mathbf{x}} \left(\frac{\partial f_z}{\partial y} - \frac{\partial f_y}{\partial z} \right) + \hat{\mathbf{y}} \left(\frac{\partial f_x}{\partial z} - \frac{\partial f_z}{\partial x} \right) + \hat{\mathbf{z}} \left(\frac{\partial f_y}{\partial x} - \frac{\partial f_x}{\partial y} \right)$$

$$\nabla^2 V = \frac{\partial^2 V}{\partial x^2} + \frac{\partial^2 V}{\partial y^2} + \frac{\partial^2 V}{\partial z^2}$$

in three-dimensional cartesian coordinates. Note that we have added the definition of the curl operator that we did not define in two dimensions.

Spherical polar coordinates (three dimensions)

Consider the three-dimensional scalar function $V(r, \theta, \varphi)$ and vector function $\mathbf{f}(r, \theta, \varphi) = \hat{\mathbf{r}}\, f_r + \hat{\boldsymbol{\theta}}\, f_\theta + \hat{\boldsymbol{\varphi}}\, f_\varphi$ expressed in spherical polar coordinates (r, θ, φ) (see Figure 5.16).[16] The displacement vector and volume element are written

$$d\mathbf{l} = \hat{\mathbf{r}}\, dr + \hat{\boldsymbol{\theta}}\, rd\theta + \hat{\boldsymbol{\varphi}}\, r\sin\theta\, \varphi = \hat{\mathbf{r}}\, dl_r + \hat{\boldsymbol{\theta}}\, dl_\theta + \hat{\boldsymbol{\varphi}}\, dl_\varphi$$

$$dV = dl_r dl_\theta dl_\varphi = r^2 \sin\theta\, dr\, d\theta\, d\varphi$$

[16] It is useful to know the identity

$$\frac{1}{r^2} \frac{\partial}{\partial r} \left(r^2 \frac{\partial V}{\partial r} \right) \equiv \frac{1}{r} \frac{\partial^2}{\partial r^2} (rV)$$

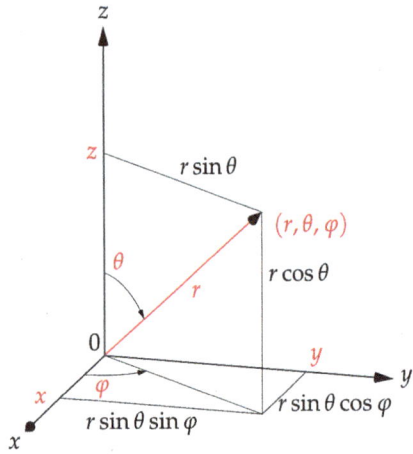

Figure 5.16: The spherical polar coordinate system in three dimensions.

The operator relations for gradient, divergence, curl, and Laplace operator are

$$\boldsymbol{\nabla} V = \hat{\mathbf{r}} \frac{\partial V}{\partial r} + \hat{\boldsymbol{\theta}} \frac{1}{r} \frac{\partial V}{\partial \theta} + \hat{\boldsymbol{\varphi}} \frac{1}{r\sin\theta} \frac{\partial V}{\partial \varphi}$$

$$\boldsymbol{\nabla} \cdot \mathbf{f} = \frac{1}{r^2} \frac{\partial}{\partial r} \left(r^2 f_r \right) + \frac{1}{r\sin\theta} \frac{\partial}{\partial \theta} \left(\sin\theta\, f_\theta \right) + \frac{1}{r\sin\theta} \frac{\partial f_\varphi}{\partial \varphi}$$

$$\boldsymbol{\nabla} \times \mathbf{f} = \hat{\mathbf{r}} \frac{1}{r\sin\theta} \left(\frac{\partial}{\partial \theta} (\sin\theta f_\varphi) - \frac{\partial f_\theta}{\partial \varphi} \right) + \hat{\boldsymbol{\theta}} \frac{1}{r} \left(\frac{1}{\sin\theta} \frac{\partial f_r}{\partial \varphi} - \frac{\partial}{\partial r} (rf_\varphi) \right)$$
$$+ \hat{\boldsymbol{\varphi}} \frac{1}{r} \left(\frac{\partial}{\partial r} (rf_\theta) - \frac{\partial f_r}{\partial \theta} \right)$$

$$\nabla^2 V = \frac{1}{r^2}\frac{\partial}{\partial r}\left(r^2\frac{\partial V}{\partial r}\right) + \frac{1}{r^2\sin\theta}\frac{\partial}{\partial\theta}\left(\sin\theta\frac{\partial V}{\partial\theta}\right) + \frac{1}{r^2\sin^2\theta}\frac{\partial^2 V}{\partial\varphi^2}$$

in three-dimensional spherical polar coordinates.

Cylindrical coordinates (three dimensions)

Consider the three-dimensional scalar function $V(r,\theta,z)$ and vector function $\mathbf{f}(r,\theta,z) = \hat{\mathbf{r}}\,f_r + \hat{\boldsymbol{\theta}}\,f_\theta + \hat{\mathbf{z}}\,f_z$ expressed in cylindrical coordinates (r,θ,z) (see Figure 5.17). The displacement vector and volume element are written

$$d\mathbf{l} = \hat{\mathbf{r}}\,dr + \hat{\boldsymbol{\theta}}\,rd\theta + \hat{\mathbf{z}}\,dz = \hat{\mathbf{r}}\,dl_r + \hat{\boldsymbol{\theta}}\,dl_\theta + \hat{\mathbf{z}}\,dl_z$$

$$dV = dl_r dl_\theta dl_z = r\,dr\,d\theta\,dz$$

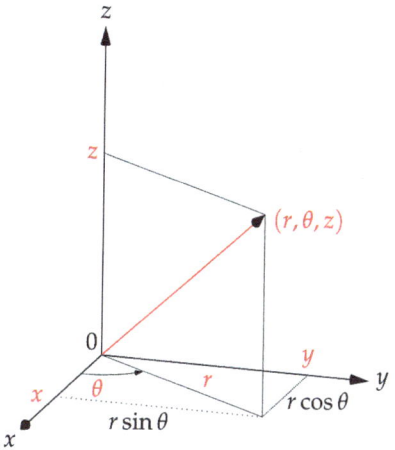

Figure 5.17: The cylindrical coordinate system in three dimensions.

The operator relations for gradient, divergence, curl, and Laplace operator are

$$\boldsymbol{\nabla} V = \hat{\mathbf{r}}\,\frac{\partial V}{\partial r} + \hat{\boldsymbol{\theta}}\,\frac{1}{r}\frac{\partial V}{\partial\theta} + \hat{\mathbf{z}}\,\frac{\partial V}{\partial z}$$

$$\boldsymbol{\nabla}\cdot\mathbf{f} = \frac{1}{r}\frac{\partial}{\partial r}\left(rf_r\right) + \frac{1}{r}\frac{\partial f_\theta}{\partial\theta} + \frac{\partial f_z}{\partial z}$$

$$\boldsymbol{\nabla}\times\mathbf{f} = \hat{\mathbf{r}}\left(\frac{1}{r}\frac{\partial f_z}{\partial\theta} - \frac{\partial f_\theta}{\partial z}\right) + \hat{\boldsymbol{\theta}}\left(\frac{\partial f_r}{\partial z} - \frac{\partial f_z}{\partial r}\right) + \hat{\mathbf{z}}\frac{1}{r}\left(\frac{\partial}{\partial r}\left(rf_\theta\right) - \frac{\partial f_r}{\partial\theta}\right)$$

$$\nabla^2 V = \frac{1}{r}\frac{\partial}{\partial r}\left(r\frac{\partial V}{\partial r}\right) + \frac{1}{r^2}\frac{\partial^2 V}{\partial\theta^2} + \frac{\partial^2 V}{\partial z^2}$$

in three-dimensional cylindrical coordinates.

We have explored vector operators in two orthogonal coordinate systems in two dimensions and three orthogonal coordinate systems in three dimensions. All five coordinate systems are commonly used in the physical sciences.

D₅ Deriving explicit forms for vector operations

The explicit forms of vector operators including the gradient, divergence, curl, and laplacian operators were presented in the previous complement. In this complement, we derive the explicit forms of the gradient, divergence, and laplacian operators in two-dimensional cartesian and plane polar coordinates, as well as three-dimensional cartesian, spherical polar, and cylindrical coordinates.

An intuitive approach utilizing the differential displacement vector is used to derive the gradient operator, which is subsequently used to derive the divergence and laplacian. This approach is more economical and significantly less complicated than direct approaches using partial differentiation.

Cartesian coordinates (two dimensions)

Consider a curve in two-dimensional cartesian coordinates where a displacement along the curve can be written in terms of the displacement vector

$$d\mathbf{l} = dx\,\hat{\mathbf{x}} + dy\,\hat{\mathbf{y}} = dl_x\hat{\mathbf{x}} + dl_y\,\hat{\mathbf{y}}$$

We can express the gradient of a function $V(x,y)$ in terms of its x and y components as

$$\boldsymbol{\nabla}V = (\boldsymbol{\nabla}V \cdot \hat{\mathbf{x}})\hat{\mathbf{x}} + (\boldsymbol{\nabla}V \cdot \hat{\mathbf{y}})\hat{\mathbf{y}} = (\boldsymbol{\nabla}V)_x\,\hat{\mathbf{x}} + (\boldsymbol{\nabla}V)_y\,\hat{\mathbf{y}}$$

Our goal is to derive an expression for the gradient operator by finding explicit expressions for $(\boldsymbol{\nabla}V)_x$ and $(\boldsymbol{\nabla}V)_y$ (see Figure 5.18).

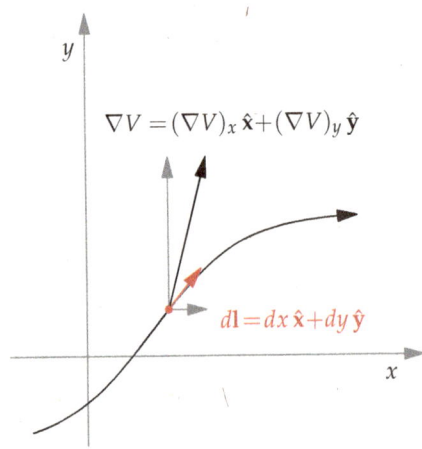

The total differential of $V(x,y)$ can be written

$$dV = \frac{\partial V}{\partial x}\,dx + \frac{\partial V}{\partial y}\,dy$$

It can also be expressed as the scalar product of the gradient $\boldsymbol{\nabla}V$ and the

Figure 5.18: The displacement vector $d\mathbf{l}$ (red) with the gradient $\boldsymbol{\nabla}V$ (black) and components of its projection onto the unit vectors $\hat{\mathbf{x}}$ and $\hat{\mathbf{y}}$ (gray).

displacement vector $d\mathbf{l}$ as

$$dV = \boldsymbol{\nabla}V \cdot d\mathbf{l} = (\boldsymbol{\nabla}V)_x\, dx + (\boldsymbol{\nabla}V)_y\, dy$$

Comparing terms in the two expressions for dV we find

$$(\boldsymbol{\nabla}V)_x = \frac{\partial V}{\partial x} \qquad (\boldsymbol{\nabla}V)_y = \frac{\partial V}{\partial y}$$

so that the gradient in two-dimensional cartesian coordinates is

$$\boldsymbol{\nabla}V = \frac{\partial V}{\partial x}\,\hat{\mathbf{x}} + \frac{\partial V}{\partial y}\,\hat{\mathbf{y}} \tag{5.13}$$

The divergence of a vector function $\mathbf{f} = f_x\,\hat{\mathbf{x}} + f_y\,\hat{\mathbf{y}}$ is defined as the scalar product of the gradient operator and the vector function

$$\boldsymbol{\nabla}\cdot\mathbf{f} = \left(\hat{\mathbf{x}}\frac{\partial}{\partial x} + \hat{\mathbf{y}}\frac{\partial}{\partial y}\right)\cdot(f_x\,\hat{\mathbf{x}} + f_y\,\hat{\mathbf{y}})$$

In cartesian coordinates, the unit vectors are independent of the coordinates. As such, the divergence is simply

$$\boldsymbol{\nabla}\cdot\mathbf{f} = \frac{\partial f_x}{\partial x} + \frac{\partial f_y}{\partial y} \tag{5.14}$$

From this result, we can evaluate the laplacian of V defined as

$$\nabla^2 V = \boldsymbol{\nabla}\cdot\boldsymbol{\nabla}V = \left(\hat{\mathbf{x}}\frac{\partial}{\partial x} + \hat{\mathbf{y}}\frac{\partial}{\partial y}\right)\cdot\left(\frac{\partial V}{\partial x}\,\hat{\mathbf{x}} + \frac{\partial V}{\partial y}\,\hat{\mathbf{y}}\right)$$

Once again, as the unit vectors are independent of the coordinates we find

$$\nabla^2 V = \frac{\partial^2 V}{\partial x^2} + \frac{\partial^2 V}{\partial y^2} \tag{5.15}$$

Plane polar coordinates (two dimensions)

Let's repeat that process in two-dimensional plane polar coordinates. We define a displacement along a curve in terms of the displacement vector

$$d\mathbf{l} = dr\,\hat{\mathbf{r}} + r d\theta\,\hat{\boldsymbol{\theta}} = dl_r\,\hat{\mathbf{r}} + dl_\theta\,\hat{\boldsymbol{\theta}}$$

We can express the gradient of a function $V(r,\theta)$ in terms of its r and θ components as

$$\boldsymbol{\nabla}V = (\boldsymbol{\nabla}V\cdot\hat{\mathbf{r}})\hat{\mathbf{r}} + (\boldsymbol{\nabla}V\cdot\hat{\boldsymbol{\theta}})\hat{\boldsymbol{\theta}} = (\boldsymbol{\nabla}V)_r\,\hat{\mathbf{r}} + (\boldsymbol{\nabla}V)_\theta\,\hat{\boldsymbol{\theta}}$$

Our goal is to derive an expression for the gradient operator by finding explicit expressions for $(\boldsymbol{\nabla}V)_r$ and $(\boldsymbol{\nabla}V)_\theta$ (see Figure 5.19). The total differential of $V(r,\theta)$ can be written

$$dV = \frac{\partial V}{\partial r}\,dr + \frac{\partial V}{\partial \theta}\,d\theta$$

It can also be expressed as the scalar product of the gradient $\boldsymbol{\nabla}V$ and the

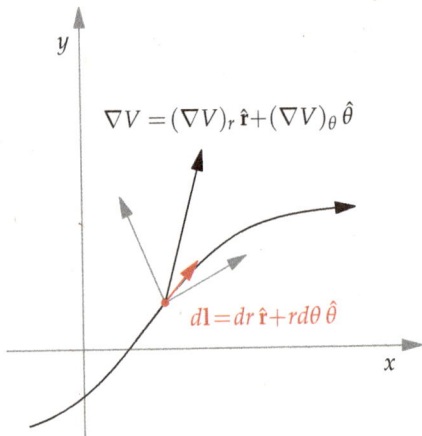

Figure 5.19: The displacement vector $d\mathbf{l}$ (red) with the gradient $\boldsymbol{\nabla}V$ (black) and components of its projection onto the unit vectors $\hat{\mathbf{r}}$ and $\hat{\boldsymbol{\theta}}$ (gray).

displacement vector $d\mathbf{l}$ as

$$dV = \boldsymbol{\nabla}V \cdot d\mathbf{l} = (\boldsymbol{\nabla}V)_r \, dr + (\boldsymbol{\nabla}V)_\theta \, r \, d\theta$$

Comparing terms in the two expressions for dV we find

$$(\boldsymbol{\nabla}V)_r = \frac{\partial V}{\partial r} \qquad (\boldsymbol{\nabla}V)_\theta = \frac{1}{r}\frac{\partial V}{\partial \theta}$$

so that the gradient in two-dimensional plane polar coordinates is

$$\boldsymbol{\nabla}V = \frac{\partial V}{\partial r}\,\hat{\mathbf{r}} + \frac{1}{r}\frac{\partial V}{\partial \theta}\,\hat{\boldsymbol{\theta}} \tag{5.16}$$

The divergence of a vector function $\mathbf{f} = f_r\,\hat{\mathbf{r}} + f_\theta\,\hat{\boldsymbol{\theta}}$ is defined as the scalar product of the gradient operator and the vector function

$$\boldsymbol{\nabla} \cdot \mathbf{f} = \left(\hat{\mathbf{r}}\frac{\partial}{\partial r} + \hat{\boldsymbol{\theta}}\frac{1}{r}\frac{\partial}{\partial \theta} \right) \cdot (f_r\,\hat{\mathbf{r}} + f_\theta\,\hat{\boldsymbol{\theta}})$$

In plane polar coordinates, the unit vectors are dependent on the coordinates as

$$\hat{\mathbf{r}} = \cos\theta\,\hat{\mathbf{x}} + \sin\theta\,\hat{\mathbf{y}} \qquad \hat{\boldsymbol{\theta}} = -\sin\theta\,\hat{\mathbf{x}} + \cos\theta\,\hat{\mathbf{y}}$$

leading to the relations

$$\frac{\partial \hat{\mathbf{r}}}{\partial r} = 0 \qquad \frac{\partial \hat{\boldsymbol{\theta}}}{\partial r} = 0 \qquad \frac{\partial \hat{\mathbf{r}}}{\partial \theta} = \hat{\boldsymbol{\theta}} \qquad \frac{\partial \hat{\boldsymbol{\theta}}}{\partial \theta} = -\hat{\mathbf{r}}$$

As such, in evaluating $\boldsymbol{\nabla}\cdot\mathbf{f}$ we must consider derivatives of the unit vectors as well as the coefficients leading to $2 \times (2 \times 2) = 8$ terms

$$\boldsymbol{\nabla}\cdot\mathbf{f} = \hat{\mathbf{r}} \cdot \left(\hat{\mathbf{r}}\frac{\partial f_r}{\partial r} + f_r\frac{\partial \hat{\mathbf{r}}}{\partial r} + \hat{\boldsymbol{\theta}}\frac{\partial f_\theta}{\partial r} + f_\theta\frac{\partial \hat{\boldsymbol{\theta}}}{\partial r} \right) + \frac{1}{r}\hat{\boldsymbol{\theta}} \cdot \left(\hat{\mathbf{r}}\frac{\partial f_r}{\partial \theta} + f_r\frac{\partial \hat{\mathbf{r}}}{\partial \theta} + \hat{\boldsymbol{\theta}}\frac{\partial f_\theta}{\partial \theta} + f_\theta\frac{\partial \hat{\boldsymbol{\theta}}}{\partial \theta} \right)$$

$$= \hat{\mathbf{r}} \cdot \left(\hat{\mathbf{r}}\frac{\partial f_r}{\partial r} + \hat{\boldsymbol{\theta}}\frac{\partial f_\theta}{\partial r} \right) + \frac{1}{r}\hat{\boldsymbol{\theta}} \cdot \left(\hat{\mathbf{r}}\frac{\partial f_r}{\partial \theta} + \hat{\boldsymbol{\theta}}f_r + \hat{\boldsymbol{\theta}}\frac{\partial f_\theta}{\partial \theta} - \hat{\mathbf{r}}f_\theta \right)$$

As the unit vectors are orthonormal we have $\hat{\mathbf{r}}\cdot\hat{\mathbf{r}} = \hat{\boldsymbol{\theta}}\cdot\hat{\boldsymbol{\theta}} = 1$ and $\hat{\mathbf{r}}\cdot\hat{\boldsymbol{\theta}} = 0$ leading to our final expression for the divergence in plane polar coordinates

$$\boldsymbol{\nabla}\cdot\mathbf{f} = \frac{\partial f_r}{\partial r} + \frac{1}{r}\left(f_r + \frac{\partial f_\theta}{\partial \theta}\right) \tag{5.17}$$

From this result, we can evaluate the laplacian of V defined as

$$\nabla^2 V = \boldsymbol{\nabla}\cdot\boldsymbol{\nabla}V = \boldsymbol{\nabla}\cdot\mathbf{f}$$

where

$$\mathbf{f} = \frac{\partial V}{\partial r}\,\hat{\mathbf{r}} + \frac{1}{r}\frac{\partial V}{\partial \theta}\,\hat{\boldsymbol{\theta}} = f_r\,\hat{\mathbf{r}} + f_\theta\,\hat{\boldsymbol{\theta}}$$

Inserting the forms for f_r and f_θ into Equation 5.17 leads to

$$\nabla^2 V = \frac{\partial}{\partial r}\left(\frac{\partial V}{\partial r}\right) + \frac{1}{r}\left[\frac{\partial V}{\partial r} + \frac{\partial}{\partial \theta}\left(\frac{1}{r}\frac{\partial V}{\partial \theta}\right)\right]$$

$$= \frac{\partial^2 V}{\partial r^2} + \frac{1}{r}\frac{\partial V}{\partial r} + \frac{1}{r^2}\frac{\partial^2 V}{\partial \theta^2}$$

where the first two terms can be cleverly combined to form the final compact expression

$$\nabla^2 V = \frac{1}{r}\frac{\partial}{\partial r}\left(r\frac{\partial V}{\partial r}\right) + \frac{1}{r^2}\frac{\partial^2 V}{\partial \theta^2} \tag{5.18}$$

Cartesian coordinates (three dimensions)

Let's repeat that process for three-dimensional cartesian coordinates. We define a displacement along a curve in terms of the displacement vector

$$d\mathbf{l} = dx\,\hat{\mathbf{x}} + dy\,\hat{\mathbf{y}} + dz\,\hat{\mathbf{z}} = dl_x\,\hat{\mathbf{x}} + dl_y\,\hat{\mathbf{y}} + dl_z\,\hat{\mathbf{z}}$$

We can express the gradient of a function $V(x,y,z)$ in terms of its x, y, and z components as

$$\boldsymbol{\nabla}V = (\boldsymbol{\nabla}V\cdot\hat{\mathbf{x}})\hat{\mathbf{x}} + (\boldsymbol{\nabla}V\cdot\hat{\mathbf{y}})\hat{\mathbf{y}} + (\boldsymbol{\nabla}V\cdot\hat{\mathbf{z}})\hat{\mathbf{z}}$$

$$= (\boldsymbol{\nabla}V)_x\,\hat{\mathbf{x}} + (\boldsymbol{\nabla}V)_y\,\hat{\mathbf{y}} + (\boldsymbol{\nabla}V)_z\,\hat{\mathbf{z}}$$

Our goal is to derive an expression for the gradient operator by finding explicit expressions for $(\boldsymbol{\nabla}V)_x$, $(\boldsymbol{\nabla}V)_y$ and $(\boldsymbol{\nabla}V)_z$.

The total differential of $V(x,y,z)$ can be written

$$dV = \frac{\partial V}{\partial x}\,dx + \frac{\partial V}{\partial y}\,dy + \frac{\partial V}{\partial z}\,dz$$

It can also be expressed as the scalar product of the gradient $\boldsymbol{\nabla}V$ and the displacement vector $d\mathbf{l}$ as

$$dV = \boldsymbol{\nabla}V\cdot d\mathbf{l} = (\boldsymbol{\nabla}V)_x\,dx + (\boldsymbol{\nabla}V)_y\,dy + (\boldsymbol{\nabla}V)_z\,dz$$

Comparing terms in the two expressions for df we find

$$(\boldsymbol{\nabla}V)_x = \frac{\partial V}{\partial x} \qquad (\boldsymbol{\nabla}V)_y = \frac{\partial V}{\partial y} \qquad (\boldsymbol{\nabla}V)_z = \frac{\partial V}{\partial z}$$

so that the gradient in three-dimensional cartesian coordinates is

$$\nabla V = \frac{\partial V}{\partial x}\,\hat{\mathbf{x}} + \frac{\partial V}{\partial y}\,\hat{\mathbf{y}} + \frac{\partial V}{\partial z}\,\hat{\mathbf{z}} \tag{5.19}$$

The gradient of a scalar function V results in a vector field.

The divergence of a vector function $\mathbf{f} = f_x\,\hat{\mathbf{x}} + f_y\,\hat{\mathbf{y}} + f_z\,\hat{\mathbf{z}}$ is defined as the scalar product of the gradient operator and the vector function

$$\nabla \cdot \mathbf{f} = \left(\hat{\mathbf{x}}\frac{\partial}{\partial x} + \hat{\mathbf{y}}\frac{\partial}{\partial y} + \hat{\mathbf{z}}\frac{\partial}{\partial z}\right) \cdot \left(f_x\,\hat{\mathbf{x}} + f_y\,\hat{\mathbf{y}} + f_z\,\hat{\mathbf{z}}\right)$$

In cartesian coordinates, the unit vectors are independent of the coordinates. As such, the divergence is simply

$$\nabla \cdot \mathbf{f} = \frac{\partial f_x}{\partial x} + \frac{\partial f_y}{\partial y} + \frac{\partial f_z}{\partial z} \tag{5.20}$$

The divergence of a vector field $\mathbf{f}$ results in a scalar function.

From this result, we can evaluate the laplacian of V defined as

$$\nabla^2 V = \nabla \cdot \nabla V = \left(\hat{\mathbf{x}}\frac{\partial}{\partial x} + \hat{\mathbf{y}}\frac{\partial}{\partial y} + \hat{\mathbf{z}}\frac{\partial}{\partial z}\right) \cdot \left(\frac{\partial V}{\partial x}\,\hat{\mathbf{x}} + \frac{\partial V}{\partial y}\,\hat{\mathbf{y}} + \frac{\partial V}{\partial z}\,\hat{\mathbf{z}}\right)$$

Once again, as the unit vectors are independent of the coordinates we find

$$\nabla^2 V = \frac{\partial^2 V}{\partial x^2} + \frac{\partial^2 V}{\partial y^2} + \frac{\partial^2 V}{\partial z^2} \tag{5.21}$$

The laplacian of a scalar function V results in a scalar function.

Finally, the curl in three-dimensional cartesian coordinates can be expressed using the determinant as

$$\nabla \times \mathbf{f} = \begin{vmatrix} \hat{\mathbf{x}} & \hat{\mathbf{y}} & \hat{\mathbf{z}} \\ \dfrac{\partial}{\partial x} & \dfrac{\partial}{\partial y} & \dfrac{\partial}{\partial z} \\ f_x & f_y & f_z \end{vmatrix}$$

$$= \hat{\mathbf{x}}\left(\frac{\partial f_z}{\partial y} - \frac{\partial f_y}{\partial z}\right) + \hat{\mathbf{y}}\left(\frac{\partial f_x}{\partial z} - \frac{\partial f_z}{\partial x}\right) + \hat{\mathbf{z}}\left(\frac{\partial f_y}{\partial x} - \frac{\partial f_x}{\partial y}\right) \tag{5.22}$$

The curl of a vector field $\mathbf{f}$ results in a vector field.

Spherical polar coordinates (three dimensions)

Let's repeat that process for three-dimensional spherical polar coordinates. We define a displacement along a curve in terms of the displacement vector

$$d\mathbf{l} = dr\,\hat{\mathbf{r}} + r d\theta\,\hat{\boldsymbol{\theta}} + r\sin\theta\,d\varphi\,\hat{\boldsymbol{\varphi}} = dl_r\,\hat{\mathbf{r}} + dl_\theta\,\hat{\boldsymbol{\theta}} + dl_\varphi\,\hat{\boldsymbol{\varphi}}$$

We can express the gradient of a function $V(r, \theta, \varphi)$ in terms of its r, θ and φ components as

$$\nabla V = (\nabla V \cdot \hat{\mathbf{r}})\hat{\mathbf{r}} + (\nabla V \cdot \hat{\boldsymbol{\theta}})\hat{\boldsymbol{\theta}} + (\nabla V \cdot \hat{\boldsymbol{\varphi}})\hat{\boldsymbol{\varphi}}$$

$$= (\nabla V)_r\,\hat{\mathbf{r}} + (\nabla V)_\theta\,\hat{\boldsymbol{\theta}} + (\nabla V)_\varphi\,\hat{\boldsymbol{\varphi}}$$

Our goal is to derive an expression for the gradient operator by finding explicit expressions for $(\nabla V)_r$, $(\nabla V)_\theta$, and $(\nabla V)_\varphi$.

The total differential of $V(r, \theta, \varphi)$ can be written

$$dV = \frac{\partial V}{\partial r}\, dr + \frac{\partial V}{\partial \theta}\, d\theta + \frac{\partial V}{\partial \varphi}\, d\varphi$$

it can also be expressed as the scalar product of the gradient ∇V and the displacement vector $d\mathbf{l}$ as

$$dV = \nabla V \cdot d\mathbf{l} = (\nabla V)_r\, dr + (\nabla V)_\theta\, r\, d\theta + (\nabla V)_\varphi\, r\sin\theta\, d\varphi$$

Comparing terms in the two expressions for dV we find

$$(\nabla V)_r = \frac{\partial V}{\partial r} \qquad (\nabla V)_\theta = \frac{1}{r}\frac{\partial V}{\partial \theta} \qquad (\nabla V)_\varphi = \frac{1}{r\sin\theta}\frac{\partial V}{\partial \varphi}$$

so that the gradient in three-dimensional spherical polar coordinates is

$$\nabla V = \frac{\partial V}{\partial r}\,\hat{\mathbf{r}} + \frac{1}{r}\frac{\partial V}{\partial \theta}\,\hat{\boldsymbol{\theta}} + \frac{1}{r\sin\theta}\frac{\partial V}{\partial \varphi}\,\hat{\boldsymbol{\varphi}} \qquad (5.23)$$

The gradient of a scalar function V results in a vector field.

The divergence of a vector function $\mathbf{f} = f_r\,\hat{\mathbf{r}} + f_\theta\,\hat{\boldsymbol{\theta}} + f_\varphi\,\hat{\boldsymbol{\varphi}}$ is defined as the scalar product of the gradient operator and the vector function

$$\nabla \cdot \mathbf{f} = \left(\hat{\mathbf{r}}\frac{\partial}{\partial r} + \hat{\boldsymbol{\theta}}\frac{1}{r}\frac{\partial}{\partial \theta} + \hat{\boldsymbol{\varphi}}\frac{1}{r\sin\theta}\frac{\partial}{\partial \varphi}\right) \cdot \left(f_r\,\hat{\mathbf{r}} + f_\theta\,\hat{\boldsymbol{\theta}} + f_\varphi\,\hat{\boldsymbol{\varphi}}\right)$$

In spherical polar coordinates, the unit vectors are dependent on the coordinates as in Equation 4.3

$$\hat{\mathbf{r}} = \sin\theta\cos\varphi\,\hat{\mathbf{x}} + \sin\theta\sin\varphi\,\hat{\mathbf{y}} + \cos\theta\,\hat{\mathbf{z}}$$

$$\hat{\boldsymbol{\theta}} = \cos\theta\cos\varphi\,\hat{\mathbf{x}} + \cos\theta\sin\varphi\,\hat{\mathbf{y}} - \sin\theta\,\hat{\mathbf{z}}$$

$$\hat{\boldsymbol{\varphi}} = -\sin\varphi\,\hat{\mathbf{x}} + \cos\varphi\,\hat{\mathbf{y}}$$

As such, in evaluating $\nabla \cdot \mathbf{f}$ we must consider derivatives of the unit vectors as well as the coefficients leading to $2 \times (3 \times 3) = 18$ terms. Evaluating the derivatives

$$\frac{\partial \hat{\mathbf{r}}}{\partial r} = 0 \qquad\qquad \frac{\partial \hat{\boldsymbol{\theta}}}{\partial r} = 0 \qquad\qquad \frac{\partial \hat{\boldsymbol{\varphi}}}{\partial r} = 0$$

$$\frac{\partial \hat{\mathbf{r}}}{\partial \theta} = \hat{\boldsymbol{\theta}} \qquad\qquad \frac{\partial \hat{\boldsymbol{\theta}}}{\partial \theta} = -\hat{\mathbf{r}} \qquad\qquad \frac{\partial \hat{\boldsymbol{\varphi}}}{\partial \theta} = 0$$

$$\frac{\partial \hat{\mathbf{r}}}{\partial \varphi} = \sin\theta\,\hat{\boldsymbol{\varphi}} \qquad \frac{\partial \hat{\boldsymbol{\theta}}}{\partial \varphi} = \cos\theta\,\hat{\boldsymbol{\varphi}} \qquad \frac{\partial \hat{\boldsymbol{\varphi}}}{\partial \varphi} = -\cos\theta\,\hat{\boldsymbol{\theta}} - \sin\theta\,\hat{\mathbf{r}}$$

and noting that the unit vectors are orthonormal, such that

$$\hat{\mathbf{r}}\cdot\hat{\mathbf{r}} = \hat{\boldsymbol{\theta}}\cdot\hat{\boldsymbol{\theta}} = \hat{\boldsymbol{\varphi}}\cdot\hat{\boldsymbol{\varphi}} = 1$$

$$\hat{\mathbf{r}}\cdot\hat{\boldsymbol{\theta}} = \hat{\mathbf{r}}\cdot\hat{\boldsymbol{\varphi}} = \hat{\boldsymbol{\theta}}\cdot\hat{\boldsymbol{\varphi}} = 0$$

we find that 10 of the 18 terms are zero. The remaining terms are

$$\nabla \cdot \mathbf{f} = \hat{\mathbf{r}} \cdot \left(\hat{\mathbf{r}} \frac{\partial f_r}{\partial r} \right) + \frac{1}{r} \hat{\boldsymbol{\theta}} \cdot \left(\hat{\boldsymbol{\theta}} f_r + \hat{\boldsymbol{\theta}} \frac{\partial f_\theta}{\partial \theta} - \hat{\mathbf{r}} f_\theta \right)$$

$$+ \frac{1}{r \sin \theta} \hat{\boldsymbol{\varphi}} \cdot \left[\hat{\boldsymbol{\varphi}} f_r \sin \theta + \hat{\boldsymbol{\varphi}} f_\theta \cos \theta + \hat{\boldsymbol{\varphi}} \frac{\partial f_\varphi}{\partial \varphi} + f_\varphi \left(-\hat{\boldsymbol{\theta}} \cos \theta - \hat{\mathbf{r}} \sin \theta \right) \right]$$

leading to the final result

$$\nabla \cdot \mathbf{f} = \frac{\partial f_r}{\partial r} + \frac{1}{r} \left(f_r + \frac{\partial f_\theta}{\partial \theta} \right) + \frac{1}{r \sin \theta} \left(f_r \sin \theta + f_\theta \cos \theta + \frac{\partial f_\varphi}{\partial \varphi} \right)$$

$$= \frac{\partial f_r}{\partial r} + \frac{2}{r} f_r + \frac{1}{r} \frac{\partial f_\theta}{\partial \theta} + \frac{1}{r \sin \theta} f_\theta \cos \theta + \frac{1}{r \sin \theta} \frac{\partial f_\varphi}{\partial \varphi}$$

A clever combination of terms leads to the final compact expression for the divergence in spherical polar coordinates

$$\nabla \cdot \mathbf{f} = \frac{1}{r^2} \frac{\partial}{\partial r} \left(r^2 f_r \right) + \frac{1}{r \sin \theta} \frac{\partial}{\partial \theta} \left(\sin \theta f_\theta \right) + \frac{1}{r \sin \theta} \frac{\partial f_\varphi}{\partial \varphi} \qquad (5.24)$$

The divergence of a vector field $\mathbf{f}$ results in a scalar function.

From this result, we can evaluate the laplacian of V defined as

$$\nabla^2 V = \nabla \cdot \nabla V = \nabla \cdot \mathbf{f}$$

where

$$\mathbf{f} = \frac{\partial V}{\partial r} \hat{\mathbf{r}} + \frac{1}{r} \frac{\partial V}{\partial \theta} \hat{\boldsymbol{\theta}} + \frac{1}{r \sin \theta} \frac{\partial V}{\partial \varphi} \hat{\boldsymbol{\varphi}}$$

$$= f_r \hat{\mathbf{r}} + f_\theta \hat{\boldsymbol{\theta}} + f_\varphi \hat{\boldsymbol{\varphi}}$$

Inserting the forms for f_r, f_θ, and f_φ into Equation 5.24 leads to

$$\nabla^2 V = \frac{1}{r^2} \frac{\partial}{\partial r} \left(r^2 \frac{\partial V}{\partial r} \right) + \frac{1}{r \sin \theta} \frac{\partial}{\partial \theta} \left(\sin \theta \frac{1}{r} \frac{\partial V}{\partial \theta} \right) + \frac{1}{r \sin \theta} \frac{\partial}{\partial \varphi} \left(\frac{1}{r \sin \theta} \frac{\partial V}{\partial \varphi} \right)$$

and our final expression for the laplacian in spherical polar coordinates

$$\nabla^2 V = \frac{1}{r^2} \frac{\partial}{\partial r} \left(r^2 \frac{\partial V}{\partial r} \right) + \frac{1}{r^2 \sin \theta} \frac{\partial}{\partial \theta} \left(\sin \theta \frac{\partial V}{\partial \theta} \right) + \frac{1}{r^2 \sin^2 \theta} \frac{\partial^2 V}{\partial \varphi^2} \qquad (5.25)$$

The laplacian of a scalar function V results in a scalar function.

Finally, the curl in three-dimensional spherical polar coordinates can be expressed as

$$\nabla \times \mathbf{f} = \frac{1}{r^2 \sin \theta} \begin{vmatrix} \hat{\mathbf{r}} & r \hat{\boldsymbol{\theta}} & r \sin \theta \, \hat{\boldsymbol{\varphi}} \\ \frac{\partial}{\partial r} & \frac{\partial}{\partial \theta} & \frac{\partial}{\partial \varphi} \\ f_r & r f_\theta & r \sin \theta f_\varphi \end{vmatrix}$$

Note that in this expression for the curl, the top line of the determinant multiplied by the prefactor is not a literal translation of the gradient operator.

Expanding the determinant, we find the expression for the curl to be

$$\nabla \times \mathbf{f} = \hat{\mathbf{r}}\frac{1}{r\sin\theta}\left(\frac{\partial}{\partial\theta}(\sin\theta f_\varphi) - \frac{\partial f_\theta}{\partial\varphi}\right) + \hat{\boldsymbol{\theta}}\frac{1}{r}\left(\frac{1}{\sin\theta}\frac{\partial f_r}{\partial\varphi} - \frac{\partial}{\partial r}(rf_\varphi)\right)$$

$$+ \hat{\boldsymbol{\varphi}}\frac{1}{r}\left(\frac{\partial}{\partial r}(rf_\theta) - \frac{\partial f_r}{\partial\theta}\right) \tag{5.26}$$

The curl of a vector field $\mathbf{f}$ results in a vector field.

Cylindrical coordinates (three dimensions)

Let's repeat that process for three-dimensional cylindrical coordinates. We define a displacement along a curve in terms of the displacement vector

$$d\mathbf{l} = dr\,\hat{\mathbf{r}} + rd\theta\,\hat{\boldsymbol{\theta}} + dz\,\hat{\mathbf{z}} = dl_r\,\hat{\mathbf{r}} + dl_\theta\,\hat{\boldsymbol{\theta}} + dl_z\,\hat{\mathbf{z}}$$

We can express the gradient of a function $V(r,\theta,z)$ in terms of its r, θ, and z components as

$$\nabla V = (\nabla V \cdot \hat{\mathbf{r}})\hat{\mathbf{r}} + (\nabla V \cdot \hat{\boldsymbol{\theta}})\hat{\boldsymbol{\theta}} + \nabla V \cdot \hat{\mathbf{z}})\hat{\mathbf{z}}$$

$$= (\nabla V)_r\,\hat{\mathbf{r}} + (\nabla V)_\theta\,\hat{\boldsymbol{\theta}} + (\nabla V)_z\,\hat{\mathbf{z}}$$

Our goal is to derive an expression for the gradient operator by finding explicit expressions for $(\nabla V)_r$, $(\nabla V)_\theta$, and $(\nabla V)_z$.

The total differential of $V(r,\theta,z)$ can be written

$$dV = \frac{\partial V}{\partial r}\,dr + \frac{\partial V}{\partial\theta}\,d\theta + \frac{\partial V}{\partial z}\,dz$$

It can also be expressed as the scalar product of the gradient ∇V and the displacement vector $d\mathbf{l}$ as

$$dV = \nabla V \cdot d\mathbf{l} = (\nabla V)_r\,dr + (\nabla V)_\theta\,r\,d\theta + (\nabla V)_z\,dz$$

Comparing terms in the two expressions for dV we find

$$(\nabla V)_r = \frac{\partial V}{\partial r} \qquad (\nabla V)_\theta = \frac{1}{r}\frac{\partial V}{\partial\theta} \qquad (\nabla V)_z = \frac{\partial V}{\partial z}$$

so that the gradient in three-dimensional cylindrical coordinates is

$$\nabla V = \frac{\partial V}{\partial r}\,\hat{\mathbf{r}} + \frac{1}{r}\frac{\partial V}{\partial\theta}\,\hat{\boldsymbol{\theta}} + \frac{\partial V}{\partial z}\,\hat{\mathbf{z}} \tag{5.27}$$

The gradient of a scalar function V results in a vector field.

The divergence of a vector function $\mathbf{f} = f_r\,\hat{\mathbf{r}} + f_\theta\,\hat{\boldsymbol{\theta}} + f_z\,\hat{\mathbf{z}}$ is defined as the scalar product of the gradient operator and the vector function

$$\nabla \cdot \mathbf{f} = \left(\hat{\mathbf{r}}\frac{\partial V}{\partial r} + \hat{\boldsymbol{\theta}}\frac{1}{r}\frac{\partial V}{\partial\theta} + \hat{\mathbf{z}}\frac{\partial V}{\partial z}\right) \cdot (f_r\,\hat{\mathbf{r}} + f_\theta\,\hat{\boldsymbol{\theta}} + f_z\,\hat{\mathbf{z}})$$

In cylindrical coordinates, the unit vectors are dependent on the coordinates as in Equation 4.4

$$\hat{\mathbf{r}} = \cos\theta\,\hat{\mathbf{x}} + \sin\theta\,\hat{\mathbf{y}} \qquad \hat{\boldsymbol{\theta}} = -\sin\theta\,\hat{\mathbf{x}} + \cos\theta\,\hat{\mathbf{y}} \qquad \hat{\mathbf{z}} = \hat{\mathbf{z}}$$

As such, in evaluating $\nabla\cdot\mathbf{f}$ we must consider derivatives of the unit vectors as well as the coefficients leading to $2 \times (3 \times 3) = 18$ terms. Evaluating the derivatives

$$\frac{\partial \hat{\mathbf{r}}}{\partial r} = 0 \qquad\qquad \frac{\partial \hat{\theta}}{\partial r} = 0 \qquad\qquad \frac{\partial \hat{\mathbf{z}}}{\partial r} = 0$$

$$\frac{\partial \hat{\mathbf{r}}}{\partial \theta} = \hat{\theta} \qquad\qquad \frac{\partial \hat{\theta}}{\partial \theta} = -\hat{\mathbf{r}} \qquad\qquad \frac{\partial \hat{\mathbf{z}}}{\partial \theta} = 0$$

$$\frac{\partial \hat{\mathbf{r}}}{\partial z} = 0 \qquad\qquad \frac{\partial \hat{\theta}}{\partial z} = 0 \qquad\qquad \frac{\partial \hat{\mathbf{z}}}{\partial z} = 0$$

and noting that the unit vectors are orthonormal, such that

$$\hat{\mathbf{r}}\cdot\hat{\mathbf{r}} = \hat{\theta}\cdot\hat{\theta} = \hat{\mathbf{z}}\cdot\hat{\mathbf{z}} = 1$$

$$\hat{\mathbf{r}}\cdot\hat{\theta} = \hat{\mathbf{r}}\cdot\hat{\mathbf{z}} = \hat{\theta}\cdot\hat{\mathbf{z}} = 0$$

we find that 14 of the 18 terms are zero. The remaining terms form the final expression for the divergence in cylindrical coordinates

$$\nabla\cdot\mathbf{f} = \frac{\partial f_r}{\partial r} + \frac{1}{r}\left(f_r + \frac{\partial f_\theta}{\partial \theta} \right) + \frac{\partial f_z}{\partial z} \tag{5.28}$$

The divergence of a vector field $\mathbf{f}$ results in a scalar function.

From this result, we can evaluate the laplacian of V defined as

$$\nabla^2 V = \nabla\cdot\nabla V = \nabla\cdot\mathbf{f}$$

where

$$\mathbf{f} = \frac{\partial V}{\partial r}\hat{\mathbf{r}} + \frac{1}{r}\frac{\partial V}{\partial \theta}\hat{\theta} + \frac{\partial V}{\partial z} = f_r\hat{\mathbf{r}} + f_\theta\hat{\theta} + f_z\hat{\mathbf{z}}$$

Inserting the forms for f_r, f_θ, and f_z into Equation 5.28 leads to

$$\nabla^2 V = \frac{\partial}{\partial r}\left(\frac{\partial V}{\partial r} \right) + \frac{1}{r}\left[\frac{\partial V}{\partial r} + \frac{\partial}{\partial \theta}\left(\frac{1}{r}\frac{\partial V}{\partial \theta} \right) \right] + \frac{\partial}{\partial z}\left(\frac{\partial V}{\partial z} \right)$$

where the first two terms can be cleverly combined to form the final expression

$$\nabla^2 V = \frac{1}{r}\frac{\partial}{\partial r}\left(r\frac{\partial V}{\partial r} \right) + \frac{1}{r^2}\frac{\partial^2 V}{\partial \theta^2} + \frac{\partial^2 V}{\partial z^2} \tag{5.29}$$

The laplacian of a scalar function V results in a scalar function.

Finally, the curl in three-dimensional cylindrical coordinates can be written

$$\nabla\times\mathbf{f} = \frac{1}{r}\begin{vmatrix} \hat{\mathbf{r}} & r\hat{\theta} & \hat{\mathbf{z}} \\ \frac{\partial}{\partial r} & \frac{\partial}{\partial \theta} & \frac{\partial}{\partial z} \\ f_r & rf_\theta & f_z \end{vmatrix}$$

$$= \hat{\mathbf{r}}\left(\frac{1}{r}\frac{\partial f_z}{\partial \theta} - \frac{\partial f_\theta}{\partial z} \right) + \hat{\theta}\left(\frac{\partial f_r}{\partial z} - \frac{\partial f_z}{\partial r} \right) + \hat{\mathbf{z}}\frac{1}{r}\left(\frac{\partial}{\partial r}(rf_\theta) - \frac{\partial f_r}{\partial \theta} \right) \tag{5.30}$$

The curl of a vector field $\mathbf{f}$ results in a vector field. These expressions define the gradient, divergence, laplacian, and curl operators in coordinate systems having square, cubic, circular, spherical, or cylindrical symmetry.

E5 End-of-chapter problems

In mathematics you don't understand
things. You just get used to them.

John von Neumann

Warm-ups

5.1 Consider the scalar total differential operator

$$\hat{D}_x \equiv \frac{d}{dx}$$

For the function $f(x) = ax^2 + bx + c$, determine $(\hat{D}_x)^2 f(x)$.

5.2 Consider the scalar partial differential operators

$$\hat{d}_x \equiv \frac{\partial}{\partial x} \qquad \hat{d}_y \equiv \frac{\partial}{\partial y}$$

For the function $g(x, y) = x^2 y + xy + 3$, show that the commutator $[\hat{d}_x, \hat{d}_y]g(x, y) = 0$.

5.3 Consider the scalar differential operator

$$\hat{p} \equiv -i\hbar \frac{d}{dx}$$

where $\hbar$ is a constant. For the function $\psi(x) = e^{ikx}$, show that $\hat{p}\psi(x) = \hbar k\ \psi(x)$.

5.4 Consider the scalar differential operator

$$\hat{H} \equiv -\frac{h^2}{8\pi^2 m}\frac{d^2}{dx^2}$$

For the function $\psi(x) = \sin\left(\frac{\pi x}{L}\right)$, show that

$$\hat{H}\psi(x) = \frac{h^2}{8mL^2}\psi(x)$$

5.5 Prove the following commutator relations where $[\hat{A}, \hat{B}] = \hat{A}\hat{B} - \hat{B}\hat{A}$ and a and b are constants.

(a) $[\hat{A}, \hat{A}] = 0$

(b) $[\hat{A}, \hat{B}] = -[\hat{B}, \hat{A}]$

(c) $[\hat{A}, \hat{B} + \hat{C}] = [\hat{A}, \hat{B}] + [\hat{A}, \hat{C}]$

(d) $[a + \hat{A}, b + \hat{B}] = [\hat{A}, \hat{B}]$

(e) $[\hat{A}, \hat{B}\hat{C}] = [\hat{A}, \hat{B}]\hat{C} + \hat{B}[\hat{A}, \hat{C}]$

(f) $[\hat{A}\hat{B}, \hat{C}] = [\hat{A}, \hat{C}]\hat{B} + \hat{A}[\hat{B}, \hat{C}]$

5.6 Show that $\nabla \cdot \nabla \varphi = \dfrac{\partial^2 \varphi}{\partial x^2} + \dfrac{\partial^2 \varphi}{\partial y^2} + \dfrac{\partial^2 \varphi}{\partial z^2}$.

5.7 Show that $\nabla(\psi\varphi) = \psi\nabla\varphi + \varphi\nabla\psi$.

5.8 For a scalar potential energy, $V(x, y, z) = x^2 + y^2 + z^2$, derive the force

$$\mathbf{F} = -\nabla V$$

defined as the negative gradient of the potential energy.

Homework exercises

5.9 Perform the following operations involving differential operators

(a) $\hat{D}_x x^3$

(b) $\hat{D}_x^2 \cos 2x$

(c) $\hat{D}_x e^{ax} \sin bx$

(d) $\hat{d}_x (x^3 y)$

(e) $\hat{d}_x^2 (x^2 y^3)$

(f) $\hat{d}_y \hat{d}_x (x^4 y^3)$

5.10 An interpretation of the *Heisenberg uncertainty principle* is that the operator for linear momentum in the x-direction does not commute with the operator for position along the x-axis. Let

$$\hat{p} = -i\hbar \frac{d}{dx} \quad \text{and} \quad \hat{x} = x$$

where $i = \sqrt{-1}$ and $\hbar$ is a constant. Evaluate the commutator

$$[\hat{p}, \hat{x}] = \hat{p}\hat{x} - \hat{x}\hat{p}$$

and show that it does not equal zero. HINT: Apply the operators $\hat{x}$ and $\hat{p}$ to an arbitrary function $\varphi(x)$ and appreciate that $x\varphi(x)$ must be differentiated as a product.

5.11 Show that the two-dimensional scalar function $V(x, y) = \ln\left(\sqrt{x^2 + y^2}\right)$ satisfies Laplace's equation

$$\nabla^2 V(x, y) = 0$$

First determine the gradient $\nabla V(x, y) = \mathbf{f}(x, y)$. Further determine the divergence $\nabla \cdot \mathbf{f}(x, y) = \nabla^2 V(x, y)$. Show that your result $\nabla^2 V = 0$.

5.12 Show that the two-dimensional scalar function $V(r, \theta) = \ln r$ satisfies Laplace's equation

$$\nabla^2 V(r, \theta) = 0$$

First determine the gradient $\nabla V(r, \theta) = \mathbf{f}(r, \theta)$. Further determine the divergence $\nabla \cdot \mathbf{f}(r, \theta) = \nabla^2 V(r, \theta)$. Show that your result $\nabla^2 V = 0$.

5.13 Show that the three-dimensional scalar function $V(x, y, z) = (x^2 + y^2 + z^2)^{-\frac{1}{2}}$ satisfies Laplace's equation

$$\nabla^2 V(x, y, z) = 0$$

First determine the gradient $\nabla V(x, y, z) = \mathbf{f}(x, y, z)$. Further determine the divergence $\nabla \cdot \mathbf{f}(x, y, z) = \nabla^2 V(x, y, z)$. Show that your result $\nabla^2 V = 0$.

5.14 Show that the three-dimensional scalar function $V(r, \theta, \varphi) = \dfrac{1}{r}$ satisfies Laplace's equation

$$\nabla^2 V(r, \theta, \varphi) = 0$$

First determine the gradient $\nabla V(r, \theta, \varphi) = \mathbf{f}(r, \theta, \varphi)$. Further determine the divergence $\nabla \cdot \mathbf{f}(r, \theta, \varphi) = \nabla^2 V(r, \theta, \varphi)$. Show that your result $\nabla^2 V = 0$.

5.15[*] The differential operator for angular momentum is given by the expression

$$\hat{\mathbf{M}} = -i\hbar(\mathbf{r} \times \nabla)$$

where $\hbar$ is a constant, $\mathbf{r} = x\,\hat{\mathbf{x}} + y\,\hat{\mathbf{y}} + z\,\hat{\mathbf{z}}$, and

$$\nabla = \hat{\mathbf{x}}\frac{\partial}{\partial x} + \hat{\mathbf{y}}\frac{\partial}{\partial y} + \hat{\mathbf{z}}\frac{\partial}{\partial z}$$

Assuming that $\hat{\mathbf{M}} = \hat{\mathbf{x}}\,\hat{M}_x + \hat{\mathbf{y}}\,\hat{M}_y + \hat{\mathbf{z}}\,\hat{M}_z$, find the components $\hat{M}_x$, $\hat{M}_y$, and $\hat{M}_z$ of $\hat{\mathbf{M}}$.

5.16[*] The electric potential, V, produced by a point electric dipole is

$$V(\mathbf{r}) = \frac{\boldsymbol{\mu} \cdot \mathbf{r}}{r^3}$$

where the dipole moment is written $\boldsymbol{\mu} = \mu_x\,\hat{\mathbf{i}} + \mu_y\,\hat{\mathbf{j}} + \mu_z\,\hat{\mathbf{k}}$, the components μ_x, μ_y, μ_z are constants, and the vector $\mathbf{r} = x\,\hat{\mathbf{i}} + y\,\hat{\mathbf{j}} + z\,\hat{\mathbf{k}}$. Shown below is the electric dipole potential produced by a point dipole located at the origin and oriented along the positive x-axis, with contours denoting $V > 0$ (solid lines) and $V < 0$ (dashed lines).

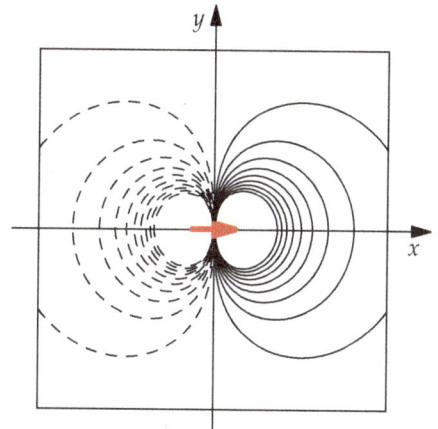

The electric field, $\mathbf{E}$, is defined as the negative gradient of the electric potential

$$\mathbf{E} = -\nabla V = \frac{3(\boldsymbol{\mu} \cdot \mathbf{r})\,\mathbf{r}}{r^5} - \frac{\boldsymbol{\mu}}{r^3}$$

Derive the equation above using cartesian coordinates in which the electric potential can be written

$$V = \frac{\boldsymbol{\mu} \cdot \mathbf{r}}{r^3} = \frac{\mu_x\,x + \mu_y\,y + \mu_z\,z}{(x^2 + y^2 + z^2)^{3/2}}$$

Note that

$$\nabla V = \left(\hat{\mathbf{i}}\frac{\partial V}{\partial x} + \hat{\mathbf{j}}\frac{\partial V}{\partial y} + \hat{\mathbf{k}}\frac{\partial V}{\partial z} \right)$$

6 EXTREMIZING FUNCTIONS OF MANY VARIABLES

6.1 Extremizing functions of one and many variables

IN MODELING PHYSICAL SYSTEMS, we often want to determine the maximum or minimum value of a function of one or many variables. Examples include characterizing the minima, maxima, and saddle points on a potential energy surface or achieving the best fit of a function to a set of data. This section explores the general problem of determining the extrema of functions of one and many variables.

6.1.1 Extremizing functions of one variable and the max/min problem

Derivatives can be used to identify extrema (maxima or minima) of functions. This is useful in characterizing the form of the function and parameterizing the function as a model of a physical property such as the potential energy. Consider the function $f(x)$ shown in Figure 6.1. This function has local and global maxima and minima. A local extremum is a maximum or minimum. A global extremum is the highest maximum or lowest minimum for that function over the allowed range of x.

At an extremum located at $x = x^*$, the slope of the function is zero and

$$\left.\frac{df}{dx}\right|_{x=x^*} = 0 \tag{6.1}$$

The type of extremum at $x = x^*$ can be determined by the second derivative test:

$$\left.\frac{d^2f(x)}{dx^2}\right|_{x=x^*} \begin{cases} < 0, & f(x^*) \text{ is a maximum} \\ > 0, & f(x^*) \text{ is a minimum} \\ = 0, & \text{the test is inconclusive} \end{cases}$$

When the second derivative is negative, the function's curvature is downward and $f(x^*)$ is a maximum. When the second derivative is positive, the function's curvature is upward and $f(x^*)$ is a minimum. When the second derivative is zero, the test is inconclusive.

Let's see how this works. Consider the polynomial function

$$y(x) = 2x^2 - x^3$$

shown in Figure 6.2. Asymptotically, the function approaches ∞ as x approaches $-\infty$ and $-\infty$ as x approaches ∞. There are two local extrema. Let's find the location of the two extrema and determine their nature.

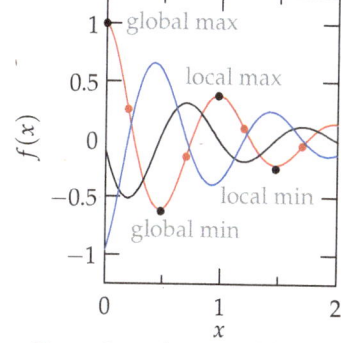

Figure 6.1: A function $f(x)$ (red line) and its extrema. Black dots mark global and local maxima and minima where the first derivative (black line) is zero. Red dots mark points of inflection where the second derivative (blue line) changes its sign passing through zero.

To find the location of the extrema, we follow Equation 6.1 and set the first derivative of the function to zero:

$$\frac{dy(x)}{dx}\bigg|_{x=x^*} = 4x^* - 3x^{*2} = 0$$

The extrema are located at positions satisfying $3x^{*2} = 4x^*$, which occur at $x^* = 0$, where $y(0) = 0$, and $x^* = \frac{4}{3}$, where $y(\frac{4}{3}) = \frac{32}{27}$.

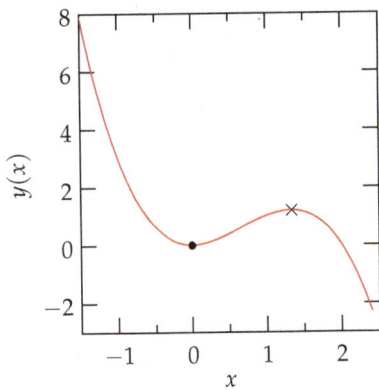

Figure 6.2: Extrema of the polynomial function $y(x) = 2x^2 - x^3$ with local minimum $(0,0)$ (black dot) and local maximum $(\frac{4}{3}, \frac{32}{27})$ (black cross).

We can use Equation 6.2 and the second derivative test to determine the nature of the extrema. The second derivative of the function is

$$\frac{d^2y(x)}{dx^2} = 4 - 6x$$

At $x^* = 0$, we find $y''(0) = 4 > 0$ and the extremum is a minimum. At $x^* = \frac{4}{3}$, we find $y''(\frac{4}{3}) = -4 < 0$ and the extremum is a maximum.

6.1.2 *Extremizing functions of many variables*

Derivatives can also be used in a similar way to identify extrema of functions of many variables. Consider a function $f(x, y)$ of two variables x and y. At an extremum, we expect that the total differential will be zero:

$$df = \frac{\partial f}{\partial x}\, dx + \frac{\partial f}{\partial y}\, dy = 0$$

The small variations in x and y, written dx and dy, are independent. As such, for $df = 0$ to be satisfied in all cases, it must be that

$$\frac{\partial f}{\partial x} = 0 \qquad \frac{\partial f}{\partial y} = 0$$

At point (x^*, y^*) at which $f_x(x^*, y^*) = f_y(x^*, y^*) = 0$ is called a *critical point*.

What type of extremum do we have at the *critical point* (x^*, y^*)? This is often determined by a generalization of the second derivative test. We define a scalar quantity called the *hessian*[1] to be

$$D = \left(\frac{\partial^2 f}{\partial x^2}\right)\left(\frac{\partial^2 f}{\partial y^2}\right) - \left(\frac{\partial^2 f}{\partial x \partial y}\right)\left(\frac{\partial^2 f}{\partial y \partial x}\right) \qquad (6.2)$$

[1] Named for German mathematician *Ludwig Otto Hesse* (1811-1874). For the functions of interest in this chapter we can assume df is an exact differential and $f_{xy} = f_{yx}$.

The hessian D is the determinant of the *hessian matrix* of second derivatives of $f(x,y)$.[2] The type of extremum at the critical point (x^*, y^*) is determined by

$$\frac{\partial^2 f}{\partial x^2}\bigg|_{(x^*,y^*)} = \begin{cases} < 0 & D > 0 \quad \text{maximum} \\ > 0 & D > 0 \quad \text{minimum} \end{cases}$$

For the case $D < 0$, we expect a *saddle point* for which the curvature is positive in some directions and negative in others. When $D = 0$, the test is inconclusive.

[2] The *hessian matrix* of second derivatives of a function $f(x,y)$ is

$$\begin{bmatrix} f_{xx} & f_{xy} \\ f_{yx} & f_{yy} \end{bmatrix}$$

The hessian

$$D = f_{xx}f_{yy} - f_{xy}f_{yx}$$

is the *determinant* of the hessian matrix.

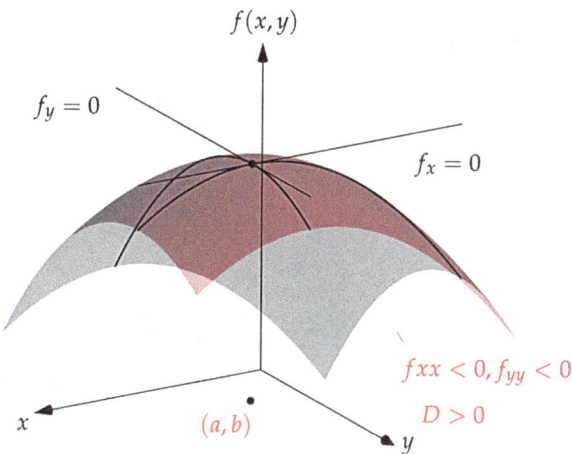

Figure 6.3: The inverted parabolic function $f(x,y)$ shown as a red surface over the xy-plane. The black point on the surface (a, b) marks the maximum of the function $f(a, b)$. Thick black lines represent one-dimensional curves $f(a, y)$ and $f(x, b)$. The slopes of the tangent lines at the point (a, b) are zero.

For example, consider the function

$$f(x,y) = f_0 - \frac{1}{2}(x - a)^2 - \frac{1}{2}(y - b)^2$$

shown in Figure 6.3. The partial derivatives are

$$\frac{\partial f}{\partial x} = -(x - a) = 0 \qquad \frac{\partial f}{\partial y} = -(y - b) = 0$$

Solving the two linear equations we find an extremum located at the critical point $x^* = a, y^* = b$, at which $f_x = f_y = 0$. Evaluating the hessian where $f_{xx} = -1$, $f_{yy} = -1$, and $f_{xy} = 0$, we find

$$D = \left(\frac{\partial^2 f}{\partial x^2}\right)\left(\frac{\partial^2 f}{\partial y^2}\right) - \left(\frac{\partial^2 f}{\partial x \partial y}\right)^2 = 1$$

for all x and y. As $D > 0$, the extremum at $(x^*, y^*) = (a, b)$ is a maximum, as shown in Figure 6.3.

An important application in the physical sciences of this method for locating extrema is identifying minima, maxima, and saddle points on potential energy surfaces. Consider the model potential energy function

$$V(x,y) = (x^2 - 1)^2 + y^2$$

shown in Figure 6.4. This is an example of a double well potential used to model chemical reactions and phase transitions.

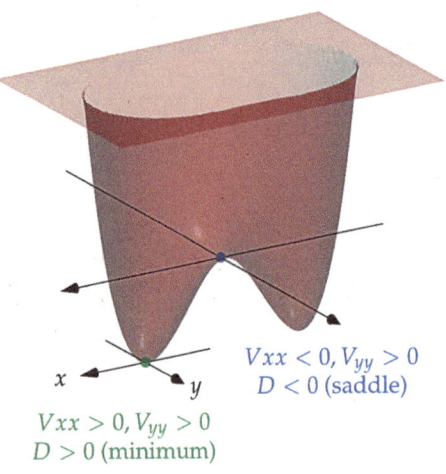

Figure 6.4: The function $V(x, y) = (x^2 - 1)^2 + y^2$. The thick horizontal lines show the derivatives $f_x = 0$ and $f_y = 0$ at a minimum (green dot) and saddle point (blue dot). The second derivatives and hessian are noted at each of these points.

$$Vxx < 0, V_{yy} > 0$$
$$D < 0 \text{ (saddle)}$$

$$Vxx > 0, V_{yy} > 0$$
$$D > 0 \text{ (minimum)}$$

We recognize two local minima on the potential energy surface. The locations of the minima are defined by

$$\frac{\partial V}{\partial x} = 0 = 4x^3 - 4x \qquad \frac{\partial V}{\partial y} = 0 = 2y$$

There are three solutions located at the critical point $(x^*, y^*) = (1, 0)$, $(0, 0)$, and $(-1, 0)$. We can use the hessian to determine the type of extremum at each point. Recognizing that

$$\frac{\partial^2 V}{\partial x^2} = 12x^2 - 4 \qquad \frac{\partial^2 V}{\partial y^2} = 2 \qquad \frac{\partial^2 V}{\partial x \partial y} = 0$$

we find that

$$D = \left(\frac{\partial^2 f}{\partial x^2}\right)\left(\frac{\partial^2 f}{\partial y^2}\right) - \left(\frac{\partial^2 f}{\partial x \partial y}\right)^2 = 24x^2 - 8$$

where the value of $D(x, y)$ depends on the position of the extremum. We find $D(-1, 0) = D(1, 0) = 16$. Given that

$$\frac{\partial^2 V}{\partial x^2} = 8 > 0 \qquad D = 16 > 0$$

the positions represent potential energy minima. We also find that $D(0, 0) = -8$ at which point

$$\frac{\partial^2 V}{\partial x^2} = -4 < 0 \qquad D = -8 < 0$$

As such, the position $(0, 0)$ is a saddle point as shown in Figure 6.4. This demonstrates how the method to identify and characterize extrema can be applied to characterize potential energy surfaces.

6.2 The method of Lagrange undetermined multipliers

THE SEARCH FOR EXTREMA of a function of many variables often occurs in the presence of constraints. For example, we might want to know the lowest possible total energy of a negative charge orbiting a positive charge, subject to the constraint that the angular moment of the system is a constant. To solve such problems, Lagrange developed the method of undetermined multipliers. This section presents the general theory of undetermined multipliers and explores applications to a number of systems.

6.2.1 The method of undetermined multipliers

We have an effective approach for extremizing functions of multiple variables in the absence of constraints. What do we do in the case of constrained optimization of a function of multiple variables, a situation that often arises in the physical sciences? Lagrange proposed one approach known as the method of undetermined multipliers.[3]

Suppose we have a function $f(x, y)$ and constraint equation

$$g(x, y) = 0$$

At an extremum of the function we know that $df = 0$ so that

$$df = \frac{\partial f}{\partial x}\, dx + \frac{\partial f}{\partial y}\, dy = 0$$

As $g(x, y) = 0$, it follows that $dg = 0$ so that

$$dg = \frac{\partial g}{\partial x}\, dx + \frac{\partial g}{\partial y}\, dy = 0$$

From the constraint equation, we can express dy in terms of dx as

$$dy = -\left(\frac{\partial g/\partial x}{\partial g/\partial y} \right) dx$$

so that

$$df = \left[\frac{\partial f}{\partial x} - \left(\frac{\partial g/\partial x}{\partial g/\partial y} \right) \frac{\partial f}{\partial y} \right] dx = 0$$

Given this result, we recognize that an extremum will be found at a point where the total derivative is

$$\frac{df}{dx} = 0$$

which is satisfied when

$$\frac{\partial f}{\partial x} \left(\frac{\partial g}{\partial x} \right)^{-1} = \frac{\partial f}{\partial y} \left(\frac{\partial g}{\partial y} \right)^{-1} = \lambda$$

where λ is a constant. We now have two equations and a constraint equation through which we can determine the undetermined multiplier and the optimal values of x and y.

[3] The method of undetermined multipliers was developed by the Italian mathematician and astronomer *Joseph-Louis Lagrange* (1736-1813). Born *Giuseppe Lodovico Lagrangia*, Lagrange's treatise *Mécanique Analytique* (1788) provided the foundation for the development of mathematical physics in the 19th century.

This result can be written as set of simultaneous equations

$$\frac{\partial f}{\partial x} - \lambda \frac{\partial g}{\partial x} = 0$$

$$\frac{\partial f}{\partial y} - \lambda \frac{\partial g}{\partial y} = 0 \tag{6.3}$$

These equations can be solved to define the location of the extremum. In practice, we determine the derivatives of the functions $f(x,y)$ and $g(x,y)$, set up the simultaneous equations, and solve for the undetermined multiplier λ. We then return to the constraint equations $g(x,y) = 0$ and substitute for λ to determine the position of the extremum in terms of $x = x^*$ and $y = y^*$.

In the next section, we explore an alternative derivation of the key equations of the method of undetermined multipliers. We will then practice the application of this powerful method of constrained optimization on a few classic examples from geometry and the physical sciences.

6.2.2 An alternative derivation

Consider the function $f(x,y)$ presented in Figure 6.5 as a red parabolic surface. The constraint $g(x,y) = 0$ is represented by a gray plane. We seek the extremum of the function $f(x,y)$ on the line defined by the intersection of the red surface defined by $f(x,y)$ and the gray plane defined by the constraint $g(x,y) = 0$.

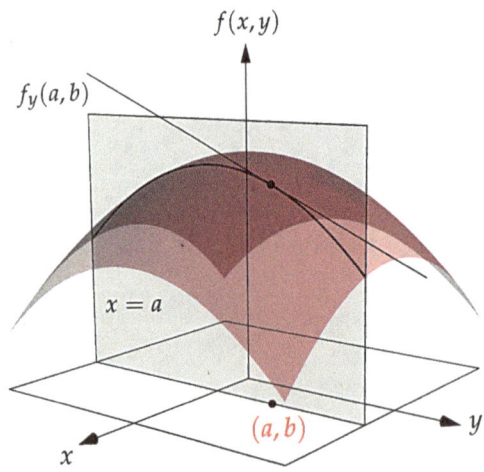

Figure 6.5: The inverted parabolic function $f(x,y)$ shown as a red surface over the xy-plane. The constraint function $g(x,y) = 0$ is represented by the gray surface. The black point on the surface is $f(a,b)$ and the thick black line is the one-dimensional curve $f(a,y)$ where x is constrained by $g(x,y) = x - a = 0$ to be $x = a$. The slope of the tangent line at the point (a,b) is $f_y(a,b)$.

The gradient of the constraint $g(x,y)$ is confined to the plane as is the gradient of the function $f(x,y)$ consistent with the constraint. This means the gradient of $f(x,y)$ and $g(x,y)$ must be parallel, differing only in magnitude. We can express this as

$$\nabla f(x,y) = \lambda \nabla g(x,y) \tag{6.4}$$

where λ is a constant. This observation provides an intuitive and elegant proof of the method of Lagrange multipliers.

We can demonstrate equivalence of Equation 6.3 and Equation 6.4. We write the gradient of $f(x,y)$ as

$$\nabla f(x,y) = \frac{\partial f}{\partial x}\,\hat{\mathbf{x}} + \frac{\partial f}{\partial y}\,\hat{\mathbf{y}}$$

and the gradient of $g(x,y)$ as

$$\nabla g(x,y) = \frac{\partial g}{\partial x}\,\hat{\mathbf{x}} + \frac{\partial g}{\partial y}\,\hat{\mathbf{y}}$$

It follows that Equation 6.3 can be rewritten

$$\left(\frac{\partial f}{\partial x} - \lambda\frac{\partial g}{\partial x}\right)\hat{\mathbf{x}} + \left(\frac{\partial f}{\partial y} - \lambda\frac{\partial g}{\partial y}\right)\hat{\mathbf{y}} = 0$$

Since the terms in the x-direction and y-direction are independent, each term in parentheses must be zero. It follows that

$$\nabla f(x,y) - \lambda\nabla g(x,y) = 0$$

6.2.3 *Application to optimizing geometric functions with constraints*

To practice the application of the method of undetermined multipliers, we consider the rectangular area in Figure 6.6 with sides of length x and y. The rectangle's perimeter has a fixed length $C = 2x + 2y$. For a rectangle of fixed perimeter C, what are the length $x = x^*$ and width $y = y^*$ that maximize the rectangle's area $A(x,y) = xy$?

We can answer this question by extremizing the area subject to a constraint on the circumference, imposed using the Lagrange multiplier. The function to be maximized is the area

$$A(x,y) = xy$$

and the constraint is expressed as

$$g(x,y) = C - (2x + 2y) = 0$$

where the circumference has been set to a constant C. Following the procedure defined by Equation 6.3, we have

$$\frac{\partial A}{\partial x} - \lambda\frac{\partial g}{\partial x} = y + 2\lambda = 0$$
$$\frac{\partial A}{\partial y} - \lambda\frac{\partial g}{\partial y} = x + 2\lambda = 0$$

Therefore, the largest possible rectangle is a square with sides $x = -2\lambda$ and $y = -2\lambda$. Since

$$C = (2x + 2y) = -8\lambda$$

we find $\lambda = -\frac{C}{8}$ and

$$x^* = y^* = \frac{C}{4}$$

Inserting back into our original equations, we find $A = \frac{C^2}{16}$. The rectangle with the largest possible area for a given circumference is a square.

Figure 6.6: A rectangle of circumference $C = 2x + 2y$ and area $A = xy$.

Let's explore another constrained extremization problem using Lagrange multipliers. Consider a cylinder of height z with circular base of radius r shown in Figure 6.7.

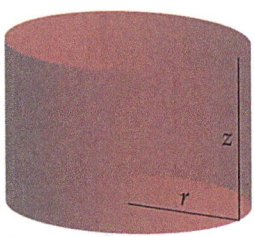

Figure 6.7: A cylinder of radius r and height z.

The volume of the cylinder is the circular area of the cap times the height:

$$V(r,z) = \pi r^2 z$$

The surface area of the cylinder is

$$A(r,z) = 2\pi r (z + r)$$

which is the sum of the area of the side of the cylinder, $2\pi r z$, and the areas of the top and bottom circular caps, each having area πr^2. Suppose we hold the surface area to be a constant A_0. For a cylinder of fixed surface area A_0, what are the values of $r = r^*$ and $z = z^*$ that maximize the cylinder's volume?

Following the procedure defined by Equation 6.3 we extremize the volume $V(r,z)$ subject to the constraint

$$g(r,z) = A_0 - [2\pi r (z + r)] = 0 \qquad (6.5)$$

so that

$$\frac{\partial V}{\partial r} - \lambda \frac{\partial g}{\partial r} = 2\pi r z + \lambda (2\pi z + 4\pi r) = 0$$
$$\frac{\partial V}{\partial z} - \lambda \frac{\partial g}{\partial z} = \pi r^2 + \lambda 2\pi r = 0$$

The second equation yields $\lambda = -\frac{r}{2}$. Substituting into the first equation yields

$$2\pi r z - \frac{r}{2}(2\pi z + 4\pi r) = 0$$

or $\pi r z = 2\pi r^2$, so that $z = 2r$. Substituting into the constraint defined by Equation 6.5 results in $A_0 = 6\pi r^2$, or equivalently $z^* = 2r^* = 2\sqrt{\frac{A_0}{6\pi}}$. As our final result, we find the maximum volume of a cylinder of surface area A_0 is

$$V(r^*, z^*) = \pi r^{*2} z^* = \frac{1}{3} A_0 \sqrt{\frac{A_0}{6\pi}}$$

Conversely, for a cylinder of fixed volume V_0 we can use the method of undetermined multipliers to determine the dimensions of height and radius that minimize the surface area of the cylinder.

A₆ Variational calculation of the energy of a one-electron atom

One interesting application of Lagrange's method of undetermined multipliers involves Bohr's model of a one-electron atom. In the *Bohr model*, an electron of charge $-e$ and mass m_e moves in a circular orbit around a central charge Ze centered at the origin (see Figure 6.8).[4]

[4] Named for the Danish mathematical physicist *Neils Bohr* (1885-1962) who developed the Correspondence Principles linking quantum and classical mechanics. He received the 1922 Nobel Prize in Physics.
Figure 6.8: Schematic of the Bohr model of a one-electron atom or ion. An electron of charge $-e$ moves in a circular orbit (red line) about the nucleus of charge Ze. A wave of wavelength $\lambda = 2\pi r/n$ is shown for $n = 4$ (gray).

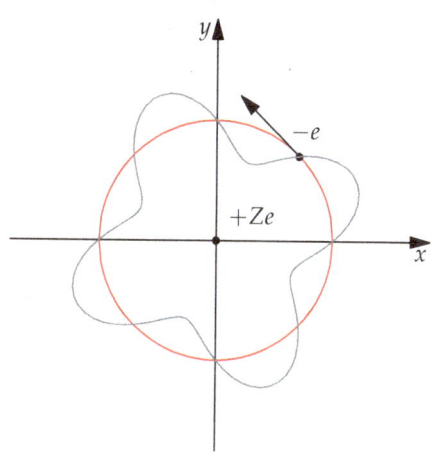

The total energy of the orbiting electron, given by the sum of the kinetic energy and potential energy, will be a function of the radius of the orbit r and the velocity of the electron v

$$E(r,v) = \frac{1}{2}m_e v^2 - \frac{Ze^2}{r}$$

where the first term is the kinetic energy of the electron and the second term is the potential energy between the negatively charged electron and the central positive charge. Bohr assumed only certain orbits were allowed.[5] In those allowed orbits, the electron's angular momentum was constrained to be

[5] As the angular momentum may assume only certain discrete values, we say that the angular momentum is *quantized*.

$$l = m_e v r = \frac{nh}{2\pi} = n\hbar$$

where n is a positive integer, h is Planck's constant, and $\hbar = \frac{h}{2\pi}$.

If we consider that the electron can be modeled as a wave with de Broglie wavelength[6] defined

[6] Named for French physicist *Louis de Broglie* (1892-1987) who proposed that electrons and all matter have properties of a wave.

$$\lambda = \frac{h}{m_e v}$$

we see that the constraint on the angular momentum is equivalent to the assumption that there are a fixed number of de Broglie wavelengths λ in the circumference of the orbit $2\pi r$. As a result

$$n\lambda = 2\pi r = \frac{nh}{m_e v}$$

This constraint is demonstrated in Figure 6.8, which shows a wave with $n = 4$.

We can find the allowed orbits using the *variational principle* where we minimize the total energy $E(r, v)$ as a function of r and v subject to the constraint on the angular momentum, written as

$$g(r, v) = m_e v r - n\hbar = 0 \qquad (6.6)$$

We follow the procedure defined by Equation 6.3, taking our undetermined multiplier to be a constant α so that

$$\frac{\partial E}{\partial r} - \alpha \frac{\partial g}{\partial r} = 0$$

$$\frac{\partial E}{\partial v} - \alpha \frac{\partial g}{\partial v} = 0$$

Computing the necessary derivatives, we find

$$\frac{\partial E}{\partial r} = \frac{Ze^2}{r^2} \qquad \frac{\partial E}{\partial v} = m_e v$$

and

$$\frac{\partial g}{\partial r} = m_e v \qquad \frac{\partial g}{\partial v} = m_e r$$

Combining these results, we find

$$\frac{\partial E}{\partial r} - \alpha \frac{\partial g}{\partial r} = \frac{Ze^2}{r^2} - \alpha m_e v = 0$$

$$\frac{\partial E}{\partial v} - \alpha \frac{\partial g}{\partial v} = m_e v - \alpha m_e r = 0$$

The second equation yields

$$\alpha = \frac{v}{r}$$

so that the first equation can be written

$$\frac{Ze^2}{r^2} = \alpha m_e v = \frac{m_e v^2}{r} \qquad (6.7)$$

This is the condition to form a stable orbit. The inward pointing *centripetal force* due to the coulomb attraction is exactly balanced by the outward pointing *centrifugal force*.

Given the constraint on the angular moment given by Equation 6.6, we can find the velocity:

$$v = \frac{n\hbar}{m_e r}$$

Substituting into Equation 6.7, we find that the allowed values of the radius and velocity are

$$r^* = \frac{n^2 \hbar^2}{Z m_e e^2} \qquad v^* = \frac{Ze^2}{n\hbar} \qquad n = 1, 2, 3, \ldots$$

With these results for the radius and velocity, we can solve for the total energy

$$E(r^*, v^*) = \frac{1}{2} m_e v^{*2} - \frac{Ze^2}{r^*}$$

$$= \frac{1}{2} m_e \left(\frac{Z^2 e^4}{n^2 \hbar^2} \right) - Ze^2 \left(\frac{Z m_e e^2}{n^2 \hbar^2} \right)$$

$$= -\frac{1}{2} m_e \frac{Z^2 e^4}{n^2 \hbar^2} \qquad n = 1, 2, 3, \dots$$

which we find to be a function of counting number n. This equation defines the infinity of allowed energies for the electron orbiting the central nuclear charge.

We can write the potential energy, $V(r)$, in the alternative units

$$V(r) = -\frac{Ze^2}{4\pi\varepsilon_0 r}$$

where ε_0 is the permittivity of a vacuum. In this system of units, we find the allowed total energies

$$E(r, v) = -\frac{Z^2}{n^2} \left(\frac{m_e e^4}{8 \varepsilon_0^2 h^2} \right) \qquad n = 1, 2, 3, \dots$$

where the complicated ratio of fundamental physical constants, including the mass and charge of the electron, the permittivity of a vacuum, Planck's constant, and the cube of two, form the expression

$$\frac{m_e e^4}{8 \varepsilon_0^2 h^2} = h\mathcal{R} = \mathcal{H}$$

that defines the *Rydberg constant*, $\mathcal{R} = 3.29 \times 10^{15}$ Hz, and the *Hartree constant*, $\mathcal{H} = 2.18 \times 10^{-18}$ Joules.[7]

[7] The lowest allowed total energy of the Bohr model for $Z = 1$ is coincidentally the ground state energy of the hydrogen atom. These fundamental constants are named for Swedish physicist *Johannes Rydberg* (1854-1919) and English mathematician and physicist *Douglas Hartree* (1897-1958).

B₆ Extremizing the multiplicity subject to constraints

Let's consider a system with N observations, each of which has $k = 2$ distinguishable and equally probable outcomes. For example, this could be a series of N coin flips, each of which might be a head or tail. In a given series of N observations there will be n_1 outcomes of the first type and n_2 outcomes of the second type where $n_1 + n_2 = N$. The total number of outcomes is given by the binomial coefficient

$$W(n_1, n_2; N) = \frac{N!}{n_1! n_2!} = \frac{N!}{n_1!(N - n_1)!}$$

where $n!$ is pronounced n *factorial* and defined

$$n! = n \times (n - 1) \times (n - 2) \times \dots \times 1$$

Suppose we want to know the most probable outcome $\{n_1^*, n_2^*\}$ that corresponds to the maximum multiplicity $W(n_1^*, n_2^*; N)$ subject to the constraint that $N = n_1^* + n_2^*$. The numerator is fixed to be $N!$ so we can maximize the multiplicity by making the denominator $n_1! n_2! = n_1!(N - n_1)!$ as small as possible. That will occur when we have an even distribution with an equal number of each type of object,[8] so that $n_1^* = n_2^* = N/2$ and

$$W(n_1^*, n_2^*; N) = \frac{N!}{\left(\frac{N}{2}\right)! \left(\frac{N}{2}\right)!}$$

[8] Note that the function

$$\frac{1}{x(1-x)}$$

has a maximum at the point $x = 1/2$ for $0 < x < 1$.

Now let's consider the slightly more complicated example of a system with N total objects of $k = 3$ distinguishable types. Suppose we want to know the most probable outcome $\{n_1^*, n_2^*, n_3^*\}$ corresponding to the maximum multiplicity $W(n_1^*, n_2^*, n_3^*; N)$, and subject to the constraint that $N = n_1^* + n_2^* + n_3^*$. To do this, we will use the method of Lagrange multipliers. The multiplicity is given by the multinomial coefficient

$$W(n_1, n_2, n_3; N) = \frac{N!}{n_1! n_2! n_3!}$$

Since the multiplicity can be a very large number, it is convenient to work with the logarithm of the multiplicity[9]

$$f(n_1, n_2, n_3) = \ln\left[W(n_1, n_2, n_3; N)\right]$$

[9] Note that as $f(n_1, n_2, n_3)$ is a monotonically increasing function of $W(n_1, n_2, n_3; N)$ any extremum of $W(n_1, n_2, n_3; N)$ will also be an extremum of $f(n_1, n_2, n_3)$.

Let's maximize the function $f(n_1, n_2, n_3)$, subject to the constraint

$$g(n_1, n_2, n_3) = N - (n_1 + n_2 + n_3) = 0$$

We assume that N as well as n_1, n_2, and n_3 are much greater than 1. As such, we can use Stirling's approximation, defined by Equation 2.21, so that

$$f(n_1, n_2, n_3) = \ln\left[W(n_1, n_2, n_3; N)\right]$$
$$\approx N \ln N - (n_1 \ln n_1 + n_2 \ln n_2 + n_3 \ln n_3)$$

Following the procedure defined by Equation 6.3 we find

$$\frac{\partial f}{\partial n_1} - \lambda \frac{\partial g}{\partial n_1} = -\ln n_1 - 1 + \lambda = 0$$
$$\frac{\partial f}{\partial n_2} - \lambda \frac{\partial g}{\partial n_2} = -\ln n_2 - 1 + \lambda = 0$$
$$\frac{\partial f}{\partial n_2} - \lambda \frac{\partial g}{\partial n_3} = -\ln n_3 - 1 + \lambda = 0$$

This leaves

$$n_1 = n_2 = n_3 = e^{\lambda - 1} = \text{constant}$$

which is a uniform distribution. Returning to our constraint equation we find

$$N = n_1 + n_2 + n_3 = \text{constant}$$

so that $n_1^* = n_2^* = n_3^* = N/3$ and the maximum multiplicity[10] is found to be

[10] Note that we were able to solve this problem without explicitly determining the value of λ.

$$W(n_1^*, n_2^*, n_3^*; N) = \frac{N!}{\left(\frac{N}{3}\right)! \left(\frac{N}{3}\right)! \left(\frac{N}{3}\right)!}$$

C$_6$ End-of-chapter problems

> Nothing takes place in the world whose meaning is not that of some maximum or minimum.
>
> Leonhard Euler

Warm-ups

6.1 For each function determine whether it reaches a maximum value, a minimum value, or both. Evaluate each function that does reach a maximum or minimum value at each extremum. NOTE: The parameters A, ε, σ, e, π, m, k, and T are positive constants.

(a) $y(x) = 3x^2 - 5x + 2$ $\quad x \in (-\infty, \infty)$

(b) $p(v) = \left(\frac{m}{2\pi kT}\right)^{\frac{3}{2}} v^2 e^{-\frac{1}{2kT}mv^2}$ $\quad v \in [0, \infty)$

(c) $V(r) = 4\varepsilon \left[\left(\frac{\sigma}{r}\right)^{12} - \left(\frac{\sigma}{r}\right)^6\right]$ $\quad r \in [0, \infty)$

(d) $P(E) = 2\pi^{-1/2}(kT)^{-3/2}e^{-E/kT}E^{1/2}$ $\quad E \in [0, \infty)$

6.2 The electron density distribution function for an electron in a one-dimensional box in its ground state is

$$p(x) = \left(\frac{2}{L}\right) \sin^2 \frac{\pi x}{L} \quad x \in [0, L]$$

where L is the length of the box. Find the value of $x = x^*$ for which $p(x)$ is a maximum.

6.3 The electron density distribution function for an electron in the ground state of the hydrogen atom is

$$p(r) = 4 \left(\frac{1}{a_0}\right)^3 r^2 e^{-\frac{2r}{a_0}}$$

Find the value of $r = r^*$ for which $p(r)$ is a maximum.

6.4 Consider a rectangular box of dimensions x, y, and z. The volume of the rectangular box is

$$V(x, y, z) = xyz$$

and the surface area of the rectangular box is

$$A(x, y, z) = 2(x^2 + y^2 + z^2).$$

The volume of the box is constrained to be a constant V_0. Using the method of undetermined multipliers, find the dimensions $x = x^*$, $y = y^*$, and $z = z^*$ that minimize the surface area of the box. Determine the value of $A(x^*, y^*, z^*)$ in terms of V_0.

6.5 Consider the function

$$f(x, y) = x^2 - y^2$$

Find the critical point at which $f_x = 0$ and $f_y = 0$. Evaluate the hessian $D = f_{xx}f_{yy} - f_{xy}f_{yx}$ at the critical point. If the second derivative test is conclusive ($D \neq 0$), classify the critical point as a minimum, maximum, or saddle point.

Homework exercises

6.6 The potential energy, $V(r)$, of a diatomic molecule as a function of its bond length, r, can be approximated by the Morse function

$$V(r) = A(1 - e^{-B(r-r_0)})^2$$

where A, B, and r_0 are positive constants. Find the bond length $r = r^*$ for which $V(r)$ is a minimum.

6.7 Consider the function

$$f(x,y) = xy$$

defining a hyperbolic saddle in three-dimensional cartesian space depicted in the figure below as the red surface. Suppose you impose the constraint

$$g(x,y) = x^2 + y^2 - 2 = 0$$

Using the method of undetermined multipliers, determine the four critical points (x^*, y^*). Evaluate the function $f(x^*, y^*)$ at each critical point and determine if that point represents a maximum or a minimum of the constrained function.

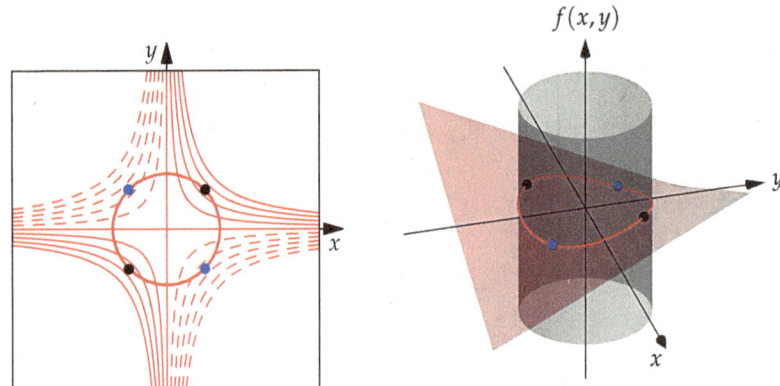

6.8 Consider the function

$$f(x,y,z) = x + y + 2z$$

in three-dimensional cartesian space. Suppose you impose the constraint

$$g(x,y,z) = x^2 + y^2 + z^2 - 3 = 0$$

Using undetermined multipliers determine the two critical points (x^*, y^*, z^*). Evaluate the function $f(x^*, y^*, z^*)$ at each critical point and determine if that point represents a maximum or a minimum.

6.9 Consider a cone of height, h, and circular base of radius, r. The volume of the cone is

$$V(r,h) = \frac{1}{3}\pi r^2 h$$

and the lateral surface area of the cone (the surface area of the top of the cone, excluding the base) is

$$A(r,h) = \pi r \sqrt{r^2 + h^2}$$

where $\sqrt{r^2 + h^2}$ is the length of the side of the cone. The volume of the cone is constrained to be a constant V_0. Using the method of undetermined multipliers, find the radius $r = r^*$ and height $h = h^*$ that minimize the lateral surface area of the cone. Determine the minimal area $A(r^*, h^*)$ in terms of the fixed volume V_0.

6.10* Consider the *butterfly potential* energy function

$$V(x, y) = ((x - y)^2 - 1)^2 + 10x^2y^2$$

shown in the figure below. There are seven critical points at which $V_x = 0$ and $V_y = 0$. Evaluate the hessian $D = V_{xx}V_{yy} - V_{xy}V_{yx}$ at the critical points $(1, 0)$, $(0, 0)$, and $(1/3, -1/3)$. If the second derivative test is conclusive $(D \neq 0)$, classify the critical point as a minimum, maximum, or saddle point.

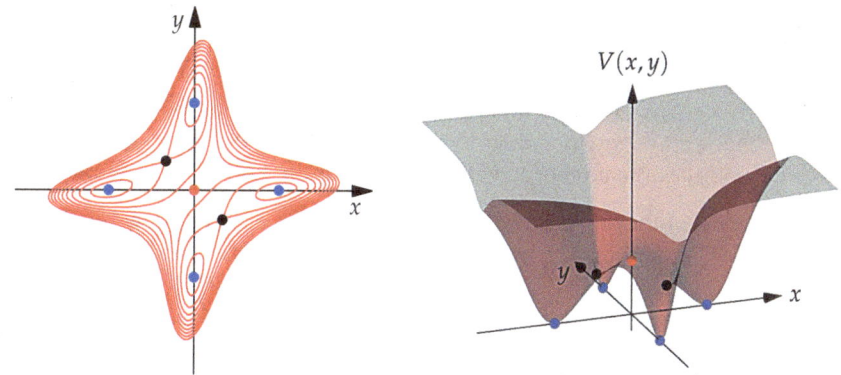

6.11* Consider the function

$$W(n_1, n_2, \ldots, n_M) = \frac{N!}{n_1! \, n_2! \ldots n_M!}$$

where

$$N = \sum_{k=1}^{M} n_k$$

(a) Using Stirling's approximation, derive an approximate form of $\ln W(n)$ that has no factorials and is valid for large values of N and n_k for all k.

(b) Using the method of undetermined multipliers, find the values of all n_k for which the function

$$S(n_1, n_2, \ldots, n_M) = \ln W(n_1, n_2, \ldots, n_M)$$

is a maximum subject to the constraint that $\sum_{k=1}^{M} n_k - N = 0$.

6.12* Find the value of λ for which the function $\rho(\lambda)$ is a maximum where

$$\rho(\lambda) = \frac{8\pi hc}{\lambda^5} \left[\frac{1}{\exp(\beta hc/\lambda) - 1} \right]$$

HINT: Introduce the variable $x = \beta hc/\lambda$. Substitute λ in terms of x to find the simplified function $\rho(x)$. Determine the maximum of the function by solving the equation

$$\left. \frac{d\rho(x)}{dx} \right|_{x = x_{max}} = 0$$

The maximum of $\rho(\lambda)$ will occur at $\lambda_{max} = \beta hc/x_{max}$ (see figure below).

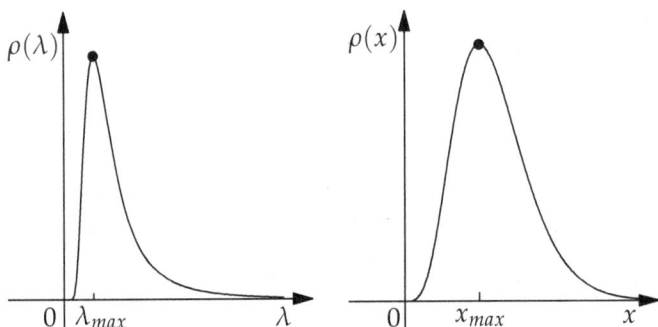

6.13*   The variational method in quantum mechanics provides a means of finding the lowest energy configuration of nuclei and electrons. A trial wave function is assumed, depending on one or more parameters, and the values of the parameters are varied to find the lowest energy possible. In this problem, you will apply the variational method to find the allowed energies of the one-electron atom or ion given by the Bohr model.

Consider the classical Bohr model of the one-electron hydrogen atom that assumes the electron moves in a circular orbit about the nucleus (see Figure 6.8). The energy of the hydrogen atom in this model is

$$E(r,v) = \frac{1}{2}m_e v^2 - \frac{e^2}{4\pi\varepsilon_0 r}$$

where ε_0 is the permittivity of a vacuum. The nucleus of charge $+e$ is centered at the origin. The electron of charge $-e$ and mass m_e moves in a circular orbit of radius, r, and linear velocity, v.

Bohr assumed only certain orbits were possible and that those orbits had values of the angular momentum

$$L = m_e v r = \frac{nh}{2\pi} \qquad n = 1, 2, 3, \ldots$$

where h is Planck's constant and n is the *quantum number*. Use the method of undetermined multipliers to minimize the energy of the atom, $E(r,v)$, subject to the constraint that

$$\frac{nh}{2\pi} - m_e v r = 0$$

(a) Show that the radius of an allowed orbits is given by

$$r = \left(\frac{4\pi\varepsilon_0 \hbar^2}{m_e e^2}\right) n^2 = a_0 n^2 \qquad n = 1, 2, 3, \ldots$$

where the constant $\hbar = h/2\pi$ and $a_0 = 0.529 \times 10^{-10} m$ is known as the *Bohr radius*.

(b) Show that the allowed energies for the orbits are given by

$$E = -\frac{1}{n^2}\left(\frac{m_e e^4}{8\varepsilon_0^2 h^2}\right) \qquad n = 1, 2, 3, \ldots$$

7 INTEGRATION IN ONE AND MANY DIMENSIONS

7.1 Integrating functions of one variable

A FOUNDATION IDEA IN CALCULUS is the concept of integration. The integral measures the area under the curve defined by the variation in a function over a range of its variable. In modeling physical processes, we are often interested in summing the value of one property (the given function) times the incremental change in another property (the variable). Pressure-volume work done by a gas is calculated by summing the pressure (the function) times the incremental change in volume (the variable) as the volume is varied. The computed pressure-volume work is the area under the pressure curve over a range of volume. In this section, we explore the definition of the integral and review the most commonly applied methods of integrating a function of one variable.

7.1.1 *The concept of the antiderivative and definition of the integral*

We can think of an integral as an *antiderivative*. While differentiation is an operation that transforms a function into its derivative, integration is an operation that transforms the derivative of a function back to the original function. Consider the derivative of a function $f(x)$ with respect to the variable x defined

$$\frac{d}{dx}f(x) = f'(x)$$

Rearranging this result, we find an expression for the total differential:

$$df(x) = f'(x)dx$$

This expression relates the change in the function $f(x)$ resulting from an incremental change in the variable dx. We can then define the *integral* as the operation that transforms the incremental change $df(x)$ into the function $f(x)$ itself:

$$f(x) = \int df(x) = \int f'(x)dx \tag{7.1}$$

The integral is the antiderivative because it transforms the derivative of a function into the function itself.

We developed a good working knowledge of differentiation in Chapter 3. That means we already know something about integrals as antiderivatives. For example, for x raised to a power n, we know that

$$f(x) = \frac{1}{n}x^n \quad \Rightarrow \quad f'(x) = x^{n-1}$$

which implies that[1]

$$\int f'(x)dx = \int x^{n-1}\, dx = \frac{1}{n}x^n = f(x)$$

[1] This form of integral, for which the range of x is unspecified, is known as an *indefinite integral*.

Similarly, for the exponential function, we know that

$$f(x) = \frac{1}{a}\, e^{ax} \quad \Rightarrow \quad f'(x) = e^{ax}$$

so that

$$f(x) = \int f'(x)dx = \int e^{ax}\, dx = \frac{1}{a}\, e^{ax}$$

For the natural logarithm, we find

$$f(x) = \ln(x) \quad \Rightarrow \quad f'(x) = \frac{1}{x}$$

so that

$$f(x) = \int f'(x)dx = \int \frac{1}{x}\, dx = \ln(x)$$

Finally, for the sine and cosine functions, we find

$$f(x) = -\frac{1}{a}\cos(ax) \quad \Rightarrow \quad f'(x) = \sin(ax)$$

so that

$$f(x) = \int f'(x)dx = \int \sin(ax)\, dx = -\frac{1}{a}\, \cos(ax)$$

and

$$f(x) = \frac{1}{a}\sin(ax) \quad \Rightarrow \quad f'(x) = \cos(ax)$$

so that

$$f(x) = \int f'(x)dx = \int \cos(ax)\, dx = \frac{1}{a}\, \sin(ax)$$

Thinking of integration as antidifferentiation, our knowledge of differentiation immediately translates into a knowledge of integration. Taking any total derivative relation, we have a relation for an indefinite integral.

7.1.2 *Interpretation of the integral*

The derivative is interpreted as the rate of change in a function for a given value of its variable. The integral sign $\int$ is an elongated S standing for summation. The process of integration is the process of summing the area under the curve of the function being integrated.[2]

Let's explore the interpretation of the integral through the example of determining the pressure-volume work done by an ideal gas at constant temperature, defined as

$$w = \int_{V_A}^{V_B} p(V)dV = \int_{V_A}^{V_B} \frac{nRT}{V}\, dV$$

where we have used the ideal gas equation of state $pV = nRT$ to define $p(V) = \frac{nRT}{V}$.[3] The function to be integrated, in this case $p(V)$, is known as the *integrand*. The factor nRT is a constant independent of V and can be pulled

[2] The fact that the definite integral of a function over an interval is equal to the difference between the antiderivatives of the function evaluated at the endpoints of the interval is known as the *fundamental theorem of calculus*. The theorem was first presented by Scottish mathematician and astronomer *James Gregory* (1638-1675) and English theologian and mathematician *Isaac Barrow* (1630-1677). The theorem was built upon by Barrow's student *Isaac Newton* (1642-1727) and systematized by *Gottfried Leibniz* (1646-1716) who introduced the notation commonly used today.

[3] This form of integral, for which the range of x is specified, is known as a *definite integral*.

out of the integral, leaving

$$w = nRT \int_{V_A}^{V_B} \frac{1}{V} \, dV$$

From our antiderivative relations in the previous section, we know that

$$\int \frac{1}{x} \, dx = \ln(x)$$

so that[4]

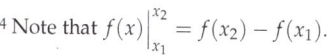

[4] Note that $f(x)\Big|_{x_1}^{x_2} = f(x_2) - f(x_1)$.

$$w = nRT \int_{V_A}^{V_B} \frac{1}{V} \, dV = nRT \, \ln(V)\Big|_{V_A}^{V_B} = nRT \, \ln\left(\frac{V_B}{V_A}\right)$$

where we have used the fact that $\ln(V_B) - \ln(V_A) = \ln\left(\frac{V_B}{V_A}\right)$.

What is the physical interpretation of this result? We interpret the integral as the antiderivative. As such, the integral represents the summing up of many small changes in the work $dw = p(V)dV$ that take place over many small changes in volume

$$w = \int dw = \int p(V) \, dV$$

Each increment of the work dw is the area formed by the height of the function $p(V)$ times the width resulting from a change in volume dV.

Figure 7.1 shows $p(V)$ as a function of volume V between the volumes V_A and V_B. Suppose we divide the interval $[V_A, V_B]$ into N subintervals $[V_1, V_2], [V_2, V_3], ..., [V_N, V_{N+1}]$ where $V_1 = V_A$, $V_{N+1} = V_B$ and $V_n = V_A + (n-1)\Delta V$. The width of each interval is $\Delta V = (V_B - V_A)/N$. We can approximate the integral defined as the area under the curve using the sum

$$w \approx p(V_1)\Delta V + p(V_2)\Delta V + \ldots + p(V_N)\Delta V$$
$$= \sum_{n=1}^{N} p(V_n)\Delta V$$

which is the sum of the areas of N rectangles each of width ΔV and height $p(V_n)$. Each shaded area is an increment of work dw. The integral is the sum of many of these areas adding up to the total change w.

As N increases, the number of rectangles increases and the accuracy of the estimate to the area under the curve improves. In the limit $N \to \infty$, the sum converges to the exact result

$$w = \lim_{N \to \infty} \sum_{n=1}^{N} p(V_n)\Delta V = \int_{V_A}^{V_B} p(V)dV \tag{7.2}$$

The area under the curve between the lower limit of integration V_A and the upper limit of integration V_B represents the definite integral of $p(V)$ taken between V_A and V_B.

Our choice of the height of each rectangle as the leftmost point in the interval was arbitrary. We could just as well take the rightmost point in the

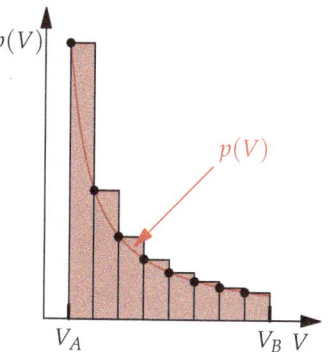

Figure 7.1: The integral of $p(V)$ between V_A and V_B is the area under the curve. The integral can be approximated by the sum of the shaded rectangular areas

$$\sum_{n=1}^{N} p(V_n)\Delta V$$

where $V_n = V_A + (n-1)\Delta V$. For finite ΔV this approximation overestimates the exact integral.

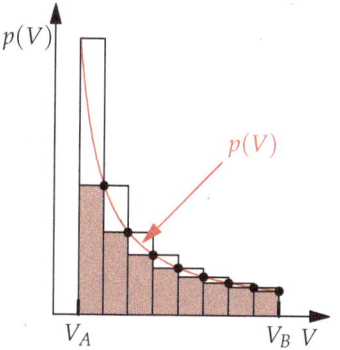

Figure 7.2: Alternatively, the integral can be approximated by the sum of the shaded rectangular areas $p(V_n)\Delta V$ where $V_n = V_A + n\Delta V$. For finite ΔV this approximation underestimates the exact integral.

interval to be the height of the rectangle. In that case, the sum approximating the integral is

$$w = \lim_{N \to \infty} \sum_{n=1}^{N} p(V_{n+1}) \Delta V = \int_{V_A}^{V_B} p(V)dV \tag{7.3}$$

which represents the shaded area in Figure 7.2. Note that for finite $\Delta V > 0$, the first approximation using the leftmost point in the interval leads to an overestimate of the area under the curve. In contrast, using the rightmost point in the interval leads to an underestimate of the area under the curve. Nevertheless, in the limit $\Delta V \to 0$ both approximations lead to the same exact result.

With these insights, we can develop a general definition of the integral in terms of the *Riemann sum*. Suppose we partition the interval $[V_A, V_B]$ into N subintervals $[V_1, V_2], [V_2, V_3], ..., [V_N, V_{N+1}]$ where $V_A = V_1 < V_2 < ... < V_N < V_{N+1} = V_B$. The integral of $p(V)$ between V_A and V_B can be approximated by the sum

$$w = \lim_{N \to \infty} \sum_{n=1}^{N} p(V_n^*) \Delta V_n = \int_{V_A}^{V_B} p(V)dV \tag{7.4}$$

where $V_n^* \in [V_n, V_{n+1}]$ and $\Delta V_n = V_{n+1} - V_n$. An example of a Riemann sum having subintervals of varying width ΔV_n and randomly chosen points V_n^* within each subinterval is shown in Figure 7.3.

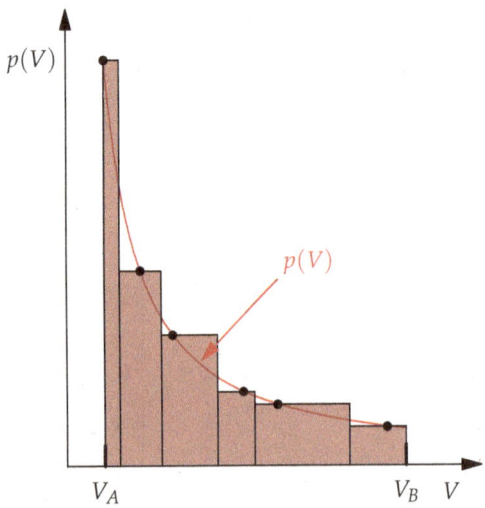

Figure 7.3: The integral of $p(V)$ between V_A and V_B can be approximated by the Riemann sum

$$\sum_{n=1}^{N} p(V_n^*) \Delta V_n$$

where $V_n^* \in [V_n, V_{n+1}]$ and $\Delta V_n = V_{n+1} - V_n$. As $N \to \infty$ the sum converges to the exact result.

In our previous sums, the subintervals ΔV_n were of uniform width. In this Riemann sum, the intervals V_n may be of varying width. Moreover, in our previous sums the height of each rectangle was taken to be the value of $p(V)$ for the leftmost or rightmost point in the interval. In the Riemann sum, the height of the rectangle is taken to be $p(V_n^*)$ where V_n^* is any point in the interval $[V_n, V_{n+1}]$.[5]

[5] The results in Equations 7.2 and 7.3 are known as *left method* and *right method* Riemann sums, respectively. Alternative definitions include the *maximum method* and *minimum method* for which the height of each rectangle is defined by the endpoint with the larger or smaller value, respectively.

We can now combine our definition of the derivative given by Equation 3.2

$$\lim_{\Delta x \to 0} \frac{f(x + \Delta x) - f(x)}{\Delta x} = \frac{df}{dx}$$

with our equivalent definition of the integral

$$\lim_{N \to \infty} \sum_{n=1}^{N} f(x_n)\Delta x = \int f(x)dx \qquad (7.5)$$

These geometric interpretations of the derivative and integral in terms of *finite differences* Δx in the variable x are useful in numerically estimating derivatives and integrals when the indefinite integrals or exact derivatives are not known. By performing the sum for increasing values of N and decreasing values of Δx, we find an increasingly accurate approximation of the integral.

7.1.3 Revisiting the rules for integrating functions of one variable

There are a number of useful rules for integrating functions of one variable. We introduce these rules by exploring the integration of a variety of commonly appearing functional forms. It is important to practice and master these rules. They are essential to our successful application of calculus to problems in the physical sciences.

1. $df(x)$. Integration of the exact differential of a function leads to the function itself plus an *integration constant*[6] written as C:

$$\int df(x) = f(x) + C \qquad (7.6)$$

For example, suppose $df(x) = dx$, then

$$\int df(x) = \int dx = x + C$$

In evaluating indefinite integrals, it is important to remember to include the integration constant C.

2. a. Integration of a constant produces a linear function plus an integration constant:

$$\int a\, dx = a \int dx = ax + C \qquad (7.7)$$

Differentiation of the result produces a. Note that when we have a constant times a function of x, we can pull the constant out of the integral as shown above. Pulling the constant out in front of the integral can be helpful in simplifying the integration of complicated expressions.

3. x^n. Using our knowledge of the *power rule* for derivatives of powers of x, we find

$$\int x^n\, dx = \frac{1}{n+1} x^{n+1} + C \qquad n \neq -1 \qquad (7.8)$$

Let's use this result to evaluate the integral

$$\int v_0\, t\, dt$$

[6] The derivative of a function is equal to the derivative of the same function plus a constant

$$\frac{d}{dx} f(x) = \frac{d}{dx}\left(f(x) + C\right)$$

As such, the antiderivative of a function equals the integral of that function plus a constant

$$\int df(x) = f(x) + C$$

The *integration constant* C is undetermined for an indefinite integral. It can be evaluated once the limits of integration are defined.

We find

$$\int v_0\, t\, dt = v_0 \int t\, dt = \frac{1}{2} v_0\, t^2 + C$$

Further consider the integral

$$\int \frac{\Delta H}{RT^2}\, dT$$

Using our result we find

$$\int \frac{\Delta H}{RT^2}\, dT = \frac{\Delta H}{R} \int \frac{1}{T^2}\, dT = -\frac{\Delta H}{R} \frac{1}{T} + C$$

Differential and integral relations of this kind are commonly encountered in thermodynamics.

4. x^{-1}. Using our knowledge of the derivative of the natural logarithm, we find the indefinite integral

$$\int \frac{1}{x}\, dx = \ln x + C \qquad (7.9)$$

The definite integral of $\frac{1}{x}$ over the range $x \in [a,b]$ can be written

$$\int_a^b \frac{1}{x}\, dx = \ln x \Big|_a^b = \ln(b) - \ln(a) = \ln\left(\frac{b}{a}\right)$$

This result is shown graphically in Figure 7.4. The integral is equal to the shaded area under the curve. The black line shows the accumulating area under the curve, which increases according to $\ln\left(\frac{x}{a}\right)$.

In the kinetics of certain reactions, the rate of change in the concentration $c(t)$ is proportional to the concentration itself. As the rate of change in $c(t)$ is dc/dt, this proportional relationship can be written

$$\frac{dc}{dt} = -kc \qquad (7.10)$$

By separating dc from dt and dividing each side by c, this equation can be reformed as

$$\frac{1}{c}\, dc = -k\, dt$$

where k is a constant.[7] We can solve this equation for $c(t)$. Using our knowledge of integrals, we can transform the result

$$\int \frac{1}{c}\, dc = -\int k\, dt$$

into a relation for the concentration as a function of time

$$\ln c(t) = -kt + C$$

By exponentiating each side of this equation, the result can also be expressed as

$$c(t) = Ae^{-kt}$$

where $A = e^C$.

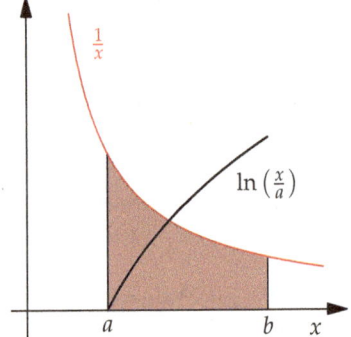

Figure 7.4: The function $\frac{1}{x}$ over a range of x (red line). The integral of $\frac{1}{x}$ between $x = a$ and $x = b$ is the shaded area under the curve. The accumulating area under the curve $\ln\left(\frac{x}{a}\right)$ (black line) is 0 at $x = a$ and grows to the final value of the integral $\ln\left(\frac{b}{a}\right)$ at $x = b$.

[7] Here we treat the derivative $\frac{dc}{dt}$ as a fraction, which allows us to separate the incremental change in concentration, dc, from the incremental change in time, dt.

As another example, consider the following fundamental relation[8] in thermodynamics:

$$\frac{dp}{dT} = \frac{\Delta H}{RT^2}\, p \qquad (7.11)$$

By separating dp from dT and dividing each side by p, this equation can be reformed as

$$\frac{1}{p}\, dp = \frac{\Delta H}{RT^2}\, dT$$

We solve the latter equation for $p(T)$ using our knowledge of integrals as

$$\int \frac{1}{p}\, dp = \frac{\Delta H}{R} \int \frac{1}{T^2} dT$$

and

$$\ln p = -\frac{\Delta H}{RT} + C$$

This can be rewritten in the familiar form

$$p(T) = A \exp\left(-\frac{\Delta H}{RT}\right)$$

where $\ln A = C$. Inserting our result in Equation 7.11 demonstrates the validity of our solution.

5. $f(x) + g(x)$. Integration is a linear operation, like multiplication or differentiation. The integral of the sum of two functions is the sum of two integrals:

$$\int [f(x) + g(x)]\, dx = \int f(x) dx + \int g(x) dx \qquad (7.12)$$

For example, consider the integral of the function $C_p(T) = a + bT + cT^2$ over T, where a, b, and c are constants. We find

$$\int C_p(T) dT = \int \left(a + bT + cT^2\right) dT = a \int dT + b \int T dT + c \int T^2 dT$$

$$= aT + \frac{1}{2}bT^2 + \frac{1}{3}bT^3 + C$$

The integration of each term in $C_p(T)$ increases the power of T. As we are taking an indefinite integral, it is important to include the integration constant C in our final result.

6. e^{ax}. Using our knowledge of the derivative of the exponential function, we find

$$\int e^{ax} dx = \frac{1}{a} e^{ax} + C \qquad (7.13)$$

For example, consider the indefinite integral over a function $c(t)$ with respect to t, where $c(t) = C_0\, e^{-t/\tau}$ and C_0 and τ are constants. We find

$$\int C_0\, e^{-t/\tau} dt = C_0 \int e^{-t/\tau} dt = -C_0 \tau\, e^{-t/\tau} + C$$

Further consider the definite integral of e^{-x} over the range $x \in [a,b]$. We

[8] Known as the *Clausius-Clapeyron equation*, this relation describes the dependence of the vapor pressure on temperature. Note that we treat $\frac{dp}{dT}$ as an algebraic ratio, which allows us to separate the incremental changes dp and dT.

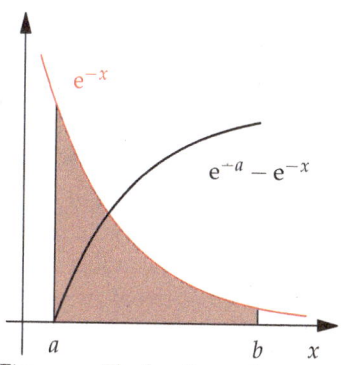

Figure 7.5: The function e^{-x} over a range of x (red line). The integral of e^{-x} between $x = a$ and $x = b$ is the shaded area under the curve. The accumulating area under the curve $e^{-a} - e^{-x}$ (black line) is 0 at $x = a$ and grows to the final value of the integral $\left(e^{-a} - e^{-b}\right)$ at $x = b$.

find

$$\int_a^b e^{-x}\, dx = -e^{-x}\Big|_a^b = e^{-a} - e^{-b}$$

This result is shown graphically in Figure 7.5. The decaying exponential is commonly encountered in physical kinetics, where it is used to model a decreasing population of species.

7. $\sin(ax)$ and $\cos(ax)$. Using our knowledge of derivatives of sine and cosine functions, we find

$$\int \sin(ax)\, dx = -\frac{1}{a}\cos(ax) + C \tag{7.14}$$

$$\int \cos(ax)\, dx = \frac{1}{a}\sin(ax) + C \tag{7.15}$$

For example, consider the integral over a sinusoidal function

$$\int_0^\pi \sin(x)dx = -\cos(x)\Big|_a^b = \cos(a) - \cos(b)$$

This result is shown graphically in Figure 7.6.

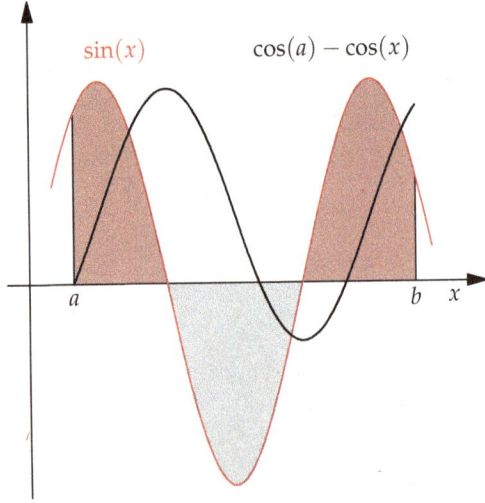

Figure 7.6: The integral of $\sin(x)$ between $x = a$ and $x = b$ has both positive (shaded red) and negative (shaded gray) contributions. The accumulating area under the curve $\cos(a) - \cos(x)$ (black line) is 0 at $x = a$ and grows to the final value of the integral $(\cos(a) - \cos(b))$ at $x = b$.

The sinusoidal function takes on positive and negative values which add and subtract from the total area under the curve. The result is an accumulating area that oscillates as the contributions from positive and negative regions of the sine are summed.

8. $\delta(x - x_0)$. The *Dirac delta function* (or δ function) was discovered by Paul Dirac[9] and plays an important role in the mathematics of quantum mechanics. While the function may seem strange, in some ways, the integral over the δ function could not be simpler. The function $\delta(x - x_0)$ is zero everywhere except at the point $x = x_0$, where it is infinite:

$$\delta(x - x_0) = \begin{cases} \infty & x = x_0 \\ 0 & x \neq x_0 \end{cases} \tag{7.16}$$

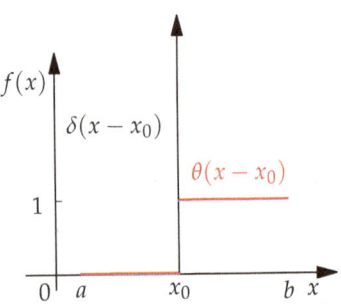

Figure 7.7: The integral of $\delta(x - x_0)$ between $x = a$ and $x = b$ is the accumulating area under the curve (red line) represented by the *Heaviside step function* $\theta(x - x_0)$, which is 0 for $x < x_0$ and unity for $x \geq x_0$.

[9] The English physicist *Paul Dirac* (1902-1984) made fundamental contributions to quantum theory.

The area under the function when integrated over a range of x containing x_0 is unity, as

$$\int_{-\infty}^{\infty} \delta(x - x_0) \, dx = 1$$

An attempt to graphically depict the Dirac delta function is shown in Figure 7.7.[10]

Consider the integration over all space of the delta function multiplying a function $f(x)$. The product of $f(x)\delta(x - x_0)$ is zero everywhere except at $x = x_0$, so that

$$\int_{-\infty}^{\infty} f(x)\delta(x - x_0) \, dx = f(x_0) \int_{-\infty}^{\infty} \delta(x - x_0) \, dx = f(x_0)$$

Note that we can pull the term $f(x_0)$ out of the integral as it is a constant. We will explore the utility of the delta function later in this chapter.

9. e^{-ax^2}. The gaussian function has many important applications in the physical sciences. There is a simple closed form solution for the definite integral of the gaussian function over all space:

$$\int_{-\infty}^{\infty} e^{-ax^2} \, dx = \sqrt{\frac{\pi}{a}} \tag{7.17}$$

This fundamentally important result is widely used in applications of integration in the physical sciences. The identity is proven in the complements.

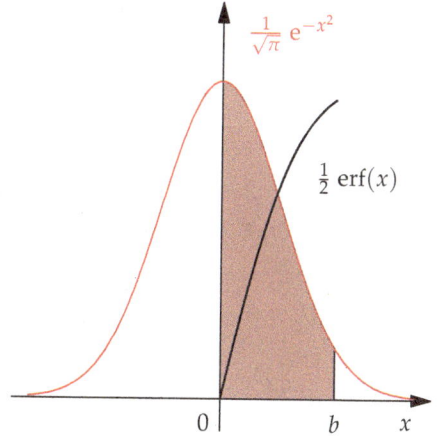

The integral over a gaussian function occurs frequently in the physical sciences. It is called the *error function* and is defined

$$\text{erf}(x) = \frac{2}{\sqrt{\pi}} \int_0^x e^{-t^2} \, dt \tag{7.18}$$

The error function is positive for positive values of x, representing the area under the gaussian function over the range $[0, x]$ (see Figure 7.8). There is

[10] Note that the Heaviside step function is related to the integral over the delta function:

$$\theta(x - x_0) = \int_{-\infty}^{x} \delta(x' - x_0) dx'$$

As such, it follows that the derivative of the Heaviside step function

$$\frac{d\theta(x - x_0)}{dx} = \delta(x - x_0)$$

is the delta function.

Figure 7.8: The function $\frac{1}{\sqrt{\pi}} e^{-x^2}$ over a range of x (red line). The integral of e^{-x^2} between $x = 0$ and $x = b$ is the shaded area under the curve. The accumulating area under the curve $\frac{1}{2}\text{erf}(x)$ (black line) is zero at $x = 0$ and grows to the final value $\frac{1}{2}\text{erf}(b)$ at $x = b$.

also a *complementary error function* that is defined

$$\text{erfc}(x) = \frac{2}{\sqrt{\pi}} \int_x^\infty e^{-t^2} \, dt = 1 - \text{erf}(x) \qquad (7.19)$$

and represents the area under the gaussian function not included in the error function. As such, $\text{erf}(x) + \text{erfc}(x) = 1$ (see Figure 7.8). Numerical values of the error function and complementary error function are tabulated in Supplement S_7.

The rules of integration above provide us with the foundation of knowledge we need to address problems in the physical sciences that involve the integration of functions.

7.1.4 Useful tricks for integrating functions of one variable

In addition to using the fundamental rules of integration previously introduced, we can add to our toolbox a few basic tricks that will prove useful in performing integrals on more complicated functions.

1. Algebraic substitution. In differentiation we encountered complicated expressions that were simplified using an *algebraic substitution* of variables. For example, consider the function

$$f(x) = e^{-\frac{1}{x}}$$

Taking the derivative of this function with respect to x can be simplified by introducing the variable

$$y = \frac{1}{x}$$

where

$$\frac{dy}{dx} = -\frac{1}{x^2}$$

Using the chain rule defined

$$\frac{df}{dx} = \frac{df}{dy}\frac{dy}{dx}$$

the derivative of $f(x)$ with respect to x can be expressed as

$$\frac{d}{dx}\left(e^{-\frac{1}{x}}\right) = \frac{d}{dy}e^{-y}\frac{dy}{dx} = -e^{-y}\left(-\frac{1}{x^2}\right) = \frac{1}{x^2}e^{-1/x}$$

We can use a similar trick in integration. Suppose we want to integrate the function

$$f(x) = \frac{1}{x^2}e^{-\frac{1}{x}}$$

This looks hard! We first simplify the function with the substitution[11]

$$y = \frac{1}{x} \qquad dy = -\frac{1}{x^2}dx$$

This allows us to write

$$\int \frac{1}{x^2}e^{-\frac{1}{x}}\,dx = -\int e^{-y}\,dy = e^{-y} + C = e^{-\frac{1}{x}} + C$$

[11] Using the fact that

$$\frac{dy}{dx} = -\frac{1}{x^2}$$

we multiply both sides of the expression by dx to find

$$dy = -\frac{1}{x^2}dx$$

where as a final step we convert back to the original variable x. This example demonstrates how a clever substitution of variables turns an apparently challenging integral into a problem with a familiar solution.[12]

Further consider the example

$$\int \frac{1}{V - nb}\, dV \tag{7.20}$$

We make the algebraic substitution commonly referred to as *u-substitution*

$$u = V - nb \qquad \frac{du}{dV} = 1 \qquad du = dV$$

Inserting these results in our integral, we find

$$\int \frac{1}{V - nb}\, dV = \int \frac{1}{u}\, du = \ln u + C = \ln(V - nb) + C$$

While the proper substitution may be apparent in this example, in other cases careful consideration and trial and error may be required to identify the optimal *u*-substitution.

Finally, consider the example

$$\int \cos^2 x \sin x \, dx$$

We make the algebraic substitution

$$u = \cos x \qquad du = -\sin x \, dx$$

and find

$$\int \cos^2 x \sin x \, dx = -\int u^2 \, du = -\frac{1}{3}u^3 + C = -\frac{1}{3}\cos^3 x + C$$

A wide range of integrals can be solved using algebraic substitution, including integrals involving polynomial, exponential, and sinusoidal integrands. Algebraic substitution is particularly valuable in the physical sciences, as the proper substitution of variables can convert a complicated integrand into a simplified and manageable function.

2. Trigonometric transformation. Consider the integral

$$\int \sin^2 x \, dx$$

We know how to integrate $\sin x$ but not $\sin^2 x$. Fortunately, we can get rid of the square by transforming the expression using the *trigonometric transformation*

$$\sin^2 x = \frac{1}{2}\left(1 - \cos(2x)\right)$$

so that

$$\int \sin^2 x \, dx = \int \frac{1}{2}\left(1 - \cos(2x)\right)\, dx = \frac{1}{2}\int dx - \frac{1}{2}\int \cos(2x)\, dx$$
$$= \frac{1}{2}x - \frac{1}{4}\sin(2x) + C$$

[12] Using this trick we can show that the integral

$$\int e^{-\frac{\Delta E}{k_B T}}\left(\frac{\Delta E}{k_B T^2}\right) dT$$

with the substitution of variables $u = -\frac{\Delta E}{k_B T}$ is equal to

$$e^{-\frac{\Delta E}{k_B T}} + C$$

When you encounter an integral over a trigonometric function that you don't recognize, consider trigonometric identities that might transform the integral into a solvable form.

3. **Partial fractions.** For an integral like Equation 7.20, the numerator dV is a derivative of the denominator $(V - nb)$. In that case, the integral can be solved using *algebraic substitution*. However, for the integral

$$\int \frac{1}{(a - x)(b - x)} \, dx$$

the numerator dx is not the derivative of the denominator $(a - x)(b - x)$. In this case, we can reform our integral using *partial fractions*.[13]

We can make use of a clever identity relating the product of the reciprocal differences to the difference of the reciprocal differences.[14] It is written as

$$\frac{1}{(a - x)(b - x)} = \frac{1}{(b - a)} \left(\frac{1}{a - x} - \frac{1}{b - x} \right)$$

and leads to the equality

$$\int \frac{1}{(a - x)(b - x)} \, dx = \int \frac{1}{(b - a)} \left(\frac{1}{a - x} - \frac{1}{b - x} \right) dx$$

As such, we can evaluate these integrals using our knowledge of the integral identity

$$\int \frac{1}{(x + a)} \, dx = \ln(x + a) + C$$

It follows that

$$\int \frac{1}{(b - a)} \left(\frac{1}{a - x} \right) dx = -\frac{1}{b - a} \ln(a - x) + C$$

and

$$\int \frac{1}{(b - a)} \left(\frac{1}{b - x} \right) dx = -\frac{1}{b - a} \ln(b - x) + C$$

resulting in

$$\int \frac{1}{(a - x)(b - x)} \, dx = \frac{1}{b - a} \ln \left(\frac{b - x}{a - x} \right) + C$$

The transformation of $\frac{1}{(a-x)(b-x)}$ using partial fractions is graphically depicted in Figure 7.9. It is somewhat remarkable that the product of the reciprocal functions $(a - x)^{-1}$ and $(b - x)^{-1}$ is proportional to the difference of the two functions.

4. **Integration by parts.** A final trick for evaluating integrals found in the physical sciences is known as *integration by parts*. Recall the *product rule*

$$\frac{d}{dx} [u(x)v(x)] = u(x)\frac{dv}{dx} + v(x)\frac{du}{dx}$$

Multiplying both sides of the equation by dx results in

$$d(uv) = udv + vdu$$

[13] The method of partial fractions is a special case of the more general *Heaviside cover-up method* named for English physicist and engineer *Oliver Heaviside* (1850-1925).

[14] We can prove this

$$\frac{1}{(b - a)} \left(\frac{1}{a - x} - \frac{1}{b - x} \right)$$
$$= \frac{1}{(b-a)} \left[\frac{b - x}{(a - x)(b - x)} - \frac{a - x}{(a - x)(b - x)} \right]$$

that can be simplified as

$$\frac{1}{(b - a)} \left[\frac{b - x - (a - x)}{(a - x)(b - x)} \right] = \frac{1}{(a - x)(b - x)}$$

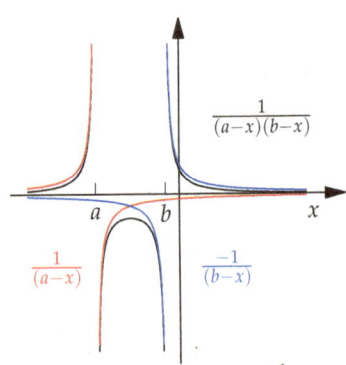

Figure 7.9: The function $\frac{1}{(a-x)(b-x)}$ (black line) alongside the partial fractions $\frac{1}{a-x}$ (red line) and $\frac{-1}{b-x}$ (blue line) over a range of x.

which can be reformed as

$$u\,dv = d\,(uv) - v\,du$$

Integrating both sides of this equation leads to[15]

$$\int u\,dv = uv - \int v\,du \qquad (7.21)$$

This relation expresses one integral $\int u\,dv$, which may be challenging to evaluate, in terms of another integral $\int v\,du$, which may be easier to solve, and a surface term uv, which is often zero. Equation 7.21 serves as our starting point for integration by parts.[16]

Let's see how this works. Consider the integral

$$\int_0^\infty x e^{-ax}\,dx = \int u\,dv$$

We decompose $x e^{-ax} dx = u\,dv$ as[17]

$$u = x \qquad\qquad dv = e^{-ax}\,dx$$

$$du = dx \qquad\qquad v = -\frac{1}{a} e^{-ax} \qquad (7.22)$$

Combining these results with Equation 7.21, we find[18]

$$\int_0^\infty x e^{-ax}\,dx = \int u\,dv = uv - \int v\,du$$

$$= -\frac{1}{a} x e^{-ax}\Big|_0^\infty - \int_0^\infty \left(-\frac{1}{a} e^{-ax}\right) dx = -\frac{1}{a^2} e^{-ax}\Big|_0^\infty = \frac{1}{a^2}$$

As a final example of integration by parts, consider the indefinite integral

$$\int x \sin x\,dx = \int u\,dv$$

We decompose $x \sin x\,dx = u\,dv$ as

$$u = x \qquad\qquad dv = \sin x\,dx$$

$$du = dx \qquad\qquad v = -\cos x \qquad (7.23)$$

Combining these results with Equation 7.21, we find

$$\int x\,\sin x\,dx = \int u\,dv = uv - \int v\,du$$

$$= -x\,\cos x + \int \cos x\,dx$$

$$= -x\,\cos x + \sin x\,dx + C$$

where C is an undetermined integration constant.

This concludes our review of useful rules for integrating functions of one variable. In the physical sciences, we often encounter functions of many variables. In the next section, we apply the rules for integrating functions of one variable to the integration of functions of many variables.

[15] Integration by parts was first introduced by English mathematician *Brook Taylor* (1685-1731).

[16] We have used the identity

$$\int df(x) = f(x)$$

to write

$$\int d(uv) = uv$$

[17] To find v we integrate dv as

$$v = \int dv = \int e^{-ax}\,dx = -\frac{1}{a}\,e^{-ax}$$

while du is the total differential dx.

[18] We use the result

$$\frac{1}{a} x e^{-ax}\Big|_0^\infty = 0$$

At the lower limit $x e^{-ax}$ is zero when $x = 0$. For the upper limit we find

$$\lim_{x\to\infty} x e^{-ax} = 0$$

This can be shown using l'Hôpital's rule as

$$\lim_{x\to\infty} \frac{x}{e^{ax}} = \lim_{x\to\infty} \frac{1}{a e^{ax}} = 0$$

7.2 Integrating functions of many variables

FUNCTIONS OF MORE THAN ONE VARIABLE are commonly encountered in modeling the properties of physical systems. The concentration of a substance in three dimensions is a function of the position in space defined by the values of three cartesian coordinates. The total amount of substance in a given volume of solution is the multiple integral of the concentration taken over all three coordinates. In this section, we explore the definition of multiple integrals and survey common methods for evaluating multiple integrals over functions of many variables.

7.2.1 Double integrals

We represent the integral of a function of one variable $f(x)$, taken between points $x = a$ and $x = b$, as

$$\int_a^b f(x)\, dx$$

We interpret this integral to be the area under the curve formed by $f(x)$ over $x \in [a, b]$. An example is shown in Figure 7.10, where the shaded area is the value of the integral. This concept can be generalized to integrals of functions of many variables.

In two dimensions, we represent the *double integral* of a function $f(x, y)$, taken over a rectangular area on the xy-plane defined by $x \in [a, b]$ and $y \in [c, d]$, as

$$\int_c^d \int_a^b f(x, y)\, dxdy$$

We interpret this double integral to be the volume under the surface formed by $f(x, y)$ over the rectangular area. An example is shown in Figure 7.11, where the shaded volume is the value of the integral.

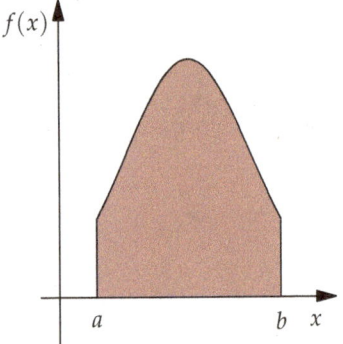

Figure 7.10: The function $f(x)$ shown over a range of x from $[a, b]$. The integral $\int_a^b f(x)\, dx$ equals the shaded area under the curve.

Figure 7.11: The function $f(x, y)$ shown as a surface over a rectangular region on the xy-plane formed by $x \in [a, b]$ and $y \in [c, d]$. The integral $\int_c^d \int_a^b f(x, y)\, dxdy$ equals the shaded volume under the surface.

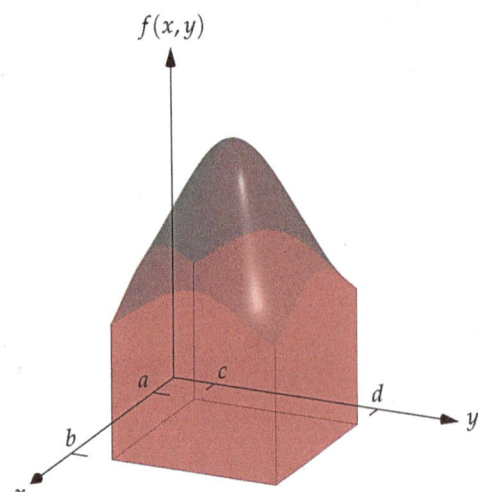

It is important to understand the notation used for a double integral. For a double integral, the order of integration is indicated by the order of the increments dx and dy in the expression. Consider the integral

$$\int_c^d \int_a^b f(x,y)\,dxdy = \int_c^d \left[\int_a^b f(x,y)\,dx \right] dy$$

Integration over x is performed first, followed by integration over y. Note that the *antiderivative* is defined strictly for functions of one variable (see Equation 7.1). As such, the definition of the *indefinite integral* does not immediately extend to functions of many variables.

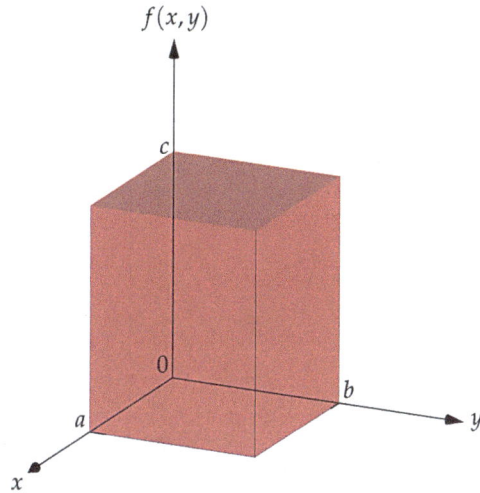

Figure 7.12: The function $f(x,y) = c$ over a rectangular region on the xy-plane formed by $x \in [0,a]$ and $y \in [0,b]$. The integral $\int_0^b \int_0^a f(x,y)\,dxdy$ equals the shaded volume $V = abc$.

Let's consider a simple example to see how the double integral works. Suppose we wish to integrate the constant function $f(x,y) = c$ over an area defined by $x \in [0,a]$ and $y \in [0,b]$ (see Figure 7.12). Since the function is a constant, we can write

$$\int_0^b \int_0^a f(x,y)\,dxdy = c \int_0^b \int_0^a dxdy$$

Since the limits of integration over x do not depend on y, we can separate the integral over x from the integral over y as

$$c \int_0^b \int_0^a dxdy = c \int_0^b dy \times \int_0^a dx$$

We have now reduced our two-dimensional integral to two one-dimensional integrals that we can readily evaluate to find

$$c \int_0^b dy \times \int_0^a dx = cb \int_0^a dx = cba$$

The value of the double integral is the height of the function, c, times the area over which the integral is performed on the xy-plane, ab, which is the volume abc.

7.2.2 *Double integrals and the order of integration*

Not all integrals on the xy-plane are as simple as our previous example. Consider the challenge of finding the area between a straight line and intersecting parabola shown in Figure 7.13. We define the area as the double integral over the xy-plane restricted to the area between the curves.

Suppose we wish to first integrate over x. As the integral is restricted to the area between the curves, the lower and upper bounds of the integral in x will depend on the particular value of y. The lower bound for integration over x is defined by the straight line so that $x = \frac{y}{2}$ while the upper bound for integration over x is defined by the parabola so that $x = \sqrt{y}$. Integration over y follows integration over x with the lower and upper bounds of $y = 0$ and $y = 4$, respectively.

With this understanding we can write a formula for the area between the curves in terms of the double integral[19]

$$\int_0^4 \int_{y/2}^{\sqrt{y}} dxdy = \int_0^4 \left[\int_{y/2}^{\sqrt{y}} dx \right] dy$$

Evaluating the inner integral, we find

$$\int_0^4 \left[\int_{y/2}^{\sqrt{y}} dx \right] dy = \int_0^4 \left[x \right]_{y/2}^{\sqrt{y}} dy = \int_0^4 \left(\sqrt{y} - \frac{y}{2} \right) dy$$

We now have two definite integrals to complete, as

$$\int_0^4 \left(\sqrt{y} - \frac{y}{2} \right) dy = \left[\frac{2}{3} y^{3/2} - \frac{1}{4} y^2 \right]_0^4 = \frac{2}{3} 4^{3/2} - \frac{1}{4} 4^2 = \frac{4}{3}$$

Conversely, this integral can also be evaluated by first integrating over y followed by integration over x:

$$\int_0^2 \int_{x^2}^{2x} dydx = \int_0^2 \left[\int_{x^2}^{2x} dy \right] dx = \int_0^2 \left(2x - x^2 \right) dx = \frac{4}{3}$$

So we see that the order of integration does not matter. This will be true for most of the integrals of interest to us.

7.2.3 *Double integrals of separable functions*

The examples above demonstrate that integration over x and y can be coupled by integrating over areas that make the limits of integration over x dependent on y (and *vice versa*). However, integration over x and y can also be coupled when $f(x,y)$ is a non-separable function of x and y and $f(x,y) \neq g(x)h(y)$.

Fortunately, most of the integrals of interest in the physical sciences are taken over *separable functions*. When $f(x,y) = g(x)h(y)$ we can write

$$\int_c^d \int_a^b f(x,y) \, dxdy = \int_c^d \int_a^b g(x)h(y) \, dxdy$$

$$= \int_c^d h(y) \, dy \times \int_a^b g(x) \, dx$$

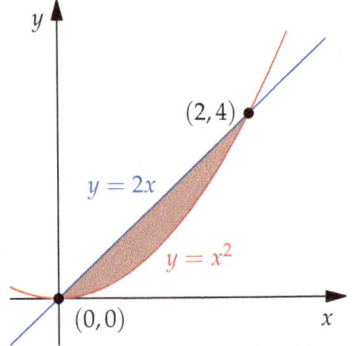

Figure 7.13: The curves defined by $y = 2x$ (blue line) and $y = x^2$ (red line) intersecting at the points $(0,0)$ and $(2,4)$ with the area between the curves shaded red.

[19] Note the we have taken the integrand to be unity.

This reduces the two-dimensional integral over the xy-plane to a product of two one-dimensional integrals. Separability will often depend on the right choice of coordinate systems. To better understand this critical idea, we will explore in detail a variety of examples involving the integration of separable functions of many variables.

7.2.4 Double integrals and tables of integrals

A valuable application of integration in many dimensions involves determination of the area or volume of a geometric object. The area or volume can be determined by integration over the area or volume of the object where the integrand is unity.

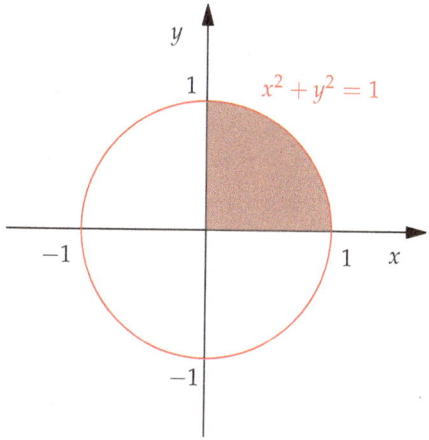

Figure 7.14: One quadrant of the unit circle with area equal to $\frac{\pi}{4}$.

Consider the evaluation of the area of a circle shown in Figure 7.14. We can divide and conquer this integral by first performing the integration over one quadrant. We then multiply that result by four. In that case, we wish to determine the shaded area of the quarter circle. The range of integration of x depends on the value of y as along the perimeter of the circle $x = \sqrt{1 - y^2}$. As such, the range of integration in x will have a lower bound of 0 and upper bound of $x = \sqrt{1 - y^2}$. The range of integration of y is from 0 to 1. With this understanding, we can write the integral over the quadrant as

$$\int_0^1 \int_0^{\sqrt{1-y^2}} dxdy = \int_0^1 \sqrt{1 - y^2}\, dy$$

What do we do now? While we have reviewed methods for performing integrals of a single variable, nothing in our survey prepared us for this integral.

We will find that for many integrals of interest, the best way to solve them is to look up the result in a *table of integrals*.[20] Tables of definite and indefinite integrals sufficient to perform the problems in this text are included in Supplements S_5 and S_6. These tables include the result

$$\int_0^a \sqrt{a^2 - x^2}\, dx = \left[\frac{1}{2} \left[x\sqrt{a^2 - x^2} + a^2 \sin^{-1}\left(\frac{x}{a} \right) \right] \right]_0^a$$

[20] The classic text is known as *Gradshteyn and Ryzhik* or simply *GR*, which still serves as a valuable compendium of results for definite and indefinite integrals.

Izrail S. Gradshteyn and Iosif M. Ryzhik. *Table of Integrals, Series, and Products [1943]*. Academic Press, eighth edition, 2014. ISBN 978-0123849335

Using this result for our case of $a = 1$, we find

$$\int_0^1 \sqrt{1^2 - x^2} \, dx = \frac{1}{2} \sin^{-1}(1) = \frac{1}{4}\pi$$

This is the area over one quadrant of the unit circle. Four times this result is our answer to the original problem, to determine the area of the unit circle. The area of the unit circle is π.

7.2.5 Double integrals and the choice of coordinate system

The evaluation of the integral representing the area within the unit circle was somewhat complicated in cartesian coordinates. Cartesian coordinates have a natural square symmetry while the unit circle has round symmetry. Let's evaluate the integral again in plane polar coordinates, which better reflect the round symmetry of the unit circle.

Using the coordinate transformations explored in Chapter 1 and depicted in Figure 7.15, we can transform the integral over a quarter of the unit circle in two-dimensional cartesian coordinates to plane polar coordinates as

$$\int_0^1 \int_0^{\sqrt{1-y^2}} dxdy = \int_0^{\pi/2} \int_0^1 r \, drd\theta = \int_0^{\pi/2} d\theta \times \int_0^1 r \, dr$$

$$= \frac{\pi}{2} \times \int_0^1 r \, dr = \frac{\pi}{2} \times \frac{1}{2} = \frac{\pi}{4}$$

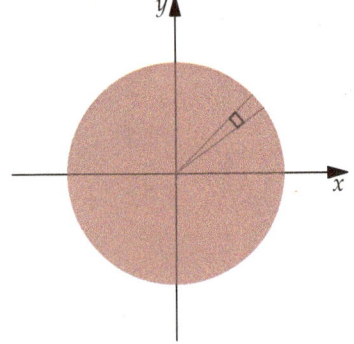

Figure 7.15: The plane polar coordinate system in two dimensions showing the volume element $dA = r \, drd\theta$ (bold). The range of variables is defined by $0 \leq r < \infty$ and $0 \leq \theta < 2\pi$.

Multiplying this answer by four leads to the result that the area of the unit circle is π. However, in evaluating the integral in plane polar coordinates there is no need to break the integral into integrals over quarter circles as was required in cartesian coordinates. We can perform the integral at once over the whole unit circle as

$$\int_0^{2\pi} \int_0^1 r \, drd\theta = \int_0^{2\pi} d\theta \times \int_0^1 r \, dr = 2\pi \times \int_0^1 r \, dr = 2\pi \times \frac{1}{2} = \pi$$

Much simpler! Note that in cartesian coordinates, the integrals over x and y were coupled through the limits of integration. In plane polar coordinates, the integrals over r and θ are separable. The take-home lesson is that the right choice of coordinate system, in which the variables reflect the natural symmetry of the function, can make the work of integration much easier.

7.2.6 Double integrals and the delta function

Suppose we want to determine the length of the circumference of a circle using an integral along the line forming the circle. We can accomplish this using the delta function. Let's start with the double integral over the xy-plane expressed in plane polar coordinates

$$\int_0^{2\pi} \int_0^\infty r \, drd\theta$$

Now suppose we insert in the integrand $\delta(r - a)$. This makes the integrand zero unless $r = a$, which defines the perimeter of a circle. In that case, the integral over the xy-plane will be reduced to an integral over the perimeter of the circle only. This gives us a formula for the circumference of the circle:

$$C = \int_0^{2\pi} \int_0^\infty \delta(r-a) r \, dr d\theta$$

Performing the integral over r first,[21] followed by the integral over θ, we find

$$\int_0^{2\pi} \int_0^\infty \delta(r-a) r \, dr d\theta = \int_0^{2\pi} d\theta \times \int_0^\infty \delta(r-a) r \, dr$$

$$= 2\pi \times \int_0^\infty \delta(r-a) r \, dr = 2\pi a$$

which is the circumference of a circle of radius a. Inserting the delta function reduces the two-dimensional integral over the xy-plane to a one-dimensional integral over the perimeter of the circle.

[21] Here we use the identity

$$\int_0^\infty \delta(r-a) r \, dr = a$$

where the Dirac delta function is zero everywhere other than the perimeter of the circle where $r = a$.

7.2.7 Triple integrals

The evaluation of the volume of a solid is of fundamental importance in the physical sciences.[22] In order to extend our understanding of multiple integrals, let's consider the evaluation of the volume of a cylinder of radius a and height h (see Figure 7.16). Defining the volume of a cylinder in cartesian coordinates is complicated. The limits of integration in x and y are coupled just as we found in the integration over the unit circle. However, the integral over the volume of a cylinder can be naturally and simply expressed in using cylindrical coordinates.

[22] *Archimedes (circa 287-212 BCE)* determined geometric volumes using the theory of *infinitesimals* before the concept of the integral was developed. The theory of infinitesimals was further applied to the calculation of areas and volumes by *Johannes Kepler (1571-1630)* in his 1615 treatise *New Solid Geometry of Wine Barrels.*

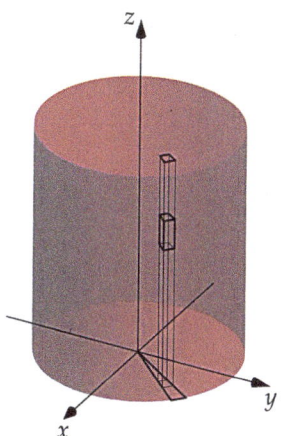

Figure 7.16: A cylinder of radius a and height h. Superimposed is the volume element $dV = r dr d\theta dz$ (bold) with sides dr, $rd\theta$, and dz. The range of variables is defined by $0 \le r < \infty$, $0 \le \theta < 2\pi$, and $-\infty < z < \infty$.

The triple integral representing the volume of the cylinder is written as

$$V = \int_0^h \int_0^{2\pi} \int_0^a r \, dr d\theta dz = \int_0^h dz \times \int_0^{2\pi} d\theta \times \int_0^a r \, dr$$

$$= h \times 2\pi \times \frac{a^2}{2} = \pi a^2 h$$

This is an intuitive result. The volume of the cylinder is equal to the circular area of the cylinder, πa^2, times the height h.

Now suppose we want to determine the volume of a sphere. We want to select a coordinate system that shares the natural symmetry of a sphere. In cartesian coordinates, the limits of integration over x, y, and z are coupled. However, the spherical polar coordinate system depicted in Figure 1.22 has the natural symmetry of a sphere. If we transform our function into spherical polar coordinates we find the integral is separable in r, θ, and φ:

$$V = \int_0^{2\pi} \int_0^{\pi} \int_0^a r^2 \sin\theta \, drd\theta d\varphi = \int_0^{2\pi} d\varphi \times \int_0^{\pi} \sin\theta d\theta \times \int_0^a r^2 \, dr$$

$$= 2\pi \times 2 \times \frac{a^3}{3} = \frac{4}{3}\pi a^3$$

This is the well-known result for the volume of a sphere of radius a.[23]

These examples show how integration in three dimensions can be used to derive important identities for volumes of shapes such as cylinders and spheres. Moreover, the examples show how the proper choice of coordinate system can convert a many dimensional integral into a series of independent one-dimensional integrations. For the integrals of greatest interest to us in the physical sciences, we will often find that multiple integrals can be simplified in this way.[24]

7.2.8 Triple integrals and the delta function

Suppose we want to determine the surface area of a sphere of radius a. We can accomplish this using the delta function. Let's start with the triple integral over all space in spherical polar coordinates

$$\int_0^{2\pi} \int_0^{\pi} \int_0^{\infty} f(r,\theta,\varphi)r^2 \sin\theta \, drd\theta d\varphi$$

We insert into the integrand a function $f(r,\theta,\varphi)$ that is zero everywhere in three-dimensional space other than on the surface of the sphere. We do that using the delta function

$$f(r,\theta,\varphi) = \delta(r-a)$$

which is zero everywhere other than on the spherical surface defined by $r = a$. As a result, we find the area of the surface of a sphere defined by the separable integral

$$A = \int_0^{2\pi} \int_0^{\pi} \int_0^{\infty} \delta(r-a)r^2 \sin\theta \, drd\theta d\varphi$$

$$= \int_0^{2\pi} d\varphi \times \int_0^{\pi} \sin\theta \, d\theta \times \int_0^{\infty} \delta(r-a)r^2 dr$$

Performing the integral over r first, followed by the integrals over θ and φ, we find

$$A = a^2 \times \int_0^{2\pi} d\varphi \times \int_0^{\pi} \sin\theta \, d\theta = a^2 \times 2\pi \times 2 = 4\pi a^2$$

which is the area of the surface of a sphere of radius a. Inserting the delta function reduces the three-dimensional integral over all space to a two-dimensional integral over the surface of the sphere.

[23] We used the identity

$$\int_0^{\pi} \sin(x)dx = -[\cos x]_0^{\pi} = -(-1-1) = 2$$

to complete the integral over θ.

[24] It is possible to extend the concept of the double and triple integrals to integrals over arbitrarily large numbers of variables. As large, in fact, as *Avogadro's number*, and larger still.

A$_7$ An alternative to integration by parts for exponential integrals

In exploring useful tricks for integrating functions of one variable, we performed integrals over an exponential function e^{ax} times a power of x of the form

$$\int xe^{ax}\,dx$$

To evaluate this integral, we used integration by parts where

$$u = x \qquad\qquad dv = e^{ax}\,dx$$

$$du = dx \qquad\qquad v = \frac{1}{a}e^{ax} \qquad\qquad (7.24)$$

This led to the result

$$\int xe^{ax}\,dx = \frac{1}{a}xe^{ax} - \int \frac{1}{a}e^{ax}dx = \frac{1}{a}xe^{ax} - \frac{1}{a^2}e^{ax}$$

$$= \frac{1}{a^2}\left(ax-1\right)e^{ax} + C$$

That was not too bad. However, consider the related integral

$$\int x^4 e^{ax}\,dx$$

To evaluate this integral using integration by parts, we need to apply the rule of integration by parts four times! Let's explore a way to improve our efficiency in performing integrals of this kind.

Consider the identity

$$xe^{ax} = \frac{d}{da}e^{ax}$$

where we have used the derivative with respect to the constant a to pull down a power of x. With this identity, we can rewrite the integral of interest as[25]

$$\int xe^{ax}\,dx = \frac{d}{da}\int e^{ax}\,dx$$

Using our knowledge of the exponential integral, this integral can be evaluated as

$$\int xe^{ax}\,dx = \frac{d}{da}\int e^{ax}\,dx = \frac{d}{da}\left(\frac{1}{a}e^{ax}\right) = -\frac{1}{a^2}e^{ax} + \frac{1}{a}xe^{ax}$$

$$= \frac{1}{a^2}\left(ax-1\right)e^{ax} + C \qquad\qquad (7.25)$$

just as we found using integration by parts.

The real power of this approach is evident in evaluating definite integrals over an exponential function e^{-ax} times a power of x. For example, we can readily evaluate the integral of xe^{-ax}:

$$\int_0^\infty xe^{-ax}\,dx = -\frac{d}{da}\int_0^\infty e^{-ax}\,dx = -\frac{d}{da}\frac{1}{a} = \frac{1}{a^2}$$

[25] We use the fact that

$$\frac{d}{da}\int f(ax)\,dx = \int \frac{df}{da}\,dx$$

This result is consistent with the result derived with greater effort using integration by parts, where

$$u = x \qquad dv = e^{-ax}\, dx$$

$$du = dx \qquad v = -\frac{1}{a}e^{ax} \tag{7.26}$$

so that

$$\int_0^\infty x e^{-ax}\, dx = -\frac{1}{a} x e^{-ax}\Big|_0^\infty - \int_0^\infty \left(-\frac{1}{a}e^{-ax}\right) dx$$

$$= \frac{1}{a^2}e^{-ax}\Big|_0^\infty = \frac{1}{a^2}$$

We see how our trick of differentiating with respect to a is more efficient than integration by parts for these particular definite integrals over the range $0 \le x \le \infty$. This is even more true for integrals such as

$$\int_0^\infty x^2 e^{-ax}\, dx = \frac{d^2}{da^2}\int_0^\infty e^{-ax}\, dx = \frac{d^2}{da^2}\frac{1}{a} = -\frac{d}{da}\frac{1}{a^2} = 2\frac{1}{a^3}$$

This integral would require two rounds of integration by parts to be solved. Using our trick, we can solve the problem by simply differentiating the exponential twice with respect to a.

B₇ Evaluating the definite integral of a gaussian function

Consider the integral over the gaussian function

$$I = \int_{-\infty}^\infty e^{-ax^2}\, dx$$

There is no obvious way to evaluate this integral using our standard methods, such as algebraic substitution or integration by parts. Consider the algebraic substitution $y = x^2$. This leads to $dy = 2xdx$, introducing a factor of x. Similar problems arise in integration by parts, where the derivative of the gaussian function introduces a factor of x. To evaluate this integral, we will use an inspired trick imagined by the German mathematician *Carl Friedrich Gauss* for whom the gaussian function is now named.

Consider the product of two equivalent one-dimensional integrals

$$\int_{-\infty}^\infty e^{-ax^2}\, dx \times \int_{-\infty}^\infty e^{-ay^2}\, dy = I \times I = I^2$$

This product can be thought of as one two-dimensional integral over the xy-plane

$$I^2 = \int_{-\infty}^\infty \int_{-\infty}^\infty e^{-a(x^2+y^2)}dxdy$$

This integral is most simply evaluated in plane polar coordinates, where

$$r^2 = x^2 + y^2 \qquad \int_{-\infty}^\infty \int_{-\infty}^\infty dxdy \rightarrow \int_0^\infty \int_0^{2\pi} r\, d\theta dr$$

With this transformation of variables, we arrive at the expression

$$I^2 = \int_{-\infty}^{\infty}\int_{-\infty}^{\infty} e^{-a(x^2+y^2)}dxdy = \int_0^{\infty}\int_0^{2\pi} e^{-ar^2} r\, d\theta dr$$

$$= 2\pi \int_0^{\infty} e^{-ar^2} r\, dr$$

Defining $u = r^2$, so that $du = 2r\, dr$, results in

$$I^2 = 2\pi \int_0^{\infty} e^{-ar^2} r\, dr = \pi \int_0^{\infty} e^{-au}\, du = \pi \left[-\frac{1}{a} e^{-au} \right]_0^{\infty} = \frac{\pi}{a}$$

leading to our final result

$$I = \int_{-\infty}^{\infty} e^{-ax^2}\, dx = \sqrt{\frac{\pi}{a}}$$

This result is recorded in the Table of Definite Integrals in Supplement S_5 #2. We commonly encounter integrals over gaussian functions in the physical sciences. As such, it is worthwhile to memorize this fundamental result.

C_7 An alternative to integration by parts for gaussian integrals

We can extend our trick for efficiently integrating over integrands of the form $x^n\, e^{-\alpha x}$ to integrands of the form $x^n\, e^{-\alpha x^2}$ containing a gaussian function.

Consider the gaussian integral

$$\int_0^{\infty} e^{-ax^2}\, dx = \frac{1}{2} \int_{-\infty}^{\infty} e^{-ax^2}\, dx = \frac{1}{2}\sqrt{\frac{\pi}{a}}$$

Differentiating the gaussian function with respect to the parameter a pulls down a factor of $-x^2$. We find

$$\frac{d}{da} \int_0^{\infty} e^{-ax^2}\, dx = \int_0^{\infty} \frac{d}{da} e^{-ax^2}\, dx = -\int_0^{\infty} x^2 e^{-ax^2}\, dx$$

That means that

$$\int_0^{\infty} x^2 e^{-ax^2}\, dx = -\frac{d}{da} \int_0^{\infty} e^{-ax^2}\, dx = -\frac{d}{da}\left(\frac{1}{2}\sqrt{\frac{\pi}{a}} \right)$$

$$= \frac{1}{4a^{3/2}}\sqrt{\pi}$$

For the odd powers of x, we can start from the integral identity[26]

$$\int_0^{\infty} x e^{-ax^2}\, dx = \frac{1}{2a}$$

and use our trick to show

$$\int_0^{\infty} x^3 e^{-ax^2}\, dx = -\frac{d}{da} \int_0^{\infty} x e^{-ax^2}\, dx = -\frac{d}{da}\left(\frac{1}{2a} \right) = \frac{1}{2a^2}$$

While integrals of this form can be evaluated using integration by parts, the

[26] We can evaluate the integral

$$I = \int_0^{\infty} x e^{-ax^2}\, dx$$

with the algebraic substitution $y = x^2$ so that $dy = 2x\, dx$ and

$$I = \frac{1}{2} \int_0^{\infty} e^{-ay}\, dy = \frac{1}{2a}$$

effort required is significant. Using our trick, we can most efficiently evaluate integrals over gaussian functions that commonly occur in the physical sciences.

D$_7$ Properties of delta functions

Delta functions can be used to select specific terms in a series or points in a function. Here we explore the properties of two delta functions that are commonly used in the physical sciences, the Kronecker delta function and the Dirac delta function.

Kronecker delta function

The *Kronecker delta function*, δ_{nm}, is a function of two discrete indices n and m. It is defined to be zero except when $n = m$:

$$\delta_{nm} = \begin{cases} 1 & n = m \\ 0 & n \neq m \end{cases}$$

This function is used in sums to select those terms for which two indices are equal (see Figure 7.17).[27] For example, consider the sum

$$\sum_{n=-\infty}^{\infty} a_n \delta_{nm} = a_m$$

where δ_{nm} is used to select only the elements a_n for which $n = m$. This is known as the *sifting property*.

The Kronecker delta function is normalized as

$$\sum_{n=-\infty}^{\infty} \sum_{m=-\infty}^{\infty} \delta_{nm} = 1$$

These examples demonstrate how the Kronecker delta function can be used to sift through and select specific terms in a finite or infinite series.

Dirac delta function

The *Dirac delta function*, $\delta(x - x_0)$, is zero everywhere except at the point $x = x_0$, where it is infinite:

$$\delta(x - x_0) = \begin{cases} \infty & x = x_0 \\ 0 & x \neq x_0 \end{cases}$$

This function is used in integrals to select a particular value of the variable x (see Figure 7.18). For example, consider the integral

$$\int_{-\infty}^{\infty} f(x)\delta(x - x_0) = f(x_0)$$

where $\delta(x - x_0)$ is used to select the single value of the function $f(x)$ for which $x = x_0$. This is the *sifting property* observed for the Kronecker delta function.

[27] The German mathematician *Leopold Kronecker* (1902-1984) made important contributions to number theory and logic.

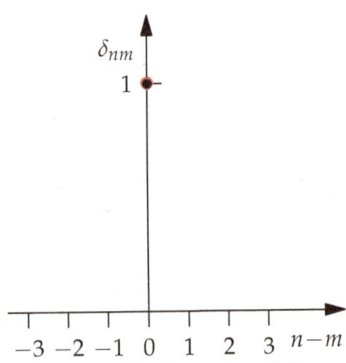

Figure 7.17: The Kronecker delta function δ_{nm} as a function of the difference between the indices n and m.

The Dirac delta function is normalized as

$$\int_{-\infty}^{\infty} \delta(x - x_0) = 1$$

and can be defined in terms of the first derivative of the *Heaviside step function*, $\theta(x - x_0)$, which is 0 for $x < x_0$ and unity for $x \geq x_0$ (see Figure 7.5).

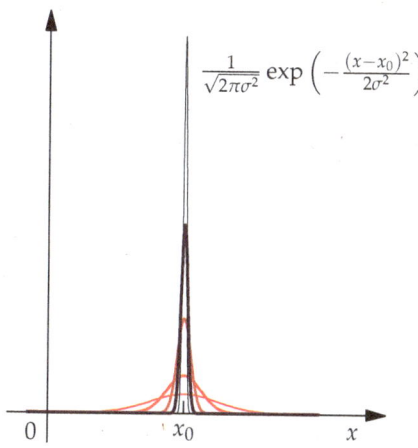

Figure 7.18: The normalized gaussian function centered at x_0 with decreasing values of σ ranging from large (pale red line) to small (black line). In the limit that $\sigma \to 0$, the function approaches the Dirac delta function $\delta(x - x_0)$.

There is a rich variety of ways to represent the Dirac delta function. In problem solving in the physical sciences, one representation of the delta function might lead to an easier solution than another representation. One popular representation of the delta function is defined in terms of a normalized gaussian function with its standard deviation σ approaching zero, represented by the limit

$$\delta(x - x_0) = \lim_{\sigma \to 0} \frac{1}{\sqrt{2\pi\sigma^2}} e^{-\frac{(x-x_0)^2}{2\sigma^2}} \tag{7.27}$$

This function is continuous and differentiable with a maximum at $x = x_0$. As $\sigma \to 0$, the function's height increases toward infinity, according to $1/\sigma$, while its width decreases to zero, according to σ, preserving the normalization

$$\int_{-\infty}^{\infty} \frac{1}{\sqrt{2\pi\sigma^2}} e^{-\frac{(x-x_0)^2}{2\sigma^2}} dx = 1$$

for all values of σ. This gaussian function is shown in Figure 7.18 for values of the parameter σ ranging from 0.5 to 0.025.

These observations provide a guide to designing other representations of the Dirac delta function. We need a function that is normalized with a single parameter that controls the width and height of the function such that in some limit the function becomes infinitely tall and infinitesimally wide.

Let's consider the *lorentzian function*

$$\delta(x - x_0) = \frac{1}{\pi} \lim_{\sigma \to 0} \frac{\sigma}{(x - x_0)^2 + \sigma^2} \tag{7.28}$$

Like the gaussian function, it is continuous and differentiable, with a height proportional to $1/\sigma$ and width proportional to σ so that the area normalized for all values of σ (see Supplement S_5 #13). This lorentzian function is shown in Figure 7.19 for values of the parameter σ ranging from 1 to 0.05.

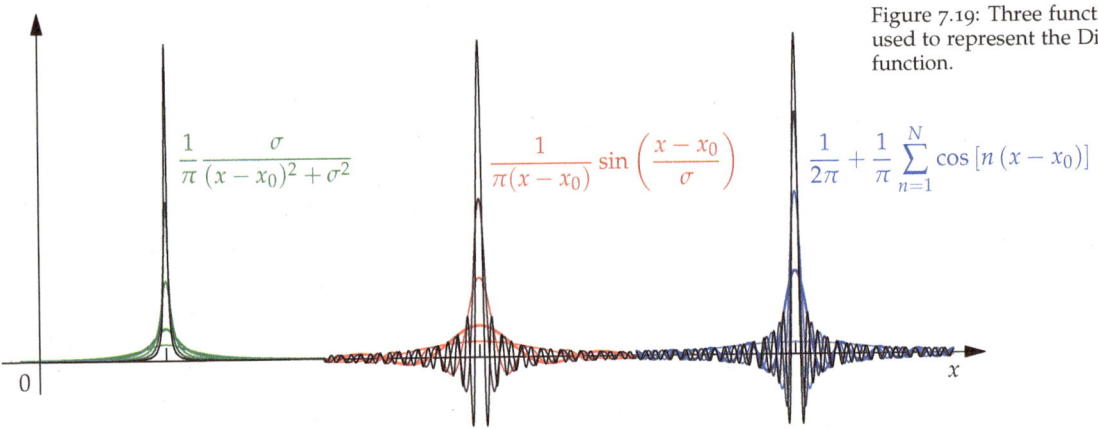

Figure 7.19: Three functional forms used to represent the Dirac delta function.

Less obvious functional representations of the Dirac delta function make use of sine or cosine terms. Consider the sinusoidal function

$$\delta(x - x_0) = \frac{1}{\pi} \lim_{\sigma \to 0} \frac{1}{(x - x_0)} \sin\left(\frac{x - x_0}{\sigma}\right) \tag{7.29}$$

This function is normalized, since

$$\int_{-\infty}^{\infty} \frac{1}{(x - x_0)} \sin\left(\frac{x - x_0}{\sigma}\right) dx = \pi$$

independent of the parameter σ (see the Table of Definite Integrals in Supplement S_5). This is surely one of the more remarkable definitions of the mathematical constant π. The height of the function at $x = x_0$ increases according to $1/\sigma$ as σ decreases, and the width of the function decreases in proportion to σ. This function is shown in Figure 7.19 for values of the parameter σ ranging from 1 to 0.05.

The Dirac delta function can also be expressed as an infinite cosine series:

$$\delta(x - x_0) = \frac{1}{2\pi} + \lim_{N \to \infty} \frac{1}{\pi} \sum_{n=1}^{N} \cos\left[n\left(x - x_0\right)\right] \tag{7.30}$$

Note that this formulation is independent of a parameter σ controlling the height and width of the function. The increased height and decreased width are realized through the constructive and destructive interference between terms of the cosine series. This function is shown in Figure 7.19 for values of the parameter N ranging from 1 to 40.

The Dirac delta function is useful in modeling a point charge, point mass, or electron. These examples show a few of the many ways the Dirac delta function can be represented in terms of continuous and differentiable functions of one variable. Many other examples exist and some can be found in Chapter 13, which explores Fourier transforms and harmonic analysis.

E₇ End-of-chapter problems

> I consider that I understand an equation
> when I can predict the properties of its
> solutions, without actually solving it.
>
> Paul Dirac

Warm-ups

7.1 Evaluate the following indefinite integrals. Consider italicized letters other than the integration variable to be constants. Include the integration constant C in your result.

(a) $\displaystyle\int mv\,dv$

(b) $\displaystyle\int \frac{1}{x^3}\,dx$

(c) $\displaystyle\int \sin 3x\,dx$

(d) $\displaystyle\int (3x+5)^2 4x\,dx$

(e) $\displaystyle\int e^{-\varepsilon/k_B T}\,d\varepsilon$

(f) $\displaystyle\int \cos(2\pi v t)\,dt$

(g) $\displaystyle\int \frac{RT}{p}\,dp$

(h) $\displaystyle\int \frac{1}{2}\kappa x^2\,dx$

(i) $\displaystyle\int \frac{q^2}{4\pi\varepsilon_0 r^2}\,dr$

7.2 Evaluate the following integrals, using various methods of integration and the tables of integrals found in Supplements S₅ and S₆ as needed. Consider italicized letters other than the integration variable to be constants. Include the integration constant C in your result.

(a) $\displaystyle\int \cos(5x)\cos(3x)\,dx$

(b) $\displaystyle\int x^4 e^{-\alpha x}\,dx$

(c) $\displaystyle\int \sin^2\left(\frac{n\pi x}{L}\right)\,dx$

(d) $\displaystyle\int \frac{dx}{(\alpha-x)^n}$

(e) $\displaystyle\int y e^{-y^2/2\sigma^2}\,dy$

(f) $\displaystyle\int e^{-\gamma x}\sin x\,dx$

(g) $\displaystyle\int \frac{dx}{(4-x)(3-x)}$

(h) $\displaystyle\int \cos^3\varphi \sin\varphi\,d\varphi$

(i) $\displaystyle\int x^3 \cos 2x\,dx$

7.3 Evaluate the following definite integrals, using the tables of integrals found in Supplements S₅ and S₆ as needed. Consider italicized letters other than the integration variable to be constants.

(a) $\displaystyle\int_{-\infty}^{\infty} x^3 e^{-\alpha x^2}\,dx$

(b) $\displaystyle\int_{p_1}^{p_2} \frac{RT}{p}\,dp$

(c) $\displaystyle\int_{V_1}^{V_2}\left(\frac{nRT}{V-nb}-\frac{n^2 a}{V^2}\right)dV$

(d) $\displaystyle\int_{0}^{L} x^2 \sin^2\left(\frac{n\pi x}{L}\right)dx$

(e) $\displaystyle\int_{T_1}^{T_2} \frac{\Delta H}{RT^2}\,dT$

(f) $\displaystyle\int_{T_1}^{T_2}\left(a+bT+cT^2+\frac{d}{T}\right)dT$

(g) $\displaystyle\int_{0}^{\infty} e^{-2r/a_0}\,r\,dr$

(h) $\displaystyle\int_{0}^{\infty} e^{-mv^2/2kT}\,v^3\,dv$

(i) $\displaystyle\int_{0}^{\infty} (2J+1)\,e^{-a(J^2+J)}\,dJ$

7.4 Evaluate the following multiple integrals, using the tables of integrals found in Supplements S₅ and S₆ as needed. For indefinite integrals be sure to include an integration constant.

(a) $\displaystyle\int\int yx^2\,dx\,dy$

(b) $\displaystyle\int\int (x^2+y^2)\,dx\,dy$

(c) $\displaystyle\int\int y \ln x \, dy \, dx$

(d) $\displaystyle\int\int\int x^2 \ln y \, e^{2x} \, dx \, dy \, dz$

(e) $\displaystyle\int_0^{\pi/2}\int_0^2 r \cos\theta \, dr \, d\theta$

(f) $\displaystyle\int_0^{2\pi}\int_0^\pi\int_0^V v^2 \sin\theta \, dv \, d\theta \, d\varphi$

(g) $\displaystyle\int_0^\infty\int_0^\infty \left(x^2+y^2\right) e^{-\frac{1}{2}\left(x^2+y^2\right)} \, dx \, dy$

(h) $\displaystyle\int_{-\infty}^\infty\int_{-\infty}^\infty\int_{-\infty}^\infty e^{-\frac{1}{2k_BT}\left(mv_x^2+mv_y^2+mv_z^2\right)} \, dv_x \, dv_y \, dv_z$

Homework exercises

7.5 Evaluate the following integral, which commonly appears in quantum mechanics and statistical thermodynamics

$$I = \int_0^\pi (3\cos^2\theta - 1)^2 \sin\theta \, d\theta$$

7.6 The allowed energies of a quantum mechanical particle in a one-dimensional box of length L can be written

$$E_n = -\frac{\hbar^2}{2m}\frac{2}{L}\int_0^L \sin\left(\frac{n\pi x}{L}\right)\left[\frac{d^2}{dx^2}\sin\left(\frac{n\pi x}{L}\right)\right] dx$$

where $n = 1, 2, 3, \cdots$ and $\hbar = h/2\pi$ is the constant h *bar*. Complete the integral to show that

$$E_n = \frac{n^2 h^2}{8mL^2}$$

7.7 Consider the integral

$$\int\int\int z^2 \, dx \, dy \, dz$$

taken over the volume of a sphere of radius a. Using spherical polar coordinates where $z = r\cos\theta$ the integral takes the form

$$\int_0^a\int_0^\pi\int_0^{2\pi} (r\cos\theta)^2 \, r^2 \sin\theta \, d\varphi \, d\theta \, dr$$

Evaluate this integral.

7.8 Consider the integral over all space

$$\int_0^\infty\int_0^\pi\int_0^{2\pi} e^{-2r}\cos^2\theta \, r^2 \sin\theta \, d\varphi \, d\theta \, dr$$

Evaluate this integral.

7.9 Consider the integral

$$\int\int\int xyz \, dx \, dy \, dz$$

taken over the volume defined by $0 \leq z < a$ and $x^2 + y^2 \leq b^2$ where $x \geq 0$ and $y \geq 0$. This integral can be expressed in cylindrical coordinates as

$$\int_0^a\int_0^{\frac{\pi}{2}}\int_0^b (r^2 \cos\theta \, \sin\theta) \, z \, r \, dr \, d\theta \, dz$$

Evaluate this integral.

7.10 Consider a gas of atoms interacting with a pairwise *square well potential* energy function dependent on the distance r between a pair of atoms

$$V(r) = \begin{cases} \infty & 0 < r \leq \sigma \\ -\varepsilon & \sigma < r \leq \lambda\sigma \\ 0 & r > \lambda\sigma \end{cases}$$

where $\lambda > 1$ controls the width of the well of depth ε. The potential function is shown below (red line) and compared with the smooth Lennard-Jones potential energy function (blue line) examined in Chapter 5 (see Figure 5.12). The *compressibility factor* of the gas can be written as a Taylor's series expansion in powers of the number of density n/V as

$$Z = \frac{PV}{nRT} = 1 + B(T)\frac{n}{V} + C(T)\left(\frac{n}{V}\right)^2 + \text{higher order terms}$$

This is known as the *virial expansion* where the coefficients $B(T)$ and $C(T)$ are the second and third virial coefficients, respectively.

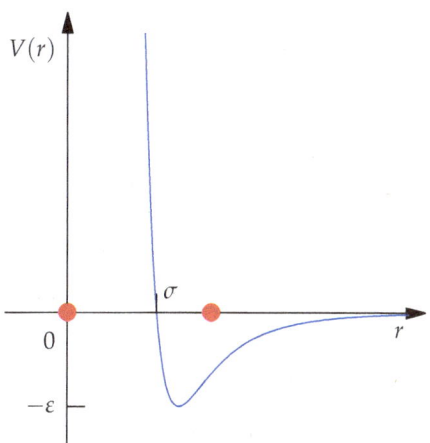

The second virial coefficient is defined as

$$B(T) = -\frac{1}{2}\int_0^\infty \int_0^\pi \int_0^{2\pi} \left(e^{-\beta V(r)} - 1\right) r^2 \sin\theta \, d\varphi \, d\theta \, dr$$

where $\beta = 1/k_B T$.

(a) Plot the potential energy $V(r)$ as a function of r over the range $0 < r < 2\lambda\sigma$ taking $\lambda = 2$.

(b) Show that $B(T)$ can be written

$$B(T) = -2\pi\int_0^\infty \left(e^{-\beta V(r)} - 1\right) r^2 \, dr$$

(c) Evaluate the integral in (b) to show that

$$B(T) = \frac{2\pi\sigma^3}{3}\left[1 - (\lambda^3 - 1)(e^{\beta\varepsilon} - 1)\right]$$

(d) Take the limit of your result as $\lambda \to 1$ and provide a physical interpretation of that limit.

7.11 Prove that

$$I_n = \int_0^\infty r^n e^{-\beta r}\, dr = \frac{n!}{\beta^{n+1}}$$

This result is recorded in the Table of Definite Integrals in Supplement S5 #1.

7.12 Starting from the result that

$$I_0 = \int_0^\infty e^{-\alpha x^2}\, dx = \frac{1}{2}\sqrt{\frac{\pi}{\alpha}}$$

show that

$$I_{2n} = \int_0^\infty x^{2n} e^{-\alpha x^2}\, dx = \frac{(2n-1)!!}{2^{n+1}\alpha^n}\sqrt{\frac{\pi}{\alpha}}$$

where the *double factorial* is defined

$$(2n-1)!! = (2n-1)(2n-3)\ldots 3\cdot 1$$

This result is recorded in the Table of Definite Integrals in Supplement S5 #4.

7.13 Starting from the result that

$$I_1 = \int_0^\infty x\, e^{-\alpha x^2}\, dx = \frac{1}{2\alpha}$$

show that

$$I_{2n+1} = \int_0^\infty x^{2n+1} e^{-\alpha x^2}\, dx = \frac{n!}{2\alpha^{n+1}}$$

This result is recorded in the Table of Definite Integrals in Supplement S5 #7.

7.14* The Dirac delta function is defined as

$$\delta(x - x_0) = \begin{cases} 0 & x \neq x_0 \\ \infty & x = x_0 \end{cases}$$

such that

$$\int_{-\infty}^\infty \delta(x - x_0)\, dx = 1$$

and

$$\int_{-\infty}^\infty \delta(x - x_0) f(x)\, dx = f(x_0)$$

Evaluate the following integrals involving the Dirac delta function.

(a) $I = \int_0^{2\pi} \delta(\theta - \pi) \cos\theta\, d\theta$ (b) $I = \int_0^\infty \delta(t-1) e^{-t}\, dt$ (c) $I = \int_0^\infty \int_0^{2\pi} \delta(r - a)\, r\, d\theta dr$

(d) $I = \int_{-2L}^{2L} \int_{-2L}^{2L} \delta(x - L)\, dx\, dy$ (e) $I = \int_{-2L}^{2L} \int_{-2L}^{2L} \int_{-2L}^{2L} \delta(y + L)\, dx\, dy\, dz$ (f) $I = \int_{-\infty}^\infty \delta(x - 1)(x^2 - 1)^2\, dx$

7.15* In Chapter 1, we considered the transformation of a volume element from one coordinate system to another. For example, using geometric arguments we found the volume element in two-dimensional cartesian coordinates, $dA = dxdy$ in Figure 1.14, is transformed to $dA = r\, drd\theta$ in two-dimensional plane polar coordinates (see Figure 1.18). This exercise explores a systematic approach for transforming a volume element between coordinate systems that draws on our knowledge of partial derivatives and the determinant.

Suppose we start from the volume element $dA = dxdy$ in two-dimensional cartesian coordinates. That volume element can be transformed to plane polar coordinates using the formula

$$dxdy = |\mathbf{J}| \, dr d\theta$$

where $|\mathbf{J}|$ is the determinant of the *jacobian matrix* defined

$$\mathbf{J} = \begin{pmatrix} \dfrac{\partial x}{\partial r} & \dfrac{\partial x}{\partial \theta} \\ \dfrac{\partial y}{\partial r} & \dfrac{\partial y}{\partial \theta} \end{pmatrix} = \dfrac{\partial(x,y)}{\partial(r,\theta)}$$

The mapping between (x, y) and (r, θ) is defined by

$$x = r\cos\theta \qquad y = r\sin\theta$$

so that the determinant of the jacobian matrix is

$$|\mathbf{J}| = \begin{vmatrix} \cos\theta & -r\sin\theta \\ \sin\theta & r\cos\theta \end{vmatrix} = r\cos^2\theta + r\sin^2\theta = r$$

resulting in

$$dxdy = |\mathbf{J}| \, dr d\theta = r \, dr d\theta$$

So you see how it works.

(a) Write the jacobian matrix for the transformation from three-dimensional cartesian coordinates (x, y, z) to spherical polar coordinates (r, θ, φ) defined by

$$dxdydz = |\mathbf{J}| \, dr d\theta d\varphi = \left| \frac{\partial(x, y, z)}{\partial(r, \theta, \varphi)} \right| dr d\theta d\varphi$$

Evaluate the determinant of the jacobian matrix to show that the volume element $dV = r^2 \sin\theta \, dr d\theta d\varphi$.

(b) Write the jacobian matrix for the transformation from three-dimensional cartesian coordinates (x, y, z) to cylindrical coordinates (r, θ, z) defined by

$$dxdydz = |\mathbf{J}| \, dr d\theta dz = \left| \frac{\partial(x, y, z)}{\partial(r, \theta, z)} \right| dr d\theta dz$$

Evaluate the determinant of the jacobian matrix to show that the volume element $dV = r \, dr d\theta dz$.

8 SEQUENCES, SERIES, AND EXPANSIONS

8.1 Series, convergence, and limits

MANY PHYSICAL PROCESSES ARE CONVENIENTLY MODELED MATHEMATICALLY USING A FINITE OR INFINITE SERIES. For example, the divergent harmonic series captures overtones of a vibrating string and harmonics in music. The convergent geometric series appears in Einstein's theory of the heat capacity of solids. In this section, we explore basic definitions of sequences and series and examine how we can determine their limits.

8.1.1 Infinite series and partial sums

Consider the general form of an *infinite series*

$$\sum_{n=0}^{\infty} u_n = u_0 + u_1 + u_2 + u_3 + \dots$$

where n is a counting number that serves as an index.[1] In forming sums of this kind, the characters i, j, k, l, m, or n are most commonly used to represent *discrete variables*.[2] Now consider the *partial sum* formed by the first $N + 1$ terms in the infinite series:

$$S_N = \sum_{n=0}^{N} u_n = u_0 + u_1 + \dots + u_N$$

The infinite series will be convergent if

$$\lim_{N \to \infty} S_N = S$$

where S is finite.[3]

Let's examine the *geometric series*

$$\sum_{n=0}^{\infty} x^n = 1 + x + x^2 + x^3 + \dots$$

This series is important in the physical sciences and arises in many problems in thermodynamic equilibria and quantum molecular science.[4] The partial sum is

$$S_N = \sum_{n=0}^{N} x^n = 1 + x + x^2 + x^3 + \dots + x^N$$

[1] The choice of the initial value of the index is arbitrary. However, a specific value typically appears to be most natural for a given series.

[2] Characters most commonly used to represent *continuous variables* include r, s, t, u, v, w, x, y, and z.

[3] The question of how a sum of an infinite series of numbers can be finite is known as *Zeno's paradox* after *Zeno of Elea* (*circa* 490-430 BCE).

[4] The geometric series is defined as a series with a constant ratio between terms. It was mentioned by *Euclid* (*circa* 300 BCE) who employed partial sums to explore its properties.

We make the clever observation that

$$S_N - xS_N = 1 + x + \ldots + x^N - \left(x + x^2 + \ldots + x^{N+1}\right)$$
$$= 1 - x^{N+1}$$

Solving for S_N we find

$$S_N = \frac{1 - x^{N+1}}{1 - x}$$

This partial sum S_N is shown in Figure 8.1 for the specific value of $x = \frac{1}{2}$ and $N = 0$ to 10. As N increases the partial sum approaches the limit $S = 2$.

Let's consider the convergence of the geometric series more generally as a function of x. Since $\lim_{N \to \infty} x^N = 0$ for $|x| < 1$ we find

$$\lim_{N \to \infty} S_N = \frac{1}{1 - x} \qquad |x| < 1$$

leading to the result that the geometric series converges for $-1 < x < 1$:

$$\sum_{n=0}^{\infty} x^n = 1 + x + x^2 + x^3 + \ldots = \frac{1}{1 - x} \qquad |x| < 1 \qquad (8.1)$$

For $|x| \geq 1$ the geometric series diverges.

As an example application of the geometric series, consider the unit square shown in Figure 8.2. We start by shading one quarter of the unit square, and then one quarter of a quarter, and then one quarter of a quarter of a quarter, and so on. Each new shaded area is $\frac{1}{4}$ the previous shaded area. After an infinite number of steps, what is the total shaded area?

This series representing the shaded area can be written as

$$S = \frac{1}{4} + \frac{1}{16} + \frac{1}{64} + \ldots = \sum_{n=1}^{\infty} \left(\frac{1}{4}\right)^n$$

This series can be transformed into a geometric series by adding one since

$$1 + S = 1 + \frac{1}{4} + \frac{1}{16} + \frac{1}{64} + \ldots = \sum_{n=0}^{\infty} x^n$$

where $x = \frac{1}{4}$. The series sums to

$$1 + S = \frac{1}{1 - x} = \frac{4}{3}$$

leading to the final result that the total shaded area $S = \frac{1}{3}$.

The area of each square in the series is reduced by $\frac{1}{4}$ relative to the area of the preceding square. As such, any set of consecutive squares is identical in form to any other set of consecutive squares, the only difference being that the size is scaled. This property is known as *self-similarity* and it is observed in nature in the form of snowflakes and shore lines. Self-similar objects are also known as *fractals*. Fractal objects can have unusual properties. For example, a fractal object might have a finite area, but an infinite perimeter. The geometric properties of fractal objects are explored in the complements.

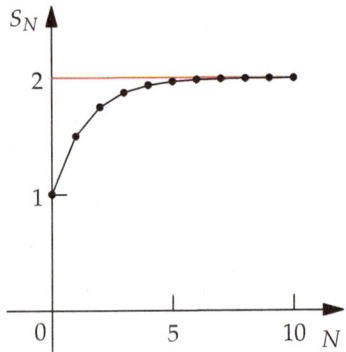

Figure 8.1: The partial sum S_N for the geometric series $S = \sum_{n=0}^{\infty} x^n$ as a function of N for the case $x = \frac{1}{2}$. The series converges to $S = 2$.

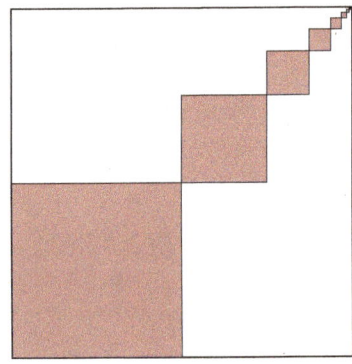

Figure 8.2: A representation of the geometric series

$$\frac{1}{4} + \frac{1}{16} + \frac{1}{64} + \ldots = \frac{1}{3}$$

The shaded area covers $\frac{1}{3}$ the total area of the bounding square.

8.1.2 Checking for convergence with the ratio test

In general, we can express an infinite series in terms of the infinite sum of terms in the series u_n:

$$\sum_{n=0}^{\infty} u_n = u_1 + u_2 + u_3 + \ldots$$

One specific example is the *harmonic series* written as

$$\sum_{n=0}^{\infty} u_n = 1 + \frac{1}{2} + \frac{1}{3} + \ldots$$

The terms in the series $u_n = \frac{1}{n}$ diminish with increasing n.

Every infinite series will converge, to a finite number, or diverge, to positive or negative infinity. How do we know if a given series, such as the harmonic series, will converge or diverge? The fact is, we can't always know! However, in many cases, we can prove that a series is convergent or divergent. Let's explore several approaches that can be used to address this question.

One approach is the *ratio test*, which assesses the relative magnitude or ratio of neighboring terms u_n and u_{n+1} as n approaches ∞.[5] We write the limit of this ratio as

$$r = \lim_{n \to \infty} \left| \frac{u_{n+1}}{u_n} \right|$$

If $r < 1$, the series converges, if $r > 1$, the series diverges, and if $r = 1$, the test is inconclusive.

Let's apply the ratio test to the geometric series where $u_n = x^n$ and $u_{n+1} = x^{n+1}$, so that

$$r = \lim_{n \to \infty} \left| \frac{u_{n+1}}{u_n} \right| = \lim_{n \to \infty} \left| \frac{x^{n+1}}{x^n} \right| = |x|$$

According to the ratio test, if $|x| < 1$ the series converges and if $|x| > 1$, the series diverges. This result agrees with our prior conclusion drawn from an analysis of the partial sum (see Equation 8.1).

Let's look at another example. Consider the exponential function

$$e^x = \sum_{n=0}^{\infty} \frac{1}{n!} x^n = 1 + x + \frac{1}{2!} x^2 + \frac{1}{3!} x^3 + \ldots$$

We can define the partial sum

$$S_N(x) = \sum_{n=0}^{N} \frac{1}{n!} x^n$$

so that $S_0(x) = 1$, $S_1(x) = 1 + x$, $S_2(x) = 1 + x + \frac{1}{2}x^2$, and so on.

As N increases, the approximation of the partial sum to the exact exponential function becomes increasingly accurate. This behavior is shown in Figure 8.3 for partial sums from $N = 0$ to 4. The partial sum is most accurate near $x = 0$. As the magnitude of x increases, an increasingly large value of N is required to accurately approximate the exponential function.

[5] The ratio test was proposed by French mathematician, physicist, philosopher, and music theorist Jean le Rond d'Alembert (1717-1783).

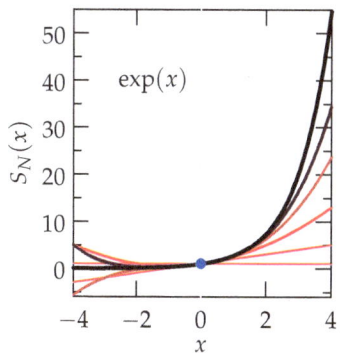

Figure 8.3: The exponential function (black) shown with partial sums $S_N(x)$ for $N = 0$ to 4 in increasingly dark shades of red.

Applying the ratio test to our infinite series where $u_n = \frac{1}{n!}x^n$ and $u_{n+1} = \frac{1}{(n+1)!}x^{n+1}$, we find

$$r = \left|\frac{u_{n+1}}{u_n}\right| = \lim_{n\to\infty}\left|\frac{n!\,x^{n+1}}{(n+1)!\,x^n}\right| = \lim_{n\to\infty}\left|\frac{x}{n+1}\right| = 0$$

This demonstrates that the series converges for all values of x. This is what we expect for the exponential function as e^x is finite for any finite value of x.

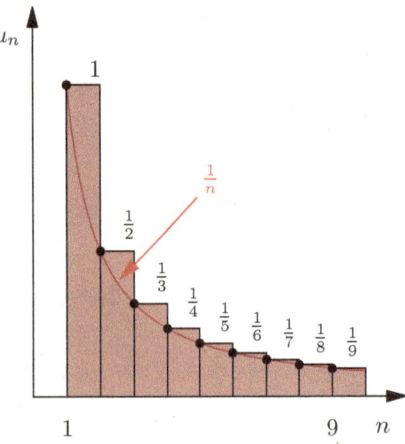

Figure 8.4: The shaded rectangles vary in area as $\frac{1}{n}$ where the total shared area is defined by

$$1 + \frac{1}{2} + \frac{1}{3} + \frac{1}{4} + \ldots = \infty$$

known as the harmonic series.

As a final example, let's consider the convergence of the *harmonic series*[6]

$$\sum_{n=1}^{\infty}\frac{1}{n} = 1 + \frac{1}{2} + \frac{1}{3} + \frac{1}{4} + \ldots$$

which is depicted graphically in Figure 8.4. Applying the ratio test we find $u_n = \frac{1}{n}$ and $u_{n+1} = \frac{1}{n+1}$, so that

$$r = \left|\frac{u_{n+1}}{u_n}\right| = \lim_{n\to\infty}\left|\frac{n}{n+1}\right| = 1$$

For the harmonic series the ratio test is inconclusive! To further test the convergence of the harmonic series, let's use the *comparison test*.[7] Consider that the series

$$S = 1 + \frac{1}{2} + \frac{1}{4} + \frac{1}{4} + \frac{1}{8} + \frac{1}{8} + \frac{1}{8} + \frac{1}{8} + \ldots$$

is divergent, since

$$S = 1 + \frac{1}{2} + \left(\frac{1}{4} + \frac{1}{4}\right) + \left(\frac{1}{8} + \frac{1}{8} + \frac{1}{8} + \frac{1}{8}\right)\ldots$$

$$= 1 + \frac{1}{2} + \frac{1}{2} + \frac{1}{2} + \ldots = \infty$$

Now note that every term in the divergent series S is less than or equal to the corresponding term in the harmonic series. As such, we conclude that the harmonic series also diverges.[8]

[6] Harmonic sequences are often associated with the Greek mathematician *Pythagoras* (570-495 BCE) and his followers. The concept of the harmonic series has inspired theories of mathematics, music, art, and architecture.

[7] This elegant proof of the divergence of the *harmonic series* was presented by French philosopher *Nicole Oresme* (*circa* 1320-1382). It is considered by many to be the high point of mathematical reasoning in medieval Europe.

[8] The comparison test is a powerful method for determining convergence properties of series. Suppose we have two series, $\sum a_n$ and $\sum b_n$. If $a_n > b_n$ for all n, then if $\sum a_n$ converges, so does $\sum b_n$. Similarly, if $\sum b_n$ diverges, so does $\sum a_n$.

8.2 Power series

POWER SERIES MAY BE USED to represent special mathematical constants such as e and π and functions commonly used in modeling physical processes such as e^x and $\ln x$. In this section, we explore the definition of a power series and how we can determine its limits.

8.2.1 Testing convergence of infinite power series

Many functions $S(x)$ of interest in the physical sciences can be represented as an infinite series of powers of x known as a *power series*

$$S(x) = \sum_{n=0}^{\infty} a_n x^n = a_0 + a_1 x + a_2 x^2 + \dots \tag{8.2}$$

We have already seen one example in the exponential function

$$e^x = \sum_{n=0}^{\infty} \frac{1}{n!} x^n = 1 + x + \frac{1}{2!} x^2 + \frac{1}{3!} x^3 + \dots$$

where the coefficients $a_n = 1/n!$. Applying the ratio test to the general case, we see that the power series will converge if

$$r = \left| \frac{u_{n+1}}{u_n} \right| = \lim_{n \to \infty} \left| \frac{a_{n+1}\, x^{n+1}}{a_n x^n} \right| = |x| \lim_{n \to \infty} \left| \frac{a_{n+1}}{a_n} \right| < 1$$

This can be expressed as a condition on the magnitude of x as

$$|x| < \frac{1}{\lim_{n \to \infty} \left| \frac{a_{n+1}}{a_n} \right|} = \lim_{n \to \infty} \left| \frac{a_n}{a_{n+1}} \right| = R$$

where R is the *radius of convergence*.[9] In general, the power series converges in the interval $-R < x < R$.

Let's consider the example

$$S(x) = \sum_{n=0} a_n x^n = \sum_{n=1}^{\infty} \frac{1}{n} \left(\frac{x}{2} \right)^n$$

where the coefficient $a_0 = 0$. The partial sum can be written

$$S_N(x) = \sum_{n=1}^{N} \frac{1}{n} \left(\frac{x}{2} \right)^n$$

For this series, the coefficients are $a_n = \frac{1}{n\, 2^n}$ and the radius of convergence is

$$R = \lim_{n \to \infty} \left| \frac{a_n}{a_{n+1}} \right| = \lim_{n \to \infty} \frac{(n+1)\, 2^{n+1}}{n\, 2^n} = 2$$

We conclude that the power series converges in the interval $-2 < x < 2$. Figure 8.5 shows the partial sum $S_N(x)$ as a function of x for $N = 1, 5$, and 10. The infinite series $S(x = 2)$ is the *harmonic series* that we found to be divergent. However, the infinite series $S(x = -2)$ is the *alternating harmonic*

[9] The radius of convergence was formulated by French mathematician and physicist *Augustin-Louis Cauchy* (1789-1857).

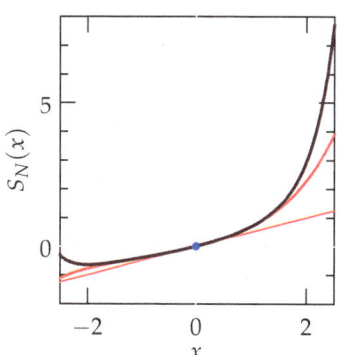

Figure 8.5: The partial sum $S_N(x)$ of the infinite power series with coefficients $a_n = \frac{1}{n\, 2^n}$ shown for $N = 1, 5$, and 10 in increasingly dark shades of red.

series. That alternating series converges to $\ln(2)$, as we will demonstrate in the next section. Therefore, we can refine our interval of convergence to be $-2 \leq x < 2$.

8.2.2 *Differentiating and integrating power series*

If a power series converges over an interval $-R < x < R$ it is also a continuous function of x over that interval. Moreover, if the power series

$$f(t) = \sum_{n=0}^{\infty} a_n t^n$$

converges, so does its integral[10]

$$S(x) = \int_0^x f(t)dt = \int_0^x \left(\sum_{n=0}^{\infty} a_n t^n \right) dt = \sum_{n=0}^{\infty} \left(\int_0^x a_n t^n dt \right)$$

$$= \sum_{n=0}^{\infty} \frac{1}{n+1} a_n x^{n+1}$$

This results from the fact that the radius of convergence of $S(x)$ is the same as for the power series itself:

$$R = \lim_{n \to \infty} \left| \frac{(n+2)\, a_n}{(n+1)\, a_{n+1}} \right| = \lim_{n \to \infty} \left| \frac{a_n}{a_{n+1}} \right|$$

Similarly, the derivative of a convergent power series[11]

$$S(x) = \frac{d}{dt} f(t) \Big|_{t=x} = \frac{d}{dt} \sum_{n=0}^{\infty} a_n t^n \Big|_{t=x} = \sum_{n=0}^{\infty} \left(\frac{d}{dt} a_n t^n \Big|_{t=x} \right)$$

$$= \sum_{n=0}^{\infty} a_n\, n x^{n-1}$$

is itself convergent, since the radius of convergence of $S(x)$ is the same as for the power series itself:

$$R = \lim_{n \to \infty} \left| \frac{(n+1)\, a_n}{n\, a_{n+1}} \right| = \lim_{n \to \infty} \left| \frac{a_n}{a_{n+1}} \right|$$

Let's explore these properties of convergent power series for the geometric series

$$f(t) = \sum_{n=0}^{\infty} t^n = 1 + t + t^2 + \ldots = \frac{1}{1-t} \qquad |t| < 1$$

Figure 8.6 shows the geometric series along with its derivative and integral as a function of x.

Integrating the geometric series, we find

$$S(x) = \int_0^x f(t)dt = \sum_{n=0}^{\infty} \left(\int_0^x t^n dt \right)$$

Using the identity

$$\int_0^x t^n dt = \frac{1}{n+1} t^{n+1} \Big|_0^x = \frac{1}{n+1} x^{n+1}$$

[10] Note that order of integration and summation may be reversed as

$$\int \left(\sum_{n=0}^{\infty} f_n(x) \right) dx$$

can be written

$$\int f_0(x)dx + \int f_1(x)dx + \ldots$$

which is equal to

$$\sum_{n=0}^{\infty} \int f_n(x)dx$$

[11] Note that order of differentiation and summation may be reversed as

$$\frac{d}{dx} \left(\sum_{n=0}^{\infty} f_n(x) \right)$$

can be written

$$\frac{d}{dx} f_0(x) + \frac{d}{dx} f_1(x) + \ldots$$

which is equal to

$$\sum_{n=0}^{\infty} \frac{d}{dx} f_n(x)$$

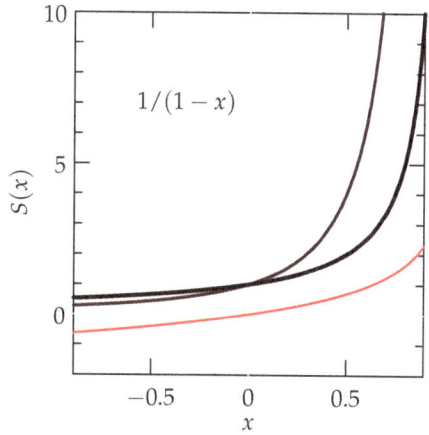

Figure 8.6: The geometric series $\frac{1}{1-x}$ (black) shown with its derivative $\frac{1}{(1-x)^2}$ (light red) and its integral $-\ln(1-x)$ (dark red).

leads to the result

$$S(x) = \sum_{n=0}^{\infty} \frac{1}{n+1} x^{n+1} = x + \frac{1}{2}x^2 + \frac{1}{3}x^3 + \dots$$

However, since $f(t) = \frac{1}{1-t}$, we can also write[12]

$$S(x) = \int_0^x f(t)dt = \int_0^x \frac{1}{1-t}\,dt = -\ln(1-x)$$

[12] Here we use the identity

$$\int_0^x \frac{1}{1-t}\,dt = -\ln(1-t)\Big|_0^x = -\ln(1-x)$$

as $\ln(1) = 0$.

Combining these results leads to an expression for the power series of the natural logarithm with a radius of convergence $R = 1$:

$$S(x) = \ln(1-x) = -\sum_{n=0}^{\infty} \frac{1}{n+1} x^{n+1}$$

This is the radius of convergence of the original geometric series.

Differentiating the geometric series, we find

$$S(x) = \frac{d}{dt} f(t)\Big|_{t=x} = \sum_{n=0}^{\infty} \left(\frac{d}{dt} t^n \Big|_{t=x} \right) = \sum_{n=0}^{\infty} nx^{n-1} = 1 + 2x + 3x^2 + \dots$$

However, we can also write

$$S(x) = \frac{d}{dt} f(t)\Big|_{t=x} = \frac{d}{dt} \frac{1}{1-t}\Big|_{t=x} = \frac{1}{(1-x)^2}$$

Combining these results leads to the identity

$$S(x) = \frac{1}{(1-x)^2} = \sum_{n=1}^{\infty} nx^{n-1}$$

with a radius of convergence $R = 1$.

These examples demonstrate how differentiating or integrating a known convergent power series can generate new power series identities. In the next section, we explore a more general approach to defining power series of functions commonly encountered in the physical sciences.

8.3 Expanding functions as Maclaurin and Taylor series

MATHEMATICAL FUNCTIONS THAT ARE CONTINUOUS AND INFINITELY DIFFERENTIABLE may be systematically represented in terms of the function's derivatives at a single point. The Maclaurin series represents a function as a power series in x with coefficients depending on the derivatives of the function at $x = 0$. The Maclaurin series is a special case of the Taylor series, which represents a function in terms of derivatives of the function at an arbitrary point x_0 and using powers of the displacement $x - x_0$. In this section, we explore the basic properties of Maclaurin and Taylor series and their application to problems in the physical sciences.

8.3.1 Maclaurin power series expansion

In our explorations of power series, we have examined power series representations of basic functions and used the ratio test to determine their associated radii of convergence. We can also ask the complementary question, for a given function $f(x)$ with the power series representation

$$f(x) = \sum_{n=0}^{\infty} a_n x^n = a_0 + a_1 x + a_2 x^2 + \ldots$$

What are the coefficients a_n? To start, we note that the value of the function at $x = 0$ is

$$f(0) = a_0$$

We further note that the first derivative of the function

$$\frac{d}{dx} f(x) = a_1 + 2 a_2 x + 3 a_3 x^2 + \ldots$$

results in an identity for the coefficient a_1 when evaluated at $x = 0$:

$$\frac{d}{dx} f(x) \Big|_{x=0} = a_1$$

Repeating this process for the second derivative of the function, we find that

$$\frac{d^2}{dx^2} f(x) = 2 a_2 + 2 \cdot 3 \, a_3 x + 3 \cdot 4 \, a_4 x^2 + \ldots$$

results in an identity for a_2 evaluated at $x = 0$:

$$\frac{d^2}{dx^2} f(x) \Big|_{x=0} = 2 a_2$$

Evaluating the third derivative of the function at $x = 0$, we find

$$\frac{d^3}{dx^3} f(x) \Big|_{x=0} = 2 \cdot 3 \, a_3$$

In general, we can show that

$$\frac{d^n}{dx^n} f(x) \Big|_{x=0} = 1 \cdot 2 \cdot 3 \cdot \ldots \cdot n \, a_n = n! \, a_n$$

where we have used the factorial $n! = 1 \cdot 2 \cdot 3 \cdot$.

This provides a general result for the identity of the coefficients:

$$a_n = \frac{1}{n!} \left.\frac{d^n}{dx^n} f(x)\right|_{x=0}$$

Returning to our original power series for $f(x)$ we can write

$$f(x) = f(0) + \left.\frac{d}{dx} f(x)\right|_{x=0} x + \frac{1}{2!} \left.\frac{d^2}{dx^2} f(x)\right|_{x=0} x^2 + \dots$$

$$= \sum_{n=0}^{\infty} \frac{1}{n!} \left.\frac{d^n}{dx^n} f(x)\right|_{x=0} x^n$$

This is known as the *Maclaurin series*.[13]

Expressing a function $f(x)$ as a Maclaurin series may at first seem rather complicated. So many derivatives! However, we often require only the first few terms in the series. We will find that the power series expansions of common functions have great utility in the physical sciences.

8.3.2 Applications of Maclaurin power series

Let's look at a few examples of how Maclaurin series can be used to derive power series representations of functions that commonly occur in the physical sciences. Take the exponential function

$$f(x) = e^x$$

for which

$$\frac{d}{dx} e^x = e^x$$

and

$$\left.\frac{d^n}{dx^n} f(x)\right|_{x=0} = \left.\frac{d^n}{dx^n} e^x\right|_{x=0} = \left.e^x\right|_{x=0} = 1$$

As a result, we find

$$a_n = \frac{1}{n!} \left.\frac{d^n}{dx^n} f(x)\right|_{x=0} = \frac{1}{n!}$$

The Maclaurin series of the exponential function is[14]

$$e^x = \sum_{n=0}^{\infty} \frac{1}{n!} x^n = 1 + x + \frac{1}{2} x^2 + \dots$$

This example demonstrates how Maclaurin series can be used to derive power series representations of functions commonly used in the physical sciences.

Now let's use a Maclaurin series to derive a power series representation of

$$f(x) = \sin x$$

To start, let's take the first few derivatives of $\sin(x)$:

$$\left.\frac{d}{dx} \sin(x)\right|_{x=0} = \cos(0) = 1 \qquad \left.\frac{d^2}{dx^2} \sin(x)\right|_{x=0} = -\sin(0) = 0$$

$$\left.\frac{d^3}{dx^3} \sin(x)\right|_{x=0} = -\cos(0) = -1 \qquad \left.\frac{d^4}{dx^4} \sin(x)\right|_{x=0} = \sin(0) = 0$$

[13] Named for Scottish mathematician *Colin Maclaurin* (1698-1746).

[14] Evaluating this expression at $x = 1$ we find an identity for the constant $e = \sum_{n=0}^{\infty} \frac{1}{n!}$.

From this, we recognize that when n is odd

$$\frac{d^n}{dx^n} f(x)\Big|_{x=0} = (-1)^{\frac{n-1}{2}} \cos x\Big|_{x=0} = (-1)^{\frac{n-1}{2}}$$

Meanwhile, when n is even

$$\frac{d^n}{dx^n} f(x)\Big|_{x=0} = (-1)^{\frac{n}{2}} \sin x\Big|_{x=0} = 0$$

As a result, the Maclaurin series of the $\sin(x)$ contains only odd powers of x.[15] Writing the first few terms we find

$$\sin x = x - \frac{1}{3!} x^3 + \frac{1}{5!} x^5 - \ldots$$

Recalling that

$$\sin x = \sum_{n=0}^{\infty} a_n x^n = a_0 + a_1 x + a_2 x^2 + \ldots$$

We determine that $a_0 = 0$, $a_1 = -1$, $a_2 = 0$, $a_3 = \frac{1}{3!}$, and so on.

To simplify our expression, let's define a new counter $k = \frac{n-1}{2}$ so that $n = 2k + 1$. Then we can write[16]

$$\sin x = x - \frac{1}{3!} x^3 + \frac{1}{5!} x^5 - \ldots = \sum_{k=0}^{\infty} (-1)^k \frac{1}{(2k+1)!} x^{2k+1}$$

We can derive the same result more deliberately by evaluating the expressions term by term:

$$\sin x = f(0) + \frac{df}{dx}\Big|_{x=0} x + \frac{1}{2!} \frac{d^2 f}{dx^2}\Big|_{x=0} x^2 + \frac{1}{3!} \frac{d^3 f}{dx^3}\Big|_{x=0} x^3 + \ldots$$

$$= \sin(0) + \cos(0)x + \frac{1}{2!}(-\sin(0))x^2 + \frac{1}{3!}(-\cos(0))x^3 + \ldots$$

$$= x - \frac{1}{3!}x^3 + \ldots \tag{8.3}$$

In Figure 8.7, the partial sum for the Maclaurin series of $\sin(x)$ written

[15] As $\sin(x)$ is an odd function of x, we expect that the Maclaurin series only contains terms with odd powers of x.

[16] We start from the restricted series

$$\sum_{n=1}^{\infty}{}' (-1)^{\frac{n-1}{2}} \frac{1}{n!} x^n$$

where the prime indicates that the sum is over odd values of n only. Defining $k = \frac{n-1}{2}$ we find $n = 2k + 1$. As such, $n = 1, 3, 5, \ldots$ corresponds to $k = 0, 1, 2, \ldots$. We can then rewrite the series in terms of the unrestricted sum $\sum_{k=0}^{\infty} (-1)^{2k+1} \frac{1}{(2k+1)!} x^{2k+1}$

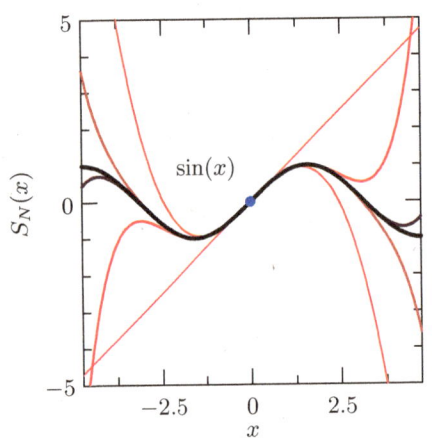

Figure 8.7: The sine function (black) shown with partial sums $S_N(x)$ for $N = 0$ to 4 in increasingly dark shades of red.

$$S_N(x) = \sum_{k=0}^{N}(-1)^k \frac{1}{(2k+1)!} x^{2k+1}$$

is shown for values of $N = 0$ to 4. For a given value of N, the partial sum $S_N(x)$ most accurately represents the function $\sin(x)$ for values of x near $x = 0$. As N increases the partial sum provides an increasingly accurate approximation of $\sin x$ over a growing range of x. Remarkably, in the limit $N \to \infty$ the Maclaurin series provides an exact representation of the infinity of oscillations of this periodic function over $-\infty < x < \infty$.

In the same way we can derive the Maclaurin series expansion of $\cos x$ as

$$\cos x = f(0) + \frac{df}{dx}\Big|_{x=0} x + \frac{1}{2!} \frac{d^2 f}{dx^2}\Big|_{x=0} x^2 + \frac{1}{3!} \frac{d^3 f}{dx^3}\Big|_{x=0} x^3 + \dots$$

$$= \cos(0) - \sin(0)x + \frac{1}{2!}\left(-\cos(0)\right)x^2 + \frac{1}{3!}\sin(0)x^3 + \dots$$

$$= 1 - \frac{1}{2!}x^2 + \dots \tag{8.4}$$

It follows that the partial sum for the Maclaurin series of $\cos(x)$ is

$$S_N(x) = \sum_{k=0}^{N}(-1)^k \frac{1}{(2k)!} x^{2k}$$

Only even powers of x appear in this Maclaurin series expansion reflecting that $\cos(x)$ is an even function of x.

We can use our power series expansions of $\sin(x)$ and $\cos(x)$ to derive approximate expressions for these functions that are valid for small values of x. Note that for small x, we find that

$$\sin(x) \approx x$$

and

$$\cos(x) \approx 1 - \frac{1}{2}x^2$$

as shown in Figure 8.8. Graphical comparison of the approximation and the exact function demonstrates the substantial range of x over which the approximation is valid.

Truncated power series are commonly applied when seeking approximate solutions to problems in the physical sciences. A number of applications of truncated power series to problems arising in molecular science are explored in the complements.

8.3.3 Taylor power series expansion

The Maclaurin series provides a general expression for the function $f(x)$ in terms of its derivatives evaluated at $x = 0$. We observed that for a finite number of terms the Maclaurin series provided the most accurate representation of the function near $x = 0$. However, at times we may be interested in representing a function most accurately near an arbitrary point $x = x_0$. For that purpose, we appeal to the Taylor series.

Suppose we want to express the function in terms of its derivatives evalu-

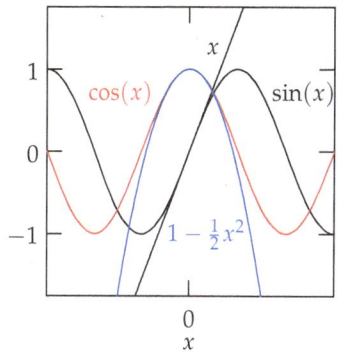

Figure 8.8: The cosine (red) and sine (black) functions compared with the first order truncated Mclaurin series approximations x (gray) and $1 - \frac{1}{2}x^2$ (blue), respectively.

ated at an arbitrary point $x = x_0$ as

$$f(x) = \sum_{n=0}^{\infty} a_n(x - x_0)^n = a_0 + a_1(x - x_0) + a_2(x - x_0)^2 + \ldots$$

Evaluating the function at $x = x_0$, we find

$$f(x_0) = a_0$$

The first derivative of the function evaluated at $x = x_0$ is

$$\left. \frac{d}{dx} f(x) \right|_{x=x_0} = a_1$$

and the second derivative of the function evaluated at $x = x_0$ is

$$\left. \frac{d^2}{dx^2} f(x) \right|_{x=x_0} = 2a_2$$

In general, we can show that

$$\left. \frac{d^n}{dx^n} f(x) \right|_{x=x_0} = 1 \cdot 2 \cdot 3 \cdot \ldots \cdot n \, a_n = n! \, a_n$$

so that

$$a_n = \frac{1}{n!} \left. \frac{d^n}{dx^n} f(x) \right|_{x=x_0}$$

With this set of coefficients we can express the function $f(x)$ with the power series

$$f(x) = f(x_0) + \left. \frac{d}{dx} f(x) \right|_{x=x_0} (x - x_0) + \frac{1}{2!} \left. \frac{d^2}{dx^2} f(x) \right|_{x=x_0} (x - x_0)^2 + \ldots$$

$$= \sum_{n=0}^{\infty} \frac{1}{n!} \left. \frac{d^n}{dx^n} f(x) \right|_{x=x_0} (x - x_0)^n$$

This is known as the *Taylor series.*[17] Note that the Maclaurin series is a special case of the Taylor series for $x_0 = 0$. Taylor series have wide-ranging applications in the physical sciences. A variety of applications of Taylor series are explored in the next section.

[17] Named for English mathematician *Brook Taylor* (1685-1731).

8.3.4 *Applications of Taylor power series*

Let's look at a few examples of how Taylor series can be used to derive power series representations of functions that commonly occur in the physical sciences. Take the natural logarithm

$$f(x) = \ln x$$

Suppose we wish to expand this function around the point $x_0 = 1$, where the function $f(1) = \ln(1) = 0$. We write

$$\ln(x) = \ln(1) + \left. \frac{d}{dx} \ln(x) \right|_{x=1} (x - 1) + \frac{1}{2!} \left. \frac{d^2}{dx^2} \ln(x) \right|_{x=1} (x - 1)^2 + \ldots$$

$$= 0 + \left. \frac{1}{x} \right|_{x=1} (x - 1) + \frac{1}{2!} \left[\frac{-1}{x^2} \right]_{x=1} (x - 1)^2 + \frac{1}{3!} \left[\frac{2}{x^3} \right]_{x=1} (x - 1)^3 + \ldots$$

When evaluated at $x_0 = 1$, this results in

$$\ln(x) = (x-1) - \frac{1}{2!}(x-1)^2 + \frac{2}{3!}(x-1)^3 - \frac{2\cdot 3}{4!}(x-1)^4 + \ldots$$

$$= (x-1) - \frac{1}{2}(x-1)^2 + \frac{1}{3}(x-1)^3 - \frac{1}{4}(x-1)^4 + \ldots \quad (8.5)$$

Now suppose we write this series in terms of $\ln(1+x)$ where we are interested in values of x near $x_0 = 1$, or small values of the deviation $(x - x_0) = (x - 1)$. Substituting $1 + x$ for x in Equation 8.5 we find

$$\ln(1+x) = x - \frac{1}{2}x^2 + \frac{1}{3}x^3 - \frac{1}{4}x^4 + \cdots = \sum_{n=1}^{\infty}(-1)^{n+1}\frac{1}{n}\,x^n$$

We can write the partial sum for this infinite series

$$S_N(x) = \sum_{n=1}^{N}(-1)^{n+1}\frac{1}{n}\,x^n$$

The convergence of the partial sum $S_N(x)$ for increasing values of N from 1 to 4 is shown in Figure 8.9 along with the exact function $\ln(1+x)$. Note that for small x we find

$$\ln(1+x) \approx x$$

which is equivalent to our previous result that near $x = 0$

$$\exp(x) \approx 1 + x$$

These approximations are useful tools that will allow us to determine limiting forms of complicated functions in an efficient and intuitive manner. A few examples of the use of expansions in determining limits are explored in the complements.

Now consider the gaussian function

$$f(x) = \exp\left[-\frac{(x-\bar{x})^2}{2\sigma^2}\right]$$

We encounter this function in the kinetic theory of gases, and in the quantum mechanics of a vibrating bond. The gaussian function is also widely applied in probability and statistics, a topic explored in Chapter 9.

Let's expand this function in a Taylor series about the point $x = \bar{x}$ as

$$f(x) = \exp\left[-\frac{(x-\bar{x})^2}{2\sigma^2}\right]$$

$$= f(\bar{x}) + \frac{df}{dx}\Big|_{x=\bar{x}}(x-\bar{x}) + \frac{1}{2!}\frac{d^2f}{dx^2}\Big|_{x=\bar{x}}(x-\bar{x})^2 + \ldots$$

The function evaluated at $x = \bar{x}$ is

$$f(\bar{x}) = 1$$

The first derivative evaluated at $x = \bar{x}$ is

$$\frac{df}{dx}\Big|_{x=\bar{x}} = -\frac{(x-\bar{x})}{\sigma^2}e^{-\frac{(x-\bar{x})^2}{2\sigma^2}}\Big|_{x=\bar{x}} = 0$$

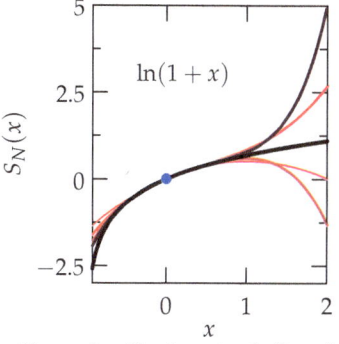

Figure 8.9: The function $\ln(1+x)$ (black) shown with partial sums $S_N(x)$ for $N = 1$ to 4 in increasingly dark shades of red.

The second derivative evaluated at $x = \bar{x}$ is

$$\frac{d^2 f}{dx^2}\bigg|_{x=\bar{x}} = \left[-\frac{1}{\sigma^2} + \frac{(x-\bar{x})^2}{\sigma^4}\right] e^{-\frac{(x-\bar{x})^2}{2\sigma^2}}\bigg|_{x=\bar{x}} = -\frac{1}{\sigma^2}$$

The third derivative evaluated at $x = \bar{x}$ is

$$\frac{d^3 f}{dx^3}\bigg|_{x=\bar{x}} = \left[\frac{3(x-\bar{x})}{\sigma^4} - \frac{(x-\bar{x})^3}{\sigma^6}\right] e^{-\frac{(x-\bar{x})^2}{2\sigma^2}}\bigg|_{x=\bar{x}} = 0$$

And so on.

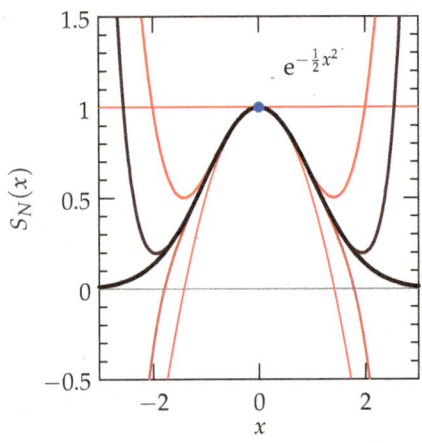

Figure 8.10: The gaussian function (black) shown with partial sums $S_N(x)$ for $N = 0$ to 4 in increasingly dark shades of red.

The final result for the Taylor series representing the gaussian function is[18]

$$e^{-\frac{(x-\bar{x})^2}{2\sigma^2}} = 1 - \frac{1}{2\sigma^2}(x-\bar{x})^2 + \frac{1}{8\sigma^4}(x-\bar{x})^4 - \frac{1}{48\sigma^6}(x-\bar{x})^6 + \dots$$

$$= \sum_{n=0}^{\infty} (-1)^n \frac{1}{n!} \left[\frac{(x-\bar{x})^2}{2\sigma^2}\right]^n$$

[18] Note that we can also find this result directly from our Maclaurin series expansion of the exponential function

$$e^{-\alpha} = \sum_{n=0}^{\infty} (-1)^n \frac{1}{n!} \alpha^n$$

$$= 1 - \alpha + \frac{1}{2}\alpha^2 - + \dots$$

where $\alpha = (x-\bar{x})^2/2\sigma^2$.

Only even powers of $(x - x_0)$ appear in the power series representation, reflecting the fact that the gaussian function is an even function of $(x - x_0)$. With this result, we can define the partial sum known as the *Taylor polynomial*

$$S_N(x) = \sum_{n=0}^{N} (-1)^n \frac{1}{n!} \left[\frac{(x-\bar{x})^2}{2\sigma^2}\right]^n$$

which is shown in Figure 8.10 for $N = 0$ to 4.

Our Taylor series expansion of the gaussian function provides us with a practically useful approximation of the gaussian function near its maximum value:

$$\exp\left[-\frac{(x-\bar{x})^2}{2\sigma^2}\right] \approx 1 - \frac{1}{2}(x - x_0)^2$$

The expansion of the gaussian function in powers of $(x - x_0)$ has great utility in addressing problems in probability and statistics, quantum theory, and statistical thermodynamics.

A8 Taylor series expansions of potential energy functions

Recall that we were able to identify minima (stable points) and saddle points (transition states) of potential energy functions by looking for those points at which the first derivatives of the potential energy were zero. For example, consider the Morse oscillator potential energy function

$$V(r) = \varepsilon \left[1 - e^{-\beta(r-r_0)} \right]^2 = \varepsilon \left[1 - 2e^{-\beta(r-r_0)} + e^{-2\beta(r-r_0)} \right]$$

This function can be used to model a bond between two atoms that can fluctuate at low energy and dissociate for energy greater than ε. This function has an extremum at $r = r_0$, since

$$\left. \frac{dV}{dr} \right|_{r=r_0} = \varepsilon \left[2\beta \, e^{-\beta(r-r_0)} - 2\beta \, e^{-2\beta(r-r_0)} \right]_{r=r_0} = 0$$

as shown in Figure 8.11. The second derivative of the potential energy function evaluated at $r = r_0$ is

$$\left. \frac{d^2V}{dr^2} \right|_{r=r_0} = \varepsilon \left[-2\beta^2 e^{-\beta(r-r_0)} + 4\beta^2 e^{-2\beta(r-r_0)} \right]_{r=r_0} = 2\varepsilon\beta^2$$

which is positive indicating that the extremum at $r = r_0$ is a potential energy minimum.

Using a Taylor series we can expand the Morse oscillator potential function about $r = r_0$ as

$$V(r) = V(r_0) + \left. \frac{dV}{dr} \right|_{r=r_0} (r - r_0) + \frac{1}{2!} \left. \frac{d^2V}{dr^2} \right|_{r=r_0} (r - r_0)^2 + \dots$$

where

$$V(r_0) = 0 \qquad \left. \frac{dV}{dr} \right|_{r=r_0} = 0 \qquad \left. \frac{d^2V}{dr^2} \right|_{r=r_0} = 2\varepsilon\beta^2$$

so that

$$V(r) = \frac{1}{2!} 2\varepsilon\beta^2 (r - r_0)^2 + \mathcal{O}\left[\beta^3 (r - r_0)^3 \right]$$

where $\mathcal{O}[x^n]$ means *on the order of* x^n.[19]

For small deviations of r from r_0, so that $\beta(r - r_0)$ is close to zero, we expect

$$\beta^2 (r - r_0)^2 \gg \beta^3 (r - r_0)^3$$

As such, we can ignore the higher order terms and employ a *harmonic approximation* of the potential energy function

$$V(r) \approx \frac{1}{2!} 2\varepsilon\beta^2 (r - r_0)^2 = \frac{1}{2}\kappa \, (r - r_0)^2$$

where κ is the harmonic oscillator force constant.[20]

Many chemical bonds are modeled using the Morse potential. At low energies, vibrations consist of small harmonic oscillations near the potential energy minimum. At larger energies, vibrations are anharmonic. When

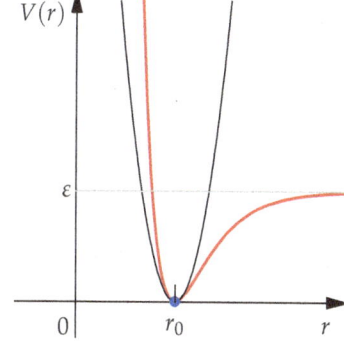

Figure 8.11: The Morse potential energy function $V(r)$ (red) for a vibrating bond with dissociation energy ε and energy minimum $r = r_0$. The harmonic approximation (black) is shown for comparison.

[19] The notation $\mathcal{O}[x^n]$ refers to terms in the full Taylor series, of order x^n and higher, which are not included in the truncated series.

[20] Recall that for a harmonic oscillator with potential energy $V(r) = \frac{1}{2}\kappa(r - r_0)^2$ the force is given by $F(r) = -\frac{dV}{dr} = -\kappa(r - r_0)$ and is proportional to the force constant κ and the displacement $r - r_0$ from the potential energy minimum at $r = r_0$.

the energy exceeds the well depth ε, the bond length r can increase toward infinity and the bond can break.

Let's consider another potential energy function that is useful for modeling chemical reactions and thermodynamic phase transitions

$$V(x) = V_0(x^2 - 1)^2 = V_0\left(x^4 - 2x^2 + 1\right) \tag{8.6}$$

This is a quartic double well potential with minima at $x = \pm 1$ and a transition state at $x = 0$.

We can expand the function $V(x)$ near the potential energy minimum at $x_0 = 1$ using a Taylor series for the special case $V_0 = 1$. We find

$$\begin{aligned} V(x) &= V(x = 1) + \left.\frac{dV}{dx}\right|_{x=1}(x-1) + \frac{1}{2!}\left.\frac{d^2V}{dx^2}\right|_{x=1}(x-1)^2 + \dots \\ &= \left.(4x^3 - 4x)\right|_{x=1}(x-1) + \frac{1}{2}\left.(12x^2 - 4)\right|_{x=1}(x-1)^2 + \dots \\ &= 4(x-1)^2 + \mathcal{O}\left[(x-1)^3\right] \end{aligned}$$

This truncated Taylor series takes the form of a quadratic equation approximation of the full potential energy function[21]

$$V(x) \approx 4(x-1)^2$$

We expect this approximation to be accurate for small displacements from the minimum near $x_0 = 1$. Locally, the potential energy is an upward facing parabolic potential well.[22] This quadratic approximation of the potential energy function takes the form of the *harmonic oscillator* potential energy function

$$V(x) = \frac{1}{2}\kappa(x - x_0)^2$$

where the minimum is $x_0 = 1$ and the force constant is $\kappa = 8$ (see Figure 8.12).

What is the nature of $V(x)$ near the transition state at $x = 0$? In that case

$$\begin{aligned} V(x) &= V(x = 0) + \left.\frac{dV}{dx}\right|_{x=0}x + \frac{1}{2!}\left.\frac{d^2V}{dx^2}\right|_{x=0}x^2 + \dots \\ &= 1 + \left.(4x^3 - 4x)\right|_{x=0}x + \frac{1}{2}\left.(12x^2 - 4)\right|_{x=0}x^2 + \dots \\ &= 1 - 2x^2 + \mathcal{O}\left[x^4\right] \end{aligned}$$

Fitting the truncated Taylor series to the harmonic oscillator potential energy function, we find

$$V(x) = 1 + \frac{1}{2}\kappa x^2 + \mathcal{O}\left[x^4\right]$$

with force constant $\kappa = -4$. Near the transition state between the two energy minima, the potential energy function is a downward facing parabola (see Figure 8.12).

Finally, let's explore the convergence of the Taylor series expansion of the quartic double well potential energy function in Equation 8.6. As the potential function is a polynomial, the Taylor series consists of a finite number of

[21] For a potential energy minimum located at $x = x_0$ the force is equal to zero making the quadratic term the first non-zero term in $(x - x_0)$ in the Taylor series.

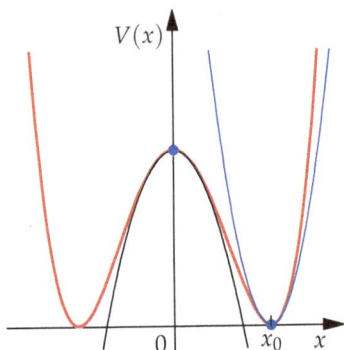

Figure 8.12: The quartic double well potential energy function $V(r)$ (red) shown with the harmonic approximation of the barrier (black) and of the well (blue).

[22] The harmonic approximation of potential energy functions near minimum energy values is a general method that finds wide application in the physical sciences.

terms. For $x_0 = 0$, the series expansion can be written

$$V(x) = 1 + \frac{1}{2!}(12x^2 - 4)\Big|_{x=0} x^2 + \frac{1}{4!} 24\Big|_{x=0} x^4$$
$$= 1 - 2x^2 + x^4$$

Note that the odd power terms in x are zero reflecting the fact that $V(x)$ is an even function of x. Furthermore, for this fourth degree polynomial function the Taylor series has a finite number of terms as derivatives higher than the fifth order in x are equal to zero.

Now consider the Taylor series expansion about $x_0 = 1$:

$$V(x) = (4x^3 - 4x)\Big|_{x=1}(x-1) + \frac{1}{2}(12x^2 - 4)\Big|_{x=1}(x-1)^2$$
$$+ \frac{1}{6} 24x\Big|_{x=1}(x-1)^3 + \frac{1}{24} 24\Big|_{x=1}(x-1)^4$$
$$= 4(x-1)^2 + 4(x-1)^3 + (x-1)^4$$

This equation sums to the exact potential energy function $V(x) = x^4 - 2x^2 + 1$. Figure 8.13 shows the convergence of the partial sums $S_2(x) = 4(x-1)^2$, $S_3(x) = S_2(x) + 4(x-1)^3$, and $S_4(x) = S_3(x) + (x-1)^4$ to the exact function.

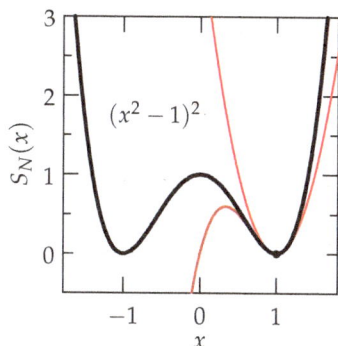

Figure 8.13: The quartic double well potential energy function $V(x)$ (black) compared with the partial sums $S_N(x)$ from $N = 2$ to 4, derived from a Taylor series expansion about $x_0 = 1$, shown in increasingly dark shades of red. Note that $S_4(x) = V(x)$.

B$_8$ Useful approximations to functions based on power series

There are many examples of physical processes in which an extreme condition justifies the use of a simpler, limiting form of a more complicated, general result. In considering those problems, the use of series solutions and *truncated series* is a powerful tool.

Partial dissociation of a weak acid at equilibrium

Consider the chemical equilibrium for the dissociation of a weak acid
$$HA(aq) + H_2O(l) \rightleftharpoons H_3O^+(aq) + A^-(aq)$$
where the concentration dependent equilibrium constant is

$$K_c = \frac{[H_3O^+][A^-]}{[HA]}$$

Suppose the initial concentration of acid is $[HA]_0$. Further suppose that a small amount of acid dissociates, creating the concentrations $[A^-] = [H_3O^+] = x$. It follows that the amount of undissociated acid remaining is $[HA] = [HA]_0 - x$.

Substituting back into our expression for the equilibrium constant, we find

$$K_c = \frac{x^2}{[HA]_0 - x}$$

This result can be written as a quadratic equation in x

$$x^2 + K_c x = K_c[HA]_0$$

The roots of this equation can be solved exactly using the quadratic formula (see Equation 1.4). The result is shown in Figure 8.14 where $x^2 + K_c x$ is plotted as a function of x for $[HA]_0 = 1.0$ and $K_c = 0.001$. The intersection of $x^2 + K_c x$ with the line at $K_c[HA]_0$ marks the equilibrium concentration.

Suppose a small fraction of the initial concentration of weak acid has dissociated. This means that $[HA]_0 \gg x$. As such, we can make the approximation that

$$K_c = \frac{x^2}{[HA]_0 - x} \approx \frac{x^2}{[HA]_0}$$

With this equation, we can immediately solve for

$$x = A^- = H_3O^+ = \sqrt{K_c[HA]_0} \tag{8.7}$$

This approximation is plotted in Figure 8.14 as a red line. For $[HA]_0 = 1.0$ and $K_c = 0.001$ the approximation is almost indistinguishable from the exact result.[23]

If we want to know the next order correction we can write

$$K_c = \frac{x^2}{[HA]_0 - x} = \frac{x^2}{[HA]_0\left(1 - \frac{x}{[HA]_0}\right)} \approx \frac{x^2}{[HA]_0}\left(1 + \frac{x}{[HA]_0}\right)$$

where we have used the property of the geometric series stating that[24]

$$\frac{1}{1-x} = 1 + x + \mathcal{O}\left[x^2\right]$$

This expression provides us with a way to evaluate the accuracy of our original approximation that led to Equation 8.7. For the approximation to be valid, it must be that

$$\frac{x}{[HA]_0} \ll 1$$

This is identical to the physical assumption that motivated our initial approximation that little of the weak acid dissociates and $[HA]_0 \gg x$.

Statistical probability distributions in the high-temperature limit

The truncated expansion is also useful in problems involving the Boltzmann probability

$$p(T) \propto \exp\left(-\frac{\varepsilon}{k_B T}\right)$$

where for sufficiently high temperatures, such that

$$\frac{\varepsilon}{k_B T} \ll 1$$

or equivalently

$$T \gg \frac{\varepsilon}{k_B}$$

We can approximate $p(T)$ as

$$p(T) \propto \exp\left(-\frac{\varepsilon}{k_B T}\right) \approx 1 - \frac{\varepsilon}{k_B T}$$

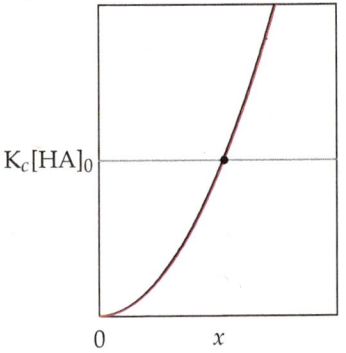

$K_c[HA]_0$

$0 \qquad x$

Figure 8.14: The concentration $x = [A^-] = [H_3O^+]$ using the exact quadratic expression (black) and the $x \ll 1$ approximation (red) for $[HA]_0 = 1.0$ and $K_c = 0.001$. The dot marks the equilibrium concentration.

[23] While it may seem that this approximate method is comparably demanding when compared with the exact solution, in cases of more complex equilibria an approximate solution can provide great efficiency and insight.

[24] Note that $\mathcal{O}[x^2]$ includes terms in the full Taylor series, of order x^2 and higher, which are not included in the truncated series.

where we have used the property of an exponential series expansion

$$e^{-x} = 1 - x + \mathcal{O}\left[x^2\right]$$

and the observation that higher order terms in the small parameter $\frac{\varepsilon}{k_B T}$ do not contribute significantly to the sum.

Let's apply this result to an expression appearing in Planck's theory of *blackbody radiation*

$$f(\nu) = \frac{1}{1 - \exp(-\beta h \nu)}$$

where $\beta \equiv \frac{1}{k_B T}$ and $h\nu$ is the energy of a photon of light of frequency ν. When the temperature is high enough that

$$T \gg \frac{h\nu}{k_B} \qquad \text{or} \qquad \beta h \nu \ll 1$$

we can approximate the exponential using a Maclaurin series truncated at first order in the small parameter $\beta h \nu$ where $\exp(-\beta h \nu) \approx 1 - \beta h \nu$. We find

$$f(\nu) = \frac{1}{1 - \exp(-\beta h \nu)} \approx \frac{1}{1 - (1 - \beta h \nu)} = \frac{1}{\beta h \nu}$$

Now consider an expression taken from Einstein's theory of the *heat capacity of solids*

$$p(\beta) = (\beta h \nu)^2 \frac{e^{-\beta h \nu}}{(1 - e^{-\beta h \nu})^2} \tag{8.8}$$

shown in Figure 8.15. We wish to evaluate this expression in the limit as $T \to \infty$ or equivalently $\beta \to 0$. Unfortunately, both the numerator and denominator go to zero as β goes to zero!

There are two ways to address this. We can apply *l'Hôpital's rule* by differentiating the numerator and denominator and retaking the limit. First we simplify the expression by substituting $x = \beta h \nu$ and noting that as $\beta \to 0$, $x \to 0$. Further noting that as $x \to 0$, the exponential in the numerator approaches unity, our limit becomes

$$\lim_{x \to 0} \frac{x^2}{(1 - e^{-x})^2}$$

Finally, we apply l'Hôpital's rule three times to find

$$\lim_{x \to 0} \frac{x^2}{(1 - e^{-x})^2} = \lim_{x \to 0} \left[\frac{2x}{2e^{-x}(1 - e^{-x})}\right] = \lim_{x \to 0} \left(\frac{x}{e^{-x} - e^{-2x}}\right)$$

$$= \lim_{x \to 0} \left(\frac{1}{2e^{-2x} - e^{-x}}\right) = 1$$

A second and more efficient approach is to use our high-temperature expansion of the Boltzmann probability

$$e^{-\beta h \nu} \approx 1 - \beta h \nu + \mathcal{O}\left[(\beta h \nu)^2\right]$$

where we maintain only the first order term in our small parameter $\beta h \nu$.

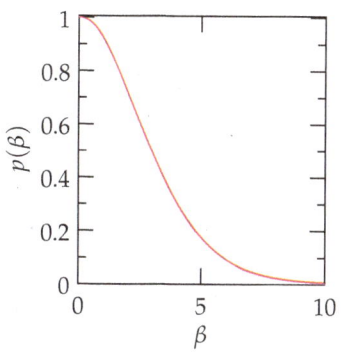

Figure 8.15: The function

$$p(\beta) = (\beta h \nu)^2 \frac{e^{-\beta h \nu}}{(1 - e^{-\beta h \nu})^2}$$

of Equation 8.8.

Taking the limit of Equation 8.8, we find

$$\lim_{\beta \to 0} (\beta h \nu)^2 \frac{e^{-\beta h \nu}}{(1 - e^{-\beta h \nu})^2} = \lim_{\beta \to 0} (\beta h \nu)^2 \left[\frac{1}{1 - (1 - \beta h \nu)} \right]^2$$

$$= \lim_{\beta \to 0} (\beta h \nu)^2 \left(\frac{1}{\beta h \nu} \right)^2 = 1$$

This demonstrates how truncated series expansions can be used to make controlled approximations that simplify expressions and facilitate the identification of limits of complicated expressions common to the physical sciences.

C$_8$ Self-similarity and fractal structures

Many objects in nature have a *self-similar* structure. The shape of part of the object is similar to the shape of the object as a whole. For example, trees display a self-similar structure on many length scales. Thick branches divide into many thin branches that divide into many still thinner branches. Consider the *Pythagoras tree* depicted in Figure 8.16. The pattern of growth repeats itself on many length scales.

Figure 8.16: This Pythagorean cherry tree is a *self-similar* object formed from squares arranged such that the area enclosed by contact between three squares forms a right triangle.

Self-similar structure is also observed in the dendritic arms of snowflakes. When a portion of a snowflake is magnified again and again, the branching of arms is found to repeat itself on many length scales. Consider the *Koch snowflake* shown in Figure 8.17.[25] One starts from an equilateral triangle. Each face of the triangle is divided into three equal segments. Smaller equilateral triangles are placed atop the center third of each face pointing outward. The line forming the base of the new triangle is then removed. At each iteration the side of a triangle is reduced by a factor of $\frac{1}{3}$ and the area by a factor of $\frac{1}{9}$. The number of triangles added is equal to the number of faces, which grows by a factor of 4 in each iteration. This process is repeated *ad infinitum* on all

[25] Discovered by Swedish mathematician *Helge von Koch* (1870-1924).

faces to form a self-similar snowflake structure that has a finite area but a perimeter of infinite length.

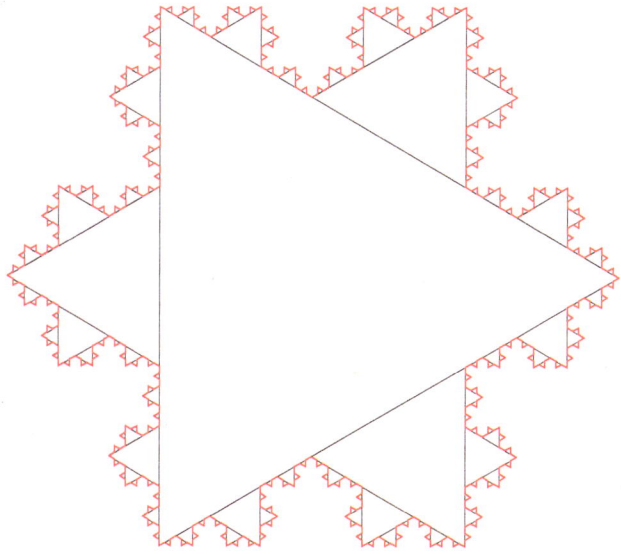

Figure 8.17: The construction of the *Koch snowflake* begins with a single equilateral triangle. An equilateral triangle having sides $\frac{1}{3}$ the length of those in the original triangle is added to each face. This process is repeated *ad infinitum*. The outer perimeter of this Koch snowflake formed from triangles on five length scales is shown in red.

Objects possessing self-similar structure are called *fractals*.[26] Fractal objects are important to the physical sciences as self-similarity occurs throughout nature. Fractal objects are found in the structure of growing crystals, agglomerated soot particles, and dendritic molecular assemblies.

[26] The term *fractal* was introduced by Polish-born mathematician *Benoit Mandelbrot* (1924-2010).

Consider the construction of a fractal object known as the *Sierpiński carpet*.[27] One starts from a solid unit square, with sides of length 1, that we call our carpet. Initially, a single square hole with side $\frac{1}{3}$ and area $\frac{1}{9}$ is added to the center of the carpet. Afterwards, in each of the surrounding 8 squares of area $\frac{1}{9}$, one forms additional holes of side $\frac{1}{9}$ and area $\frac{1}{81}$. This process is repeated once again to add 64 holes with sides of length $\frac{1}{27}$ and area $\frac{1}{729}$. An example of a Sierpiński carpet is shown in Figure 8.18.

[27] Named for Polish mathematician *Wacław Sierpiński* (1882-1969).

Rather than thinking of the area of the holes, we can equivalently consider the area of carpet remaining after n iterations of the process of forming holes. Initially, the carpet has area $A_0 = 1$. After $n = 1$ iteration, the first square hole is formed with side of length $L_1 = \frac{1}{3}$ and area $L_1^2 = \frac{1}{9}$. The area of the remaining carpet is

$$A_1 = 1 - \frac{1}{9} = \frac{8}{9}$$

We can think of the area of the remaining carpet being $N_1 L_1^2 = \frac{8}{9}$ where $N_1 = 8$ is the number of squares of carpet with area equal to that of the smallest hole L_1^2.

After $n = 2$ iterations, 8 new square holes of side $L_2 = \frac{1}{9}$ and area $L_2^2 = \frac{1}{81}$ are added. The area of the remaining carpet is

$$A_2 = 1 - \frac{1}{9} - \frac{8}{81} = \frac{64}{81} = \left(\frac{8}{9}\right)^2$$

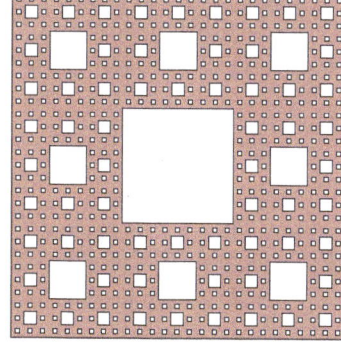

Figure 8.18: The Sierpiński carpet drawn in a unit square to $n = 4$ where the smallest holes have sides of length $\frac{1}{3^4} = \frac{1}{81}$ and number $8^3 = 512$.

This area can be thought of as consisting of $N_2 = 64$ squares of carpet each of area $\frac{1}{81}$. After n iterations, the area of carpet remaining will be

$$A_n = N_n L_n^2$$

The total area replaced by holes is $1 - A_n$.

Figure 8.19: The Sierpiński gasket formed in an equilateral triangle to $n = 5$.

Repeating this process starting from an equilateral triangle as opposed to a square results in the *Sierpiński gasket* shown in Figure 8.19. The Sierpiński carpet and Sierpiński gasket have self-similar fractal structures. A small square of carpet repeats the structure of the carpet as a whole. The pattern of holes in the gasket is repeated at every length scale.

We can define the *fractal dimension* in terms of the limit[28]

$$d \equiv - \lim_{n \to \infty} \frac{\ln N_n}{\ln L_n}$$

where N_n is the number of elements of area L_n^2 needed to cover the fractal surface. The fractal dimension can be thought of as the capacity of a particular pattern to fill space. For this reason, the fractal dimension is also referred to as the *capacity dimension*.

For example, suppose we cover a unit square surface with $N = n^2$ solid elements of area $L_n^2 = 1/n^2$. The fractal dimension is calculated as

$$d = - \lim_{n \to \infty} \frac{\ln N_n}{\ln L_n} = - \lim_{n \to \infty} \frac{\ln \left(n^2 \right)}{\ln \left(n^{-1} \right)} = \lim_{n \to \infty} \frac{2 \ln n}{\ln n} = 2$$

The repeating pattern of the solid square has the capacity to cover the surface completely. As such, the fractal dimension of the solid two-dimensional square is $d = 2$.

Now let's consider the Sierpiński carpet where $N_n = 8^n$ and $L_n = 3^{-n}$. Returning to our definition of the fractal dimension we find

$$d = - \lim_{n \to \infty} \frac{\ln 8^n}{\ln 3^{-n}} = \frac{\ln 8}{\ln 3} = \log_3(8) = 1.8927 \ldots$$

The repeating pattern of the porous Sierpiński carpet does not fill the two-dimensional space. As such, the fractal dimension is less than 2.

[28] The definition of the fractal dimension suggests the scaling relation
$$N \propto L^{-d}$$
where the number of elements, N, required to cover the fractal surface grows exponentially as the length scale of the elements, L, diminishes.

D₈ End-of-chapter problems

All things are number.

Pythagoras

Warm-ups

8.1 Determine whether the following series are convergent or divergent. If the series is convergent, determine the sum.

(a) $1 + 3 + 5 + 7 + 9 + \ldots$

(b) $\frac{1}{1!} + \frac{1}{2!} + \frac{1}{3!} + \frac{1}{4!} + \ldots$

(c) $\frac{1}{1} - \frac{1}{2} + \frac{1}{3} - \frac{1}{4} - + \ldots$

(d) $\frac{3}{2} + \frac{3}{4} + \frac{3}{6} + \frac{3}{8} + \ldots$

(e) $4 + \frac{4}{3} + \frac{4}{9} + \frac{4}{27} + \frac{4}{81} + \ldots$

(f) $\frac{1}{4} + \frac{2}{7} + \frac{3}{10} + \frac{4}{13} + \ldots$

8.2 Use the ratio test to determine whether the following series are convergent, divergent, or if your evaluation is inconclusive.

(a) $\frac{1}{2} + \frac{1}{2^2} + \frac{1}{2^3} + \frac{1}{2^4} + \ldots$

(b) $3 + \frac{3^2}{2} + \frac{3^3}{3} + \frac{3^4}{4} + \ldots$

(c) $\frac{1}{2} + \frac{2}{3} + \frac{3}{4} + \frac{4}{5} + \ldots$

(d) $2 + \frac{2^2}{2^2} + \frac{2^3}{3^2} + \frac{2^4}{4^2} + \ldots$

(e) $\frac{1}{2} + \frac{2^2}{2^2} + \frac{3^2}{2^3} + \frac{4^2}{2^4} + \ldots$

(f) $3 + \frac{3^2}{2!} + \frac{3^3}{3!} + \frac{3^4}{4!} + \ldots$

8.3 Determine the interval of convergence for the following power series.

(a) $1 + x + x^2 + x^3 + \ldots$

(b) $1 - x^2 + \frac{x^4}{2!} - \frac{x^6}{3!} + - \ldots$

(c) $1 + x + \frac{x^2}{2!} + \frac{x^3}{3!} + \ldots$

(d) $x - \frac{x^2}{2!} + \frac{x^3}{3!} - \frac{x^4}{4!} + - \ldots$

(e) $1 - 2x + 3x^2 - 4x^3 + - \ldots$

(f) $x - \frac{1}{3}x^3 + \frac{1}{5}x^5 - + \ldots$

8.4 Expand the following functions in a Maclaurin series.

(a) $\frac{1}{1+x}$

(b) $\frac{1}{(1+x)^2}$

(c) $(1+x)^{1/2}$

(d) $\ln(1-x)$

(e) e^{-x^2}

(f) a^x

(g) $\cos x$

(h) $(1+x)^3$

8.5 Show that for small values of X_B the function $\ln(1 - X_B) \cong -X_B$.

8.6 Show that for small values of θ the function $\sin \theta \cong \theta$.

Homework exercises

8.7 Use your knowledge of power series to derive the following expressions for fundamental mathematical constants defined in terms of infinite series

(a) $e = \sum_{n=0}^{\infty} \frac{1}{n!}$

(b) $\ln 2 = \sum_{n=1}^{\infty} (-1)^{n+1} \frac{1}{n}$

(c) $\frac{\pi}{4} = \sum_{n=0}^{\infty} (-1)^n \frac{1}{2n+1}$

This infinite series representation of the mathematical constant e was used by *Leonhard Euler* (1707-1783) to provide an approximate value of e = 2.718281828459045235 to 18 decimal places.

8.8 Consider the power series

$$\sum_{n=1}^{\infty} nx^n$$

(a) Show that the interval of convergence of this series is $|x| < 1$.

(b) Find the function, $f(x)$, that corresponds to the above power series. To do so, differentiate the geometric series

$$\frac{1}{1-x} = \sum_{n=0}^{\infty} x^n$$

with respect to x and then multiply by x. Use your result to show that

$$f(x) = \sum_{n=1}^{\infty} nx^n = \frac{x}{(1-x)^2} \qquad |x| < 1$$

8.9 Find the function, $f(x)$, that corresponds to the series

$$\sum_{n=1}^{\infty} n^2 x^n$$

HINT: Start from the fact that

$$\frac{x}{(1-x)^2} = \sum_{n=1}^{\infty} nx^n$$

8.10 Show that the Maclaurin series of the hyperbolic sine function

$$\sinh(x) = x + \frac{x^3}{3!} + \frac{x^5}{5!} + \cdots$$

for all x where

$$\sinh(x) = \frac{e^x - e^{-x}}{2}$$

8.11 Expand the function $\sin x$ in powers of $(x - a)$ using a Taylor series including terms to the fourth order in $(x - a)$.

8.12 Consider the integral below where the upper limit of integration is a variable

$$I(x) = \int_0^x e^{-t^2} \, dt$$

and the function $I(x)$ is related to the *error function* as $\mathrm{erf}(x) = \frac{2}{\sqrt{\pi}} I(x)$.

(a) Derive a formula for $I(x)$ by expanding the integrand in a Maclaurin series, keeping only the first four terms.

(b) Determine the numerical value of $I(\frac{1}{3})$. Compare your result to the value derived using the error function tabulated in Supplement S_7.

(c) Determine the numerical value of $I(\frac{1}{2})$. Compare your result to the value derived using the error function tabulated in Supplement S_7.

8.13 Consider the expansion of $\ln(x)$ near $x = 1$.

(a) Show that the Maclaurin series of $\ln(1 + x)$ expanded about $x = 0$ is

$$\ln(1 + x) = x - \frac{x^2}{2} + \frac{x^3}{3} - + \dots$$

(b) Show that the Taylor series of $\ln(x)$ expanded about $x = 1$ is

$$\ln(x) = (x - 1) - \frac{1}{2}(x - 1)^2 + \frac{1}{3}(x - 1)^3 - + \dots$$

(c) Show that the series derived in (a) and (b) have equivalent intervals of convergence $|x| < 1$ and $|x - 1| < 1$, respectively.

8.14 The distribution of frequencies, ν, for *blackbody radiation* was derived by Planck and found to be

$$\rho(\nu, T) \cong \frac{8\pi h}{c^3} \frac{\nu^3}{e^{\beta h \nu} - 1}$$

where $\rho(\nu, T)$ is the intensity of radiation from the blackbody at frequency ν and temperature T where $\beta = \frac{1}{k_B T}$ (see below). Show that for low frequencies or high temperatures, such that $\beta \nu h \ll 1$, the distribution can be written

$$\rho(\nu, T) \cong \frac{8\pi k_B T}{c^3} \nu^2$$

In the figure on the right below, the exact distribution $\rho(\nu)$ is shown in black. The approximation good at low frequencies or high temperatures appears in red.

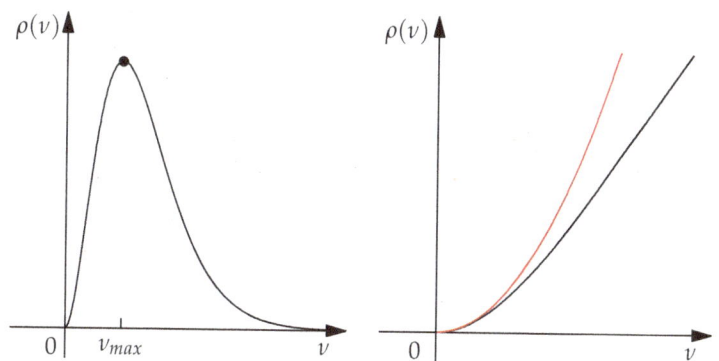

8.15 Consider the quartic double well potential energy function

$$V(x) = (x^2 - 1)^2$$

with minima at $x = \pm 1$ and local maximum at $x = 0$ (see Figure 8.12).

(a) What is the height of the barrier separating the two minima?

(b) For small displacements around $x = 1$, the potential energy can be approximated as the *harmonic oscillator* potential energy function

$$V(x) = \frac{1}{2}\kappa(x - 1)^2$$

Expand the potential in a Taylor series about the point $x = 1$ to second order in the displacement $(x - 1)$ to determine the value of κ.

(c) For small displacements near $x = 0$, the potential can be approximated as a downward facing parabola

$$V(x) = 1 - \frac{1}{2}\kappa x^2$$

Expand the potential in a Taylor series about the point $x = 0$ to second order in the displacement x to determine the value of κ.

8.16 The Einstein model for the temperature dependence of the constant volume heat capacity of a crystal can be written

$$C_V(T) = 3R \left(\frac{\Theta_E}{T}\right)^2 \frac{e^{-\Theta_E/T}}{(1 - e^{-\Theta_E/T})^2}$$

where Θ_E is a constant known as the Einstein temperature. The temperature dependence of $C_V(T)$ is shown below.

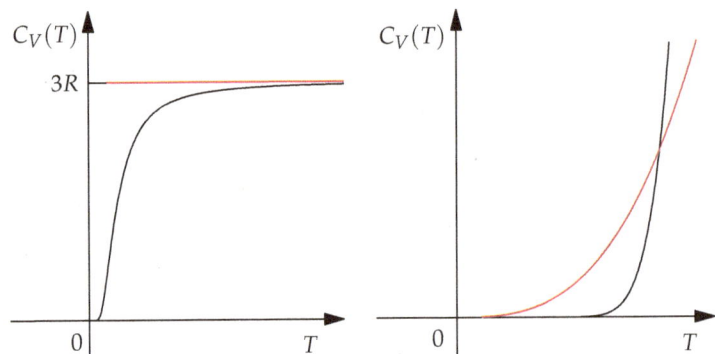

Show that for high temperatures $T \gg \Theta_E$ that

$$C_V(T) \cong 3R$$

which is the classic law of Dulong and Petit.

8.17* According to one theory of electrolyte (salt) solutions the thermodynamic energy of the solution can be written

$$U(\kappa, a) = \frac{(\kappa a)^2 + (\kappa a) - \kappa a(1 + 2\kappa a)^{1/2}}{4\pi\beta a^3}$$

where κ is related to the concentration of the salt solution, $\beta = \frac{1}{k_B T}$, and a is the average radius of a positive or negative ion found in the solution.

Show that as $\kappa a \to 0$, $U(\kappa, a)$ varies as

$$U(\kappa, a) \cong \frac{\kappa^3}{8\pi\beta} + \mathcal{O}[\kappa a]$$

8.18 A chemical bond modeled as a harmonic oscillator has quantum mechanically allowed vibrational energies given by

$$\varepsilon_n = \left(n + \frac{1}{2}\right)h\nu \qquad n = 0, 1, 2, 3, \ldots$$

where ν is the frequency of the bond oscillation. The average energy of the bond is defined

$$\langle \varepsilon \rangle = \frac{1}{q} \sum_{n=0}^{\infty} \left(n + \frac{1}{2} \right) h\nu \, e^{-\beta\left(n+\frac{1}{2}\right)h\nu}$$

where

$$q = \sum_{n=0}^{\infty} e^{-\beta\left(n+\frac{1}{2}\right)h\nu}$$

(a) Show that

$$q = \sum_{n=0}^{\infty} e^{-\beta\left(n+\frac{1}{2}\right)h\nu} = \frac{e^{-\frac{\beta h\nu}{2}}}{1 - e^{-\beta h\nu}}$$

HINT: Pull out a factor of $e^{-\frac{\beta h\nu}{2}}$ and substitute $x = e^{-\beta h\nu}$. You will recognize a familiar series.

(b) Return to the definition of the average energy, $\langle \varepsilon \rangle$. Use the result in (a) to show that

$$\langle \varepsilon \rangle = \frac{h\nu}{2} + \frac{h\nu e^{-\beta h\nu}}{1 - e^{-\beta h\nu}}$$

8.19* Consider the Koch snowflake shown in Figure 8.17. The initial area of the equilateral triangle at the center of the snowflake is $A_0 = 1$.

(a) In the $n = 1$ iteration, three new triangles of area $\frac{1}{9} A_0$ are added. The overall area of the snowflake becomes

$$A_1 = A_0 \left(1 + 3 \frac{1}{9} \right)$$

Use your knowledge of the geometric sequence to derive a formula for the area of the snowflake A_n after n iterations.

(b) Use your knowledge of the geometric series to show that in the limit that the number of iterations $n \rightarrow \infty$ the area of the snowflake is

$$\lim_{n \to \infty} A_n = A_0 \frac{8}{5}$$

(c) Consider the initial equilateral triangle at center of the snowflake. The length of a side of the triangle is S_0. There are $N_0 = 3$ faces so that the perimeter of the triangle is $P_0 = N_0 S_0 = 3 S_0$. After $n = 1$ iteration of adding additional triangles there are $N_1 = 12$ faces of the snowflake of equal length $S_1 = \frac{1}{3} S_0$. As such, the length of the perimeter of the snowflake is

$$P_1 = N_1 S_1 = \frac{12}{3} S_0$$

Derive a formula for the perimeter of the snowflake $P_n = N_n S_n$ after n iterations.

(d) Take the limit that the number of iterations $n \rightarrow \infty$ to determine

$$\lim_{n \to \infty} P_n = P_\infty$$

8.20* Consider the *Sierpiński carpet*, shown in Figure 8.18, formed from a square of area $A_0 = 1$ in the absence of holes. A square hole of area $H_1 = \frac{1}{9}$ is formed in the center of the carpet. The remaining area of the carpet is $A_1 = 1 - H_1 = 1 - \frac{1}{9} = \frac{8}{9}$.

(a) Derive the geometric sequence that equals the total area of the carpet occupied by holes $H_n = 1 - A_n$ after n iterations.

(b) Use your knowledge of the geometric series to determine

$$\lim_{n\to\infty} H_n = H_\infty$$

where H_∞ is total area occupied by holes once the number of iterations $n \to \infty$.

8.21* Consider the *Menger sponge*, shown below. It is formed from a volume $V_0 = 1$ in the absence of holes. Each face of the cube is divided into nine squares, subdividing the cube into 27 smaller cubes (as for a *Rubik's cube*). The central cube of each face and the cube at the center of the larger cube (7 smaller cubes in all) are removed. This creates a sponge with a hole of volume $H_1 = \frac{7}{27}$. The remaining volume of the sponge is $V_1 = 1 - H_1 = 1 - \frac{7}{27} = \frac{20}{27}$. The process is then repeated *ad infinitum* for each of the remaining cubes.

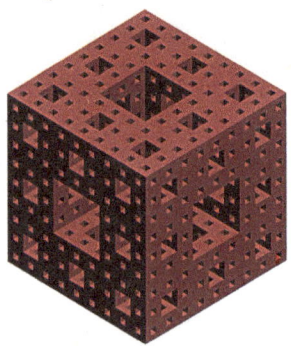

(a) Show that the total volume of the sponge remaining after n iterations is

$$V_n = \left(\frac{20}{27}\right)^n$$

(b) The total volume of the sponge remaining after n iterations derived in (a) can be interpreted as $V_n = N_n L_n^3$ where N_n is the number of cubes of volume L_n^3. Determine N_n and L_n for the Menger sponge.

(c) The fractal dimension is defined in terms of N_n and L_n as

$$d \equiv -\lim_{n\to\infty} \frac{\ln N_n}{\ln L_n}$$

Show that the fractal dimension of the Menger sponge is $\log_3(20) = 2.7268\ldots$ which is a number greater than the dimension of a square, $d = 2$, and less than the dimension of a filled cube, $d = 3$.

8.22* Consider the *Sierpiński gasket*, shown in Figure 8.19, formed from an equilateral triangle with area $A_0 = 1$ in the absence of holes. A triangular hole of area $H_1 = \frac{1}{4}$ is formed in the center of the gasket. The remaining area of the gasket is

$$A_1 = 1 - H_1 = 1 - \frac{1}{4} = \frac{3}{4}$$

After $n = 2$ iterations the remaining area of the gasket is

$$A_2 = 1 - \frac{1}{4} - 3\left(\frac{1}{4}\right)^2 = 1 - \frac{1}{4} - \frac{3}{16} = 1 - H_2$$

(a) Show that the total area of the gasket remaining after n iterations is

$$A_n = \left(\frac{3}{4}\right)^n$$

(b) The total area of the gasket remaining after n iterations derived in (a) can be interpreted as $A_n = N_n L_n^2$ where N_n is the number of triangles of area L_n^2. Determine N_n and L_n for the Sierpiński gasket.

(c) The fractal dimension is defined in terms of N_n and L_n as

$$d \equiv -\lim_{n \to \infty} \frac{\ln N_n}{\ln L_n}$$

Show that the fractal dimension of the Sierpiński gasket is $\log_2(3) = 1.5849\ldots$ which is a number greater than the dimension of a line, $d = 1$, and less than the dimension of a filled triangle, $d = 2$.

9 FUNDAMENTALS OF PROBABILITY AND STATISTICS

9.1 Probability distributions of discrete variables

FUNCTIONS OF DISCRETE VARIABLES arise in problems in which a number of observations are made of a property of the system consisting of individually separate and distinct states. Matter at the atomic level is discrete. Energy levels in quantum mechanical systems are discrete. Simple models of classical phenomena such as gases, liquids, and solids often assume a discrete set of possible states of the system. This section explores functions of discrete variables commonly used in the physical sciences.

9.1.1 Factorials

Consider the set of numbers $\{1, 2, 3, 4\}$. The number of ways to order the four numbers is

$$4 \times 3 \times 2 \times 1$$

since there are four possible choices for the first number in the sequence, leaving three choices for the second number, two choices for the third number, and one choice for the fourth number. This sequential product is known as a *factorial* and defined as

$$n! = n \times (n-1) \times (n-2) \times \ldots \times 1$$

where we pronounce $n!$ as *n factorial*. We say that there are $4! = 4 \times 3 \times 2 \times 1 = 24$ ways to order the four numbers.[1] The dependence of $n!$ on n is displayed in Figure 9.1 on a semi-log plot.

Let's explore a few examples to appreciate the size of factorials. Suppose we have a peptide consisting of a sequence of 10 unique amino acids. There are $10! = 3.6 \times 10^6$ possible unique peptide sequences. Consider a deck of 52 playing cards. There are $52! \approx 8.1 \times 10^{67}$ ways of ordering the deck of cards. To appreciate how large this number is, consider that there may be 10^{22} stars in the observable Universe and 10^{50} atoms composing the Earth. If we shuffled our deck of cards once every four seconds, we would explore a meager 10^{17} shuffles over the age of the Universe. Now consider a crystal consisting of $N_0 = 6 \times 10^{23}$ indistinguishable atoms. There are $N_0!$ ways of arranging the atoms in the crystal, a number that seems too large to imagine.

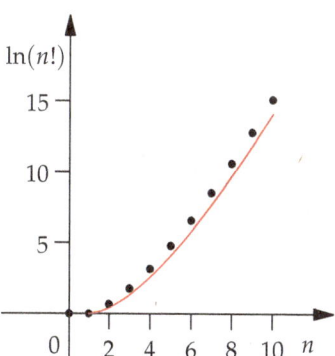

Figure 9.1: The natural logarithm of $n!$ (black dots) for $n \in [0, 10]$. Shown for comparison is Sterling's approximation $\ln(n!) \approx n \ln n - n + 1$ (red line).

[1] Note that as $n! = n \times (n-1)!$ it must be that $1! = 1 \times 0! = 1$ so that $0! = 1$.

9.1.2 *The binomial probability distribution*

Consider an event with two possible outcomes, such as the flip of a coin. Further consider the probability of observing an outcome of interest, referred to as a favorable outcome, such as observing two heads in two tosses of a coin.[2] For a fair coin, for each coin flip the probability of observing a *head* is $\frac{1}{2}$ and the probability of observing a *tail* is $\frac{1}{2}$.

We assume that the probability of observing a head or tail is independent of whether a head or tail was observed the prior flip. The outcome of the coin toss is treated as an *independent random variable*. When an independent random variable is sampled (say by observing the flip of a coin), the result of the next observation (a second flip of the coin) is independent of all prior observations.[3]

It follows that if we have two coin flips, the probability of a specific outcome such as a head followed by a tail (written HT) is the *joint probability* of flipping a head first followed by a tail:

$$\frac{1}{2} \times \frac{1}{2} = \frac{1}{4}$$

In general, for N coin flips each with one of two possible outcomes, the total number of possible outcomes is 2^N and the probability of observing any one specific outcome is

$$\frac{1}{2^N}$$

Now suppose that we don't care about the order of outcomes. We only care how many heads and how many tails occur in N coin flips. Note that for two coin flips there is a total of $2^2 = 4$ possible outcomes:

HH   HT   TH   TT

If we ignore the order, there are three unique outcomes, one with two heads, one with a head and a tail, and one with two tails. There is one way of observing two heads (HH), two ways of observing one head and one tail (HT, TH), and one way of observing two tails (TT). The results are distributed according to a *binomial distribution*, a distribution composed of two terms H and T.

The number of ways to order N outcomes is $N!$. For $N = 2$ coin flips, there will be $N! = 2! = 2$ ways of ordering any outcome. However, if we flip $n_1 = 2$ heads, there is really only one unique sequence. So we must divide $2!$ by the number of ways of permuting the 2 heads, which is $n_1! = 2! = 2$, times the number of ways of permuting 0 tails, which is $n_2! = 0! = 1$. We say there is one way of flipping two heads:

$$\frac{N!}{n_1! n_2!} = \frac{2!}{2!\,0!} = 1$$

Similarly, if we flip two tails there are $n_1 = 0$ heads, $n_2 = 2$ tails, and $\frac{N!}{n_1! n_2!} = \frac{2!}{0!\,2!} = 1$ way of flipping two tails. The number of ways to flip one head and one tail is the total number of ways $N!$ divided by the number of ways of permuting one head, $n_1! = 1$, times the number of ways of permuting

[2] The *probability* of an event is a measure of the likelihood of the event happening. Mathematically, the probability of a favorable event is the ratio of the number of favorable outcomes to the total number of possible outcomes. As such, the probability is expressed as a number between 0 and 1, with 0 representing impossibility and 1 representing certainty. The larger the probability of an event, the more likely the event is to occur.

[3] A *random process* is a series of actions having no defined method and leading to no discernable pattern of outcomes.

one tail, $n_2! = 1$, or $\frac{N!}{n_1!n_2!} = \frac{2!}{1!\,1!} = 2$ ways to flip a head and a tail.

If we have three coin flips, there are $2^3 = 8$ possible outcomes:

HHH   HHT   HTH   HTT   THH   THT   TTH   TTT

The number of ways of flipping three heads is $\frac{N!}{n_1!} = \frac{3!}{3!} = 1$, the number of ways of flipping two heads and one tail is $\frac{N!}{n_1!n_2!} = \frac{3!}{2!1!} = 3$, the number of ways of flipping one head and two tails is $\frac{N!}{n_1!n_2!} = \frac{3!}{1!2!} = 3$, and the number of ways of flipping three tails is $\frac{N!}{n_2!} = \frac{3!}{3!} = 1$. The resulting distribution is shown in Figure 9.2.

In general, for N coin flips each having two possible outcomes, the number of ways of having a specific outcome with n_1 heads and n_2 tails is

$$W(n_1, n_2; N) = \frac{N!}{n_1!n_2!} = \frac{N!}{n_1!(N - n_1)!} \tag{9.1}$$

where we recognize that $N = n_1 + n_2$. The factor $W(n_1, n_2; N)$ is known as the *binomial coefficient*.

We can now consider the probability of observing a given outcome of interest in a series of coin flips. Recall that for N coin flips there are 2^N possible outcomes with each individual outcome occurring with a probability of $\frac{1}{2^N}$. As such, the probability of having N coin flips with n_1 heads and $n_2 = N - n_1$ tails is the probability $\frac{1}{2^N}$ times the number of outcomes having n_1 heads and n_2 tails:

$$p(n_1, n_2; N) = \frac{1}{2^N} \, W(n_1, n_2; N) = \frac{1}{2^N} \, \frac{N!}{n!(N - n)!} \tag{9.2}$$

In the final form of the equation we have set $n = n_1$. This result is known as the *binomial probability distribution*.[4]

The resulting distribution of probabilities for the case $N = 6$ is shown in Figure 9.3. Note that the mean value of the distribution is $\frac{N}{2} = 3$. Further note that in comparison to the distribution for $N = 3$ (see Figure 9.2) the distribution for $N = 6$ is somewhat more sharply peaked.

Let's evaluate the probabilities defined by Equation 9.2 for a few specific cases. For $N = 2$ events, the probabilities for the three possible outcomes (HH), (HT or TH), and (TT) are

$$\frac{1}{4} \quad \frac{2}{4} \quad \frac{1}{4}$$

Each probability is equal to $\frac{1}{2^2}$ times the binomial coefficient, 1, 2, or 1, expressing the number of ways of observing a given outcome (HH), (HT or TH), or (TT). The binomial coefficients form a series $1, 2, 1$.

For $N = 3$ events, the probabilities for the four possible outcomes (HHH), (HHT or HTH or THH), (TTH or THT or HTT), and (TTT) are

$$\frac{1}{8} \quad \frac{3}{8} \quad \frac{3}{8} \quad \frac{1}{8}$$

where the binomial coefficients form a series $1, 3, 3, 1$. For $N = 4$ events, the

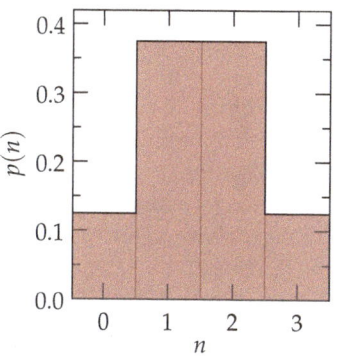

Figure 9.2: The binomial distribution for the relative probability of the four possible outcomes of $N = 3$ flips of a coin. Here we assume the order of events does not matter. As such, TTH, THT, and HTT are identical outcomes as each has $n = 1$ head and 2 tails. The same holds for THH, HTH, and HHH, which are three identical outcomes each having $n = 2$ heads and 1 tail.

[4] The binomial theorem was first mentioned by Greek mathematician *Euclid* (*circa* 300 BCE).

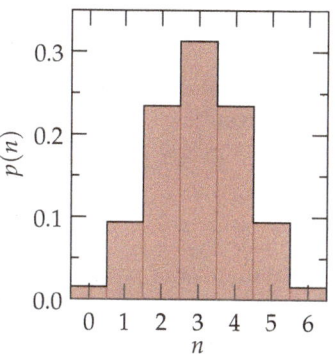

Figure 9.3: The binomial distribution for the relative probability of seven possible outcomes from $N = 6$ flips of a coin, where n is the total number of heads.

probabilities for the five possible outcomes are

$$\frac{1}{16} \quad \frac{4}{16} \quad \frac{6}{16} \quad \frac{4}{16} \quad \frac{1}{16}$$

where the binomial coefficients form a series 1, 4, 6, 4, 1. And so on.

The pattern of binomial coefficients can be captured graphically in the form of *Pascal's triangle* shown in Figure 9.4. The pattern starts with 1 at the top. We then form the next line by summing the number above to the left and the number above to the right.

1
1 1
1 2 1
1 3 3 1
1 4 6 4 1
1 5 10 10 5 1
1 6 15 20 15 6 1
1 7 21 35 35 21 7 1
1 8 28 56 70 56 28 8 1
1 9 36 84 126 126 84 36 9 1
1 10 45 120 210 256 210 120 45 10 1

Figure 9.4: Pascal's triangle is a convenient source of binomial coefficients. It also has a number of additional properties. Consider the red triplets. In each case, the two numbers above sum to the number below ($3+3=6$ and $15+6=21$). In addition, the numbers along a diagonal sum to a *Fibonacci number*. For example, the series of boxed numbers sums to 13. Finally and remarkably, coloring all even numbers white and all odd numbers black results in a *Sierpiński gasket* (see Figure 8.19).

Almost magically, each row in Pascal's triangle forms a series of numbers equal to the binomial coefficients for a value of N.[5]

9.1.3 *Applications of the binomial probability distribution*

Let's use the binomial distribution to analyze the outcome of a coin flipped six times, resulting in four heads and two tails. The number of ways to flip a coin $N=6$ times with a total of $n_1=2$ heads and $n_2=4$ tails is

$$W(2,4;6) = \frac{6!}{2!4!} = 15$$

The total number of possible outcomes for $N=6$ coin flips is $2^6=64$. The probability of flipping the coin $N=6$ times with a total of $n_1=2$ heads and $n_2=4$ tails is

$$p(2,4;6) = \frac{1}{2^6}\,W(2,4;6) = \frac{15}{64} \approx 0.234$$

The computed probability at $n=2$ is $p(n=2) \approx 0.234$ as shown in Figure 9.3.[6]

Now suppose we have $N=10$ electrons, each with the property of spin that can take on one of two states said to be *up* or *down*. If each spin is

[5] Named for French mathematician, physicist, and inventor *Blaise Pascal* (1623-1662). A child prodigy, as a teenager he carried out fundamental work on the properties of computing machines and the properties of fluids. He is known beyond the sciences for proposing *Pascal's wager*.

[6] Once N is fixed, the binomial probability depends on the single variable n. We write $p(n_1,n_2;N) = p(n,N-n;N) = p(n)$ where it is implied that $N=6$.

distinguishable, there are $2^N = 2^{10} = 1024$ possible ways of ordering the 10 spins.

And what if the spins are indistinguishable, rather than distinguishable? If 9 spins are up and 1 spin is down, the 1 down spin could be any of the 10 spins. In that case, the total number of ways of ordering $n_1 = 9$ up spins and $n_2 = 1$ down spin is $\frac{N!}{n_1!n_2!} = \frac{10!}{9!1!} = 10$. It follows that the total probability of observing 9 up spins and 1 down spin is $p(9,1;10) = \frac{10}{1024} = 0.00976$ or just less than 1%.[7]

9.1.4 The multinomial probability distribution

Suppose we randomly choose a number from a set $\{1,2,3,4\}$. There are $k = 4$ numbers in the set and the probability of choosing any one number is $\frac{1}{4}$, which is 1 specific outcome divided by 4 possible outcomes. If we return to the full set $\{1,2,3,4\}$ and again choose a number at random, the probability of choosing any one number is again $\frac{1}{4}$. For the two choices, there are 4^2 possible outcomes so the probability of observing 1 specific outcome will be $\frac{1}{4^2} = \frac{1}{16} \approx 6.2\%$.

If we return to the full set $\{1,2,3,4\}$ a third time, the probability of choosing any one number is $\frac{1}{4}$. It follows that the probability of choosing any particular sequence of three numbers such as $2,3,4$ or $1,4,2$ or $3,1,1$ will be $\frac{1}{4^3} \approx 1.5\%$. In general, the probability of observing a particular sequence of N choices each with k possible outcomes is

$$\frac{1}{k^N}$$

Now let's ask what the probability of choosing a sequence of N numbers will be if we ignore the order in which the numbers are chosen. For example, for $N = 3$ choices of $k = 4$ numbers from the set $\{1,2,3,4\}$ the outcomes $2,3,4$ and $1,3,4$ are unique but $2,3,4$ and $3,2,4$ are taken to be the same. The probability of choosing a unique sequence of numbers will be the probability of choosing that sequence times the multiplicity of ways that the sequence can be chosen.

For example, there is one way a sequence with 3 twos can be chosen $(2,2,2)$, but there are three ways a sequence with 1 one and 2 twos can be chosen:

$$1,2,2 \qquad 2,1,2 \qquad 2,2,1$$

For a sequence in which each number is different such as $3,2,4$ there are six unique ways the numbers can be ordered:

$$3,2,4 \qquad 3,4,2 \qquad 2,3,4 \qquad 2,4,3 \qquad 4,3,2 \qquad 4,2,3$$

Let's generalize these observations. For a sequence of N choices from a set $\{1,2,3,4\}$ having n_1 ones, n_2 twos, n_3 threes, and n_4 fours, there will be

$$n_1!n_2!n_3!n_4!$$

ways of ordering the sequence where $n_1 + n_2 + n_3 + n_4 = N$. It follows that the number of unique sequences that can be chosen, having n_1 ones, n_2 twos, n_3 threes, and n_4 fours, is the total number of possible sequences $N!$ divided

[7] Consider this alternative derivation. There are 10 ways of ordering the 1 down spin. For a given down spin, the order of the 9 indistinguishable up spins does not matter. As a result, there is a total of 10 unique ways to order the up and down spins.

by the number of non-unique ways of ordering the sequence $n_1!n_2!n_3!n_4!$, or

$$\frac{N!}{n_1!n_2!n_3!n_4!}$$

In general, for N choices from k distinguishable outcomes with n_j choices of the jth specific outcome, the number of ways a unique sequence can be chosen is

$$W(n_1, n_2, n_3, \ldots, n_k; N) = \frac{N!}{n_1!n_2!n_3!\ldots n_k!} \tag{9.3}$$

where $N = n_1 + n_2 + n_3 + \ldots + n_k$. The function $W(n_1, n_2, n_3, \ldots, n_k; N)$ is known as the *multinomial coefficient*.

The probability of observing N choices leading to an outcome defined by $n_1, n_2, n_3, \ldots, n_k$ will be

$$p(n_1, n_2, n_3, \ldots, n_k; N) = \frac{1}{k^N} W(n_1, n_2, n_3, \ldots, n_k; N)$$

which is $\frac{1}{k^N}$ times the multiplicity of ways the sequence can be chosen.

9.1.5 Applications of the multinomial probability distribution

The multinomial coefficient can be used to determine the number of ways one can roll a six-sided die N times with the outcome n_1 ones, n_2 twos, n_3 threes, n_4 fours, n_5 fives, and n_6 sixes. For example, the number of ways to roll the die $N = 5$ times, resulting in $n_1 = 2$ ones and $n_6 = 3$ sixes, is

$$W(2, 0, 0, 0, 0, 3; 5) = \frac{5!}{2!\,0!\,0!\,0!\,0!\,3!} = 10$$

The total number of possible outcomes rolling the die $N = 5$ times is $k^N = 6^5 = 7,776$. Therefore, the probability of rolling the die $N = 5$ times and having $n_1 = 2$ ones and $n_6 = 3$ sixes[8] is

$$p(2, 0, 0, 0, 0, 3; 5) = \frac{1}{6^5} W(2, 0, 0, 0, 0, 3; 5)$$

$$= \frac{10}{7,776} \approx 1.28 \times 10^{-3}$$

For another example, consider $k = 6$ different levels of varying energy, ε, that can be occupied by $N = 24$ indistinguishable particles. The total number of ways the particles can be arranged over the six energy levels is given by the multinomial coefficient

$$W(n_1, n_2, n_3, \ldots, n_6; N) = \frac{N!}{n_1!n_2!n_3!\ldots n_6!}$$

where n_j is the number of particles occupying the jth energy level. Consider the distribution shown in Figure 9.5. For this case, $n_1 = 10$, $n_2 = 7$, $n_3 = 4$, $n_4 = 2$, $n_5 = 1$, and $n_6 = 0$. The number of ways the N indistinguishable particles can be arranged is

$$W(10, 7, 4, 2, 1, 0; 24) = \frac{24!}{10!\,7!\,4!\,2!\,1!\,0!} \approx 7.0 \times 10^{11} \tag{9.4}$$

[8] There are $6 \times 5 = 30$ different pairs that can be formed from the numbers $\{1, 2, 3, 4, 5, 6\}$. As such, the probability of rolling a *full house* of 2 of one kind and 3 of another in 5 rolls is $30\,(1.28 \times 10^{-3}) = 3.85\%$.

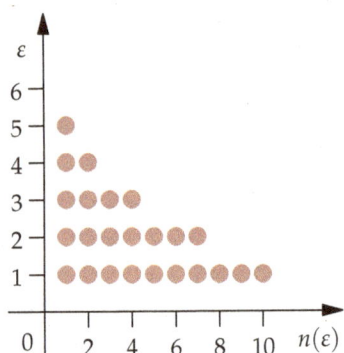

Figure 9.5: An arrangement of $N = 24$ particles (red balls) occupying $k = 6$ energy levels of varying energy ε. $n(\varepsilon)$ represents the number of particles occupying a level of energy ε.

9.1.6 Moments of the binomial and multinomial distributions

Suppose there is a measurement on an observed property of a system. There are n possible outcomes sampled over N trials. The probability of observing a specific outcome is

$$p_k = \frac{N_k}{N} \qquad k = 1, 2, \ldots, n$$

where N_k is the number of times that the kth possible outcome is observed in N trials. We know that the total number of observations must equal the number of trials

$$\sum_{k=1}^{n} N_k = N$$

and the sum

$$\sum_{k=1}^{n} p_k = \sum_{k=1}^{n} \left(\frac{N_k}{N} \right) = \frac{1}{N} \sum_{k=1}^{n} N_k = \frac{1}{N} N = 1$$

Now suppose there is a property x that has a specific value x_k for each of the n possible outcomes. We can write the average or the *mean value* of x as the probability of a specific outcome p_k times the measured value of x for that outcome x_k summed over all possible outcomes. This can be written

$$\bar{x} = \sum_{k=1}^{n} x_k p_k$$

where $\bar{x}$ is the mean value of x.

We can also define the average of the square of the observable x, known as the *mean square value* of x and written[9]

$$\overline{x^2} = \sum_{k=1}^{n} (x_k)^2 p_k$$

[9] The averages $\bar{x}$ and $\overline{x^2}$ are also known as *moments* and provide insight into the shape of the distribution.

Now consider the mean square deviation of x from its average value $\bar{x}$:

$$\overline{(x - \bar{x})^2} = \sum_{k=1}^{n} (x_k - \bar{x})^2 p_k$$

This is known as the *variance*, and is often written

$$\sigma_x^2 \equiv \overline{(x - \bar{x})^2}$$

The variance provides a measure of the size of fluctuations in observations of x around its mean value $\bar{x}$.

If we expand the square, we find

$$\overline{(x - \bar{x})^2} = \sum_{k=1}^{n} \left((x_k)^2 - 2\bar{x}x_k + \bar{x}^2 \right) p_k$$

$$= \sum_{k=1}^{n} (x_k)^2 p_k - 2\bar{x} \sum_{k=1}^{n} x_k p_k + \bar{x}^2 \sum_{k=1}^{n} p_k$$

$$= \overline{x^2} - 2\bar{x}^2 + \bar{x}^2 = \overline{x^2} - \bar{x}^2 \geq 0$$

This expression relates the mean-square difference to the averages $\overline{x^2}$ and $\bar{x}$.

The resulting identity for the variance is

$$\sigma_x^2 = \overline{(x - \overline{x})^2} = \overline{x^2} - \overline{x}^2 \tag{9.5}$$

This is a fundamental result that we will return to in many contexts in the physical sciences, including quantum theory and thermodynamics.

An interesting quantity is the square root of the variance, σ_x, known as the *standard deviation*. The ratio of the standard deviation to the mean value of the distribution

$$\frac{\sigma_x}{\overline{x}} = \frac{\sqrt{\overline{(x - \overline{x})^2}}}{\overline{x}}$$

provides a useful measure of the size of the fluctuations about the mean value relative to the mean value itself.

Consider the binomial probability distribution. There are two outcomes, each assumed to have probability $\frac{1}{2}$, so that

$$p(n; N) = \frac{1}{2^N} \frac{N!}{n!(N-n)!}$$

where

$$\sum_{n=0}^{N} p(n; N) = \frac{1}{2^N} \sum_{n=0}^{N} \frac{N!}{n!(N-n)!} = \frac{1}{2^N} 2^N = 1 \tag{9.6}$$

The mean value of n for the binomial distribution is

$$\overline{n} = \sum_{n=0}^{N} n\, p(n; N) = \frac{1}{2^N} \sum_{n=0}^{N} n \frac{N!}{n!(N-n)!}$$

$$= \frac{1}{2^N} \sum_{n=1}^{N} n \frac{N!}{n!(N-n)!} \tag{9.7}$$

where we note that the $n = 0$ term in the sum is zero. Rewriting the coefficient as

$$n \frac{N!}{n!(N-n)!} = \frac{N!}{(n-1)!(N-n)!}$$

$$= N \frac{(N-1)!}{(n-1)!((N-1)-(n-1))!}$$

and substituting $r = n - 1$, we find

$$\overline{n} = \sum_{n=0}^{N} n\, p(n; N) = \frac{1}{2^N} \sum_{r=0}^{N-1} N \frac{(N-1)!}{r!((N-1)-r)!} \tag{9.8}$$

In a final step, we note that the sum appearing in Equation 9.6 can alternatively be written in terms of the index r, with the sum extending from $r = 0$ to $r = N - 1$:

$$\sum_{r=0}^{N-1} \frac{(N-1)!}{r!((N-1)-r)!} = 2^{N-1}$$

Substituting this relation in Equation 9.8, we arrive at our final result

$$\bar{n} = \frac{N}{2^N} \sum_{r=0}^{N-1} \frac{(N-1)!}{r!((N-1)-r)!} = \frac{N}{2^N} 2^{N-1} = \frac{N}{2} \tag{9.9}$$

This result makes sense. The binomial distribution has a single peak, and the average $\bar{n}$ is the maximum value and midpoint $N/2$.

With additional effort we can also prove that the mean square value of n is

$$\overline{n^2} = \sum_{n=0}^{N} n^2 p(n; N) = \frac{1}{2^N} \sum_{n=0}^{N} n^2 \frac{N!}{n!(N-n)!} = \frac{N^2}{4} + \frac{N}{4}$$

so that the variance is

$$\sigma_n^2 = \overline{(n-\bar{n})^2} = \left(\frac{N^2}{4} + \frac{N}{4}\right) - \left(\frac{N}{2}\right)^2 = \frac{N}{4} \tag{9.10}$$

As a result, the standard deviation is

$$\sigma_n = \frac{1}{2}\sqrt{N}$$

and the ratio of the standard deviation to the mean is[10]

$$\frac{\sigma_n}{\bar{n}} = \frac{\sqrt{N}}{N} = \frac{1}{\sqrt{N}}$$

The mean and standard deviation are depicted for a binomial distribution with $N = 10$ in Figure 9.6.

This final result states that as the number of observations N increases, the magnitude of fluctuations in the average value of n from the mean value $\bar{n}$ will decrease according to $N^{-1/2}$. This is a fundamental property of the binomial probability distribution that has important implications in the kinetic theory of matter.

Finally, consider the multinomial distribution in which there are k possible outcomes each with probability p_k. When there are N observations, the expected value of the kth outcome is

$$n_k = N p_k$$

The variance, defined as the mean square fluctuation in the kth outcome relative to the average, is

$$\sigma_k^2 = N p_k (1 - p_k)$$

Let's consider the binomial distribution as a special case of the multinomial distribution. We take the probability of one outcome to be $p_1 = p$ and that of the second outcome to be $p_2 = 1 - p = q$. In that case, the mean value of observing the first outcome is $Np_1 = Np$ and the associated variance is $\sigma_1^2 = Np_1(1-p_1) = Np(1-p) = Npq$. These results agree with our prior results in Equation 9.9 and Equation 9.10 for the binomial distribution. For equal probabilities $p_1 = p_2 = \frac{1}{2}$, we found a mean of $\frac{N}{2}$ and variance $\frac{N}{4}$.

In the next section, we explore the properties of probability distributions of continuous variables that are commonly used in the physical sciences.

[10] For the general case of the binomial process in which one outcome has probability p and the other probability $(1 - p)$ we find the mean $\bar{n} = Np$ and variance $\sigma_n^2 = Np(1-p)$.

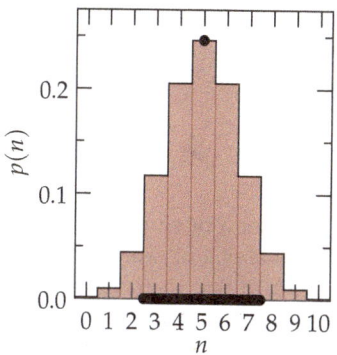

Figure 9.6: The binomial distribution for the relative probability of 11 possible outcomes resulting from $N = 10$ flips of a coin, where n is the total number of heads. The black dot marks the mean $\bar{n}$ and the thick black bar marks the range $\bar{n} - \sigma_n$ to $\bar{n} + \sigma_n$.

9.2 Probability distributions of continuous variables

FUNCTIONS OF CONTINUOUS VARIABLES arise in problems in which a number of observations are made of a property of the system consisting of a continuity of states. The position of atoms in space is defined in terms of continuous variables. The kinetic energy of atoms modeled as classical particles is taken to be continuous. The variables defining the pressure and temperature of a system are taken to be continuous. This section introduces probability distributions for continuous variables and explores a few functions of continuous variables commonly encountered in the physical sciences.

9.2.1 Properties of probability distributions of a continuous variable

For discrete variables such as a number of coin flips, the probability of a favorable event is the ratio of the number of favorable outcomes to the total number of possible outcomes. A sum over all possible outcomes must equal unity, the probability of any event happening. For a continuous variable, such as a particle's position or momentum, we need to generalize our concept of probability. The probability of observing a discrete state, p_k, is replaced by the probability of finding the variable within a range of values, $x \in [x, x + dx]$,[11] such that

$$p_k \to p(x)dx$$

[11] Note that p_k is dimensionless while $p(x)$ has units of $1/x$ making the product $p(x)dx$ dimensionless.

where $p(x)$ is a *probability distribution function*. The sum over the probabilities of all discrete states is replaced by an integral

$$\sum_k p_k \to \int p(x)dx$$

Let's see how this works for the specific case of the *uniform probability distribution*.

Consider a random variable x. There is a constant probability C of observing a value of x between $x = a$ and $x = b$ and zero probability of observing x outside of that region. We define the corresponding uniform probability distribution function as

$$p(x) = \begin{cases} 0 & x < a \\ C & x \in [a, b] \\ 0 & x > b \end{cases}$$

The probability of selecting a value of x in the region $[x, x + dx]$ at random is $p(x)\,dx$, where dx represents a small increment of x.[12] The probability of randomly choosing a value of x somewhere between $x = a$ and $x = b$ must be unity. To reflect this, we say that the probability distribution is a *normalized distribution*, meaning that

[12] Note that the increment dx has the units of x while the probability distribution function $p(x)$ has units of $1/x$. The product $p(x)dx$ is dimensionless.

$$\text{probability of finding } x \in [a, b] = \int_a^b p(x)\,dx = 1$$

The sum over all the probabilities of all possible outcomes must equal one.

This normalization constant can be used to determine the value of $\mathcal{C}$ as

$$\int_a^b p(x)\, dx = \int_a^b \mathcal{C}\, dx = \mathcal{C} \int_a^b dx = \mathcal{C}(b-a) = 1$$

From this result, we find $\mathcal{C} = 1/(b-a)$. We call $\mathcal{C}$ the *normalization constant*. The resulting normalized uniform probability distribution function is

$$p(x) = \begin{cases} 0 & x < a \\ \dfrac{1}{b-a} & x \in [a,b] \\ 0 & x > b \end{cases}$$

This function is depicted in Figure 9.7.

The general definition of the mean value or average of x is

$$\overline{x} = \int_a^b x p(x)\, dx$$

For the uniform distribution, we evaluate $\overline{x}$ as

$$\overline{x} = \frac{1}{(b-a)} \int_a^b x\, dx = \frac{1}{(b-a)} \left[\frac{1}{2} x^2 \right]_a^b$$
$$= \frac{1}{2} \frac{1}{(b-a)} (b^2 - a^2) = \frac{1}{2}(b+a)$$

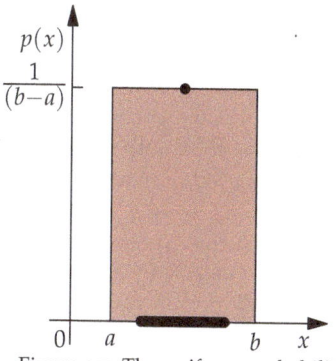

Figure 9.7: The uniform probability distribution function $p(x)$ as a function of x. The distribution function is normalized so that the total shaded area is unity. The black dot marks the mean $\overline{x}$ and the black bar marks the range $\overline{x} - \sigma_x$ to $\overline{x} + \sigma_x$.

We can reform this result as

$$\overline{x} = a + \frac{1}{2}(b-a)$$

This is a very intuitive result. The average value of x is the midpoint of the range $[a, b]$. For example, for the specific values $a = 0$ and $b = 1$ the mean, $\overline{x} = \frac{1}{2}$.

The average value of x^2 is defined as

$$\overline{x^2} = \int_a^b x^2 p(x)\, dx$$

which we evaluate for the uniform distribution as

$$\overline{x^2} = \frac{1}{(b-a)} \int_a^b x^2\, dx = \frac{1}{(b-a)} \left[\frac{1}{3} x^3 \right]_a^b = \frac{1}{3} \frac{1}{(b-a)} (b^3 - a^3)$$
$$= \frac{1}{3}(b^2 + ab + a^2)$$

For the specific values $a = 0$ and $b = 1$, we find $\overline{x^2} = \frac{1}{3}$.

With this result we can evaluate the variance of x as

$$\sigma_x^2 = \overline{(x - \overline{x})^2} = \overline{x^2} - \overline{x}^2 = \frac{1}{3}(b^2 + ab + a^2) - \left(\frac{1}{2}(b+a) \right)^2$$
$$= \frac{1}{12}(b-a)^2$$

For the specific values $a = 0$ and $b = 1$ we find $\sigma_x^2 = \overline{x^2} - \overline{x}^2 = \frac{1}{12}$. Finally, the ratio of the root-mean square value of $(x - \bar{x})$ and the mean $\bar{x}$ is

$$\frac{\sigma_x}{\bar{x}} = \frac{\sqrt{\overline{x^2} - \bar{x}^2}}{\bar{x}} = \frac{1}{\sqrt{3}}\left(\frac{b-a}{b+a}\right)$$

For the specific values $a = 0$ and $b = 1$ we find the ratio $\frac{1}{\sqrt{3}}$.

9.2.2 *Properties of exponential probability distributions*

Many measured observables in the physical sciences are well described by an exponential probability distribution function. Examples include the Boltzmann probability distribution, which is an exponential function of the energy used in modeling chemical equilibria and the rates of chemical reactions.

Consider the exponential distribution function $p(x)$, characterized by a maximum value at $x = 0$ and an exponential decrease with increasing x:

$$p(x) = C \, \exp\left(-\gamma x\right)$$

where $\gamma > 0$. The probability distribution function is normalized, so that

$$\int_0^\infty p(x) \, dx = 1$$

which means that

$$C \int_0^\infty \exp\left(-\gamma x\right) dx = C\left[-\frac{1}{\gamma}e^{-\gamma x}\right]_0^\infty = \frac{1}{\gamma}C = 1$$

This results in a normalization constant

$$C = \gamma$$

With this result, we find the normalized exponential probability distribution function to be

$$p(x) \, dx = \gamma \, \exp\left(-\gamma x\right) \, dx$$

which is shown in Figure 9.8. We can define the probability of randomly selecting a value of x between $x = a$ and $x = b$:

$$\text{probability of finding } x \in [a, b] = \int_a^b p(x) \, dx$$

This is depicted as the shaded area in Figure 9.8.

The general definition of the mean value or average of x is

$$\bar{x} = \int_0^\infty x p(x) \, dx$$

For the exponential probability distribution we find

$$\bar{x} = \gamma \int_0^\infty x \exp\left(-\gamma x\right) dx$$

We can evaluate this integral using either integration by parts or our trick of exploiting differentiation with respect to the constant γ. In either case, we

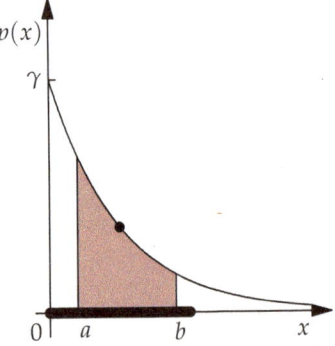

Figure 9.8: The exponential probability distribution function $p(x)$ as a function of x. The distribution function is normalized so that the total area under the curve is unity. The probability of randomly choosing a value of $x \in [a, b]$ is equal to the shaded area.

find the mean:

$$\overline{x} = \gamma \int_0^\infty x \exp\left(-\gamma x\right) dx = \gamma \frac{1}{\gamma^2} = \frac{1}{\gamma}$$

The average value of x^2 is defined as

$$\overline{x^2} = \int_0^\infty x^2 p(x)\, dx$$

For the exponential probability distribution function we find

$$\overline{x^2} = \gamma \int_0^\infty x^2 \exp\left(-\gamma x\right) dx$$

We can evaluate this integral using either integration by parts or our trick of exploiting differentiation with respect to the constant γ, discussed in Chapter 7 in the complements. This leads to

$$\overline{x^2} = \gamma \int_0^\infty x^2 \exp\left(-\gamma x\right) dx = \gamma \frac{2}{\gamma^3} = \frac{2}{\gamma^2}$$

With this result, we can evaluate the variance

$$\sigma_x^2 = \overline{(x - \overline{x})^2} = \overline{x^2} - \overline{x}^2 = \frac{2}{\gamma^2} - \left(\frac{1}{\gamma}\right)^2 = \frac{1}{\gamma^2}$$

The ratio of the root-mean square value of x to the mean of x is

$$\frac{\sigma_x}{\overline{x}} = \frac{\sqrt{\overline{x^2} - \overline{x}^2}}{\overline{x}} = \frac{\sqrt{1/\gamma^2}}{1/\gamma} = 1$$

The exponential probability distribution has the special property that the root-mean square width of the distribution is equal to the mean.

9.2.3 Applications of the exponential probability distribution

Consider the case of a sample of N_0 *radioactive atomic nuclei*. Observations of the system demonstrate that the rate of radioactive decay of the nuclei varies with time[13] according to the relation

$$N(t)\, dt = \frac{N_0}{\tau} \exp(-t/\tau)\, dt$$

where $N(t = 0) = N_0/\tau$ is the initial rate of decay at $t = 0$ and $N(t)dt$ is the number of nuclei that decay between times t and $t + dt$. Over all time, the number of nuclei that will decay is

$$\int_0^\infty N(t)\, dt = \frac{N_0}{\tau} \int_0^\infty \exp(-t/\tau)\, dt = N_0$$

which is the total number of nuclei in the sample. The form of $N(t)$ is shown in Figure 9.9.

Suppose we would like to know how many nuclei decay between a time t_a and t_b. We can calculate this as

$$\int_{t_a}^{t_b} N(t)\, dt = \frac{N_0}{\tau} \int_{t_a}^{t_b} \exp(-t/\tau)\, dt = N_0 \left[e^{-t_a/\tau} - e^{-t_b/\tau} \right]$$

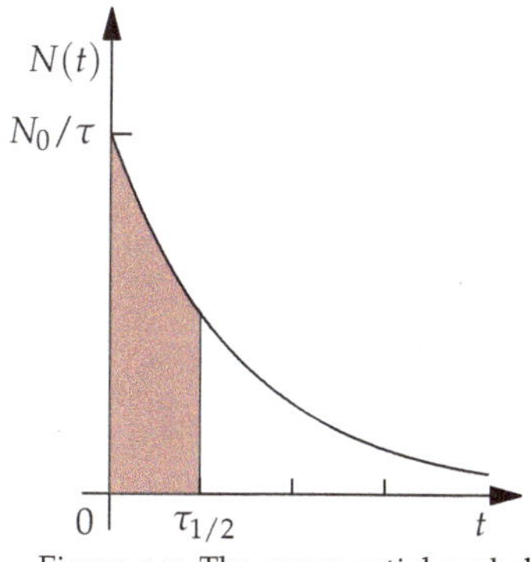

[13] Note that $N(t)\, dt$ has units of number while $N(t)$ has units of number per unit time.

Figure 9.9: The exponential probability distribution function

$$N(t) = \frac{N_0}{\tau} \exp(-t/\tau)$$

as a function of time t. The shaded area represents half of the total area under the curve reached at the half-life time $t = \ln(2)\tau$.

If we set $t_a = 0$ and $t_b = t$, we find

$$\text{number of nuclei decayed} = N_0 \left[1 - e^{-t/\tau} \right]$$

It follows that the *survival probability*, which is the probability that a nucleus has not decayed after time t, is

$$\text{survival probability} = \frac{N_0 - N(t)}{N_0} = e^{-t/\tau}$$

For example, after a time $t = \ln(2)\tau = 0.693\tau$, the survival probability is $\frac{1}{2}$. We call this time the *half-life* of the exponential distribution:

$$\tau_{1/2} = \ln(2)\tau$$

After a time $\tau_{1/2}$, one-half of the nuclei have decayed and one-half have *survived*.

Let's consider another application involving an exponential distribution of energy:[14]

$$p(E) = \mathcal{C} \exp(-E/k_B T)$$

[14] Note that $p(E)\, dE$ has units of number while $p(E)$ has units of number per unit energy.

To normalize this probability distribution, we demand that

$$\int_0^\infty p(E)\, dE = \mathcal{C} \int_0^\infty \exp(-E/k_B T)\, dE = \mathcal{C} k_B T = 1$$

so that $\mathcal{C} = 1/k_B T$, and

$$p(E)\, dE = \frac{1}{k_B T} \exp(-E/k_B T)\, dE$$

What is the probabilty of having energy less than or equal to the thermal energy $k_B T$? We can evaluate this as

$$\frac{1}{k_B T} \int_0^{k_B T} \exp(-E/k_B T)\, dE = -\frac{1}{k_B T} k_B T \left[\exp(-E/k_B T) \right]_0^{k_B T}$$
$$= 1 - 1/e = 0.632$$

We can further ask for the fraction having energy less than $2k_B T$. That turns out to be

$$\frac{1}{k_B T} \int_0^{2k_B T} \exp(-E/k_B T)\, dE = 1 - \frac{1}{e^2} = 0.865$$

or for the fraction having energy less than $3k_B T$, which is $1 - \frac{1}{e^3} = 0.950$. This result is shown graphically in Figure 9.10.

9.2.4 *Properties of gaussian probability distributions*

In the physical sciences, we commonly employ the gaussian function to represent a probability distribution function that is centered at $\overline{x} = x_0$ with mean square width σ^2, such as

$$p(x) = \mathcal{C} \exp\left[-\frac{(x - x_0)^2}{2\sigma^2} \right]$$

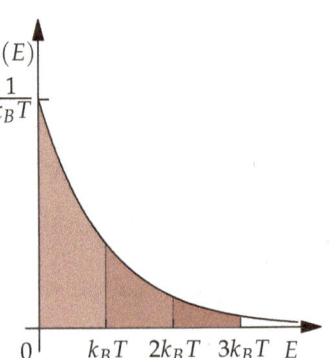

Figure 9.10: The exponential probability distribution function

$$p(E) = \frac{1}{k_B T} \exp(-E/k_B T)$$

as a function of energy E. The shaded areas represent the fraction of atoms having energy E less than $k_B T$, $2k_B T$, or $3k_B T$.

where the distribution function is normalized as

$$\int_{-\infty}^{\infty} p(x)\, dx = 1$$

Imposing the normalization condition, we find

$$\int_{-\infty}^{\infty} p(x)\, dx = C \int_{-\infty}^{\infty} \exp\left[-\frac{(x-x_0)^2}{2\sigma^2}\right] dx = 1$$

To perform this integral, we want to center the gaussian at the origin. We do this with the algebraic substitution $y = (x - x_0)$, so that $dy = dx$. We find

$$C \int_{-\infty}^{\infty} \exp\left[-\frac{(x-x_0)^2}{2\sigma^2}\right] dx = C \int_{-\infty}^{\infty} \exp\left(-\frac{y^2}{2\sigma^2}\right) dy$$

Using the identity

$$\int_{-\infty}^{\infty} e^{-\alpha y^2}\, dy = \sqrt{\frac{\pi}{\alpha}}$$

we find

$$C \int_{-\infty}^{\infty} \exp\left(-\frac{y^2}{2\sigma^2}\right) dy = C\,\sqrt{2\pi\sigma^2} = 1$$

so that the normalization constant $C = 1/\sqrt{2\pi\sigma^2}$. The resulting normalized gaussian probability distribution function is

$$p(x)\, dx = \frac{1}{\sqrt{2\pi\sigma^2}} \exp\left[-\frac{(x-x_0)^2}{2\sigma^2}\right] dx$$

This result is shown in Figure 9.11.

Note that the probability of randomly selecting a value of x between $x_0 - \sigma$ and $x_0 + \sigma$ is

$$\text{probability of finding } x \in [x_0 - \sigma, x_0 + \sigma] = \int_{x_0-\sigma}^{x_0+\sigma} p(x)\, dx \approx 0.6827$$

The total probability of observing x within one standard deviation σ from the mean x_0 is depicted as the shaded area in Figure 9.11.

The probability of randomly selecting a value of x within two standard deviations of the mean, $x \in [x_0 - 2\sigma, x_0 + 2\sigma]$, is roughly 0.9545, and within three standard deviations of the mean, $x \in [x_0 - 3\sigma, x_0 + 3\sigma]$, is approximately 0.9973. This is the origin of the *68-95-99.7 rule* that is often evoked in discussions of the probability and statistics of the gaussian distribution.

The mean of x is defined as

$$\overline{x} = \int_{-\infty}^{\infty} x p(x)\, dx$$

For the gaussian distribution centered at $x = x_0$ with variance σ^2, we find

$$\overline{x} = \frac{1}{\sqrt{2\pi\sigma^2}} \int_{-\infty}^{\infty} x \exp\left[-\frac{(x-x_0)^2}{2\sigma^2}\right] dx$$

With the algebraic substitution $y = (x - x_0)$ so that $x = y + x_0$ and $dy = dx$,

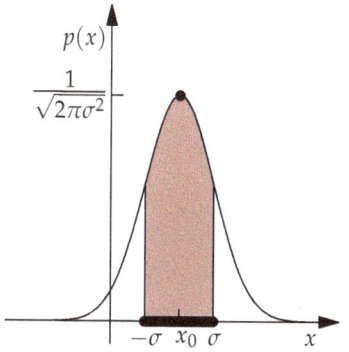

Figure 9.11: The normalized gaussian probability distribution function $p(x)$ as a function of x. The probability of randomly choosing a value of $x \in [-\sigma, \sigma]$ is equal to the shaded area.

we find

$$\overline{x} = \frac{1}{\sqrt{2\pi\sigma^2}} \int_{-\infty}^{\infty} (y + x_0) \exp\left(-\frac{y^2}{2\sigma^2}\right) dy$$

$$= \frac{1}{\sqrt{2\pi\sigma^2}} \left[\int_{-\infty}^{\infty} y \exp\left(-\frac{y^2}{2\sigma^2}\right) dy + x_0 \int_{-\infty}^{\infty} \exp\left(-\frac{y^2}{2\sigma^2}\right) dy \right]$$

$$= x_0$$

The first integral is zero

$$\int_{-\infty}^{\infty} y \exp(-ay^2)\, dy = 0$$

as the integrand $y \exp(-ay^2)$ is an odd function of y. As such, the integral from $-\infty$ to 0 is equal in magnitude and opposite in sign to the integral from 0 to ∞.

The average value of x^2 is defined as

$$\overline{x^2} = \int_{-\infty}^{\infty} x^2 p(x)\, dx$$

For the gaussian distribution centered at $x = x_0$ with variance σ^2, we find

$$\overline{x^2} = \frac{1}{\sqrt{2\pi\sigma^2}} \int_{-\infty}^{\infty} x^2 \exp\left[-\frac{(x - x_0)^2}{2\sigma^2}\right] dx$$

With the algebraic substitution $y = (x - x_0)$ so that $x^2 = y^2 + 2yx_0 + x_0^2$ and $dy = dx$, we find

$$\overline{x^2} = \frac{1}{\sqrt{2\pi\sigma^2}} \int_{-\infty}^{\infty} (y^2 + 2yx_0 + x_0^2) \exp\left(-\frac{y^2}{2\sigma^2}\right) dy$$

The first integral is

$$\frac{1}{\sqrt{2\pi\sigma^2}} \int_{-\infty}^{\infty} y^2 \exp\left(-\frac{y^2}{2\sigma^2}\right) dy = \sigma^2$$

The second integral is zero by symmetry since the integrand is an odd function of y. The third integral is equal to x_0^2. This leads to the final result

$$\overline{x^2} = x_0^2 + \sigma^2$$

Using this result we can determine the average of the square of the deviation in x from the mean value $\overline{x} = x_0$ as

$$\overline{(x - x_0)^2} = \int_{-\infty}^{\infty} (x - x_0)^2 p(x)\, dx$$

$$= \overline{x^2} - 2x_0\overline{x} + x_0^2$$

$$= \overline{x^2} - x_0^2 = \sigma^2$$

This demonstrates that σ^2 is the variance of the gaussian distribution. In the next section, we explore a variety of applications of the gaussian distribution in the physical sciences.

9.2.5 Applications of the gaussian probability distribution

Consider the kinetic energy of an ideal gas of atoms. In one dimension, the kinetic energy of an atom is

$$E = \frac{1}{2}mv^2$$

where m is the mass of the atom and v is the velocity. The distribution of kinetic energy is given by the Boltzmann exponential probability distribution

$$p(E)\, dE = \frac{1}{k_B T} \exp\left(-\frac{E}{k_B T}\right)\, dE$$

Since the energy varies in proportion to the square of the velocity, the distribution of velocities is the gaussian probability distribution function

$$p(v)\, dv = \sqrt{\frac{m}{2\pi k_B T}}\, \exp\left(-\frac{mv^2}{2k_B T}\right)\, dv$$

This distribution is shown graphically in Figure 9.12. The mean velocity is $\bar{v} = 0$, and the variance in the distribution of velocities is $\sigma_v^2 = \frac{k_B T}{m}$.

Suppose we want to know the fraction of atoms with kinetic energy less than or equal to $\frac{k_B T}{2}$, so that $\frac{1}{2}mv^2 \leq \frac{k_B T}{2}$. The velocities of atoms in this fraction are found in the range $v \in [-\sigma_v, \sigma_v]$ where $\sigma_v = \sqrt{\frac{k_B T}{m}}$. That magnitude of that fraction can be quantified using the integral

$$\sqrt{\frac{1}{2\pi\sigma_v^2}} \int_{-\sigma_v}^{\sigma_v} \exp\left(-\frac{v^2}{2\sigma_v^2}\right)\, dv = \mathrm{erf}\left(\frac{1}{\sqrt{2}}\right) = 0.683$$

where $\mathrm{erf}(x)$ is the *error function* introduced in Chapter 7 through Equation 7.18 and Figure 7.6.[15] It follows that the fraction of atoms having kinetic energy greater than $\frac{k_B T}{2}$ is $1 - \mathrm{erf}\left(\frac{1}{\sqrt{2}}\right) = \mathrm{erfc}\left(\frac{1}{\sqrt{2}}\right)$, where $\mathrm{erf}(x)$ is the *error function* and $\mathrm{erfc}(x)$ is the *complementary error function*.

9.2.6 Probability distributions in many dimensions

The concept of the probability distribution can be extended to many dimensions. Consider two random variables, x and y, where the probability of measuring a value of $x \in [x, x + dx]$ and $y \in [y, y + dy]$ is written

$$p(x, y)\, dx dy$$

where $p(x, y)$ is a two-dimensional probability distribution function. An example is shown graphically in Figure 9.13.

The integral over all x and y is the total probability of observing an outcome (x, y) somewhere on the xy-plane, which must equal unity:

$$\int_{-\infty}^{\infty} \int_{-\infty}^{\infty} p(x, y)\, dx dy = 1$$

This is our normalization condition. The probability of measuring $x \in [x, x + dx]$ and $y \in [y, y + dy]$ is the volume of the column over the area $dx\, dy$ under the surface defined by $p(x, y)$.

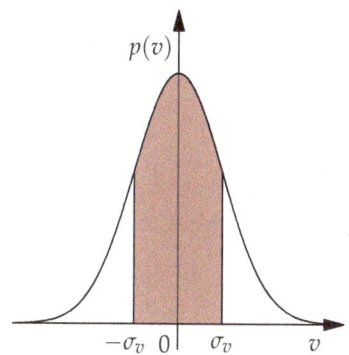

$p(v)$

$-\sigma_v \quad 0 \quad \sigma_v \qquad v$

Figure 9.12: The integral over $p(v)$ between $-\sigma_v$ and σ_v is the fraction of atoms with kinetic energy $E \leq \frac{k_B T}{2}$, shown as the shaded area under the curve.

[15] The error function is defined as

$$\mathrm{erf}(x) = \frac{1}{\sqrt{\pi}} \int_{-x}^{x} e^{-t^2}\, dt$$

and is positive for positive values of x. In this case, $t^2 = \frac{v^2}{2\sigma_v^2}$. Since $v = \sigma_v$ corresponds to $t^2 = \frac{1}{2}$, the limit of integration $x = \frac{1}{\sqrt{2}}$.

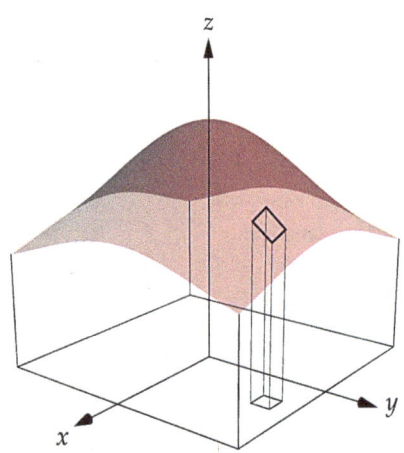

Figure 9.13: A two-dimensional probability distribution $p(x,y)$ shown as a surface over the xy-plane. The probability of observing a value of $x \in [x, x + dx]$ and $y \in [y, y + dy]$ is given by the volume of the vertical column over the area $dx \, dy$ and beneath the surface defined by $p(x,y)$.

We can further define the mean values of the x and y as

$$\overline{x} = \int_{-\infty}^{\infty} \int_{-\infty}^{\infty} x \, p(x,y) \, dxdy \qquad \overline{y} = \int_{=\infty}^{\infty} \int_{-\infty}^{\infty} y \, p(x,y) \, dxdy$$

In defining the variances of this two-dimensional probability distribution function, we encounter a new concept. We define the variances σ_x^2 and σ_y^2 as

$$\sigma_x^2 = \overline{x^2} - \overline{x}^2 \qquad \sigma_y^2 = \overline{y^2} - \overline{y}^2$$

where the mean square values of x and y are defined as

$$\overline{x^2} = \int_{-\infty}^{\infty} \int_{-\infty}^{\infty} x^2 \, p(x,y) \, dxdy \qquad \overline{y^2} = \int_{-\infty}^{\infty} \int_{-\infty}^{\infty} y^2 \, p(x,y) \, dxdy$$

We can also define a measure of the *cross correlation* between a deviation from the mean in x, $(x - \overline{x})$, and a deviation from the mean in y, $(y - \overline{y})$, as

$$\sigma_{xy} = \int_{-\infty}^{\infty} \int_{-\infty}^{\infty} (x - \overline{x})(y - \overline{y}) \, p(x,y) \, dxdy$$

This is known as the *covariance*.

The value of σ_{xy} provides a measure of the correlation between observations of the two variables. We can define a useful measure of the magnitude of the correlation as $\sigma_{xy} = \rho_{xy} \sigma_x \sigma_y$ where ρ_{xy} is the *correlation coefficient*. When measurements of x and y are uncorrelated, the probability distribution $p(x,y)$ is separable and can be written as a product of the probability of observing a value of x times the independent probability of observing a value of y:

$$p(x,y) \, dxdy = p_x(x) \, dx \times p_y(y) \, dy$$

In that case, we find

$$\sigma_{xy} = \int_{-\infty}^{\infty} \int_{-\infty}^{\infty} (x - \overline{x})(y - \overline{y}) \, p(x,y) \, dxdy$$

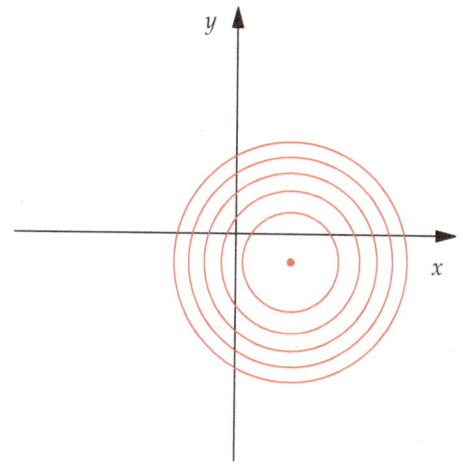

Figure 9.14: A two-dimensional probability distribution $p(x, y)$ over the xy-plane with correlation coefficient $\rho_{xy} = 0$. Contours define constant values of the probability and the dot marks the mean value of the distribution.

which can be evaluated as

$$\sigma_{xy} = \int_{-\infty}^{\infty} (x - \overline{x}) p_x(x) \, dx \times \int_{-\infty}^{\infty} (y - \overline{y}) \, p_y(y) \, dy = 0$$

and the correlation coefficient $\rho_{xy} = 0$.

An example of a distribution with $\rho_{xy} = 0$ is shown in Figure 9.14. In this case, we say that the observations of x and y are uncorrelated. In general, the correlation coefficient will vary over the range: $-1 \leq \rho_{xy} \leq 1$ where positive values of ρ_{xy} indicate positive correlation and negative values of ρ_{xy} indicate negative correlation. When $\rho_{xy} = 0$, the variables x and y are independent. Examples of distributions with positive and negative values of ρ_{xy} are provided in Figure 9.15.

Consider the case of two fair coins. We expect the outcome of the flip of the first coin to be independent of the outcome of a flip of the second coin. In that case, the value of $\rho_{xy} = 0$ and the observations are uncorrelated. Now consider the case of an ideal gas for which the pressure can be written

$$p = \frac{nRT}{V}$$

Suppose the gas is at constant volume. The average pressure is measured to be $\overline{p}$ and the average temperature $\overline{T}$. If the pressure is higher than its mean value, so that $(p - \overline{p}) > 0$, we expect the temperature to be higher than its mean value, so that $(T - \overline{T}) > 0$. In this case we expect $\sigma_{pT} > 0$ and $\rho_{pT} > 0$. In such a case, we say that the pressure and temperature are positively correlated.

Now suppose the gas is at constant temperature. The average pressure is measured to be $\overline{p}$ and the average volume $\overline{V}$. If the pressure is higher than its mean value, then $(p - \overline{p}) > 0$. Given the inverse dependence of volume on pressure, if the pressure is greater than its mean value, the volume will be lower than its mean value and $(V - \overline{V}) < 0$. In this case we expect $\sigma_{pV} < 0$ and $\rho_{pV} < 0$. In such a case, we say that the pressure and volume are negatively correlated.

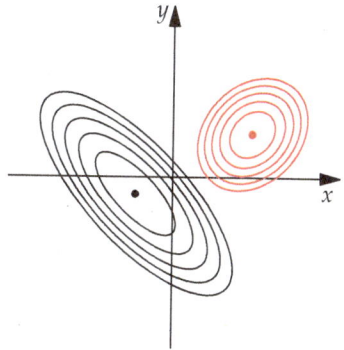

Figure 9.15: A two-dimensional probability distribution $p(x, y)$ over the xy-plane with correlation coefficient $\rho_{xy} = 1$ (red) and $\rho_{xy} = -1$ (black). Contours define constant values of the probability and dots mark the mean values of the distributions.

9.3 Probability distributions in the physical sciences

THE MOST COMMONLY ENCOUNTERED MULTIPLE INTEGRALS in the physical sciences involve integration over multivariate probability distributions. Examples include the Maxwell-Boltzmann probability distribution in kinetic theory, the Boltzmann probability distribution in thermodynamics, and square of the modulus of the wave function in quantum theory. The functions are typically complicated by the presence of many physical constants. However, careful simplification of the expressions by algebraic substitution followed by the right choice of coordinate system can convert an intimidating multiple integral into a readily evaluated expression. This section will explore some of the more common functional forms encountered in the physical sciences.

9.3.1 Properties of the Maxwell-Boltzmann probability distribution

Probability distributions are common in the physical sciences. We are often challenged to describe the properties of a very large number of particles, so large that it is hopeless to track the energy or velocity of each particle. Instead, we hope to describe the average properties of the system using statistical assumptions and statistical probability distributions.

The probability of observing a state of given energy can be written in terms of the Boltzmann exponential probability distribution

$$\exp\left(-\frac{E}{k_B T}\right)$$

Suppose E is the kinetic energy of a particle with velocity $\mathbf{v} = u_x \hat{\mathbf{x}} + u_y \hat{\mathbf{y}} + u_z \hat{\mathbf{z}}$. We can express the kinetic energy in vector notation as

$$E = \frac{1}{2} m \, \mathbf{v} \cdot \mathbf{v} = \frac{1}{2} m (u_x^2 + u_y^2 + u_z^2)$$

Inserting this energy into the Boltzmann probability distribution leads to the Maxwell-Boltzmann distribution

$$p(u_x, u_y, u_z) = C \exp\left[-\frac{m}{2k_B T}(u_x^2 + u_y^2 + u_z^2)\right]$$

If we demand that the distribution is normalized such that

$$\int_{-\infty}^{\infty}\int_{-\infty}^{\infty}\int_{-\infty}^{\infty} p(u_x, u_y, u_z)\,du_x du_y du_z = 1$$

then

$$\int_{-\infty}^{\infty}\int_{-\infty}^{\infty}\int_{-\infty}^{\infty} C \exp\left[-\frac{m}{2k_B T}(u_x^2 + u_y^2 + u_z^2)\right]\,du_x du_y du_z = 1$$

We note that this triple integral is separable and can be represented as a product of three one-dimensional integrals[16]

$$C \int_{-\infty}^{\infty} e^{-\frac{m}{2k_B T}u_x^2}\,du_x \times \int_{-\infty}^{\infty} e^{-\frac{m}{2k_B T}u_y^2}\,du_y \times \int_{-\infty}^{\infty} e^{-\frac{m}{2k_B T}u_z^2}\,du_z$$

$$= C\sqrt{\frac{2\pi k_B T}{m}} \times \sqrt{\frac{2\pi k_B T}{m}} \times \sqrt{\frac{2\pi k_B T}{m}} = C\left(\frac{2\pi k_B T}{m}\right)^{\frac{3}{2}} = 1$$

[16] We use the identity

$$\int_{\infty}^{\infty} e^{-\alpha x^2}\,dx = \sqrt{\frac{\pi}{\alpha}}$$

where $x = u_x$ and $\alpha = \frac{m}{2k_B T}$.

As a result, the normalization constant $\mathcal{C} = \left(\frac{m}{2\pi k_B T}\right)^{\frac{3}{2}}$. The final result for the statistical distribution of velocities is

$$p(u_x, u_y, u_z) = \left(\frac{m}{2\pi k_B T}\right)^{\frac{3}{2}} \exp\left[-\frac{m}{2k_B T}(u_x^2 + u_y^2 + u_z^2)\right]$$

Note that

$$p(u_x, u_y, u_z)\, du_x du_y du_z$$

expresses the probability of finding a particle with velocity components in the range

$$[u_x, u_x + du_x] \times [u_y, u_y + du_y] \times [u_z, u_z + du_z]$$

which is a small increment of volume $du_x du_y du_z$ in the space of all possible velocities.

9.3.2 Mean and variance of the Maxwell-Boltzmann distribution

We can compute the mean and variance of the Maxwell-Boltzmann distribution using a similar approach. Consider first the mean of the velocity defined as

$$\overline{\mathbf{v}} = \int_{-\infty}^{\infty} \int_{-\infty}^{\infty} \int_{-\infty}^{\infty} \mathbf{v}\, p(u_x, u_y, u_z)\, du_x du_y du_z$$

$$= \int_{-\infty}^{\infty} \int_{-\infty}^{\infty} \int_{-\infty}^{\infty} (u_x \hat{\mathbf{x}} + u_y \hat{\mathbf{y}} + u_z \hat{\mathbf{z}})\, p(u_x, u_y, u_z)\, du_x du_y du_z$$

where integration is carried out in cartesian coordinates, defined in Figure 9.16. This integral is a sum of three triple integrals, each corresponding to the average of the velocity in one of the three principal directions x, y, and z depicted in Figure 9.17. As such, we express the average velocity as

$$\overline{\mathbf{v}} = \overline{u_x}\, \hat{\mathbf{x}} + \overline{u_y}\, \hat{\mathbf{y}} + \overline{u_z}\, \hat{\mathbf{z}}$$

where the averages are defined

$$\overline{u_x} = \int_{-\infty}^{\infty} \int_{-\infty}^{\infty} \int_{-\infty}^{\infty} u_x\, p(u_x, u_y, u_z)\, du_x du_y du_z$$

$$\overline{u_y} = \int_{-\infty}^{\infty} \int_{-\infty}^{\infty} \int_{-\infty}^{\infty} u_y\, p(u_x, u_y, u_z)\, du_x du_y du_z$$

$$\overline{u_z} = \int_{-\infty}^{\infty} \int_{-\infty}^{\infty} \int_{-\infty}^{\infty} u_z\, p(u_x, u_y, u_z)\, du_x du_y du_z$$

Each of the averages is a three-dimensional integral that we recognize is separable and can be evaluated as a product of three one-dimensional integrals.

The average $\overline{u_x}$ can be expressed as

$$\overline{u_x} = \int_{-\infty}^{\infty} \int_{-\infty}^{\infty} \int_{-\infty}^{\infty} u_x\, p(u_x, u_y, u_z)\, du_x du_y du_z$$

$$= \left(\frac{m}{2\pi k_B T}\right)^{\frac{3}{2}} \int_{-\infty}^{\infty} u_x e^{-\frac{mu_x^2}{2k_B T}}\, du_x \times \int_{-\infty}^{\infty} e^{-\frac{mu_y^2}{2k_B T}}\, du_y \times \int_{-\infty}^{\infty} e^{-\frac{mu_z^2}{2k_B T}}\, du_z$$

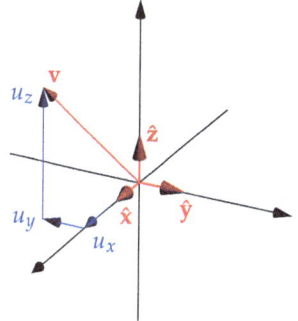

Figure 9.16: The decomposition of the velocity vector $\mathbf{v} = u_x\, \hat{\mathbf{x}} + u_y\, \hat{\mathbf{y}} + u_z\, \hat{\mathbf{z}}$ in terms of its cartesian components.

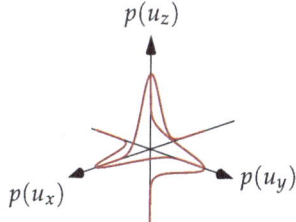

Figure 9.17: Projections of the three one-dimensional Maxwell-Boltzmann velocity probability distributions $p(u_x)$, $p(u_y)$, and $p(u_z)$. The product of these three one-dimensional distributions forms the separable three-dimensional velocity probability distribution $p(u_x, u_y, u_z)$.

Evaluating the integrals, we find

$$\overline{u_x} = \left(\frac{m}{2\pi k_B T}\right)^{\frac{3}{2}} \int_{-\infty}^{\infty} u_x e^{-\frac{m}{2k_B T}u_x^2} \, du_x \times \sqrt{\frac{2\pi k_B T}{m}} \times \sqrt{\frac{2\pi k_B T}{m}}$$

$$= \left(\frac{m}{2\pi k_B T}\right)^{\frac{1}{2}} \int_{-\infty}^{\infty} u_x e^{-\frac{m}{2k_B T}u_x^2} \, du_x$$

where we have evaluated the integrals over u_y and u_z as we did in normalizing the distribution. We are left with an integral over u_x in which the integrand is an odd function of u_x. As such, we find

$$\overline{u_x} = \int_{-\infty}^{\infty} u_x e^{-\frac{m}{2k_B T}u_x^2} \, du_x = 0$$

Repeating the process for the averages of u_y and u_z leads to $\overline{u_y} = \overline{u_z} = 0$. The final result is that the mean velocity $\overline{\mathbf{v}} = 0$.[17]

What about the variance of the velocity? We can define the variance in terms of the mean and mean square values of $\mathbf{v}$ as

$$\sigma_v^2 = \overline{|\mathbf{v}|^2} - |\overline{\mathbf{v}}|^2 = \left(\overline{u_x^2} + \overline{u_y^2} + \overline{u_z^2}\right) - \left(\overline{u_x}^2 + \overline{u_y}^2 + \overline{u_z}^2\right)$$

Grouping terms we find

$$\sigma_v^2 = \left(\overline{u_x^2} - \overline{u_x}^2\right) + \left(\overline{u_y^2} - \overline{u_y}^2\right) + \left(\overline{u_z^2} - \overline{u_z}^2\right)$$

$$= \sigma_{u_x}^2 + \sigma_{u_y}^2 + \sigma_{u_z}^2$$

The total variance in the distribution of velocities is the sum of the variances in the components of the velocities u_x, u_y, and u_z. Therefore, to determine σ_v^2 we should determine $\sigma_{u_x}^2$, $\sigma_{u_y}^2$, and $\sigma_{u_z}^2$. Noting that $\sigma_{u_x}^2 = \overline{u_x^2} - \overline{u_x}^2 = \overline{u_x^2}$, as we determined that $\overline{u_x} = 0$, we find

$$\overline{u_x^2} = \int_{-\infty}^{\infty}\int_{-\infty}^{\infty}\int_{-\infty}^{\infty} u_x^2 \, p(u_x, u_y, u_z) \, du_x du_y du_z$$

$$= \left(\frac{m}{2\pi k_B T}\right)^{\frac{3}{2}} \int_{-\infty}^{\infty} u_x^2 \, e^{-\frac{mu_x^2}{2k_B T}} du_x \times \int_{-\infty}^{\infty} e^{-\frac{mu_y^2}{2k_B T}} du_y \times \int_{-\infty}^{\infty} e^{-\frac{mu_z^2}{2k_B T}} du_z$$

$$= \left(\frac{m}{2\pi k_B T}\right)^{\frac{3}{2}} \int_{-\infty}^{\infty} u_x^2 \, e^{-\frac{m}{2k_B T}u_x^2} du_x \times \sqrt{\frac{2\pi k_B T}{m}} \times \sqrt{\frac{2\pi k_B T}{m}}$$

$$= \left(\frac{m}{2\pi k_B T}\right)^{\frac{1}{2}} \int_{-\infty}^{\infty} u_x^2 \, e^{-\frac{m}{2k_B T}u_x^2} \, du_x$$

The remaining integral can be performed by integration by parts or by using our differentiation with respect to a parameter trick explored in the complements to Chapter 7. The result is[18]

$$\overline{u_x^2} = \frac{k_B T}{m}\sqrt{\frac{2\pi k_B T}{m}}\sqrt{\frac{m}{2\pi k_B T}} = \frac{k_B T}{m}$$

By symmetry, we recognize the result will be the same for $\overline{u_y^2}$ and $\overline{u_z^2}$. With

[17] This result makes physical sense. It is equally likely that the velocity will be oriented in the positive or negative direction in a given cartesian dimension. This makes the average velocity in each cartesian dimension zero.

[18] We use the identity

$$\int_{-\infty}^{\infty} x^2 \, e^{-\alpha x^2} \, dx = -\frac{d}{d\alpha}\int_{-\infty}^{\infty} e^{-\alpha x^2} \, dx$$

$$= -\frac{d}{d\alpha}\sqrt{\frac{\pi}{\alpha}} = \sqrt{\frac{\pi}{4\alpha^3}}$$

where $x = u_x$ and $\alpha = m/2k_B T$.

this result, we can write

$$\sigma_v^2 = \sigma_{u_x}^2 + \sigma_{u_y}^2 + \sigma_{u_z}^2 = \frac{k_B T}{m} + \frac{k_B T}{m} + \frac{k_B T}{m} = \frac{3k_B T}{m}$$

for the variance in the velocity in three dimensions. These examples demonstrate the challenge of multiple integration over complicated functions resulting from models in the physical sciences. It is not that the integrals require great invention. The integral identities used in the derivation are familiar. The principal challenge comes from managing the complicated algebraic expressions without error.

9.3.3 Properties of the electron density probability distribution

The state of an electron in a one-electron atom is described in terms of a *wave function*. The square of the modulus of that function describes the electron's spatial distribution (or shape) in three-dimensional space.

Consider the form of the wave function for four possible states of an electron in a one-electron atom. These functions are most naturally expressed in spherical polar coordinates. For the electron in the lowest energy state, referred to as the 1s *ground state*, the wave function is

$$\psi_{1s}(r, \theta, \varphi) = N_1 \, e^{-r/a_0}$$

where N_1 is a normalization constant, r is the distance of the electron from the nucleus, and a_0 is known as the *Bohr radius*.

The wave functions for the electron in the four excited states are

$$\psi_{2s}(r, \theta, \varphi) = N_2 \left(2 - \frac{r}{a_0}\right) e^{-r/2a_0}$$

$$\psi_{2p_x}(r, \theta, \varphi) = N_2 \left(\frac{r}{a_0}\right) e^{-r/2a_0} \sin\theta \cos\varphi$$

$$\psi_{2p_y}(r, \theta, \varphi) = N_2 \left(\frac{r}{a_0}\right) e^{-r/2a_0} \sin\theta \sin\varphi$$

$$\psi_{2p_z}(r, \theta, \varphi) = N_2 \left(\frac{r}{a_0}\right) e^{-r/2a_0} \cos\theta$$

where N_2 is a normalization constant. Figure 9.18 depicts the radial dependence of the $\psi_{1s}(r, \theta, \varphi)$ and $\psi_{2s}(r, \theta, \varphi)$ wave functions.

We can determine the normalization constants N_1 and N_2 by taking the square of the wave function and integrating the resulting electron distribution over all space. Consider the ψ_{1s} wave function where the square of the wave function is

$$|\psi_{1s}(r, \theta, \varphi)|^2 = \left(N_1 \, e^{-r/a_0}\right)^2 = N_1^2 \, e^{-2r/a_0}$$

The normalization constant can be determined from the definition

$$\int_0^\infty \int_0^\pi \int_0^{2\pi} |\psi_{1s}(r, \theta, \varphi)|^2 \ r^2 \sin\theta \, d\varphi d\theta dr = 1$$

where we chose to work in spherical polar coordinates reflecting the fact that

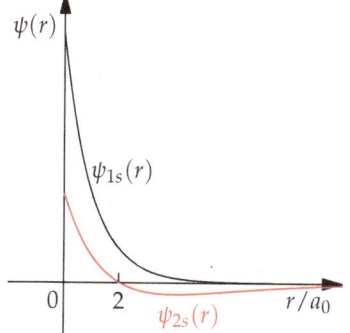

Figure 9.18: The wave functions $\psi_{1s}(r)$ (black line) and $\psi_{2s}(r)$ (red line) as a function of the distance expressed in reduced units r/a_0.

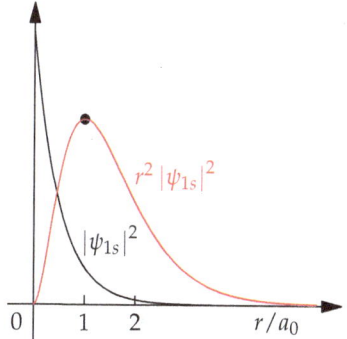

Figure 9.19: The functions $|\psi_{1s}(r)|^2$ (black line) and $r^2|\psi_{1s}(r)|^2$ (red line) over $r \in [0, 6a_0]$. The peak of the distribution occurs at the Bohr radius $r = a_0$ (black dot). The volume element $dV = r^2 \sin\theta \, drd\theta d\varphi$ contributes a factor of r^2 to the overall radial dependence of the distribution.

the wave functions are natural functions of r, θ, and φ. It follows that

$$\int_0^\infty \int_0^\pi \int_0^{2\pi} N_1^2 \, e^{-2r/a_0} \, r^2 \sin\theta \, d\varphi d\theta dr = 1$$

The volume element for integration in spherical polar coordinates $dV = r^2 \sin\theta dr d\theta d\varphi$ contributes a factor of r^2 to the radial dependence of the integrand (see Figure 9.19).

This three-dimensional integral is separable in spherical polar coordinates and can be expressed in terms of the product of three one-dimensional integrals:

$$N_1^2 \int_0^\infty e^{-2r/a_0} \, r^2 \, dr \times \int_0^\pi \sin\theta \, d\theta \times \int_0^{2\pi} d\varphi = 1$$

We have encountered each of these integrals before. Integration over θ and φ results in a familiar factor of 4π:

$$\int_0^\pi \sin\theta \, d\theta \times \int_0^{2\pi} d\varphi = 2 \times 2\pi = 4\pi$$

This means that the original triple integral can be simplified to

$$4\pi N_1^2 \int_0^\infty e^{-2r/a_0} \, r^2 \, dr = 1$$

where the r^2 term scales $|\psi(r)|^2$ in proportion to the growing area to be integrated over as r increases (see Figure 9.19). The final integral over an exponential function can be evaluated using integration by parts or differentiation with respect to a parameter explored in the complements to Chapter 7. The result is

$$\int_0^\infty x^2 e^{-ax} dx = \frac{2}{a^3}$$

where $x = r$ and $a = 2/a_0$. The final result is

$$4\pi N_1^2 \left(\frac{a_0^3}{4}\right) = N_1^2 \pi a_0^3 = 1$$

so that $N_1 = (\pi a_0^3)^{-1/2}$. A similar calculation can be used to show that $N_2 = (32\pi a_0^3)^{-1/2}$ for each of these four excited states. Examples of the three-dimensional forms of the wave functions noted above are provided in Figure 9.20 and Figure 9.21.

9.3.4 Mean and variance of the electron density probability distribution

Let's compute the mean and variance of the position of the electron in the 1s ground state. The mean position of the electron is defined as the position r averaged over the electron density distribution $|\psi_{1s}(r,\theta,\varphi)|^2$ and is written

$$\bar{r} = \int_0^\infty \int_0^\pi \int_0^{2\pi} r \, |\psi_{1s}(r,\theta,\varphi)|^2 \, r^2 \sin\theta \, d\varphi d\theta dr$$

$$= \frac{1}{\pi a_0^3} \int_0^\infty \int_0^\pi \int_0^{2\pi} r^3 \, e^{-2r/a_0} \sin\theta \, d\varphi d\theta dr$$

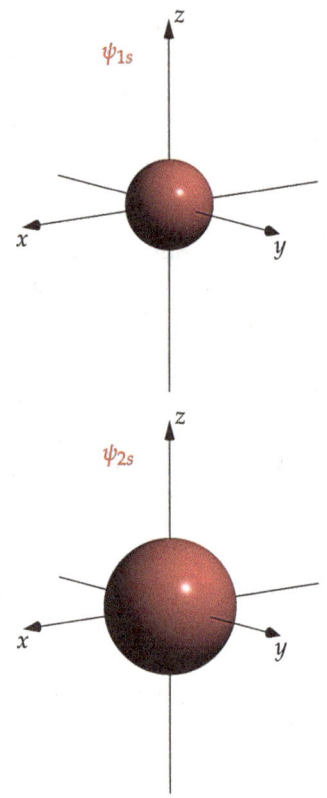

Figure 9.20: Surface of the modulus squared of the ψ_{1s} and ψ_{2s} wave functions for a one-electron atom.

This three-dimensional integral over all space is separable and can be written as a product of three one-dimensional integrals:

$$\bar{r} = \frac{1}{\pi a_0^3} \int_0^\infty r^3 \, e^{-2r/a_0} \, dr \times \int_0^\pi \sin\theta \, d\theta \times \int_0^{2\pi} d\varphi$$

$$= 4\pi \frac{1}{\pi a_0^3} \int_0^\infty r^3 \, e^{-2r/a_0} \, dr$$

The integral over r can be performed using integration by parts, but only with many steps. The most straightforward way to determine the integral is to use the identity

$$\int_0^\infty x^3 e^{-ax} dx = -\frac{d^3}{da^3} \int_0^\infty e^{-ax} dx = -\frac{d^3}{da^3} \frac{1}{a} = \frac{6}{a^4}$$

Applying this identity to our integral of interest with $x = r$ and $a = 2/a_0$ results in

$$\bar{r} = 4\pi \frac{1}{\pi a_0^3} \int_0^\infty r^3 \, e^{-2r/a_0} \, dr = 4\frac{1}{a_0^3}\left(\frac{6a_0^4}{16}\right) = \frac{3}{2}a_0$$

We can use the same approach to determine the variance in the position of the electron, defined as

$$\sigma_r^2 = \overline{r^2} - \bar{r}^2$$

We determine $\overline{r^2}$ to be

$$\overline{r^2} = \int_0^\infty \int_0^\pi \int_0^{2\pi} r^2 \, |\psi_{1s}(r,\theta,\varphi)|^2 \, r^2 \sin\theta \, d\varphi d\theta dr$$

$$= \frac{1}{\pi a_0^3} \int_0^\infty \int_0^\pi \int_0^{2\pi} r^4 \, e^{-2r/a_0} \sin\theta \, d\varphi d\theta dr$$

$$= 4\pi \frac{1}{\pi a_0^3} \int_0^\infty r^4 \, e^{-2r/a_0} \, dr$$

Using the integral identity

$$\int_0^\infty x^4 e^{-ax} dx = \frac{d^4}{da^4} \int_0^\infty e^{-ax} dx = \frac{d^4}{da^4} \frac{1}{a} = \frac{24}{a^5}$$

where $x = r$ and $a = 2/a_0$, leads to

$$\overline{r^2} = 4\pi \frac{1}{\pi a_0^3} \int_0^\infty r^4 \, e^{-2r/a_0} \, dr = 4\frac{1}{a_0^3}\left(\frac{24a_0^5}{32}\right) = 3a_0^2$$

The final result for the variance in the position of the electron is

$$\sigma_r^2 = \overline{r^2} - \bar{r}^2 = 3a_0^2 - \left(\frac{3a_0}{2}\right)^2 = \frac{3}{4}a_0^2$$

We find that an apparently complicated set of three-dimensional integrals can be reduced to a series of one-dimensional integrations of familiar functions.

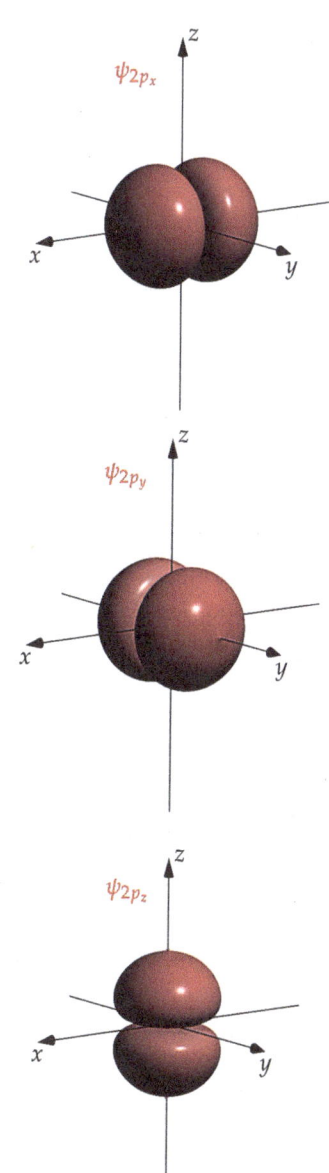

Figure 9.21: Surface of the modulus squared of the $\psi_{2p_x}(r,\theta,\varphi)$, $\psi_{2p_y}(r,\theta,\varphi)$, and $\psi_{2p_z}(r,\theta,\varphi)$ wave functions for a one-electron atom.

A$_9$ Connecting the gaussian and binomial probability distributions

Let's examine the binomial probability distribution when the number of outcomes is $N = 20$. The distribution of probabilities is shown in Figure 9.22. Note that the mean of the distribution is $\bar{n} = \frac{N}{2} = 10$ and the variance is $\sigma_n^2 = \frac{N}{4} = 5$. Suppose we use these values of the mean and variance to parameterize a gaussian distribution. We find

$$p(n;N) = \frac{1}{\sqrt{2\pi\sigma_n^2}}\,\exp\left[-\frac{(n-\bar{n})^2}{2\sigma_n^2}\right]$$

so that

$$p(n;N) = \frac{1}{\sqrt{\pi N/2}}\,\exp\left[-\frac{(n-N/2)^2}{N/2}\right] \tag{9.11}$$

The result is shown in Figure 9.22, where the continuous gaussian probability distribution function is compared with the discrete binomial probability distribution.

For the binomial distribution, we found that the ratio of the width of the distribution σ_n relative to the mean $\bar{n}$ varies as

$$\frac{\sigma_n}{\bar{n}} = \frac{\sqrt{N}}{N} = \frac{1}{\sqrt{N}}$$

As N increases the mean increases according to $\bar{n} = \frac{N}{2}$ and standard deviation increases according to $\sigma_n = \frac{\sqrt{N}}{2}$. As N increases, the distribution appears to be more and more narrow. Another way to interpret this is to think of σ_n as the uncertainty in the measurement of the mean $\bar{n}$, where we might report our measurement as

$$\bar{n} \pm \sigma_n$$

The relative uncertainty, σ_n, in the measurement of the mean, $\bar{n}$, decreases with increasing N according to $N^{-1/2}$. For the distribution in Figure 9.2 with $N = 3$, we find $\frac{\sigma_n}{\bar{n}} = 0.577$, for the distribution in Figure 9.3 with $N = 6$, we find $\frac{\sigma_n}{\bar{n}} = 0.408$, and for the distribution in Figure 9.13 with $N = 20$, we find $\frac{\sigma_n}{\bar{n}} = 0.224$. As N increases, the ratio $\frac{\sigma_n}{\bar{n}}$ decreases, making the variance a smaller and smaller fraction of the mean. This demonstrates the increasing certainty in our measurement of the mean with increasing N.

This point is made more clearly in Figure 9.23 in which the gaussian approximation of the binomial distribution provided by Equation 9.11 is used to compare $Np(n;N)$ for varying $N = 10, 50$, and 150. Both the mean and standard deviation increase with N. The mean increases according to

$$\bar{n} \propto N$$

while the standard deviation increases more slowly according to

$$\sigma \propto \sqrt{N}$$

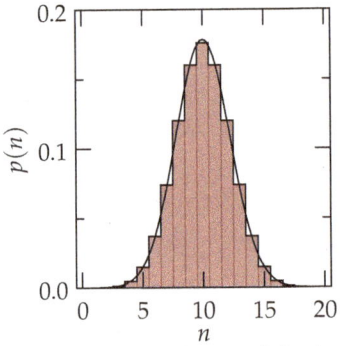

Figure 9.22: The binomial distribution for the relative probability of 21 possible outcomes resulting from $N = 20$ flips of a coin. Shown for comparison is the gaussian distribution.

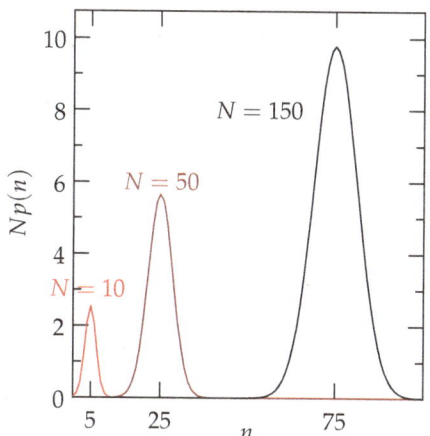

Figure 9.23: The gaussian approximation of the binomial distribution $p(n; N)$ for $N = 10, 50$, and 150.

As N increases, the probability becomes more narrowly distributed about the mean. This relative narrowing is best exposed in Figure 9.24 in which the distributions are plotted as a function of the normalized index $\frac{n}{N}$ for $N = 10, 50$, and 150. As N increases, the mean increases according to $\bar{n} = \frac{N}{2}$ and standard deviation increases according to $\sigma_n = \frac{\sqrt{N}}{2}$.

Finally, let's consider the random sampling of an *independent random variable*. We assume that when we sample the variable once, the outcome of the next sampling is independent of the prior result. We make a series of N observations from which we compute the mean. As N is increased, the normalized sum of the measured averages tends to a gaussian distribution. Remarkably this is true even if the variables themselves (such as the coin flips) are not distributed according to a gaussian distribution. This important result is known as the *central limit theorem*.[19] Let's return to the coin flip. Suppose we flip a coin many times. As the number of flips is increased, the probability of observing a certain number of heads in a series of flips will approach a gaussian probability distribution also referred to as the *normal distribution*. This remarkable result explains the ubiquitous use of the normal distribution in modeling phenomena in the physical and social sciences.

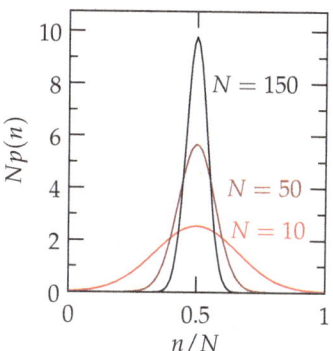

Figure 9.24: The gaussian approximation of the binomial distribution $p(n; N)$ for N, including $10, 50$, and 150.

B9 Uniform distributions of independent random variables

Flipping a coin provides a means of generating independent *random numbers*. For example, suppose we take a head to be 0 and a tail to be 1. The outcome of $N = 20$ random coin flips can be written

$$00110 \ 11110 \ 11101 \ 00101$$

forming a series of independent random variables. In this case there are 8 heads and 12 tails, an outcome with probability $p(n; N) = p(8; 20) = 0.120$, or 12 percent. Other physical methods for generating random variables include choosing numbered balls from a bin or measuring decay events from a source of radioactive matter.

[19] This observation and its application to the study of coin flips was first reported by French mathematician *Abraham de Moivre* (1667-1754).

Now suppose we want to generate 10,000 independent random variables. We won't want to do that by flipping a coin! The challenge of generating independent random variables using mathematics alone is not at all trivial and has captured the imagination of mathematicians for decades. Consider the *middle square method*.[20] One takes a large odd integer, squares it, and then selects the middle set of numbers as the random variable.

For example, we start with the *random seed* 579,203, square it to form 335,476,115,209, and select the middle digits to arrive at 476,115. Repeating the process leads to 685,493. Repeating again yields 900,653 and 175,826 and 914,782. And so on. We can visualize this process as

$$(579,203)^2$$
$$335,476,115,209$$
$$(476,115)^2$$
$$226,685,493,225$$
$$(685,493)^2$$
$$469,900,653,049$$
$$(900,653)^2$$
$$811,175,826,409$$
$$(175,826)^2$$
$$030,914,782,276$$
$$914,782$$

Note that the last operation produced an 11-digit number. We created a 12-digit number by padding the left side with a 0.

This method is not perfect. It has a relatively short *recurrence time* after which the series repeats itself. It can also converge to zero. As such, we say that the series is formed of *pseudorandom numbers*. If we wish to have the generated series of pseudorandom numbers occur over the interval $[0,1]$, we simply divide by one million.

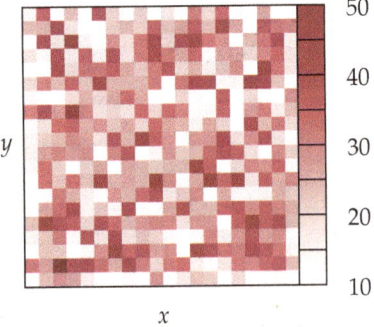

Suppose we have a process for generating random numbers $x \in [0,1]$ from a *uniform probability distribution*. The result of sampling 10,000 independent random numbers $x \in [0,1]$ is shown in Figure 9.25 and compared with a uniform distribution. The randomly sampled numbers approximate the uniform distribution over the range of $x \in [0,1]$. The computed mean value $\bar{x} = 0.497$ is slightly smaller than the exact value $\bar{x} = 0.5$, and the computed

[20] This method was developed by Hungarian-American mathematician and physicist *John von Neumann* (1903-1957). Von Neumann invented and built the first modern computer, developed *game theory* into a powerful tool in the social sciences, and made critical contributions to the development of the first atomic bomb.

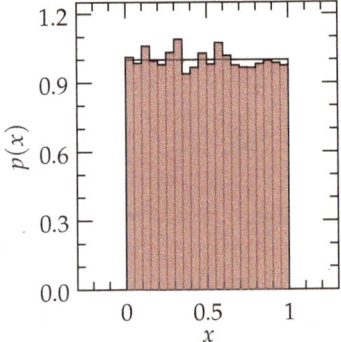

Figure 9.25: A distribution derived from 10,000 independent uniform random variables $x \in [0,1]$ (shaded red). The black straight line is the exact uniform distribution.

Figure 9.26: A two-dimensional probability distribution $p(x,y)$ for a sample of 10,000 independent uniform random variables x and y (shaded red). The color bar shows the number of observations of a pair (x,y) in a given region.

standard deviation $\sigma_x = 0.2883$ is slightly smaller than the exact value $\sigma_x = 12^{-\frac{1}{2}} = 0.2887\ldots$ for $N = 10,000$.

We can use the same method to generate a two-dimensional uniform probability distribution of independent random variables $x \in [0,1]$ and $y \in [0,1]$. To do this we can use the middle square method to generate x and repeat the process to generate y. The result for sampling $10,000$ independent random variables from a uniform distribution in x and y is shown in Figure 9.26.

Note the significant variation in the number of observations in a given increment $dx\,dy$ across the distribution, even for a sampling of $10,000$ numbers. As the number of dimensions increases, the number of samples must increase to achieve the same level of convergence. These examples convey the challenges associated with generating random samples, and the substantial data set required to represent an adequately converged random distribution.

C₉ Gaussian distributions of independent random variables

The middle square method provides a way to generate independent random variables corresponding to a uniform probability distribution. How can we sample independent random variables from a gaussian distribution? It turns out that there is a clever method for transforming independent random variables sampled from a uniform distribution to independent random variables sampled from a gaussian distribution.

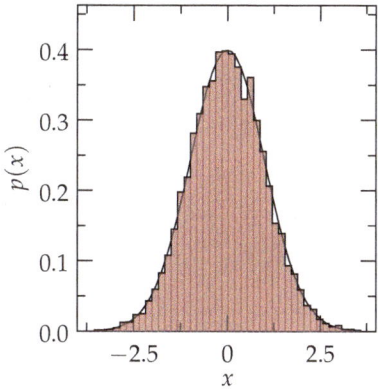

Figure 9.27: A distribution derived from $10,000$ independent gaussian random variables (shaded red). The black line is the fit to a gaussian probability distribution function.

Suppose we generate independent random numbers uniformly over the interval $x \in [0,1]$. We generate two such numbers x_1 and x_2. We can transform these numbers from two independent random variables sampled from a uniform distribution, x_1 and x_2, into two independent random variables sampled from a gaussian distribution, z_1 and z_2, using the relations

$$z_1 = \sqrt{-2\ln x_1}\,\cos(2\pi x_2)$$
$$z_2 = \sqrt{-2\ln x_1}\,\sin(2\pi x_2)$$

This is known as the *Box-Mueller method*.[21]

[21] This method was originally presented by American mathematician and philosopher *Norbert Wiener* (1894-1964). Wiener was a famous child prodigy who made seminal contributions to the fields of cybernetics, robotics, and control systems.

Take two independent pseudorandom variables sampled from a uniform distribution in the interval $[0, 1]$:

$$x_1 = 0.579203 \qquad x_2 = 0.476115$$

Using the Box-Mueller method they can be transformed to two independent random variables sampled from a gaussian distribution:

$$z_1 = \sqrt{-2\ln(0.579203)}\cos\left[2\pi(0.476115)\right] = -0.680981$$

$$z_2 = \sqrt{-2\ln(0.579203)}\sin\left[2\pi(0.476115)\right] = 0.102972$$

Figure 9.27 shows the result of a sampling of 10,000 independent random variables compared with a gaussian probability distribution. We see that the Box-Mueller method provides a simple and effective way to generate independent random gaussian variables.

It is also possible to extend this method to sample a two-dimensional gaussian probability distribution in the variables x and y. Figure 9.28 shows a two-dimensional histogram over the xy-plane, which depicts the frequency of sampling two independent gaussian random variables, x and y, for a sample of 50,000 points. For a large sampling of points the distribution will be well approximated by a two-dimensional gaussian probability distribution function.

The ability to sample independent random variables from a gaussian distribution is useful to model physical processes such as the random motion of atoms in an ideal gas or the molecular diffusion of molecules in a liquid. It also provides a means of modeling statistical error. Ultimately, our success in generating independent gaussian random variables relies on our ability to effectively generate uniform random numbers.

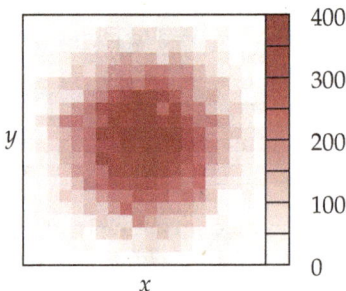

Figure 9.28: A two-dimensional probability distribution $p(x, y)$ for a sample of 50,000 independent gaussian random variables x and y (shaded red). The color bar shows the number of observations of a pair (x, y) in a given region.

D9 Three definitions of Pythagorean means

There are several ways to define the mean value of a series of numbers. The most familiar is the *arithmetic mean*, defined as

$$A_M = \frac{1}{n}\left(x_1 + \ldots + x_n\right)$$

This mean is useful for averaging *reaction times*. The arithmetic mean of a and b is $A_M = \frac{1}{2}(a + b)$.

However, we can also consider the *geometric mean*, defined as

$$G_M = \sqrt[n]{x_1 \times \ldots \times x_n}$$

The geometric mean of a and b is $G_M = \sqrt{ab}$. In contrast, the *harmonic mean* is defined as

$$H_M = n\left(\frac{1}{x_1} + \ldots + \frac{1}{x_n}\right)^{-1}$$

This mean is useful for averaging *reaction rates*. The harmonic mean of a and b is $H_M = 2(\frac{1}{a} + \frac{1}{b})^{-1} = 2\frac{ab}{a+b}$. The arithmetic mean, geometric mean, and

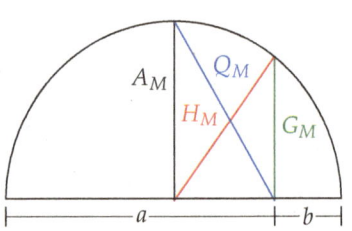

Figure 9.29: The geometric relationship between the arithmetic mean, A_M, geometric mean, G_M, harmonic mean, H_M, and quadratic mean, Q_M.

harmonic mean are known as *Pythagorean means*.

In addition, we have the *quadratic mean,* also known as the *root mean square* value, and defined as

$$Q_M = \frac{1}{n}\sqrt{x_1^2 + \ldots + x_n^2}$$

This mean is useful in computing the root mean square speed. The quadratic mean of a and b is $Q_M = \frac{1}{2}\sqrt{a^2 + b^2}$.

The geometric relationships between these four definitions of the mean are displayed in Figure 9.29 for the case of $a = 0.8$ and $b = 1 - a = 0.2$, for which $A_M = 0.5$, $G_M = 0.4$, $H_M = 0.32$, and $Q_M = 0.41$. Note that only the case $a = b$ leads to the same mean value $A_M = G_M = H_M = Q_M$ for all four definitions of the mean.

E₉ Propagation of error through total differentials and Taylor series

---

Suppose we have a function $f(x, y)$ of two independent variables x and y. Let's see how the error in our estimate of the function f is related to the known error in the averages of the variables x and y.

We define the mean value of the function, $\overline{f}$, and the mean values of the variables, $\overline{x}$ and $\overline{y}$. We further define the deviation of the observed values from the mean values as the errors

$$\varepsilon_x = x - \overline{x} \qquad \varepsilon_y = y - \overline{y} \qquad \varepsilon_f = f - \overline{f}$$

Since $f(x, y)$ is a function of x and y, we can ask how the error in our estimate of the function, ε_f, is related to the error in our measurements of the variables, ε_x and ε_y. We start from the Taylor series

$$f(x, y) = \overline{f} + \frac{\partial f}{\partial x}\Big|_{\overline{x}, \overline{y}}(x - \overline{x}) + \frac{\partial f}{\partial y}\Big|_{\overline{x}, \overline{y}}(y - \overline{y}) + \frac{1}{2}\frac{\partial^2 f}{\partial x^2}\Big|_{\overline{x}, \overline{y}}(x - \overline{x})^2$$
$$+ \frac{1}{2}\frac{\partial^2 f}{\partial y^2}\Big|_{\overline{x}, \overline{y}}(y - \overline{y})^2 + \frac{\partial^2 f}{\partial y \partial x}\Big|_{\overline{x}, \overline{y}}(x - \overline{x})(y - \overline{y}) + \ldots$$

which we can rewrite as

$$\varepsilon_f = f(x, y) - \overline{f} = \frac{\partial f}{\partial x}\Big|_{\overline{x}, \overline{y}}\varepsilon_x + \frac{\partial f}{\partial y}\Big|_{\overline{x}, \overline{y}}\varepsilon_y + \frac{1}{2}\frac{\partial^2 f}{\partial x^2}\Big|_{\overline{x}, \overline{y}}\varepsilon_x^2$$
$$+ \frac{1}{2}\frac{\partial^2 f}{\partial y^2}\Big|_{\overline{x}, \overline{y}}\varepsilon_y^2 + \frac{\partial^2 f}{\partial y \partial x}\Big|_{\overline{x}, \overline{y}}\varepsilon_x \varepsilon_y + \ldots$$

If the errors ε_x and ε_y are small enough, we can truncate the series and ignore higher order terms to find

$$\varepsilon_f \approx \frac{\partial f}{\partial x}\Big|_{\overline{x}, \overline{y}}\varepsilon_x + \frac{\partial f}{\partial y}\Big|_{\overline{x}, \overline{y}}\varepsilon_y$$

If the errors ε_x and ε_y are large, we need to retain higher order terms in the Taylor series to determine an accurate measure of the error in our estimate of f. We want to relate the mean square error in our estimate of the function

$f(x,y)$, defined as

$$\overline{\varepsilon_f^2} = \overline{(f - \overline{f})^2}$$

to the mean square error in our measurements of the variables x and y, defined $\overline{\varepsilon_x^2}$ and $\overline{\varepsilon_y^2}$.[22] Using the Taylor series approximation above, we find

$$\varepsilon_f^2 \approx \left(\frac{\partial f}{\partial x}\Big|_{\overline{x},\overline{y}} \varepsilon_x + \frac{\partial f}{\partial y}\Big|_{\overline{x},\overline{y}} \varepsilon_y \right)^2$$

$$= \left(\frac{\partial f}{\partial x} \right)^2_{\overline{x},\overline{y}} \varepsilon_x^2 + \left(\frac{\partial f}{\partial y} \right)^2_{\overline{x},\overline{y}} \varepsilon_y^2 + 2 \left(\frac{\partial f}{\partial x} \frac{\partial f}{\partial y} \right)_{\overline{x},\overline{y}} \varepsilon_x \varepsilon_y$$

Now we must average over each side of the equation.

On the left-hand side, ε_f^2 becomes $\overline{\varepsilon_f^2}$. On the right-hand side, the partial derivatives are constant coefficients. We average the squared error so that ε_x^2 and ε_y^2 become $\overline{\varepsilon_x^2}$ and $\overline{\varepsilon_y^2}$. Finally, we note that since x and y are independent variables, $\overline{\varepsilon_x \varepsilon_y} = 0$. This leads to our final result[23]

$$\overline{\varepsilon_f^2} \approx \left(\frac{\partial f}{\partial x} \right)^2_{\overline{x},\overline{y}} \overline{\varepsilon_x^2} + \left(\frac{\partial f}{\partial y} \right)^2_{\overline{x},\overline{y}} \overline{\varepsilon_y^2}$$

We interpret the mean square error in our estimate of f as the variance

$$\sigma_f^2 = \overline{\varepsilon_f^2} = \overline{(f - \overline{f})^2}$$

Similarly $\sigma_x^2 = \overline{\varepsilon_x^2}$ and $\sigma_y^2 = \overline{\varepsilon_y^2}$. We can now write our final result relating the mean square error in our measurement of the variables x and y to the mean square error in our estimate of the function f as

$$\sigma_f^2 \approx \left(\frac{\partial f}{\partial x} \right)^2_{\overline{x},\overline{y}} \sigma_x^2 + \left(\frac{\partial f}{\partial y} \right)^2_{\overline{x},\overline{y}} \sigma_y^2 \qquad (9.12)$$

Let's apply this result to the measurement of the pressure of a gas. Suppose we measure the temperature, T, and volume, V, of the gas. Using the ideal gas law, the pressure of the gas can be written as a function of the temperature and volume:

$$p(T,V) = \frac{RT}{V}$$

Let's see how the error in our estimate of p is related to our error in the measurements of T and V. Taking $p \equiv f$, $T \equiv x$, and $V \equiv y$, the resulting expression for the mean square error in our estimate of the pressure σ_p^2 is

$$\sigma_p^2 \approx \left(\frac{\partial p}{\partial T} \right)^2_{\overline{T},\overline{V}} \sigma_T^2 + \left(\frac{\partial p}{\partial V} \right)^2_{\overline{T},\overline{V}} \sigma_V^2 \qquad (9.13)$$

where σ_T^2 and σ_V^2 are the mean square errors in the temperature and volume, respectively. This distribution is depicted in Figure 9.30.

Suppose there is 20% random error in $\overline{T}$ and $\overline{V}$. Interpreting the random error as the root-mean square error or standard deviation, we can write

$$\sigma_T = 0.2\,\overline{T} \qquad \sigma_V = 0.2\,\overline{V}$$

[22] Note that $\overline{\varepsilon_f} = \overline{(f - \overline{f})} = \overline{f} - \overline{f} = 0$. To measure the error σ_f we must average the square of the error and take the square root

$$\sigma_f = \sqrt{\overline{\varepsilon_f^2}} = \sqrt{\overline{(f - \overline{f})^2}}$$

[23] Suppose we have a function of one variable $f(x)$. We define the mean values $\overline{f}$ and $\overline{x}$ and associated errors $\varepsilon_f = f - \overline{f}$ and $\varepsilon_x = x - \overline{x}$. Suppose we take N measurements, each uncorrelated in error. We expect that $\overline{\varepsilon_f^2} = N f_x^2 \, \overline{\varepsilon_x^2}$, which we can also write as $\sigma_f^2 = N \sigma_x^2$. As such

$$\sigma_f \propto \sqrt{N} \sigma_x$$

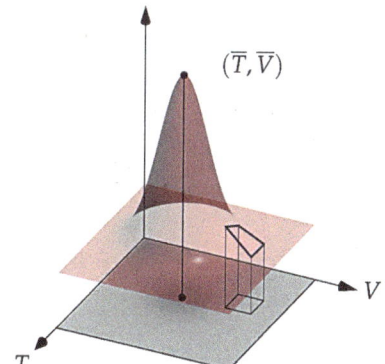

Figure 9.30: A two-dimensional probability distribution for the variables T and V. The distribution is centered around the average $(\overline{T}, \overline{V})$. The width of the distribution represents the standard deviations σ_T and σ_V in the T and V dimensions, respectively. The probability of observing a value of $T \in [T, T + dT]$ and $V \in [V, V + dV]$ is given by the volume of the vertical column over the area $dT\, dV$.

so that $\sigma_T^2 = (0.2\overline{T})^2$ and $\sigma_V^2 = (0.2\overline{V})^2$. We further know that

$$\frac{\partial p}{\partial T} = \frac{R}{V} \qquad \frac{\partial p}{\partial V} = -\frac{RT}{V^2}$$

Combining these results with Equation 9.13, we find

$$\sigma_p^2 \approx \left(\frac{R}{\overline{V}}\right)^2 \times (0.2\overline{T})^2 + \left(-\frac{R\overline{T}}{\overline{V}^2}\right)^2 \times (0.2\overline{V})^2$$

$$= 0.08 \left(\frac{R\overline{T}}{\overline{V}}\right)^2 = 0.08\, p^2$$

Let's look at one more example involving the average of small differences between large numbers. Consider the function

$$f(x, y) = x - y$$

where $\overline{x} \pm \sigma_x = 30 \pm 1$ and $\overline{y} \pm \sigma_y = 29 \pm 1$. We can return to our earlier result for the mean square error, defined as

$$\sigma_f^2 \approx \left(\frac{\partial f}{\partial x}\right)_{\overline{x},\overline{y}}^2 \sigma_x^2 + \left(\frac{\partial f}{\partial y}\right)_{\overline{x},\overline{y}}^2 \sigma_y^2$$

where

$$\frac{\partial f}{\partial x} = 1 \qquad \frac{\partial f}{\partial y} = -1$$

so that $\sigma_f^2 \approx \sigma_x^2 + \sigma_y^2 = 1 + 1 = 2$. It follows that

$$\sigma_f = \sqrt{2} = 1.41$$

From our original expression for $f(x, y)$, we find $\overline{f} = \overline{x} - \overline{y} = 30 - 29 = 1$ so that $\overline{f} \pm \sigma_f = 1 \pm 1.41$. The error is larger than the difference itself! This example demonstrates the importance of considering not only the relative error in variables, but how that relative error translates into the total error in the function itself.

F₉ End-of-chapter problems

> Life's most important questions are, for the most part, nothing but probability problems.
>
> Pierre-Simon Laplace

Warm-ups

9.1 Calculate the probability of getting (a) exactly 5 heads in 10 tosses of a fair coin, (b) exactly 1 head in 5 tosses of a fair coin, and (c) 5 heads in the order of THHTHTTTHH.

HINT: A coin toss has two possible outcomes, head or tail. The binomial distribution gives the probability of observing n heads and $N - n$ tails in N flips of a fair coin as

$$p(n; N) = \frac{N!}{(N-n)! \, n!} \left(\frac{1}{2}\right)^N$$

where the binomial coefficient

$$\binom{N}{n} = \frac{N!}{(N-n)! \, n!}$$

is the number of ways that n heads and $N - n$ tails can be observed in N flips out of 2^N possible outcomes.

9.2 Determine the relative error in *Stirling's approximation*

$$\ln N! \cong N \ln N - N$$

in calculations of $\ln N!$ for $N = 10, 50$, and 100. The relative error is defined as the difference between the estimated value and the exact value, divided by the exact value.

9.3 In the card game *bridge* you can distribute 52 cards having 4 suits (hearts, diamonds, clubs, and spades) of 13 cards each. A bridge hand consists of 13 cards. What is the total number of possible bridge hands?

9.4 Find the number of different permutations for the letters in the words Euler, Laplace, and Lagrange.

HINT: An event has k possible outcomes. The multinomial coefficient gives the number of ways of observing k possible outcomes, defined as a set of $\{n_1, n_2, n_3, \ldots, n_k\}$ where n_l is the number of times the lth outcome was observed in N trials

$$\binom{N}{n_1, n_2, n_3, \cdots, n_k} = \frac{(n_1 + n_2 + n_3 + \cdots n_k)!}{n_1! n_2! n_3! \cdots n_k!} = \frac{N!}{n_1! n_2! n_3! \cdots n_k!}$$

where

$$\sum_{l=1}^{k} n_l = N$$

9.5 Consider the simple uniform distribution

$$p(x) = \begin{cases} C & 0 \leq x \leq 5 \\ 0 & \text{otherwise} \end{cases}$$

(a) If $p(x)$ is normalized, what is the normalization constant C?

(b) What is the mean $\bar{x}$?

(c) What is the standard deviation σ_x defined in terms of the variance $\sigma_x^2 = \overline{(x - \bar{x})^2}$?

9.6 Consider the exponential distribution

$$p(r)\, dr = Ce^{-\lambda r} dr \qquad 0 \leq r \leq \infty$$

(a) If $p(r)$ is normalized, what is the normalization constant C?

(b) What is the mean $\bar{r}$?

(c) What is standard deviation σ_r defined in terms of the variance $\sigma_r^2 = \overline{(r - \bar{r})^2}$

Homework exercises

9.7 Consider the probability distribution of the one-dimensional *particle in a box*

$$p(x)\, dx = |\psi_n(x)|^2\, dx = \frac{2}{L} \sin^2\left(\frac{n\pi x}{L}\right) dx \qquad n = 1, 2, 3, \dots$$

where $0 \leq x \leq L$.

(a) Show that $p(x)$ is normalized.

(b) Show that the mean is

$$\bar{x} = \frac{L}{2}$$

(c) Show that the variance is

$$\overline{(x - \bar{x})^2} = \frac{L^2}{12}\left(1 - \frac{6}{n^2\pi^2}\right)$$

9.8 Consider the distribution of energies of an ideal gas

$$p(\varepsilon)\, d\varepsilon = \frac{2\pi}{(\pi k_B T)^{3/2}} \varepsilon^{1/2} \exp\left(-\frac{\varepsilon}{k_B T}\right) d\varepsilon$$

where $0 \leq \varepsilon < \infty$.

(a) Show that $p(\varepsilon)$ is normalized.

(b) Determine the mean $\bar{\varepsilon}$. Compare your result to the average kinetic energy of an ideal gas particle

$$\frac{1}{2} m\overline{v^2} = \frac{3}{2} k_B T$$

(c) Show that the variance is

$$\sigma_\varepsilon^2 = \overline{(\varepsilon - \bar{\varepsilon})^2} = \frac{3}{2}(k_B T)^2$$

9.9 The mean value $\bar{n}$ and variance $\sigma_n^2 = \overline{(n - \bar{n})^2}$ of a binomial distribution are defined by Equations 9.9 and 9.10 to be

$$\bar{n} = \frac{N}{2} \qquad \sigma_n^2 = \overline{(n - \bar{n})^2} = \frac{N}{4}$$

when each outcome is assumed to have probability $\frac{1}{2}$. Consider the general case in which one outcome has probability p and the other outcome has probability $1 - p = q$. The binomial probability of observing n events each having probability p and $N - n$ events each having probability $1 - p = q$ takes the form

$$p(n; N) = \frac{N!}{n!(N-n)!} \, p^n \, (1-p)^{N-n} = \frac{N!}{n!(N-n)!} \, p^n \, q^{N-n}$$

(a) Show that the mean value is given by

$$\bar{n} = \sum_{n=0}^{N} np(n; N) = Np$$

which reduces to the result $\bar{n} = N/2$ for the case $p = q = \frac{1}{2}$.

(b) Show that the variance is given by

$$\sigma_n^2 = \overline{(n - \bar{n})^2} = Np(1 - p) = Npq$$

which reduces to the result $\sigma_n^2 = \overline{(n - \bar{n})^2} = N/4$ for the case $p = q = \frac{1}{2}$.

9.10 Find the mean value

$$\bar{r} = 4 \left(\frac{1}{a_0} \right)^3 \int_0^\infty e^{-2r/a_0} r^3 \, dr$$

for an electron in the 1s state of the hydrogen atom.

9.11 The probability distribution of a quantum mechanical particle in a three-dimensional rectangular box of volume $V = L_x L_y L_z$ can be written as

$$I = \int \int \int |\psi(x, y, z)|^2 \, dx \, dy \, dz$$
$$= \int_0^{L_z} \int_0^{L_y} \int_0^{L_x} C \sin^2 \left(\frac{n_x \pi x}{L_x} \right) \sin^2 \left(\frac{n_y \pi y}{L_y} \right) \sin^2 \left(\frac{n_z \pi z}{L_z} \right) dx \, dy \, dz = 1$$

where $n_x = 1, 2, 3, \cdots$, $n_y = 1, 2, 3, \cdots$, and $n_z = 1, 2, 3, \cdots$.

(a) Carry out the multiple integral to determine the normalization constant C.

(b) Determine the average position of the particle in the box

$$\bar{\mathbf{r}} = \int \int \int \mathbf{r} |\psi(x, y, z)|^2 \, dx \, dy \, dz$$

where $\mathbf{r} = x \, \hat{\mathbf{x}} + y \, \hat{\mathbf{y}} + z \, \hat{\mathbf{z}}$.

9.12 Consider the Maxwell distribution of velocities defined as

$$p(v)dv = C \exp \left[-\frac{\beta m}{2} \left(u_x^2 + u_y^2 + u_z^2 \right) \right] du_x \, du_y \, du_z$$

where $\beta = \dfrac{1}{k_B T}$, the velocity $v = u_x \mathbf{i} + u_y \mathbf{j} + u_z \mathbf{k}$, and the speed u is defined as the magnitude of the velocity

$$u = |\mathbf{v}| = \sqrt{u_x^2 + u_y^2 + u_z^2}$$

(a) Carry out the following integral

$$I = \iiint p(\boldsymbol{v})d\boldsymbol{v} = \int_{-\infty}^{\infty}\int_{-\infty}^{\infty}\int_{-\infty}^{\infty} p(u_x, u_y, u_z)\, du_x\, du_y\, du_z = 1$$

to determine the normalization constant C.

(b) Show that the distribution in (a) may be written in spherical polar coordinates as

$$p(\boldsymbol{v})d\boldsymbol{v} = C\exp\left(-\frac{\beta m}{2}u^2\right) u^2 \sin\theta\, du\, d\theta\, d\varphi$$

where $u^2 = u_x^2 + u_y^2 + u_z^2$.

(c) Determine the average speed defined by the integral

$$I = \iiint u\, p(\boldsymbol{v})d\boldsymbol{v} = \int_0^{\infty}\int_0^{\pi}\int_0^{2\pi} C\, u\exp\left(-\frac{\beta m}{2}u^2\right) u^2 \sin\theta\, d\varphi\, d\theta\, du$$

(d) Determine the mean square speed defined by the integral

$$I = \iiint u^2 p(\boldsymbol{v})d\boldsymbol{v} = \int_0^{\infty}\int_0^{\pi}\int_0^{2\pi} C\, u^2\exp\left(-\frac{\beta m}{2}u^2\right) u^2 \sin\theta\, d\varphi\, d\theta\, du$$

9.13 Consider a bivariate (two variable) probability distribution $p(x,y)$ with variances $\sigma_x^2 = \overline{(x-\bar{x})^2}$ and $\sigma_y^2 = \overline{(y-\bar{y})^2}$. The cross correlation between x and y can be measured using

$$\rho_{xy} = \frac{1}{\sigma_x\sigma_y}\overline{(x-\bar{x})(y-\bar{y})}$$

where ρ_{xy} is defined as the correlation coefficient. The distribution is shown in the figure below for three values of the correlation coefficient ρ_{xy}.

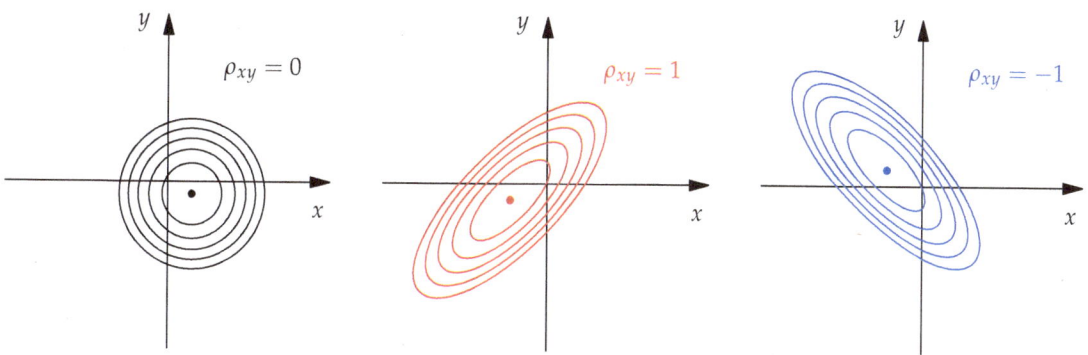

Assume that x and y are independent random variables with probability distribution given by

$$p(x,y)\, dx\, dy = \frac{\alpha}{\pi}e^{-\alpha\left[(x-\bar{x})^2+(y-\bar{y})^2\right]} dx\, dy$$

Show that $\rho_{xy} = 0$.

9.14 Consider the bivariate normal probability distribution

$$p(x,y)\,dx\,dy = \frac{1}{2\pi\sqrt{1-\rho_{xy}^2}}\exp\left[-\frac{1}{2(1-\rho_{xy}^2)}(x^2+y^2-2\rho_{xy}\,xy)\right]dx\,dy$$

The distribution is shown in the figure above for three values of the correlation coefficient ρ_{xy}.

(a) Show that the *marginal probability* distributions take the form

$$p(x) = \int_{-\infty}^{\infty} p(x,y)\,dy = \frac{1}{\sqrt{2\pi}}e^{-x^2/2}$$

and

$$p(y) = \int_{-\infty}^{\infty} p(x,y)\,dx = \frac{1}{\sqrt{2\pi}}e^{-y^2/2}$$

by integrating over x (or y) leading to a normal distribution in y (or x).

(b) Perform the integral

$$I = \int_{-\infty}^{\infty}\int_{-\infty}^{\infty} p(x,y)\,dx\,dy$$

and demonstate that the distribution is normalized.

9.15* The volume of a cylindrical capillary tube is given by the expression $V = \pi r^2 h$, where r is the radius of the tube and h is the length of the tube. If the radius of the capillary tube is found to be 0.030 cm with a error of ± 0.002 cm and the length of the tube is found to be 4.0 cm with a error of ± 0.1 cm, what is the volume of the capillary tube and its error? What measurement must be made to a higher precision to decrease the error in the volume?

9.16* The molar mass of a vapor is determined by filling a bulb of known volume with the vapor at a known temperature and pressure and measuring the mass of the bulb. The vapor is assumed to be an ideal gas so that molar mass, M, is given by

$$M = \frac{mRT}{pV}$$

where m is the mass of the vapor, R is the gas constant, T is the absolute temperature, p is the pressure, and V is the volume. Measurements are summarized as $m = 1.0339 \pm 0.0007$ g, $T = 274.0 \pm 0.5$ K, $p = 1.050 \pm 0.001$ bar, $V = 0.1993 \pm 0.0001$ liters, and $R = 0.08314\,L\cdot bar\cdot mol^{-1}\cdot K^{-1}$ (no error as the value is taken to be exact). Determine the error in the molar mass M.

9.17* Suppose you measure independent variables x and y. You have a function $f(x,y)$ that depends on those variables. Average quantities are defined $\bar{x}$, $\bar{y}$, and $\bar{f}$. We can further define the error in x, y, and $f(x,y)$ as

$$\varepsilon_x = x - \bar{x} \qquad \varepsilon_y = y - \bar{y} \qquad \varepsilon_f = f - \bar{f}$$

(a) Use a Taylor series expansion to express the measured error in $f(x,y)$, ε_f, as a function of errors ε_x and ε_y in x and y. HINT: Expand $f(x,y)$ about the point $(\bar{x},\bar{y})$ to first order in $x - \bar{x}$ and $y - \bar{y}$.

(b) Derive a formula for the mean square error $\overline{\varepsilon_f^2}$ as a function of $\overline{\varepsilon_x^2}$ and $\overline{\varepsilon_y^2}$.

(c) Consider the ideal gas law with $f \equiv p = \dfrac{RT}{V}$, $x \equiv T$, and $y \equiv V$. If T and V have 10% errors, how large is the error in p?

(d) You add N measurements with uncorrelated error in x, ε_x. How will the error in ε_f increase with increasing N?

10 ORDINARY DIFFERENTIAL EQUATIONS

10.1 First order ordinary differential equations

THE FUNDAMENTAL EQUATIONS of thermodynamics, kinetics, and quantum theory are expressed in the form of differential equations. Differential equations are used to relate the change in a property, such as pressure, concentration, or position, to the change in a variable, such as temperature, position, or time. This section explores the general form and properties of first order ordinary differential equations.

10.1.1 General features of first order ordinary differential equations

In order to become familiar with the general features of differential equations, we will begin by looking at a few specific examples. We will then generalize our observations and discuss approaches to analyzing and solving differential equations.

Consider the concentration of some species, $c(t)$, that varies as a function of time, t. The initial concentration is $c(t = 0) = c_0$ and the rate of change in the concentration with changing time is defined by the derivative

$$\frac{dc}{dt}$$

which is written in terms of the change in concentration, dc, divided by the change in time, dt. Suppose the concentration is observed to decrease with time at a rate proportional to the concentration itself. We can record all of this information in the simple equation

$$\frac{d}{dt}c(t) = -k\,c(t) \tag{10.1}$$

where k is a positive proportionality constant known as the *rate constant* with units of inverse time. The negative sign captures the fact that $dc < 0$ and the concentration is decreasing. This is known as a linear ordinary first order differential equation. It is *linear* since the rate of change in $c(t)$ is a linear function of $c(t)$. It is *ordinary* since it involves a function $c(t)$ of a single variable t. It is *first order* since the highest order derivative in the equation is a first derivative.

We seek a function $c(t)$ that is proportional to its own derivative, which must be the exponential function $\exp(\alpha t)$. With this insight, we can guess the functional form for $c(t)$ to be

$$c(t) = Ae^{\alpha t} \tag{10.2}$$

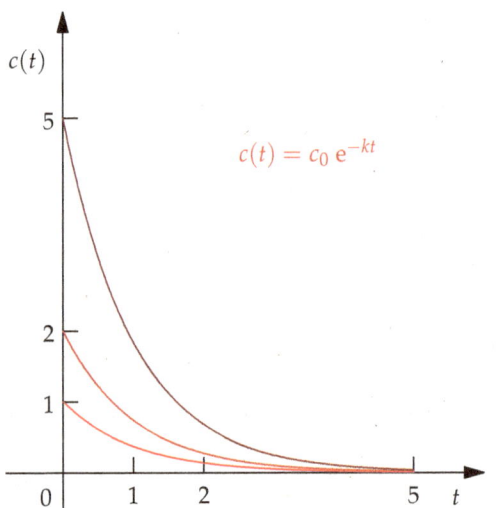

Figure 10.1: The concentration $c(t)$ as a function of time for $k = 1$ and three initial conditions $c_0 = 5, 2,$ and 1.

where A is a proportionality constant. We call this guess our *ansatz*.[1] Inserting Equation 10.2 into Equation 10.1, we find

$$\frac{d}{dt}c(t) = \frac{d}{dt}Ae^{\alpha t} = A\frac{d}{dt}e^{\alpha t} = A\alpha e^{\alpha t} = \alpha c(t) = -k\,c(t)$$

so that $\alpha = -k$. We can determine the value of the constant A using the *initial condition* for the concentration $c(t = 0) = c_0$ to be

$$c(t = 0) = Ae^{\alpha t}\Big|_{t=0} = A = c_0$$

so that the final form for the concentration as a function of time is

$$c(t) = c_0\,e^{-kt} \tag{10.3}$$

This result is plotted in Figure 10.1.

As another example, suppose we seek to model the population of some species, $p(t)$, as it changes in time, t. One simple model is that each unit in the population reproduces itself at a constant rate

$$\text{birth rate} = b$$

and dies at a constant rate

$$\text{death rate} = d$$

We model this *birth-death process* mathematically with the linear first order ordinary differential equation:

$$\frac{d}{dt}p(t) = bp(t) - dp(t) = (b - d)p(t)$$

When the birth rate exceeds the death rate, $(b - d) > 0$ and the population increases. When the death rate exceeds the birth rate, $(b - d) < 0$ and the population decreases. When $b = d$, the population is constant. The behavior of $p(t)$ in the domains of growth, decline, and stasis is shown in Figure 10.2.

[1] This educated guess is often called an *ansatz*, the German word for approach or attempt.

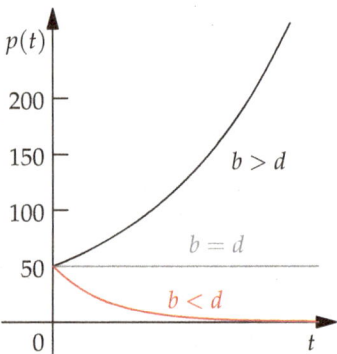

Figure 10.2: The population $p(t)$ as a function of time for the conditions of growth $b > d$, stasis $b = d$, and decline $b < d$.

Consider a slightly different model for the birth-death process. In this model, each species has a birth rate b, but as the population increases the birth rate decreases in proportion to the population as a result of competition for limited resources. This can be written

$$\text{birth rate} = b - cp(t)$$

where c is a constant. The birth rate leads to a rate of change in the population defined by the differential equation

$$\frac{d}{dt}p(t) = (b - cp(t))\, p(t) - d\, p(t)$$
$$= (b - d)p(t) - cp(t)^2$$

This is a *non-linear* first order ordinary differential equation as the rate of change in $p(t)$ depends on both $p(t)$ and $p(t)^2$.[2]

Non-linear differential equations can be challenging to solve. Moreover, the solutions can present qualitatively new properties not observed in linear differential equations. The study of non-linear complex phenomena is also known as *chaos theory*. Fortunately, many of the physical processes of interest to us can be modeled using linear differential equations.

10.1.2 *General approach to solve linear first order differential equations*

Consider the linear first order ordinary differential equation

$$\frac{d}{dx}y(x) + q(x)y(x) = r(x) \tag{10.4}$$

where the goal is to solve for $y(x)$. Both $q(x)$ and $r(x)$ are coefficients that may depend on x. This is a *non-homogeneous* differential equation because there is a term $r(x)$ that is not proportional to $y(x)$. Our first step toward solving this non-homogeneous differential equation is to solve the *reduced equation*[3]

$$\frac{d}{dx}y(x) + q(x)y(x) = 0 \tag{10.5}$$

in which $r(x)$ is taken to be zero. This is known as a *homogeneous equation* as each term is proportional to $y(x)$.[4]

To solve this equation, we employ a trick familiar from solving elementary kinetic equations such as

$$\frac{d}{dt}c(t) = -k\, c(t)$$

with the initial condition $c(t = 0) = c_0$. We separate the numerator and denominator of the derivative and rearrange the equation to be

$$\frac{1}{c(t)}\, dc(t) = -k\, dt$$

so that the dependence on $c(t)$ is on the left-hand side and the explicit dependence on t is on the right-hand side. Integrating each side of the equation from the initial time (or concentration) to the final time (or concentration) leads to[5]

[2] Note if there are two solutions, $p_1(t)$ and $p_2(t)$, each satisfying the differential equation, then the sum or *superposition* of the solutions $p_1(t) + p_2(t)$ is also a solution to the differential equation. Suppose

$$\frac{dp}{dt} = ap = 0$$

then

$$\frac{d}{dt}(p_1 + p_2) = \frac{dp_1}{dt} + \frac{dp_2}{dt} = a(p_1 + p_2)$$

Since

$$\frac{dp_1}{dt} = ap_1 = 0 \qquad \frac{dp_2}{dt} = ap_2 = 0$$

it follows that $p_1 + p_2$ is also a solution.

[3] How do we know that solving Equation 10.5 is a reasonable starting point on our way to solving Equation 10.4? Simply put, because someone tried it, and it worked!

[4] Consider the homogeneous differential equation

$$y' + q(x)y = 0$$

If $y(x)$ is a solution, then so is $cy(x)$ where c is a constant. The equation

$$y' + q(x)y = r(x)$$

is a *non-homogeneous equation*. When $y(x)$ is a solution, $cy(x)$ will not be (even for the case $c = 0$).

[5] We introduced the *dummy variable* t' to distinguish the upper bound in integration, t, from the integration variable, t'.

$$\int_{c(0)}^{c(t)} \frac{1}{c}\, dc = \ln c(t') \Big|_{c(0)}^{c(t)} = -\int_{0}^{t} k\, dt' = -k\, t' \Big|_{0}^{t}$$

with the solution for $c(t)$ being

$$\ln[c(t)] - \ln[c(0)] = \ln\left(\frac{c(t)}{c(0)}\right) = -kt$$

We can use this approach to solve our original reduced equation, Equation 10.5. We find

$$\frac{1}{y}\, dy = -q(x)dx$$

so that

$$\int \frac{1}{y}\, dy = -\int q(x)dx$$

This leaves us with

$$\ln y(x) = -\int q(x)dx + \mathcal{C}$$

where $\mathcal{C}$ is an integration constant that will be defined by the *boundary condition*.[6] Exponentiating each side results in

$$y(x) = \exp\left[-\int q(x)dx + \mathcal{C}\right] = A\exp\left[-\int q(x)dx\right]$$

where $A = e^{\mathcal{C}}$. This is a general solution of our reduced equation for $y(x)$.

The form of this equation may be unfamiliar and appear quite complex. There is an exponential in which the argument is an integral. To better understand this expression, let's apply it to the solution of Equation 10.1. We recognize that

$$\frac{d}{dx}y(x) = -q(x)y(x) \;\rightarrow\; \frac{d}{dt}c(t) = -k\, c(t)$$

so that $y(x) \equiv c(t)$, $x \equiv t$, and $q(x) \equiv k$. According to our general solution, the function $y(x)$ can be written

$$y(x) = A\exp\left[-\int q(x)dx\right] \;\rightarrow\; c(t) = A\exp\left[-\int k\, dt\right]$$

The integral in the exponential is evaluated as $\int k dt = kt$, leading to our final result

$$c(t) = A\exp\left(-\int k\, dt\right) = A\exp(-kt)$$

To evaluate the undetermined constant A, we exploit the initial condition $c(t = 0) = c_0$ so that

$$c(t = 0) = A = c_0$$

leading to the final result

$$c(t) = c_0 \exp(-kt)$$

which agrees with our earlier solution (see Equation 10.3). This example shows how we can use the general solution to the reduced equation to address a familiar problem.

[6] An *initial condition* specifies the value of a function of time at the initial time. A *boundary condition* specifies the value of a function of space at some point in space for all time. In determining the undefined integration constant, one is as good as the other.

Let's return to the original differential equation, Equation 10.4, written

$$\frac{d}{dx}y(x) + q(x)y(x) = r(x)$$

where we employ the *ansatz*

$$y(x) = u(x) \, \exp\left[-\int q(x)dx\right] \tag{10.6}$$

This functional form might not immediately come to mind, but let's see where it takes us.[7] Inserting this form for $y(x)$ into Equation 10.4, we find

[7] Many differential equations are solved by qualitatively analyzing the expression and proposing an *ansatz* based on experience and intuition.

$$\frac{d}{dx}y(x) + q(x)y(x) = \frac{d}{dx}u(x) \times \exp\left[-\int q(x)dx\right]$$

$$- u(x) \times q(x)\exp\left[-\int q(x)dx\right] + q(x)u(x)\exp\left[-\int q(x)dx\right]$$

$$= r(x)$$

or stated more simply

$$\frac{d}{dx}u(x) \times \exp\left[-\int q(x)dx\right] = r(x)$$

Using our trick of separating the numerator and denominator of the derivative and the fact that

$$\exp\left[-\int q(x)dx\right] = \frac{1}{\exp\left[\int q(x)dx\right]}$$

leads to

$$du(x) = r(x)\exp\left[\int q(x)dx\right]dx$$

Integrating each side of this relation as

$$\int du(x) = \int r(x)\exp\left[\int q(x)dx\right]dx$$

results in

$$u(x) = \int r(x)\exp\left[\int q(x)dx\right]dx + C$$

Substituting this expression for $u(x)$ into our *ansatz* for $y(x)$ leads to a general solution to Equation 10.4

$$y(x) = u(x)e^{-\int q(x)dx} = e^{-\int q(x)dx}\left[\int r(x)e^{\int q(x)dx}dx + C\right] \tag{10.7}$$

where the integration constant, C, is determined by specifying an initial or boundary condition.

Equation 10.7 provides a general solution to any first order differential equation of the form defined by Equation 10.4. In the next section, we apply this general result to a number of problems of interest in the physical sciences.

10.2 Applications of first order differential equations

THE PRINCIPAL APPLICATION OF FIRST ORDER ORDINARY DIFFERENTIAL EQUATIONS in physical science is *physical kinetics*. The rate of change in time in the amount or concentration of a species is modeled as the first derivative of the concentration set equal to a function of concentration. In this section we explore the application of our general method for solving first order differential equations to two important models of physical kinetics.

10.2.1 *Application to reversible first order kinetic equations*

Let's explore the application of our general solution to linear first order ordinary differential equations for the case of reversible kinetics

$$A \underset{k_b}{\overset{k_f}{\rightleftharpoons}} B$$

where A is the reactant, B is the product, and k_f and k_b are the forward and backward reaction rate constants, respectively. The initial conditions of the concentrations of each species at time $t = 0$ are defined as $A(0) = A_0$ and $B(0) = 0$. Moreover, the sum $A(t) + B(t)$ is a constant since when one A is lost, one B is gained, and *vice versa*.

We interpret this kinetic scheme in the form of two coupled linear first order differential equations

$$\frac{dA}{dt} = -k_f A + k_b B \qquad \frac{dB}{dt} = k_f A - k_b B \qquad (10.8)$$

While there are two variables $A(t)$ and $B(t)$ there is really only one independent variable since

$$A(t) + B(t) = A(0) + B(0) = A_0$$

and so $B(t) = A_0 - A(t)$. Inserting this result in our equation for the rate of change in A, we find

$$\frac{dA}{dt} = -k_f A + k_b(A_0 - A(t)) = -(k_f + k_b)A(t) + k_b A_0$$

Comparing this specific equation with the general form of our linear first order ordinary differential equation defined by Equation 10.4, we find

$$\frac{d}{dx}y(x) + q(x)y(x) = r(x) \rightarrow \frac{dA}{dt} + (k_f + k_b)A(t) = k_b A_0$$

where $y \equiv A$, $x \equiv t$, $q(x) \equiv (k_f + k_b)$, and $r(x) \equiv k_b A_0$. It follows that the solution for $A(t)$ can be written

$$A(t) = e^{-\int (k_f + k_b)dt}\left[\int k_b A_0 e^{\int (k_f + k_b)dt} dt + C\right]$$

Since $\int (k_f + k_b)dt = (k_f + k_b)t$, we find

$$A(t) = e^{-(k_f + k_b)t}\left[\int k_b A_0 e^{(k_f + k_b)t} dt + C\right]$$

Performing the final integral leads to

$$A(t) = e^{-(k_f+k_b)t} \left[A_0 \frac{k_b}{k_f+k_b} e^{(k_f+k_b)t} + C \right] = \left[A_0 \frac{k_b}{k_f+k_b} + Ce^{-(k_f+k_b)t} \right]$$

Applying the initial condition $A(0) = A_0$ results in

$$A(0) = A_0 \frac{k_b}{k_f+k_b} + C = A_0$$

so that $C = A_0 \frac{k_f}{k_f+k_b}$. We arrive at the final result

$$A(t) = A_0 \frac{k_b}{k_f+k_b} \left(1 + \frac{k_f}{k_b} e^{-(k_f+k_b)t} \right) \tag{10.9}$$

We know that

$$A(t) + B(t) = A_0$$

for all time. As such, we can say that $B(t) = A_0 - A(t)$, so that

$$B(t) = A_0 \frac{k_f}{k_f+k_b} \left(1 - e^{-(k_f+k_b)t} \right)$$

The behavior of $A(t)$ and $B(t)$ as a function of t is shown in Figure 10.3.

The rate of change in the populations of A and B is defined by the sum of the elementary forward and backward rate constants k_f and k_b. Note that as $t \to \infty$ we find the equilibrium populations

$$A_{eq} = A_0 \frac{k_b}{k_f+k_b} \qquad B_{eq} = A_0 \frac{k_f}{k_f+k_b}$$

where the ratio of the equilibrium populations defines the *equilibrium constant*

$$K_{eq} = \frac{B_{eq}}{A_{eq}} = \frac{k_f}{k_b}$$

This reflects the underlying condition of *detailed balance*

$$k_f A_{eq} = k_b B_{eq}$$

This condition states that the forward rate of reaction is equal to the backward rate of reaction. While reactions continue to occur, the populations of reactant and product are kept macroscopically constant in a state of microscopic dynamic equilibrium.

10.2.2 *Application to coupled first order kinetic equations*

Let's explore the application of our general solution to linear first order ordinary differential equations for the kinetic scheme

$$A \xrightarrow{k_1} B \xrightarrow{k_2} C \tag{10.10}$$

The initial conditions define the concentration of each species at time $t = 0$ as $A(0) = A_0$ and $B(0) = C(0) = 0$. We interpret this kinetic scheme in the form

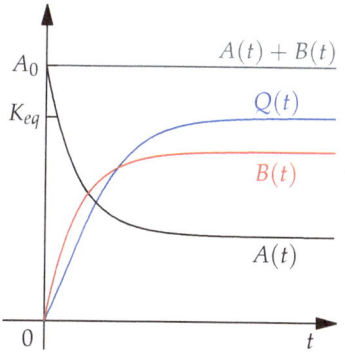

Figure 10.3: The variation in $A(t)$ and $B(t)$ as a function of time for $k_f = 3$ and $k_b = 1.5$. The reaction quotient $Q(t) = B(t)/A(t)$ converges to the equilibrium constant K_{eq} at long times. The gray line shows the sum $A(t) + B(t)$ which is constant in time.

of three coupled linear first order ordinary differential equations

$$\frac{dA}{dt} = -k_1 A \qquad \frac{dB}{dt} = k_1 A - k_2 B \qquad \frac{dC}{dt} = k_2 B \qquad (10.11)$$

These equations will be solved one at a time using our general approach to the solution of first order ordinary differential equations.

Comparing the general linear first order ordinary differential equation with the specific form of our differential equation for the rate of change in A we find

$$\frac{d}{dx} y(x) + q(x)y(x) = r(x) \rightarrow \frac{dA}{dt} + k_1 A = 0$$

where $y \equiv A$, $x \equiv t$, $q(x) \equiv k_1$, and $r(x) \equiv 0$. It follows that the solution for $A(t)$ can be expressed

$$y(x) = A \exp\left[-\int q(x)dx\right] \rightarrow A(t) = A_0 \exp\left[-\int k_1 dt\right]$$

Since $\int k_1 dt = k_1 t$ we arrive at the result

$$A(t) = A_0 e^{-k_1 t} \qquad (10.12)$$

Returning to our equation for the rate of change in $B(t)$, Equation 10.13, we insert our result for $A(t)$ to find

$$\frac{dB}{dt} = k_1 A - k_2 B = k_1 A_0 e^{-k_1 t} - k_2 B \qquad (10.13)$$

Let's compare this equation with the general form of a linear first order ordinary differential equation in Equation 10.4. We find

$$\frac{d}{dx} y(x) + q(x)y(x) = r(x) \rightarrow \frac{dB}{dt} + k_2 B = k_1 A_0 e^{-k_1 t}$$

From this we find $y \equiv B$, $x \equiv t$, $q(x) \equiv k_2$, and $r(x) \equiv k_1 A_0 \exp(-k_1 t)$. Since the solution for $y(x)$ is given by Equation 10.7, we find

$$y(x) = e^{-\int q(x)dx}\left[\int r(x)e^{\int q(x)dx}dx + C\right]$$

It follows that the solution for $B(t)$ is

$$B(t) = e^{-\int k_2 dt}\left[\int k_1 A_0 e^{-k_1 t} e^{\int k_2 dt}dt + C\right]$$

Since $\int k_2 dt = k_2 t$, we arrive at the result

$$B(t) = e^{-k_2 t}\left[\int k_1 A_0 e^{-k_1 t} e^{k_2 t}dt + C\right]$$

We are left with an integral over an exponential function

$$\int k_1 A_0 e^{-k_1 t} e^{k_2 t}dt = k_1 A_0 \int e^{-(k_1 - k_2)t}dt = -\frac{k_1 A_0}{k_1 - k_2} e^{-(k_1 - k_2)t}$$

so that

$$B(t) = e^{-k_2 t} \left[-\frac{k_1 A_0}{k_1 - k_2} \, e^{-(k_1 - k_2)t} + C \right] = -\frac{k_1 A_0}{k_1 - k_2} e^{-k_1 t} + C e^{-k_2 t}$$

Applying our initial condition at $t = 0$, we find

$$B(0) = 0 = -\frac{k_1 A_0}{k_1 - k_2} + C$$

so that

$$C = \frac{k_1 A_0}{k_1 - k_2}$$

leading to the solution

$$B(t) = \frac{k_1 A_0}{k_2 - k_1} \left[e^{-k_1 t} - e^{-k_2 t} \right] = \frac{k_1 A_0}{k_2 - k_1} e^{-k_1 t} \left[1 - e^{-(k_2 - k_1)t} \right] \qquad (10.14)$$

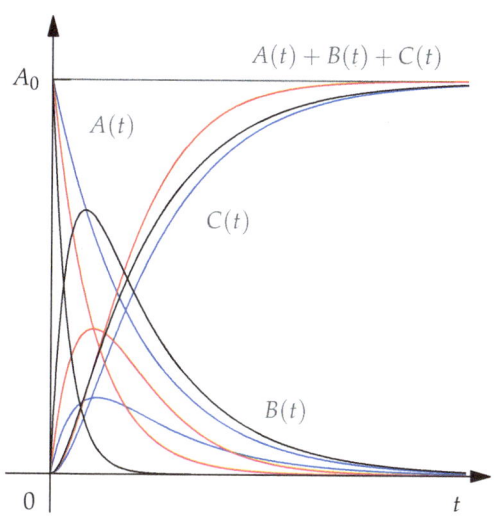

Figure 10.4: The general solution for the amounts $A(t)$, $B(t)$, and $C(t)$ as a function of time for three cases of $(k_1, k_2) = (1, 3)$ (blue), $(2, 2)$ (red), and $(5, 1)$ (black).

With general solutions for $A(t)$ and $B(t)$, we are left to solve for $C(t)$. We could approach this using our general solution for linear first order ordinary differential equations defined by Equation 10.7. However, we can also note that the total amount of $A(t) + B(t) + C(t) = $ constant. From the initial conditions $A(0) = A_0$ and $B(0) = C(0) = 0$, we can say that

$$C(t) = A_0 - (A(t) + B(t)) \qquad (10.15)$$

or that

$$C(t) = A_0 \left(1 - e^{-k_1 t} \right) - \frac{k_1 A_0}{k_2 - k_1} e^{-k_1 t} \left(1 - e^{-(k_2 - k_1)t} \right)$$

$$= A_0 \left(1 - \frac{k_2}{k_2 - k_1} e^{-k_1 t} + \frac{k_1}{k_2 - k_1} e^{-k_2 t} \right) \qquad (10.16)$$

$A(t)$, $B(t)$, and $C(t)$ are shown in Figure 10.4 for three sets of elementary rate constants (k_1, k_2).

10.2.3 Analyzing solutions using limiting laws and approximations

Let's explore our solution in three domains of rate constant parameter space, defined as

$$k_1 \ll k_2 \qquad k_1 = k_2 \qquad k_1 \gg k_2$$

We will find that in each regime our general solution takes on a specific form that provides insight into the underlying kinetics.[8]

[8] When a process occurs at a rate k, we say that the characteristic time scale for that process is $1/k$.

Case #1: $k_1 \gg k_2$

In the domain $k_1 \gg k_2$, we expect the rapid production of B from A with slower conversion of B to C. In this case, starting from the general solution

$$B(t) = \frac{k_1 A_0}{k_2 - k_1} \left(e^{-k_1 t} - e^{-k_2 t} \right)$$

we can approximate

$$\frac{k_1 A_0}{k_2 - k_1} \approx -\frac{k_1 A_0}{k_1} = -A_0$$

and

$$e^{-k_1 t} - e^{-k_2 t} \approx -e^{-k_2 t}$$

Substituting our approximate expression into our general solution, we find

$$B(t) \approx A_0\, e^{-k_2 t} \tag{10.17}$$

This result is identical to the result found in Equation 10.3 and for the

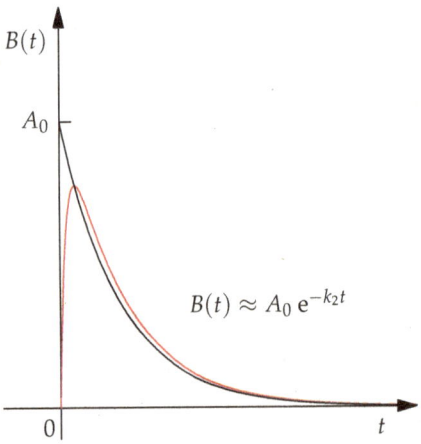

Figure 10.5: The amount of $B(t)$ as a function of time for $k_1 \gg k_2$. The exact solution (red) is compared to the approximate solution (black).

reaction A $\rightleftharpoons$ B in Equation 10.9 for the case that $k_f = k_2$ and $k_b = 0$. Our result reflects the fact that A is rapidly converted to B which is subsequently converted to C at a rate k_2. This behavior is depicted in Figure 10.5. While the approximate solution (black) deviates from the exact result at short times, it provides an accurate representation of the exact result (red) for longer times on the order of $t \gg \frac{1}{k_1}$.

Case #2: $k_1 \ll k_2$

In the domain $k_1 \ll k_2$, we expect that once B is made it is rapidly converted to C. Starting from the general solution

$$B(t) = \frac{k_1 A_0}{k_2 - k_1} \left(e^{-k_1 t} - e^{-k_2 t} \right)$$

we can approximate

$$\frac{k_1 A_0}{k_2 - k_1} \approx \frac{k_1 A_0}{k_2}$$

and $e^{-k_1 t} - e^{-k_2 t} \approx e^{-k_1 t}$. Substituting our approximate expressions into our general solution, we find

$$B(t) \approx \frac{k_1 A_0}{k_2} e^{-k_1 t} \tag{10.18}$$

The exponential decay at a rate k_1 reflects the fact that the rate of production of B is governed by the slow conversion of A to B at a rate k_1. The prefactor is an effective initial condition for the amount of B (see Figure 10.6).

Let's return to the original kinetic scheme and Equation 10.13 for the rate of change in the concentration of B, written

$$\frac{dB}{dt} + k_2 B = k_1 A_0 e^{-k_1 t}$$

Suppose we assume that the system is in a *steady state* condition in which the concentration of B is constant. As such

$$\frac{dB}{dt} = 0$$

Inserting and solving for $B(t)$ results in

$$B(t) \approx \frac{k_1 A_0}{k_2} e^{-k_1 t}$$

Our approximate expression is equivalent to assuming a steady state condition for the concentration of B. When $k_1 \ll k_2$, the amount of B diminishes slowly at a rate k_1 from an effective initial amount given by $B(0) \approx A_0 k_1 / k_2$. This behavior is depicted in Figure 10.6.

Case #3: $k_1 = k_2$

For the special case of $k_1 = k_2$, we note that

$$B(t) = \frac{k_1 A_0}{k_2 - k_1} \left(e^{-k_1 t} - e^{-k_2 t} \right)$$

which can also be written

$$B(t) = \frac{k_1 A_0}{k_2 - k_1} e^{-k_1 t} \left(1 - e^{-(k_2 - k_1)t} \right)$$

As $k_1 \to k_2$, the difference in the numerator

$$1 - e^{-(k_2 - k_1)t}$$

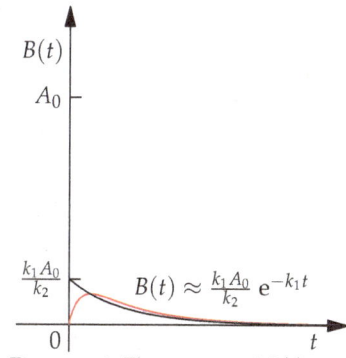

Figure 10.6: The amount of $B(t)$ as a function of time for $k_1 \ll k_2$. The exact solution (red) is compared to the approximate solution (black).

goes to zero. However, the difference in the denominator, $(k_2 - k_1)$, also goes to zero! We need to take care in evaluating $B(t)$ when $k_1 = k_2$.

We can evaluate this limit using l'Hôpital's rule or alternatively use the trick of expanding the exponential as a Taylor series, such that

$$e^{-(k_2 - k_1)t} = 1 - (k_2 - k_1)t + \dots$$

Retaining only the leading order term $(k_2 - k_1)t$, we find

$$B(t) = \frac{k_1 A_0}{k_2 - k_1} e^{-k_1 t} \left[1 - [1 - (k_2 - k_1)t] \right] = k_1 A_0 \, t e^{-k_1 t} \qquad (10.19)$$

Evaluating this solution, we see that initially B is produced from A at a rate k_1. This leads to a linear increase in $B(t) \approx k_1 A_0 t$ at short times where $t \ll 1/k_1$ and the exponential term is near unity. As B is produced, it is converted to C. At longer times, the conversion of B to C exceeds the production of A from B and the overall amount of B decreases. The amount of B reaches a maximum value when

$$\frac{d}{dt} B(t) = 0$$

which is determined by

$$\frac{d}{dt} B(t) = \frac{d}{dt} \, k_1 A_0 \, t e^{-k_1 t} = k_1 A_0 \, (1 - k_1 t) \, e^{-k_1 t} = 0$$

with the solution $t = 1/k_1$. This behavior is depicted in Figure 10.7.

Let's return to the original equation kinetic scheme, Equation 10.10, where we set $k_2 = k_1$, so that

$$\text{A} \xrightarrow{k_1} \text{B} \xrightarrow{k_1} \text{C}$$

With the initial conditions $A(0) = A_0$ and $B(0) = C(0) = 0$, we find

$$\frac{dB}{dt} + k_1 B = k_1 A_0 e^{-k_1 t}$$

with the solution

$$B(t) = e^{-\int k_1 dt} \left[\int k_1 A_0 e^{-k_1 t} e^{\int k_1 dt} dt + C \right]$$

$$= e^{-k_1 t} \, (k_1 A_0 t + C)$$

Evaluating this expression at $t = 0$, we find $B(0) = C = 0$, and the final result

$$B(t) = k_1 A_0 t e^{-k_1 t} \qquad (10.20)$$

As expected, this agrees with our result in Equation 10.19, found by evaluating the general solution in the limit $k_1 \to k_2$.

First order ordinary differential equations describing physical kinetics can be used to model population dynamics, the spread of infectious disease, predator-prey relations, and the reactions of atoms and molecules. These applications involving first order physical kinetics demonstrate the power of our general method for the solution of first order ordinary differential equations.

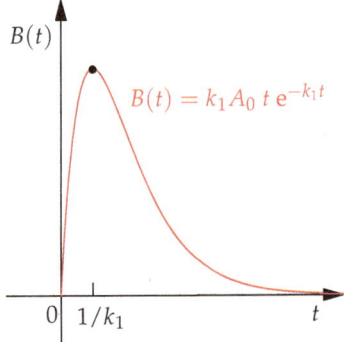

Figure 10.7: The amount of $B(t)$ as a function of time for $k_1 = k_2$. The dot marks the maximum value that occurs at $t = 1/k_1$.

A$_{10}$ Functions derived from exact differentials and integrating factors

Consider the *exact differential* of an unknown function $f(x,y)$

$$df = a(x,y)\, dx + b(x,y)\, dy$$

which satisfies Euler's test

$$\frac{\partial}{\partial y}a(x,y) = \frac{\partial}{\partial x}b(x,y)$$

How can we determine the function $f(x,y)$ for this exact differential? We approach this problem by first noting that

$$\frac{\partial f}{\partial x} = a(x,y)$$

so that

$$f(x,y) = \int df = \int a(x,y)dx + c(y) \qquad (10.21)$$

Equation 10.21 provides an expression for $f(x,y)$. If we can determine $c(y)$, we can define $f(x,y)$.[9]

Let's differentiate the proposed function $f(x,y)$ with respect to y as

$$\frac{\partial f}{\partial y} = \frac{\partial}{\partial y}\int a(x,y)dx + \frac{\partial}{\partial y}c(y) = b(x,y)$$

Rearranging this expression, we find

$$\frac{\partial}{\partial y}c(y) = b(x,y) - \frac{\partial}{\partial y}\int a(x,y)dx$$

We can then integrate this expression to determine $c(y)$:

$$c(y) = \int \left[b(x,y) - \frac{\partial}{\partial y}\int a(x,y)dx \right] dy$$

This results in our final expression for $f(x,y)$:

$$f(x,y) = \int a(x,y)dx + \int \left[b(x,y) - \frac{\partial}{\partial y}\int a(x,y)dx \right] dy \qquad (10.22)$$

This expression is an integral equation with a rather intimidating form. However, a few examples will demonstrate how to apply this method.

Suppose we have the total differential

$$dV = \left(\frac{\partial V}{\partial r}\right) dr + \left(\frac{\partial V}{\partial h}\right) dh = 2\pi rh\, dr + \pi r^2\, dh$$

and want to determine the function $V(r,h)$. Using our general result in Equation 10.22, we find

$$V(r,h) = \int 2\pi rh\, dr + \int \left(\pi r^2 - \frac{\partial}{\partial h}\int 2\pi rh\, dr \right) dh$$

[9] We can prove the last relation by noting that

$$\frac{\partial f}{\partial x} = \frac{\partial}{\partial x}\left[\int a(x,y)dx + c(y) \right] = a(x,y)$$

We can evaluate this expression as

$$V(r,h) = \pi r^2 h + \int \left[\pi r^2 - \frac{\partial}{\partial h} \left(\pi r^2 h \right) \right] dh + \mathcal{C}$$

$$= \pi r^2 h + \int \left(\pi r^2 - \pi r^2 \right) dh + \mathcal{C} = \pi r^2 h + \mathcal{C}$$

providing a form for the function $V(r,h)$. To determine the integration constant $\mathcal{C}$, we note that when $h = 0$, the volume should be zero, therefore $\mathcal{C} = 0$.

Now suppose we have an *inexact differential*

$$q = nC_V \, dT + \frac{nRT}{V} \, dV$$

that fails Euler's test

$$\frac{\partial}{\partial V} (nC_V) = 0 \qquad \frac{\partial}{\partial T} \left(\frac{nRT}{V} \right) = \frac{nR}{V}$$

It is possible to turn this inexact differential into an exact differential by applying an *integrating factor*. If we divide by T, we have a new total differential

$$\frac{q}{T} = \frac{nC_V}{T} \, dT + \frac{nR}{V} \, dV$$

which passes Euler's test

$$\frac{\partial}{\partial V} \left(\frac{nC_V}{T} \right) = 0 \qquad \frac{\partial}{\partial T} \left(\frac{nR}{V} \right) = 0$$

and is an exact differential. We can integrate this exact differential to determine a function that we will call

$$S(T,V) = \frac{q}{T}$$

We find that

$$S(T,V) = \int \frac{nC_V}{T} \, dT + \int \left(\frac{nR}{V} - \frac{\partial}{\partial V} \int \frac{nC_V}{T} dT \right) dV$$

$$= nC_V \ln(T) + \int \left(\frac{nR}{V} - \frac{\partial}{\partial V} nC_V \ln(T) \right) dV + \mathcal{C}$$

$$= nC_V \ln(T) + \int \frac{nR}{V} dV + \mathcal{C} = nC_V \ln(T) + nR \ln(V) + \mathcal{C}$$

To determine the integration constant $\mathcal{C}$, we note that when $n = 0$, the function $S = 0$, therefore $\mathcal{C} = 0$. Finally, we can prove our result by taking the total differential of $S(T,V)$ to find[10]

$$dS(T,V) = \left(\frac{\partial S}{\partial T} \right) dT + \left(\frac{\partial S}{\partial V} \right) dV = \frac{nC_V}{T} dT + \frac{nR}{V} dV$$

as desired. The function $S(V,T)$ and its first derivatives $S_V(V,T)$ (gray) and $S_T(V,T)$ are shown graphically in Figure 10.8.

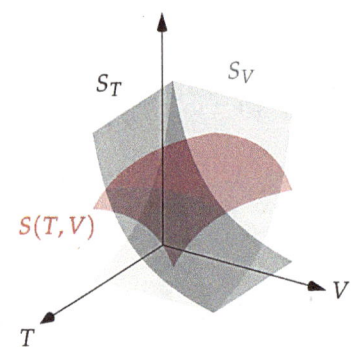

Figure 10.8: The function $S(V,T) = nC_V \ln(T) + nR \ln(V)$ (red) plotted with the first derivatives $S_T(T,V)$ (black) and $S_V(T,V)$ (gray).

[10] This recovered function can be recognized as the thermodynamic entropy $S(T,V)$ as a function of temperature T and volume V.

B$_{10}$ End-of-chapter problems

> It is still an unending source of surprise for me to see how a few scribbles on a blackboard or on a sheet of paper could change the course of human affairs.
>
> Stanislaw Ulam

Warm-ups

10.1 Determine the general solution, $y(x)$, for the following first order linear differential equations. Take k_1, k_2, and a to be constants.

(a) $\dfrac{dy(x)}{dx} + 3y(x) = 0$

(b) $x\dfrac{dy(x)}{dx} + y(x) = x^2 + 1$

(c) $\dfrac{dx(t)}{dt} = k_1(a - x(t)) - k_2 x(t)$

(d) $\dfrac{df(r)}{dr} + 4rf(r) = r$

10.2 Consider the differential equation

$$\frac{dy(x)}{dx} = x^2 - 3x^2 y(x)$$

(a) Find the general solution, $y(x)$.

(b) Prove that your solution is correct by inserting your result for $y(x)$ into the differential equation and confirming the equality.

10.3 Consider the differential equation

$$\frac{dy(x)}{dx} + \frac{2}{x}y(x) = x^2 + 2$$

(a) Find the general solution, $y(x)$.

(b) Prove that your solution is correct by inserting your result for $y(x)$ into the differential equation and confirming the equality.

10.4 Find the general solution, $f(r)$, to the following differential equation

$$r\frac{df(r)}{dr} + f(r) = 2r$$

with boundary condition $f(2) = 2$.

10.5 Find the general solution, $s(t)$, to the following differential equation

$$t\frac{ds(t)}{dt} = (3t + 1)s(t) + t^3 e^{3t}$$

with boundary condition $\left.\dfrac{ds}{dt}\right|_{t=0} = 1$.

Homework exercises

10.6 Find the general solution, $m(t)$, to the differential equation

$$\frac{dm(t)}{dt} + \frac{4m(t)}{20+t} = 2$$

with initial condition $m(0) = 20$.

10.7 Consider the Clapeyron equation, a first order differential equation describing the change in pressure, $p(T)$, with changing temperature, T

$$\frac{dp(T)}{dT} = \frac{\Delta H}{T(V_g - V_l)}$$

where ΔH is the enthalpy of vaporization and V_g and V_l are the volumes of the bulk substance in the gas and liquid phases, respectively. Since $V_g \gg V_l$, we find that

$$\frac{dp(T)}{dT} = \frac{\Delta H}{T(V_g - V_l)} \cong \frac{\Delta H}{T V_g}$$

Using the ideal gas equation of state, we arrive at the differential equation

$$\frac{dp(T)}{dT} = \frac{\Delta H}{RT^2} p(T)$$

(a) Solve this equation for $p(T)$. Your solution will include an undetermined constant C.

(b) Consider the log-lin plot below where $p(T_0) = p_0$ is the reference pressure at the reference temperature T_0 (red dot). Express the constant C in terms of p_0 and T_0.

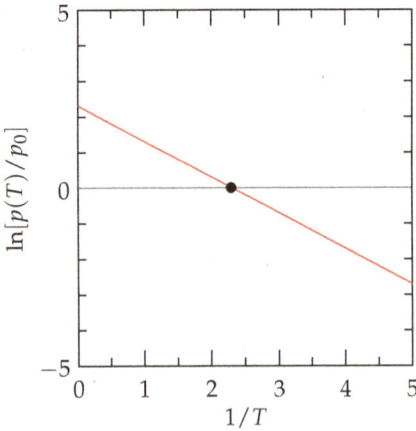

10.8 In physical kinetics, we frequently encounter the reaction

$$A \underset{k_b}{\overset{k_f}{\rightleftharpoons}} B$$

The corresponding rate equation is

$$\frac{dA}{dt} = -k_f A + k_b B$$

where k_f and k_b are the forward and backward rate constants. By conservation of mass, we know that

$$A(t) + B(t) = A_0 + B_0$$

with the initial conditions $A(0) = A_0$ and $B(0) = B_0$.

(a) Solve for $A(t)$ and $B(t)$.

(b) The reaction quotient $Q(t) = B(t)/A(t)$. At sufficiently long times we find

$$\lim_{t \to \infty} Q(t) = \frac{B(\infty)}{A(\infty)} = K_{eq}$$

Show that $K_{eq} = k_f/k_b$.

10.9* A container holds 50.0 L of a 1.00 mol/L solution. Pure water is added to the container at a rate of 1.00 L/s. The resulting solution is pumped out of the container at a rate of 0.50 L/s. How long will it take for the solution to reach a concentration of 0.05 mol/L?

10.10 Consider the following two total differentials for the pressure, p, as a function of temperature, T, and volume, V.

$$(a) \ \frac{2RT}{(V-b)^2} \, dV + \frac{R(V-b)}{b^2} \, dT \qquad (b) \ -\frac{RT}{(V-b)^2} \, dV + \frac{R}{(V-b)} \, dT$$

Determine if either or both are exact differentials.

10.11* Consider the exact differential for a function of two variables

$$dp(T, V) = \frac{nR}{V} \, dT - \frac{nRT}{V^2} \, dV$$

Use the method of integrating factors and Equation 10.22 to find the function $p(T, V)$. To validate your result, show that the function $p(V, T)$ has the correct total differential.

10.12* Consider the exact differential for a function of two variables

$$dE(T, V) = nC_V \, dT + \frac{n^2 a}{V^2} \, dV$$

Use the method of integrating factors and Equation 10.22 to find the function $E(T, V)$. To validate your result, show that the function $E(T, V)$ has the correct total differential.

10.13* Use Euler's test to prove that the total differential $f'(x)g(y)dx + f(x)g'(y)dy$ is an exact differential. Use the method of integrating factors and Equation 10.22 to find the corresponding function $h(x, y)$.

10.14* Consider the two coupled first order differential equations

$$\frac{dx(t)}{dt} = \frac{p(t)}{m} \qquad \frac{dp(t)}{dt} = -kx(t)$$

where k and m are constants. The functions $x(t)$ and $p(t)$ are related by the equation

$$2E = kx(t)^2 + \frac{1}{m}p(t)^2 = \text{constant}$$

which forms an ellipse in the xp-plane (see figure below).

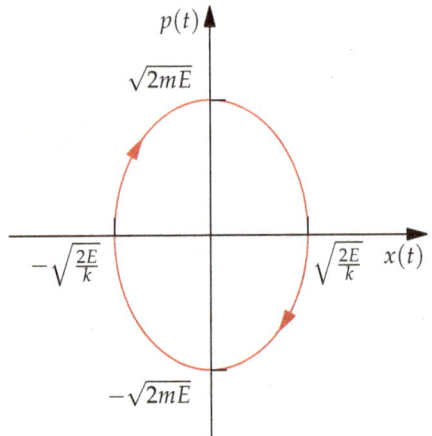

(a) Define the complex variable

$$\rho(t) = \sqrt{k}x(t) + i\frac{1}{\sqrt{m}}p(t)$$

where $x(t) = \frac{1}{\sqrt{k}}\,\text{Re}[\rho(t)]$ and $p(t) = \sqrt{m}\,\text{Im}[\rho(t)]$. Show that $|\rho|^2 = \rho^*\rho = 2E$.

(b) Show that the first order differential equation for the complex variable $\rho(t)$ written

$$\frac{d\rho(t)}{dt} = -i\omega\rho(t) \qquad \omega^2 = \frac{k}{m}$$

is equivalent to the two first order differential equations above for the real variables $x(t)$ and $p(t)$.

(c) Solve this equation to find

$$\rho(t) = \left(\sqrt{k}x_0 + i\frac{1}{\sqrt{m}}p_0\right)e^{-i\omega t}$$

where $x(0) = x_0$ and $p(0) = p_0$.

(d) Show that

$$x(t) = \frac{1}{\sqrt{k}}\,\text{Re}[\rho(t)] = x_0\cos(\omega t) + \frac{p_0}{m\omega}\sin(\omega t)$$
$$p(t) = \sqrt{m}\,\text{Im}[\rho(t)] = p_0\cos(\omega t) - m\omega x_0\sin(\omega t)$$

11 MORE ORDINARY DIFFERENTIAL EQUATIONS

11.1 Second order ordinary differential equations

SECOND ORDER DIFFERENTIAL EQUATIONS are central to the modeling of physical systems. While first order ordinary differential equations model processes with exponentially growing or diminishing observables, second order ordinary differential equations can be used to model a variety of phenomena including oscillatory motion. This section explores the general form of second order ordinary differential equations and diverse nature of the solutions.

11.1.1 General features of second order ordinary differential equations

In Chapter 10, we considered the first order ordinary differential equation

$$\frac{d}{dx}y(x) = ay(x)$$

We found the solution $y(x)$ must be a function of the form $y(x) = c_1 e^{ax}$ where c_1 is a constant, since the first derivative of the exponential function is proportional to the function itself. Similarly, the equation

$$\frac{d}{dx}y(x) = -ay(x)$$

will have a solution of the form $y(x) = c_1 e^{-ax}$. In each case, the constant coefficient c_1 is determined by a boundary condition.

Let's apply the same approach to the solution of the second order ordinary differential equation

$$\frac{d^2}{dx^2}y(x) = a^2 y(x) \tag{11.1}$$

where $a > 0$ and real. In this case, we seek a function $y(x)$ with second derivative proportional to a positive constant times the function itself. The exponential functions $c_1 e^{ax}$ and $c_2 e^{-ax}$ with real arguments possess that essential property, yielding the solution

$$y(x) = c_1 e^{ax} + c_2 e^{-ax}$$

where c_1 and c_2 are constant coefficients.[1]

[1] Since $y_1(x) = c_1 e^{ax}$ and $y_2(x) = c_2 e^{-ax}$ are each solutions to the linear second order differential equation, Equation 11.1, it follows that the linear *superposition* $y(x) = y_1(x) + y_2(x)$ is also a solution.

Now consider the equation

$$\frac{d^2}{dx^2}y(x) = -a^2\,y(x) \tag{11.2}$$

We seek a function $y(x)$ with a second derivative equal to a negative constant times the function itself. The exponential functions $c_1\,e^{iax}$ and $c_2\,e^{-iax}$ with imaginary arguments have that property.[2] As such, we can write the solution $y(x) = c_1\,e^{iax} + c_2\,e^{-iax}$ where the coefficients c_1 and c_2 are determined by the boundary conditions.

[2] We could equivalently propose solutions of the form $c_1'\cos(ax)$ and $c_2'\sin(ax)$.

Now consider the homogeneous linear second order ordinary differential equation

$$\frac{d^2}{dx^2}y(x) + b\,\frac{d}{dx}y(x) + c\,y(x) = 0 \tag{11.3}$$

with constant coefficients b and c. We propose the *ansatz*

$$y(x) = e^{\alpha x}$$

where α is a *complex number*. Inserting this form for $y(x)$, we find

$$\frac{d^2}{dx^2}e^{\alpha x} + b\,\frac{d}{dx}e^{\alpha x} + c\,e^{\alpha x} = \alpha^2 e^{\alpha x} + b\,\alpha e^{\alpha x} + c\,e^{\alpha x} = 0$$

Ignoring the *trivial solution* $e^{\alpha x} = 0$, we arrive at the result[3]

$$\alpha^2 + b\alpha + c = 0 \tag{11.4}$$

[3] A *trivial solution* is a formal solution to the problem from which we learn nothing.

This is known as the *auxiliary equation*. This quadratic equation in the variable α has two roots, $\alpha_\pm$, defined by the *quadratic formula*

$$\alpha_\pm = \frac{-b \pm \sqrt{b^2 - 4c}}{2}$$

Returning to our *ansatz*, we observe that both $c_1 e^{\alpha_+ x}$ and $c_2 e^{\alpha_- x}$ are solutions to Equation 11.3, where c_1 and c_2 are constant coefficients. As such, we write the general solution to Equation 11.3 as

$$y(x) = c_1\,e^{\alpha_+ x} + c_2\,e^{\alpha_- x} \tag{11.5}$$

where the constant coefficients c_1 and c_2 are determined by the *boundary conditions*, which are typically defined in terms of the value of the function, $y(0)$, and its derivative, $y'(0)$. However, knowledge of the function at two values of x can serve as well.

11.1.2 Qualitative theory of second order ordinary differential equations

Our observations above can be summarized as follows. For the second order differential equation

$$\frac{d^2}{dx^2}y(x) + b\,\frac{d}{dx}y(x) + c\,y(x) = 0$$

there is a general solution of the form

$$y(x) = c_1\,e^{\alpha_+ x} + c_2\,e^{\alpha_- x}$$

where the constants $\alpha_\pm$ are the roots of the auxiliary equation

$$\alpha^2 + b\alpha + c = 0$$

defined by the quadratic formula

$$\alpha_\pm = \frac{-b \pm \sqrt{b^2 - 4c}}{2}$$

The nature of the solution $y(x)$ is determined by the roots $\alpha_\pm$, which may be real, imaginary, complex, or degenerate (double) roots, depending on the particular values of the parameters b and c. Figure 11.1 shows the bc-parameter space divided into the four regimes.

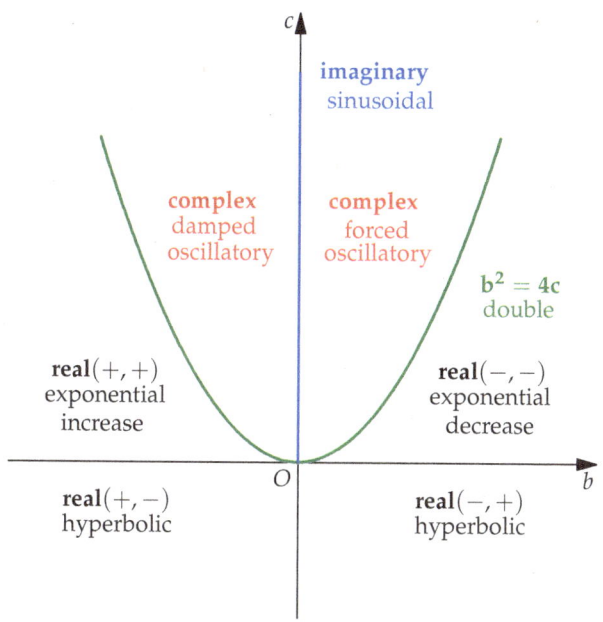

Figure 11.1: The nature of roots of the auxiliary equation as a function of the parameters b and c. Double roots are found on the parabola defined by $b^2 = 4c$ (thick green). Complex roots are found atop the parabolic curve for $c > b^2/4$. Purely imaginary roots occupy a surface normal to the bc-plane (thick blue). In this projection, imaginary roots are restricted to the positive c-axis. Below the parabolic curve the roots are real, forming exponential or hyperbolic functions.

When $b^2 > 4c$, the roots $\alpha_\pm$ are real. The solution $y(x)$ is exponentially increasing or decreasing with increasing x. Real roots support solutions involving exponentially decaying or growing functions of the kind found in physical kinetics, capturing the behavior of populations or concentrations that grow or diminish exponentially in time.

When $b = 0$, the roots $\alpha_\pm$ are imaginary. The solution $y(x)$ is a purely oscillatory function of x. Imaginary roots support solutions involving oscillatory functions of the kind used to model a vibrating mass on a spring and waves found in the study of quantum theory.

When $b^2 < 4c$, the roots $\alpha_\pm$ are complex. The solution $y(x)$ oscillates with an amplitude that exponentially increases or decreases with increasing x. Complex roots capture damped oscillatory motion as for found an oscillator feeling the effects of friction that gradually attenuate the amplitude of motion. Solutions with complex roots are also used to model propagating electromagnetic waves.

When $b^2 = 4c$, we find the special case of double roots where $\alpha_+ = \alpha_- = -b/2$. Double roots are a special coincidence in the values of the b and c parameters that we do not expect to find in nature. Examples of the various kinds of qualitative behavior resulting from Equation 11.3 are presented below in Figure 11.2.

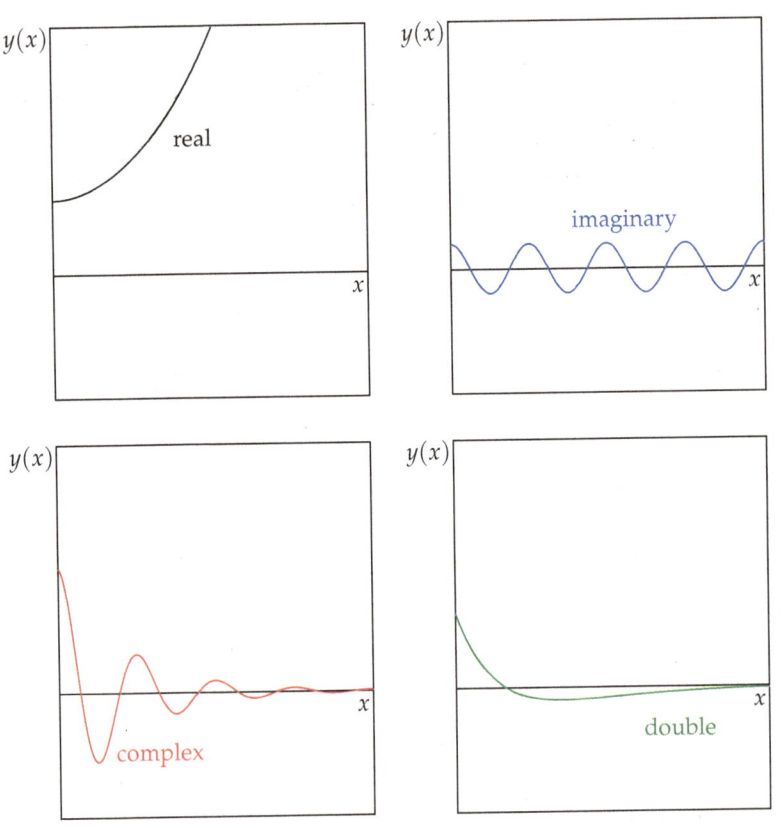

Figure 11.2: The nature of the roots of the auxiliary equation, Equation 11.4, determine the character of the solution to Equation 11.3. Examples of solutions with real roots (upper left), imaginary roots (upper right), complex roots (lower left), and double roots (lower right).

In summary, our analysis demonstrates that Equation 11.3 can be used to model a wide variety of physical phenomena.

$$\frac{d^2}{dx^2}y(x) + b\frac{d}{dx}y(x) + c\,y(x) = 0$$

The particular choice of parameters b and c lead to roots

$$\alpha_\pm = \frac{-b \pm \sqrt{b^2 - 4c}}{2}$$

that determine the nature of the solution

$$y(x) = c_1\,e^{\alpha_+ x} + c_2\,e^{\alpha_- x}$$

In the next section, we explore the solution of the homogeneous linear second order ordinary differential equation, Equation 11.3, for a variety of examples relevant to modeling processes in the physical sciences.

11.2 Applications of second order differential equations

SECOND ORDER DIFFERENTIAL EQUATIONS with constant coefficients yield solutions exhibiting a wide range of behavior, including exponential growth or decay, sinusoidal oscillation, and exponentially attenuated oscillation. This section explores applications exhibiting the variety of behavior found in solutions of second order ordinary differential equations.

11.2.1 Survey of second order ordinary differential equations

In this section, we explore specific examples of differential equations representing domains with real roots, imaginary roots, complex roots, and double roots. In doing so, we will appreciate how this variety of solutions can be used to model a remarkable diversity of physical systems.

Case #1: real roots

When the coefficients b and c are such that $b^2 > 4c$, the roots

$$\alpha_{\pm} = \frac{-b \pm \sqrt{b^2 - 4c}}{2}$$

are real. For the case $c = 0$, there will be one zero root and one non-zero root equal to $-b$, leading to a solution of the form

$$y(x) = c_1 + c_2\, e^{-bx}$$

For the more general case of $c \neq 0$ and $b^2 > 4c$, the roots of the auxiliary equation $\alpha_{\pm}$ are real and the solution $y(x)$ will have exponentially divergent behavior as a function of x.

When the coefficients c_1 and c_2 are of the same sign, the solution $y(x)$ has the character of a *hyperbolic cosine* (see Figure 11.3, red line). In contrast, when the coefficients c_1 and c_2 are of opposite sign, the general solution has the character of a *hyperbolic sine* (see Figure 11.3, black line). In each case, the solution $y(x)$ diverges with increasing x. Exponential growth of this kind is observed in population growth as observed in Chapter 10.

Let's explore the solution of the second order ordinary differential equation

$$\frac{d^2}{dx^2}y(x) + \frac{d}{dx}y(x) - 2\,y(x) = 0 \qquad (11.6)$$

with the boundary conditions $y(0) = 3$ and $y'(0) = 0$. We take the *ansatz*

$$y(x) = e^{\alpha x}$$

and insert it into Equation 11.6 with the result

$$\frac{d^2}{dx^2}e^{\alpha x} + \frac{d}{dx}e^{\alpha x} - 2\,e^{\alpha x} = \alpha^2\,e^{\alpha x} + \alpha\,e^{\alpha x} - 2\,e^{\alpha x} = 0$$

leading to the auxiliary equation

$$\alpha^2 + \alpha - 2 = (\alpha - 1)(\alpha + 2) = 0$$

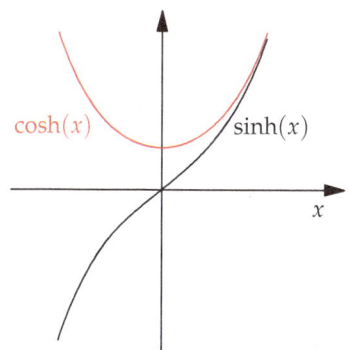

Figure 11.3: The hyperbolic functions $\sinh(x) = (e^x - e^{-x})/2$ (black line) and $\cosh(x) = (e^x + e^{-x})/2$ (red line).

This quadratic equation has real roots

$$\alpha_\pm = \begin{cases} 1 \\ -2 \end{cases}$$

leading to the general solution defined by Equation 11.5 as

$$y(x) = c_1 e^x + c_2 e^{-2x}$$

The coefficients c_1 and c_2 are determined by applying the boundary conditions

$$y(0) = c_1 + c_2 = 3$$

and

$$\left. \frac{dy}{dx} \right|_{x=0} = \left[c_1 e^x - 2c_2 e^{-2x} \right]_{x=0} = c_1 - 2c_2 = 0$$

We find $c_1 = 3 - c_2$ and $c_1 = 2c_2$ so that $c_1 = 2$ and $c_2 = 1$. The final solution to Equation 11.6 is

$$y(x) = 2 e^x + e^{-2x} \tag{11.7}$$

This result is plotted in Figure 11.4. Note that we can validate our solution, Equation 11.7, by inserting it into the original differential equation, Equation 11.6, and proving the equality.

Case #2: imaginary roots

When the coefficients b and c are such that $b^2 < 4c$ and $b = 0$, the roots

$$\alpha_\pm = \pm i \sqrt{c}$$

are purely imaginary, leading to solutions of the form

$$y(x) = c_1 e^{i\sqrt{c}x} + c_2 e^{-i\sqrt{c}x} \tag{11.8}$$

This solution appears to have real and imaginary parts. However, we expect the solution to be purely real. We will find the boundary conditions result in coefficients c_1 and c_2 that make the solution $y(x)$ purely real.

Recall that Euler's formula provides a connection between exponentials of imaginary arguments and the cosine and sine functions:

$$e^{ix} = \cos(x) + i\sin(x)$$

Using this identity, we can reformulate Equation 11.8 as

$$y(x) = c_1' \cos(\sqrt{c}x) + c_2' \sin(\sqrt{c}x)$$

where $c_1' = c_1 + c_2$ and $c_2' = i(c_1 - c_2)$. When the roots of the auxiliary equation are imaginary and the coefficients c_1 and c_2 are real, the resulting solution $y(x)$ is a real sinusoidal function of x.

Let's apply this method to solve the second order ordinary differential equation

$$\frac{d^2}{dx^2} y(x) + 9 \, y(x) = 0 \tag{11.9}$$

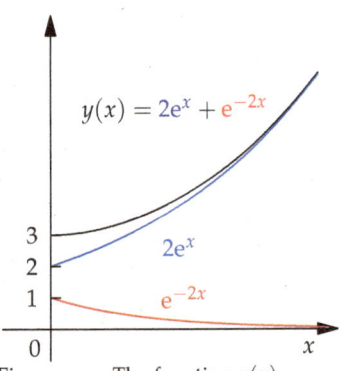

Figure 11.4: The function $y(x) = 2\exp(x) + \exp(-2x)$ (black) as a function of x. Shown for comparison are the two contributions forming the superposition $y(x)$, each of which is a solution to Equation 11.6.

with the boundary conditions $y(0) = 1$ and $y'(0) = 6$. We take the *ansatz*

$$y(x) = e^{\alpha x}$$

and insert it into Equation 11.9 as

$$\frac{d^2}{dx^2}e^{\alpha x} + 9\,e^{\alpha x} = \alpha^2\,e^{\alpha x} + 9\,e^{\alpha x} = 0$$

leading to the auxiliary equation

$$\alpha^2 + 9 = (\alpha - 3i)(\alpha + 3i) = 0$$

This quadratic equation has imaginary roots

$$\alpha_\pm = \begin{cases} 3i \\ -3i \end{cases}$$

leading to the general solution defined by Equation 11.5:

$$y(x) = c_1'\,\cos(3x) + c_2'\,\sin(3x)$$

The coefficients c_1' and c_2' can be determined by applying the boundary conditions

$$y(0) = c_1' = 1$$

and

$$\left.\frac{dy}{dx}\right|_{x=0} = [-3c_1'\,\sin(3x) + 3c_2'\,\cos(3x)]_{x=0} = 3c_2' = 6$$

We find $c_1' = 1$ and $c_2' = 2$. The final solution to Equation 11.9 is

$$y(x) = \cos(3x) + 2\sin(3x) \tag{11.10}$$

This result is plotted in Figure 11.5. Note that we can validate our solution, Equation 11.10, by inserting it into the original differential equation, Equation 11.9, and proving the equality.

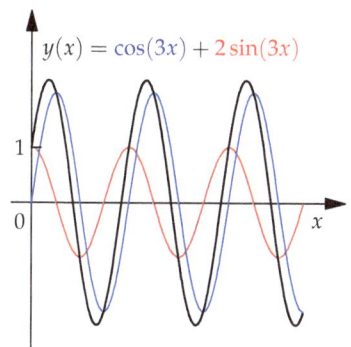

Figure 11.5: The solution to Equation 11.9 is the sinusoidal function $y(x) = 2\cos(3x) + \sin(3x)$ characterized by imaginary roots of the auxiliary equation and satisfying the boundary conditions $y(0) = 1$ and $y'(0) = 6$.

Case #3: complex roots

When the coefficients b and c are such that $b^2 < 4c$, the radical $\sqrt{b^2 - 4c}$ is imaginary. When $b \neq 0$, the roots

$$\alpha_\pm = \frac{-b \pm \sqrt{b^2 - 4c}}{2} = \frac{-b \pm \sqrt{c'}}{2}$$

are complex with real and imaginary parts. As a result, solutions are of the form

$$y(x) = c_1\,\exp\left[-\frac{1}{2}\left(b - i\sqrt{c'}\right)x\right] + c_2\,\exp\left[-\frac{1}{2}\left(b + i\sqrt{c'}\right)x\right]$$

where $c' = b^2 - 4c$. Pulling out the common exponential factor $e^{-bx/2}$ leads to

$$y(x) = e^{-bx/2}\left(c_1\,e^{i\sqrt{c'}x/2} + c_2\,e^{-i\sqrt{c'}x/2}\right)$$

Using Euler's formula, we can convert the sum of exponentials with imaginary arguments to sinusoidal functions with the result

$$y(x) = e^{-bx/2} \left[c_1' \cos\left(\sqrt{c'}x/2\right) + c_2' \sin\left(\sqrt{c'}x/2\right) \right]$$

where again $c_1' = c_1 + c_2$ and $c_2' = i(c_1 - c_2)$. The solutions $y(x)$ are exponentially decreasing (when $b > 0$) or increasing (when $b < 0$) sinusoidal functions.

Let's apply this method to solve the second order ordinary differential equation

$$\frac{d^2}{dx^2}y(x) + 2\frac{d}{dx}y(x) + 10y(x) = 0 \tag{11.11}$$

with the boundary conditions $y(0) = 1$ and $y'(0) = 0$. We take the *ansatz*

$$y(x) = e^{\alpha x}$$

and insert it into Equation 11.11 as

$$\frac{d^2}{dx^2}e^{\alpha x} + 2\frac{d}{dx}e^{\alpha x} + 10e^{\alpha x} = \alpha^2 e^{\alpha x} + 2\alpha e^{\alpha x} + 10e^{\alpha x} = 0$$

leading to the auxiliary equation

$$\alpha^2 + 2\alpha + 10 = 0$$

This quadratic equation has complex roots[4]

$$\alpha_{\pm} = \begin{cases} -1 + 3i \\ -1 - 3i \end{cases}$$

[4] Note that α_+ and α_- are complex conjugates so that $\alpha_+ = \alpha_-^*$.

leading to the general solution defined by Equation 11.5 as

$$y(x) = e^{-x}\left(c_1 e^{3ix} + c_2 e^{-3ix}\right)$$

and more conveniently written as

$$y(x) = e^{-x}\left[c_1' \cos(3x) + c_2' \sin(3x)\right]$$

The coefficients c_1' and c_2' can be determined by applying the boundary conditions

$$y(0) = c_1' = 1$$

and

$$\begin{aligned} \frac{dy}{dx}\Big|_{x=0} &= \Big[-e^{-x}\left[c_1' \cos(3x) + c_2' \sin(3x)\right] \\ &\quad + e^{-x}\left[-3c_1' \sin(3x) + 3c_2' \cos(3x)\right] \Big]_{x=0} \\ &= -c_1' + 3c_2' = 0 \end{aligned}$$

We find $c_1' = 1$ and $c_2' = \frac{1}{3}$. The final solution to Equation 11.11 is

$$y(x) = e^{-x}\left[\cos(3x) + \frac{1}{3}\sin(3x)\right] \tag{11.12}$$

This result is plotted in Figure 11.6 (black line). The solution exhibits oscillations that exponentially decay with increasing x. Note that we can validate our solution, Equation 11.12, by inserting it into the original differential equation, Equation 11.11, and proving the equality.

Now consider the second order ordinary differential equation

$$\frac{d^2}{dx^2}y(x) - 2\frac{d}{dx}y(x) + 10y(x) = 0 \qquad (11.13)$$

where we have changed the coefficient of the middle term from $b = 2$ in Equation 11.11 to $b = -2$. The roots of the auxiliary equation become

$$\alpha_{\pm} = \begin{cases} 1 + 3i \\ 1 - 3i \end{cases}$$

leading to the general solution

$$y(x) = e^x \left[c_1' \cos(3x) + c_2' \sin(3x) \right]$$

The decaying exponential in Equation 11.12 has been replaced with a growing exponential. Applying the boundary conditions $y(0) = 0$ and $y'(0) = 1/200$, the final result is

$$y(x) = \frac{1}{600} e^x \sin(3x)$$

as shown in Figure 11.6 (red line). The solution exhibits oscillations that exponentially grow in amplitude with increasing x.

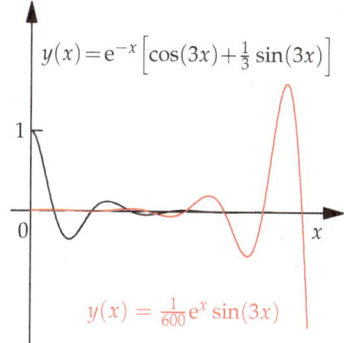

Figure 11.6: The solutions to Equations 11.11 and 11.13 are characterized by complex roots of the auxiliary equation resulting in sinusoidal functions with amplitudes that diminish (black line) or grow (red line) exponentially with increasing x.

Case #4: double roots

When the coefficients b and c are such that $b^2 = 4c$, the radical $\sqrt{b^2 - 4c}$ is zero and the two roots of the auxiliary equation are degenerate:

$$\alpha_{\pm} = \alpha = -\frac{b}{2}$$

This case is known as *double roots*. In the case of double roots, the solution takes the special form

$$y(x) = c_1 e^{\alpha x} + c_2 x e^{\alpha x}$$

Due to the degeneracy in the roots, each exponential varies at the same rate.

Let's prove this is true for the second order ordinary differential equation

$$\frac{d^2}{dx^2}y(x) + 2\frac{d}{dx}y(x) + y(x) = 0 \qquad (11.14)$$

with boundary conditions $y(0) = 1$ and $y'(0) = -3$. The solution to the corresponding auxiliary equation

$$\alpha^2 + 2\alpha + 1 = (\alpha + 1)(\alpha + 1) = 0$$

is the double root $\alpha = -1$ leading to the special solution

$$y(x) = c_1 e^{-x} + c_2 x e^{-x}$$

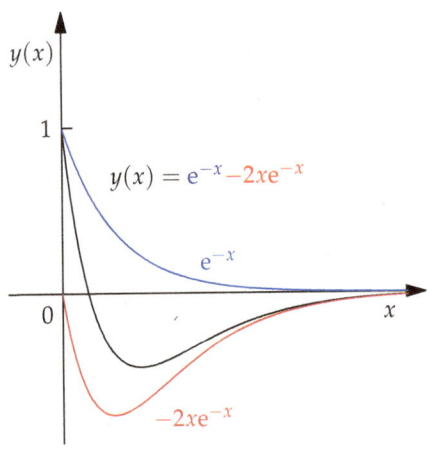

Figure 11.7: The solution for double roots $y(x) = \exp(-x) - 2x \exp(-x)$ (black) as a function of x. Shown for comparison are the two contributions forming $y(x)$, each of which is a solution to Equation 11.14.

Applying the boundary conditions

$$y(0) = c_1 = 1$$

and

$$\left.\frac{dy}{dx}\right|_{x=0} = \left[-c_1 e^{-x} + c_2 \left(e^{-x} - xe^{-x}\right)\right]_{x=0} = -c_1 + c_2 = -3$$

we find that $c_1 = 1$ and $c_2 = -2$. The final result is

$$y(x) = e^{-x} - 2xe^{-x} \tag{11.15}$$

as is shown in Figure 11.7. Checking our solution, we use $y'(x) = -3e^{-x} + 2xe^{-x}$ and $y''(x) = 5e^{-x} - 2xe^{-x}$. Inserting these relations into our original Equation 11.14 yields

$$\frac{d^2}{dx^2}y(x) + 2\frac{d}{dx}y(x) + y(x) = \left(5e^{-x} - 2xe^{-x}\right) + 2\left(-3e^{-x} + 2xe^{-x}\right)$$
$$+ \left(e^{-x} - 2xe^{-x}\right) = 0$$

validating our solution.

When we model a system using Equation 11.3 written as

$$\frac{d^2}{dx^2}y(x) + b\frac{d}{dx}y(x) + cy(x) = 0$$

We expect the coefficients b and c to be real numbers defined by fundamental physical constants. The coincidence of having an auxiliary equation with the exact equality $b^2 = 4c$ is unlikely. As such, we do not expect to encounter double roots when solving accurately parameterized models of physical systems.

In this section, we have developed a general method for the solution of linear second order ordinary differential equations with constant coefficients b and c, as in Equation 11.3. In the section that follows, we explore a more general class of second order ordinary differential equations, which will be applicable to cases where the coefficients $b(x)$ and $c(x)$ are functions of x.

11.3 Power series solutions to differential equations

MANY FIRST AND SECOND ORDER ORDINARY DIFFERENTIAL EQUATIONS yield solutions represented by common functions such as the exponential, sine and cosine, and hyperbolic sine and cosine. Our approach has been to make an educated guess, or *ansatz*, plug that in, and find a set of parameters representing a solution. However, it is possible to employ a more general approach to the solution of ordinary differential equations in which our *ansatz* is a power series representing continuous and differentiable functions normally encountered in modeling physical systems. In this section, we will explore this general and powerful approach to the solution of first and second order ordinary differential equations.

11.3.1 *Solving ordinary differential equations using power series*

Let's review our current ability to solve first and second order ordinary differential equations. We developed a general solution for homogeneous and non-homogeneous linear first order ordinary differential equations of the form

$$\frac{d}{dx}y(x) + q(x)y(x) = r(x)$$

and homogeneous linear second order ordinary differential equations with constant coefficients[5]

$$\frac{d^2}{dx^2}y(x) + b\frac{d}{dx}y(x) + c\,y(x) = 0$$

[5] While it is possible to form ordinary differential equations of arbitrary order, in the study of thermodynamics, quantum theory, and kinetics, we typically encounter first and second order equations.

However, we have no general solution for non-homogeneous linear second order ordinary differential equations of the form

$$\frac{d^2}{dx^2}y(x) + b(x)\frac{d}{dx}y(x) + c(x)\,y(x) + d(x) = 0$$

We also have no general solution for homogeneous linear second order ordinary differential equations where $d(x) = 0$ but the coefficients $b(x)$ and $c(x)$ depend on x.

How can we solve this more general differential equation when we cannot anticipate the specific form of our solution? We recall that most functions used to model physical properties can be expressed as a *power series*

$$y(x) = \sum_{n=0}^{\infty} a_n x^n$$

Depending on the specific set of coefficients $\{a_n\}$, the function $y(x)$ may be an exponential, sinusoid, damped sinusoid, polynomial, logarithm, or a variety of other functions. With this insight, we propose the power series as our *ansatz*, representing a general continuous and differentiable function. Inserting our proposed solution into the differential equation of interest, we solve for the coefficients $\{a_n\}$ defining our solution. Let's see how this works by solving two familiar ordinary differential equations using the power series method.

11.3.2 Power series solutions for first order differential equations

Let's apply this approach to solve the familiar first order ordinary differential equation

$$\frac{d}{dx}y(x) + y(x) = 0 \tag{11.16}$$

with the boundary condition $y(0) = 2$. As our *ansatz*, we propose the power series

$$y(x) = \sum_{n=0}^{\infty} a_n x^n \tag{11.17}$$

Inserting this proposed solution in Equation 11.16, we must evaluate the first derivative of $y(x)$ with respect to x as

$$\frac{d}{dx}y(x) = \frac{d}{dx}\sum_{n=0}^{\infty} a_n x^n = \sum_{n=0}^{\infty} a_n \left(\frac{d}{dx}x^n\right) = \sum_{n=0}^{\infty} a_n n x^{n-1}$$

Using this result in Equation 11.16, we find

$$\frac{d}{dx}y(x) + y(x) = \sum_{n=0}^{\infty} a_n n x^{n-1} + \sum_{n=0}^{\infty} a_n x^n = 0$$

To simplify this result, we must add the two power series together. To do that, we must form each series over the same range of n, with terms of a given n having the same power of x.

We are left to combine the two series. In the first series, the coefficient is proportional to n so that the $n = 0$ term is always zero. Removing the $n = 0$ term from the first series results in

$$\sum_{n=0}^{\infty} a_n n x^{n-1} \rightarrow \sum_{n=1}^{\infty} a_n n x^{n-1} = a_1 + 2a_2 x + 3a_3 x^2 + \ldots$$

where we have increased the lower end of the index n from 0 to 1. We are left to add the two series

$$\sum_{n=1}^{\infty} a_n n x^{n-1} + \sum_{n=0}^{\infty} a_n x^n = 0$$

This first power series starts from $n = 1$, but we would like this series to start from $n = 0$. To accomplish this, we shift the index $n \rightarrow n + 1$ with the result

$$\sum_{n=1}^{\infty} a_n n x^{n-1} \rightarrow \sum_{n=0}^{\infty} a_{n+1}(n+1)x^n = a_1 + 2a_2 x + 3a_3 x^2 + \ldots$$

Properly shifting the index in a power series can be tricky, but we will learn to do this with practice.[6]

Returning to our initial Equation 11.16, we insert our final form for the derivative of $y(x)$. We find

$$\frac{d}{dx}y(x) + y(x) = \sum_{n=0}^{\infty} a_{n+1}(n+1)x^n + \sum_{n=0}^{\infty} a_n x^n$$

$$= \sum_{n=0}^{\infty} \left[a_{n+1}(n+1) + a_n\right] x^n = 0 \tag{11.18}$$

[6] Equivalently, you can introduce a new index $n' = n - 1$, substituting $n \rightarrow n' + 1$. You can then rewrite the new index n' as n before combining the sums. Try both approaches and see what works best for you.

For Equation 11.18 to be true for all x, it must be that

$$a_{n+1}(n+1) + a_n = 0$$

for all values of n. This leads to the *recursion relation* that is written as

$$a_{n+1} = -\frac{a_n}{(n+1)} \qquad n = 0, 1, 2, \ldots$$

This expression can be used to recursively generate our coefficients (see Figure 11.8).[7]

[7] The negative sign in the recursion relation signals the fact that the coefficients lead to an alternating series.

Figure 11.8: The power series coefficients $a_n = (-1)^n a_0/n!$ (red dots) for $a_0 = 1$.

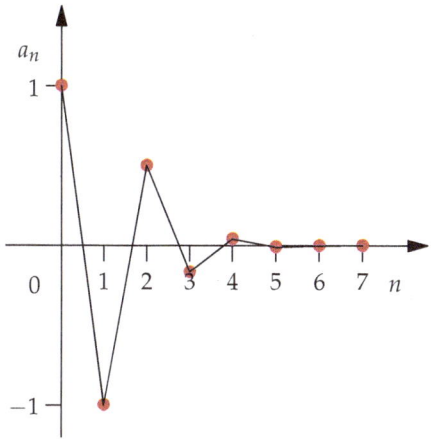

We take our first coefficient to be a_0, a value we will determine later using our boundary conditions. We can generate the coefficient a_1 as

$$a_1 = -\frac{a_0}{0+1} = -a_0$$

where the negative sign in the recursion relation tells us that our solution for $y(x)$ will be an alternating series. We can continue to generate coefficients through recursion, including

$$a_2 = -\frac{a_1}{1+1} = -\frac{1}{2}a_1 = -\frac{1}{2}(-a_0) = \frac{1}{2}a_0$$

and

$$a_3 = -\frac{a_2}{2+1} = -\frac{1}{3}a_2 = -\frac{1}{3}\left(\frac{1}{2}a_0\right) = -\frac{1}{3 \cdot 2}a_0$$

and so on. In general, we recognize that

$$a_n = (-1)^n \frac{1}{n!}a_0 \qquad n = 0, 1, 2, \ldots$$

Inserting this result for our coefficients in our original power series from Equation 11.17, we find

$$y(x) = \sum_{n=0}^{\infty} a_n x^n = \sum_{n=0}^{\infty}(-1)^n \frac{1}{n!}a_0 x^n = a_0 \sum_{n=0}^{\infty}(-1)^n \frac{1}{n!}x^n$$

We recognize this result as the power series of an exponential

$$e^{-x} = \sum_{n=0}^{\infty} (-1)^n \frac{1}{n!} x^n$$

so that

$$y(x) = a_0 e^{-x}$$

Applying our initial condition that $y(0) = 2$, we find $y(0) = a_0 = 2$, leading to our final solution to Equation 11.16

$$y(x) = 2e^{-x}$$

Our solution can be validated by inserting our result for $y(x)$ into Equation 11.16 and proving the equality

$$\frac{d}{dx} y(x) + y(x) = \frac{d}{dx} 2e^{-x} + 2e^{-x} = -2e^{-x} + 2e^{-x} = 0$$

11.3.3 *Power series solutions for second order differential equations*

Consider the second order ordinary differential equation

$$\frac{d^2}{dx^2} y(x) + y(x) = 0 \tag{11.19}$$

with the initial conditions $y(0) = 3$ and $y'(0) = 1$. As our *ansatz*, we propose

$$y(x) = \sum_{n=0}^{\infty} a_n x^n \tag{11.20}$$

Inserting this proposed solution in Equation 11.19, we must evaluate the second derivative of $y(x)$ with respect to x as

$$\frac{d^2}{dx^2} y(x) = \frac{d^2}{dx^2} \sum_{n=0}^{\infty} a_n x^n = \sum_{n=0}^{\infty} a_n \left(\frac{d^2}{dx^2} x^n \right) = \sum_{n=0}^{\infty} a_n n(n-1) x^{n-2}$$

Inserting this result in Equation 11.19, we find

$$\frac{d^2}{dx^2} y(x) + y(x) = \sum_{n=0}^{\infty} a_n n(n-1) x^{n-2} + \sum_{n=0}^{\infty} a_n x^n = 0$$

We are left to combine the two series. In the first series, the coefficient is proportional to $n(n-1)$ so that the $n = 0$ and $n = 1$ terms are always zero. Removing the $n = 0$ and $n = 1$ terms from the first series results in

$$\sum_{n=0}^{\infty} a_n n(n-1) x^{n-2} \to \sum_{n=2}^{\infty} a_n n(n-1) x^{n-2}$$

This leaves us with

$$\sum_{n=2}^{\infty} a_n n(n-1) x^{n-2} + \sum_{n=0}^{\infty} a_n x^n = 0$$

The first power series starts from $n = 2$, but we would like this series to start from $n = 0$ as we must eventually combine the first series with the second series that starts from $n = 0$. To accomplish this, we shift the index $n \to n + 2$ by substituting $n \to n + 2$, $n - 1 \to n + 1$, and $n - 2 \to n$, with the result[8]

$$\sum_{n=2}^{\infty} a_n n(n-1) x^{n-2} \to \sum_{n=0}^{\infty} a_{n+2}(n+2)(n+1) x^n$$

Returning to our initial equation, we find

$$\frac{d^2}{dx^2} y(x) + y(x) = \sum_{n=0}^{\infty} a_{n+2}(n+2)(n+1) x^n + \sum_{n=0}^{\infty} a_n x^n$$

$$= \sum_{n=0}^{\infty} \left[a_{n+2}(n+2)(n+1) + a_n \right] x^n = 0$$

This leads to the condition

$$a_{n+2}(n+2)(n+1) + a_n = 0$$

and the recursion relation

$$a_{n+2} = -\frac{1}{(n+2)(n+1)} a_n \qquad n = 0, 1, 2, \ldots$$

which we will use to generate the set of power series coefficients (see Figure 11.9).[9] The coefficients a_0 and a_1 will be determined through our boundary conditions. All other coefficients will be determined in terms of a_0 and a_1 through the recursion relation.

The coefficients with even indices are

$$a_0 \qquad a_2 = -\frac{a_0}{2 \cdot 1} \qquad a_4 = -\frac{a_2}{4 \cdot 3} = \frac{a_0}{4!} \qquad a_6 = -\frac{a_4}{6 \cdot 5} = -\frac{a_0}{6!}$$

and so on. As such, we find

$$a_{2n} = (-1)^n \frac{1}{(2n)!} a_0 \qquad n = 0, 1, 2, \ldots \qquad (11.21)$$

The coefficients with odd indices are

$$a_1 \qquad a_3 = -\frac{a_1}{3 \cdot 2} \qquad a_5 = -\frac{a_3}{5 \cdot 4} = \frac{a_1}{5!} \qquad a_7 = -\frac{a_5}{7 \cdot 6} = -\frac{a_1}{7!}$$

and so on, such that

$$a_{2n+1} = (-1)^n \frac{1}{(2n+1)!} a_1 \qquad n = 0, 1, 2, \ldots \qquad (11.22)$$

As we have identified the coefficients, we can form a series solution for $y(x)$.

We will form two series, one for even powers of x and one for odd powers of x, and combine those results. Inserting Equation 11.21 for even powers of x in our original power series Equation 11.20 results in[10]

$$a_0 \sum_{n=0}^{\infty} (-1)^n \frac{1}{(2n)!} x^{2n}$$

Similarly, inserting Equation 11.22 for odd powers of x into our original

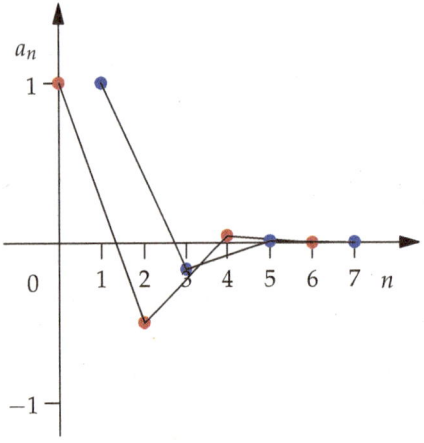

Figure 11.9: The power series coefficients $a_{2n} = (-1)^n a_0 / (2n)!$ for even n and $a_0 = 1$ (red dots) and $a_{2n+1} = (-1)^n a_1 / (2n+1)!$ for odd n and $a_1 = 1$ (blue dots).

power series Equation 11.20 results in the following power series

$$a_1 \sum_{n=0}^{\infty} (-1)^n \frac{1}{(2n+1)!} x^{2n+1}$$

Combining these results we arrive at the result

$$y(x) = a_0 \sum_{n=0}^{\infty} (-1)^n \frac{1}{(2n)!} x^{2n} + a_1 \sum_{n=0}^{\infty} (-1)^n \frac{1}{(2n+1)!} x^{2n+1}$$

Referring to our results for power series expansions of familiar functions explored in Chapter 8, including Equation 8.4 and Equation 8.3, we recognize that[11]

$$\cos(x) = \sum_{n=0}^{\infty} \frac{(-1)^n}{(2n)!} x^{2n} \qquad \sin(x) = \sum_{n=0}^{\infty} \frac{(-1)^n}{(2n+1)!} x^{2n+1}$$

We can reformulate our result as

$$y(x) = a_0 \cos(x) + a_1 \sin(x)$$

Applying the boundary condition $y(0) = 3$ leads to $y(0) = a_0 = 3$, while the boundary condition $y'(0) = 1$ leads to $a_1 = 1$. Our final solution is

$$y(x) = 3\cos(x) + \sin(x)$$

which can be validated by inserting our result for $y(x)$ into Equation 11.19 and proving the equality.

We have used the power series method to again derive familiar solutions to first and second order ordinary differential equations. However, the power series method also provides a general means to solve more complicated ordinary differential equations that arise in the physical sciences. Specific examples of differential equations encountered in quantum theory, when modeling the translations of atoms and molecules, the vibrations of chemical bonds, the electronic properties of atoms, and the rotations of molecules, are explored in the complements.

[11] Note that the sum

$$\sum_{n=0}^{\infty} x^{2n+1}$$

contains only odd powers of x.

A$_{11}$ Quantum theory of a particle in a box

Consider the time-independent Schrödinger equation for a particle in one dimension:

$$-\frac{\hbar^2}{2m}\frac{d^2}{dx^2}\psi(x) = E\psi(x) \tag{11.23}$$

where $\hbar = h/2\pi$, h is Planck's constant, m is the mass of the particle, E is the particle's energy, and $\psi(x)$ is the wave function describing the extent of the quantum particle along x. The particle is confined to a one-dimensional box defined by $0 \leq x \leq L$. The *boundary conditions* define the value of the wave function at the two ends of the box:

$$\psi(0) = \psi(L) = 0$$

We require two boundary conditions for any second order differential equation. Since we interpret the square of the wave function, $|\psi(x)|^2$, as a probability distribution, once we solve Equation 11.23 for $\psi(x)$, we will also impose the normalization condition

$$\int_0^L |\psi(x)|^2 dx = 1$$

This is a second order ordinary differential equation, and we can solve it using the methods developed in this chapter. We start by inserting $e^{\alpha x}$ for $\psi(x)$ in Equation 11.23 to find the auxiliary equation

$$-\frac{\hbar^2}{2m}\alpha^2 = E$$

As $E \geq 0$, we find the pure imaginary roots

$$\alpha_{\pm} = \pm i\sqrt{\frac{2mE}{\hbar^2}}$$

and expect an oscillatory solution (see Figure 11.5). The general solution can be written as

$$\psi(x) = c_1 \exp\left(i\sqrt{\frac{2mE}{\hbar^2}}x\right) + c_2 \exp\left(-i\sqrt{\frac{2mE}{\hbar^2}}x\right)$$

We can determine the values of the coefficients c_1 and c_2 by applying the boundary conditions. Knowing that

$$\psi(0) = c_1 + c_2 = 0$$

tells us that $c_1 = -c_2$, so that

$$\psi(x) = c_1 \exp\left(i\sqrt{\frac{2mE}{\hbar^2}}x\right) - c_1 \exp\left(-i\sqrt{\frac{2mE}{\hbar^2}}x\right) = 2ic_1 \sin\left(\sqrt{\frac{2mE}{\hbar^2}}x\right)$$

where we have used the identity $\sin(x) = \frac{1}{2i}(e^{ix} - e^{-ix})$.

Knowing that

$$\psi(L) = 2ic_1 \sin\left(\sqrt{\frac{2mE}{\hbar^2}}L\right) = 0$$

it must be that

$$\sqrt{\frac{2mE}{\hbar^2}}L = n\pi \qquad n = 1,2,3,\ldots$$

This result provides us with a solution for the allowed values of the energy

$$E_n = \frac{n^2h^2}{8mL^2} \qquad n = 1,2,3,\ldots \qquad (11.24)$$

We know the wave function is normalized such that

$$\int_0^L |\psi(x)|^2 dx = 1 = -4c_1^2 \int_0^L \sin^2\left(\frac{n\pi x}{L}\right) dx = -4c_1^2\,\frac{L}{2}$$

where the integral can be taken using the trigonometric half-angle identity $\sin^2(ax) = \frac{1}{2}(1 - \cos(2ax))$, noting that the integral over the cosine term will be zero. This allows us to solve for the coefficient

$$c_1 = -\frac{i}{2}\sqrt{\frac{2}{L}}$$

As such, our final result for the solution to Equation 11.23 is

$$\psi_n(x) = \begin{cases} \sqrt{\frac{2}{L}}\sin\left(\frac{n\pi x}{L}\right) & 0 \le x \le L \\ 0 & \text{everywhere else} \end{cases}$$

for $n = 1,2,3,\ldots$

Examples of the wave function $\psi_n(x)$ and the modulus squared of the wave function $|\psi_n(x)|^2$ are provided in Figure 11.10 and Figure 11.11 for $n = 1,2$, and 3. Note that each solution $\psi_n(x)$ has n half-oscillations in the box. As n increases, the number of half-oscillations increases, reflecting the increasing energy of the particle. Returning to Equation 11.23, we note that the second derivative of $\psi(x)$ is proportional to E. As E increases, the curvature of the wave function increases, reflected in an increasing number of oscillations in the box.

Finally, inserting our result for $\psi_n(x)$ into our original differential equation, we find

$$-\frac{\hbar^2}{2m}\frac{d^2}{dx^2}\psi_n(x) = -\frac{\hbar^2}{2m}\left[-\left(\frac{n\pi}{L}\right)^2\sqrt{\frac{2}{L}}\sin\left(\frac{n\pi x}{L}\right)\right]$$

$$= \frac{\hbar^2}{2m}\left(\frac{n\pi}{L}\right)^2\psi_n(x) = E_n\psi_n(x)$$

The resulting allowed energies E_n for our particle in a box agree with those in Equation 11.24, validating our solution. This example demonstrates the effectiveness of our general approach to solving linear second order differential equations. Subsequent complements apply this approach to differential equations arising in classical theory of motion.

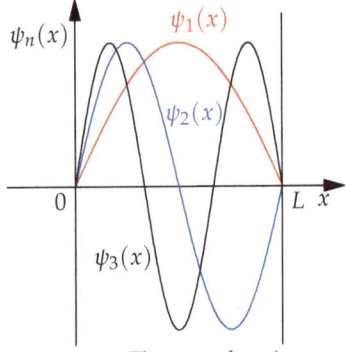

Figure 11.10: The wave function $\psi_n(x)$ for a particle confined to a box of length L. Three possible wave functions are shown for $n = 1,2$, and 3 corresponding to energies $E_1 = h^2/8mL^2$, $E_2 = 4h^2/8mL^2$, and $E_3 = 9h^2/8mL^2$.

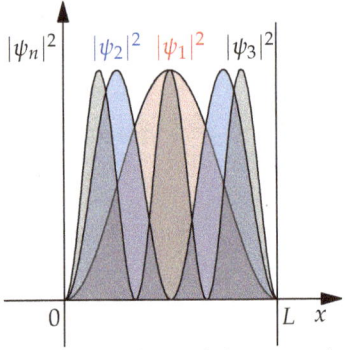

Figure 11.11: The modulus squared of the wave function $|\psi_n(x)|^2$ for a particle confined to a box of length L. Three possible wave functions are shown for $n = 1,2$, and 3. The area under each curve is unity making each $|\psi_n(x)|^2$ a normalized probability distribution.

B$_{11}$ Classical theory of motion of a harmonic oscillator

In Chapter 8, we found that a mass on a spring or a vibrating bond between two atoms can be modeled as a harmonic oscillator with potential energy

$$V(x) = \frac{1}{2}\kappa(x - x_0)^2$$

and force

$$F(x) = -\frac{dV}{dx} = -\kappa(x - x_0)$$

where x is the displacement of the oscillator from its mechanically stable position at x_0 and κ is the force constant. When $x > x_0$, the force acts in the negative direction to shorten the oscillator. When $x < x_0$, the force acts in the positive direction to extend the oscillator. And when $x = x_0$, the force is zero and the oscillator is in a state of *mechanical equilibrium*. These results are depicted in Figure 11.12.

The velocity is the rate of change in the position with respect to time:

$$v = \frac{dx}{dt}$$

The total energy of the oscillator is the sum of the kinetic energy and potential energy:

$$E = \frac{1}{2}mv^2 + \frac{1}{2}\kappa(x - x_0)^2$$

The total energy is constant in time.

The acceleration is the rate of change in the velocity with respect to time:

$$a = \frac{dv}{dt} = \frac{d^2x}{dt^2}$$

When a force acts on the oscillator, the mass accelerates according to Newton's equation of motion $ma = F$, or

$$m\frac{d^2x}{dt^2} = -\kappa(x - x_0)$$

where $\kappa > 0$ is the force constant. This is a linear second order ordinary differential equation. It can be solved to determine the position of the mass, $x(t)$, as a function of time.

We can reform the equation in terms of the variable $y(t) = x(t) - x_0$, representing the displacement of the oscillator from its equilibrium position:[12]

$$m\frac{d^2y}{dt^2} = -\kappa y$$

We would like to solve this equation to determine the position of the oscillator as a function of time $y(t)$. Substituting $e^{\alpha t}$ for $y(t)$ leads to

$$m\frac{d^2}{dt^2}e^{\alpha t} = m\alpha^2 e^{\alpha t} = -\kappa e^{\alpha t}$$

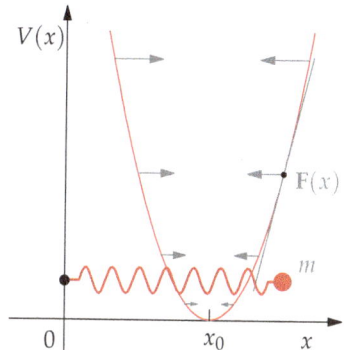

Figure 11.12: The quadratic potential energy function $V(x)$ and the corresponding force $\mathbf{F}(x)$.

[12] Note that the velocity

$$v = \frac{dx}{dt} = \frac{d}{dt}(x - x_0) = \frac{dy}{dt}$$

and

$$a = \frac{d^2x}{dt^2} = \frac{d^2y}{dt^2}$$

for the acceleration are unchanged by this transformation.

which leads to the auxiliary equation

$$m\alpha^2 = -\kappa$$

with purely imaginary roots

$$\alpha_\pm = \pm i \sqrt{\frac{\kappa}{m}}$$

and the solution

$$y(t) = c_1 \exp\left(i\sqrt{\frac{\kappa}{m}}t\right) + c_2 \exp\left(-i\sqrt{\frac{\kappa}{m}}t\right)$$

The expression describes undamped oscillatory motion.

The time scale for the dynamics of the oscillator can be defined either by the *linear frequency* of motion

$$\nu = \frac{1}{2\pi}\sqrt{\frac{\kappa}{m}}$$

or equivalently by the *angular frequency*

$$\omega = \sqrt{\frac{\kappa}{m}} = 2\pi\nu$$

The period of oscillation is

$$T = \frac{1}{\nu} = \frac{2\pi}{\omega}$$

The larger the force constant, the higher the frequency of oscillation. The heavier the mass, the lower the frequency of oscillation.

With these definitions, we can reformulate $y(t) = x(t) - x_0$ as

$$y(t) = c_1 \exp\left(i\omega t\right) + c_2 \exp\left(-i\omega t\right)$$

We assume that at $t = 0$ the oscillator is displaced to y_0 and the velocity is zero

$$y(0) = y_0 \qquad v(0) = \left.\frac{dy}{dt}\right|_{t=0} = 0$$

As such, the initial total energy of the oscillator is $E(0) = \frac{1}{2}\kappa y_0^2$. Applying the initial condition for the velocity, we find

$$\left.\frac{dy}{dt}\right|_{t=0} = \left[ic_1\omega e^{i\omega t} - ic_2\omega e^{-i\omega t}\right]_{t=0}$$

$$= ic_1\omega - ic_2\omega = 0$$

Accordingly, $c_1 = c_2$ so that

$$y(t) = c_1 \left[\exp\left(i\omega\right) + \exp\left(-i\omega\right)\right] = 2c_1 \cos\left(\omega t\right)$$

Applying the second initial condition, we find

$$y(0) = 2c_1 \cos\left(\omega t\right)\Big|_{t=0} = 2c_1 = y_0$$

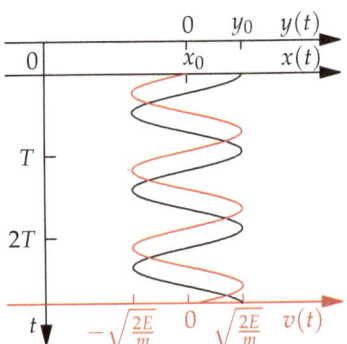

Figure 11.13: The trajectory of the harmonic oscillator shown in terms of the position, $x(t)$, the displaced coordinate, $y(t) = x(t) - x_0$, and the velocity, $v(t)$, as a function of t. The period of oscillation is $T = 2\pi/\omega$. The corresponding velocity oscillates between $\pm\sqrt{2E/m}$. Note that the velocity (red) is the derivative of the position (black) with respect to time.

so that the constant

$$c_1 = \frac{1}{2} y_0$$

Our final result for the position and velocity is

$$y(t) = y_0 \cos(\omega t) \qquad v(t) = \frac{dy}{dt} = -y_0 \omega \sin(\omega t)$$

This result is depicted in Figure 11.13 which shows the position as a function of time in terms of $x(t)$ and $y(t)$ over three periods of oscillation $T = 2\pi/\omega$.

With a knowledge of the position and velocity of the oscillator as a function of time, we can determine the total energy:

$$
\begin{aligned}
E(t) &= \frac{1}{2} m v(t)^2 + \frac{1}{2} \kappa y(t)^2 \\
&= \frac{1}{2} m \left(-y_0 \omega \sin(\omega t)\right)^2 + \frac{1}{2} \kappa \left(y_0 \cos(\omega t)\right)^2 \\
&= \frac{1}{2} m \omega^2 y_0^2 \sin^2(\omega t) + \frac{1}{2} \kappa y_0^2 \cos^2(\omega t) \\
&= \frac{1}{2} \kappa y_0^2 \sin^2(\omega t) + \frac{1}{2} \kappa y_0^2 \cos^2(\omega t) \\
&= \frac{1}{2} \kappa y_0^2
\end{aligned}
$$

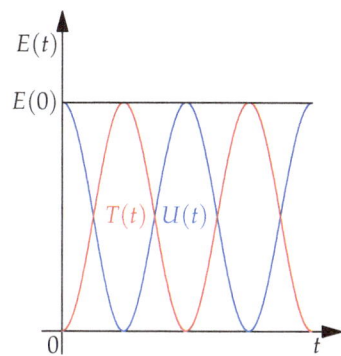

Figure 11.14: The kinetic energy, $T(t)$, potential energy, $U(t)$, and total energy $E(t) = T(t) + U(t)$ of the harmonic oscillator as a function of t. The total energy is conserved and is constant in time (black line).

During the vibration of the oscillator, the total energy is conserved and equals the initial energy:

$$E(t) = E(0) = \frac{1}{2} \kappa y_0^2$$

Figure 11.14 shows the complementary oscillations in the kinetic and potential energies. Potential energy is transformed to kinetic energy, and kinetic energy is transformed to potential energy in a repeating cycle, while the total energy remains constant.

It is interesting to consider the conservation of total energy

$$\frac{dE}{dt} = 0 = \frac{d}{dt} \left(\frac{1}{2} m v^2 + \frac{1}{2} \kappa (x - x_0)^2 \right)$$

Evaluating the total time derivative leads to

$$
\begin{aligned}
\frac{d}{dt} \left(\frac{1}{2} m v^2 + \frac{1}{2} \kappa (x - x_0)^2 \right) &= m v \frac{dv}{dt} + \kappa (x - x_0) \frac{dx}{dt} \\
&= m v a + v \kappa (x - x_0) = 0
\end{aligned}
$$

Canceling the common value of v, we find

$$m a + \kappa (x - x_0) = 0$$

or

$$m a = -\kappa (x - x_0) = F(x)$$

which is Newton's equation of motion. In this case of energy-conserving dynamics in one dimension, the conservation of energy implies Newton's equation of motion. This can be thought of as a special case of *Noether's theorem*.[13] The principle of conservation of energy (that the energy is constant

[13] Named for German mathematician *Emmy Noether* (1882-1935) whose work illuminated the relationship between conserved quantities, such as energy and angular momentum, and underlying symmetries of the system, such as time and rotation.

in time) is a consequence of an underlying symmetry, the invariance under translation in time. That is, when the time changes, the energy remains constant.

Finally, note that the position and velocity can also be written as

$$y(t) = \sqrt{\frac{2E}{\kappa}} \cos(\omega t) \qquad v(t) = \frac{dy}{dt} = -\sqrt{\frac{2E}{m}} \sin(\omega t)$$

A slightly different set of equations defining the oscillator's motion are written in terms of the position and linear momentum $p(t) = mv(t)$:

$$y(t) = \sqrt{\frac{2E}{\kappa}} \cos(\omega t) \qquad p(t) = m\frac{dy}{dt} = -\sqrt{2mE} \sin(\omega t)$$

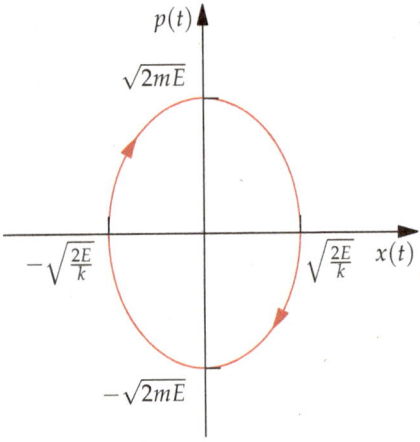

Figure 11.15: The motion of the harmonic oscillator defined by the position, $y(t)$, and momentum, $p(t)$, as a function of t. The elliptical curve is the phase portrait of the harmonic oscillator.

These equations define a trajectory on the yp-plane formed by the position, y, and the momentum, p. The yp-plane is called *phase space*. All possible states of the oscillator's dynamics are represented in this phase space of position and momentum. The constant total energy restricts the position and momentum to move in a cycle on an elliptical curve defined by[14]

$$E = \frac{\kappa}{2}y^2 + \frac{1}{2m}p^2$$

The elliptical curve defining the states visited in the oscillatory dynamics of the oscillator is called the *phase portrait*. Each total energy results in a unique phase portrait, an elliptical phase portrait that grows larger as the energy grows. This result is shown in Figure 11.15.

[14] The total energy of our oscillator is
$$E = \frac{\kappa}{2}y^2 + \frac{1}{2m}p^2$$
Dividing by E we find
$$1 = \frac{\kappa}{2E}y^2 + \frac{1}{2mE}p^2 = \frac{y^2}{a^2} + \frac{p^2}{b^2}$$
which is the equation of an ellipse where $a = \sqrt{\frac{2E}{\kappa}}$ and $b = \sqrt{2mE}$ (see Supplement S₂).

C_{11} Classical theory of a damped harmonic oscillator

We found that a mass on a spring or a vibrating bond between two atoms can be modeled as a harmonic oscillator with potential energy

$$V(x) = \frac{1}{2}\kappa(x - x_0)^2$$

and corresponding force

$$F_{\text{spring}} = -\frac{dV}{dx} = -\kappa(x - x_0)$$

where x is the displacement of the oscillator from its mechanically stable position at x_0 and κ is the force constant. The corresponding velocity is the rate of change in the position with time, or

$$v = \frac{dx}{dt}$$

and the linear momentum is $p = mv$.

The motion of this model oscillator is undamped in time. Once it begins to oscillate, it continues to oscillate with fixed frequency of motion and constant energy. However, physical oscillators such as a mass on a spring in the air or a vibrating molecule in a liquid will lose energy to the surroundings over time, damping the motion of the oscillator. To model this phenomenon, we introduce an additional force acting on the mass in the form of a frictional damping force

$$F_{\text{friction}} = -\gamma\frac{dx}{dt} = -\gamma v$$

where $\gamma \geq 0$ and the frictional force is proportional to the velocity. If the velocity is positive, the damping force is in the negative direction, slowing the oscillator. If the velocity is negative, the damping force is in the positive direction, slowing the oscillator. As such, the frictional force acts to reduce the speed, removing energy from the system and slowing the oscillator until the speed eventually reaches zero. This behavior is demonstrated in Figure 11.16, which can be compared with the motion of an undamped harmonic oscillator shown in Figure 11.13.

The acceleration is the rate of change in the velocity with time

$$a = \frac{dv}{dt} = \frac{d^2x}{dt^2}$$

The equation of motion for the oscillator is given by Newton's equation $F = ma$ where m is the mass and a is the acceleration:

$$m\frac{d^2x}{dt^2} = F_{\text{spring}} + F_{\text{friction}} = -\kappa(x - x_0) - \gamma\frac{dx}{dt}$$

We can reform the equation in terms of the variable $y = x - x_0$, representing the displacement of the oscillator from its equilibrium position

$$m\frac{d^2y}{dt^2} = -\kappa y - \gamma\frac{dy}{dt}$$

We would like to solve this equation to determine the displacement of the oscillator, $y(t)$, as a function of time. Substituting $e^{\alpha t}$ for $y(t)$ leads to

$$m\frac{d^2}{dt^2}e^{\alpha t} = -\kappa e^{\alpha t} - \gamma\frac{d}{dt}e^{\alpha t}$$

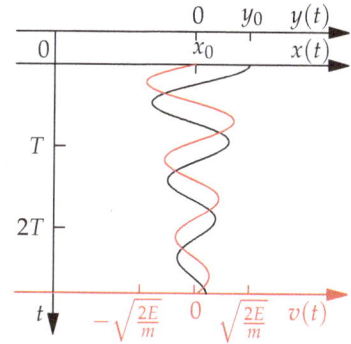

Figure 11.16: The trajectory of the damped harmonic oscillator shown in terms of the position, $x(t)$, the displaced coordinate, $y(t) = x(t) - x_0$, and the velocity, $v(t)$, as a function of t. The period of the undamped oscillator is $T = 2\pi/\omega$.

Evaluating the derivatives, we find

$$m\alpha^2 e^{\alpha t} = -\kappa e^{\alpha t} - \alpha\gamma e^{\alpha t}$$

which reduces to the auxiliary equation

$$\alpha^2 + \alpha\frac{\gamma}{m} + \frac{\kappa}{m} = 0$$

with roots

$$\alpha_\pm = \frac{1}{2}\left[-\frac{\gamma}{m} \pm \sqrt{\left(\frac{\gamma}{m}\right)^2 - \frac{4\kappa}{m}}\right] = -\frac{\gamma}{2m} \pm \sqrt{\left(\frac{\gamma}{2m}\right)^2 - \frac{\kappa}{m}}$$

This leads to the general solution for the displacement as a function of time, written

$$y(t) = c_1 e^{\alpha_+ t} + c_2 e^{\alpha_- t}$$

Let's consider the nature of the roots, $\alpha_\pm$, that can be purely imaginary, complex, or real depending on the magnitude of γ. In the case that $\gamma = 0$, there is no friction and we recover the pure imaginary roots

$$\alpha_\pm = \pm i\sqrt{\frac{\kappa}{m}} = \pm i\omega$$

associated with undamped oscillatory motion. In the case that $0 < \gamma < 2\kappa$, there is low friction and the roots will be complex, with real and imaginary terms

$$\alpha_\pm = -\frac{\gamma}{2m} \pm \sqrt{\left(\frac{\gamma}{2m}\right)^2 - \omega^2} = -\frac{\gamma}{2m} \pm i\omega'$$

where

$$\omega' = \sqrt{\omega^2 - \left(\frac{\gamma}{2m}\right)^2} = \omega\sqrt{1 - \left(\frac{\gamma}{2m\omega}\right)^2}$$

For these complex roots, there is an exponential damping term, with rate of damping $\gamma/2m$, and an oscillatory term, with frequency ω' shifted relative to the frequency of undamped motion. In this case, the motion of the oscillator is *underdamped* (see Figure 1.12). Finally, in the case that $\gamma \geq 2\kappa$, there is high friction and the roots are purely real:

$$\alpha_\pm = -\frac{\gamma}{2m} \pm \sqrt{\left(\frac{\gamma}{2m}\right)^2 - \omega^2}$$

In this case, the motion of the oscillator is *overdamped* (see Figure 1.12).

In each of the three cases, we can write

$$\alpha_\pm = \alpha_0 \pm \Delta\alpha$$

where

$$\alpha_0 = -\frac{\gamma}{2m} \qquad \Delta\alpha = \sqrt{\left(\frac{\gamma}{2m}\right)^2 - \omega^2}$$

and $\Delta\alpha$ may be real or imaginary.[15] With this notation, our solution takes the form

$$y(t) = c_1 e^{\alpha_+ t} + c_2 e^{\alpha_- t} = e^{\alpha_0 t}\left(c_1 e^{\Delta\alpha t} + c_2 e^{-\Delta\alpha t}\right)$$

[15] In the special case $\gamma = 2m\omega$ we find $\Delta\alpha = 0$, the double root $\alpha = -\gamma/2m$, and the solution

$$y(t) = c_1 e^{\alpha t} + c_2 t e^{\alpha t}$$

In the modeling of physical systems, it is possible but very improbable that the friction, frequency, and mass will exactly satisfy the equality $\gamma = 2m\omega$. As such, this special case will not be discussed in detail.

Considering the initial conditions

$$y(0) = y_0 \qquad \left.\frac{dy}{dt}\right|_{t=0} = 0$$

we find

$$y(0) = c_1 + c_2 = y_0$$

and

$$
\begin{aligned}
\left.\frac{dy}{dt}\right|_{t=0} &= \frac{d}{dt}\left[e^{\alpha_0 t}\left(c_1 e^{\Delta\alpha t} + c_2 e^{-\Delta\alpha t}\right)\right] \\
&= \left[\alpha_0 e^{\alpha_0 t}\left(c_1 e^{\Delta\alpha t} + c_2 e^{-\Delta\alpha t}\right) + \Delta\alpha\, e^{\alpha_0 t}\left(c_1 e^{\Delta\alpha t} - c_2 e^{-\Delta\alpha t}\right)\right]_{t=0} \\
&= \alpha_0 \left(c_1 + c_2\right) + \Delta\alpha \left(c_1 - c_2\right) = 0
\end{aligned}
$$

As such, $c_2 = y_0 - c_1$ and

$$c_1 = -\frac{(\alpha_0 - \Delta\alpha)y_0}{2\Delta\alpha} \qquad c_2 = \frac{(\alpha_0 + \Delta\alpha)y_0}{2\Delta\alpha}$$

Note that when $\gamma = 0$, we find that $\alpha_0 = 0$, and we recover the coefficients expected for the harmonic oscillator in the absence of friction, where $c_1 = c_2 = y_0/2$. For the general case, we find

$$y(t) = y_0 e^{\alpha_0 t}\left[\cosh\left(\Delta\alpha t\right) - \frac{\alpha_0}{\Delta\alpha}\sinh\left(\Delta\alpha t\right)\right]$$

$$v(t) = y_0 e^{\alpha_0 t}\left[-\frac{\omega^2}{\Delta\alpha}\sinh\left(\Delta\alpha t\right)\right]$$

where high friction leads to a real $\Delta\alpha$, real roots, and exponential damping. In contrast, low friction leads to an imaginary $\Delta\alpha$, complex roots, and an exponentially damped oscillator. We can also write the linear momentum as

$$p(t) = y_0 e^{\alpha_0 t}\left[-\frac{m\omega^2}{\Delta\alpha}\sinh\left(\Delta\alpha t\right)\right]$$

The solution is shown in Figure 11.16.

The total energy of the damped harmonic oscillator is defined as

$$E(t) = \frac{1}{2}mv(t)^2 + \frac{1}{2}\kappa y(t)$$

An example of the time-dependence of $E(t)$ is shown in Figure 11.17 for an underdamped harmonic oscillator for which $0 < \gamma < 2m\omega$. For $\gamma > 0$, the vibration of the oscillator is damped and the total energy decreases monotonically from its initial value

$$E(0) = \frac{1}{2}\kappa y_0^2$$

When the kinetic energy reaches a maximum, the velocity is maximized as is the rate of frictional damping. This is reflected in the variation in the total energy, where periods of higher kinetic energy are also periods of greater energy loss. This leads to the stepping behavior observed in the total

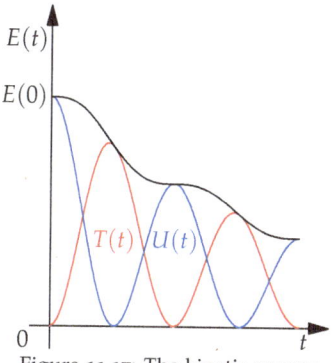

Figure 11.17: The kinetic energy, $T(t)$, potential energy, $U(t)$, and total energy $E(t) = T(t) + U(t)$ of the damped harmonic oscillator. The total energy is monotonically decreasing as a function of time (black line).

energy as a function of time. Compare this behavior to that of the undamped harmonic oscillator shown in Figure 11.14 where the trajectory performs a repeating elliptical orbit in phase space.

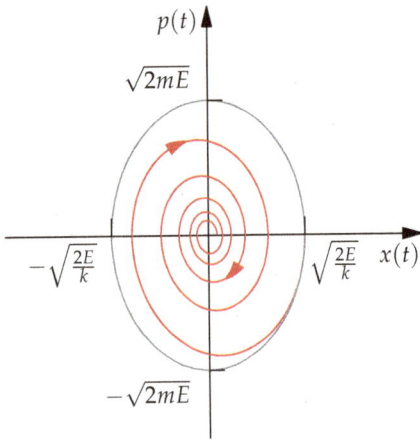

Figure 11.18: The phase portrait of the damped harmonic oscillator (red). The gray ellipse is formed by the values of position and momentum consistent with the initial energy $E(0)$. The continuous energy loss of the damped oscillator is reflected in the narrowing of the inward spiral with increasing time.

All possible states of the oscillator's dynamics are represented in this *phase space* of position and momentum. Our solutions for $y(t)$ and $p(t)$ define a trajectory on the yp-plane called the *phase portrait*. For an undamped harmonic oscillator, the position and momentum move in a repeating cycle restricted to an elliptical curve defined by the constant total energy

$$E(t) = \frac{\kappa}{2}y(t)^2 + \frac{1}{2m}p(t)^2 = E(0)$$

For a damped harmonic oscillator, the total energy of the oscillator decreases with time. As a result, the breadth of the elliptical path decreases with time leading to a phase portrait with the form of an inward elliptical spiral. This behavior is depicted in Figure 11.18 for a harmonic oscillator for which $0 < \gamma < 2m\omega$ and the motion is underdamped. This behavior can be compared to that of the undamped harmonic oscillator shown in Figure 11.15.

D₁₁ Power series solutions to special equations in quantum theory

A number of special ordinary differential equations arise in the quantum theory of matter when solving the Schrödinger equation for electronic energy of a one-electron atom or the rotations and vibrations of a diatomic molecule. These special differential equations may be solved using power series. In each case, the resulting power series represents a type of *special function* that finds wide use in the physical sciences.[16]

Special functions explored in this complement include the Hermite polynomials, found in the quantum theory of vibrational motion, the Laguerre polynomials, found in the quantum theory of the hydrogen atom, and the Legendre polynomials, found in the quantum theory of rotational motion and multipole expansions in the classical theory of electricity and magnetism.

[16] A classic reference for the properties of special functions is *Abramowitz and Stegun* or simply *AS*, which still serves as a valuable compendium of special functions and their properties.

Milton Abramowitz and Irene A. Stegun, editors. *Handbook of Mathematical Functions with Formulas, Graphs, and Mathematical Tables [1964]*. Dover Publications, first edition, 1972. ISBN 978-0486612720

Hermite's equation and Hermite polynomials

In the quantum theory of the harmonic oscillator, a special differential equation appears:

$$\frac{d^2}{dx^2}y(x) - 2x\frac{d}{dx}y(x) + 2\alpha y(x) = 0 \qquad (11.25)$$

where α is a constant. This is known as *Hermite's equation.*[17] Proposing a power series solution of the form

$$y(x) = \sum_{n=0}^{\infty} a_n x^n$$

we find that Equation 11.25 can be written

$$\sum_{n=0}^{\infty} a_n n(n-1)x^{n-2} - 2x\sum_{n=0}^{\infty} a_n n x^{n-1} + 2\alpha\sum_{n=0}^{\infty} a_n x^n = 0 \qquad (11.26)$$

In order to identify the recursion relation that defines the coefficients in our series, we must transform the first two series so that they can be combined with the third.

In the first series, the coefficients are proportional to $n(n-1)$, so that the $n = 0$ and $n = 1$ terms are always zero. Removing the $n = 0$ and $n = 1$ terms from the first series results in

$$\sum_{n=0}^{\infty} a_n n(n-1)x^{n-2} \rightarrow \sum_{n=2}^{\infty} a_n n(n-1)x^{n-2}$$

Now the series starts with $n = 2$, but we would like it to start from $n = 0$. As such, we shift the index by $n \rightarrow n+2$ so that[18]

$$\sum_{n=2}^{\infty} a_n n(n-1)x^{n-2} \rightarrow \sum_{n=0}^{\infty} a_{n+2}(n+2)(n+1)x^n$$

For the second series, we simply absorb the factor of x in the sum, so that

$$-2x\sum_{n=0}^{\infty} a_n n x^{n-1} = -2\sum_{n=0}^{\infty} a_n n x^n$$

Inserting these results into Equation 11.26 leads to

$$\sum_{n=0}^{\infty} a_{n+2}(n+2)(n+1)x^n - 2\sum_{n=0}^{\infty} a_n n x^n + 2\alpha\sum_{n=0}^{\infty} a_n x^n = 0$$

Combining the sums, we find

$$\sum_{n=0}^{\infty} \left[a_{n+2}(n+2)(n+1) - 2a_n n + 2\alpha a_n\right] x^n = 0$$

and the condition

$$a_{n+2}(n+2)(n+1) - 2a_n n + 2\alpha a_n = 0$$

This condition can be used to derive a recursion relation relating a_{n+2} to a_n.

[17] Named for French mathematician *Charles Hermite* (1822-1901).

[18] Equivalently, you can introduce a new index $n' = n - 2$, substituting $n \rightarrow n' + 2$. You can then rewrite the new index n' as n before combining the sums.

In particular, we find the recursion relation

$$a_{n+2} = -\frac{2(\alpha - n)}{(n+2)(n+1)} a_n$$

The even indices can be generated from a_0 as[19]

$$a_2 = -\frac{2\alpha}{2} a_0 \qquad a_4 = -\frac{2(\alpha - 2)}{4 \cdot 3} a_2 = \frac{2^2 \alpha(\alpha - 2)}{4!} a_0$$

leading to the following power series in even powers of x:

$$y_0(x) = a_0 \left[1 - \frac{2\alpha}{2!} x^2 + \frac{2^2 \alpha(\alpha - 2)}{4!} x^4 - + \dots \right]$$

The odd indices can be generated from a_1 as

$$a_3 = -\frac{2(\alpha - 1)}{3 \cdot 2} a_1 \qquad a_5 = -\frac{2(\alpha - 3)}{5 \cdot 4} a_3 = \frac{2^2 (\alpha - 3)(\alpha - 1)}{5!} a_1$$

leading to the following power series in odd powers of x:

$$y_1(x) = a_1 \left[x - \frac{2(\alpha - 1)}{3!} x^3 + \frac{2^2 (\alpha - 3)(\alpha - 1)}{5!} x^5 - + \dots \right]$$

The overall solution is written as a linear *superposition* of the two independent solutions $y_0(x)$ and $y_1(x)$. The remaining coefficients a_0 and a_1 are defined by the boundary conditions.

Let's examine the results for various values of the parameter α. We will find that the solutions to Hermite's equation form a series of polynomial equations known as the *Hermite polynomials*.[20] In defining the Hermite polynomials, it is helpful to consider the even power series, $y_0(x)$, and odd power series, $y_1(x)$, separately. When $\alpha = 0$, Hermite's equation is

[20] The Hermite polynomials were originally described by *Pierre-Simon Laplace* (1749-1827).

$$\frac{d^2}{dx^2} y(x) - 2x \frac{d}{dx} y(x) = 0$$

with a solution defined by the even power series

$$y_0(x) = a_0$$

as all higher order terms contain a multiplicative factor of α and are therefore zero. This is the Hermite polynomial $H_0(x) = 1$ for $a_0 = 1$. For $\alpha = 1$, Hermite's equation is

$$\frac{d^2}{dx^2} y(x) - 2x \frac{d}{dx} y(x) + 2y(x) = 0$$

with a solution defined by the odd power series

$$y_1(x) = a_1 x$$

as all higher order terms in the series contain a multiplicative factor of $(\alpha - 1)$ and are zero. This is the Hermite polynomial $H_1(x) = 2x$ when $a_1 = 2$. When

$\alpha = 2$, Hermite's equation is

$$\frac{d^2}{dx^2}y(x) - 2x\frac{d}{dx}y(x) + 4y(x) = 0$$

with a solution defined by the even power series

$$y_0(x) = a_0\left(1 - 2x^2\right)$$

which is the Hermite polynomial $H_2(x) = 4x - 2$ when $a_0 = -2$.

Repeating this process, we find that when $\alpha = 3$, the Hermite polynomial $H_3(x) = 8x^3 - 12x$ when $a_1 = -12$. When $\alpha = 4$, the Hermite polynomial $H_4(x) = 16x^4 - 48x^2 + 12$ when $a_0 = 12$. The first five Hermite polynomials are presented in Figure 11.19.

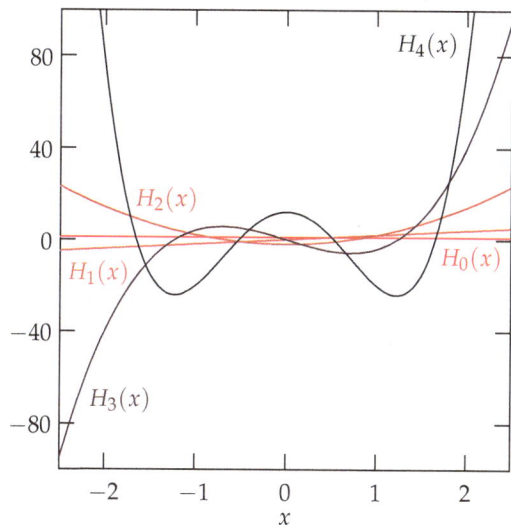

Figure 11.19: Variation in the first five Hermite polynomials $H_n(x)$ over the range $x \in [-2.5, 2.5]$.

Our success in solving Hermite's equation demonstrates the applicability of the power series method to the solution of linear second order differential equations with non-constant coefficients. The resulting solutions to Hermite's equation consist of an infinite number of Hermite polynomials. The Hermite polynomials possess special properties that will be explored in Chapter 13 in the complements.

Laguerre's equation and Laguerre polynomials

In the quantum theory of the one-electron atom, a special differential equation describes the radial dependence of the electron's wave function:

$$x\frac{d^2}{dx^2}y(x) - (1-x)\frac{d}{dx}y(x) + \alpha y(x) = 0 \tag{11.27}$$

where α is a constant. This is known as *Laguerre's equation*.[21] Proposing a power series solution

$$y(x) = \sum_{n=0}^{\infty} a_n x^n$$

[21] Named for French mathematician *Edmond Laguerre* (1834-1886).

we find that Equation 11.27 can be written

$$x \sum_{n=0}^{\infty} a_n n(n-1)x^{n-2} - (1-x) \sum_{n=0}^{\infty} a_n n x^{n-1} + \alpha \sum_{n=0}^{\infty} a_n x^n = 0 \qquad (11.28)$$

We must transform the first two series so that they can be combined with the third.

In the first series, the coefficients are proportional to n, so the $n = 0$ term is always zero. Removing the $n = 0$ term from the first series and absorbing the factor of x leads to

$$x \sum_{n=0}^{\infty} a_n n(n-1)x^{n-2} \to \sum_{n=1}^{\infty} a_n n(n-1)x^{n-1}$$

Shifting the index by $n \to n+1$ results in[22]

$$\sum_{n=1}^{\infty} a_n n(n-1)x^{n-1} \to \sum_{n=0}^{\infty} a_{n+1}(n+1)n x^n$$

which is in the desired form in terms of the range of the index n and the sum over x^n, weighted by constant coefficients. We transform the second series by distributing the factor of $(1-x)$, resulting in two series

$$(1-x) \sum_{n=0}^{\infty} a_n n x^{n-1} = \sum_{n=0}^{\infty} a_n n x^{n-1} - \sum_{n=0}^{\infty} a_n n x^n$$

In the first series the coefficients are proportional to n so the $n = 0$ term is always zero. Removing the $n = 0$ term from the first series leads to

$$\sum_{n=0}^{\infty} a_n n x^{n-1} \to \sum_{n=1}^{\infty} a_n n x^{n-1}$$

Finally, shifting the index in the first series by $n \to n+1$ results in

$$\sum_{n=1}^{\infty} a_n n x^{n-1} \to \sum_{n=0}^{\infty} a_{n+1}(n+1)x^n$$

Inserting these results in Equation 11.28 leads to

$$\sum_{n=0}^{\infty} [a_{n+1}(n+1)n + a_{n+1}(n+1) - a_n n + \alpha a_n] x^n = 0$$

and the condition

$$a_{n+1}(n+1)n + a_{n+1}(n+1) - a_n n + \alpha a_n = 0$$

This leads to the recursion relation[23]

$$a_{n+1} = -\frac{\alpha - n}{(n+1)^2} a_n$$

The coefficients can be generated starting from a_0 as

$$a_1 = -\alpha a_0 \qquad a_2 = \frac{(\alpha-1)\alpha}{(2!)^2} a_0 \qquad a_3 = -\frac{(\alpha-2)(\alpha-1)\alpha}{(3!)^2} a_0$$

[22] Equivalently, you can introduce a new index $n' = n - 1$, substituting $n \to n' + 1$. You can then rewrite the new index n' as n before combining the sums.

[23] The negative sign in the recursion relation signals the fact that the coefficients lead to an alternating series.

leading to the power series solution

$$y(x) = a_0 \left[1 - \alpha x + \frac{(\alpha-1)\alpha}{(2!)^2} x^2 - \frac{(\alpha-2)(\alpha-1)\alpha}{(3!)^2} x^3 + - \ldots \right]$$

Let's examine our results for various values of the parameter α. We will find that the solutions to Laguerre's equation form a series of polynomial equations known as the *Laguerre polynomials*. When $\alpha = 0$, Laguerre's equation is

$$x \frac{d^2}{dx^2} y(x) - (1-x) \frac{d}{dx} y(x) = 0$$

with the solution

$$y(x) = a_0$$

as all higher order terms contain a multiplicative factor of α and are therefore zero. This is the Laguerre polynomial $L_0(x) = 1$ for $a_0 = 1$. For $\alpha = 1$, Laguerre's equation is

$$x \frac{d^2}{dx^2} y(x) - (1-x) \frac{d}{dx} y(x) + y(x) = 0$$

and the solution is

$$y(x) = a_0(1-x)$$

as all higher order terms in the series contain a multiplicative factor of $(\alpha - 1)$ and are zero. This is the Laguerre polynomial $L_1(x) = 1 - x$ when $a_0 = -1$.

Repeating this process for $\alpha = 2$, we find the Laguerre polynomial $L_2(x) = \frac{1}{2} (x^2 - 4x + 2)$ when $a_0 = 1$. When $\alpha = 3$, we find the Laguerre polynomial $L_3(x) = \frac{1}{6} (-x^3 + 9x^2 - 18x + 6)$ when $a_0 = 1$. The first four Laguerre polynomials are depicted in Figure 11.20.

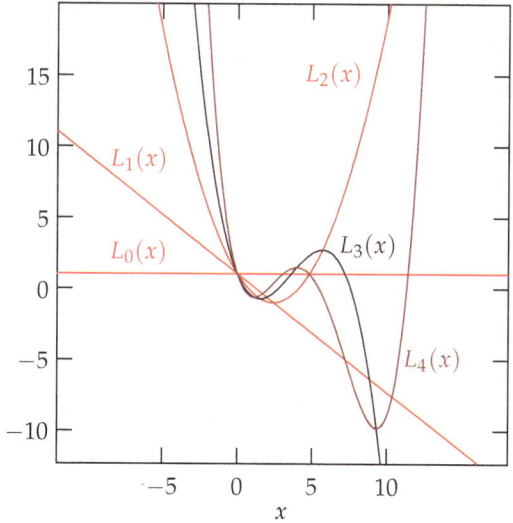

Figure 11.20: Variation in the first five Laguerre polynomials $L_n(x)$ over the range $x \in [-10, 15]$.

The Laguerre polynomials possess special properties that will be explored in Chapter 13 in the complements.

Legendre's equation and Legendre polynomials

In the quantum theory of rotational motion of a diatomic molecule, a special differential equation describes the rotational wave function:

$$(1 - x^2)\frac{d^2}{dx^2}y(x) - 2x\frac{d}{dx}y(x) + l(l+1)y(x) = 0 \qquad (11.29)$$

where l is a constant. This is known as *Legendre's equation*.[24] Proposing a power series solution

$$y(x) = \sum_{n=0}^{\infty} a_n x^n$$

we find that Equation 11.29 can be written

$$(1 - x^2)\sum_{n=0}^{\infty} a_n n(n-1)x^{n-2} - 2x\sum_{n=0}^{\infty} a_n n x^{n-1} + l(l+1)\sum_{n=0}^{\infty} a_n x^n = 0 \quad (11.30)$$

We must transform the first two series so that they can be combined with the third.

We transform the first series by distributing the factor of $(1 - x^2)$, resulting in two series

$$(1 - x^2)\sum_{n=0}^{\infty} a_n n(n-1)x^{n-2} \to \sum_{n=0}^{\infty} a_n n(n-1)x^{n-2} - \sum_{n=0}^{\infty} a_n n(n-1)x^n$$

In the first series the coefficients are proportional to $n(n-1)$, so that the $n = 0$ and $n = 1$ terms are always zero. Removing the $n = 0$ and $n = 1$ terms from the first series results in

$$\sum_{n=0}^{\infty} a_n n(n-1)x^{n-2} \to \sum_{n=2}^{\infty} a_n n(n-1)x^{n-2}$$

Now the series starts with $n = 2$ but we would like it to start from $n = 0$. As such, we shift the index by $n \to n+2$ so that[25]

$$\sum_{n=2}^{\infty} a_n n(n-1)x^{n-2} \to \sum_{n=0}^{\infty} a_{n+2}(n+2)(n+1)x^n$$

For the second series in Equation 11.30, we absorb the factor of $-2x$, resulting in

$$-2x\sum_{n=0}^{\infty} a_n n x^{n-1} = -2\sum_{n=0}^{\infty} a_n n x^n$$

Inserting these results in Equation 11.30 leads to

$$\sum_{n=0}^{\infty} \left[a_{n+2}(n+2)(n+1) - a_n n(n-1) - 2a_n n \right.$$

$$\left. + l(l+1)a_n\right]x^n = 0$$

and the condition

$$a_{n+2}(n+2)(n+1) - a_n n(n-1) - 2a_n n + l(l+1)a_n = 0$$

[24] Named for French mathematician *Adrien-Marie Legendre* (1752-1833).

[25] Equivalently, you can introduce a new index $n' = n - 2$, substituting $n \to n' + 2$. You can then rewrite the new index n' as n before combining the sums.

We find the recursion relation

$$a_{n+2} = \frac{(n+1)n - l(l+1)}{(n+2)(n+1)} a_n = -\frac{[l + (n+1)](l-n)}{(n+2)(n+1)} a_n$$

where the negative sign leads to an alternating series.

The even indices can be generated from a_0 as

$$a_2 = -\frac{(l+1)l}{2} a_0 = -\frac{(l+1)l}{2!} a_0$$

$$a_4 = -\frac{(l+3)(l-2)}{4 \cdot 3} a_2 = \frac{(l+3)(l-2)(l+1)l}{4!} a_0$$

and so on, leading to the following power series in even powers of x:

$$y_0(x) = a_0 \left[1 - \frac{(l+1)l}{2!} x^2 \right.$$
$$\left. + \frac{(l+3)(l-2)(l+1)l}{4!} x^4 - + \dots \right]$$

The odd indices can be generated from a_1 as

$$a_3 = -\frac{(l+2)(l-1)}{3 \cdot 2} a_1 = -\frac{(l+2)(l-1)}{3!} a_1$$

$$a_5 = -\frac{(l+4)(l-3)}{5 \cdot 4} a_3 = \frac{(l+4)(l-3)(l+2)(l-1)}{5!} a_1$$

and so on, leading to the following power series in odd powers of x:

$$y_1(x) = a_1 \left[x - \frac{(l+2)(l-1)}{3!} x^3 + \frac{(l+4)(l-3)(l+2)(l-1)}{5!} x^5 - + \dots \right]$$

The overall solution is written as a linear *superposition* of the two independent solutions $y_0(x)$ and $y_1(x)$. The remaining coefficients a_0 and a_1 are defined by the boundary conditions.

Let's examine our results for various values of the parameter l. We will find that the solutions to Legendre's equation form a series of polynomial equations known as the *Legendre polynomials*. In defining the Legendre polynomials, it is helpful to consider the even power series, $y_0(x)$, and odd power series, $y_1(x)$, separately. When $l = 0$, Legendre's equation is

$$(1 - x^2) \frac{d^2}{dx^2} y(x) - 2x \frac{d}{dx} y(x) = 0$$

and the solution is

$$y(x) = a_0$$

as all higher order terms contain a multiplicative factor of l and are therefore zero. This is the Legendre polynomial $P_0(x) = 1$ for $a_0 = 1$. For $l = 1$, Legendre's equation is

$$(1 - x^2) \frac{d^2}{dx^2} y(x) - 2x \frac{d}{dx} y(x) + 2y(x) = 0$$

and the solution is

$$y(x) = a_1 x$$

as all higher order terms in the series contain a multiplicative factor of $(l-1)$ and are therefore zero. This the Legendre polynomial $P_1(x) = x$ when $a_1 = 1$. When $l = 2$, Legendre's equation is

$$(1 - x^2)\frac{d^2}{dx^2}y(x) - 2x\frac{d}{dx}y(x) + 6y(x) = 0$$

and the solution is

$$y(x) = a_0\left(1 - 3x^2\right)$$

which is the Legendre polynomial $P_2(x) = \frac{1}{2}\left(3x^2 - 1\right)$ when $a_0 = -\frac{1}{2}$. When $l = 3$, Legendre's equation is

$$(1 - x^2)\frac{d^2}{dx^2}y(x) - 2x\frac{d}{dx}y(x) + 12y(x) = 0$$

and the solution is

$$y(x) = a_1\left(x - \frac{5}{3}x^3\right)$$

which is the Legendre polynomial $P_3(x) = \frac{1}{2}\left(5x^3 - 3x\right)$ when $a_1 = -3$. The first four Legendre polynomials are depicted in Figure 11.21. The Legendre polynomials possess special properties that will be explored in Chapter 13 in the complements.

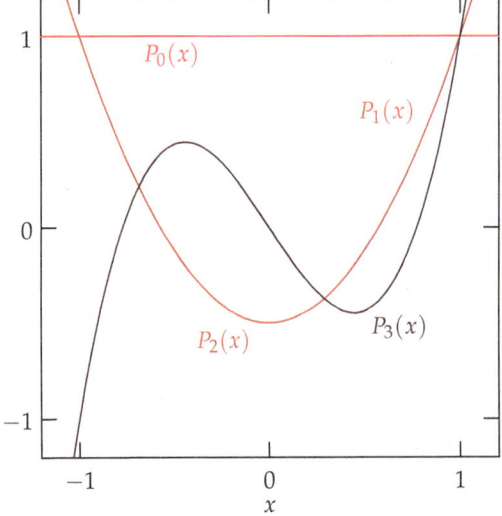

Figure 11.21: Variation in the first four Legendre polynomials $P_l(x)$ over the range $x \in [-1, 1]$.

The differential equations discussed in this complement play a special role in the quantum theory of atoms and molecules. We have found that differential equations can be solved using one or more *common functions* such as the exponentials and sinusoids.

The solution of these three differential equations using the power series method led to the discovery of three new special functions: the Hermite, Laguerre, and Legendre polynomials. It is good to remember that at one time the exponential function was a *special function* as well.[26] Through regular use, we come to regard special functions as common.

[26] The constant e, also known as *Euler's number* or *Napier's constant*, was not well defined until the late 17th century.

E_{11} End-of-chapter problems

> How can it be that mathematics, being after all a product of human thought independent of experience, is so admirably adapted to the objects of reality?
>
> Albert Einstein

Warm-ups

11.1 Consider the following second order ordinary differential equations.

(a) $\dfrac{d^2y(x)}{dx^2} - 4y(x) = 0$

(b) $\dfrac{d^2y(x)}{dx^2} + 2\dfrac{dy(x)}{dx} + 4y(x) = 0$

(c) $\dfrac{d^2y(x)}{dx^2} + 9y(x) = 0$

(d) $\dfrac{d^2y(x)}{dx^2} + 6\dfrac{dy(x)}{dx} = 0$

In each case, assume the solution $y(x) = e^{\alpha x}$. Find the auxiliary equation and determine the two roots $\alpha_\pm$. Compose the overall solution

$$y(x) = c_1 e^{\alpha_+ x} + c_2 e^{\alpha_- x}$$

Apply the boundary conditions $y(0) = 0$ and $\left.\dfrac{dy(x)}{dx}\right|_{x=0} = 1$ to determine the coefficients c_1 and c_2.

11.2 Consider the linear second order differential equation

$$\frac{d^2y(x)}{dx^2} + 2\frac{dy(x)}{dx} + y(x) = 0$$

(a) Assume the solution $y(x) = e^{\alpha x}$. Find the auxiliary equation and determine the roots $\alpha_\pm$. You should find double roots where $\alpha_+ = \alpha_- = \alpha$.

(b) Prove that

$$y(x) = c_1 e^{\alpha x} + c_2\, x\, e^{\alpha x}$$

is a solution to the differential equation.

11.3 Consider the general second order ordinary differential equation

$$\frac{d^2y(x)}{dx^2} + a(x)\frac{dy(x)}{dx} + b(x)\, y(x) = 0$$

Prove that if $y_1(x)$ and $y_2(x)$ are each solutions to the differential equation, so is the linear superposition

$$y(x) = c_1\, y_1(x) + c_2\, y_2(x)$$

11.4 Reform the following sums so that the first term in the sum is $n = 0$

(a) $\displaystyle\sum_{n=2}^{\infty} (n-2)a_{n-2}x^n$

(b) $\displaystyle\sum_{n=2}^{\infty} n(n-1)a_n x^{n-2}$

(c) $\displaystyle\sum_{n=1}^{\infty} na_n x^{n-1}$

11.5 Consider the following recursion relations where $n \geq 0$. In each case, derive a general expression for the coefficient a_n in terms of a_0.

(a) $\quad a_{n+1} = -\dfrac{a_n}{(n+1)^2}$
(b) $\quad a_{n+1} = \dfrac{n+2}{2(n+1)} a_n$
(c) $\quad a_{n+1} = -\dfrac{2a_n}{n+1}$

11.6 Consider the second order differential equation

$$\frac{d^2}{dx^2} y(x) - y(x) = 0$$

(a) Find two linearly independent power series solutions of the form

$$\sum_{n=0}^{\infty} a_n x^n$$

Start by determining recursion relations for the coefficients a_n. You should find two recursion relations, one for n odd and one for n even.

(b) In each case, derive a general expression for a_n in terms of a_0 or a_1.

(c) Identify the common functions representing each series.

Homework exercises

11.7 The general solution for a homogeneous second order ordinary differential equation of the form

$$y(x) = c_1 e^{ibx} + c_2 e^{-ibx}$$

can be rewritten

$$y(x) = c_1' \cos(bx) + c_2' \sin(bx)$$

Derive expressions for c_1' and c_2' in terms of c_1 and c_2.

11.8 The position of an oscillator, $x(t)$, as a function of time, t, satisfies the second order differential equation

$$\frac{d^2 x(t)}{dt^2} + \omega^2 x(t) = 0$$

where ω is a constant. Solve this equation for $x(t)$ given the following initial conditions.

(a) $\quad x(0) = 0$ and $\dfrac{dx}{dt}\Big|_{t=0} = u_0$
(b) $\quad x(0) = x_0$ and $\dfrac{dx}{dt}\Big|_{t=0} = u_0$

In each case, prove that $x(t)$ oscillates with frequency $v = \dfrac{\omega}{2\pi}$ and period $T = \dfrac{2\pi}{\omega} = \dfrac{1}{v}$.

11.9 The second order differential equation

$$\frac{d^2 f(x)}{dx^2} + \frac{25\pi^2}{L^2} f(x) = 0$$

models the displacement, $f(x)$, of a plucked string that is fixed at each end leading to the boundary conditions $f(0) = f(L) = 0$ and $\frac{df}{dx}\big|_{x=0} = 5$.

(a) Determine the general solution of the form

$$f(x) = c_1 e^{\alpha_+ x} + c_2 e^{\alpha_- x}$$

(b) Simplify your result by expressing your solution in terms of a sinusoidal function.

11.10 Consider the second order inhomogeneous ordinary differential equation describing the height, $z(t)$, of a mass, m, falling under the force of gravity with constant rate of acceleration, g, and experiencing a frictional drag force proportional to the speed, $u(t) = \dfrac{dz(t)}{dt}$, written as

$$m \frac{d^2 z(t)}{dt^2} = -\gamma \frac{dz(t)}{dt} + mg$$

where γ is the friction constant.

(a) Show that this equation is equivalent to the first order ordinary differential equation for the speed of the particle

$$\frac{du(t)}{dt} + \frac{\gamma}{m} u(t) = g$$

(b) Solve the differential equation for $u(t)$ given that $u(0) = 0$.

(c) Show that as $t \to \infty$, the speed of the falling mass approaches a constant terminal speed $u_T = \dfrac{mg}{\gamma}$ as shown in the figure below.

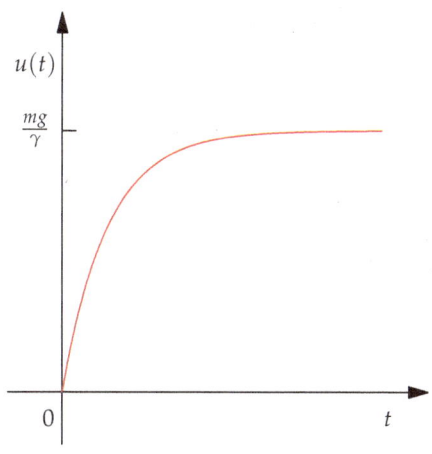

11.11 Determine the coefficients a_n for which the equation

$$\sum_{n=1}^{\infty} n a_n x^{n-1} + 2 \sum_{n=0}^{\infty} a_n x^n = 0$$

is satisfied. Substitue the resulting coefficients a_n in the power series

$$\sum_{n=0}^{\infty} a_n x^n$$

and identify the corresponding function.

11.12* Consider the differential equation

$$(x^2 + 1)\frac{d^2}{dx^2}y(x) - 4x\frac{d}{dx}y(x) + 6y(x) = 0$$

with the boundary conditions $y(0) = 1$ and $y'(0) = 2$.

(a) Propose a solution in the form of the power series

$$y(x) = \sum_{n=0}^{\infty} a_n x^n$$

Derive the recursion relation

$$a_{n+2} = -\frac{(n-2)(n-3)}{(n+2)(n+1)} a_n$$

(b) Using the recursion relation above, determine the four coefficients a_n that are non-zero in terms of a_0 and a_1.

(c) Apply the boundary conditions to determine the solution for $y(x)$.

11.13* Show that

$$c_1 \cos(\omega t) + c_2 \sin(\omega t) = A \sin(\omega t + \varphi)$$

where

$$A = \sqrt{c_1^2 + c_2^2} \qquad \varphi = \tan^{-1}\left(\frac{c_1}{c_2}\right)$$

or equivalently

$$c_1 \cos(\omega t) + c_2 \sin(\omega t) = A \cos(\omega t + \varphi)$$

where

$$A = \sqrt{c_1^2 + c_2^2} \qquad \varphi = \tan^{-1}\left(-\frac{c_2}{c_1}\right)$$

HINT: Make use of the trigonometric identities provided in Supplement S_3. Consider working backward from the identity $\cos(\alpha + \beta) = \cos(\alpha)\cos(\beta) - \sin(\alpha)\sin(\beta)$. Also note that $\sin\left(\tan^{-1}(x)\right) = \frac{x}{\sqrt{x^2+1}}$ and $\cos\left(\tan^{-1}(x)\right) = \frac{1}{\sqrt{x^2+1}}$.

11.14* Consider the differential equation

$$\frac{d^2}{dx^2}y(x) - \frac{d}{dx}y(x) = y(x)$$

with the boundary conditions $y(0) = 0$ and $y'(0) = 1$.

(a) Derive the corresponding auxiliary equation and show that the two roots are $\alpha_+ = \varphi$ and $\alpha_- = 1 - \varphi$ where

$$\varphi = \frac{1}{2}\left(1 + \sqrt{5}\right)$$

is the *golden ratio*.

(b) Using the result from (a) and the boundary conditions, determine the solution for $y(x)$.

(c) Propose a solution to the original differential equation in the form of the power series

$$y(x) = \sum_{n=0}^{\infty} a_n x^n$$

and show that the coefficients

$$a_n = \frac{1}{n!} f_n$$

where f_n are the *Fibonacci numbers* and $f_0 = 0$, $f_1 = 1$, $f_2 = 1$, $f_3 = 2$, $f_4 = 3$, $f_5 = 5$, $f_6 = 8$, and so on.

(d) Expand your result from (b) in a Maclaurin series. Compare that series term-by-term with the result from (c). Show that

$$f_n = \frac{1}{\sqrt{5}} \left[\left(\frac{1+\sqrt{5}}{2} \right)^n - \left(\frac{1-\sqrt{5}}{2} \right)^n \right] = \frac{1}{\sqrt{5}} \left[\varphi^n - (1-\varphi)^n \right]$$

This equation relating the Fibonacci numbers to the golden mean is known as *Binet's formula*.

11.15* Consider the second order differential equation known as *Bessel's equation*

$$x^2 \frac{d^2 y(x)}{dx^2} + x \frac{dy(x)}{dx} + (x^2 - c^2) y(x) = 0$$

where c is a constant. The method we have developed in this chapter cannot be used to solve this problem as the term $x^2 y(x)$ leads to terms of order x^{n+2} in addition to the usual terms of order x^n.

(a) Propose a solution of the form $y(x) = x^r \sum_{n=0}^{\infty} a_n x^n$. Inserting into Bessel's equation will lead to a sum of two power series. Assuming that $a_0 \neq 0$, show that a consistent solution can be found if $(n+r)(n+r-1)+(n+r)-c^2 = 0$ for $n = 0$. Solve that equation to identify the two allowed values of $r = c$ and $r = -c$.

(b) For $r = c$, show that $a_1 = 0$ and therefore $a_{2n+1} = 0$ for $n \geq 1$.

(c) For $r = c$, determine the recursion relation for the coefficients a_{2n} in terms of a_0. Use your coefficients to determine the solution $y_+(x)$ to Bessel's equation.

(d) Repeat steps (b) and (c) for $r = -c$ to arrive at a second solution $y_-(x)$ to Bessel's equation.

12 Partial differential equations

12.1 The classical heat equation

THE PROCESS OF THERMAL TRANSPORT involves energy moving from regions of high temperature to regions of low temperature. Microscopically, this occurs through the random motions of atoms and molecules. At a larger scale, the process can be modeled in terms of changes in temperature as a function of space and time described by the classical heat equation. This section explores the properties and solution of the classical heat equation in one dimension.

12.1.1 The classical heat equation in one dimension (finite bounds)

Variation in the temperature of an object as a function of position and time can be modeled by the *classical heat equation*

$$\frac{\partial u(x,t)}{\partial t} = \kappa \frac{\partial^2 u(x,t)}{\partial x^2} \qquad (12.1)$$

where $u(x,t) \geq 0$ is the temperature profile that is dependent on position x and time t (see Figure 12.1).[1] The classical heat equation is a *partial differential equation* as it describes changes in a function of many variables in terms of partial derivatives of the function.

> [1] The classical heat equation may also be written $u_t = \kappa u_{xx}$ using our compact notation for partial derivatives.

The equation states that the rate of change in the temperature $u(x,t)$, given by the first derivative of $u(x,t)$ with respect to t, is proportional to the second derivative of the temperature $u(x,t)$ with respect to x. The parameter κ is the *thermal diffusivity* that modulates the proportionality between the partial derivatives in time and position and has units of length squared divided by time.

Our goal is to solve this partial differential equation for the temperature, $u(x,t)$. We assume that $u(x,t)$ is a separable function of x and t, so that it can be written as

$$u(x,t) = X(x)T(t)$$

Inserting this function into Equation 12.1, we find

$$X(x)\frac{dT(t)}{dt} = \kappa T(t)\frac{d^2 X(x)}{dx^2}$$

Note that as $X(x)$ and $T(t)$ are functions of one variable, the derivatives of those functions are represented as total derivatives rather than partial derivatives. Rearranging the equation so that the time dependence is on the

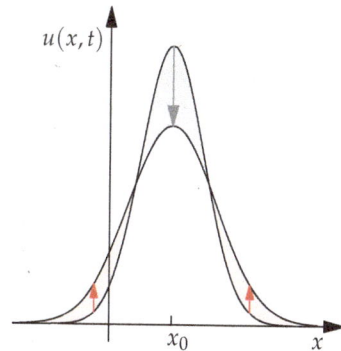

Figure 12.1: Curves representing the temperature profile, $u(x,t)$, of a *hot spot* as a function of position x taken at two points in time t. Regions of higher temperature fall while regions of lower temperature rise. The cooling in the center (shaded gray) equals the heating in the wings (shaded red).

left and the position dependence is on the right leads to

$$\frac{1}{\kappa T(t)}\frac{dT(t)}{dt} = \frac{1}{X(x)}\frac{d^2 X(x)}{dx^2}$$

Since x and t are independent variables, for this equality to hold each side of the equation must be proportional to a constant, such that

$$\frac{1}{\kappa T(t)}\frac{dT(t)}{dt} = -\alpha^2 = \frac{1}{X(x)}\frac{d^2 X(x)}{dx^2} \tag{12.2}$$

Note that α is a real constant that can be determined using initial conditions or boundary conditions for $u(x,t)$. We expect the constant $-\alpha^2$ will be negative, as we expect the functions $X(x)$ and $T(t)$ will be oscillatory.

By assuming a separable solution $u(x,t) = X(x)T(t)$, we have converted a single partial differential equation into two ordinary differential equations. We can now use techniques to solve first and second order ordinary differential equations of a single variable, developed in Chapters 10 and 11, to determine solutions for $X(x)$ and $T(t)$. Those solutions can then be combined to determine the overall solution for $u(x,t) = X(x)T(t)$.

Let's determine the solution for $X(x)$ by returning to Equation 12.2 and solving the second order ordinary differential equation[2]

$$\frac{d^2 X(x)}{dx^2} = -\alpha^2 X(x)$$

The boundary conditions can take on many forms. We will consider the case of *reflecting boundary conditions* at $x = 0$ and $x = L$ for which the gradient of the temperature in space is zero:

$$\frac{\partial u}{\partial x}(0,t) = \frac{\partial u}{\partial x}(L,t) = 0$$

The reflecting boundary condition models a perfectly insulating wall. A perfectly insulating wall will not induce change in the temperature profile. As such, at the wall the gradient of the the temperature profile will be zero.

We take the constant $\alpha^2 \geq 0$ and draw on our experience in solving second order ordinary differential equations in Chapter 11 to propose a general solution of the form[3]

$$X(x) = a_1 \cos(\alpha x) + b_1 \sin(\alpha x)$$

Let's apply the boundary conditions to this proposed solution.[4] Since

$$\frac{dX}{dx}(0) = b_1 \alpha = 0$$

it must be that $b_1 = 0$. Furthermore, since

$$\frac{dX}{dx}(L) = -a_1 \alpha \sin(\alpha L) = 0$$

it must be that

$$\alpha L = n\pi \qquad n = 0,1,2,\ldots$$

[2] As this equation depends solely on x, we have replaced the partial derivative with respect to x in Equation 12.2 with a total derivative.

[3] Suppose $\alpha^2 < 0$. Defining $k^2 = -\alpha^2 > 0$ then $X(x) = a_1 e^{kx} + b_1 e^{-kx}$. But since $X'(0) = a_1 k - b_1 k = 0$ and $X'(L) = a_1 k e^{kL} - b_1 k e^{-kL} = 0$ it must be that $a_1 = b_1 = 0$. As such, results for $\alpha^2 < 0$ are considered to be *trivial solutions*.

[4] Here we use the fact that if $\frac{\partial u}{\partial x} = 0$ then $\frac{dX}{dx} = 0$.

This provides us with an infinite number of possible values of our constant

$$\alpha^2 = \frac{n^2\pi^2}{L^2} \qquad n = 0, 1, 2, \ldots \qquad (12.3)$$

and an infinite number of potential solutions

$$X_n(x) = a_n \cos\left(\frac{n\pi}{L}x\right) \qquad n = 0, 1, 2, \ldots$$

The constant coefficients a_n must be determined by additional initial or boundary conditions. Note that we include the constant solution $X_0(x) = a_0$. This is a *non-trivial solution* that will prove to be critical to our overall solution for $u(x,t)$. Since $u(x,t)$ is a sum of cosines that can take on positive and negative values, the constant term a_0 is essential to maintain the condition that the temperature profile $u(x,t) > 0$ for all x and t.

Now let's return to Equation 12.2 and determine the solution for $T(t)$ by solving the first order ordinary differential equation[5]

$$\frac{dT(t)}{dt} = -\kappa\alpha^2 T(t)$$

[5] We have replaced the partial derivative with respect to t in Equation 12.2 with a total derivative as t is the sole variable.

Having determined acceptable values of α^2 in Equation 12.3, we can rewrite this first order ordinary differential equation for $T(t)$ as

$$\frac{dT(t)}{dt} = -\kappa\frac{n^2\pi^2}{L^2}T(t) \qquad n = 0, 1, 2, \ldots$$

Drawing on our experience solving first order ordinary differential equations in Chapter 10, we propose an infinite number of potential solutions

$$T_n(t) = c_n \exp\left(-\frac{n^2\pi^2}{L^2}\kappa t\right)$$

where c_n are constant coefficients that must be determined by additional initial or boundary conditions.

For a given value of n we find that the product $X_n(x)T_n(t)$ is a solution of Equation 12.1. It follows that a sum of all possible solutions $X_n(x)T_n(t)$ represents a general solution for the time and position dependence of $u(x,t)$, which can be written as the series[6]

$$u(x,t) = \sum_{n=0}^{\infty} X_n(x)T_n(t)$$

$$= \sum_{n=0}^{\infty} a_n \cos\left(\frac{n\pi}{L}x\right) \times c_n \exp\left(-\frac{n^2\pi^2}{L^2}\kappa t\right)$$

[6] This general solution to the classical heat equation was first presented by French mathematician and physicist *Joseph Fourier* (1768-1830). Fourier is also credited with the discovery of the *greenhouse effect*.

where the constant coefficients $u_n = a_n c_n$. Separating the first term in the series, we write the general solution

$$u(x,t) = u_0 + \sum_{n=1}^{\infty} u_n \cos\left(\frac{n\pi}{L}x\right) \exp\left(-\frac{n^2\pi^2}{L^2}\kappa t\right) \qquad (12.4)$$

To determine the values of the constant coefficients u_n, we typically appeal to an initial condition at $t = 0$, for which our general solution takes the form

$$u(x,0) = u_0 + \sum_{n=1}^{\infty} u_n \cos\left(\frac{n\pi}{L}x\right) \tag{12.5}$$

Our result is a general solution for the heat equation with reflecting boundary conditions. Using this result, the parameters u_n are fit to match the given initial temperature profile, $u(x,0)$, providing a complete solution for $u(x,t)$.

12.1.2 *Application of the classical heat equation (reflecting boundaries)*

Let's consider the case of an initial condition for the temperature profile described by

$$u(x,0) = \sin^2\left(\frac{\pi}{L}x\right) \tag{12.6}$$

and shown in Figure 12.2. The *reflecting boundary conditions* for the temperature profile are

$$\frac{\partial u}{\partial x}(0,t) = \frac{\partial u}{\partial x}(L,t) = 0$$

We return to our general solution for $u(x,t)$, Equation 12.4, appropriate for reflecting boundary conditions

$$u(x,t) = u_0 + \sum_{n=1}^{\infty} u_n \cos\left(\frac{n\pi}{L}x\right) \exp\left(-\frac{n^2\pi^2}{L^2}\kappa t\right)$$

Taking the general solution at $t = 0$ and setting it equal to the initial distribution, Equation 12.6, we find

$$u(x,0) = u_0 + \sum_{n=1}^{\infty} u_n \cos\left(\frac{n\pi}{L}x\right) = \sin^2\left(\frac{\pi}{L}x\right)$$

Using our knowledge of trigonometric identities, we recognize that $\sin^2(x) = (1 - \cos(2x))/2$ so that

$$u(x,0) = \sin^2\left(\frac{\pi}{L}x\right) = \frac{1}{2} - \frac{1}{2}\cos\left(\frac{2\pi}{L}x\right)$$

$$= u_0 + u_2 \cos\left(\frac{2\pi}{L}x\right)$$

where $u_0 = 1/2$, $u_2 = -1/2$, and all other coefficients are zero. Having identified the coefficients of the series, we can write the general solution as a function of position and time

$$u(x,t) = \frac{1}{2} - \frac{1}{2}\cos\left(\frac{2\pi}{L}x\right) \exp\left(-\frac{4\pi^2}{L^2}\kappa t\right) \tag{12.7}$$

Figure 12.2 shows the time evolution of $u(x,t)$ described by Equation 12.7.

Initially, the temperature distribution is peaked in the center of the box. This maximum in the distribution of temperature represents a hot spot. The temperature at the edges of the box is initially zero. Over time, the temperature in the center decreases, the temperature in the wings increases, and $u(x,t)$ approaches a final uniform temperature distribution, $u(x,\infty)$.

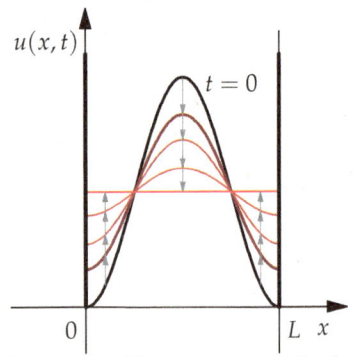

Figure 12.2: The temperature $u(x,t)$ over a box of length L. The initial distribution $u(x,0)$ (black) evolves in time toward a constant distribution, $u(x,\infty)$ (light red line).

12.2 The classical diffusion equation

THE PROCESS OF DIFFUSION IS FUNDAMENTAL TO STUDIES OF KINETICS AND TRANSPORT in the physical sciences. At the microscopic level, diffusion occurs through the random motions of atoms and molecules. At a larger scale, the process of diffusion can be modeled in terms of changes in the concentration profile as a function of space and time described by the classical diffusion equation. This section explores the classical diffusion equation and its solution in one and many dimensions.

12.2.1 The classical diffusion equation in one dimension (free diffusion)

Variation in the concentration of particles as a function of position and time can be modeled by the *classical diffusion equation*[7]

$$\frac{\partial c(x,t)}{\partial t} = D\frac{\partial^2 c(x,t)}{\partial x^2} \tag{12.8}$$

[7] The diffusion equation was discovered by German physician and physiologist *Adolf Fick* (1829-1901).

where $c(x,t)$ describes the physical density or concentration of particles that is dependent on position x and time t.[8] The equation states that the rate of change in the distribution $c(x,t)$, given by the first derivative of $c(x,t)$ with respect to t, is proportional to the second derivative of the distribution $c(x,t)$ with respect to x. The parameter D is the *diffusion coefficient* that modulates the proportionality between the partial derivatives in time and position and has units of length squared divided by time.

[8] The classical diffusion equation may also be written $c_t = D\,c_{xx}$ using our compact notation for partial derivatives.

For free diffusion, the total amount of diffusing particles is conserved so that the integral of the distribution $c(x,t)$ over all positions x is constant for all time:

$$\int_{-\infty}^{\infty} c(x,t)\,dx = c_0$$

An initial condition $c(x,0)$ describes the distribution of our particles in space at $t = 0$. The distribution of our particles at some time $t > 0$ in the future is defined by the solution of Equation 12.8.

Let's consider the case of diffusion in one dimension with no spatial boundary conditions. We call this *free diffusion*. Assume that our diffusing particles are initially concentrated at a single point in space, $x = x_0$, providing the initial condition

$$c(x,0) = c_0\,\delta(x - x_0)$$

We propose an *ansatz* for the dependence of the distribution on position x and time t, written as a gaussian function[9]

$$c(x,t) = \frac{c_0}{\sqrt{4\pi D t}} \exp\left[-\frac{(x - x_0)^2}{4Dt}\right] \tag{12.9}$$

[9] We propose this *ansatz* based on the qualitative observation that the solution $c(x,t)$ must be normalized and have $\partial_t c$ proportional to $\partial_{xx} c$.

We note several important properties of this proposed distribution. The integral of the distribution over all positions x is a constant

$$\int_{-\infty}^{\infty} c(x,t)dx = \frac{c_0}{\sqrt{4\pi D t}} \int_{-\infty}^{\infty} \exp\left[-\frac{(x - x_0)^2}{4Dt}\right] dx = c_0$$

[10] Here we use the identity

$$\int_{-\infty}^{\infty} e^{-\frac{1}{2\sigma^2}(x-x_0)^2}\,dx = \sqrt{2\pi\sigma^2}$$

where $\sigma^2 = 2Dt$.

for all time t.[10] Furthermore, at $t = 0$, the proposed distribution $c(x,t)$ is concentrated at a single point in space, $x = x_0$, which can be represented by

$$c(x,0) = \lim_{t \to 0} \frac{c_0}{\sqrt{4\pi Dt}} \exp\left[-\frac{(x-x_0)^2}{4Dt}\right] = c_0\delta(x-x_0)$$

Here we have used the definition of the Dirac delta function

$$\delta(x-x_0) = \lim_{\sigma \to 0} \frac{1}{\sqrt{2\pi\sigma^2}} \exp\left[-\frac{(x-x_0)^2}{2\sigma^2}\right]$$

where $\sigma^2 = 2Dt$. The normalized gaussian distribution function approaches a Dirac delta function as the width of the distribution, $\sigma = \sqrt{2Dt}$, approaches zero. As time evolves the width of the distribution increases according to

$$\sigma = \sqrt{2Dt} \tag{12.10}$$

at a rate proportional to $t^{1/2}$. These properties of the distribution $c(x,t)$ are depicted in Figure 12.3.

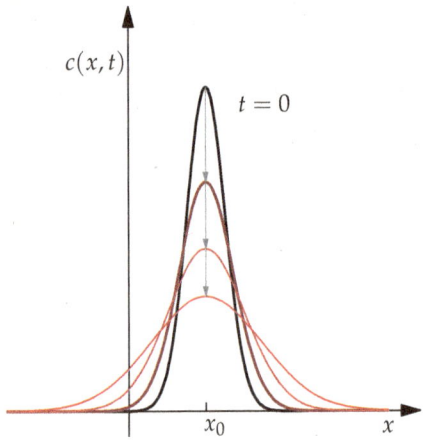

Figure 12.3: The normalized distribution $c(x,t)$ as a function of position x for several values of time t. The width of the distribution grows with time according to $t^{1/2}$ (lighter and lighter shades of red).

Now let's prove that our *ansatz* for $c(x,t)$ is a solution to the classical diffusion equation by substituting Equation 12.9 into Equation 12.8 and demonstrating the equality. We will do this in steps, by evaluating the partial derivatives with respect to time and position separately before combining them in Equation 12.8.

For the partial derivative with respect to time, we find

$$\frac{\partial c(x,t)}{\partial t} = \left[\frac{(x-x_0)^2}{4Dt^2} - \frac{1}{2t}\right]c(x,t)$$

For the first partial derivative with respect to position, we find

$$\frac{\partial c(x,t)}{\partial x} = -\frac{2(x-x_0)}{4Dt}c(x,t)$$

so that the second partial derivative with respect to position is

$$\frac{\partial^2 c(x,t)}{\partial x^2} = \left[\frac{4(x-x_0)^2}{(4Dt)^2} - \frac{2}{4Dt}\right]c(x,t)$$

Combining our results, we find

$$\frac{\partial c(x,t)}{\partial t} = D\frac{\partial^2 c(x,t)}{\partial x^2}$$

$$\left[\frac{(x-x_0)^2}{4Dt^2} - \frac{1}{2t}\right]c(x,t) = D\left[\frac{4(x-x_0)^2}{(4Dt)^2} - \frac{2}{4Dt}\right]c(x,t)$$

Canceling $c(x,t)$ on each side and multiplying through by D in the right-hand expression, we demonstrate the equality. This proves that our *ansatz* in the form of Equation 12.9 satisfies the diffusion equation. In the next section, we explore classical diffusion in the presence of a reflecting boundary.

12.2.2 *Application of the classical diffusion equation (reflecting boundary)*

Let's consider the case of an initial condition for the concentration profile $c(x,t)$ described by

$$c(x,0) = c_0\delta(x-x_0) \tag{12.11}$$

where $x_0 > 0$. Assume there is a reflecting boundary at the origin $x = 0$. Any particles diffusing to $x = 0$ are reflected back and maintained in the range $0 \leq x < \infty$. The *reflecting boundary condition* for the distribution is written as

$$\frac{\partial c}{\partial x}(0,t) = 0$$

In this case, the solution to the diffusion equation for $x \geq 0$ is

$$c(x,t) = \frac{c_0}{\sqrt{4\pi Dt}}\exp\left[-\frac{(x-x_0)^2}{4Dt}\right] + \frac{c_0}{\sqrt{4\pi Dt}}\exp\left[-\frac{(x+x_0)^2}{4Dt}\right] \tag{12.12}$$

where we have added a gaussian centered at $x = x_0$ to an *image* of that gaussian function,[11] a second gaussian function centered at $x = -x_0$. With the reflecting boundary, the total area under our distribution is constant in time:

$$\int_0^\infty c(x,t)dx = c_0$$

The function is shown in Figure 12.4.

[11] We refer to the complementary gaussian function as an image function, as it is the reflection of the primary gaussian function across the y-axis, as if the y-axis were a mirror.

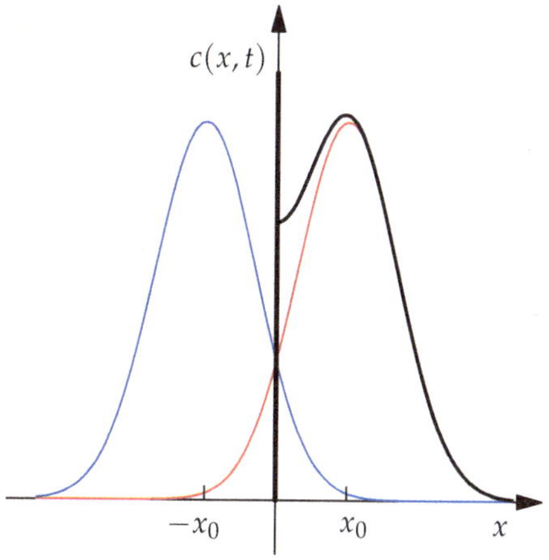

Figure 12.4: The concentration profile $c(x,t)$ for particles diffusing on $0 \leq x < \infty$ with a reflecting boundary at $x = 0$ (thick black line). The total distribution (black line) is composed of the sum of two gaussians, one centered at $x_0 > 0$ (red line) and an image function centered at $-x_0$ (blue line). The sum of the two functions meets the boundary condition that $c_x(0,t) = 0$.

How can we see that this function represents a valid solution to the diffusion equation over $0 \leq x < \infty$ with a reflecting boundary at $x = 0$? First, we demonstrated earlier that gaussian functions of the form in Equation 12.12 satisfy the diffusion equation $c_t(x,t) = D c_{xx}(x,t)$. As such, the sum of the two functions will also be a solution to the diffusion equation. Second, we can see that the derivative will be

$$\frac{\partial c}{\partial x}(0,t) = 0$$

since the derivative of the first gaussian at $x = 0$ will be equal in magnitude and opposite in sign to the derivative of the second gaussian at $x = 0$. Third, at $t = 0$, each gaussian becomes a Dirac delta function, so that

$$c(x,0) = c_0 \delta(x - x_0) + c_0 \delta(x + x_0)$$

As our solution is defined only for $0 \leq x < \infty$, only the first Dirac delta function contributes. As such, our solution satisfies the initial condition that

$$c(x,0) = c_0\, \delta(x - x_0)$$

Finally, we are left to show that the total area under our distribution is constant in time. We can do this by evaluating

$$\int_0^\infty c(x,t)dx = \frac{c_0}{\sqrt{4\pi Dt}} \int_0^\infty e^{-\frac{(x-x_0)^2}{4Dt}}\, dx + \frac{c_0}{\sqrt{4\pi Dt}} \int_0^\infty e^{-\frac{(x+x_0)^2}{4Dt}}\, dx$$

Defining $u = (x - x_0)/\sqrt{4Dt}$, we note that for the first integral is

$$\frac{1}{\sqrt{4\pi Dt}} \int_0^\infty e^{-\frac{(x-x_0)^2}{4Dt}}\, dx = \frac{1}{\sqrt{\pi}} \int_{\frac{-x_0}{\sqrt{4Dt}}}^\infty e^{-u^2}\, du$$

$$= 1 - \frac{1}{2}\mathrm{erfc}\left(\frac{x_0}{\sqrt{4Dt}}\right)$$

where $\mathrm{erfc}(x)$ is the *complementary error function*. The error function and complementary error function, discussed in Chapters 7 and 9, are special functions that represent partial integration over a gaussian function.[12]
Defining $u = (x + x_0)/\sqrt{4Dt}$, we note that for the second integral is

$$\frac{1}{\sqrt{4\pi Dt}} \int_0^\infty e^{-\frac{(x+x_0)^2}{4Dt}}\, dx = \frac{1}{\sqrt{\pi}} \int_{\frac{x_0}{\sqrt{4Dt}}}^\infty e^{-u^2}\, du$$

$$= \frac{1}{2}\mathrm{erfc}\left(\frac{x_0}{\sqrt{4Dt}}\right)$$

with the final result that

$$\int_0^\infty c(x,t)\, dx = c_0 \left[1 - \frac{1}{2}\mathrm{erfc}\left(\frac{x_0}{\sqrt{4Dt}}\right) + \frac{1}{2}\mathrm{erfc}\left(\frac{x_0}{\sqrt{4Dt}}\right)\right] = c_0$$

We've demonstrated that Equation 12.12 satisfies the diffusion equation, the reflecting boundary condition, initial condition, and the constraint that the

[12] The *error function* is defined as

$$\mathrm{erf}(x) = \frac{2}{\sqrt{\pi}} \int_0^x e^{-t^2}\, dt$$

while the *complementary error function* is defined as

$$\mathrm{erfc}(x) = \frac{2}{\sqrt{\pi}} \int_x^\infty e^{-t^2}\, dt$$

or $\mathrm{erfc}(x) = 1 - \mathrm{erf}(x)$.

total area under the distribution is conserved. This validates our solution for $c(x,t)$. The final result is shown in Figure 12.5.

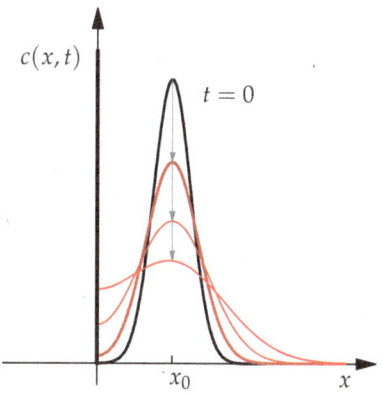

Figure 12.5: The distribution $c(x,t)$ evolving over $0 \leq x < \infty$ with a reflecting boundary at $x = 0$. The distribution is shown as a function of x for several values of t. As time evolves, solutions are shown in lighter and lighter shades of red.

Note the effect of the reflecting boundary condition on $c(x,t)$. Initially, near $x = 0$ there is growth in the distribution $c(x,t)$ as particles diffusing to the left are reflected back to $x > 0$. Over time, the distribution broadens as a function of x. Eventually, particles will diffuse to form a uniform concentration profile.

12.2.3 The classical diffusion equation in two dimensions (free diffusion)

For the classical diffusion equation in one dimension

$$\frac{\partial c(x,t)}{\partial t} = D\frac{\partial^2 c(x,t)}{\partial x^2}$$

we proposed the *ansatz* that

$$c(x,t) = \frac{c_0}{\sqrt{4\pi Dt}} \exp\left[-\frac{(x-x_0)^2}{4Dt}\right]$$

where the variance of the distribution

$$\sigma^2 = 2Dt$$

We proved this to be valid for free diffusion with the initial condition

$$c(x,0) = c_0\, \delta(x - x_0)$$

Now consider the classical diffusion equation in two dimensions expressed in cartesian coordinates:

$$\frac{\partial c(x,y,t)}{\partial t} = D\left[\frac{\partial^2 c(x,y,t)}{\partial x^2} + \frac{\partial^2 c(x,y,t)}{\partial y^2}\right] \tag{12.13}$$

Here $c(x,y,t)$ is the concentration profile in position, defined by x and y, and time t. Let's examine the case of free diffusion where there are no boundary conditions on our concentration profile. The initial condition is

$$c(x,y,0) = c_0\, \delta(x - x_0)\delta(y - y_0)$$

As time evolves, the distribution broadens with growing variances

$$\sigma_x^2 = \sigma_y^2 = 2Dt$$

Note that we have assumed isotropic diffusion, where the rates of diffusion in the x-direction and y-direction are identical.[13] In general, it is possible to model anisotropic diffusion, in which the rate of diffusion in the x-direction and y-direction differ.

With these observations in mind and following our solution in one dimension, we propose the following *ansatz* for the concentration profile

$$c(x,y,t) = \frac{1}{\sqrt{4\pi Dt}} \exp\left[-\frac{(x-x_0)^2}{4Dt}\right] \times \frac{1}{\sqrt{4\pi Dt}} \exp\left[-\frac{(y-y_0)^2}{4Dt}\right]$$

This can be further simplified as

$$c(x,y,t) = \frac{1}{4\pi Dt} \exp\left[-\frac{1}{4Dt}\left[(x-x_0)^2 + (y-y_0)^2\right]\right] \qquad (12.14)$$

This represents the solution for free diffusion starting at a point (x_0, y_0) and spreading at a rate defined by the diffusion coefficient, D. The concentration profile is initially defined by a delta function at $t = 0$. As time evolves, the variance of the distribution grows according to t and the width grows according to

$$\text{distribution width} \propto t^{1/2}$$

Let's examine the properties of our solution to the two-dimensional diffusion equation given by Equation 12.14. The concentration profile $c(x,y,t)$ is shown in Figure 12.6 at five times $t = 1, 2, 4, 8,$ and ∞. At $t = 1$, the distribution remains peaked about the point (x_0, y_0) with a width equal to $\sqrt{2D}$. As the distribution evolves in time to $t = 2$ and then $t = 4$, the width grows by a factor of $\sqrt{4} = 2$. When time has evolved to $t = 8$, the width of the distribution has increased by a factor of $\sqrt{8} = 2\sqrt{2}$ from the distribution at $t = 1$. As time evolves still further, the distribution will continue to spread as it approaches a uniform distribution over the xy-plane at $t = \infty$.

While we have assumed free diffusion in two dimensions, it is also possible to impose boundary conditions that are absorbing or reflecting, as was done in the case of the classical heat equation. Moreover, while we have expressed Equation 12.13 in two-dimensional cartesian coordinates, the equation can be written more generally as

$$\frac{\partial c}{\partial t} = D\nabla^2 c \qquad (12.15)$$

where ∇^2 is the *Laplace operator* (see Chapter 5). As such, we can readily express the diffusion equation in two-dimensional plane polar coordinates for $c(r, \theta, t)$, three-dimensional cartesian coordinates for $c(x, y, z, t)$, cylindrical coordinates for $c(r, \theta, z, t)$, or spherical polar coordinates for $c(r, \theta, \varphi, t)$, by simply using the appropriate form of the operator ∇^2 (as provided in Complement C$_5$). The best choice of coordinates will be those that reflect the natural symmetries of our system, be they square, circular, cubic, spherical, or cylindrical.

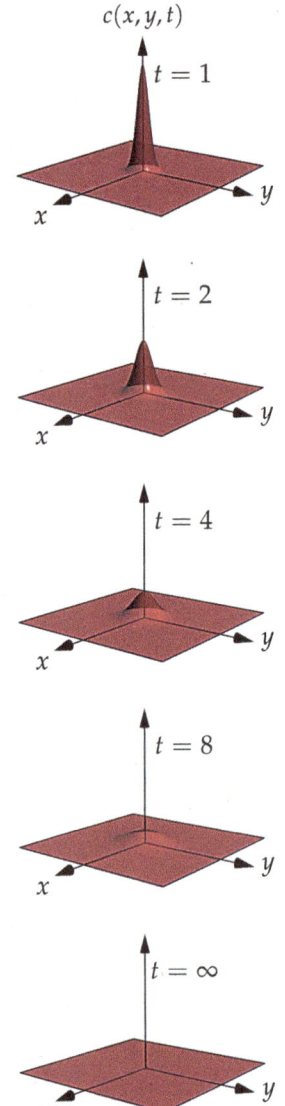

$c(x,y,t)$

$t = 1$

$t = 2$

$t = 4$

$t = 8$

$t = \infty$

Figure 12.6: A concentration profile in the xy-plane represented by a two-dimensional gaussian distribution at times $t = 1, 2, 4, 8,$ and ∞.

[13] As we assume diffusion in x and y is uncorrelated, the correlation coefficient $\rho_{xy} = 0$ and the covariance $\sigma_{xy} = 0$.

12.3 The classical wave equation

MANY SYSTEMS INVOLVE INTERFACES SEPARATING ONE DOMAIN FROM ANOTHER. Common examples include the surface of a liquid exposed to a gas, the interface between two immiscible liquids, and the surface of a membrane exposed to water. In each case, fluctuations of the interface involve the formation and propagation of wave forms that can be modeled using the classical wave equation. This section explores the classical wave equation and its solution in one and two dimensions.

12.3.1 The classical wave equation in one dimension

The classical wave equation for the displacement $h(x,t)$ of a string as a function of position x and time t is[14]

$$\frac{\partial^2 h(x,t)}{\partial t^2} = v^2 \frac{\partial^2 h(x,t)}{\partial x^2} \tag{12.16}$$

where v is a constant with units of speed, or length over time.[15] We assume that the ends of the string are fixed so that the displacement is zero at $x = 0$ and $x = L$. As such, we have the boundary conditions

$$h(0,t) = h(L,t) = 0$$

as shown in Figure 12.7 below.

[14] The wave equation was discovered by French mathematician, physicist, philosopher, and music theorist *Jean le Rond d'Alembert* (1717-1783).

[15] The classical wave equation may also be written $h_{tt} = v^2 h_{xx}$ using our compact notation for partial derivatives.

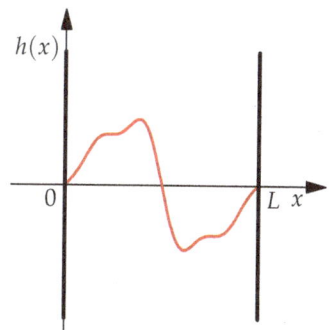

Figure 12.7: The displacement $h(x,t)$ of a classical wave over a length L.

To solve the wave equation, we follow the procedure used to solve the classical heat equation, Equation 12.1, by proposing a solution for the displacement $h(x,t)$ that is separable in its dependence on position and time:

$$h(x,t) = X(x)T(t)$$

Inserting this function into Equation 12.16 we find

$$X(x)\frac{d^2 T(t)}{dt^2} = v^2 T(t)\frac{d^2 X(x)}{dx^2}.$$

Note that as $X(x)$ and $T(t)$ are functions of one variable, the derivatives of those functions are represented as total derivatives rather than partial derivatives.

Rearranging this equation so that the time dependence is on the left and the position dependence is on the right, we find

$$\frac{1}{v^2 T(t)} \frac{d^2 T(t)}{dt^2} = \frac{1}{X(x)} \frac{d^2 X(x)}{dx^2}$$

Since we expect x and t to be independent, the only way for this equality to hold is if each side of the equation is proportional to a constant:

$$\frac{1}{v^2 T(t)} \frac{d^2 T(t)}{dt^2} = -k^2 = \frac{1}{X(x)} \frac{d^2 X(x)}{dx^2} \qquad (12.17)$$

Note that k is a real constant that can be determined by initial or boundary conditions. We expect the constant $-k^2$ will be negative, as we expect the functions $X(x)$ and $T(t)$ will be oscillatory. The use of a negative constant $-k^2$ will simplify the steps leading to the overall solution.

By assuming a separable solution, we have converted one second order partial differential equation into two second order ordinary differential equations. We can now use the techniques we developed to solve second order ordinary differential equations of a single variable to determine solutions for $T(t)$ and $X(x)$. Those solutions can then be combined to determine the overall solution for $h(x,t) = X(x)T(t)$.

Let's determine the solution for $X(x)$ by returning to Equation 12.17 and solving the second order ordinary differential equation[16]

$$\frac{d^2 X(x)}{dx^2} = -k^2 X(x)$$

As the constant $k^2 > 0$, we draw on our experience in solving second order ordinary differential equations in Chapter 11 to propose a general solution of the form[17]

$$X(x) = a_1 \cos(kx) + b_1 \sin(kx)$$

The constants a_1 and b_1 can be determined through the boundary conditions. We find

$$X(0) = a_1 = 0$$

Furthermore, since

$$X(L) = b_1 \sin(kL) = 0$$

it follows that

$$kL = n\pi \qquad n = 1,2,\ldots$$

providing an infinite number of possible values of our constant

$$k^2 = \frac{n^2 \pi^2}{L^2} \qquad n = 1,2,\ldots \qquad (12.18)$$

This provides us with an infinite number of potential solutions

$$X_n(x) = b_n \sin\left(\frac{n\pi}{L}x\right) \qquad n = 1,2,\ldots$$

The constant coefficients b_n must be determined by additional initial or boundary conditions.

[16] As x is the sole variable, we have replaced the partial derivative with respect to x in Equation 12.17 with a total derivative.

[17] If $k = 0$ then $X(x) = a_1 + b_1 x$. Since $X(0) = 0$ and $X(L) = 0$ it follows that $a_1 = b_1 = 0$. Now suppose $k^2 < 0$. Defining $\alpha^2 = -k^2 > 0$ then $X(x) = a_1 e^{\alpha x} + b_1 e^{-\alpha x}$. But since $X(0) = a_1 + b_1 = 0$ and $X(L) = a_1 e^{\alpha L} + b_1 e^{-\alpha L} = 0$ it must be that $a_1 = b_1 = 0$. As such, results for $k^2 \leq 0$ are considered to be *trivial solutions*.

Now let's return to Equation 12.17 and determine the solution for $T(t)$ by solving the second order ordinary differential equation

$$\frac{d^2 T(t)}{dt^2} = -v^2 k^2 T(t)$$

Having determined acceptable values of k^2 in Equation 12.18, we can rewrite this second order ordinary differential equation for $T(t)$ as

$$\frac{d^2 T(t)}{dt^2} = -v^2 \frac{n^2 \pi^2}{L^2} T(t) \qquad n = 1, 2, \ldots$$

Drawing on our experience solving second order ordinary differential equations in Chapter 11, we propose an infinite number of potential solutions

$$T_n(t) = c_n \cos\left(\frac{n\pi}{L} vt\right) + d_n \sin\left(\frac{n\pi}{L} vt\right)$$

where c_n and d_n are constant coefficients that must be determined by additional initial or boundary conditions.

For a given value of n, we find that the product $X_n(x) T_n(t)$ is a solution of Equation 12.16. It follows that a sum of all possible solutions $X_n(x) T_n(t)$ represents a general solution for the time and position dependence of $h(x, t)$, which can be written

$$h(x, t) = \sum_{n=1}^{\infty} X_n(x) T_n(t)$$

$$= \sum_{n=1}^{\infty} b_n \sin\left(\frac{n\pi}{L} x\right) \times \left(c_n \cos\left(\frac{n\pi}{L} vt\right) + d_n \sin\left(\frac{n\pi}{L} vt\right)\right)$$

$$= \sum_{n=1}^{\infty} \sin\left(\frac{n\pi}{L} x\right) \left(s_n \cos\left(\omega_n t\right) + t_n \sin\left(\omega_n t\right)\right)$$

where $\omega_n = n\pi v / L$ and the constant coefficients are $s_n = b_n c_n$ and $t_n = b_n d_n$.

To determine the values of the constant coefficients s_n and t_n, we typically appeal to the initial conditions at $t = 0$ for our general solution

$$h(x, 0) = \sum_{n=1}^{\infty} s_n \sin\left(\frac{n\pi}{L} x\right)$$

and its derivative in time

$$\frac{\partial h}{\partial t}(x, 0) = \sum_{n=1}^{\infty} t_n \omega_n \sin\left(\frac{n\pi}{L} x\right)$$

Given a knowledge of $h(x, 0)$ and $h_t(x, 0)$, the parameters s_n and t_n are fit to provide an overall solution to $h(x, t)$.

12.3.2 *Application of the classical wave equation (plucked string)*

Let's consider the specific case of a plucked string. The ends of the string located at $x = 0$ and $x = L$ are fixed, imposing the boundary conditions

$$h(0, t) = 0 \qquad h(L, t) = 0$$

Suppose that the initial condition for the displacement of the string is described by

$$h(x,0) = \sin\left(\frac{2\pi}{L}x\right)\cos\left(\frac{2\pi}{L}x\right) \tag{12.19}$$

with

$$\frac{\partial h}{\partial t}(x,0) = 0 \tag{12.20}$$

The initial displacement profile $h(x,0)$ differs from the time-independent resting profile $h(x,0) = 0$. Note that the first initial condition determines the initial displacement of the string $h(x,0)$, while the second initial condition indicates that at $t = 0$ the string is drawn but not moving (heavy black line in Figure 12.8).

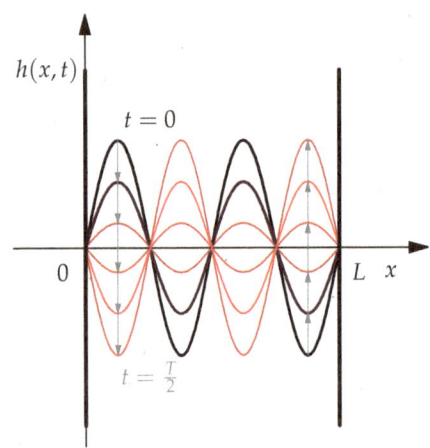

Figure 12.8: The displacement of a string $h(x,t)$ over a box of length L, shown for the initial displacement (black) and at several times in the future (lighter and lighter shades of red).

Once released, the string moves through sinusoidal repeating oscillations.

We return to our general solution appropriate for these boundary conditions:

$$h(x,t) = \sum_{n=1}^{\infty} \sin\left(\frac{n\pi}{L}x\right)\left[s_n\cos\left(\omega_n t\right) + t_n\sin\left(\omega_n t\right)\right]$$

Taking the general solution at $t = 0$ and setting it equal to the initial distribution defined by Equation 12.19, we find

$$h(x,0) = \sum_{n=1}^{\infty} s_n \sin\left(\frac{n\pi}{L}x\right) = \sin\left(\frac{2\pi}{L}x\right)\cos\left(\frac{2\pi}{L}x\right)$$

Using our knowledge of trigonometric identities, we recognize that $\sin(x)\cos(x) = \sin(2x)/2$, so that

$$h(x,0) = \sin\left(\frac{2\pi}{L}x\right)\cos\left(\frac{2\pi}{L}x\right) = \frac{1}{2}\sin\left(\frac{4\pi}{L}x\right)$$

$$= s_4 \sin\left(\frac{4\pi}{L}x\right)$$

As such, $s_4 = 1/2$ and all other coefficients $s_{n\neq4} = 0$. Applying the second initial condition, Equation 12.20, we find

$$\frac{\partial h}{\partial t}(x,0) = \sum_{n=1}^{\infty} t_n \omega_n \sin\left(\frac{n\pi}{L}x\right) = 0$$

As such $t_n = 0$ for all n. Having identified the coefficients of the series, we can write the general solution as a function of position and time:

$$h(x,t) = \frac{1}{2}\sin\left(\frac{4\pi}{L}x\right)\cos\left(\frac{4\pi}{L}vt\right)$$

This final result for the displacement of the string and its variation in time is shown in Figure 12.8. The solution is periodic in time with a period

$$T = \frac{L}{2v}$$

which is proportional to L and inversely proportional to the speed of the wave v.

12.3.3 *The classical wave equation in two dimensions*

The classical wave equation in one dimension describes the displacement $h(x,t)$ of a *string* in position x and time t as

$$\frac{\partial^2 h(x,t)}{\partial x^2} = \frac{1}{v^2}\frac{\partial^2 h(x,t)}{\partial t^2}$$

Assuming that the ends of the string are fixed so that

$$h(0,t) = 0 \qquad h(L,t) = 0$$

we identified the general solution

$$h(x,t) = \sum_{n=1}^{\infty} \sin\left(\frac{n\pi}{L}x\right)\left(s_n \cos\left(\omega_n t\right) + t_n \sin\left(\omega_n t\right)\right)$$

where $\omega_n = n\pi v/L$, and the constant coefficients s_n and t_n are defined by appealing to initial or boundary conditions.

Now consider the classical wave equation in two dimensions expressed in cartesian coordinates:

$$\frac{\partial^2 h(x,y,t)}{\partial x^2} + \frac{\partial^2 h(x,y,t)}{\partial y^2} = \frac{1}{v^2}\frac{\partial^2 h(x,y,t)}{\partial t^2} \qquad (12.21)$$

Here $h(x,y,t)$ describes the displacement of a two-dimensional *sheet* in position, defined by x and y, and time t.

Note that this equation can be written more generally as

$$\nabla^2 h = \frac{1}{v^2}\frac{\partial^2 h}{\partial t^2} \qquad (12.22)$$

where ∇^2 is the *Laplace operator* (see Chapter 5). As such, we can readily express the wave equation in two-dimensional plane polar coordinates for $c(r,\theta,t)$, three-dimensional cartesian coordinates for $c(x,y,z,t)$, cylindrical coordinates for $c(r,\theta,z,t)$, or spherical polar coordinates for $c(r,\theta,\varphi,t)$, by simply using the appropriate form of the operator ∇^2 (as provided in Complement C_5).

Our string was held fixed at two ends, leading to the boundary conditions $h(0,t) = h(L,t) = 0$. We take our square sheet to be held fixed along each of its four sides, imposing the boundary conditions

$$h(0,y,t) = h(L,y,t) = 0 \qquad h(x,0,t) = h(x,L,t) = 0$$

An example of a wave form in our two-dimensional square sheet is shown in Figure 12.9.

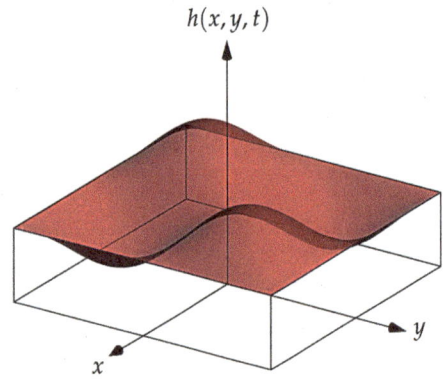

$h(x,y,t)$

Figure 12.9: A two-dimensional classical wave over the xy-plane with fixed boundaries.

Following our solution of the classical wave equation in one dimension, we propose an *ansatz* for the displacement $h(x,y,t)$ that is separable in position and time

$$h(x,y,t) = X(x)Y(y)T(t)$$

Inserting this function into Equation 12.21, we find

$$Y(y)T(t)\frac{d^2X(x)}{dx^2} + X(y)T(t)\frac{d^2Y(y)}{dy^2} = X(x)Y(y)\frac{1}{v^2}\frac{d^2T(t)}{dt^2}$$

Dividing each term by $h(x,y,t) = X(x)Y(y)T(t)$ leads to

$$\frac{1}{X(x)}\frac{d^2X(x)}{dx^2} + \frac{1}{Y(y)}\frac{d^2Y(y)}{dy^2} = \frac{1}{v^2T(t)}\frac{d^2T(t)}{dt^2}$$

Each of the terms depends solely on x, y, or t and can be solved independently. Our solutions for $X(x)$ and $Y(y)$ are

$$X_n(x) = a_n \sin\left(\frac{n\pi}{L}x\right) \qquad n = 1,2,\ldots$$

$$Y_m(y) = b_m \sin\left(\frac{m\pi}{L}y\right) \qquad m = 1,2,\ldots$$

where we have introduced indices n and m that independently track the number of half oscillations in the x- and y-directions. As such, our equation in time is

$$\frac{\partial^2 T(t)}{\partial t^2} = -v^2 \frac{\pi^2}{L^2}\left(n^2 + m^2\right) T(t) = -\omega_{nm}^2 T(t)$$

with the solution

$$T_{nm}(t) = c_{nm}\cos\left(\omega_{nm}t\right) + d_{nm}\sin\left(\omega_{nm}t\right)$$

where the frequencies are determined by the indices n and m:

$$\omega_{nm} = v\frac{\pi}{L}\left(n^2 + m^2\right)^{1/2}$$

As v has units of speed, or length over time, the ratio L/v has units of time.

We can simplify our result as

$$c_{nm}\cos\left(\omega_{nm}t\right) + d_{nm}\sin\left(\omega_{nm}t\right) = e_{nm}\sin\left(\omega_{nm}t + \varphi_{nm}\right)$$

where

$$e_{nm} = \sqrt{c_{nm}^2 + d_{nm}^2} \qquad \varphi_{nm} = \tan^{-1}\left(c_{nm}/d_{nm}\right)$$

Here we have used the relation

$$c_1\cos(\omega t) + c_2\sin(\omega t) = A\sin(\omega t + \varphi)$$

where $A = \sqrt{c_1^2 + c_2^2}$ and $\varphi = \tan^{-1}(c_1/c_2)$. Combining these results leads to the overall solution

$$h(x,y,t) = \sum_{n=1}^{\infty}\sum_{m=1}^{\infty} X_n(x)Y_m(y)T_{nm}(t)$$

$$= \sum_{n=1}^{\infty}\sum_{m=1}^{\infty} u_{nm}\sin\left(\frac{n\pi}{L}x\right)\sin\left[\frac{m\pi}{L}y\right]\sin\left(\omega_{nm}t + \varphi_{nm}\right)$$

where the coefficient $u_{nm} = a_n b_m e_{nm}$.

Let's examine the properties of our solution to the two-dimensional wave equation by exploring the time evolution of the displacement $h(x,y,t)$. Consider the wave form $h(x,y,t)$ shown in Figure 12.10 for $n = 2$, $m = 4$, and $\omega_{nm} = \omega_{24}$. The period of oscillation is defined as

$$T = \frac{2\pi}{\omega_{24}}$$

The time evolution of the two-dimensional wave shown in Figure 12.10 represents one-half a period of oscillation. The maxima become minima and the minima become maxima.[18]

At the initial time $t = 0$, the wave form is at maximum amplitude. As time evolves to $t = T/8$, the extrema in the wave are diminished while the nodal lines along which $h(x,y,t) = 0$ are maintained. At $t = T/4$, the wave form is flat. Further evolution to $t = 3T/8$ reverses the phase of the initial oscillations while maintaining the diminished amplitudes. As time evolves further to $t = T/2$, the full amplitude of the initial oscillation in the wave is restored. However, the wave is now inverted. Further evolution of the wave over an additional time $T/2$ will restore the original wave form. This demonstrates the periodic nature of the general solution to the classical wave equation.

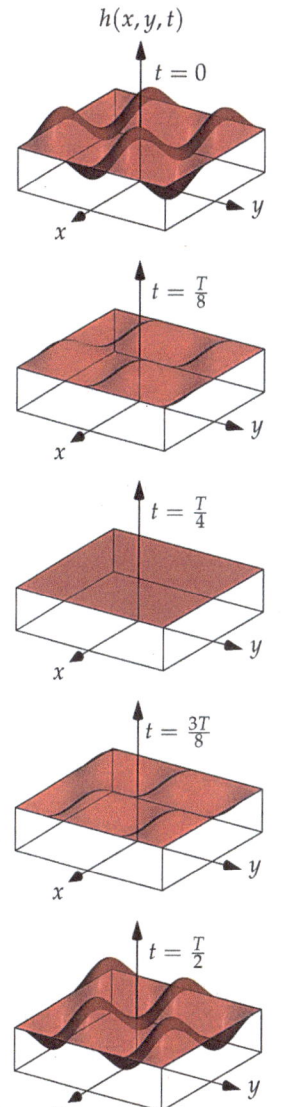

$h(x,y,t)$

Figure 12.10: The displacement $h(x,y,t)$ for a classical wave in two dimensions for $n = 2$ and $m = 4$. The wave is shown at five distinct times during one-half a period of oscillation.

[18] Along the nodal lines defined by $x = 0$ and $y = 0$ the amplitude of the wave is zero at all times.

A_{12} Survey of partial differential equations in the physical sciences

Partial differential equations appear in all corners of the physical sciences including thermodynamics, kinetics, mechanics, and quantum theory. Examples include the classical heat equation, classical diffusion equation, classical wave equation, and quantum Schrödinger equation. In each case, a partial differential equation describes the variation in a function of two or more variables in terms of partial derivatives with respect to those variables. Here we explore the physical derivation of the classical heat equation, classical diffusion equation, and classical wave equation.

The classical heat equation in one dimension

In the physical sciences, we are often interested in knowing how a distribution in space changes with time. For example, suppose we have a metal bar. The two ends of the bar are kept cold. The center of the bar is initially heated at time $t = 0$, creating a distribution of temperature that evolves in time as shown in Figure 12.11. The temperature in the center of the distribution decreases over time as heat flows toward the cold ends of the bar. Eventually, the entire bar is as cold as the ends.

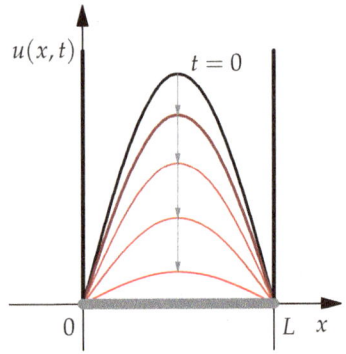

Figure 12.11: The temperature profile $u(x,t)$ is shown as a function of distance x over a metal bar (thick gray line) of length L. At time $t = 0$, the bar is heated in the center while the ends are kept cold. Over time, the bar cools as depicted by profiles of temperature $u(x,t)$ that decrease with increasing time.

How can we model this process as the temperature evolves in space and time? We expect that the rate of heat transfer is proportional to the temperature gradient:

$$\text{temperature gradient} = \frac{\partial u(x,t)}{\partial x}$$

If the temperature gradient is zero, there will be no net heat flow and no change in temperature. If the temperature gradient is large, the temperature profile is expected to change as heat flows from the region of higher temperature to the region of lower temperature. There is a driving force toward a uniform distribution of temperature.

The heat energy in a length Δx at position x and time t will be

$$\text{heat energy} = \Delta x \, u(x,t)$$

The flux of heat energy passing any point x in time Δt will be

$$\text{heat energy flux} = -\kappa\,\Delta t\,\frac{\partial u(x,t)}{\partial x}$$

where κ is the *thermal diffusion coefficient*. Note that if the gradient is positive, the heat energy flux is in the negative direction, and *vice versa*. This is known as *Fourier's law of thermal conduction*.

The change in heat energy in a given segment of the bar of length Δx in time Δt will be proportional to the difference between the heat entering at one side, x, and the other, $x + \Delta x$, as

$$\text{change in heat energy} = \text{difference in heat energy flux}$$

or

$$\Delta x\, u(x,t+\Delta t) - \Delta x\, u(x,t) = -\kappa\Delta t\,\frac{\partial u(x,t)}{\partial x}\bigg|_x - \left(-\kappa\Delta t\,\frac{\partial u(x,t)}{\partial x}\bigg|_{x+\Delta x}\right)$$

which we can rearrange as

$$\frac{1}{\Delta t}\left[u(x,t+\Delta t) - u(x,t)\right] = \kappa\frac{1}{\Delta x}\left(\frac{\partial u(x,t)}{\partial x}\bigg|_{x+\Delta x} - \frac{\partial u(x,t)}{\partial x}\bigg|_x\right)$$

This takes the form of a *finite difference* approximation of the first derivative. When the width of the segment Δx and the period of time Δt are small enough, this equation reduces to the form

$$\frac{\partial u(x,t)}{\partial t} = \kappa\frac{\partial^2 u(x,t)}{\partial x^2} \tag{12.23}$$

This is the *classical heat equation*.

The classical diffusion equation in one dimension

Suppose we wish to model the transport of particles in a gas, a fluid, or a solid. We start by defining a function $c(x,t)$, which describes the concentration of the particles in position x and time t. Suppose an initial concentration profile is centered at $x = x_0$. At $x = 0$ there is an *absorbing boundary*. If a particle reaches the absorbing boundary, it is removed from the system. As particles diffuse in space, the concentration profile $c(x,t)$ broadens as shown in Figure 12.12. Over time, some fraction of the diffusing particles are absorbed while the remainder diffuses to larger values of x.[19]

How can we model the process of particle diffusion in space and time? We note that the rate of diffusion of particles is proportional to the gradient of the concentration profile

$$\text{concentration gradient} = \frac{\partial c(x,t)}{\partial x}$$

If the gradient is zero, there will be no net diffusion of particles and no change in the concentration profile. If the concentration gradient is large, the concentration profile is expected to change as particles diffuse from regions of higher concentration to lower concentration.

[19] The question of how much of the initial concentration is absorbed and how much diffuses away is equivalent to the problem known as the *gambler's ruin*.

The number of particles over a length Δx at position x and time t will be

$$\text{number of particles} = \Delta x \, c(x, t)$$

The flux of particles passing any point x in time Δt will be

$$\text{particle flux} = -D \, \Delta t \frac{\partial c(x, t)}{\partial x}$$

where D is the *particle diffusion coefficient*. Note that if the gradient is positive, the flux is in the negative direction, and *vice versa*. This is known as *Fick's law of diffusion*.

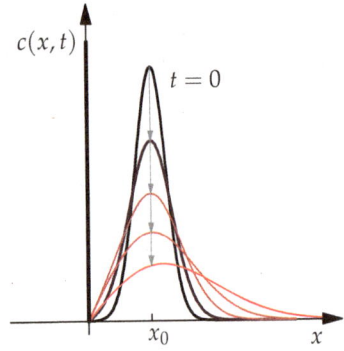

Figure 12.12: The concentration profile $c(x, t)$ for particles diffusing on $0 < x < \infty$ with an absorbing boundary at $x = 0$. The distribution is shown as a function of x and t. As time evolves, solutions are depicted in lighter and lighter shades of red.

The change in the distribution in a given length Δx in time Δt will be proportional to the difference between the number of particles entering at one side, x, and the other, $x + \Delta x$, as

$$\text{change in number of particles} = \text{difference in particle flux}$$

or

$$\Delta x \, c(x, t + \Delta t) - \Delta x \, c(x, t) = -D \, \Delta t \left. \frac{\partial c(x, t)}{\partial x} \right|_x - \left(-D \Delta t \left. \frac{\partial c(x, t)}{\partial x} \right|_{x + \Delta x} \right)$$

which we can rearrange as

$$\frac{1}{\Delta t} \left[c(x, t + \Delta t) - c(x, t) \right] = D \frac{1}{\Delta x} \left(\left. \frac{\partial c(x, t)}{\partial x} \right|_{x + \Delta x} - \left. \frac{\partial c(x, t)}{\partial x} \right|_x \right)$$

where we recognize *finite difference* approximations to the first derivatives with respect to x and t. When the width of the segment Δx and the period of time Δt are small enough, this equation reduces to

$$\frac{\partial c(x, t)}{\partial t} = D \frac{\partial^2 c(x, t)}{\partial x^2} \tag{12.24}$$

This is the *classical diffusion equation*.

The classical wave equation in one dimension

Now let's consider the motion of a wave in a string. The wave is described by a displacement in the string $h(x, t)$ as a function of position x and time t over a length L (see Figure 12.13). We consider the string to consist of a series of masses at positions $x - \Delta x$, x, $x + \Delta x$, and so on. Each mass is connected to the nearest neighboring masses by harmonic springs. How will the string move in space and time?

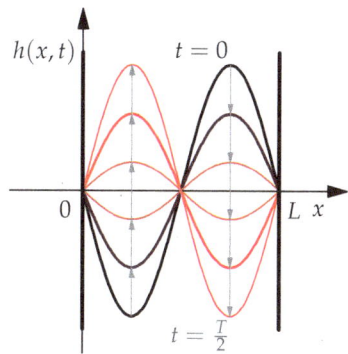

Figure 12.13: The initial displacement $h(x,0)$ (black) over a box of length L is shown with the displacement $h(x,t)$ at several times in the future (lighter and lighter shades of red).

The motion of each mass is defined by Newton's equation of motion. For a mass at point x and time t, the acceleration in the displacement $h(x,t)$ is equal to the force divided by the mass:

$$\frac{\partial^2 h(x,t)}{\partial t^2} = \frac{1}{m} F(x)$$

where $F(x)$ is the force and m is the mass. The force resulting from the displacement of the mass at x relative to the displacements of its nearest neighboring masses at $x + \Delta x$ and $x - \Delta x$ can be written

$$F(x) = k \ (h(x + \Delta x) - h(x)) - k \ (h(x) - h(x - \Delta x))$$

where k is the force constant. Combining these results leads to

$$\frac{\partial^2 h(x,t)}{\partial t^2} = \frac{k}{m} \ (h(x + \Delta x) - 2h(x) + h(x - \Delta x))$$

The total length of the string is divided into $N = L/\Delta x$ segments so that the total mass of the string is $M = Nm$ while the total force constant is $K = k/N$. As such, we can write

$$\frac{1}{m}k = \frac{N}{M}KN = \frac{K}{M}N^2 = \frac{K}{M}\left(\frac{L}{\Delta x}\right)^2$$

This leads to the result

$$\frac{\partial^2 h(x,t)}{\partial t^2} = \frac{KL^2}{M} \frac{(h(x + \Delta x) - 2h(x) + h(x - \Delta x))}{\Delta x^2}$$

with the form of a *finite difference* approximation of the second derivative.

When the width of the segment Δx is small enough, this equation reduces to

$$\frac{\partial^2 h(x,t)}{\partial t^2} = v^2 \frac{\partial^2 h(x,t)}{\partial x^2} \tag{12.25}$$

where $v^2 = KL^2/M$. This is the *classical wave equation*.

B$_{12}$ End-of-chapter problems

> I have deeply regretted that I did not proceed far enough at least to understand something of the great leading principles of mathematics; for [those] thus endowed seem to have an extra sense.
>
> Charles Darwin

Warm-ups

12.1 The one-dimensional classical heat equation

$$\frac{\partial u(x,t)}{\partial t} = \kappa \frac{\partial^2 u(x,t)}{\partial x^2}$$

describes the dependence of the temperature profile $u(x,t)$ as a function of position x and time t. Note this equation can be written in the compact form $u_t(x,t) = \kappa u_{xx}(x,t)$.

The general solution for $u(x,t)$ was found to be

$$u(x,t) = u_0 + \sum_{n=1}^{\infty} u_n \cos\left(\frac{n\pi}{L}x\right) \exp\left(-\frac{n^2\pi^2}{L^2}\kappa t\right)$$

for the temperature profile subject to the *reflecting boundary conditions* $u_t(0,t) = u_t(L,t) = 0$.

(a) Consider the *absorbing boundary conditions*

$$u(0,t) = 0 \qquad u(L,t) = 0$$

representing cold walls at $x = 0$ and $x = L$ where the temperature is always zero. Show that the general solution to the classical heat equation can be written as

$$u(x,t) = \sum_{n=1}^{\infty} u_n \sin\left(\frac{n\pi}{L}x\right) \exp\left(-\frac{n^2\pi^2}{L^2}\kappa t\right)$$

(b) Consider the initial condition for the temperature profile

$$u(x,0) = 4\cos\left(\frac{2\pi}{L}x\right) \sin\left(\frac{2\pi}{L}x\right)$$

Determine the values of the coefficients $\{u_n\}$ for all n and express your final solution for $u(x,t)$.

12.2 The one-dimensional classical wave equation

$$\frac{\partial^2 h(x,t)}{\partial t^2} = v^2 \frac{\partial^2 h(x,t)}{\partial x^2}$$

describes the dependence of the displacement $h(x,t)$ as a function of position x and time t. Note this equation can be written in the compact form $h_{tt}(x,t) = v^2 h_{xx}(x,t)$.

The general solution for $h(x,t)$ was found to be

$$h(x,t) = \sum_{n=1}^{\infty} \sin\left(\frac{n\pi}{L}x\right)\left(a_n \cos(\omega_n t) + b_n \sin(\omega_n t)\right)$$

where $\omega_n = \frac{n\pi v}{L}$ and the displacement $h(x,t)$ satisfies the boundary conditions $h(0,t) = h(L,t) = 0$. Solve the one-dimensional classical wave equation for the displacement $h(x,t)$ subject to the boundary condition $h(0,t) = h(L,t) = 0$ and initial conditions (see figure below)

$$h(x,0) = 5\sin\left(\frac{3\pi x}{L}\right)$$

and

$$\frac{\partial h(x,0)}{\partial t} = 0$$

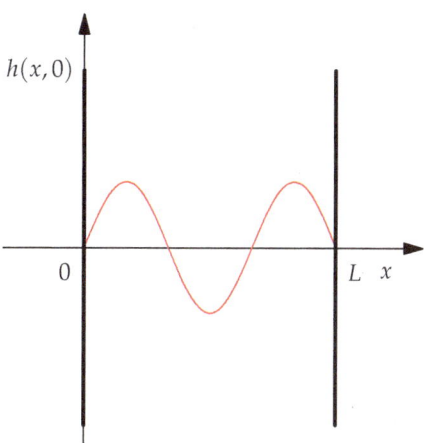

12.3 Consider the one-dimensional classical diffusion equation

$$\frac{\partial c(x,t)}{\partial t} = D\frac{\partial^2 c(x,t)}{\partial x^2}$$

describing the dependence of the distribution, $c(x,t)$, as a function of position x and time t. Note this equation can be written in the compact form $c_t(x,t) = Dc_{xx}(x,t)$.

(a) Show that the distribution

$$c(x,t) = \frac{1}{\sqrt{4\pi Dt}}\exp\left[-\frac{(x-x_0)^2}{4Dt}\right]$$

is a solution to the diffusion equation.

(b) Show that the distribution is normalized by proving

$$\int_{-\infty}^{\infty} c(x,t)dx = 1$$

for all values of time.

(c) What is the functional form of the distribution at $t = 0$?

(d) Plot the distribution $c(x,t)$ for $t = 1/4D$, $2/(4D)$, and $4/(4D)$ with $x_0 = 1$.

Homework exercises

12.4 In Chapter 3, we found that $\varphi(x,y,z) = (x^2 + y^2 + z^2)^{-1/2}$ satisfies

$$\frac{\partial^2 \varphi}{\partial x^2} + \frac{\partial^2 \varphi}{\partial y^2} + \frac{\partial^2 \varphi}{\partial z^2} = 0$$

which is known as *Laplace's equation*. In doing so, we solved a second order partial differential equation.

A simple extension of Laplace's equation is

$$\frac{\partial^2 \varphi}{\partial x^2} + \frac{\partial^2 \varphi}{\partial y^2} + \frac{\partial^2 \varphi}{\partial z^2} + \frac{8\pi^2 mE}{h^2}\varphi = 0$$

which is the time-independent Schrödinger equation of quantum theory and $\varphi(x,y,z)$ is the wave function for a free particle of mass m. Show that

$$\varphi_{n_x n_y n_z}(x,y,z) = \sqrt{\frac{8}{L^3}}\sin\left(\frac{n_x \pi}{L}x\right)\sin\left(\frac{n_y \pi}{L}y\right)\sin\left(\frac{n_z \pi}{L}z\right)$$

is a solution to this differential equation with the boundary conditions

$$\varphi_{n_x n_y n_z}(x,y,z) = 0$$

for $x = 0, L$, $y = 0, L$, or $z = 0, L$ where $n_x = 1,2,3,\ldots$, $n_y = 1,2,3,\ldots$ and $n_z = 1,2,3,\ldots$. Find a relation for E in terms of the constants m, h, n_x, n_y, and n_z.

12.5 Consider the one-dimensional classical diffusion equation

$$\frac{\partial c(x,t)}{\partial t} = D\frac{\partial^2 c(x,t)}{\partial x^2}$$

describing the dependence of the distribution $c(x,t)$ as a function of position x and time t (see figure below).

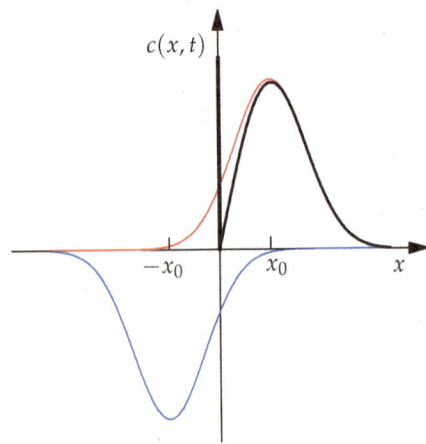

(a) Suppose that diffusion takes place on a semi-infinite plane $0 \le x < \infty$ where there is an *absorbing boundary condition* at $x = 0$ defined by

$$c(0,t) = 0$$

Further assume the initial condition for $x_0 > 0$

$$c(x,0) = \delta(x - x_0)$$

Consider the concentration profile

$$c(x,t) = \frac{1}{\sqrt{4\pi Dt}}\exp\left[-\frac{(x-x_0)^2}{4Dt}\right] - \frac{1}{\sqrt{4\pi Dt}}\exp\left[-\frac{(x+x_0)^2}{4Dt}\right]$$

shown in the figure above. The total distribution (black line) is composed of the difference of two gaussians, one centered at $x_0 > 0$ (red line) and an image function centered at $-x_0$ (blue line). Demonstrate that the sum of the two functions meets the absorbing boundary condition that $c(0,t) = 0$ and initial condition stated above.

(b) We can define the surviving population of the diffusing particles in terms of the *survival probability*

$$p(t) = \int_0^\infty c(x,t)dx$$

which decreases with time. Determine a formula for $p(t)$ in terms of the error function.

(c) Plot the survival probability $p(t)$ as a function of t taking $D = 1$ for $x_0 = 0.2, 1.0$, and 2.0.

12.6 Consider the second order partial differential equation

$$i\hbar\frac{\partial\psi(x,t)}{\partial t} = -\frac{\hbar^2}{2m}\frac{\partial^2\psi(x,t)}{\partial x^2} = E\psi(x,t)$$

which is the time-dependent Schrödinger equation of quantum theory and $\psi(x,t)$ is the wave function for a free particle of mass m. Note that $\hbar = h/2\pi$.

(a) Show that

$$\psi(x,t) = A\exp\left[2\pi i\left(\frac{x}{\lambda} - \nu t\right)\right]$$

is a solution to this differential equation if

$$h\nu = \frac{h^2}{2m\lambda^2}$$

(b) For this free particle, the total energy E is the kinetic energy

$$E = \frac{p^2}{2m}$$

where p is the particle's momentum. Prove that the solution provided in (a) implies

$$\lambda = \frac{h}{p}$$

which is the de Broglie relation between the particle's momentum and wavelength.

12.7[*] Consider the time-dependent Schrödinger equation

$$i\hbar\frac{\partial\varphi(x,t)}{\partial t} = -\frac{\hbar^2}{2m}\frac{\partial^2\varphi(x,t)}{\partial x^2} + V(x)\varphi(x,t) = E\varphi(x,t)$$

for a particle or mass m experiencing a potential energy $V(x)$.

(a) Consider the *imaginary time*, τ, defined as

$$\tau \equiv \frac{it}{\hbar}$$

Substitute $t = -i\hbar\tau$ into the time-dependent Schrödinger equation above to prove that

$$\frac{\partial \varphi(x,\tau)}{\partial \tau} = \frac{\hbar^2}{2m}\frac{\partial^2 \varphi(x,\tau)}{\partial x^2} - V(x)\varphi(x,\tau) = -E\varphi(x,\tau)$$

(b) For a particle in a harmonic potential

$$V(x) = \frac{1}{2}m\omega^2 x^2$$

the imaginary-time Schrödinger equation can be written

$$\frac{\partial \varphi(x,\tau)}{\partial \tau} = \frac{\hbar^2}{2m}\frac{\partial^2 \varphi(x,\tau)}{\partial x^2} - \frac{1}{2}m\omega^2 x^2 \varphi(x,\tau) = -E\varphi(x,\tau)$$

Prove that the function

$$\varphi(x,\tau) = \left(\frac{m\omega}{\pi\hbar}\right)^{1/4} \exp\left(-\frac{m\omega}{2\hbar}x^2\right) \exp(-E\tau)$$

is a solution to this imaginary-time Schrödinger equation (see figure below). Find a relation for E in terms of ω.

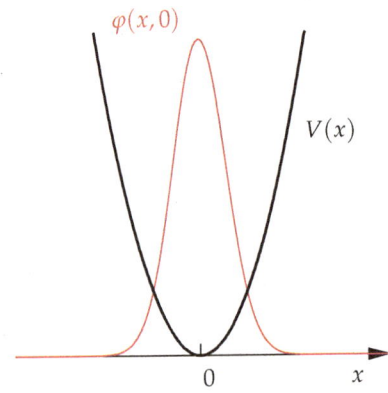

(c) Show that imaginary-time Schrödinger equation for a free particle, for which $V(x) = 0$, is *isomorphic* (of the same form) with the classical diffusion equation

$$\frac{\partial c(x,t)}{\partial t} = D\frac{\partial^2 c(x,t)}{\partial x^2}$$

where the diffusion constant is

$$D = \frac{\hbar^2}{2m}$$

(d) Consider the solution to the time-dependent Schrödinger equation for a free particle

$$\psi(x,t) = A\exp\left[2\pi i\left(\frac{x}{\lambda} - \nu t\right)\right]$$

Substitute $t = -i\hbar\tau$ and show that the resulting function is a solution to the imaginary-time Schrödinger equation for a free particle, derived in (c), where the energy is

$$E = h\nu = \frac{h^2}{2m\lambda^2}$$

13 FOURIER SERIES, FOURIER TRANSFORMS, AND HARMONIC ANALYSIS

13.1 Fourier series

AN INFINITE SERIES OF SINE AND COSINE FUNCTIONS emerged as a general solution to a partial differential equation, such as the classical heat equation and classical wave equation, applied over a finite range. In this section, we expand upon this idea and explore ways to represent any periodic function as an infinite weighted sum of sines and cosines known as a Fourier series. The general study of Fourier series is known as harmonic analysis.

13.1.1 Periodic functions and Fourier series

In Chapter 12, we explored the general solution of partial differential equations, describing the flow of heat, the diffusion of particles, or the motion of waves over some region $0 \leq x \leq L$, in terms of an infinite series of sine and cosine functions.

We explored the general solution to the *classical heat equation*, which describes the temperature $u(x)$ of some substance as a function of position x as a weighted sum of sine and cosine terms. For example, the temperature profile of some substance bounded by insulated ends[1] at $x = 0$ and $x = L$ might take the following form:

[1] This example appears in Chapter 12 as Figure 12.2.

$$u(x) = \frac{1}{2} - \frac{1}{2} \cos\left(\frac{2\pi x}{L}\right)$$

As shown in Figure 13.1, the temperature is never negative so that $u(x) \geq 0$.

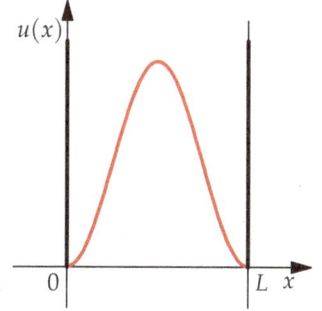

Figure 13.1: The temperature profile $u(x)$ over a length L with insulated ends.

In addition, the fact that the temperature is reflected from the insulated ends means the slope of the temperature profile is zero at $x = 0$ and $x = L$.

We also explored the general solution to the *classical wave equation*, which describes the displacement of the wave $h(x)$ as a function of position x using a weighted sum of sine and cosine terms. For example, the displacement of a string with fixed ends[2] at $x = 0$ and $x = L$ might take the following form:

[2] This example appears in Chapter 12 as Figure 12.8.

$$h(x) = \frac{1}{2} \sin\left(\frac{4\pi}{L}x\right)$$

The displacement of the string varies, oscillating between positive and negative values, while being fixed at the end points $x = 0$ and $x = L$. This is reflected in the shape of the displacement shown in Figure 13.2.

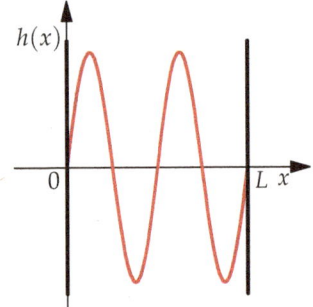

Figure 13.2: The displacement $h(x)$ of a classical wave over length L.

The solutions above are defined over a finite interval $0 \leq x \leq L$. Moreover, they have the property that the solution is zero at the boundaries $x = 0$ and $x = L$. Now suppose that we take our solutions, defined over the interval $0 \leq x < L$, and repeat them periodically over an infinite series of intervals

$$(n-1)L \leq x < nL$$

where $n = \ldots, -2, -1, 0, 1, 2, \ldots$. For example, the function $u(x)$ describing the temperature profile, shown in Figure 13.1 over the range $[0, L]$, can be repeated to form a periodic function $f(x)$ over $(-\infty, \infty)$, as shown in Figure 13.3.

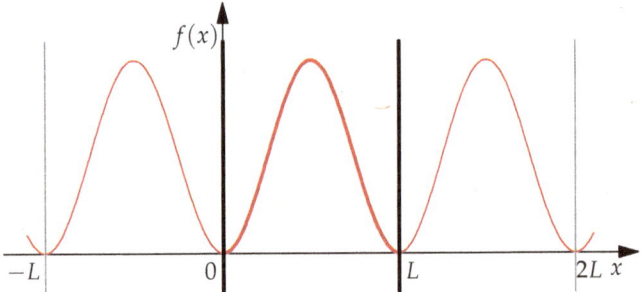

Figure 13.3: The temperature profile $f(x)$ for a substance over a box of length L repeated to form an infinite periodic wave.

The result is a periodic function defined over the range $(-\infty, \infty)$. The resulting infinite periodic function is both continuous and smooth. That is, the function itself and the first derivative of the function are continuous over the full range of the function $-\infty < x < \infty$.

Similarly, the function $h(x)$ describing the displacement, shown in Figure 13.2 over the range $[0, L]$, can be repeated to form a periodic function over the range $(-\infty, \infty)$, as shown in Figure 13.4.

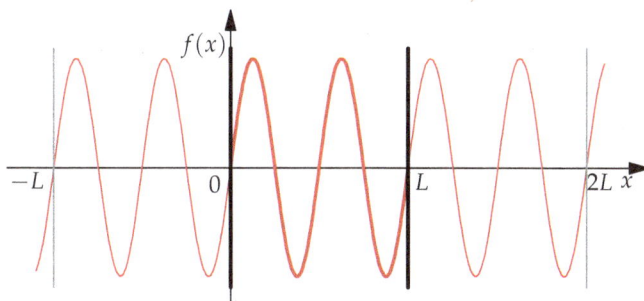

Figure 13.4: The displacement $f(x)$ for a classical wave over a box of length L repeated to form an infinite periodic wave.

It turns out that periodic functions of interest in the physical sciences can be represented as an infinite sum of sines and cosines, such that

$$f(x) = \frac{a_0}{2} + \sum_{n=1}^{\infty} a_n \cos\left(\frac{2n\pi x}{L}\right) + \sum_{n=1}^{\infty} b_n \sin\left(\frac{2n\pi x}{L}\right)$$

where the function $f(x)$ is periodic over $(-\infty, \infty)$ with period L. The coefficients are defined as follows

$$a_n = \frac{2}{L} \int_0^L f(x) \cos\left(\frac{2n\pi x}{L}\right) dx \qquad b_n = \frac{2}{L} \int_0^L f(x) \sin\left(\frac{2n\pi x}{L}\right) dx$$

Each term represents the contribution of an oscillation of wavelength $\lambda_n = \frac{L}{n}$ with amplitude defined by the coefficients a_n or b_n.[3]

In this section, we develop a general approach to determine the infinity of coefficients $\{a_n\}$ and $\{b_n\}$ for any periodic function $f(x)$. In doing so, we will derive the relations above for the Fourier series and coefficients. We then apply the resulting theory to a variety of examples, including the square wave, triangle wave, sawtooth wave, and shark fin wave.

Through those examples, we will develop an intuition for the form of the Fourier series needed to describe a given function. In addition, we will learn to interpret the patterns found in the coefficients a_n and b_n as n is varied. The coefficients define the decomposition of the function $f(x)$ into waves of varying wavelengths, and the study of the patterns found in the coefficients is known as *harmonic analysis*.

13.1.2 *Deriving the Fourier series for an arbitrary periodic function*

Consider almost any function $f(x)$ defined over the range $-\pi \le x \le \pi$ and periodic over all x.[4] To determine the Fourier series we must evaluate the coefficients a_n and b_n that weight the contributions of the cosine and sine terms, respectively. We do that by determining the overlap of the function $f(x)$ with each of the component functions $\cos(nx)$ and $\sin(nx)$ that form the basis functions of the Fourier series. The overlap determines the magnitudes of the coefficients a_n or b_n that weight each term in the series.

[3] The Fourier series is named for French mathematician and physicist *Joseph Fourier* (1768-1830) who developed this harmonic analysis to study the flow of heat in a metal bar of finite length.

[4] Here we refer to those functions $f(x)$ with a finite number of discontinuities for which

$$\int_{-\pi}^{\pi} |f(x)| dx$$

is finite.

To determine the coefficients a_n and b_n, we will exploit the special property that the cosine and sine functions are orthogonal. In the case of vectors in three dimensions, we can express any vector $\mathbf{a}$ in terms of a linear combination of *basis vectors*, such as the unit vectors $\hat{\mathbf{x}}$, $\hat{\mathbf{y}}$, and $\hat{\mathbf{z}}$, weighted by scalar coefficients a_x, a_y, and a_z as

$$\mathbf{a} = a_x\,\hat{\mathbf{x}} + a_y\,\hat{\mathbf{y}} + a_z\,\hat{\mathbf{z}}$$

In the case of the Fourier series, we can express almost any function $f(x)$ in terms of a linear combination of *basis functions* $\cos(nx)$ and $\sin(nx)$ weighted by scalar coefficients a_n and b_n as

$$f(x) = \frac{a_0}{2} + \sum_{n=1}^{\infty} a_n \cos\left(\frac{2n\pi x}{L}\right) + \sum_{n=1}^{\infty} b_n \sin\left(\frac{2n\pi x}{L}\right)$$

In the case of vectors, we determined the coefficients a_x, a_y, and a_z by projecting the vector $\mathbf{a}$ onto each basis vector, so that

$$a_x = \mathbf{a} \cdot \hat{\mathbf{x}} \qquad a_y = \mathbf{a} \cdot \hat{\mathbf{y}} \qquad a_z = \mathbf{a} \cdot \hat{\mathbf{z}}$$

In the case of Fourier series, the coefficients a_n and b_n are determined by projecting the function $f(x)$ onto each basis function $\cos(nx)$ and $\sin(nx)$ as

$$a_n = \frac{2}{L} \int_0^L f(x) \cos\left(\frac{2n\pi x}{L}\right) dx \qquad b_n = \frac{2}{L} \int_0^L f(x) \sin\left(\frac{2n\pi x}{L}\right) dx$$

Let's see how this works.

Evaluating the cosine coefficients a_n for $n \geq 1$

Let's determine the coefficients a_n for $n \geq 1$ by projecting the function $f(x)$ onto the basis function $\cos(mx)$. To do this, we multiply the function $f(x)$ by $\cos(mx)$ and integrate over one period of oscillation as[5]

[5] We have used the fact that
$$\int \sum_n f_n(x)dx = \sum_n \int f_n(x)dx$$

$$
\begin{aligned}
\frac{1}{\pi}\int_{-\pi}^{\pi} f(x)\cos(mx)dx &= \frac{1}{\pi}\int_{-\pi}^{\pi}\left[\frac{a_0}{2} + \sum_{n=1}^{\infty} a_n\cos(nx) + \sum_{n=1}^{\infty} b_n\sin(nx)\right]\cos(mx)dx \\
&= \frac{1}{\pi}\frac{a_0}{2}\int_{-\pi}^{\pi}\cos(mx)dx + \frac{1}{\pi}\sum_{n=1}^{\infty} a_n\int_{-\pi}^{\pi}\cos(nx)\cos(mx)dx \\
&\quad + \frac{1}{\pi}\sum_{n=1}^{\infty} b_n\int_{-\pi}^{\pi}\sin(nx)\cos(mx)dx \qquad (13.1)
\end{aligned}
$$

We can evaluate the three terms using our knowledge of integration and the table of indefinite integrals in Supplement S_6.

There are three integrals to evaluate.[6] The first integral is zero, as integration performed over m full periods of oscillation of the cosine is zero. The second integral can be evaluated for $n \neq m$, using S_6 #102, as

[6] This property of cosine functions is equivalent to *orthogonality* in vectors. The properties of orthogonal functions are discussed in the complements.

$$\int_{-\pi}^{\pi} \cos(nx)\cos(mx)\,dx = 0$$

and for $n = m$, using S_6 #16, as

$$\int_{-\pi}^{\pi} \cos^2(nx)\, dx = \pi$$

We can express these two results in the compact form

$$\int_{-\pi}^{\pi} \cos(nx) \cos(mx)\, dx = \pi\, \delta_{nm}$$

where δ_{nm} is the Kronecker delta function that selects the single coefficient in the infinite series for which $n = m$.[7] We say that $\cos(nx)$ and $\cos(mx)$ are mutually orthogonal functions when $n \neq m$.

The third integral can be evaluated for $n = m$ using S_6 #17 and for $n \neq m$ using S_6 #103. We find that

$$\int_{-\pi}^{\pi} \sin(nx) \cos(mx)\, dx = 0$$

We say that $\sin(nx)$ and $\cos(mx)$ are mutually orthogonal functions for all n and m. Inserting our results for the three integrals in Equation 13.1, we find

$$\frac{1}{\pi} \int_{-\pi}^{\pi} f(x) \cos(mx)\, dx = \frac{1}{\pi} \sum_{n=1}^{\infty} a_n \times \pi\, \delta_{nm} = a_m \qquad (13.2)$$

which defines the coefficients weighting the cosine terms.[8]

Evaluating the sine coefficients b_n for $n \geq 1$

Let's determine the coefficients b_n for $n \geq 1$ by projecting the function $f(x)$ onto the basis function $\sin(mx)$. To do this, we multiply the function $f(x)$ by $\sin(mx)$ and integrate over one period of oscillation as

$$
\begin{aligned}
\frac{1}{\pi} \int_{-\pi}^{\pi} f(x) \sin(mx)\, dx &= \frac{1}{\pi} \int_{-\pi}^{\pi} \left[\frac{a_0}{2} + \sum_{n=1}^{\infty} a_n \cos(nx) + \sum_{n=1}^{\infty} b_n \sin(nx) \right] \sin(mx)\, dx \\
&= \frac{1}{\pi} \frac{a_0}{2} \int_{-\pi}^{\pi} \sin(mx)\, dx + \frac{1}{\pi} \sum_{n=1}^{\infty} a_n \int_{-\pi}^{\pi} \cos(nx) \sin(mx)\, dx \\
&+ \frac{1}{\pi} \sum_{n=1}^{\infty} b_n \int_{-\pi}^{\pi} \sin(nx) \sin(mx)\, dx \qquad (13.3)
\end{aligned}
$$

The first integral is zero as integration performed over m periods of oscillation of the sine is zero.[9] The second integral can be evaluated for $n = m$ using S_6 #17 and for $n \neq m$ using S_6 #103 . We find that

$$\int_{-\pi}^{\pi} \cos(nx) \sin(mx)\, dx = 0$$

The third integral can be evaluated using for $n = m$ using S_6 #14, since

$$\int_{-\pi}^{\pi} \sin^2(nx)\, dx = \pi$$

and for $n \neq m$, using S_6 #93, as

$$\int_{-\pi}^{\pi} \sin(nx)\sin(mx)\,dx = 0$$

We can express these two results in the compact form[10]

$$\int_{-\pi}^{\pi} \sin(nx)\sin(mx)\,dx = \pi\,\delta_{nm}$$

Inserting our results for the three integrals in Equation 13.3, we find

$$\frac{1}{\pi}\int_{-\pi}^{\pi} f(x)\sin(mx)\,dx = \frac{1}{\pi}\sum_{n=1}^{\infty} b_n \times \pi\,\delta_{nm} = b_m \qquad (13.4)$$

which defines the coefficients weighting the sine terms.[11]

Evaluating the constant coefficient a_0

Finally, let's determine the coefficient a_0 by projecting the function $f(x)$ onto $\cos(0) = 1$. To do this, we evaluate

$$
\begin{aligned}
\frac{1}{\pi}\int_{-\pi}^{\pi} f(x)dx &= \frac{1}{\pi}\int_{-\pi}^{\pi}\left[\frac{a_0}{2} + \sum_{n=1}^{\infty} a_n\cos(nx) + \sum_{n=1}^{\infty} b_n\sin(nx)\right]dx \qquad (13.5)\\
&= \frac{1}{\pi}\frac{a_0}{2}\int_{-\pi}^{\pi} dx + \frac{1}{\pi}\sum_{n=1}^{\infty} a_n\int_{-\pi}^{\pi}\cos(nx)dx + \frac{1}{\pi}\sum_{n=1}^{\infty} b_n\int_{-\pi}^{\pi}\sin(nx)dx
\end{aligned}
$$

The first integral is

$$\int_{-\pi}^{\pi} dx = 2\pi$$

while the second and third integrals, involving integration of cosine or sine over one period of oscillation, are zero. This leads to

$$\frac{1}{\pi}\int_{-\pi}^{\pi} f(x)dx = \frac{1}{\pi}\frac{a_0}{2}\times 2\pi = a_0 \qquad (13.6)$$

which acts as a constant offset of the overall series, also known as the *zero frequency component*.

Let's summarize our findings. We have derived a most remarkable result. We can express any function $f(x)$ in terms of a Fourier series:[12]

$$f(x) = \frac{a_0}{2} + \sum_{n=1}^{\infty} a_n\cos(nx) + \sum_{n=1}^{\infty} b_n\sin(nx) \qquad (13.7)$$

The coefficients $\{a_n\}$ and $\{b_n\}$ defining this series are determined using the integral relations provided by Equations 13.2, 13.4, and 13.6 and collected here:[13]

$$a_n = \frac{1}{\pi}\int_{-\pi}^{\pi} f(x)\cos(nx)\,dx \qquad b_n = \frac{1}{\pi}\int_{-\pi}^{\pi} f(x)\sin(nx)\,dx \qquad (13.8)$$

Our variable x extends over the range $-\pi \leq x < \pi$ with a period of 2π.

[10] This property of sine functions is equivalent to *orthogonality* in vectors. The properties of orthogonal functions are discussed in the complements.

[11] Note that the result

$$\frac{1}{\pi}\int_{-\pi}^{\pi} f(x)\sin(mx)dx = b_m$$

can also be written

$$\frac{1}{\pi}\int_{-\pi}^{\pi} f(x)\sin(nx)dx = b_n$$

We will typically use the latter convention with the index n.

[12] Note that only multiples of the full wavelength appear in the Fourier series. Multiples of half wavelengths are excluded as they do not form periodic functions over the interval $[-\pi, \pi]$.

[13] Note that we have combined our results from Equations 13.2 and 13.6 since for $n = 0$ the coefficient

$$a_0 = \frac{1}{\pi}\int_{-\pi}^{\pi} f(x)\,dx$$

Fourier series for functions of variables with units

In the Fourier series defined by Equations 13.7 and 13.8, the variable x was dimensionless. Suppose we want to define a Fourier series representation of a function $f(x)$ where the variable x has units of length over a range $-\frac{L}{2} \leq x < \frac{L}{2}$. We are substituting the range

$$-\pi \leq x < \pi \rightarrow -\frac{L}{2} \leq x < \frac{L}{2}$$

As such, we must translate the argument of the sine and cosine functions as

$$nx \rightarrow \frac{2n\pi x}{L}$$

Note that the argument must be dimensionless as the sine and cosine functions require dimensionless arguments.

In that case, the Fourier series can be written as

$$f(x) = \frac{a_0}{2} + \sum_{n=1}^{\infty} a_n \cos\left(\frac{2n\pi x}{L}\right) + \sum_{n=1}^{\infty} b_n \sin\left(\frac{2n\pi x}{L}\right) \qquad (13.9)$$

where the coefficients are defined as

$$a_n = \frac{2}{L} \int_{-\frac{L}{2}}^{\frac{L}{2}} f(x) \cos\left(\frac{2n\pi x}{L}\right) dx \quad b_n = \frac{2}{L} \int_{-\frac{L}{2}}^{\frac{L}{2}} f(x) \sin\left(\frac{2n\pi x}{L}\right) dx \qquad (13.10)$$

Note that the coefficients are dimensionless. While the increment dx has units of length, the prefactor $\frac{2}{L}$ has units of inverse length, making the coefficients dimensionless.

We now have expressions for describing almost any periodic function in terms of a Fourier series of sine and cosine functions. The process of determining the coefficients a_n and b_n is known as a *Fourier decomposition*, because we can decompose the function $f(x)$ in terms of the contributing harmonics. As such, the Fourier decomposition is also known as *harmonic analysis*.

Let's apply these results to a number of common periodic functions to develop an intuitive feeling for how Fourier series are defined and why they are useful. To start, we will determine the Fourier series of a periodic function $f(x)$ known as the square wave.

13.1.3 *Deconstructing the Fourier series of the square wave*

Consider the periodic function of x known as the *square wave* function. The function is written

$$f(x) = 2\theta\left(\frac{x}{L}\right) - 1 \qquad -\frac{L}{2} \leq x < \frac{L}{2}$$

where $\theta\left(\frac{x}{L}\right)$ is the *Heaviside step function*, which is unity for $x \geq 0$ and zero otherwise.

How does this equation describe a square wave? Multiplying the step function by a factor of two leads to a function that is 0 over the range $-\frac{L}{2} \leq x < 0$ and 2 over the range $0 \leq x < \frac{L}{2}$. Subtracting unity then leads to the square wave that is -1 over the range $-\frac{L}{2} \leq x < 0$ and 1 over the range $0 \leq x < \frac{L}{2}$. The square wave function is shown in Figure 13.5 over two periods of oscillation with wavelength L.

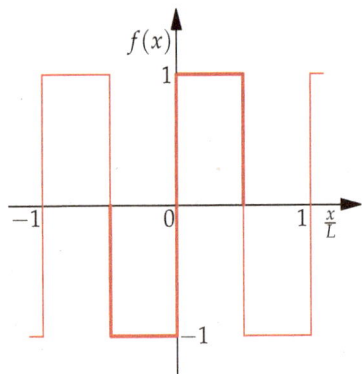

Figure 13.5: The square wave $f(x)$ (red line) is an odd periodic function of x with wavelength L.

It is one of the miracles of the Fourier series that a function composed of straight lines and right angles can be expressed as a sum of an infinite number of smooth sinusoidal functions. Let's see how that works.

To represent the square wave using the Fourier series defined in Equation 13.9 as

$$f(x) = \frac{a_0}{2} + \sum_{n=1}^{\infty} a_n \cos\left(\frac{2n\pi x}{L}\right) + \sum_{n=1}^{\infty} b_n \sin\left(\frac{2n\pi x}{L}\right)$$

we must determine the coefficients by evaluating the integrals appearing in Equation 13.10:

$$a_n = \frac{2}{L} \int_{-\frac{L}{2}}^{\frac{L}{2}} f(x) \cos\left(\frac{2n\pi x}{L}\right) dx \qquad b_n = \frac{2}{L} \int_{-\frac{L}{2}}^{\frac{L}{2}} f(x) \sin\left(\frac{2n\pi x}{L}\right) dx$$

Let's determine the Fourier coefficients for the periodic square wave.

Fourier coefficients for the square wave

For the coefficients a_n, the integrand is a product of the square wave, which is an odd function of x, and a cosine wave, which is an even function of x, making the integrand an odd function of x. This means that the integral over the range $-\infty < x \leq 0$ is equal in magnitude and opposite in sign to the integral over the range $0 \leq x < \infty$ As such, the coefficients are $a_n = 0$ for all n by symmetry.[14]

For the coefficients b_n, we note that $f(x) = -1$ for $x \in [-\frac{L}{2}, 0)$ and $f(x) = 1$ for $x \in [0, \frac{L}{2})$. As such, the coefficients are

$$b_n = -\frac{2}{L} \int_{-\frac{L}{2}}^{0} \sin\left(\frac{2n\pi x}{L}\right) dx + \frac{2}{L} \int_{0}^{\frac{L}{2}} \sin\left(\frac{2n\pi x}{L}\right) dx$$

[14] Recall that for an odd function

$$f_o(-x) = -f_o(x)$$

so that

$$\int_{0}^{\frac{L}{2}} f_o(x)\, dx = -\int_{-\frac{L}{2}}^{0} f_o(x)\, dx$$

and

$$\int_{-\frac{L}{2}}^{\frac{L}{2}} f_o(x)\, dx = 0$$

As the sine function is an odd function of x, we have

$$b_n = \frac{4}{L} \int_0^{\frac{L}{2}} \sin\left(\frac{2n\pi x}{L}\right) dx = -\frac{4}{L}\left(\frac{L}{2n\pi}\right) \cos\left(\frac{2n\pi x}{L}\right)\Big|_0^{\frac{L}{2}}$$

$$= \frac{2}{n\pi}\left(1 - \cos(n\pi)\right) \tag{13.11}$$

where $\cos(n\pi) = (-1)^n$.

When n is even in Equation 13.11, the cosine term is unity and the coefficients are

$$b_n = 0 \quad n = 2, 4, 6, \ldots$$

When n is odd, the cosine term is -1 and the coefficients are[15]

$$b_n = \frac{4}{n\pi} \quad n = 1, 3, 5, \ldots \tag{13.12}$$

The square wave function is an odd function of x. Reflecting this fact, the corresponding Fourier series is composed solely of sine functions, which are also odd functions of x. Inserting these results into Equation 13.9 leads to our final result

$$f(x) = \sum_{n=1,3,5,\ldots}^{\infty} \frac{4}{n\pi} \sin\left(\frac{2n\pi x}{L}\right) \tag{13.13}$$

for the Fourier series representation of the square wave function.

[15] In general, step functions with jump discontinuities have Fourier coefficients that vary as $\frac{1}{n}$. The slow convergence of the series with increasing n reflects the difficulty in describing a discontinuous function using an infinite series of smoothly varying and continuous sine and cosine functions.

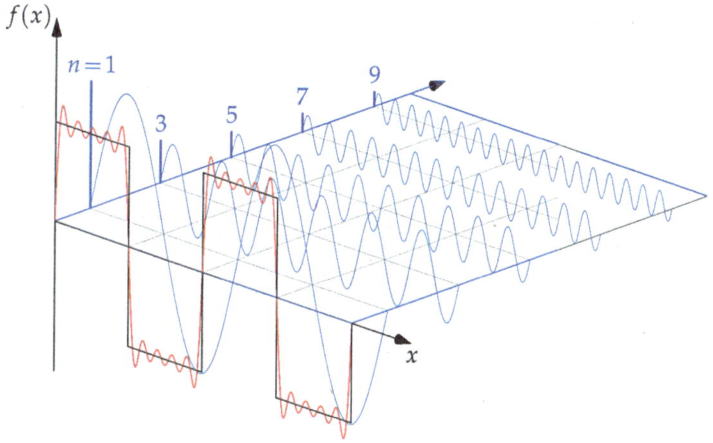

Figure 13.6: The Fourier series approximating the square wave (black line) using the first five terms in the series. The sum of the five blue curves of varying wavelength forms the composite red curve. The height of each term indicates the magnitude of the coefficients b_n (thick blue bars).

Figure 13.6 shows the square wave function (black line) and the first five contributions to the Fourier series (blue lines).

Viewed one way, we see how an infinite series of sine functions can be used to describe a square wave function. Viewed another way, we see how to decompose the square wave function into an infinite series of sinusoidal functions of varying wavelength $\frac{L}{n}$ weighted by the coefficient b_n. The latter process is known as Fourier decomposition and it forms the heart of harmonic analysis.

Harmonic analysis of the square wave

The square wave is represented as an infinite sum of *harmonics*, where each harmonic has a characteristic wavelength $\lambda_n = \frac{1}{n}L$ and amplitude $b_n \propto \frac{1}{n}$. Figure 13.6 shows the square wave function compared with the partial sum, defined as

$$f_N(x) = \sum_{n=1,3,5,\dots}^{N} \frac{4}{n\pi} \sin\left(\frac{2n\pi x}{L}\right) \qquad (13.14)$$

For $N = 1$, the partial sum is a single sine function

$$f_1(x) = \frac{4}{\pi} \sin\left(\frac{2\pi x}{L}\right)$$

representing a wavelength $\lambda_1 = L$.[16] The wavelength L of the square wave is captured exactly by this one term. However, the amplitude is too great, as the coefficient $b_1 = \frac{4}{\pi} > 1$ causes $f_1(x)$ to overshoot the bounds of the square wave that oscillates between -1 and 1. In addition, the smooth sinusoidal form lacks the rectilinear square wave form.

Adding additional terms to the sum with wavelengths L/n enhances the accuracy of the approximation. For example, including two terms, such that

$$f_3(x) = \frac{4}{\pi} \sin\left(\frac{2\pi x}{L}\right) + \frac{4}{3\pi} \sin\left(\frac{6\pi x}{L}\right)$$

reduces the maximum amplitude of oscillation from $\frac{4}{\pi} = 1.273\dots$ to $\frac{8\sqrt{2}}{3\pi} = 1.2004\dots$ through *destructive interference*, while enhancing the square symmetry at the corners, through *constructive interference*. In the limit $N \to \infty$, an accurate representation of the square wave function is achieved.

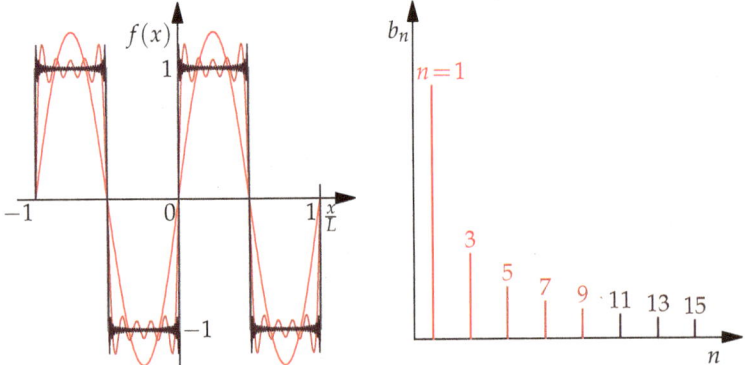

The first five terms in the partial sum are displayed individually in Figure 13.6 (blue lines). The partial sum $f_9(x)$ over the five contributions (red line) is compared with the exact square wave function (gray line). The combination of destructive and constructive interference between the sinusoidal waves of varying wavelength and amplitude leads to an increasingly accurate representation of the square wave function as N increases.[17]

Figure 13.7 shows the same information from another perspective. The amplitude of each term in the sum decreases in proportion to $1/n$ with

[16] This term is the *fundamental* or the *first harmonic* appearing in the Fourier series. It is the longest wavelength contribution with $\lambda_1 = L$. The wavelength of other contributions are a fraction $\lambda_n = \frac{1}{n}L$ of the fundamental.

Figure 13.7: The partial sum

$$f_N(x) = \sum_{n=1,3,5,\dots}^{N} \frac{4}{n\pi} \sin\left(\frac{2n\pi x}{L}\right)$$

formed from the infinite Fourier series for the square wave function (gray line). Increasing N from 1 (pale red) to 9 (red) to 60 (dark red line).

[17] Even in the limit $N \to \infty$, the Fourier series overshoots the square wave form at points of discontinuity where it is not possible to achieve uniform convergence. This is known as the *Gibbs phenomenon*.

increasing n. Note that while the square wave function is discontinuous at the points $x = \frac{nL}{2}$, those points are zeros of each term in the Fourier series. The continuous Fourier series approximates each discontinuity in the function $f(x)$ by having every sinusoidal wave that contributes to the series pass through zero at each point of discontinuity.

13.1.4 Deconstructing the Fourier series of the triangle wave

Consider the periodic function known as the *triangle wave*, shown in Figure 13.8 over two periods of oscillation with period T. It can be written over one period of oscillation as

$$f(t) = \begin{cases} 1 - \dfrac{4t}{T} & 0 \le t < \dfrac{T}{2} \\ -1 + \dfrac{4}{T}\left(t - \dfrac{T}{2}\right) & \dfrac{T}{2} \le t < T \end{cases}$$

Remarkably, this function, which is composed of straight lines and sharp angles, can be represented as a sum of an infinite number of smooth sinusoidal functions.

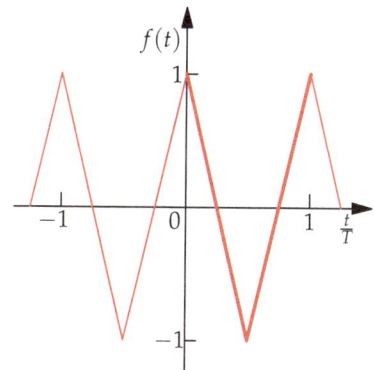

Figure 13.8: The triangle wave $f(t)$ (red line) is an even periodic function of t with period T.

We can represent the triangle wave using a reformulation of Equation 13.9 where the variable x is replaced by t and the wavelength L is replaced by the period T:

$$f(t) = \frac{a_0}{2} + \sum_{n=1}^{\infty} a_n \cos\left(\frac{2n\pi t}{T}\right) + \sum_{n=1}^{\infty} b_n \sin\left(\frac{2n\pi t}{T}\right)$$

The coefficients are determined by a reformulation of Equation 13.10

$$a_n = \frac{2}{T}\int_{-\frac{T}{2}}^{\frac{T}{2}} f(t)\cos\left(\frac{2n\pi t}{T}\right) dt \qquad b_n = \frac{2}{T}\int_{-\frac{T}{2}}^{\frac{T}{2}} f(t)\sin\left(\frac{2n\pi t}{T}\right) dt$$

Let's determine the Fourier coefficients for the periodic triangle wave.

Fourier coefficients for the triangle wave

Each coefficient b_n is an integral over a product of the triangle wave, which is an even function of t, and a sine wave, which is an odd function of t, making the integrand an odd function of t. As such, the coefficients are $b_n = 0$ for all n by symmetry.

Considering the coefficients a_n, we note the integrand $f(t) \cos\left(\frac{2n\pi t}{T}\right)$ is an even function of t about the point $t = \frac{T}{2}$. It follows that the integral over $[0, \frac{T}{2}]$ equals the integral over $[\frac{T}{2}, T]$. As such, we find[18]

$$a_n = \frac{2}{T} \int_0^{\frac{T}{2}} \left(1 - \frac{4t}{T}\right) \cos\left(\frac{2n\pi t}{T}\right) dt - \frac{2}{T} \int_{\frac{T}{2}}^T \left[-1 + \frac{4}{T}\left(t - \frac{T}{2}\right)\right] \cos\left(\frac{2n\pi t}{T}\right) dt$$

$$= \frac{4}{T} \int_0^{\frac{T}{2}} \left(1 - \frac{4t}{T}\right) \cos\left(\frac{2n\pi t}{T}\right) dt \tag{13.15}$$

The coefficients a_n can now be evaluated by performing the two integrals

$$a_n = \frac{4}{T} \int_0^{\frac{T}{2}} \cos\left(\frac{2n\pi t}{T}\right) dt - \frac{4}{T} \int_0^{\frac{T}{2}} \frac{4t}{T} \cos\left(\frac{2n\pi t}{T}\right) dt \tag{13.16}$$

The first integral of a cosine function taken over an integer multiple of half periods is zero:

$$\int_0^{\frac{T}{2}} \cos\left(\frac{2n\pi t}{T}\right) dt = 0$$

The second integral can be evaluated using the table of indefinite integrals in Supplement S_6 #96:

$$\int x \cos(ax + b)\, dx = \frac{1}{a^2} \cos(ax + b) + \frac{x}{a} \sin(ax + b) + C$$

Substituting in $a = \frac{2n\pi}{T}$ and $b = 0$, we find

$$\int_0^{\frac{T}{2}} \frac{4t}{T} \cos\left(\frac{2n\pi t}{T}\right) dt = \frac{T}{n^2 \pi^2} \left(\cos(n\pi) - 1\right)$$

Inserting our results in Equation 13.16 leads to

$$a_n = -\frac{4}{T} \times \frac{T}{n^2 \pi^2} \left(\cos(n\pi) - 1\right) = \frac{4}{n^2 \pi^2} \left(1 - (-1)^n\right)$$

where we have used the fact that $\cos(n\pi) = (-1)^n$. When n is even the coefficients $a_n = 0$, and when n is odd the coefficients are[19]

$$a_n = \frac{8}{n^2 \pi^2} \qquad n = 1, 3, 5, \ldots \tag{13.17}$$

Inserting the coefficients in Equation 13.9 leads to our final result for the Fourier series representation of the triangle wave function:

$$f(t) = \sum_{n=1,3,5,\ldots}^{\infty} \frac{8}{n^2 \pi^2} \cos\left(\frac{2n\pi t}{T}\right) \tag{13.18}$$

The triangle wave is an even function of t. Reflecting this fact, the corresponding Fourier series is composed solely of cosine functions, which are also even functions of t. The terms of the Fourier series reflect the underlying symmetries of the function itself.

[18] This can be shown by substituting $x = T - t$ and $dx = -dt$ into the integral over $\frac{T}{2} \le t < T$ such that

$$\int_{\frac{T}{2}}^T \left[-1 + \frac{4}{T}\left(t - \frac{T}{2}\right)\right] \cos\left(\frac{2n\pi t}{T}\right) dt \rightarrow$$

$$\int_0^{\frac{T}{2}} \left(1 - \frac{4x}{T}\right) \cos\left[2n\pi - \left(\frac{2n\pi x}{T}\right)x\right] dx$$

For integer values of n we know that $\cos(2n\pi - x) = \cos(-x) = \cos(x)$ so that

$$\int_0^{\frac{T}{2}} \left(1 - \frac{4x}{T}\right) \cos\left[2n\pi - \left(\frac{2n\pi x}{T}\right)\right] dx =$$

$$\int_0^{\frac{T}{2}} \left(1 - \frac{4x}{T}\right) \cos\left(\frac{2n\pi x}{T}\right) dx$$

as we proposed based on symmetry.

[19] In general, continuous *ramp functions* with discontinuities in the first derivative have Fourier coefficients that vary as $\frac{1}{n^2}$. The convergence of the Fourier series as $\frac{1}{n^2}$ with increasing n is more rapid than the convergence as $\frac{1}{n}$ characteristic of discontinuous *step functions*.

Harmonic analysis of the triangle wave

The Fourier series expresses the function $f(t)$ as an infinite series of harmonics of wavelength $\lambda = \frac{1}{n}T$ and amplitude $a_n \propto \frac{1}{n^2}$. Figure 13.9 shows the *residual difference* between the exact function $f(t)$ and the partial sum of the Fourier series representation $f_N(x)$ for $N = 1$ and 19. Note that the greatest difference between the exact function and the approximate Fourier series representation occurs at the sharp points where the derivative of the function is discontinuous. Even so, as $N \to \infty$, the residual tends to zero.

We can also express the Fourier series in terms of the frequency of oscillation, defined as

$$\omega_0 = \frac{2\pi}{T}$$

so that

$$f(t) = \sum_{n=1,3,5,\ldots}^{\infty} \frac{8}{n^2 \pi^2} \cos\left(n\omega_0 t\right) \tag{13.19}$$

The first five terms in the partial sum are displayed in Figure 13.10 (blue lines). The partial sum $f_9(t)$ over the five contributions (red line) is compared with the exact triangle wave function (gray line). The coefficients a_n for $n = 1$ through 9 are shown as blue bars. As n increases, the amplitude of each term in the sum decreases and the frequency of oscillation increases.

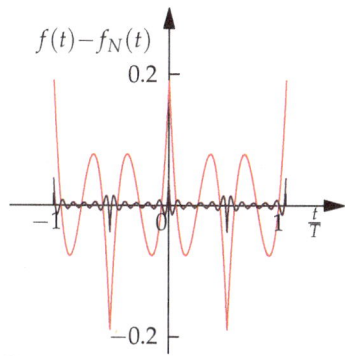

Figure 13.9: The residual difference $f(t) - f_N(t)$ between the exact triangle wave function $f(t)$ and the partial sum $f_N(t)$ for $N = 1$ (red line) and 9 (dark red line). The greatest residual difference is found at the sharp angles appearing at $t = -T, -\frac{T}{2}, 0, \frac{T}{2},$ and T.

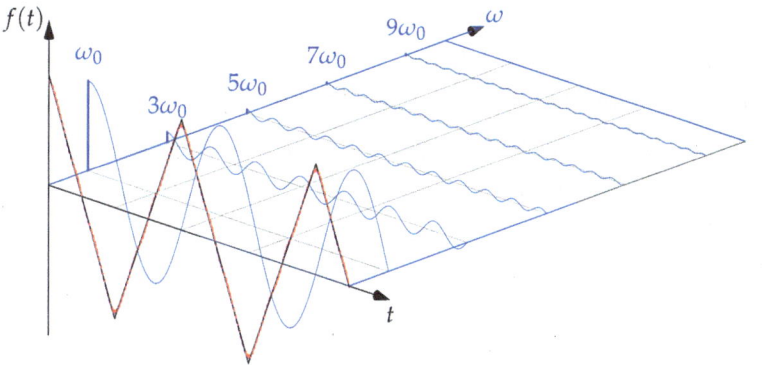

Figure 13.10: The Fourier series approximating the triangle wave (black line) using the first five terms in the series. The sum of the five blue curves of varying wavelength forms the composite red curve. The height of each term indicates the magnitude of the coefficients b_n (thick blue bars).

Note that for the triangle wave, the magnitudes of the non-zero Fourier coefficients for the continuous triangle wave diminish according to

$$\frac{a_n}{a_1} = \frac{1}{n^2} \qquad n = 1,3,5,\ldots$$

while for the discontinuous square wave the magnitudes of the Fourier coefficients diminish according to

$$\frac{b_n}{b_1} = \frac{1}{n} \qquad n = 1,3,5,\ldots$$

As such, the Fourier series for the triangle wave converges faster to an accurate representation of the actual function more rapidly than the Fourier series for the square wave.

13.1.5 *The sawtooth wave as an alternating sine series*

Consider the periodic function of x known as the *sawtooth wave*, shown in Figure 13.11 over three periods of oscillation with period L.

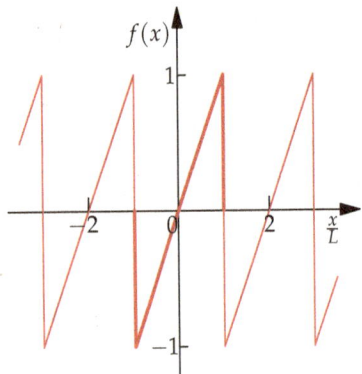

Figure 13.11: The sawtooth wave $f(x)$ (red line) is an odd periodic function of x with period $2L$.

The function can be written over one period as

$$f(x) = \frac{x}{L} \qquad -L \le x < L \tag{13.20}$$

The Fourier series is

$$f(x) = \frac{a_0}{2} + \sum_{n=1}^{\infty} a_n \cos\left(\frac{n\pi x}{L}\right) + \sum_{n=1}^{\infty} b_n \sin\left(\frac{n\pi x}{L}\right)$$

where the coefficients are

$$a_n = \frac{1}{L}\int_{-L}^{L} \frac{x}{L} \cos\left(\frac{n\pi x}{L}\right) dx \qquad b_n = \frac{1}{L}\int_{-L}^{L} \frac{x}{L} \sin\left(\frac{n\pi x}{L}\right) dx$$

Let's determine the Fourier coefficients for the periodic sawtooth wave.

Fourier coefficients for the sawtooth wave

Since $f(x)$ is an odd function of x, the Fourier series is an odd function of x and the coefficients multiplying the even cosine functions are zero for all n:

$$a_n = 0$$

As such, the coefficients for the odd sine function can be evaluated using the table of indefinite integrals in Supplement S_6 #89:

$$\int x \sin(ax+b)\, dx = \frac{1}{a^2}\sin(ax+b) - \frac{x}{a}\cos(ax+b) + C$$

Substituting in $a = \frac{n\pi}{L}$ and $b = 0$, we find

$$b_n = \frac{1}{L^2}\int_{-L}^{L} x \sin\left(\frac{n\pi x}{L}\right) dx = -\frac{1}{L^2}\left[\frac{L^2}{n\pi}\cos(n\pi) - \left(-\frac{L^2}{n\pi}\cos(-n\pi)\right)\right]$$

$$= -\frac{2}{n\pi}\cos(n\pi) = -\frac{2}{n\pi}(-1)^n = \frac{2}{n\pi}(-1)^{n+1}$$

This leads to our final result for the Fourier series of the odd sawtooth wave:[20]

$$f(x) = \frac{2}{\pi} \sum_{n=1}^{\infty} \frac{(-1)^{n+1}}{n} \sin\left(\frac{n\pi x}{L}\right) \tag{13.21}$$

The result is shown in Figure 13.12, where the exact function (gray line) is compared with the partial sum $f_N(x)$ for $N = 1$ (light red), $N = 5$ (red), and $N = 20$ terms (dark red).

Harmonic analysis of the sawtooth wave

The harmonic analysis is summarized in Figure 13.12, showing the variation of the coefficients b_n with increasing frequency.

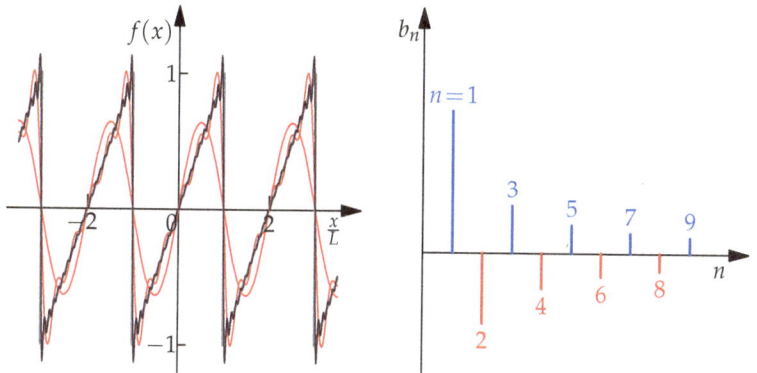

The magnitude of the Fourier coefficients diminishes according to $\frac{1}{n}$, as we found for the square wave in Equation 13.13. However, unlike the coefficients for the square wave, which were all positive, the coefficients for the sawtooth wave have alternating signs:

$$\frac{b_n}{b_1} = \frac{(-1)^{n+1}}{n} \qquad n = 1, 2, 3, \ldots$$

As was observed for the square wave, the Fourier series overshoots the exact sawtooth wave form on either side of the discontinuity. In the case of the sawtooth wave, rather than stepping down from 1 to -1 there is a step down from roughly 1.089 to -1.089, making the overshoot approximately 9% of the magnitude of the discontinuity.

13.1.6 *The shark fin wave as a series of sines and cosines*

Now consider the periodic function of x known as the *shark fin wave*. The function can be written over one period as

$$f(x) = \begin{cases} 0 & -L \leq x < 0 \\ \dfrac{x}{L} & 0 \leq x < L \end{cases} \tag{13.22}$$

The shark fin wave can be thought of as the positive half of the sawtooth wave. While the sawtooth wave is an odd function of x, the shark fin wave is

[20] The sawtooth wave has step and ramp features. The Fourier coefficients vary as $\frac{1}{n}$, characteristic of discontinuous step functions.

Figure 13.12: The partial sum

$$f_N(x) = \frac{2}{\pi} \sum_{n=1}^{N} \frac{(-1)^{n+1}}{n} \sin\left(\frac{n\pi x}{L}\right)$$

approximating the sawtooth wave (gray line) for $N = 1$ (light red), $N = 5$ (red), and $N = 20$ (dark red). Also shown are the coefficients contributing to the series, which alternate in sign.

neither an odd function of x nor an even function of x. The shark fin wave is shown in Figure 13.13 over three periods of oscillation with period $2L$.

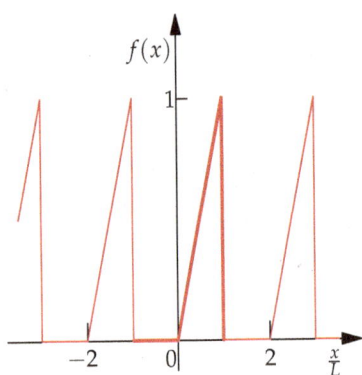

Figure 13.13: The shark fin wave (red line) is a periodic function of x with period $2L$.

Since $f(x)$ is neither an even function of x nor an odd function of x, we expect the Fourier series to have contributions from both cosine and sine terms:[21]

$$f(x) = \frac{a_0}{2} + \sum_{n=1}^{\infty} a_n \cos\left(\frac{n\pi x}{L}\right) + \sum_{n=1}^{\infty} b_n \sin\left(\frac{n\pi x}{L}\right) \qquad (13.23)$$

[21] The range of integration is $0 \le x \le L$ since $f(x) = 0$ for $-L \le x \le 0$.

The coefficients are defined by Equation 13.10

$$a_0 = \frac{1}{L} \int_0^L \frac{x}{L} \, dx \quad a_n = \frac{1}{L} \int_0^L \frac{x}{L} \cos\left(\frac{n\pi x}{L}\right) \, dx \quad b_n = \frac{1}{L} \int_0^L \frac{x}{L} \sin\left(\frac{n\pi x}{L}\right) \, dx$$

Let's determine the Fourier coefficients for the periodic shark fin wave.

Fourier coefficients for the shark fin wave

The coefficient for the constant term is

$$a_0 = \frac{1}{L^2} \int_0^L x \, dx = \frac{1}{L^2} \left[\frac{x^2}{2}\right]_0^L = \frac{1}{2} \qquad (13.24)$$

We can evaluate the cosine coefficients using the table of indefinite integrals in Supplement S_6 #96:

$$\int x \cos(ax + b) \, dx = \frac{1}{a^2} \cos(ax + b) + \frac{x}{a} \sin(ax + b) + C$$

Substituting in $a = \frac{n\pi}{L}$ and $b = 0$, we find

$$a_n = \frac{1}{L^2} \int_0^L x \cos\left(\frac{n\pi x}{L}\right) \, dx = \frac{1}{n^2 \pi^2} \left(\cos(n\pi) - 1\right) = -\frac{1}{n^2 \pi^2} \left(1 - (-1)^n\right)$$

As such, $a_n = 0$ for even n, whereas for odd n, we find

$$a_n = -\frac{2}{n^2 \pi^2} \qquad n = 1, 3, 5, \ldots \qquad (13.25)$$

We can evaluate the sine coefficients b_n using the table of indefinite integrals in Supplement S_6 #89:

$$\int x \sin(ax+b)\,dx = \frac{1}{a^2}\sin(ax+b) - \frac{x}{a}\cos(ax+b) + C$$

Substituting in $a = \frac{n\pi}{L}$ and $b = 0$, we find

$$b_n = \frac{1}{L^2}\int_0^L x \sin\left(\frac{n\pi x}{L}\right)dx = -\frac{1}{n\pi}\cos(n\pi) = -\frac{1}{n\pi}(-1)^n = \frac{1}{n\pi}(-1)^{n+1}$$

As such, we find

$$b_n = \frac{1}{n\pi}(-1)^{n+1} \qquad n = 1, 2, 3, \ldots \qquad (13.26)$$

Inserting the coefficients defined by Equations 13.24, 13.25, and 13.26 into

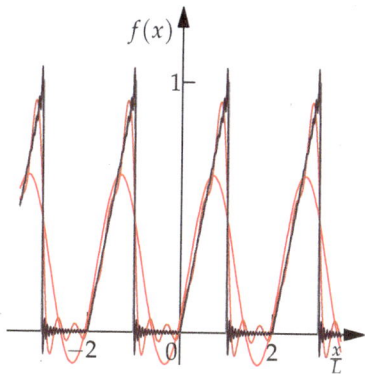

Figure 13.14: Gradual convergence of the Fourier series approximating the shark fin wave (gray line) for increasing number of terms in the series (darker shades of red).

Equation 13.23 leads to the Fourier series for the shark fin wave:[22]

$$f(x) = \frac{1}{4} - \sum_{n=1,3,5,\ldots}^{\infty}\frac{2}{n^2\pi^2}\cos\left(\frac{n\pi x}{L}\right) + \sum_{n=1}^{\infty}\frac{1}{n\pi}(-1)^{n+1}\sin\left(\frac{n\pi x}{L}\right) \qquad (13.27)$$

Three realizations of the partial sum representing the Fourier series for the shark fin wave are shown in Figure 13.14. The Gibbs phenomenon is apparent at the discontinuities.

Harmonic analysis of the shark fin wave

The behavior of the Fourier coefficients of the shark fin wave can be summarized as

$$a_0 = \frac{1}{2}, n = 0 \qquad \frac{a_n}{a_1} = \frac{1}{n^2}, n = 1, 3, 5, \ldots \qquad \frac{b_n}{b_1} = \frac{(-1)^{n+1}}{n}, n = 1, 2, 3, \ldots$$

After $a_0 = \frac{1}{4}$, the coefficients for the cosine terms are uniformly negative, decaying in magnitude according to

$$\frac{1}{n^2}$$

due to discontinuities in the derivative of the function. Coefficients for the

[22] The shark fin wave has step and ramp features. The Fourier coefficients vary as $\frac{1}{n}$, characteristic of discontinuous *step functions*, and $\frac{1}{n^2}$, characteristic of *ramp functions*. The overall convergence of the Fourier series will be limited by the terms that varying most slowly as $\frac{1}{n}$ with increasing n.

sine terms have alternating signs, decaying in magnitude more gradually according to

$$\frac{1}{n}$$

due to discontinuities in the function itself. In addition, since the coefficients for the cosine series vary according to $a_n \propto \frac{1}{n^2}$ while the coefficients of the sine series vary according to $b_n \propto \frac{1}{n}$, the sine series makes the greater overall contribution. The coefficients vary as depicted in Figure 13.15.

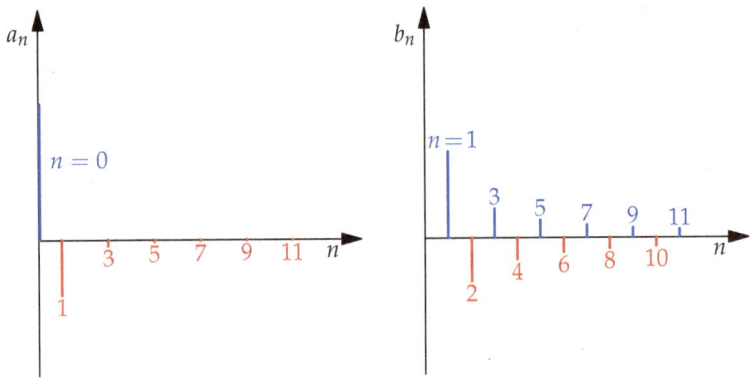

Figure 13.15: The coefficients a_n and b_n for the Fourier series approximating the shark fin wave.

In contrast to the Fourier series for the square wave and sawtooth wave, which are odd functions represented by sine series, and the triangle wave, which is an even function represented by a cosine series, the Fourier series for the shark fin wave includes both even cosine terms and odd sine terms. In fact, the shark fin wave $f(x)$ can be expressed as a linear sum of a constant, the triangle wave $f_T(x)$ and the sawtooth wave $f_S(x)$, resulting in

$$f(x) = \frac{1}{4} - \frac{1}{4}f_T(x) + \frac{1}{2}f_S(x)$$

Similarly, the Fourier series of the shark fin wave, in Equation 13.27, can be expressed as a linear sum of a constant term, the Fourier series of the triangle wave, in Equation 13.18, and the Fourier series of the sawtooth wave, in Equation 13.21.

13.1.7 *The complex Fourier series*

So far we have discussed Fourier series of real functions in terms of sine and cosine series. We will now consider a convenient compact definition of the Fourier series of real functions known as the complex Fourier series. We begin with

$$f(x) = \frac{a_0}{2} + \sum_{n=1}^{\infty} a_n \cos(nx) + \sum_{n=1}^{\infty} b_n \sin(nx)$$

Substituting $\cos(nx) = \frac{1}{2}\left(e^{inx} + e^{-inx}\right)$ and $\sin(nx) = \frac{1}{2i}\left(e^{inx} - e^{-inx}\right)$, we find

$$f(x) = \frac{1}{2}a_0 + \frac{1}{2}\sum_{n=1}^{\infty} a_n \left(e^{inx} + e^{-inx}\right) + \frac{1}{2i}\sum_{n=1}^{\infty} b_n \left(e^{inx} - e^{-inx}\right)$$

Rearranging terms, we have

$$f(x) = \frac{1}{2}a_0 + \frac{1}{2}\sum_{n=1}^{\infty}(a_n - ib_n)\,e^{inx} + \frac{1}{2}\sum_{n=1}^{\infty}(a_n + ib_n)\,e^{-inx} \qquad (13.28)$$

Let's take the second series in Equation 13.28 above, and substitute $n \to -n$:

$$\sum_{n=1}^{\infty}(a_n + ib_n)\,e^{-inx} = \sum_{n=-1}^{-\infty}(a_{-n} + ib_{-n})\,e^{inx}$$

As the cosine function is even, the coefficients $a_{-n} = a_n$ are even, and as the sine function is odd, the coefficients $b_{-n} = -b_n$ are odd. As such, we find

$$\sum_{n=-1}^{-\infty}(a_{-n} + ib_{-n})\,e^{inx} = \sum_{n=-1}^{-\infty}(a_n - ib_n)\,e^{inx}$$

Substituting this result in Equation 13.28, we find

$$f(x) = \frac{1}{2}a_0 + \frac{1}{2}\sum_{n=1}^{\infty}(a_n - ib_n)\,e^{inx} + \frac{1}{2}\sum_{n=-1}^{-\infty}(a_n - ib_n)\,e^{inx}$$

Defining the coefficient $b_0 = 0$, inserting it into the series, and reordering terms, we have

$$f(x) = \frac{1}{2}\sum_{n=-1}^{-\infty}(a_n - ib_n)\,e^{inx} + \frac{1}{2}(a_0 - ib_0) + \frac{1}{2}\sum_{n=1}^{\infty}(a_n - ib_n)\,e^{inx}$$

Combining the three sums, we arrive at a single series with an index running from $-\infty$ to ∞:

$$f(x) = \frac{1}{2}\sum_{n=-\infty}^{\infty}(a_n - ib_n)\,e^{inx} = \sum_{n=-\infty}^{\infty}c_n\,e^{inx} \qquad (13.29)$$

We have previously shown in Equation 13.8 that

$$a_n = \frac{1}{\pi}\int_{-\pi}^{\pi}f(x)\cos(nx)\,dx \qquad b_n = \frac{1}{\pi}\int_{-\pi}^{\pi}f(x)\sin(nx)\,dx \qquad (13.30)$$

Substituting these relations for a_n and b_n, we find

$$\begin{aligned}
c_n = \frac{1}{2}(a_n - ib_n) &= \frac{1}{2\pi}\int_{-\pi}^{\pi}f(x)\cos(nx)dx - i\frac{1}{2\pi}\int_{-\pi}^{\pi}f(x)\sin(nx)dx \\
&= \frac{1}{2\pi}\int_{-\pi}^{\pi}f(x)\left(\cos(nx) - i\sin(nx)\right)dx \\
&= \frac{1}{2\pi}\int_{-\pi}^{\pi}f(x)\,e^{-inx}dx
\end{aligned}$$

where we have used the identity $\cos(nx) - i\sin(nx) = e^{-inx}$.

Note that we can also define the complex Fourier series over the range $-\frac{L}{2} \le x < \frac{L}{2}$ as

$$f(x) = \sum_{n=-\infty}^{\infty}c_n\exp\left(i\frac{2n\pi x}{L}\right) \qquad (13.31)$$

where the coefficients are defined[23]

[23] Note that $c_{-n} = c_n^*$ the complex conjugate of c_n.

$$c_n = \frac{1}{2}(a_n - ib_n) = \frac{1}{L} \int_{-\frac{L}{2}}^{\frac{L}{2}} f(x) \exp\left(-i\frac{2n\pi x}{L}\right) dx \qquad (13.32)$$

The complex Fourier series provides a compact representation of the Fourier series of a periodic function. It also provides a foundation for the harmonic analysis of aperiodic functions explored in the next section.

13.1.8 Complex Fourier series of the sawtooth wave

Consider again the sawtooth wave, which is a periodic function of x. It can be written over one period as

$$f(x) = \frac{x}{L} \qquad -L \le x < L \qquad (13.33)$$

The function is shown in Figure 13.11 over three periods of oscillation.

The complex Fourier series defined by Equations 13.31 and 13.32 can be extended to treat functions with a period $2L$ over a range $-L \le x < L$ as

$$f(x) = \sum_{n=-\infty}^{\infty} c_n \exp\left(\frac{in\pi x}{L}\right) \qquad (13.34)$$

where the corresponding Fourier coefficients are

$$c_n = \frac{1}{2L} \int_{-L}^{L} \frac{x}{L} \exp\left(-\frac{in\pi x}{L}\right) dx \qquad (13.35)$$

Let's determine the coefficients for this complex Fourier series.

Coefficients for the complex Fourier series of the sawtooth wave

For c_0, we find

$$c_0 = \frac{1}{2L} \int_{-L}^{L} \frac{x}{L} dx = 0 \qquad (13.36)$$

as the integrand is an odd function of x. We can evaluate the remaining coefficients c_n for $n > 0$ using the table of indefinite integrals in Supplement S_6 #12:

$$\int x e^{bx} dx = \frac{x}{b} e^{bx} - \frac{1}{b^2} e^{bx} + C$$

Substituting in $b = -\frac{in\pi}{L}$, we find

$$c_n = \frac{1}{2L^2} \left[-\frac{xL}{in\pi} \exp\left(-\frac{in\pi x}{L}\right) + \frac{L^2}{n^2\pi^2} \exp\left(-\frac{in\pi x}{L}\right) \right]_{-L}^{L}$$

$$= \frac{1}{2L^2} \left[-\frac{L^2}{in\pi}\left(e^{-in\pi} + e^{in\pi}\right) + \frac{L^2}{n^2\pi^2}\left(e^{-in\pi} - e^{in\pi}\right) \right]$$

which can be reformed as

$$c_n = \left[-\frac{1}{in\pi}\left(\frac{e^{inx} + e^{-inx}}{2}\right) - \frac{i}{n^2\pi^2}\left(\frac{e^{inx} - e^{-inx}}{2i}\right) \right]$$

$$= -\frac{1}{in\pi}\cos(n\pi) - i\frac{1}{n^2\pi^2}\sin(n\pi)$$

for $n \neq 0$ and $c_0 = 0$. Since $\sin(n\pi) = 0$ for any integer n and $\cos(n\pi) = (-1)^n$, we find[24]

$$c_n = -\frac{1}{in\pi}(-1)^n = \frac{1}{in\pi}(-1)^{n+1}$$

With this result, we can write the complex Fourier series of the sawtooth wave as

$$f(x) = \frac{1}{i\pi} \sum_{n=-\infty}^{\infty}{}' \frac{(-1)^{n+1}}{n} \exp\left(\frac{in\pi x}{L}\right) \qquad (13.37)$$

where the primed sum is restricted to $n \neq 0$.

Harmonic analysis of the complex Fourier series of the sawtooth wave

The Fourier series expresses the function $f(x)$ as an infinite series of harmonics of wavelength $\lambda = \frac{2}{n}L$ and amplitude $a_n \propto \frac{1}{n}$. Figure 13.16 shows the *residual difference* between the exact function $f(x)$ and the partial sum of the Fourier series representation $f_N(x)$ for $N = 1$ and 20. Note that the greatest differences between the exact function and the approximate Fourier series representation occur at the sharp points where the function is discontinuous. This is an example of the *Gibbs phenomenon*.

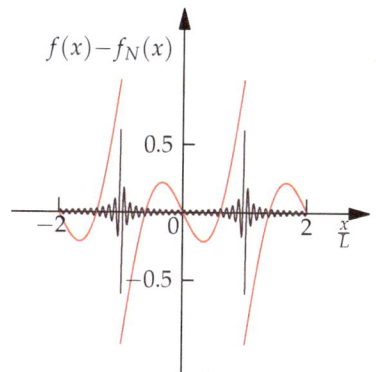

Note that for every term n in the sum there is a term $-n$. Grouping the n and $-n$ terms together, we can rewrite the sum as

$$f(x) = \frac{1}{i\pi} \sum_{n=1}^{\infty} \frac{(-1)^{n+1}}{n} \left[\exp\left(\frac{in\pi x}{L}\right) - \exp\left(-\frac{in\pi x}{L}\right)\right]$$

Given that $\frac{1}{2i}(e^{\frac{in\pi x}{L}} - e^{-\frac{in\pi x}{L}}) = \sin\left(\frac{n\pi x}{L}\right)$, the complex Fourier series becomes

$$f(x) = \frac{2}{\pi} \sum_{n=1}^{\infty} \frac{(-1)^{n+1}}{n} \sin\left(\frac{n\pi x}{L}\right) \qquad (13.38)$$

Our result is identical to that previously derived for the Fourier series representation of the sawtooth wave given by Equation 13.21. In the next section, we will extend our harmonic analysis from periodic functions, using Fourier series, to aperiodic functions, using Fourier transforms.

[24] Note that $(-1)^{n+1} = (-1)^{-n+1}$, indicating that the sign of the coefficient c_n is the same as the sign of c_{-n}.

Figure 13.16: The residual difference between the exact sawtooth wave function $f(x)$ and the partial sum $f_N(x)$ for $N = 1$ (pale red line) and 20 (dark red line). For larger values of N the greatest residual difference is found at the discontinuities appearing at $x = -L$ and L.

13.2 Fourier transforms

THE FOURIER SERIES can be used to represent an infinitely repeating periodic function in terms of an infinite sum of periodic waves. The coefficients weight the contributions of each sine and cosine component to sum to the total function. The Fourier transform extends the harmonic analysis to aperiodic functions by including a continuum of component frequencies. It has widespread application in the solution of differential equations, the formulation of quantum theory, and signal processing and analysis, including vibrational spectroscopy.

13.2.1 Fourier transforms and the connection to Fourier series

In this section, we extend our harmonic analysis from infinitely *periodic functions*, using Fourier series, to *aperiodic functions*, using Fourier transforms. We will start by introducing the definition of the Fourier transform. By comparing the Fourier transform with the complex Fourier series, we will gain a deep insight into the connection between Fourier transforms and the coefficients used in Fourier series. The specific example of g of a sawtooth wave, defined by Equation 13.21, with the corresponding Fourier series will be used to build on our understanding of Fourier series to develop an intuitive feeling for the nature of the Fourier transform.

A general function $f(x)$ can be written in the form of an *integral transform* known as the *Fourier transform*:

$$f(x) = \int_{-\infty}^{\infty} F(k)\, e^{ikx} dk \qquad (13.39)$$

The integral over the variable k extends from $-\infty$ to ∞. The integrand consists of the complex exponential function e^{ikx} weighted by a function $F(k)$. The function $F(k)$ is defined as an integral over the complex exponential function e^{-ikx}, the complex conjugate of the complex exponential appearing above, weighted by the function $f(x)$:

$$F(k) = \frac{1}{2\pi} \int_{-\infty}^{\infty} f(x)\, e^{-ikx} dx \qquad (13.40)$$

These expressions remind us of the complex Fourier series for a function $f(x)$:

$$f(x) = \sum_{n=-\infty}^{\infty} c(k_n)\, e^{ik_n x} \qquad (13.41)$$

The sum over the variable n extends from $-\infty$ to ∞. In this case, $k_n = n\pi/L$ is a function of the integer n. The summand consists of the complex exponential function $e^{ik_n x}$ weighted by the coefficients $c(k_n)$, which are functions of k_n:

$$c(k_n) = \frac{1}{2L} \int_{-L}^{L} f(x)\, e^{-ik_n x} dx \qquad (13.42)$$

Comparing Equations 13.40 and 13.42, we see that the Fourier transform $F(k)$ plays the role of the coefficients $c(k_n)$ in the Fourier series. That is, the Fourier transform of a function is related to the coefficients of the Fourier series of that function.

To get a better feeling for the relationship between Fourier transforms and Fourier series, let's perform an analysis of the sawtooth wave. We will determine the Fourier transform of one oscillation of the sawtooth wave. We will then compare that result with our prior result for the Fourier series of the periodic sawtooth wave.

Recall the general form of the odd sawtooth wave shown in Figures 13.12 and 13.17. This sawtooth wave is an infinite periodic function of x. It is also an odd function of x as $f(-x) = -f(x)$.

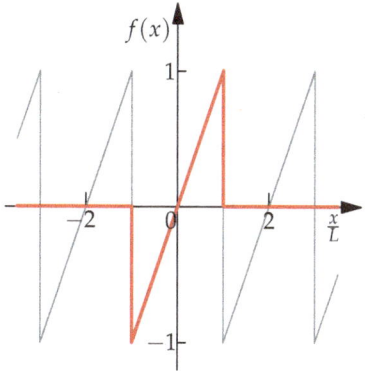

Figure 13.17: The infinite sawtooth wave (gray line) is an odd periodic function of x with period $2L$. The sawtooth function (thick red line) is a finite aperiodic function.

Using Equation 13.42, we can derive our result in Equation 13.21 for the Fourier series of the infinite periodic sawtooth wave:

$$f(x) = \frac{2}{\pi} \sum_{n=1}^{\infty} \frac{(-1)^{n+1}}{n} \sin\left(\frac{n\pi x}{L}\right) \qquad (13.43)$$

where the coefficients in the series can be written

$$c(k_n) = -\frac{i}{n\pi}(-1)^{n+1} = \frac{1}{2}(a_n + ib_n) \qquad (13.44)$$

so that $b_n = \frac{2}{n\pi}(-1)^{n+1}$. The first eight coefficients are plotted as vertical bars in Figure 13.18. Note the alternating sign of the coefficients and the diminishing magnitude as n increases according to $\frac{1}{n}$.

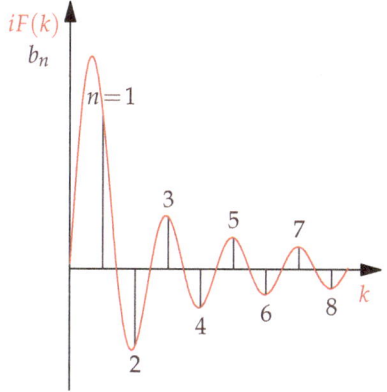

Figure 13.18: The coefficients $b_n = \frac{2}{n\pi}(-1)^{n+1}$ of the infinite Fourier series for the odd sawtooth wave function with wavelength $2L$ (dark red lines) compared with the magnitude of the Fourier transform $iF(k)$ of the aperiodic odd sawtooth wave defined over $-L \leq x < L$. Recall that

$$b_n = i\frac{\sqrt{2\pi}}{L}\, F\left(\frac{n\pi}{L}\right)$$

when relating the Fourier coefficients to the Fourier transform at specific values of the wave vector $k = \frac{n\pi}{L}$.

Using Equation 13.40, we evaluate the Fourier transform[25] of the finite aperiodic sawtooth wave:

$$F(k) = i\,\frac{1}{\sqrt{2\pi}}\,\frac{2}{k^2 L}\,(kL\cos(kL) - \sin(kL)) \tag{13.45}$$

At first, it may not appear that Equations 13.44 and 13.45 are closely related. However, if we take $k = \frac{n\pi}{L}$ and note that $\cos(n\pi) = (-1)^n$ and $\sin(n\pi) = 0$, we find $F(\frac{n\pi}{L}) = i\,\frac{1}{\sqrt{2\pi}}\,\frac{2L}{n\pi}\,(-1)^n$ and

$$b_n = \frac{2}{n\pi}(-1)^{n+1} = i\,\frac{\sqrt{2\pi}}{L}\,F\!\left(\frac{n\pi}{L}\right) \tag{13.46}$$

For a function $f(x)$, the coefficients of the Fourier series $c_n = \frac{1}{2}(a_n - ib_n)$ are proportional to the Fourier transform $F(k)$ evaluated at $k = n\pi/L$.

In essence, we have replaced one order of infinity with another. The coefficients used in Fourier series, which are functions of integers, are replaced with Fourier transforms, which are functions of real wave numbers.[26] The greater level of detail that is possible through the Fourier transform allows us to treat aperiodic functions as well as periodic functions.

13.2.2 *Conventions used for Fourier transforms and Fourier series*

In practice, Fourier transforms are applied to functions of number, length, and time. Fourier transforms are also applied in many fields of the physical and social sciences. In each case, a slightly different definition of the Fourier transform may be used to simplify the evaluation. Or the form most commonly used may be a matter of established convention in a given field. Let's explore several common conventions used to define the Fourier transform. In each case, we will define the relationship between the Fourier series of a periodic function and the Fourier transform over one period of that function.

Asymmetric prefactors with dependence on 2π: The Fourier transform is defined as

$$F(k) = \int_{-\infty}^{\infty} f(x)\,e^{-ikx}\,dx \tag{13.47}$$

Now recall that for the complex Fourier series, the coefficients c_n are defined in Equation 13.32 to be

$$c_n = \frac{1}{2}(a_n - ib_n) = \frac{1}{2\pi}\int_{-\pi}^{\pi} f(x)\,e^{-inx}\,dx$$

Comparing these results, we find that the Fourier series coefficients a_n and b_n are related to the Fourier transform $F(k)$, evaluated at discrete values of the *wave number*[27]

$$k = n$$

through the following equality:

$$c_n = \frac{1}{2}(a_n - ib_n) = \frac{1}{2\pi}F(n) \tag{13.48}$$

---

[25] Here we evaluate

$$F(k) = \int_{-L}^{L} \frac{x}{L}\,e^{-ikx}\,dx$$

using S_6 #80 which yields

$$F(k) = \frac{1}{k^2}(ikx+1)e^{-ikx}\Big|_{-L}^{L}$$

After a bit of algebra and noting that $2i\sin(kx) = e^{ikx} - e^{-ikx}$ and $2\cos(kx) = e^{ikx} + e^{-ikx}$ we arrive at the final result

$$F(k) = i\,\frac{2}{k^2 L}\,(kL\cos(kL) - \sin(kL))$$

[26] The multiple orders of infinity were discovered by German mathematician *Georg Cantor* (1845-1918). He demonstrated that integers are *countably infinite*. The rational numbers, that can be expressed as a quotient of integers, are *countably infinite* and of the same order of infinity as the integers. However, the real numbers are *uncountably infinite*, representing a higher order of infinity.

[27] The wave number

$$k = n = \frac{2\pi}{\lambda}$$

corresponds to wavelengths $\lambda = 2\pi/n$.

Symmetric prefactors with dependence on 2π: Let's divide the reciprocal of 2π evenly between the integrals as

$$F(k) = \frac{1}{\sqrt{2\pi}} \int_{-\infty}^{\infty} f(x) \, e^{-ikx} dx \tag{13.49}$$

and

$$f(x) = \frac{1}{\sqrt{2\pi}} \int_{-\infty}^{\infty} F(k) \, e^{ikx} dk \tag{13.50}$$

Using this convention, the wave number is[28]

$$k = \frac{2n\pi}{L}$$

and the coefficients are defined as

$$c_n = \frac{1}{L} \int_{-\frac{L}{2}}^{\frac{L}{2}} f(x) \, \exp\left(-i\frac{2n\pi x}{L}\right) dx \tag{13.51}$$

so that[29]

$$c_n = \frac{1}{2}(a_n - ib_n) = \frac{\sqrt{2\pi}}{L} F\left(\frac{2n\pi}{L}\right) \tag{13.52}$$

Symmetric unit prefactors (no 2π): Another convention is based on including a factor of 2π in the argument of the exponential. In that form, the Fourier transform pairs are written

$$F(k) = \int_{-\infty}^{\infty} f(x) \, e^{-i2\pi kx} dx \tag{13.53}$$

and

$$f(x) = \int_{-\infty}^{\infty} F(k) \, e^{i2\pi kx} dk \tag{13.54}$$

which is the most symmetric form of the reciprocal pair. The wave number is[30]

$$k = \frac{n}{L}$$

where[31]

$$c_n = \frac{1}{L} \int_{-\frac{L}{2}}^{\frac{L}{2}} f(x) \, \exp\left(-i\frac{2n\pi x}{L}\right) dx \tag{13.55}$$

so that

$$c_n = \frac{1}{2}(a_n - ib_n) = \frac{1}{L} F\left(\frac{n}{L}\right) \tag{13.56}$$

We find that the coefficients of the Fourier series, each corresponding to a characteristic frequency of oscillation, are defined in terms of the corresponding Fourier transform, evaluated at that characteristic frequency. The Fourier transform $F(k)$ provides a harmonic analysis of the function $f(x)$.

Harmonic analysis is commonly used in the study of molecular vibrations and the application of various spectroscopies. Let's apply the Fourier transform to the harmonic analysis of a variety of functions commonly used in the physical sciences. The results of these and other examples are tabulated in the table of Fourier transform pairs in Supplement S_9.

---

[28] The wave number

$$k = \frac{2\pi}{\lambda} = \frac{2n\pi}{L}$$

corresponds to allowed wavelengths $\lambda = L/n$.

[29] If integration is carried out over a range $-L \leq x \leq L$ then the correspondence is

$$c_n = \frac{\sqrt{2\pi}}{2L} F\left(\frac{n\pi}{L}\right)$$

where we have substituted L for $\frac{L}{2}$ in Equation 13.52.

[30] The wave number

$$k = \frac{n}{L} = \frac{2\pi}{\lambda}$$

corresponds to wavelengths $\lambda = 2\pi L/n$.

[31] If integration is carried out over a range $-L \leq x \leq L$ then the correspondence is

$$c_n = \frac{1}{2L} F\left(\frac{n}{2L}\right)$$

13.2.3 *Deconstructing the Fourier transform of the square wave*

In the previous section, we derived the Fourier series of the infinite periodic square wave function (see Figure 13.5). Let's define the aperiodic function of x consisting of one oscillation of the *square wave* defined over the range $-\frac{L}{2} < x < \frac{L}{2}$ and shown in Figure 13.19. It is written in terms of a difference:

$$f(x) = 2\theta\left(\frac{x}{L}\right) - 1 \qquad -\frac{L}{2} \leq x < \frac{L}{2} \tag{13.57}$$

The *Heaviside step function* $\theta(x)$ equals 1 for $x \geq 0$ and 0 otherwise.

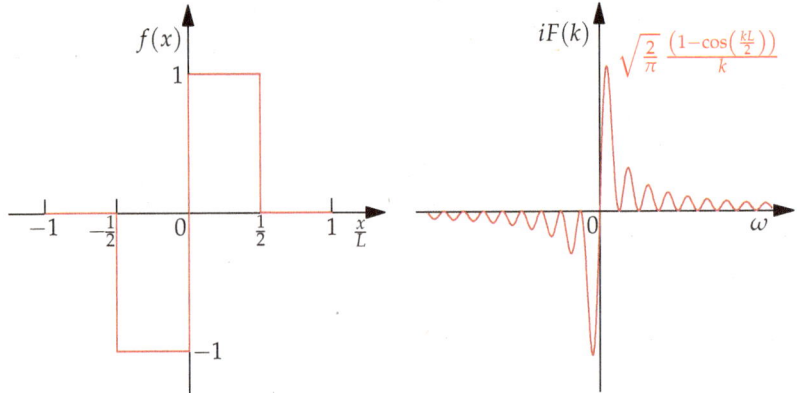

Figure 13.19: Comparison of the square wave function $f(x) = 2\theta\left(\frac{x}{L}\right) - 1$ and the imaginary Fourier transform complement $iF(k) = \sqrt{\frac{2}{\pi}} \frac{(1-\cos(kL/2))}{k}$.

Let's evaluate the Fourier transform of this aperiodic square wave.

Evaluating the Fourier transform of the aperiodic square wave

The Fourier transform of the square wave function is defined by the integral

$$F(k) = \frac{1}{\sqrt{2\pi}} \int_{-\infty}^{\infty} f(x)e^{-ikx} dx = \frac{1}{\sqrt{2\pi}} \int_{-\infty}^{\infty} \left(2\theta\left(\frac{x}{L}\right) - 1\right) e^{-ikx} dx$$

The integrand is zero other than over the interval $-\frac{L}{2} < x < \frac{L}{2}$. The function is -1 for $-\frac{L}{2} < x < 0$ and 1 for $0 < x < \frac{L}{2}$, so the integral can be divided as

$$F(k) = -\frac{1}{\sqrt{2\pi}} \int_{-\frac{L}{2}}^{0} e^{-ikx} dx + \frac{1}{\sqrt{2\pi}} \int_{0}^{\frac{L}{2}} e^{-ikx} dx$$

$$= \frac{1}{\sqrt{2\pi}} \left[\frac{1}{ik}\left(1 - e^{\frac{ikL}{2}}\right) - \frac{1}{ik}\left(e^{-\frac{ikL}{2}} - 1\right)\right]$$

$$= -i\frac{1}{\sqrt{2\pi}} \left[\frac{1}{k}\left(2 - e^{\frac{ikL}{2}} - e^{-\frac{ikL}{2}}\right)\right]$$

We then substitute $\cos(x) = (e^{ix} + e^{-ix})/2$ to form our final result

$$F(k) = -i\sqrt{\frac{2}{\pi}} \frac{\left(1 - \cos\left(\frac{kL}{2}\right)\right)}{k} \tag{13.58}$$

shown in Figure 13.19. The rapid oscillation or *ringing* is a signature of the abrupt steps in the square wave function.

Relating the Fourier transform to the Fourier series

To extend our harmonic analysis of periodic functions, using Fourier series, to aperiodic functions, we must extend the finite interval of the Fourier series to $(-\infty, \infty)$. In essence, we have replaced one order of infinity, the integer coefficients used in Fourier series, with another, the real wave numbers used in Fourier transforms.

We found the Fourier series of the periodic square wave function defined in Equation 13.13 to be

$$f(x) = \sum_{n=1,3,5,\ldots}^{\infty} \frac{4}{n\pi} \sin\left(\frac{2n\pi x}{L}\right)$$

where the coefficients were defined in Equation 13.11 as

$$b_n = \frac{2}{n\pi}\left(1 - \cos(n\pi)\right) = \frac{4}{L} \frac{\left(1 - \cos\left(\frac{kL}{2}\right)\right)}{k}$$

Here we use the definition of the wave number

$$k = \frac{2n\pi}{L}$$

The Fourier coefficients b_n and the Fourier transform $F(k)$ are related through Equation 13.52 where

$$-i\frac{1}{2} b_n = \frac{\sqrt{2\pi}}{L} F\left(\frac{2n\pi}{L}\right) \tag{13.59}$$

where we have accounted for the fact that the Fourier coefficients are scaled by $\frac{1}{L}$ while the Fourier transform is scaled by $\frac{1}{\sqrt{2\pi}}$.

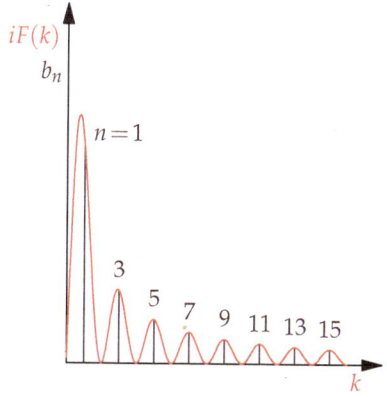

Figure 13.20: The coefficients $b_n = \frac{4}{n\pi}$ of the infinite Fourier series for the periodic square wave function with wavelength L (dark red lines) compared with the magnitude of the Fourier transform $iF(k) = \sqrt{\frac{2}{\pi}} \frac{\left(1-\cos\left(\frac{kL}{2}\right)\right)}{k}$ of the aperiodic square wave defined over $-L/2 \leq x < L/2$.

The coefficients b_n and Fourier transform $F(k)$, appropriately scaled using Equation 13.59, are compared in Figure 13.20. The Fourier coefficients represent points located near the top of the peaks of the continuously oscillating Fourier transform.

13.2.4 Fourier transform of the impulse function is the sampling function

Consider the aperiodic function of t known as the *impulse function*, which is unity in the range $-a < t < a$ and zero otherwise (see Figure 13.21). It can be written in terms of a difference:

$$f(t) = \theta(t+a) - \theta(t-a) \tag{13.60}$$

The *Heaviside step function* $\theta(x)$ equals 1 for $x \geq 0$ and 0 otherwise.

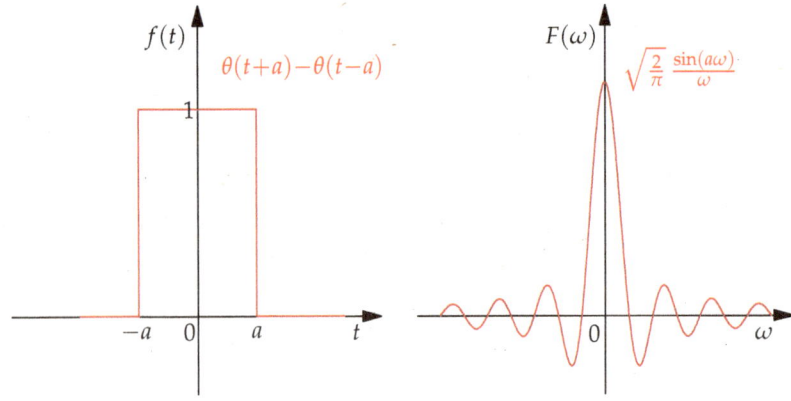

Figure 13.21: Comparison of the impulse function $f(t) = \theta(t+a) - \theta(t-a)$ formed as a difference between two Heaviside step functions and the Fourier transform complement $F(\omega) = \sqrt{\frac{2}{\pi}} \frac{\sin(a\omega)}{\omega}$.

The Fourier transform of the *impulse function* is defined by the integral

$$F(\omega) = \frac{1}{\sqrt{2\pi}} \int_{-\infty}^{\infty} f(t) e^{-i\omega t} dt = \frac{1}{\sqrt{2\pi}} \int_{-\infty}^{\infty} (\theta(t+a) - \theta(t-a)) e^{-i\omega t} dt$$

The integrand is zero other than over the interval $-a < x < a$, where it is unity. As such, the integral can be written as

$$F(\omega) = \frac{1}{\sqrt{2\pi}} \int_{-a}^{a} e^{-i\omega t} dt = \frac{1}{\sqrt{2\pi}} \left(-\frac{1}{i\omega} \right) e^{-i\omega t} \Big|_{-a}^{a} = \frac{1}{\sqrt{2\pi}} \left(\frac{e^{i\omega a} - e^{-i\omega a}}{i\omega} \right)$$

We then substitute $\sin(x) = (e^{ix} - e^{-ix})/2i$ to form our final result[32]

$$F(\omega) = \sqrt{\frac{2}{\pi}} \frac{\sin(a\omega)}{\omega} \tag{13.61}$$

which is shown in Figure 13.21 and recorded in Supplement S_9 #5. The period of oscillation is inversely proportional to the width a. The *ringing* on the flanks of the central peak is a signature of the discontinuous steps in the impulse function.

[32] The function

$$\text{sinc}(x) = \begin{cases} 1 & x = 0 \\ \dfrac{\sin x}{x} & \text{otherwise} \end{cases}$$

is called the *sinc function* and is also known as the *sampling function*.

13.2.5 Fourier transform of the gaussian is a gaussian

Consider the aperiodic function of t known as the *gaussian function*:

$$f(t) = e^{-at^2} \tag{13.62}$$

The gaussian function is used to model a wide range of phenomenon in the physical sciences.

The Fourier transform of the gaussian function is defined by the integral

$$F(\omega) = \frac{1}{\sqrt{2\pi}} \int_{-\infty}^{\infty} f(t)e^{-i\omega t} dt = \frac{1}{\sqrt{2\pi}} \int_{-\infty}^{\infty} e^{-at^2} e^{-i\omega t} \, dt$$

We can evaluate this integral using the table of definite integrals in Supplement S_5 #11, where

$$\int_{-\infty}^{\infty} e^{-ax^2 + bx} \, dx = \sqrt{\frac{\pi}{a}} \, e^{b^2/4a}$$

Conveniently, setting $a = a$ and $b = -i\omega$, we find our final result

$$F(\omega) = \frac{1}{\sqrt{2\pi}} \int_{-\infty}^{\infty} e^{-\alpha t^2} e^{-i\omega t} \, dt = \frac{1}{\sqrt{2\alpha}} e^{-\omega^2/4\alpha} \qquad (13.63)$$

The gaussian function $f(t)$ and its Fourier transform $F(\omega)$ are shown in Figure 13.22 and recorded in Supplement S_9 #2.

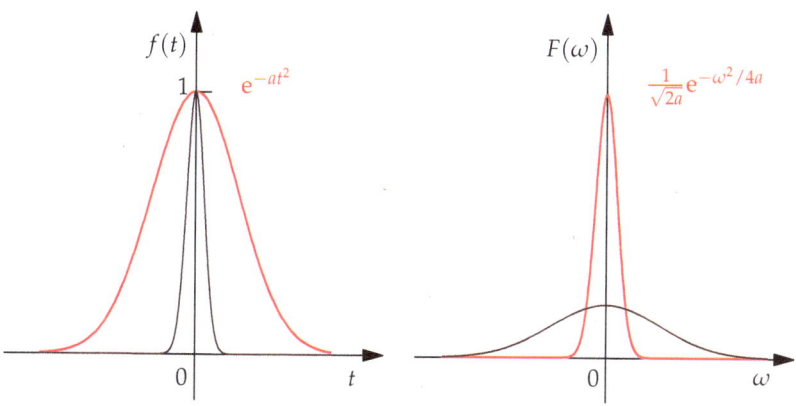

Figure 13.22: Comparison of the gaussian function $f(t) = e^{-at^2}$ and its Fourier transform complement $F(\omega) = \frac{1}{\sqrt{2a}} e^{-\omega^2/4a}$ for two values of a. A darker shade of red indicates a larger value of a.

The smooth variation of the gaussian function translates to an equally smooth variation in its Fourier transform.[33]

13.2.6 Fourier transform of an exponential is a lorentzian

Consider the aperiodic function of t known as the symmetric *exponential function* written

$$f(t) = e^{-a|t|} \qquad (13.64)$$

where the absolute value $|t|$ guarantees that the exponential function decreases to zero as $t \to \infty$ and as $t \to -\infty$. This is quite different from our usual definition of the exponential function. While the exponential function e^{-at} is neither and even nor odd function of t, the symmetric exponential function $e^{-a|t|}$ is an even function of t.

The Fourier transform of the exponential function is defined by the integral

$$F(\omega) = \frac{1}{\sqrt{2\pi}} \int_{-\infty}^{\infty} f(t)e^{-i\omega t} dt = \frac{1}{\sqrt{2\pi}} \int_{-\infty}^{\infty} e^{-a|t|} e^{-i\omega t} \, dt$$

[33] The gaussian function $f(x) = e^{-\frac{1}{2}x^2}$ is the only function for which the Fourier transform $F(k) = e^{-\frac{1}{2}k^2}$ is the function itself.

To eliminate the absolute value, we can divide the integral in two as

$$\int_{-\infty}^{\infty} e^{-a|t|}e^{-i\omega t}\, dt = \int_{-\infty}^{0} e^{at}e^{-i\omega t}\, dt + \int_{0}^{\infty} e^{-at}e^{-i\omega t}\, dt$$

$$= \int_{-\infty}^{0} e^{(a-i\omega)t}\, dt + \int_{0}^{\infty} e^{-(a+i\omega)t}\, dt$$

$$= \frac{1}{a-i\omega} + \frac{1}{a+i\omega}$$

We can simplify this result by multiplying by the complex conjugate, to form a real common denominator, and summing terms. We find

$$\frac{1}{a-i\omega}\left(\frac{a+i\omega}{a+i\omega}\right) + \frac{1}{a+i\omega}\left(\frac{a-i\omega}{a-i\omega}\right) = \frac{a+i\omega}{a^2+\omega^2} + \frac{a-i\omega}{a^2+\omega^2} = \frac{2a}{a^2+\omega^2}$$

Combining these terms leads to the final result

$$F(\omega) = \sqrt{\frac{2}{\pi}}\left(\frac{a}{a^2+\omega^2}\right) \tag{13.65}$$

The exponential function $f(t)$ and its Fourier transform $F(\omega)$ are shown in Figure 13.23 and recorded in Supplement S_9 #3.

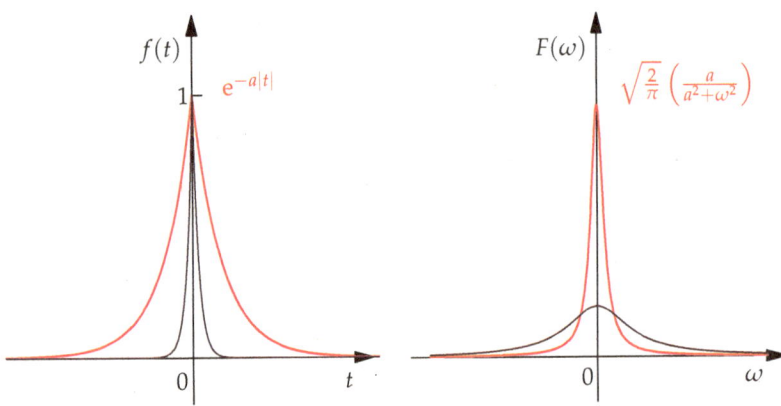

Figure 13.23: Comparison of the exponential function $f(t) = e^{-a|t|}$ and its Fourier transform complement $F(\omega) = \sqrt{\frac{2}{\pi}}\left(\frac{a}{a^2+\omega^2}\right)$ for two values of a. A darker shade of red indicates a larger value of a.

The lorentzian function is commonly used as a model of line shapes observed in spectroscopy. Note that the parameter a that controls the rate of decay of the exponential function is related to the width of the lorentzian function. Increasing the rate of decay of the exponential function broadens the width of the corresponding lorentzian function that is its Fourier transform. Decreasing the rate of decay of the exponential function narrows the width of the corresponding lorentzian function that is its Fourier transform.

13.2.7 Fourier transform of a sinusoid and Dirac delta functions

Now let's consider the periodic cosine function

$$f(t) = \cos(\omega_0 t) \tag{13.66}$$

The cosine function is an even function of t. The Fourier transform of the cosine function is defined by the integral

$$F(\omega) = \frac{1}{\sqrt{2\pi}} \int_{-\infty}^{\infty} f(t)e^{-i\omega t} dt = \frac{1}{\sqrt{2\pi}} \int_{-\infty}^{\infty} \cos(\omega_0 t)e^{-i\omega t} dt$$

Inserting the identity $\cos(x) = (e^{ix} + e^{-ix})/2$ we can simplify this integral as

$$F(\omega) = \frac{1}{\sqrt{2\pi}} \int_{-\infty}^{\infty} \left(\frac{e^{i\omega_0 t} + e^{-i\omega_0 t}}{2} \right) e^{-i\omega t} dt$$

$$= \frac{1}{\sqrt{2\pi}} \left(\frac{1}{2} \int_{-\infty}^{\infty} e^{-i(\omega - \omega_0)t} dt + \frac{1}{2} \int_{-\infty}^{\infty} e^{-i(\omega + \omega_0)t} dt \right)$$

Recall that one of many definitions of the Dirac delta function is[34]

$$\delta(\omega) = \frac{1}{2\pi} \int_{-\infty}^{\infty} e^{-i\omega t} dt$$

[34] This is the Fourier transform of unity as shown in S_9 #7.

This identity allows us to reform our result as

$$F(\omega) = \sqrt{\frac{\pi}{2}} \left(\frac{1}{2\pi} \int_{-\infty}^{\infty} e^{-i(\omega - \omega_0)t} dt + \frac{1}{2\pi} \int_{-\infty}^{\infty} e^{-i(\omega + \omega_0)t} dt \right)$$

$$= \sqrt{\frac{\pi}{2}} \left(\delta(\omega - \omega_0) + \delta(\omega + \omega_0) \right) \tag{13.67}$$

The cosine function $f(t)$ and its Fourier transform $F(\omega)$ are shown in Figure 13.24 and recorded in Supplement S_9 #10.

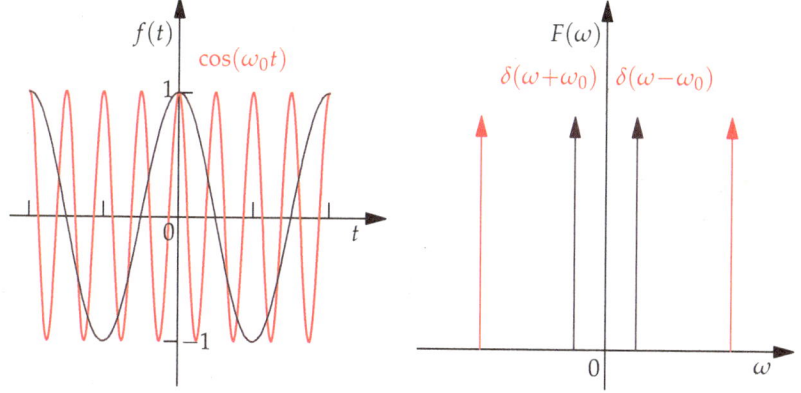

Figure 13.24: Comparison of sinusoidal function $f(t) = \cos(\omega_0 t)$ and its Fourier transform complement
$$F(\omega) = \sqrt{\frac{\pi}{2}} \left(\delta(\omega - \omega_0) + \delta(\omega + \omega_0) \right)$$
for two values of ω_0. A darker shade of red indicates a smaller value of ω.

The cosine function has a single characteristic frequency ω_0 leading to a single representative harmonic frequency in the Fourier transform. As the Fourier transform cannot differentiate between positive and negative frequencies, the Fourier transform of the cosine function with characteristic frequency ω_0 results in two delta functions forming a symmetric pair of peaks at ω_0 and $-\omega_0$. When the frequency of oscillation is high, the spacing between the delta functions is large. As the frequency is reduced, the peaks grow closer together. As the frequency approaches zero, the delta functions approach the origin.

13.2.8 Fourier transform of a damped oscillator

We have considered the Fourier transform of the exponential function, Equation 13.65, and the Fourier transform of the cosine function, Equation 13.67. Now consider the exponentially damped sinusoidal function

$$f(t) = e^{-a|t|}\cos(\omega_0 t) \tag{13.68}$$

where the Fourier transform is defined

$$F(\omega) = \frac{1}{\sqrt{2\pi}}\int_{-\infty}^{\infty} f(t)e^{-i\omega t}dt = \frac{1}{\sqrt{2\pi}}\int_{-\infty}^{\infty} e^{-a|t|}\cos(\omega_0 t)e^{-i\omega t}\,dt$$

$$= \frac{1}{\sqrt{2\pi}}\int_{-\infty}^{\infty} e^{-a|t|}\cos(\omega_0 t)e^{-i\omega t}\,dt$$

Employing the identity $\cos(x) = \frac{1}{2}\left(e^{ix} + e^{-ix}\right)$ we find

$$\int_{-\infty}^{\infty} e^{-a|t|}\cos(\omega_0 t)e^{-i\omega t}\,dt = \frac{1}{2}\int_{-\infty}^{\infty} e^{-a|t|}e^{-i(\omega-\omega_0)t}\,dt + \frac{1}{2}\int_{-\infty}^{\infty} e^{-a|t|}e^{-i(\omega+\omega_0)t}\,dt$$

We eliminate the absolute value by dividing each integral into two integrals, one from $-\infty$ to 0 and another from 0 to ∞. This results in the equation

$$\int_{-\infty}^{\infty} e^{-a|t|}\cos(\omega_0 t)e^{-i\omega t}\,dt = \frac{1}{2}\int_{-\infty}^{0} e^{at}e^{-i(\omega-\omega_0)t}\,dt + \frac{1}{2}\int_{0}^{\infty} e^{-at}e^{-i(\omega-\omega_0)t}\,dt$$

$$+ \frac{1}{2}\int_{-\infty}^{0} e^{at}e^{-i(\omega+\omega_0)t}\,dt + \frac{1}{2}\int_{0}^{\infty} e^{-at}e^{-i(\omega+\omega_0)t}\,dt$$

$$= \frac{1}{2}\int_{-\infty}^{0} e^{(a-i(\omega-\omega_0))t}\,dt + \frac{1}{2}\int_{0}^{\infty} e^{-(a+i(\omega-\omega_0))t}\,dt$$

$$+ \frac{1}{2}\int_{-\infty}^{0} e^{(a-i(\omega+\omega_0))t}\,dt + \frac{1}{2}\int_{0}^{\infty} e^{-(a+i(\omega+\omega_0))t}\,dt$$

Evaluating each of the four integrals leads to

$$\int_{-\infty}^{\infty} e^{-a|t|}\cos(\omega_0 t)e^{-i\omega t}\,dt = \frac{1}{a - i(\omega-\omega_0)} + \frac{1}{a + i(\omega-\omega_0)}$$

$$+ \frac{1}{a - i(\omega+\omega_0)} + \frac{1}{a + i(\omega+\omega_0)}$$

Note that the top two terms differ from the bottom two terms solely by the sign of the ω_0 term in the denominator. We can further simplify this result by multiplying and dividing each by its complex conjugate. This creates real common denominators, which allows us to sum terms. We find

$$\frac{1}{a-i(\omega\pm\omega_0)}\frac{a+i(\omega\pm\omega_0)}{a+i(\omega\pm\omega_0)} + \frac{1}{a+i(\omega\pm\omega_0)}\frac{a-i(\omega\pm\omega_0)}{a-i(\omega\pm\omega_0)} = \frac{2a}{a^2 + (\omega\pm\omega_0)^2}$$

Combining these terms leads to the final result

$$F(\omega) = \sqrt{\frac{2}{\pi}}\left[\left(\frac{a/2}{a^2 + (\omega+\omega_0)^2}\right) + \left(\frac{a/2}{a^2 + (\omega-\omega_0)^2}\right)\right] \tag{13.69}$$

The exponentially damped cosine function $f(t)$ and its Fourier transform are shown in Figure 13.25 and recorded in Supplement S_9 #12.

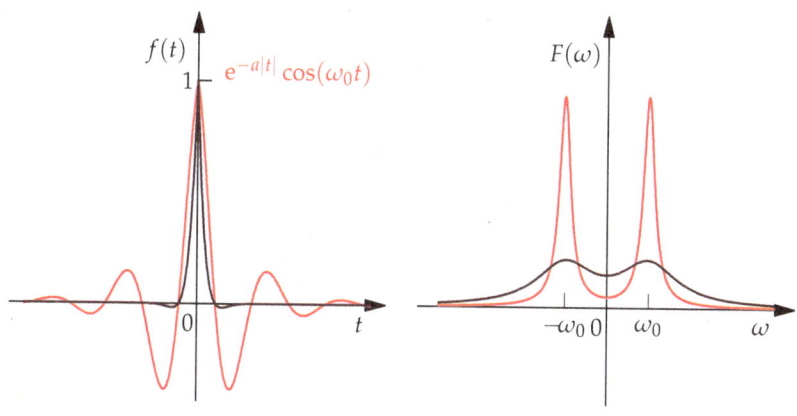

Figure 13.25: Comparison of exponentially damped sinusoidal function $f(t) = e^{-a|t|}\cos(\omega_0 t)$ and its Fourier transform complement $F(\omega) = \sqrt{\frac{2}{\pi}}\left[\left(\frac{a/2}{a^2+(\omega+\omega_0)^2}\right) + \left(\frac{a/2}{a^2+(\omega-\omega_0)^2}\right)\right]$ for two values of a. A darker shade of red indicates a larger value of a.

The cosine function has a single characteristic frequency ω_0 leading to peaks in its transform at $\pm\omega_0$. The exponential decay leads to a broadened line. The more rapid the decay, the broader the line. The result is a Fourier transform with peaks at $\pm\omega_0$ that have widths proportional to the rate of decay a.

13.2.9 Fourier transformation turns differentiation into multiplication

Consider the Fourier transform of a differentiable function $f(x)$:

$$F(k) = \int_{-\infty}^{\infty} f(x)e^{-ikx}\,dx \qquad (13.70)$$

What is the Fourier transform of the derivative of $f(x)$ with respect to x? Starting from

$$\int_{-\infty}^{\infty} \frac{d}{dx}f(x)e^{-ikx}\,dx$$

we can integrate by parts to find

$$\int_{-\infty}^{\infty} \frac{d}{dx}f(x)e^{-ikx}\,dx = f(x)e^{-ikx}\Big|_{-\infty}^{\infty} + ik\int_{-\infty}^{\infty} f(x)e^{-ikx}\,dx$$

The first term is zero if

$$\lim_{x\to\pm\infty} f(x) = 0$$

The second term is the Fourier transform of $f(x)$ times a factor ik. The final result is

$$\int_{-\infty}^{\infty} \frac{d}{dx}f(x)e^{-ikx}\,dx = ikF(k) \qquad (13.71)$$

which is recorded in Supplement S_9 #13. We see that the Fourier transform turns differentiation into multiplication.

The Fourier transform is a powerful tool for problem-solving and analysis in the physical sciences. This survey provides and strong foundation for the application of Fourier transforms in the solution of differential equations, the formulation of quantum theory, and signal processing and analysis.

A_{13} Orthogonal vectors and orthogonal functions

Any vector in three-dimensional space can be expressed as a weighted sum over orthonormal unit vectors multiplied by scalar coefficients. We call the unit vectors *basis vectors*, as they provide a basis for the representation of any vector in three-dimensional space. Functions of continuous variables can be expressed as a weighted sum over continuous orthonormal *basis functions*. The properties of basis vectors and commonly used basis functions are explored below.

Orthonormal sets of vectors

In Chapter 4 we considered the unit vectors $\hat{\mathbf{x}}$, $\hat{\mathbf{y}}$, and $\hat{\mathbf{z}}$. We found that they were *normalized*, meaning that

$$\hat{\mathbf{x}} \cdot \hat{\mathbf{x}} = 1 \qquad \hat{\mathbf{y}} \cdot \hat{\mathbf{y}} = 1 \qquad \hat{\mathbf{z}} \cdot \hat{\mathbf{z}} = 1$$

and *orthogonal*, meaning that

$$\hat{\mathbf{x}} \cdot \hat{\mathbf{y}} = \hat{\mathbf{y}} \cdot \hat{\mathbf{z}} = \hat{\mathbf{z}} \cdot \hat{\mathbf{x}} = 0$$

As such, we refer to $\hat{\mathbf{x}}$, $\hat{\mathbf{y}}$, and $\hat{\mathbf{z}}$ as orthonormal vectors.

In three-dimensional space, any vector can be written in terms of the three unit vectors $\hat{\mathbf{x}}$, $\hat{\mathbf{y}}$, and $\hat{\mathbf{z}}$ as

$$\mathbf{a} = a_x\hat{\mathbf{x}} + a_y\hat{\mathbf{y}} + a_z\hat{\mathbf{z}}$$

The scalar coefficients a_x, a_y, and a_z are defined in terms of the projection of the vector $\mathbf{a}$ on each of the unit vectors, where

$$a_x = \mathbf{a} \cdot \hat{\mathbf{x}} \qquad a_y = \mathbf{a} \cdot \hat{\mathbf{y}} \qquad a_z = \mathbf{a} \cdot \hat{\mathbf{z}}$$

Figure 13.26 depicts the relationship between a vector $\mathbf{a}$ and the scalar coefficients a_x, a_y, and a_z.

Suppose we define the unit vectors $\mathbf{e}_n$ where $\mathbf{e}_1 = \hat{\mathbf{x}}$, $\mathbf{e}_2 = \hat{\mathbf{y}}$, and $\mathbf{e}_3 = \hat{\mathbf{z}}$. Then we can state the orthonormality condition as[35]

$$\mathbf{e}_n \cdot \mathbf{e}_m = \delta_{nm}$$

Any vector $\mathbf{a}$ can be represented in terms of the unit vectors as

$$\mathbf{a} = \sum_{n=1}^{3} a_n\mathbf{e}_n$$

where the scalar coefficients are defined[36]

$$a_n = \mathbf{a} \cdot \mathbf{e}_n$$

In the sections that follow we will explore extensions of this idea to orthonormal functions of continuous variables.

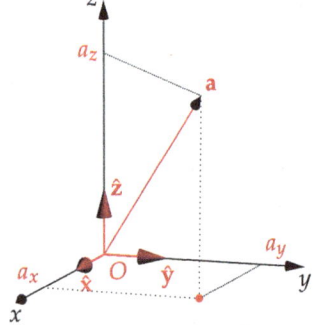

Figure 13.26: The vector **a** shown with its individual scalar components a_x, a_y, and a_z in the three-dimensional cartesian coordinate system. Also shown are the unit vectors $\hat{\mathbf{x}}$, $\hat{\mathbf{y}}$, and $\hat{\mathbf{z}}$.

[35] The Kronecker delta function is defined

$$\delta_{nm} = \begin{cases} 1 & n = m \\ 0 & n \neq m \end{cases}$$

Its properties are explored in Chapter 7.

[36] The three-dimensional space in which all points are represented by an ordered triple of real numbers (x, y, z) is a *vector space*. Any vector resulting from the addition of vectors in the vector space or the scalar multiplication of a vector in that vector space is also in the vector space.

Orthonormal sets of sinusoidal functions

In this chapter, we found that many functions can be accurately expressed in terms of a weighted sum of sine and cosine functions. The sine and cosine functions are orthogonal and normalized. The orthonormality conditions are written

$$\frac{1}{\pi} \int_{-\pi}^{\pi} \cos(nx) \cos(mx) dx = \delta_{nm} \qquad \frac{1}{\pi} \int_{-\pi}^{\pi} \sin(nx) \sin(mx) dx = \delta_{nm}$$

where δ_{nm} is the Kronecker delta function. In addition, we find

$$\frac{1}{\pi} \int_{-\pi}^{\pi} \cos(nx) \sin(mx) dx = 0$$

for any n and m. The properties of orthonormality are presented graphically in Figure 13.27.

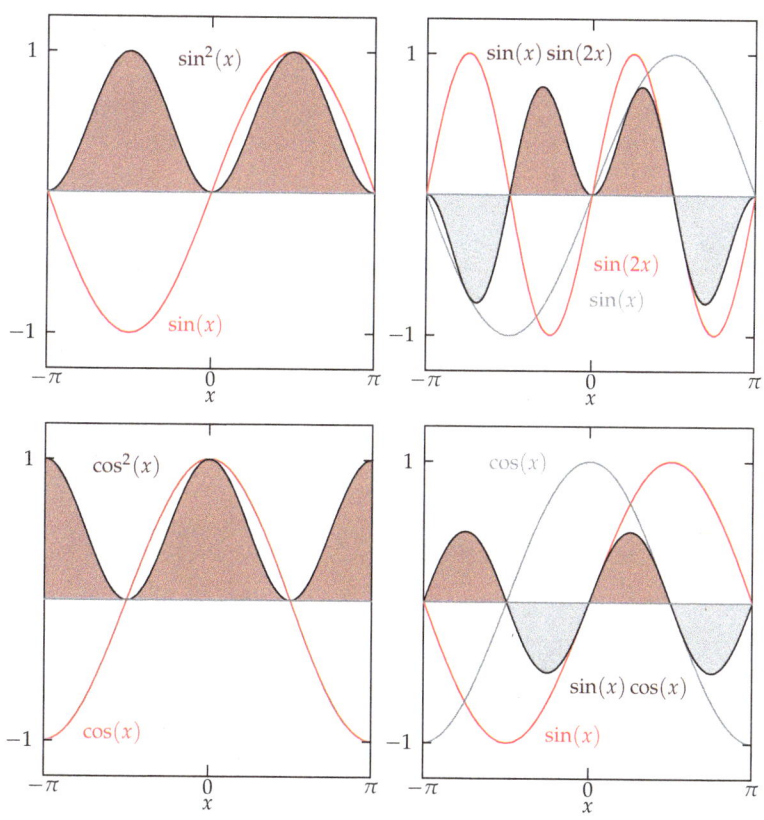

Figure 13.27: Visualizing the property of orthogonality for continuous functions. Variation in the normalized sine and cosine functions over the range $x \in [-\pi, \pi]$ (left). The integral over the product formed by two orthogonal functions equals zero (right), shown as the sum of positive (red) and negative (gray) areas.

We can express any function $f(x)$ in terms of a Fourier series

$$f(x) = \frac{a_0}{2} + \sum_{n=1}^{\infty} a_n \cos(nx) + \sum_{n=1}^{\infty} b_n \sin(nx)$$

where the coefficients are determined using the integral relations provided by

Equations 13.2, 13.4, and 13.6 and collected here

$$a_n = \frac{1}{\pi} \int_{-\pi}^{\pi} f(x) \cos(nx)\, dx \qquad b_n = \frac{1}{\pi} \int_{-\pi}^{\pi} f(x) \sin(nx)\, dx$$

In Figure 13.27, we compare the normalized sine and cosine basis functions (left) and products of two orthogonal functions (right). When vectors in three-dimensional space are orthogonal, they are perpendicular. When two continuous functions are orthogonal, the integral over their product is zero.

Orthonormal sets of Hermite, Laguerre, and Legendre polynomial functions

In Chapter 11, we found the Hermite polynomials, Laguerre polynomials, and Legendre polynomials to be solutions to particular second order ordinary differential equations. In the case of Fourier series, we expressed a given function in terms of a sum of orthonormal sine and cosine functions. In a similar manner, we can express a given function in terms of a sum of orthonormal functions based on the Hermite polynomials, Laguerre polynomials, or Legendre polynomials.

Hermite polynomials: The orthonormality condition for the *Hermite polynomials* is written

$$\frac{1}{\sqrt{\pi}\, 2^n n!} \int_{-\infty}^{\infty} H_n(x)\, H_m(x) e^{-x^2} dx = \delta_{nm}$$

where the first five Hermite polynomials are listed below.

$$H_0(x) = 1$$

$$H_1(x) = 2x$$

$$H_2(x) = 4x^2 - 2$$

$$H_3(x) = 8x^3 - 12x$$

$$H_4(x) = 16x^4 - 48x^2 + 12$$

The gaussian e^{-x^2} is the *weighting function*. Figure 13.28 shows the first five Hermite polynomials, each multiplied by the square root of the weighting function $e^{-x^2/2}$.

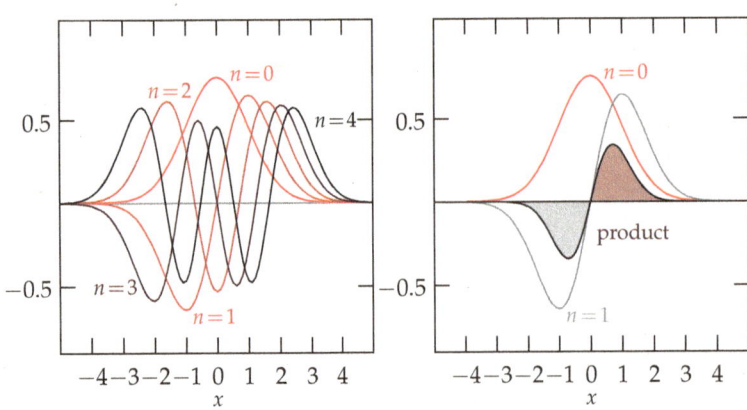

Figure 13.28: Variation in the first five weighted and normalized Hermite polynomials proportional to $e^{-\frac{1}{2}x^2} H_n(x)$ over the range $x \in [-5, 5]$ (left). The equal areas of opposite sign observed in the product of the first two polynomials demonstrates the orthogonality of those functions (right).

The integral defining orthonormality includes a product of two Hermite polynomials and the gaussian weighting function. Note that n is the number of nodes or zeros. The Hermite polynomials are used to represent the wave function of the quantum harmonic oscillator.

Laguerre polynomials: The orthonormality condition for the *Laguerre polynomials* is written

$$\int_0^\infty L_n(x)\, L_m(x) e^{-x} dx = \delta_{nm}$$

where the first five Laguerre polynomials are listed below.

$$L_0(x) = 1$$

$$L_1(x) = -x + 1$$

$$L_2(x) = \frac{1}{2}(x^2 - 4x + 2)$$

$$L_3(x) = \frac{1}{6}(-x^3 + 9x^2 - 18x + 6)$$

$$L_4(x) = \frac{1}{24}(x^4 - 16x^3 + 72x^2 - 96x + 24)$$

The exponential e^{-x} is the *weighting function*. The integral defining orthonormality includes a product of two Laguerre polynomials and the exponential weighting function. Figure 13.29 shows the first five Laguerre polynomials, each multiplied by the square root of the weighting function, $e^{-x/2}$.

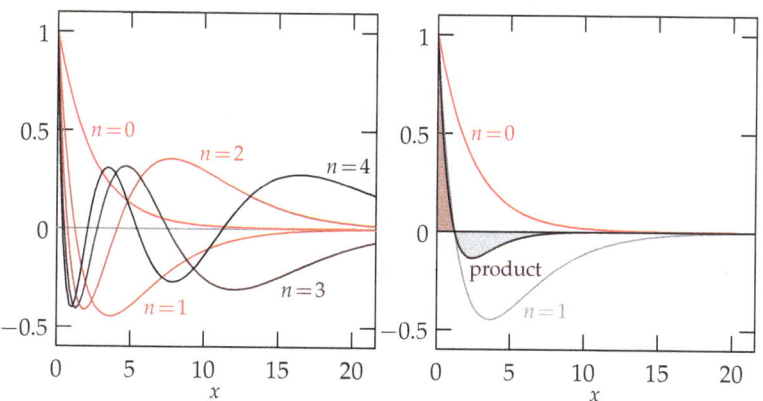

Figure 13.29: Variation in the first five weighted and normalized Laguerre polynomials proportional to $e^{-\frac{1}{2}x} L_n(x)$ over the range $x \in [0, 22]$ (left). The equal areas of opposite sign observed in the product of the first two polynomials demonstrates the orthogonality of those functions (right).

The Laguerre polynomials are used to represent the radial component of the wave function of the quantum one-electron atom.

Legendre polynomials: The orthonormality condition for the *Legendre polynomials* is written

$$\frac{2n+1}{2} \int_{-1}^1 P_n(x)\, P_m(x) dx = \delta_{nm}$$

where the first five Legendre polynomials are listed below.

$$P_0(x) = 1$$

$$P_1(x) = x$$

$$P_2(x) = \frac{1}{2}(3x^2 - 1)$$

$$P_3(x) = \frac{1}{2}(5x^3 - 3x)$$

$$P_4(x) = \frac{1}{8}(35x^4 - 30x^2 + 3)$$

The *weighting function* is unity. The integral defining orthonormality includes a product of two Legendre polynomials and a weighting function of unity. Figure 13.30 shows the first five Legendre polynomials, each multiplied by the square root of the weighting function 1.

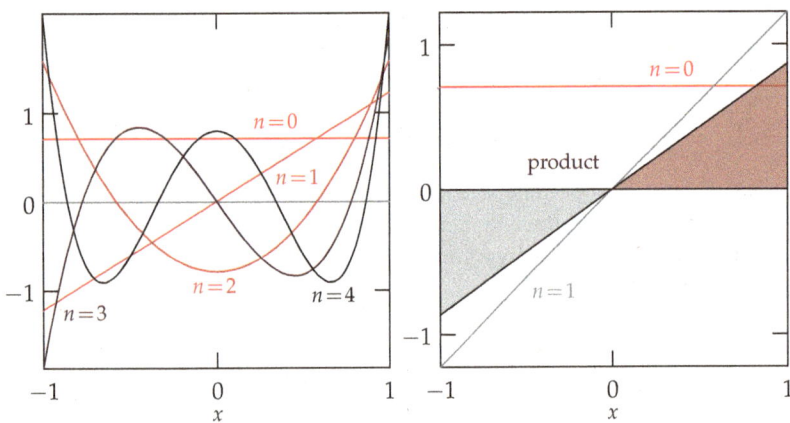

Figure 13.30: Variation in the first five normalized Legendre polynomials $P_n(x)$ over the range $x \in [-1, 1]$ (left). The equal areas of opposite sign observed in the product of the first two polynomials demonstrates the orthogonality of those functions (right).

Almost any function $f(x)$, defined over the interval $x \in [-1, 1]$, can be expanded as an infinite series known as the *Fourier-Legendre series*.[37] Using the set of orthogonal Legendre polynomial functions, we can express $f(x)$ as

$$f(x) = \sum_{n=0}^{\infty} a_n P_n(x)$$

where the coefficients a_n are defined as

$$a_n = \frac{2n+1}{2} \int_{-1}^{1} f(x)\, P_n(x)\, dx$$

The relative contribution of each Legendre polynomial is given by the coefficient a_n. Just as one carries out *harmonic analysis* using Fourier transforms, one can analyze functions $f(x)$, defined over the interval $[-1, 1]$, by decomposing the function in terms of Legendre polynomials $P_n(x)$.

The Legendre polynomials described above are also known as the unassociated Legendre polynomials. A related set of orthogonal polynomial functions are the *associated Legendre polynomials*. The associated Legendre polynomial functions can be generated from the Legendre polynomial func-

[37] As in the case of Fourier series, we refer to those functions $f(x)$ with a finite number of discontinuities for which

$$\int_{-1}^{1} |f(x)|\, dx$$

is finite.

tions through the following relation:

$$P_n^m(x) = (-1)^m (1-x^2)^{m/2} \frac{d^m}{dx^m} P_n(x) \qquad (13.72)$$

In this expression, n is a positive integer and $m = 0, 1, \ldots, n$. The first six associated Legendre polynomials are listed below.

$$P_0^0(x) = 1$$

$$P_1^0(x) = x \qquad\qquad P_1^1(x) = -\sqrt{1-x^2}$$

$$P_2^0(x) = \frac{1}{2}\left(3x^2 - 1\right) \quad P_2^1(x) = -3x\sqrt{1-x^2} \quad P_2^2(x) = 3\left(1-x^2\right)$$

The associated Legendre polynomial functions are commonly expressed in terms of $x = \cos(\theta)$ as shown below.

$$P_0^0(\cos(\theta)) = 1$$

$$P_1^0(\cos(\theta)) = \cos(\theta) \qquad\qquad P_1^1(\cos(\theta)) = -\sin(\theta)$$

$$P_2^0(\cos(\theta)) = \frac{1}{2}\left(3\cos^2(\theta) - 1\right) \quad P_2^1(\cos(\theta)) = -3\cos(\theta)\sin(\theta) \quad P_2^2(\cos(\theta)) = 3\sin^2(\theta)$$

The associated Legendre polynomials form a component of the spherical harmonic functions. The spherical harmonic functions are discussed in the next section.[38]

Orthonormal sets of spherical harmonic functions

The *spherical harmonic functions*, $Y_l^m(\theta, \varphi)$, are solutions to Laplace's equation in spherical coordinates. The associated Legendre polynomials, $P_l^m(\cos(\theta))$, are related to the spherical harmonic functions as follows:

$$Y_l^m(\theta, \varphi) = \sqrt{\frac{2l+1}{4\pi} \frac{(l-m)!}{(l+m)!}}\, P_l^m(\cos(\theta))\, e^{im\varphi} \qquad (13.73)$$

The normalization factor is chosen to satisfy the following orthonormality condition:

$$\int_0^{2\pi} \int_0^{\pi} Y_{l,m}(\theta, \varphi)\, Y_{l',m'}^*(\theta, \varphi)\sin(\theta)d\theta d\varphi = \delta_{l,l'}\delta_{m,m'}$$

Note that the spherical harmonic for specific l and m can be written equivalently as $Y_l^m(\theta, \varphi)$ or $Y_{l,m}(\theta, \varphi)$. The latter form is more convenient when representing the *complex conjugate* of the function as $Y_{l,m}^*(\theta, \varphi)$.

The first six spherical harmonic functions are shown in Figure 13.31 where the red and white surfaces indicate the positive and negative sign of the function for the specific value of (θ, φ).

[38] The associated Legendre polynomials are defined for negative values of m as

$$P_n^{-m}(x) = (-1)^m \frac{(n-m)!}{(n+m)!} P_n^m(x)$$

where $m = 0, 1, \ldots, n$.

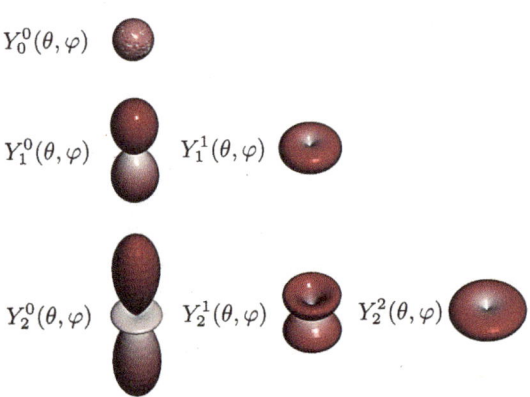

$Y_0^0(\theta, \varphi)$

$Y_1^0(\theta, \varphi)$ $Y_1^1(\theta, \varphi)$

$Y_2^0(\theta, \varphi)$ $Y_2^1(\theta, \varphi)$ $Y_2^2(\theta, \varphi)$

Figure 13.31: Variation in the normalized spherical harmonic functions $Y_l^m(\theta, \varphi)$ over the range $\theta \in [0, \pi]$ and $\varphi \in [0, 2\pi)$. Change in color indicates a change in the sign of the function.

The functions graphically depicted above are defined by the equations below. The integral defining orthonormality includes a product of one spherical harmonic function and the complex conjugate of a second spherical harmonic function. That combination ensures that the integral is real.

$$Y_0^0 = \frac{1}{2}\sqrt{\frac{1}{\pi}}$$

$$Y_1^0 = \frac{1}{2}\sqrt{\frac{3}{\pi}}\cos(\theta) \qquad Y_1^{\pm 1} = \mp\frac{1}{2}\sqrt{\frac{3}{2\pi}}\sin(\theta)e^{\pm i\varphi}$$

$$Y_2^0 = \frac{1}{4}\sqrt{\frac{5}{\pi}}(3\cos^2(\theta)-1) \quad Y_2^{\pm 1} = \mp\frac{1}{2}\sqrt{\frac{15}{2\pi}}\sin(\theta)\cos(\theta)e^{\pm i\varphi} \quad Y_2^{\pm 2} = \frac{1}{4}\sqrt{\frac{15}{2\pi}}\sin^2(\theta)e^{\pm 2i\varphi}$$

To perceive the orthogonality of these functions, imagine overlapping $Y_0^0(\theta, \varphi)$, which is even over all space, with $Y_1^0(\theta, \varphi)$, which is an odd function of z (see variation in sign between the two lobes). An integral over all space of the product of the two functions will therefore be zero, making the functions orthogonal.

Any real function $f(\theta, \varphi)$ can be expanded as an infinite series known as the *Laplace series* using the set of orthogonal spherical harmonic functions

$$f(\theta, \varphi) = \sum_{l=0}^{\infty}\sum_{m=0}^{l} f_{l,m}Y_{l,m}(\theta, \varphi)$$

where the coefficients $f_{l,m}$ are defined as

$$f_{l,m} = \int_0^{2\pi}\int_0^{\pi} f(\theta, \varphi)\, Y_{l,m}^*(\theta, \varphi)\sin(\theta)d\theta d\varphi$$

Just as one carries out *harmonic analysis* using Fourier transforms, one can carry out *spherical harmonic analysis* in decomposing an angular distribution $f(\theta, \varphi)$ in terms of spherical harmonics $Y_{l,m}(\theta, \varphi)$, where the relative contribution of each spherical harmonic is given by the coefficient $f_{l,m}$.

The spherical harmonic functions are the angular component of the solutions of *Laplace's equation*

$$\nabla^2 f(r, \theta, \varphi) = 0$$

in spherical polar coordinates.[39] In addition to forming a complete set of basis functions for the representation of angular functions $f(\theta, \varphi)$, the spherical harmonics are commonly used to represent the multipole expansion of the electrostatic potential, the fluctuations in the surface of a spherical droplet, and the distribution of electrons in atomic orbital electron configurations.

[39] The spherical harmonics were introduced by French mathematician, physicist, and astronomer Pierre-Simon Laplace (1749-1827).

B$_{13}$ Building orthogonal polynomials using Gram-Schmidt orthogonalization

We discovered sets of orthogonal functions, including the Hermite, Laguerre, and Legendre polynomials, through the solution of differential equations. Here we explore an alternative way to generate an infinite set of orthogonal polynomials, by generalizing the Gram-Schmidt orthogonalization procedure, introduced in Chapter 4 for vectors, to the orthogonalization of functions.

In this procedure, we start by defining the first polynomial function, which is often taken to be a constant

$$\varphi_0 = c_1$$

The one constant c_1 is determined using the one constraint that the function meets some normalization condition.[40] We then build a second polynomial function, often taken to be a linear function

$$\varphi_1 = c_2\, x + c_3$$

The two constants c_2 and c_3 are determined using the two constraints that the function is orthogonal to φ_0 and normalized. Then we the build a third polynomial, often taken to be a quadratic function

$$\varphi_2(x) = c_4\, x^2 + c_5\, x + c_6$$

The three constants c_4, c_5, and c_6 are determined using the three constraints that the function is orthogonal to φ_0, orthogonal to φ_1, and normalized.

The condition that our polynomial functions are normalized is written as

$$\int r(x)\, \varphi_n^2(x)\, dx = 1$$

The function $r(x)$ is an integrating factor that is naturally combined with the polynomial function. The condition that our polynomials are orthogonal is written as[41]

$$\int r(x)\, \varphi_n(x)\varphi_m(x)\, dx = 0 \qquad n \neq m$$

The general form for our polynomial functions is taken to be

$$\varphi_n(x) = \sum_{k=0}^{n} c_{i_n+k}\, x^k$$

[40] We have often considered cases where the integral over a function is set to unity. However, a more general definition of normalization is simply that the integral over the function has a well defined value, which need not be unity. Normalization is regularization.

[41] Overall, the condition that the polynomial functions are normalized and mutually orthogonal can be expressed as

$$\int r(x)\, \varphi_n(x)\varphi_m(x)\, dx = \delta_{nm}$$

where δ_{nm} is the Kronecker delta function.

where $i_n = \frac{1}{2}n^2 + \frac{1}{2}n + 1$. This formula generates a series of polynomials of increasing order:

$$\varphi_0(x) = c_1$$

$$\varphi_1(x) = c_2\, x + c_3$$

$$\varphi_2(x) = c_4\, x^2 + c_5\, x + c_6$$

$$\varphi_3(x) = c_7\, x^3 + c_8\, x^2 + c_9\, x + c_{10}$$

By imposing our orthonormality condition, we can determine the unknown constants and define the infinite set of orthogonal polynomial functions.

In building each new polynomial function $\varphi_n(x)$ of degree n, we have $n+1$ constants to determine. Fortunately, we have n constraints from orthogonality and 1 constraint from normalization, providing the $n+1$ constraints needed to determine the $n+1$ unknown constants.

What makes one set of polynomials different from another is the choice of the integrating factor $r(x)$. For example, for the Legendre polynomials, the integrating factor is a constant

$$r(x) = 1$$

For the Laguerre polynomials, the integrating factor is an exponential function

$$r(x) = e^{-x}$$

And for the Hermite polynomials, the integrating factor $r(x)$ is a gaussian function

$$r(x) = e^{-x^2}$$

Each set of orthogonal polynomials is defined by its unique integrating factor.

In practice, it is convenient to first set the constant $c_1 = 1$ so that $\varphi_0(x) = 1$. We then build the full set of orthogonal polynomials from that starting point. In addition, rather than imposing the normalization condition at each step, we sometimes demand that the polynomial function have a certain value at $x = 1$:

$$\varphi_n(1) = 1$$

Only at the end do we impose the normalization condition. This is more practical than imposing the normalization condition at each step, and leads to the same final result.

Let's see how this works. We start with our choice that $c_0 = 1$. We can write the second polynomial using the general form $\varphi_1(x) = c_2 x + c_3$. We want to chose the constants c_2 and c_3 such that $\varphi_0(x)$ and $\varphi_1(x)$ are orthogonal. As such, we demand that

$$\int r(x)\varphi_0(x)\varphi_1(x)\,dx = \int r(x)(c_2 x + c_3)\,dx = 0$$

$$= \int r(x)c_2 x\,dx + \int r(x)c_3\,dx = 0$$

For the second constraint, we take $\varphi_1(1) = 1$. Combining the two constraint equations we can solve for c_1 and c_2 and therefore define $\varphi_1(x)$. This procedure is then repeated to determine $\varphi_2(x)$, $\varphi_3(x)$, and so on.

Let's apply the Gram-Schmidt orthogonalization procedure to a specific example. Suppose we chose the integrating factor $r(x) = 1$, where the range of integration is

$$-1 \leq x \leq 1$$

In determining each new polynomial function, we impose the orthogonality condition and the condition the $\varphi_n(1) = 1$ for all n. To determine $\varphi_0(x) = c_1$, we demand that $\varphi_0(1) = c_1 = 1$. So $c_1 = 1$ and $\varphi_0(x) = 1$.

To determine $\varphi_1(x) = c_2 x + c_3$, we first impose the orthogonality condition

$$\int_{-1}^{1} \varphi_0(x)\varphi_1(x)\,dx = \int_{-1}^{1} (c_2 x + c_3)\,dx$$

$$= \int_{-1}^{1} c_2 x\,dx + \int_{-1}^{1} c_3\,dx = 2c_3 = 0$$

This fixes $c_3 = 0$ so that $\varphi_1(x) = c_2 x$. Imposing the condition $\varphi_1(1) = c_2 = 1$, we determine the second constant $c_2 = 1$. As such, $\varphi_1(x) = x$.

To determine $\varphi_2(x)$, we demand

$$\int_{-1}^{1} \varphi_0(x)\varphi_2(x)\,dx = 0 \qquad \int_{-1}^{1} \varphi_1(x)\varphi_2(x)\,dx = 0$$

The first integral is evaluated as

$$\int_{-1}^{1} \varphi_0(x)\varphi_2(x)\,dx = \int_{-1}^{1} (c_4 x^2 + c_5 x + c_6)\,dx = \frac{2}{3}c_4 + 2c_6 = 0$$

The second integral is evaluated as

$$\int_{-1}^{1} \varphi_1(x)\varphi_2(x)\,dx = \int_{-1}^{1} x(c_4 x^2 + c_5 x + c_6)\,dx$$

$$= \int_{-1}^{1} (c_4 x^3 + c_5 x^2 + c_6 x)\,dx = \frac{2}{3}c_5 = 0$$

Finally, we have the condition

$$\varphi_2(1) = c_4 + c_5 + c_6 = 1$$

We now have three equations with three unknowns. We find $c_5 = 0$. We have $c_4 + c_6 = 1$, so that $c_6 = 1 - c_4$. Substituting into the third equation we find

$$\frac{2}{3}c_4 + 2c_6 = \frac{2}{3}c_4 + 2(1 - c_4) = -\frac{4}{3}c_4 + 2 = 1$$

so that $c_4 = \frac{3}{2}$. Finally, returning to

$$\varphi_2(1) = c_4 + c_5 + c_6 = \frac{3}{2} + c_6 = 1$$

we find $c_6 = -\frac{1}{2}$, leading to our final result that

$$\varphi_2(x) = \frac{1}{2}\left(3x^2 - 1\right)$$

This polynomial function has broad application in the physical sciences.

Continuing to apply this procedure, we find that

$$\varphi_3(x) = \frac{1}{2}\left(5x^3 - 3x\right)$$

and that

$$\varphi_4(x) = \frac{1}{8}\left(35x^4 - 30x^2 + 3\right)$$

and on to infinity. We recognize this set of polynomial functions to be the Legendre polynomials:

$$\varphi_0(x) = 1$$

$$\varphi_1(x) = x$$

$$\varphi_2(x) = \frac{1}{2}\left(3x^2 - 1\right)$$

$$\varphi_3(x) = \frac{1}{2}\left(5x^3 - 3x\right)$$

$$\varphi_4(x) = \frac{1}{8}\left(35x^4 - 30x^2 + 3\right)$$

It is remarkable that these polynomial functions, which we first discovered in Chapter 11 as solutions to the second order ordinary differential equation known as Legendre's equation, can be generated using the Gram-Schmidt orthogonalization procedure by assuming that the integrating factor is

$$r(x) = 1$$

and the range of integration is $-1 \leq x \leq 1$.

Finally, let's evaluate the integrals

$$\int_{-1}^{1} \varphi_n^2(x)\, dx$$

to determine the normalization condition. We note that

$$\int_{-1}^{1} \varphi_0^2(x)\, dx = \int_{-1}^{1} dx = 2$$

Integrating over $\varphi_1(x)$, we find

$$\int_{-1}^{1} \varphi_1^2(x)\, dx = \int_{-1}^{1} x^2\, dx = \frac{1}{3}x^3\Big|_{-1}^{1} = \frac{2}{3}$$

And when integrating over $\varphi_2(x)$, we find

$$\int_{-1}^{1} \varphi_2^2(x)\, dx = \int_{-1}^{1} \frac{1}{4}\left(3x^2 - 1\right)^2 dx = \int_{-1}^{1} \frac{1}{4}\left(9x^4 - 6x^2 + 1\right) dx$$

$$= \frac{1}{4}\left[\frac{9}{5}x^5 - \frac{6}{3}x^3 + x\right]_{-1}^{1} = \frac{9}{10} - 1 + \frac{1}{2} = \frac{2}{5}$$

In general, for this set of orthogonal polynomial functions:

$$\int_{-1}^{1} \varphi_n(x)\varphi_m(x)\, dx = \frac{2}{2n+1}\delta_{nm}$$

This serves as the normalization condition for the Legendre polynomials. In general, each set of orthogonal polynomials will have its unique normalization condition.

The Gram-Schmidt orthogonalization procedure can be followed using the integrating factor $r(x) = e^{-x}$ with $0 \leq x < \infty$, leading to the Laguerre polynomials, and $r(x) = e^{-x^2}$ with $-\infty < x < \infty$, leading to the Hermite polynomials. Those examples are explored in the exercises.

C$_{13}$ Deriving Fourier integral transforms from complex Fourier series

Let's derive the Fourier transform, defined by Equations 13.39 and 13.40, from the complex Fourier series. We start from the complex Fourier series over the range $-\frac{L}{2} \leq x < \frac{L}{2}$ originally defined in Equation 13.31 as

$$f(x) = \sum_{n=-\infty}^{\infty} c_n \exp\left(i\frac{2n\pi x}{L}\right)$$

with the coefficients defined in Equation 13.32 as

$$c_n = \frac{1}{L}\int_{-\frac{L}{2}}^{\frac{L}{2}} f(x)\exp\left(-i\frac{2n\pi x}{L}\right) dx$$

where n is an integer.

Let's define a new variable called the *wave number*:

$$k_n = \frac{2n\pi}{L} \tag{13.74}$$

The wave number k is the spatial frequency of the wave. It is defined in terms of the wavelength λ as

$$k = \frac{2\pi}{\lambda}$$

The wave number is the spatial analog of the angular frequency ω that is related to the period of oscillation T in time as

$$\omega = \frac{2\pi}{T}$$

The wave number is a measure of the number of waves per unit distance.

As k_n is a real number that depends on the integer index n, we write the coefficient c_n as a function of k_n, $c(k_n)$. The result is

$$c(k_n) = \frac{1}{L}\int_{-\frac{L}{2}}^{\frac{L}{2}} f(x)e^{-ik_n x}dx$$

so that our complex Fourier series is

$$f(x) = \sum_{n=-\infty}^{\infty} c(k_n)e^{ik_n x}$$

Note that the index n in the complex Fourier series extends from $-\infty$ to ∞. We would now like to convert the sum over n to an integral over k_n. This will allow us to transform functions using a continuous function of the wave number, rather than a discrete sum.

Noting that the increment in the sum is $\Delta n = (n_i + 1) - n_i = 1$, we write

$$f(x) = \sum_{n=-\infty}^{\infty} c(k_n) e^{ik_n x} = \sum_{n=-\infty}^{\infty} c(k_n) e^{ik_n x} \Delta n$$

Inserting the formula for $c(k_n)$, we have[42]

$$f(x) = \sum_{n=-\infty}^{\infty} \left(\frac{1}{L} \int_{-\frac{L}{2}}^{\frac{L}{2}} f(x') e^{-ik_n x'} dx' \right) e^{ik_n x} \Delta n$$

where the bracketed term is a function of k_n and independent of x. Finally, let's substitute[43]

$$1 = \Delta n = \frac{L}{2\pi} \Delta k_n$$

to find[44]

$$f(x) = \sum_{n=-\infty}^{\infty} \left(\frac{1}{2\pi} \int_{-\frac{L}{2}}^{\frac{L}{2}} f(x') e^{-ik_n x'} dx' \right) e^{ik_n x} \Delta k_n$$

Suppose we take the limit that $L \to \infty$. Our steps in k_n become infinitesimally small as $L \to \infty$ and

$$\Delta k_n = \frac{2\pi}{L} \to dk$$

As such, we can convert the sum to an integral as

$$\sum_{n=-\infty}^{\infty} c(k_n) e^{ik_n x} \Delta k_n \to \int_{-\infty}^{\infty} c(k) e^{ikx} dk$$

where

$$c(k) = \frac{1}{2\pi} \int_{-\infty}^{\infty} f(x) e^{-ikx} dx$$

Combining our results, we find the overall formula

$$f(x) = \int_{-\infty}^{\infty} \left(\frac{1}{2\pi} \int_{-\infty}^{\infty} f(x') e^{-ikx'} dx' \right) e^{ikx} dk$$

$$= \frac{1}{2\pi} \int_{-\infty}^{\infty} F(k) e^{ikx} dk$$

The inner integral transforms $f(x)$ from a function of x to a function of k, which we write $F(k)$. The outer integral reverses the transformation and returns our original function $f(x)$. The result is the *Fourier transform pair* defined as follows:[45]

$$F(k) = \int_{-\infty}^{\infty} f(x) e^{-ikx} dx$$

and

$$f(x) = \frac{1}{2\pi} \int_{-\infty}^{\infty} F(k) e^{ikx} dk$$

Here $F(k)$ is the *Fourier transform* of $f(x)$ and $f(x)$ is the *inverse Fourier transform* of $F(k)$.

[42] Note the difference between x' and x. We integrate over the variable x' to determine the coefficient $c(k_n)$ used in the expression for $f(x)$.

[43] This is equivalent to the derivative

$$\frac{dk_n}{dn} = \frac{2\pi}{L}$$

using Equation 13.74.

[44] Note that

$$\int_{-\frac{L}{2}}^{\frac{L}{2}} f(x) e^{-ik_n x} dx = \int_{-\frac{L}{2}}^{\frac{L}{2}} f(x') e^{-ik_n x'} dx'$$

[45] Integral transforms were first explored by the great *Leonhard Euler* (1707-1783). They were used by *Joseph Fourier* (1768-1830) to study heat flow in an unbounded semi-infinite domain as an extension of his Fourier series.

D$_{13}$ End-of-chapter problems

> The profound study of nature is the most fertile source of mathematical discoveries.
>
> Joseph Fourier

Warm-ups

13.1 Find the Fourier series representing the *periodic step* function

$$f(x) = \begin{cases} 0 & -\pi \leq x < 0 \\ 1 & 0 \leq x < \pi \end{cases}$$

defined over the range $-\pi \leq x < \pi$ and repeating with a period of 2π as shown below.

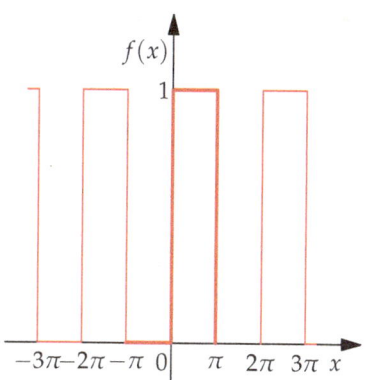

Compare your resulting coefficients with those of the square wave in Equation 13.12. Explain the observed differences in terms of the properties of the periodic step function and the square wave.

13.2 Consider the general Fourier series of Equation 13.7

$$f(x) = \frac{a_0}{2} + \sum_{n=1}^{\infty} a_n \cos(nx) + \sum_{n=1}^{\infty} b_n \sin(nx)$$

where the coefficients are defined by Equation 13.8

$$a_0 = \frac{1}{\pi} \int_{-\pi}^{\pi} f(x)\, dx$$

and

$$a_n = \frac{1}{\pi} \int_{-\pi}^{\pi} f(x) \cos(nx)\, dx \qquad b_n = \frac{1}{\pi} \int_{-\pi}^{\pi} f(x) \sin(nx)\, dx$$

Show that $a_n = 0$ for $n = 0, 1, 2, \ldots$ for any odd function $f_o(x)$ (sine transform). Similarly, show that $b_n = 0$ for $n = 1, 2, 3, \ldots$ for any even function $f_e(x)$ (cosine transform).

13.3 Find the Fourier series representing the periodic positive *triangle wave* function

$$f(x) = \begin{cases} \dfrac{x}{L}, & 0 \leq x < L \\[2ex] \dfrac{2L-x}{L}. & L \leq x < 2L \end{cases}$$

defined over the range $0 \leq x < 2L$ and repeating with a period of $2L$ as shown below.

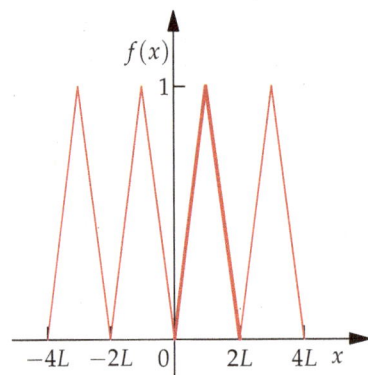

13.4 Use the Table of Fourier transform pairs in Supplement S_9 to determine the Fourier transform of the function

$$f(t) = 0.5e^{-t/10}\cos(12t) + 1.8e^{-t/4}\cos(20t) + 0.7e^{-3t/4}\cos(26t)$$

Sketch the form of the function $f(t)$ over the range $0 \leq t \leq 10$ and its Fourier transform $F(\omega)$ over the range $0 \leq \omega < 30$.

13.5 The Fourier transform of the damped cosine function $e^{-\alpha|t|}\cos(\omega_0 t)$ is a lorentzian function of the form

$$I(\omega) = \frac{\alpha}{\alpha^2 + (\omega - \omega_0)^2}$$

Determine the position of the maximum of $I(\omega)$, the maximum value of $I(\omega)$, and the width of $I(\omega)$ at half of its maximum value known as the *full width at half maximum* (FWHM).

13.6 Consider the function $f(t)$ with Fourier transform $F(\omega)$. Show that the Fourier transform of $f(t + t_0)$ is $F(\omega)e^{i\omega t_0}$.

Homework exercises

13.7 Consider the Fourier series of the periodic *sawtooth wave* function, Equation 13.21, over the range $-L \leq x < L$ where

$$f(x) = \frac{2}{\pi} \sum_{n=1}^{\infty} \frac{(-1)^{n+1}}{n} \sin\left(\frac{n\pi x}{L}\right)$$

Set $x = \frac{L}{2}$ and show that

$$\frac{\pi}{4} = 1 - \frac{1}{3} + \frac{1}{5} - \frac{1}{7} + \frac{1}{9} - \frac{1}{11} + - \cdots$$

which is *Leibniz's formula* for π.

13.8 Find the Fourier series representing the periodic *parabolic wave* function

$$f(x) = x^2 \qquad -\pi \leq x < \pi$$

defined over the range $-\pi \leq x < \pi$ and repeating with a period of 2π as shown below.

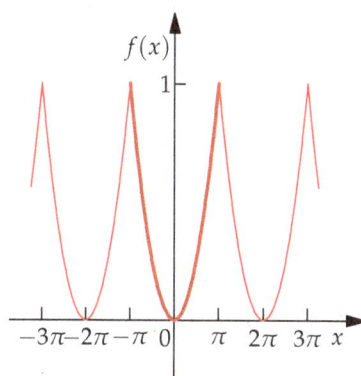

13.9 Consider the exponentially *damped cosine* function

$$f(t) = e^{-at}\cos(\omega_0 t)$$

defined over the range $0 \leq t < \infty$ with $a = \frac{1}{3}$ and $b = \frac{1}{3}$. Determine its cosine transform defined as

$$F_c(\nu) = \int_0^\infty f(t)\cos(2\pi\nu t)dt$$

Sketch the form of the function $f(t)$ over the range $0 \leq t \leq 20$ and the cosine transform $F_c(\nu)$ over the range $-1 \leq \nu \leq 1$.

13.10 Find the Fourier series representing the periodic positive *sawtooth wave* function

$$f(x) = x \qquad 0 \leq x < 2L$$

defined over the range $0 \leq x < 2L$ and repeating with a period of $2L$ as shown below.

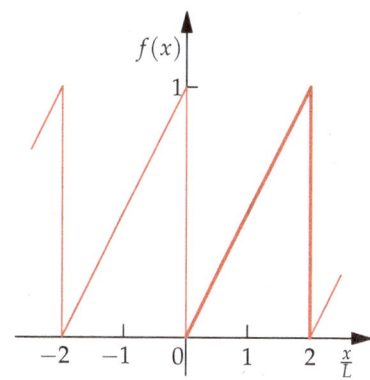

Unlike the sawtooth wave shown in Figure 13.11, that is an odd function of x, this sawtooth wave is not even or odd.

13.11 Find the Fourier series representing the periodic odd *triangle wave* function

$$f(t) = \begin{cases} \frac{2t}{T} & -\frac{T}{2} \leq t < \frac{T}{2} \\ 2\left(1 - \frac{t}{T}\right) & \frac{T}{2} \leq t < \frac{3T}{2} \end{cases}$$

defined over the range $-\frac{T}{2} \leq t < \frac{3T}{2}$ and repeating with a period of $2T$ as shown below.

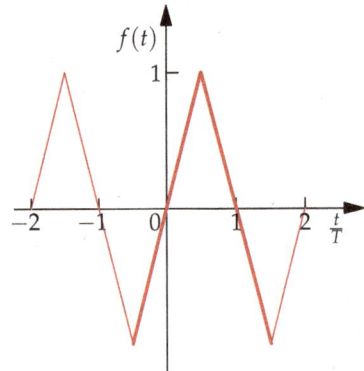

13.12 Consider the function $f(t)$ with Fourier transform $F(\omega)$. Show that the Fourier transform of $f(st)$ is $\frac{1}{s}F(\frac{\omega}{s})$ when $s > 0$.

13.13* Consider the exponential function $f(t) = e^{-a|t|}$ with Fourier transform $F(\omega)$. Starting from the definition of the Fourier transform of the derivative of $f(t)$ written

$$G(\omega) = \frac{1}{\sqrt{2\pi}} \int_{-\infty}^{\infty} \left(\frac{d}{dt} e^{-a|t|}\right) e^{-i\omega t}\, dt$$

demonstrate that $G(\omega) = i\omega F(\omega)$.

13.14 Consider the sine series solution to the *classical wave equation*

$$h(x) = \sin\left(\frac{\pi x}{L}\right)$$

defined over the range $0 \leq x < L$. We can use $h(x)$ defined over $0 \leq x < L$ to build a periodic function $f(x)$ defined over all x by repeating the behavior with a period of L as shown below.

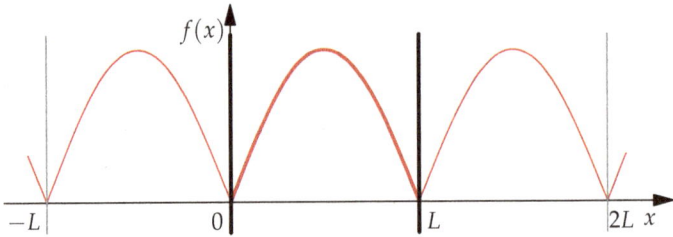

Derive the Fourier series representation of this solution to the wave equation.

13.15 Consider the cosine series solution to the *classical heat equation*

$$u(x) = \frac{1}{2} - \frac{1}{2}\cos\left(\frac{2\pi x}{L}\right)$$

defined over the range $0 \leq x < L$. We can use $u(x)$ defined over $0 \leq x < L$ to build a periodic function $f(x)$ defined over all x by repeating the behavior with a period of L as shown below.

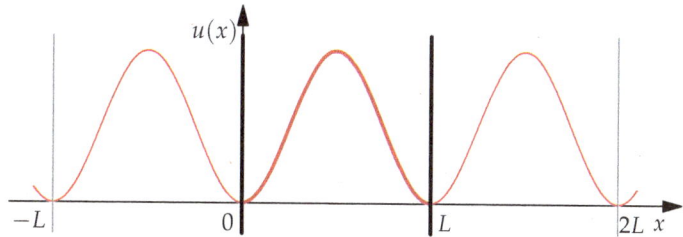

Derive the Fourier series representation of this solution to the heat equation.

13.16* Find the Fourier series representing the inverted *parabolic wave* function

$$f(x) = L^2 - (x - 2n)^2. \qquad n = \ldots, -2, -1, 0, 1, 2, \ldots$$

defined over the range $(2n - 1)L \leq x < (2n + 1)L$ and repeating with a period of $2L$ as shown below.

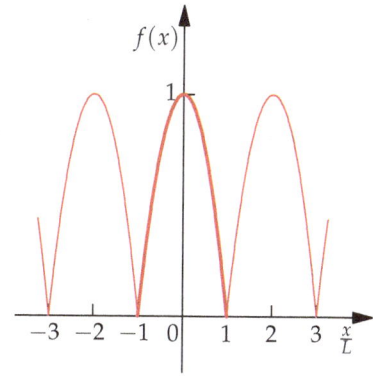

Prove that your answer satisfies the boundary conditions $f(0) = L^2$ and $f(-L) = f(L) = 0$.

13.17 Complete the following steps to derive the *energy theorem*, also known as *Parseval's theorem*, and explore its meaning.

(a) Evaluate the integral over the square of the function $f(t)$ written

$$E = \frac{1}{T}\int_{-\frac{T}{2}}^{\frac{T}{2}} (f(t))^2 dt$$

when $f(t)$ is expressed as a Fourier series

$$f(t) = \frac{a_0}{2} + \sum_{n=1}^{\infty} a_n \cos\left(\frac{2n\pi t}{T}\right) + \sum_{n=1}^{\infty} b_n \sin\left(\frac{2n\pi t}{T}\right)$$

(b) Evaluate the integral defining the mean square residual difference written

$$\Delta E = \frac{1}{T} \int_{-\frac{T}{2}}^{\frac{T}{2}} (f(t) - f_N(t))^2 \, dt$$

where $f(t)$ is expressed as a Fourier series and $f_N(t)$ is the partial sum.

(c) Evaluate E for the square wave Fourier series. Do that by evaluating the integral as well as by summing the infinite series. Demonstrate that your results are equal.

13.18 Let's consider another representation of the Fourier cosine series of a function $f(t)$ in terms of a function $g(\omega)$ defining a spectrum of frequencies ω contributing to the Fourier series. The function can be written as

$$f(t) = \sum_{n=-\infty}^{\infty} a_n \cos\left(\frac{2n\pi t}{T}\right) = \int_0^\infty g(\omega) \cos(\omega t) \, d\omega$$

Derive $g(\omega)$. HINT: Consider the possibility of a weighted sum over Dirac delta functions.

13.19* Consider the sinusoidal function $\cos(\omega_0 t)$ sampled over $N+1$ periods of oscillation and written

$$f(t) = \begin{cases} 0 & t < -\dfrac{N\pi}{\omega_0} \\ \cos(\omega_0 t) & -\dfrac{N\pi}{\omega_0} < t < \dfrac{N\pi}{\omega_0} \\ 0 & t > \dfrac{N\pi}{\omega_0} \end{cases}$$

The Fourier transform $F(\omega)$ of $f(t)$ can be written

$$F(\omega) = \frac{1}{\sqrt{2\pi}} \left[\frac{\sin\left(\frac{N\pi}{\omega_0}(\omega + \omega_0)\right)}{\omega + \omega_0} + \frac{\sin\left(\frac{N\pi}{\omega_0}(\omega - \omega_0)\right)}{\omega - \omega_0} \right]$$

(a) Plot the function $f(t)$ and its Fourier transform $F(k)$ for $N = 1$ over the range $-60 < \omega < 60$.

(b) Plot the function $f(t)$ and its Fourier transform $F(k)$ for $N = 11$ over the range $-15 < \omega < 15$.

(c) Determine the form of the Fourier transform in the limit

$$\lim_{N \to \infty} F(\omega)$$

where we have sampled an infinity of oscillations of the sinusoidal wave. HINT: Equation 7.29 will be useful.

(d) What feature of this function is responsible for the ringing oscillations observed in the Fourier transform?

13.20* Consider the two functions $f(t)$ and $g(t)$ with Fourier transforms $F(\omega)$ and $G(\omega)$, respectively. The integral convolution of the two functions $f(t)$ and $g(t)$ is

$$(f * g)(t) = \int_{-\infty}^{\infty} f(s) g(t - s) \, ds$$

where we use the notation $f * g$ to denote convolution and write $(f * g)(t)$ to emphasize that the result is a function of t. Prove that the Fourier transform of the convolution $(f * g)(t)$ is the product $\sqrt{2\pi} \, F(\omega) G(\omega)$.

13.21* The complex Fourier series is defined as

$$f(x) = \sum_{m=-\infty}^{\infty} c_m\, e^{imx}$$

where $e^{imx} = \cos(mx) + i\sin(mx)$ and the coefficients c_m are complex numbers. Prove that

$$c_n = \frac{1}{2\pi} \int_{-\pi}^{\pi} f(x)\, e^{-inx} dx$$

HINT: Do this by inserting the equation above for $f(x)$ into the proposed definition of c_n and evaluating the integrals.

13.22* The aperiodic *sawtooth function* can be written for one period of oscillation over $-L \leq x < L$ as

$$f(x) = \begin{cases} 0 & x < -L \\ \dfrac{x}{L} & -L \leq x < L \\ 0 & x > L \end{cases}$$

The function is shown below for a period of 2L.

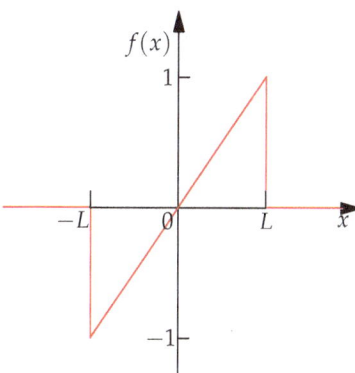

(a) Derive the Fourier transform of the aperiodic sawtooth function $f(x)$ defined

$$F(k) = \frac{1}{\sqrt{2\pi}} \int_{-\infty}^{\infty} f(x) e^{-ikx}\, dx$$

as a function of k.

(b) We derived the Fourier series of the periodic *sawtooth wave* function shown in Figure 13.11. The resulting Fourier series defined in Equation 13.21 is

$$f(x) = \sum_{n=1}^{\infty} b_n \sin\left(\frac{n\pi x}{L}\right) = \frac{2}{\pi} \sum_{n=1}^{\infty} \frac{(-1)^{n+1}}{n} \sin\left(\frac{n\pi x}{L}\right)$$

Show that the Fourier series coefficients b_n are related to the value of the Fourier transform evaluated at $k = \dfrac{n\pi}{L}$ through the relation

$$c_n = \frac{\sqrt{2\pi}}{2L} F\left(\frac{n\pi}{L}\right)$$

13.23* The orthonormality condition for the *Hermite polynomials* is

$$\frac{1}{\sqrt{\pi}\,2^n n!}\int_{-\infty}^{\infty} H_n(x)\,H_m(x)\,e^{-x^2}dx = \delta_{nm}$$

where δ_{nm} is the Kronecker delta function, which is unity when $n = m$ and zero otherwise. Show that the Hermite polynomials $H_1(x) = 2x$ and $H_2(x) = 4x^2 - 2$ are normalized and orthogonal over $-\infty < x < \infty$.

13.24* The orthonormality condition for the *Laguerre polynomials* is

$$\int_0^{\infty} L_n(x)\,L_m(x)\,e^{-x}dx = \delta_{nm}$$

Show that the Laguerre polynomials $L_1(x) = 1 - x$ and $L_2(x) = \frac{1}{2}(x^2 - 4x + 2)$ are normalized and orthogonal over $0 \le x < \infty$.

13.25* The orthonormality condition for the *Legendre polynomials* is

$$\frac{2n+1}{2}\int_{-1}^{1} P_n(x)\,P_m(x)\,dx = \delta_{nm}$$

Show that the Legendre polynomials $P_1(x) = x$ and $P_2(x) = \frac{1}{2}(3x^2 - 1)$ are normalized and orthogonal over $-1 \le x \le 1$.

13.26* The orthonormality condition for the *spherical harmonic functions* is

$$\int_0^{2\pi}\int_0^{\pi} Y_{l,m}(\theta,\varphi)\,Y_{l',m'}^{*}(\theta,\varphi)\sin(\theta)\,d\theta d\varphi = \delta_{l,l'}\delta_{m,m'}$$

where $Y_{l,m}^{*}(\theta,\varphi)$ is the complex conjugate of $Y_{l,m}(\theta,\varphi)$. Show that the spherical harmonic functions $Y_{1,0}(\theta,\varphi) = \frac{1}{2}\sqrt{\frac{3}{\pi}}\cos\theta$ and $Y_{1,1}(\theta,\varphi) = -\frac{1}{2}\sqrt{\frac{3}{2\pi}}\sin\theta\,e^{i\varphi}$ are normalized and orthogonal over $0 \le \theta < \pi$ and $0 \le \varphi < 2\pi$.

13.27* Build a set of orthogonal polynomial functions $\varphi_n(x)$ with $n = 0, 1, 2, \ldots$ where

$$\int_0^{\infty} e^{-x}\,\varphi_n(x)\varphi_m(x)\,dx = \frac{1}{n!}\Gamma(n)\,\delta_{nm}$$

and $\Gamma(n) = 1 \cdot 2 \cdot \ldots (n-2) \cdot (n-1) = (n-1)!$. Take $\varphi_n(0) = 1$ and $\varphi_0(x) = 1$. Determine $\varphi_1(x)$ and $\varphi_2(x)$.

13.28* Build a set of orthogonal polynomial functions $\varphi_n(x)$ with $n = 0, 1, 2, \ldots$ where

$$\int_{-\infty}^{\infty} e^{-x^2}\,\varphi_n(x)\varphi_m(x)\,dx = \sqrt{\pi}\,2^n\,n!\,\delta_{nm}$$

Take $\varphi_0(x) = 1$. Determine $\varphi_1(x)$ and $\varphi_2(x)$.

14 MATRICES AND MATRIX ALGEBRA

14.1 Vectors, matrices, and determinants

LINEAR ALGEBRA INVOLVES THE STUDY OF LINEAR EQUATIONS. We have seen that linear equations naturally arise in the physical sciences in the form of linear polynomial equations, linear operators, and linear differential equations. Of particular interest is the solution of systems of coupled linear equations in many variables. This section explores methods for solving systems of linear equations in a way that naturally leads to the definition of the matrix and the determinant.

14.1.1 The vector and the matrix

In exploring the properties of *vectors* and the principles of vector algebra in Chapter 4, we expressed a vector shown in Figure 14.1 as

$$\mathbf{a} = a_x\,\hat{\mathbf{x}} + a_y\,\hat{\mathbf{y}} + a_z\,\hat{\mathbf{z}}$$

An alternative notation for the vector **a** is

$$\mathbf{a} = \begin{pmatrix} a_x \\ a_y \\ a_z \end{pmatrix}$$

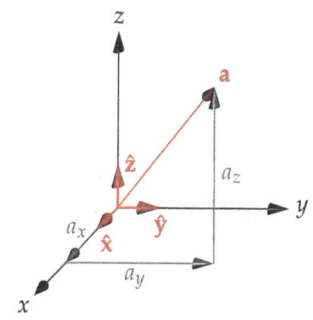

Figure 14.1: A vector $\mathbf{a} = a_x\,\hat{\mathbf{x}} + a_y\,\hat{\mathbf{y}} + a_z\,\hat{\mathbf{z}}$ represented in three-dimensional cartesian coordinates.

Using this notation the *unit vectors* are written

$$\hat{\mathbf{x}} = \begin{pmatrix} 1 \\ 0 \\ 0 \end{pmatrix} \qquad \hat{\mathbf{y}} = \begin{pmatrix} 0 \\ 1 \\ 0 \end{pmatrix} \qquad \hat{\mathbf{z}} = \begin{pmatrix} 0 \\ 0 \\ 1 \end{pmatrix}$$

Thinking of **a** as the sum $\mathbf{a} = a_x\,\hat{\mathbf{x}} + a_y\,\hat{\mathbf{y}} + a_z\,\hat{\mathbf{z}}$, we can write

$$\mathbf{a} = a_x\,\hat{\mathbf{x}} + a_y\,\hat{\mathbf{y}} + a_z\,\hat{\mathbf{z}} = a_x \begin{pmatrix} 1 \\ 0 \\ 0 \end{pmatrix} + a_y \begin{pmatrix} 0 \\ 1 \\ 0 \end{pmatrix} + a_z \begin{pmatrix} 0 \\ 0 \\ 1 \end{pmatrix} = \begin{pmatrix} a_x \\ a_y \\ a_z \end{pmatrix}$$

This demonstrates how these two distinct representations of the vector **a** are equivalent.

We can generalize the concept of the vector to form a *matrix*. A matrix is a two-dimensional array of terms forming m rows and n columns. We say that the *matrix order* is $m \times n$ for a matrix with m rows and n columns. Consider the example

$$\mathbf{A} = \begin{pmatrix} a_{11} \\ a_{21} \\ a_{31} \\ a_{41} \end{pmatrix}$$

A is a *column matrix* or, equivalently, a *column vector*, and its order is 4×1. The position of each element a_{ij} is identified by a row index i and a column index j.[1] Now consider a related example of

$$\mathbf{B} = \begin{pmatrix} b_{11} & b_{12} & b_{13} \end{pmatrix}$$

B is formed from a row rather than a column and is referred to as a *row matrix* or, equivalently, a *row vector*. Its order is 1×3. Finally, consider

$$\mathbf{C} = \begin{pmatrix} c_{11} & c_{12} & c_{13} \\ c_{21} & c_{22} & c_{23} \\ c_{31} & c_{32} & c_{33} \end{pmatrix}$$

C is a *square matrix* and its order is 3×3. **C** can be thought of as being formed from three row vectors of order 1×3. It can also be thought of as being formed from three column vectors of order 3×1.

Note that we can also form more general rectangular matrices, such as

$$\mathbf{D} = \begin{pmatrix} d_{11} & d_{12} & d_{13} \\ d_{21} & d_{22} & d_{23} \\ d_{31} & d_{32} & d_{33} \\ d_{41} & d_{42} & d_{43} \end{pmatrix}$$

which is a matrix of order 4×3, and

$$\mathbf{E} = \begin{pmatrix} e_{11} & e_{12} & e_{13} & e_{14} \\ e_{21} & e_{22} & e_{23} & e_{24} \end{pmatrix}$$

which is a matrix of order 2×4.

Over the course of this chapter, we will encounter a variety of matrices. The most common matrices in the physical sciences are row or column matrices, which act as vectors, and square matrices. Square matrices have many special properties, one of which is the matrix determinant.

14.1.2 *The matrix determinant and its geometric interpretation*

For any square matrix we can calculate a property called the *determinant*.[2] Consider the determinant $|\mathbf{A}|$ of the 2×2 square matrix **A**, defined as

$$|\mathbf{A}| = \begin{vmatrix} a_{11} & a_{12} \\ a_{21} & a_{22} \end{vmatrix} = a_{11}a_{22} - a_{12}a_{21} \tag{14.1}$$

We encountered the determinant in Chapter 4 while exploring the vector *cross product*. We learned to express the vector cross product as the determinant of a 3×3 matrix (see Equation 4.9).

[1] We write the element in the ith row and jth column of the matrix as $[\mathbf{A}]_{ij}$.

[2] Determinants were first considered in the West by German philosopher and mathematician *Gottfried Leibniz* (1646-1716). However, the determinant was discovered earlier in China and appears in *The Nine Chapters on the Mathematical Art* in 179 CE with commentary added later by *Liu Hui* [劉徽] (c. 225- c. 295).

In addition, we learned how the determinant of a 3×3 matrix can be written in terms of the determinants of three 2×2 matrices, as in this expression for the cross product:

$$\mathbf{a} \times \mathbf{b} = \begin{vmatrix} \hat{\mathbf{x}} & \hat{\mathbf{y}} & \hat{\mathbf{z}} \\ a_x & a_y & a_z \\ b_x & b_y & b_z \end{vmatrix} = \hat{\mathbf{x}} \begin{vmatrix} a_y & a_z \\ b_y & b_z \end{vmatrix} - \hat{\mathbf{y}} \begin{vmatrix} a_x & a_z \\ b_x & b_z \end{vmatrix} + \hat{\mathbf{z}} \begin{vmatrix} a_x & a_y \\ b_x & b_y \end{vmatrix}$$

$$= (a_y b_z - a_z b_y)\, \hat{\mathbf{x}} + (a_z b_x - a_x b_z)\, \hat{\mathbf{y}} + (a_x b_y - a_y b_x)\, \hat{\mathbf{z}}$$

Let's consider another use of the matrix determinant. Suppose we have a set of linear equations

$$a_{11}x + a_{12}y = d_1$$
$$a_{21}x + a_{22}y = d_2 \tag{14.2}$$

where the goal is to solve the equations for the variables x and y in terms of the various coefficients (see Figure 14.2). We can multiply the first equation by a_{22} and the second by a_{12} (noted in red), to produce

$$a_{11}a_{22}x + a_{12}a_{22}y = d_1 a_{22}$$
$$a_{21}a_{12}x + a_{22}a_{12}y = d_2 a_{12}$$

Subtracting the two equations, we find

$$(a_{11}a_{22} - a_{21}a_{12})\, x = d_1 a_{22} - d_2 a_{12}$$

Solving this equation for x leads to

$$x = \frac{d_1 a_{22} - d_2 a_{12}}{a_{11}a_{22} - a_{21}a_{12}} \tag{14.3}$$

Suppose we start again, this time multiplying the first equation by a_{21} and the second equation by a_{11}, to produce

$$a_{11}a_{21}x + a_{12}a_{21}y = d_1 a_{21}$$
$$a_{21}a_{11}x + a_{22}a_{11}y = d_2 a_{11}$$

Subtracting the two equations, we find

$$(a_{12}a_{21} - a_{22}a_{11})\, y = d_1 a_{21} - d_2 a_{11}$$

leading to

$$y = \frac{d_1 a_{21} - d_2 a_{11}}{a_{12}a_{21} - a_{22}a_{11}} = \frac{d_2 a_{11} - d_1 a_{21}}{a_{11}a_{22} - a_{12}a_{21}} \tag{14.4}$$

which completes our solution to Equation 14.2 for x and y.

The common denominator in Equations 14.3 and 14.4 is the determinant of the 2×2 square matrix

$$|\mathbf{A}| = \begin{vmatrix} a_{11} & a_{12} \\ a_{21} & a_{22} \end{vmatrix} = a_{11}a_{22} - a_{12}a_{21}$$

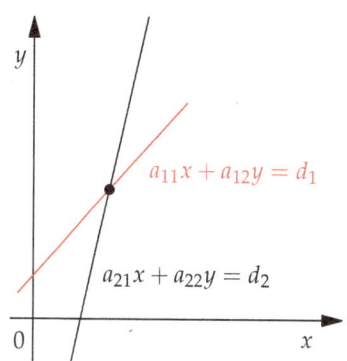

Figure 14.2: In two dimensions, points representing the solution to a linear equation form a line. The point of intersection (black dot) of two lines represents the common solution (x, y) to the two linear equations.

So we see how the algebraic form of the determinant naturally arises when solving for the unknown variables in a set of linear equations.[3]

14.1.3 Geometric interpretation of the matrix determinant

Geometrically, the determinant $|\mathbf{A}|$ can be interpreted as the area of a parallelogram, with edges defined by the column vectors forming the matrix. For example, we can consider the 2×2 square matrix $\mathbf{A}$ appearing in Equation 14.1 as being composed of the column vectors

$$\mathbf{a} = \begin{pmatrix} a_{11} \\ a_{21} \end{pmatrix} \qquad \mathbf{b} = \begin{pmatrix} a_{12} \\ a_{22} \end{pmatrix}$$

as shown in Figure 14.3.

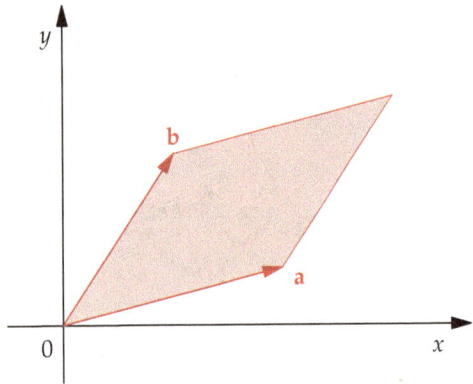

For the determinant of a 3×3 matrix, we write[4]

$$|\mathbf{A}| = \begin{vmatrix} a_{11} & a_{12} & a_{13} \\ a_{21} & a_{22} & a_{23} \\ a_{31} & a_{32} & a_{33} \end{vmatrix}$$

$$= a_{11} \begin{vmatrix} a_{22} & a_{23} \\ a_{32} & a_{33} \end{vmatrix} - a_{12} \begin{vmatrix} a_{21} & a_{23} \\ a_{31} & a_{33} \end{vmatrix} + a_{13} \begin{vmatrix} a_{21} & a_{22} \\ a_{31} & a_{32} \end{vmatrix} \qquad (14.5)$$

$$= a_{11} \left(a_{22} a_{33} - a_{23} a_{32} \right) - a_{12} \left(a_{21} a_{33} - a_{23} a_{31} \right) + a_{13} \left(a_{21} a_{32} - a_{22} a_{31} \right)$$

Geometrically, the determinant $|\mathbf{A}|$ of a 3×3 matrix $\mathbf{A}$ can be interpreted as the volume of a parallelepiped with edges defined by the three column vectors forming the matrix (see Figure 14.4). This geometric interpretation is a useful qualitative guide in the quantitative evaluation of matrix determinants.

14.1.4 Matrix determinants of arbitrary dimension

We can extend this approach to computing determinants of $n \times n$ square matrices to arbitrarily large n.[5] To do so, we introduce the *cofactor* A_{ij} of the matrix $\mathbf{A}$, where A_{ij} is the determinant of an $(n-1) \times (n-1)$ square matrix, formed by deleting the ith row and jth column of $\mathbf{A}$, multiplied by $(-1)^{i+j}$.

[3] Our approach is readily generalized to the solution of n linear algebraic equations in n unknowns through a similar treatment of $n \times n$ square matrix determinants. The resulting method will be discussed later in this chapter.

Figure 14.3: The magnitude of the determinant $|\mathbf{A}|$ of the 2×2 matrix $\mathbf{A}$ is the area of the parallelogram (shaded red) with edges defined by the column vectors $\mathbf{a}$ and $\mathbf{b}$.

[4] This looks familiar. In fact, we previously encountered this determinant in our evaluation of the vector cross product (see Equation 4.9).

[5] This method was developed by the French mathematician, physicist, and astronomer *Pierre-Simon Laplace* (1749-1827).

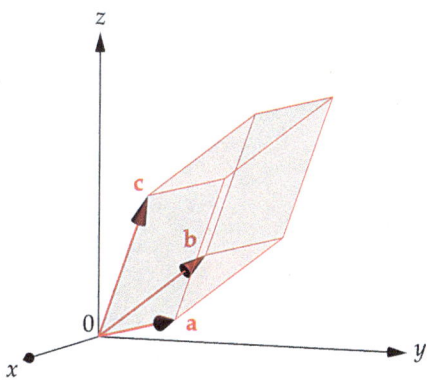

Figure 14.4: The magnitude of the determinant $|\mathbf{A}|$ of the 3×3 matrix $\mathbf{A}$ is the volume of the parallelepiped (shaded red) formed by the three vectors $\mathbf{a}$, $\mathbf{b}$, and $\mathbf{c}$.

The determinant of a $n \times n$ square matrix $\mathbf{A}$ can then be written

$$|\mathbf{A}| = a_{11}A_{11} + a_{12}A_{12} + \ldots + a_{1n}A_{1n}$$

To determine the n cofactors, we need to evaluate n determinants of $(n-1) \times (n-1)$ matrices. For each of these determinants, we need to compute $n-1$ determinants of $(n-2) \times (n-2)$ matrices. And so on.

For example, returning to the case above, we can express the determinant of the 3×3 matrix $\mathbf{A}$ in terms of three cofactors as

$$|\mathbf{A}| = \begin{vmatrix} a_{11} & a_{12} & a_{13} \\ a_{21} & a_{22} & a_{23} \\ a_{31} & a_{32} & a_{33} \end{vmatrix} = a_{11}A_{11} + a_{12}A_{12} + a_{13}A_{13}$$

where the cofactors are

$$A_{11} = (-1)^2 \begin{vmatrix} a_{22} & a_{23} \\ a_{32} & a_{33} \end{vmatrix} \quad A_{12} = (-1)^3 \begin{vmatrix} a_{21} & a_{23} \\ a_{31} & a_{33} \end{vmatrix} \quad A_{13} = (-1)^4 \begin{vmatrix} a_{21} & a_{22} \\ a_{31} & a_{32} \end{vmatrix}$$

Evaluating the 2×2 determinants, watching our signs, leads to the final result

$$\begin{aligned} |\mathbf{A}| &= a_{11}A_{11} + a_{12}A_{12} + a_{13}A_{13} \\ &= a_{11}\left[a_{22}a_{33} - a_{23}a_{32}\right] + a_{12}\left[-\left(a_{21}a_{33} - a_{23}a_{31}\right)\right] + a_{13}\left[a_{21}a_{32} - a_{22}a_{31}\right] \end{aligned}$$

which matches our earlier result in Equation 14.5, validating our method.

Consider the determinant of the 3×3 matrix

$$|\mathbf{A}| = \begin{vmatrix} 2 & -1 & -1 \\ -1 & 2 & -1 \\ -1 & -1 & 2 \end{vmatrix}$$

which can be evaluated as

$$\begin{aligned} |\mathbf{A}| &= 2(-1)^2 \begin{vmatrix} 2 & -1 \\ -1 & 2 \end{vmatrix} + (-1)(-1)^3 \begin{vmatrix} -1 & -1 \\ -1 & 2 \end{vmatrix} + (-1)(-1)^4 \begin{vmatrix} -1 & 2 \\ -1 & -1 \end{vmatrix} \\ &= 6 - 3 - 3 = 0 \end{aligned}$$

This demonstrates how the determinant of the 3×3 matrix is evaluated in terms of three 2×2 determinants. In the end, our goal is to reduce the determinant of any size of matrix down to many 2×2 determinants.

Now let's consider the determinant of a 4×4 matrix

$$|\mathbf{B}| = \begin{vmatrix} 2 & -1 & 0 & -1 \\ -1 & 2 & -1 & 0 \\ 0 & -1 & 2 & -1 \\ -1 & 0 & -1 & 2 \end{vmatrix} = b_{11}B_{11} + b_{12}B_{12} + b_{13}B_{13} + b_{14}B_{14}$$

where the cofactors are[6]

$$B_{11} = (-1)^2 \begin{vmatrix} 2 & -1 & 0 \\ -1 & 2 & -1 \\ 0 & -1 & 2 \end{vmatrix} = 4 \qquad B_{12} = (-1)^3 \begin{vmatrix} -1 & -1 & 0 \\ 0 & 2 & -1 \\ -1 & -1 & 2 \end{vmatrix} = 4$$

$$B_{13} = (-1)^4 \begin{vmatrix} -1 & 2 & 0 \\ 0 & -1 & -1 \\ -1 & 0 & 2 \end{vmatrix} = 4 \qquad B_{14} = (-1)^5 \begin{vmatrix} -1 & 2 & -1 \\ 0 & -1 & 2 \\ -1 & 0 & -1 \end{vmatrix} = 4$$

This leads to the final result

$$|\mathbf{B}| = 2(4) + (-1)(4) + 0(4) + (-1)(4) = 0$$

The determinant of this 4×4 matrix required the calculation of four 3×3 determinants, each of which required the calculation of three 2×2 determinants. Ultimately, we reduced the determinant of the 4×4 matrix to twelve 2×2 determinants.

Another approach to computing $|\mathbf{B}|$ is to simply expand the larger determinant as

$$|\mathbf{B}| = +2 \begin{vmatrix} 2 & -1 & 0 \\ -1 & 2 & -1 \\ 0 & -1 & 2 \end{vmatrix} - (-1) \begin{vmatrix} -1 & -1 & 0 \\ 0 & 2 & -1 \\ -1 & -1 & 2 \end{vmatrix} + 0 \begin{vmatrix} -1 & 2 & 0 \\ 0 & -1 & -1 \\ -1 & 0 & 2 \end{vmatrix} - (-1) \begin{vmatrix} -1 & 2 & -1 \\ 0 & -1 & 2 \\ -1 & 0 & -1 \end{vmatrix} = 0$$

This is equivalent to the approach using cofactor matrices but treats the alternating signs in the initial expansion. This makes a sign error less likely. These examples demonstrate how our method for computing matrix determinants works for 2×2, 3×3, and 4×4 matrices, and extend to any square matrix of arbitrary dimension.

The calculation of matrix determinants is involved in many applications of linear algebra in the physical sciences. As such, it is critical to become efficient in the calculation of matrix determinants and familiar with their properties. The next section surveys five fundamental properties of matrix determinants. An effort is made to explain the origin of each property through a simple example or related geometric interpretation. More than formal observations, we will find that each of these properties has real utility in problem solving in the physical sciences.

[6] In each case, we use the approach summarized by Equation 14.5 to evaluate the 3×3 determinants. For example

$$\begin{vmatrix} 2 & -1 & 0 \\ -1 & 2 & -1 \\ 0 & -1 & 2 \end{vmatrix} = 2 \begin{vmatrix} 2 & -1 \\ -1 & 2 \end{vmatrix}$$

$$-(-1) \begin{vmatrix} -1 & -1 \\ 0 & 2 \end{vmatrix}$$

$$+0 \begin{vmatrix} -1 & 2 \\ 0 & -1 \end{vmatrix}$$

$$= 2(3) + 1(-2) + 0(1) = 4$$

and so on.

14.1.5 Special properties of matrix determinants

We identify five fundamental properties of the matrix determinant. Terms in a determinant that have been transformed are highlighted in red.

1. If we change the rows of a matrix $\mathbf{A}$ to the columns of a matrix in the same order, we create the *transpose matrix* $\mathbf{A}^T$.[7] The determinant of $\mathbf{A}$ and $\mathbf{A}^T$ are identical. Compare the determinant

$$|\mathbf{A}| = \begin{vmatrix} a & b & c \\ d & e & f \\ g & h & i \end{vmatrix}$$

$$= a\,(ei - fh) - b\,(di - fg) + c\,(dh - eg)$$

to the determinant of the transpose matrix

$$\left|\mathbf{A}^T\right| = \begin{vmatrix} a & d & g \\ b & e & h \\ c & f & i \end{vmatrix}$$

$$= a\,(ei - hf) - d\,(bi - hc) + g\,(bf - ec)$$

While the terms in blue have changed order, the determinants $|\mathbf{A}|$ and $\left|\mathbf{A}^T\right|$ are identical.

2. If any row or column of a square matrix is multiplied by a constant k, then the determinant is multiplied by k. Consider that

$$|\mathbf{B}| = \begin{vmatrix} a & b & kc \\ d & e & kf \\ g & h & ki \end{vmatrix}$$

$$= a\,(eki - kfh) - b\,(dki - kfg) + kc\,(dh - eg)$$

$$= k\,[a\,(ei - fh) - b\,(di - fg) + c\,(dh - eg)]$$

$$= k\,|\mathbf{A}|$$

Geometrically, the multiplication of one column vector by a constant k corresponds to scaling one side of the parallelepiped by k. This increases the volume of the parallelepiped, equal to the magnitude of the determinant, by the same factor (see Figure 14.5).[8]

3. If any two rows or columns of a square matrix are identical, the determinant is zero. Consider the determinant with two identical columns

$$|\mathbf{C}| = \begin{vmatrix} a & b & a \\ d & e & d \\ g & h & g \end{vmatrix}$$

$$= a\,(eg - dh) - b\,(dg - dg) + a\,(dh - eg) = 0$$

where the first term is equal in magnitude and opposite in sign to the third term, and the second term is zero.

[7] Consider a matrix $\mathbf{A}$

$$\mathbf{A} = \begin{pmatrix} a & b & c \\ d & e & f \\ g & h & i \end{pmatrix}$$

By changing the rows of the matrix to columns, we create the *matrix transpose* $\mathbf{A}^T$ as

$$\mathbf{A}^T = \begin{pmatrix} a & d & g \\ b & e & h \\ c & f & i \end{pmatrix}$$

In compact notation we write $\left[\mathbf{A}^T\right]_{ij} = [\mathbf{A}]_{ji}$.

[8] Note that if each element of a $n \times n$ determinant is multiplied by a constant k, then the determinant is multiplied by k^n.

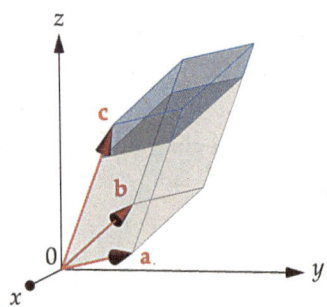

Figure 14.5: Interpreted geometrically, multiplying one column by a constant k increases the length of one side of the parallelepiped, and its volume, by the same constant k. The difference in volume is shaded blue.

Geometrically, the three-dimensional parallelepiped is reduced to a two-dimensional parallelogram of zero volume (see Figure 14.6).

4. If two rows or columns of a matrix $\mathbf{A}$ are interchanged to form a matrix $\mathbf{D}$, the sign of the determinant changes. Consider that

$$
|\mathbf{D}| = \begin{vmatrix} a & c & b \\ d & f & e \\ g & i & h \end{vmatrix}
$$
$$
= a\,(fh - ei) - c\,(dh - eg) + b\,(di - fg)
$$
$$
= -\,|\mathbf{A}|
$$

Relative to the determinant $|\mathbf{A}|$, each of the three terms in $|\mathbf{D}|$ has the sign reversed.

5. If one row or column is added to or subtracted from another row or column, the determinant is unchanged. Consider the matrix $\mathbf{E}$ formed from matrix $\mathbf{A}$ by adding the first column to the second column as

$$
|\mathbf{E}| = \begin{vmatrix} a & b+a & c \\ d & e+d & f \\ g & h+g & i \end{vmatrix}
$$
$$
= a\,((e+d)\,i - f\,(h+g)) - (b+a)\,(di - fg) + c\,(d\,(h+g) - (e+d)\,g)
$$
$$
= |\mathbf{A}|
$$

Note that the six terms involving a red coefficient sum to zero. Geometrically, this corresponds to a transformed parallelepiped of equal volume (see Figure 14.7).

Let's practice computing determinants while exploring a number of square matrices that commonly occur in the physical sciences.

14.1.6 Applications to determinants of common square matrices

Consider the set of 2×2 complex matrices known as the *Pauli matrices*:[9]

$$
\sigma_x = \begin{pmatrix} 0 & 1 \\ 1 & 0 \end{pmatrix} \qquad \sigma_y = \begin{pmatrix} 0 & -i \\ i & 0 \end{pmatrix} \qquad \sigma_z = \begin{pmatrix} 1 & 0 \\ 0 & -1 \end{pmatrix}
$$

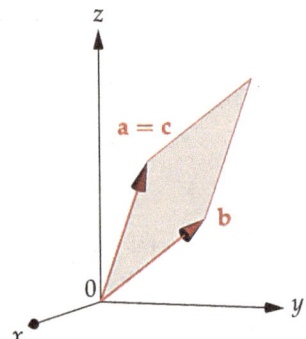

Figure 14.6: Interpreted geometrically, when two columns of the matrix are identical the three-dimensional parallelepiped is reduced to a two-dimensional parallelogram of zero volume.

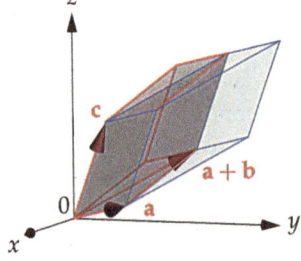

Figure 14.7: Interpreted geometrically, when the elements of one column of the matrix are added to the corresponding elements of another column, the volume of the parallelepiped and the determinant of the matrix are unchanged. The volume shaded red is equal to the volume shaded blue.

We can compute the determinant of each matrix as

$$|\sigma_x| = \begin{vmatrix} 0 & 1 \\ 1 & 0 \end{vmatrix} = |\sigma_y| = \begin{vmatrix} 0 & -i \\ i & 0 \end{vmatrix} = |\sigma_z| = \begin{vmatrix} 1 & 0 \\ 0 & -1 \end{vmatrix} = -1$$

[9] Named for Austrian physicist *Wolfgang Pauli* (1900-1958) who discovered the law of nature known as the *Pauli exclusion principle*.

Now consider another set of 2×2 square matrices known as *quaternions:*[10]

$$\mathbf{U} = \begin{pmatrix} 1 & 0 \\ 0 & 1 \end{pmatrix} \quad \mathbf{I} = \begin{pmatrix} i & 0 \\ 0 & -i \end{pmatrix}$$

$$\mathbf{J} = \begin{pmatrix} 0 & 1 \\ -1 & 0 \end{pmatrix} \quad \mathbf{K} = \begin{pmatrix} 0 & i \\ i & 0 \end{pmatrix}$$

[10] Quaternions were discovered by Irish mathematician, astronomer, and physicist *William Rowan Hamilton* (1805-1865).

The quaternions are related to the Pauli matrices through a factor of i as $\mathbf{I} = i\sigma_z$, $\mathbf{J} = i\sigma_y$, and $\mathbf{K} = i\sigma_x$. The matrix determinants of the quaternions are

$$|\mathbf{U}| = \begin{vmatrix} 1 & 0 \\ 0 & 1 \end{vmatrix} = |\mathbf{I}| = \begin{vmatrix} i & 0 \\ 0 & -i \end{vmatrix} = |\mathbf{J}| = \begin{vmatrix} 0 & 1 \\ -1 & 0 \end{vmatrix} = |\mathbf{K}| = \begin{vmatrix} 0 & i \\ i & 0 \end{vmatrix} = 1$$

Quaternions have a variety of interesting properties and can be used to model rotations in three-dimensional space.

Another group of matrices commonly employed in the physical sciences are the *rotation matrices*. In two dimensions, the rotation matrix is defined

$$\mathbf{R}(\theta) = \begin{pmatrix} \cos\theta & -\sin\theta \\ \sin\theta & \cos\theta \end{pmatrix}$$

for counter-clockwise rotation through an angle θ in the xy-plane. Note that the elements of the matrix are functions of the variable θ, making the matrix itself a *matrix function* of θ.[11] For example, depending on the specific value of θ, the rotation matrix can take on a variety of values, including

[11] The matrix function is a generalization of the concept of the *vector function* discussed in Chapter 5.

$$\mathbf{R}(0) = \begin{pmatrix} 1 & 0 \\ 0 & 1 \end{pmatrix} \qquad \mathbf{R}\left(\frac{\pi}{2}\right) = \begin{pmatrix} 0 & -1 \\ 1 & 0 \end{pmatrix}$$

$$\mathbf{R}(\pi) = \begin{pmatrix} -1 & 0 \\ 0 & -1 \end{pmatrix} \qquad \mathbf{R}\left(\frac{3\pi}{2}\right) = \begin{pmatrix} 0 & 1 \\ -1 & 0 \end{pmatrix}$$

This matrix has interesting properties that we will explore in the context of the rotation of vectors in a plane. For now, we compute the determinant of $\mathbf{R}(\theta)$ as

$$|\mathbf{R}(\theta)| = \begin{vmatrix} \cos\theta & -\sin\theta \\ \sin\theta & \cos\theta \end{vmatrix} = \cos^2\theta + \sin^2\theta = 1$$

This result is independent of the value of θ, as shown in Figure 14.8.

In this section, we have defined the matrix as a higher dimensional generalization of the concept of the vector. The determinant provides us with insight into the fundamental properties of the matrix. In the next section, we explore the rules for adding, subtracting, multiplying, and dividing matrices.

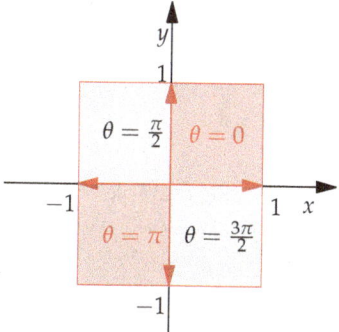

Figure 14.8: The magnitude of the determinant of the 2×2 rotation matrix $\mathbf{R}(\theta)$ for counter-clockwise rotation is the area of the unit square. Shown for for four values of θ (coloring shown for contrast).

14.2 Basic properties of matrix algebra

SYSTEMS OF LINEAR EQUATIONS MAY BE COMPACTLY EXPRESSED USING MATRICES. Matrices have an algebra all their own, and may be added, subtracted, multiplied, and divided. They may also be real or complex. Some properties of matrix algebra are shared by scalars, while other properties are distinctly different. This section explores the fundamental properties of matrix algebra.

14.2.1 Addition and subtraction of matrices

Recall the rules for adding and subtracting vectors. Two vectors $\mathbf{a}$ and $\mathbf{b}$ can be added to form a new vector $\mathbf{c}$ by adding each of the components:

$$\mathbf{a} + \mathbf{b} = \begin{pmatrix} a_x \\ a_y \\ a_z \end{pmatrix} + \begin{pmatrix} b_x \\ b_y \\ b_z \end{pmatrix} = \begin{pmatrix} a_x + b_x \\ a_y + b_y \\ a_z + b_z \end{pmatrix} = \begin{pmatrix} c_x \\ c_y \\ c_z \end{pmatrix} = \mathbf{c}$$

We can add two 2×2 matrices $\mathbf{A}$ and $\mathbf{B}$ to form a new 2×2 matrix $\mathbf{C}$ in the following way:

$$\mathbf{A} + \mathbf{B} = \begin{pmatrix} a_{11} & a_{12} \\ a_{21} & a_{22} \end{pmatrix} + \begin{pmatrix} b_{11} & b_{12} \\ b_{21} & b_{22} \end{pmatrix}$$

$$= \begin{pmatrix} a_{11} + b_{11} & a_{12} + b_{12} \\ a_{21} + b_{21} & a_{22} + b_{22} \end{pmatrix} = \begin{pmatrix} c_{11} & c_{12} \\ c_{21} & c_{22} \end{pmatrix} = \mathbf{C}$$

As we see from the terms highlighted in red, the matrix operation of addition applies to each element of the matrix individually. This process can be generalized to higher order $m \times n$ matrices. We can subtract one matrix from another matrix in the following way:[12]

[12] In compact notation we write $[\mathbf{C}]_{ij} = [\mathbf{A}]_{ij} - [\mathbf{B}]_{ij}$.

$$\mathbf{A} - \mathbf{B} = \begin{pmatrix} a_{11} & a_{12} \\ a_{21} & a_{22} \\ a_{31} & a_{32} \end{pmatrix} - \begin{pmatrix} b_{11} & b_{12} \\ b_{21} & b_{22} \\ b_{31} & b_{32} \end{pmatrix}$$

$$= \begin{pmatrix} a_{11} - b_{11} & a_{12} - b_{12} \\ a_{21} - b_{21} & a_{22} - b_{22} \\ a_{31} - b_{31} & a_{32} - b_{32} \end{pmatrix} = \begin{pmatrix} d_{11} & d_{12} \\ d_{21} & d_{22} \\ d_{31} & d_{32} \end{pmatrix} = \mathbf{D}$$

For a specific example of addition and subtraction of matrices, consider the addition of two 2×3 matrices:

$$\mathbf{E} + \mathbf{F} = \begin{pmatrix} -3 & 6 & 4 \\ 1 & 0 & 2 \end{pmatrix} + \begin{pmatrix} 2 & 1 & 1 \\ -6 & 4 & 3 \end{pmatrix}$$

$$= \begin{pmatrix} -3 + 2 & 6 + 1 & 4 + 1 \\ 1 - 6 & 0 + 4 & 2 + 3 \end{pmatrix} = \begin{pmatrix} -1 & 7 & 5 \\ -5 & 4 & 5 \end{pmatrix} = \mathbf{G}$$

Note that when adding and subtracting matrices, the two matrices must have the same order.

If we multiply or divide a matrix by a constant, we multiply or divide each of the elements of that matrix by the same constant.[13] For example, we find

$$cA = c \begin{pmatrix} a_{11} & a_{12} \\ a_{21} & a_{22} \end{pmatrix} = \begin{pmatrix} ca_{11} & ca_{12} \\ ca_{21} & ca_{22} \end{pmatrix}$$

This property along with the rules for addition and subtraction allow us to form linear equations involving matrices, just as we have formed linear equations involving constants such as a and b. Consider the equation

$$3E - 2G = 3 \begin{pmatrix} -3 & 6 & 4 \\ 1 & 0 & 2 \end{pmatrix} - 2 \begin{pmatrix} -1 & 7 & 5 \\ -5 & 4 & 5 \end{pmatrix}$$

$$= \begin{pmatrix} -9 & 18 & 12 \\ 3 & 0 & 6 \end{pmatrix} + \begin{pmatrix} 2 & -14 & -10 \\ 10 & -8 & -10 \end{pmatrix} = \begin{pmatrix} -7 & 4 & 2 \\ 13 & -8 & -4 \end{pmatrix} = H$$

These examples demonstrate how the rules of addition and subtraction of scalars and vectors can be extended to the addition and subtraction of matrices and the scaling of vectors.

14.2.2 Multiplication of matrices

The rule for multiplying two matrices is similar to the rule for multiplying two vectors using the *dot product*. Consider the dot product between two vectors **a** and **b**, which we express as

$$a \cdot b = \begin{pmatrix} a_x & a_y & a_z \end{pmatrix} \begin{pmatrix} b_x \\ b_y \\ b_z \end{pmatrix} = a_x b_x + a_y b_y + a_z b_z$$

Note that while the vector **b** is standing straight, the vector **a** is laying on its side. This is important as the number of columns in the first matrix must match the number of rows in the second matrix to take the product of the two. Recall the relationship between a matrix **A** and its transpose A^T.[14]

$$A = \begin{pmatrix} a & b \\ c & d \end{pmatrix} \qquad A^T = \begin{pmatrix} a & c \\ b & d \end{pmatrix}$$

The 3×1 matrix **B** and its transpose B^T can be written as follows:

$$B = \begin{pmatrix} a \\ d \\ g \end{pmatrix} \qquad B^T = \begin{pmatrix} a & d & g \end{pmatrix}$$

So we see that we can multiply a 3×1 matrix and its transpose just as we multiply two vectors using the dot product:

$$B^T B = \begin{pmatrix} a & d & g \end{pmatrix} \begin{pmatrix} a \\ d \\ g \end{pmatrix} = a^2 + d^2 + g^2 = c$$

This product of a 1×3 matrix and 3×1 matrix is a scalar c.

[13] In compact notation we write $c[A]_{ij} = [cA]_{ij}$.

[14] The sum of diagonal elements of a matrix
$$\sum_i [A]_{ii} = \text{Tr } A$$
is known as the *trace*. The trace of a matrix is identical for the matrix and its transpose $\text{Tr } A = \text{Tr } A^T$.

Now consider multiplying two 2×2 matrices $\mathbf{C}$ and $\mathbf{D}$ to form a third 2×2 matrix $\mathbf{E}$ in the following way:[15]

$$\mathbf{CD} = \begin{pmatrix} c_{11} & c_{12} \\ c_{21} & c_{22} \end{pmatrix} \begin{pmatrix} d_{11} & d_{12} \\ d_{21} & d_{22} \end{pmatrix}$$

$$= \begin{pmatrix} c_{11}d_{11} + c_{12}d_{21} & c_{11}d_{12} + c_{12}d_{22} \\ c_{21}d_{11} + c_{22}d_{21} & c_{21}d_{12} + c_{22}d_{22} \end{pmatrix} = \begin{pmatrix} e_{11} & e_{12} \\ e_{21} & e_{22} \end{pmatrix} = \mathbf{E}$$

The multiplication of a particular row and column, and the resulting product matrix element, are highlighted in red, demonstrating that matrix multiplication is a series of vector multiplications. Equivalently, vector multiplication is matrix multiplication that results in a 1×1 matrix or a scalar.

This approach is readily generalized to the multiplication of a matrix of order $m \times k$ by a matrix of order $k \times n$ to form a matrix of order $m \times n$. Consider the example of multiplying a 2×3 matrix by a 3×4 matrix to yield a 2×4 matrix:

$$\mathbf{FG} = \begin{pmatrix} -4 & 3 & 1 \\ 0 & -2 & 1 \end{pmatrix} \begin{pmatrix} 2 & 1 & -1 & 0 \\ 4 & -2 & 1 & -1 \\ -3 & 0 & 2 & 1 \end{pmatrix} = \begin{pmatrix} 1 & -10 & 9 & -2 \\ -11 & 4 & 0 & 3 \end{pmatrix} = \mathbf{H}$$

Note that the product $\mathbf{GF}$ is not defined using our rules of matrix multiplication explored in this chapter.

14.2.3 Properties of matrix multiplication

Now let's compare the basic properties of matrix multiplication with the familiar properties of scalar multiplication.

Associative property. The multiplication of scalars has the *associative property*, defined as follows:

$$a(bc) = (ab)c$$

The same associative property is true for matrices, where

$$\mathbf{A}(\mathbf{BC}) = (\mathbf{AB})\mathbf{C} \tag{14.6}$$

As in scalar multiplication, multiplying a square matrix $\mathbf{A}$ times itself is squaring the matrix, so that

$$\mathbf{AA} = \mathbf{A}^2 \tag{14.7}$$

In general, we can raise a matrix to arbitrary powers as $\mathbf{A}^n$.

Distributive property. The multiplication of scalars has the *distributive property*, defined as follows:

$$a(b + c) = ab + ac$$

The same distributive property is true for matrices, where

$$\mathbf{A}(\mathbf{B} + \mathbf{C}) = \mathbf{AB} + \mathbf{AC} \tag{14.8}$$

and

$$(\mathbf{A} + \mathbf{B})\mathbf{C} = \mathbf{AC} + \mathbf{BC}$$

[15] In compact notation we write

$$[\mathbf{C}]_{ij} = \sum_k [\mathbf{A}]_{ik} [\mathbf{B}]_{kj}$$

or even more compactly

$$[\mathbf{C}]_{ij} = [\mathbf{A}]_{ik} [\mathbf{B}]_{kj}$$

The lower expression uses the *Einstein summation convention* in which the repeated index k is implicitly summed over.

This is where the similarity between scalar and matrix multiplication ends.

Commutative property... or not! The multiplication of scalars has the *commutative property*, defined as follows:

$$ab = ba$$

which can also be expressed in terms of the *commutator*, defined as

$$[a, b] = ab - ba = 0$$

In this case, matrix multiplication does not share the general commutative property of scalar multiplication. Earlier we multiplied the 2×2 matrices $\mathbf{A}$ and $\mathbf{B}$ to find $\mathbf{C} = \mathbf{AB}$. Reversing the order of multiplication, we find

$$\mathbf{D} = \mathbf{BA} = \begin{pmatrix} b_{11} & b_{12} \\ b_{21} & b_{22} \end{pmatrix} \begin{pmatrix} a_{11} & a_{12} \\ a_{21} & a_{22} \end{pmatrix}$$

$$= \begin{pmatrix} b_{11}a_{11} + b_{12}a_{21} & b_{11}a_{12} + b_{12}a_{22} \\ b_{21}a_{11} + b_{22}a_{21} & b_{21}a_{12} + b_{22}a_{22} \end{pmatrix} = \begin{pmatrix} d_{11} & d_{12} \\ d_{21} & d_{22} \end{pmatrix}$$

Comparing the elements of the resulting matrices $\mathbf{C}$ and $\mathbf{D}$, it is clear that in general $\mathbf{C} \neq \mathbf{D}$, and the commutator

$$[\mathbf{A}, \mathbf{B}] = \mathbf{AB} - \mathbf{BA} \neq \mathbf{O} \tag{14.9}$$

$\mathbf{O}$ is the *zero matrix*, for which all elements are zero.

Let's examine a specific example. Consider the 2×2 matrices

$$\mathbf{A} = \begin{pmatrix} 0 & 2 \\ 1 & 0 \end{pmatrix} \qquad \mathbf{B} = \begin{pmatrix} 3 & 0 \\ 0 & -1 \end{pmatrix}$$

where

$$\mathbf{AB} = \begin{pmatrix} 0 & 2 \\ 1 & 0 \end{pmatrix} \begin{pmatrix} 3 & 0 \\ 0 & -1 \end{pmatrix} = \begin{pmatrix} 0 & -2 \\ 3 & 0 \end{pmatrix}$$

while

$$\mathbf{BA} = \begin{pmatrix} 3 & 0 \\ 0 & -1 \end{pmatrix} \begin{pmatrix} 0 & 2 \\ 1 & 0 \end{pmatrix} = \begin{pmatrix} 0 & 6 \\ -1 & 0 \end{pmatrix}$$

Clearly, $\mathbf{A}\,\mathbf{B} \neq \mathbf{B}\,\mathbf{A}$. We can also express this relation using the matrix commutator:

$$[\mathbf{A}, \mathbf{B}] = \mathbf{AB} - \mathbf{BA} = \begin{pmatrix} 0 & -2 \\ 3 & 0 \end{pmatrix} - \begin{pmatrix} 0 & 6 \\ -1 & 0 \end{pmatrix} = \begin{pmatrix} 0 & -8 \\ 4 & 0 \end{pmatrix} \neq \mathbf{O}$$

Since $[\mathbf{A}, \mathbf{B}] \neq \mathbf{O}$, the matrices $\mathbf{A}$ and $\mathbf{B}$ do not commute.

Ways to zero. Another property of matrix multiplication that differs from the multiplication of scalars is uncovered by the following example. For scalars, the equation

$$ab = 0$$

is only satisfied when a or b is equal to zero. For matrix multiplication, we can satisfy the equivalent relation

$$\mathbf{AB} = \mathbf{O} \qquad (14.10)$$

even when neither **A** nor **B** are the zero matrix. For example, consider

$$\mathbf{AB} = \begin{pmatrix} 1 & 1 \\ 2 & 2 \end{pmatrix} \begin{pmatrix} -1 & 1 \\ 1 & -1 \end{pmatrix} = \begin{pmatrix} 0 & 0 \\ 0 & 0 \end{pmatrix} = \mathbf{O}$$

These examples demonstrate key similarities and differences between scalar and matrix algebra. In the next section, we explore three types of matrices, each having special properties. As in the case of determinants, exploiting the fundamental properties of these matrices can be helpful in problem solving.

14.2.4 Survey of matrix multiplication

Let's look at an example of multiplying two 3×3 matrices:

$$\mathbf{A} = \begin{pmatrix} 1 & 2 & 1 \\ 3 & 0 & -1 \\ -1 & -1 & 2 \end{pmatrix} \qquad \mathbf{B} = \begin{pmatrix} -3 & 0 & -1 \\ 1 & 4 & 0 \\ 1 & 1 & 1 \end{pmatrix}$$

where the multiplication is carried out as follows:

$$\mathbf{AB} = \begin{pmatrix} 1 & 2 & 1 \\ 3 & 0 & -1 \\ -1 & -1 & 2 \end{pmatrix} \begin{pmatrix} -3 & 0 & -1 \\ 1 & 4 & 0 \\ 1 & 1 & 1 \end{pmatrix}$$

$$= \begin{pmatrix} -3+2+1 & 0+8+1 & -1+0+1 \\ -9+0-1 & 0+0-1 & -3+0-1 \\ 3-1+2 & 0-4+2 & 1+0+2 \end{pmatrix} = \begin{pmatrix} 0 & 9 & 0 \\ -10 & -1 & -4 \\ 4 & -2 & 3 \end{pmatrix}$$

We have used red to track the center 1×3 row vector in A and center 3×1 column vector in **B** to show how they are multiplied as a dot product to form the central element of the resulting matrix. Now let's explore the properties of matrix multiplication for three special classes of matrices.

Identity matrices

Consider the *identity matrix*, in which all diagonal elements are unity and all off-diagonal elements are zero:

$$\mathbf{I} = \begin{pmatrix} 1 & 0 & 0 \\ 0 & 1 & 0 \\ 0 & 0 & 1 \end{pmatrix}$$

The identity matrix derives its name from the fact that any matrix **A** multiplied by the identity matrix **I** equals the original matrix, such that

$$\mathbf{AI} = \begin{pmatrix} a_{11} & a_{12} & a_{13} \\ a_{21} & a_{22} & a_{23} \\ a_{31} & a_{32} & a_{33} \end{pmatrix} \begin{pmatrix} 1 & 0 & 0 \\ 0 & 1 & 0 \\ 0 & 0 & 1 \end{pmatrix} = \begin{pmatrix} a_{11} & a_{12} & a_{13} \\ a_{21} & a_{22} & a_{23} \\ a_{31} & a_{32} & a_{33} \end{pmatrix}$$

and

$$\mathbf{IA} = \begin{pmatrix} 1 & 0 & 0 \\ 0 & 1 & 0 \\ 0 & 0 & 1 \end{pmatrix} \begin{pmatrix} a_{11} & a_{12} & a_{13} \\ a_{21} & a_{22} & a_{23} \\ a_{31} & a_{32} & a_{33} \end{pmatrix} = \begin{pmatrix} a_{11} & a_{12} & a_{13} \\ a_{21} & a_{22} & a_{23} \\ a_{31} & a_{32} & a_{33} \end{pmatrix}$$

The identity matrix commutes with all matrices. The geometric interpretation of the determinant of the identity matrix is shown in Figure 14.9.

Block matrices

First consider a square *block matrix* that can be divided into blocks of largely non-zero elements and blocks in which all elements are zero. For example, consider the block matrices

$$\mathbf{A} = \begin{pmatrix} -2 & 1 & 0 & 0 \\ 6 & 5 & 0 & 0 \\ 0 & 0 & 9 & 2 \\ 0 & 0 & 1 & 4 \end{pmatrix} \qquad \mathbf{B} = \begin{pmatrix} 1 & 3 & 0 & 0 \\ 6 & -2 & 0 & 0 \\ 0 & 0 & 4 & 1 \\ 0 & 0 & 9 & -2 \end{pmatrix}$$

In the product matrix **AB**, the only non-zero elements will be those that are non-zero in the block matrices. The non-zero elements may be efficiently found by multiplying the two upper submatrices

$$\begin{pmatrix} -2 & 1 \\ 6 & 5 \end{pmatrix} \begin{pmatrix} 1 & 3 \\ 6 & -2 \end{pmatrix} = \begin{pmatrix} 4 & -8 \\ 36 & 8 \end{pmatrix}$$

and two lower submatrices

$$\begin{pmatrix} 9 & 2 \\ 1 & 4 \end{pmatrix} \begin{pmatrix} 4 & 1 \\ 9 & -2 \end{pmatrix} = \begin{pmatrix} 54 & 5 \\ 40 & -7 \end{pmatrix}$$

The resulting products of submatrices can be combined to form the product matrix

$$\mathbf{AB} = \begin{pmatrix} -2 & 1 & 0 & 0 \\ 6 & 5 & 0 & 0 \\ 0 & 0 & 9 & 2 \\ 0 & 0 & 1 & 4 \end{pmatrix} \begin{pmatrix} 1 & 3 & 0 & 0 \\ 6 & -2 & 0 & 0 \\ 0 & 0 & 4 & 1 \\ 0 & 0 & 9 & -2 \end{pmatrix} = \begin{pmatrix} 4 & -8 & 0 & 0 \\ 36 & 8 & 0 & 0 \\ 0 & 0 & 54 & 5 \\ 0 & 0 & 40 & -7 \end{pmatrix}$$

This efficiency can be extended to block matrices that are not square, as long as the submatrix products are defined by the normal rules for matrix multiplication.

Orthogonal matrices

Finally, let's explore the properties of *orthogonal matrices*, in which each column of the matrix, viewed as a vector, is orthogonal to every other column in the matrix.[16] For example, consider the rotation matrix

$$\mathbf{R}(\theta) = \begin{pmatrix} \cos\theta & -\sin\theta \\ \sin\theta & \cos\theta \end{pmatrix}$$

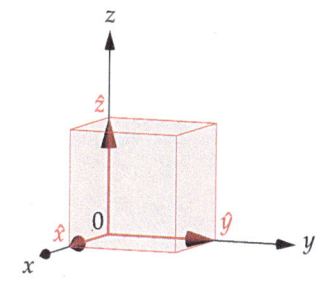

Figure 14.9: The magnitude of the determinant of the 3×3 identity matrix, $|\mathbf{I}| = 1$, is the volume of the unit cube (shaded red).

[16] Recall that two vectors **a** and **b** are orthogonal vectors when their dot product $\mathbf{a} \cdot \mathbf{b} = 0$. This property was explored in Chapter 4.

We can view this matrix as being composed of two vectors

$$\mathbf{a} = \begin{pmatrix} \cos\theta \\ \sin\theta \end{pmatrix} \qquad \mathbf{b} = \begin{pmatrix} -\sin\theta \\ \cos\theta \end{pmatrix}$$

If we take the dot product of the two vectors, we see that

$$\mathbf{a} \cdot \mathbf{b} = \begin{pmatrix} \cos\theta & \sin\theta \end{pmatrix} \begin{pmatrix} -\sin\theta \\ \cos\theta \end{pmatrix} = -\cos\theta\sin\theta + \sin\theta\cos\theta = 0$$

This indicates that the vectors $\mathbf{a}$ and $\mathbf{b}$ are orthogonal. In addition, we find

$$\mathbf{a} \cdot \mathbf{a} = \begin{pmatrix} \cos\theta & \sin\theta \end{pmatrix} \begin{pmatrix} \cos\theta \\ \sin\theta \end{pmatrix} = \cos^2\theta + \sin^2\theta = 1$$

and

$$\mathbf{b} \cdot \mathbf{b} = \begin{pmatrix} -\sin\theta & \cos\theta \end{pmatrix} \begin{pmatrix} -\sin\theta \\ \cos\theta \end{pmatrix} = \sin^2\theta + \cos^2\theta = 1$$

This indicates that the vectors $\mathbf{a}$ and $\mathbf{b}$ are normalized.

As the vectors $\mathbf{a}$ and $\mathbf{b}$ are mutually orthogonal and individually normalized, they are *orthonormal vectors* (see Chapter 4). It follows that the matrix $\mathbf{R}(\theta)$ multiplied by its transpose leads to the identity matrix, such that

$$\mathbf{R}(\theta)^T\mathbf{R}(\theta) = \begin{pmatrix} \cos\theta & \sin\theta \\ -\sin\theta & \cos\theta \end{pmatrix} \begin{pmatrix} \cos\theta & -\sin\theta \\ \sin\theta & \cos\theta \end{pmatrix} = \begin{pmatrix} 1 & 0 \\ 0 & 1 \end{pmatrix} = \mathbf{I}$$

and

$$\mathbf{R}(\theta)\mathbf{R}(\theta)^T = \begin{pmatrix} \cos\theta & -\sin\theta \\ \sin\theta & \cos\theta \end{pmatrix} \begin{pmatrix} \cos\theta & \sin\theta \\ -\sin\theta & \cos\theta \end{pmatrix} = \begin{pmatrix} 1 & 0 \\ 0 & 1 \end{pmatrix} = \mathbf{I}$$

In fact, an orthonormal matrix times its transpose is always the identity matrix. Given that $\mathbf{R}(\theta)^T\mathbf{R}(\theta) = \mathbf{R}(\theta)\mathbf{R}(\theta)^T = \mathbf{I}$, it follows that the commutator is

$$\left[\mathbf{R}(\theta)^T, \mathbf{R}(\theta)\right] = \mathbf{R}(\theta)^T\mathbf{R}(\theta) - \mathbf{R}(\theta)\mathbf{R}(\theta)^T = \mathbf{I} - \mathbf{I} = \mathbf{O}$$

so that the orthogonal matrices $\mathbf{R}(\theta)$ and $\mathbf{R}(\theta)^T$ commute.

Consider one more example involving the matrix

$$\mathbf{A} = \begin{pmatrix} \dfrac{1}{\sqrt{2}} & \dfrac{1}{\sqrt{2}} \\ \dfrac{1}{\sqrt{2}} & -\dfrac{1}{\sqrt{2}} \end{pmatrix}$$

We can demonstrate that $\mathbf{A}$ is an orthogonal matrix, since

$$\begin{pmatrix} \dfrac{1}{\sqrt{2}} & \dfrac{1}{\sqrt{2}} \end{pmatrix} \begin{pmatrix} \dfrac{1}{\sqrt{2}} \\ \dfrac{1}{\sqrt{2}} \end{pmatrix} = \begin{pmatrix} \dfrac{1}{\sqrt{2}} & -\dfrac{1}{\sqrt{2}} \end{pmatrix} \begin{pmatrix} \dfrac{1}{\sqrt{2}} \\ -\dfrac{1}{\sqrt{2}} \end{pmatrix} = \dfrac{1}{2} + \dfrac{1}{2} = 1$$

and

$$\left(\frac{1}{\sqrt{2}} \quad \frac{1}{\sqrt{2}} \right) \begin{pmatrix} \frac{1}{\sqrt{2}} \\ -\frac{1}{\sqrt{2}} \end{pmatrix} = \left(\frac{1}{\sqrt{2}} \quad -\frac{1}{\sqrt{2}} \right) \begin{pmatrix} \frac{1}{\sqrt{2}} \\ \frac{1}{\sqrt{2}} \end{pmatrix} = \frac{1}{2} - \frac{1}{2} = 0$$

As expected for an orthogonal matrix, we find that

$$\mathbf{A}^T \mathbf{A} = \begin{pmatrix} \frac{1}{\sqrt{2}} & \frac{1}{\sqrt{2}} \\ \frac{1}{\sqrt{2}} & -\frac{1}{\sqrt{2}} \end{pmatrix} \begin{pmatrix} \frac{1}{\sqrt{2}} & \frac{1}{\sqrt{2}} \\ \frac{1}{\sqrt{2}} & -\frac{1}{\sqrt{2}} \end{pmatrix} = \begin{pmatrix} 1 & 0 \\ 0 & 1 \end{pmatrix}$$

and

$$\mathbf{A}\mathbf{A}^T = \begin{pmatrix} \frac{1}{\sqrt{2}} & \frac{1}{\sqrt{2}} \\ \frac{1}{\sqrt{2}} & -\frac{1}{\sqrt{2}} \end{pmatrix} \begin{pmatrix} \frac{1}{\sqrt{2}} & \frac{1}{\sqrt{2}} \\ \frac{1}{\sqrt{2}} & -\frac{1}{\sqrt{2}} \end{pmatrix} = \begin{pmatrix} 1 & 0 \\ 0 & 1 \end{pmatrix}$$

The geometric interpretation of the determinant of the orthogonal matrix **A** is shown in Figure 14.10. This is a special case of the general observation that the determinant $|\mathbf{A}|$ of the 2×2 matrix **A** is the area of a parallelogram (see Figure 14.3). Orthogonal vectors form a right angle. In addition, the vectors are normalized. As such, the determinant is the area of a unit square. Orthogonal matrices commonly arise in problems in the physical sciences, including quantum theory and the vibrational dynamics of molecules.

14.2.5 Matrix division and the matrix inverse

In considering the division of scalars we recognize that

$$a\frac{1}{a} = aa^{-1} = 1$$

where we say that a^{-1} is the inverse of a, and the product of a scalar and its inverse is unity. For matrices, the equivalent expression is

$$\mathbf{A}\mathbf{A}^{-1} = \mathbf{I}$$

where **I** is the identity matrix and $\mathbf{A}^{-1}$ is known as the *matrix inverse* of **A**.

While it is a simple matter to determine the inverse of a scalar a, as long as $a \neq 0$, finding the inverse of a matrix takes more effort. Fortunately, as long as the matrix **A** is *non-singular* meaning that

$$|\mathbf{A}| \neq 0$$

there is a systematic method for determining the matrix inverse, $\mathbf{A}^{-1}$. Recall the definition of a matrix **A** and its transpose $\mathbf{A}^T$

$$\mathbf{A} = \begin{pmatrix} a_{11} & a_{12} \\ a_{21} & a_{22} \end{pmatrix} \qquad \mathbf{A}^T = \begin{pmatrix} a_{11} & a_{21} \\ a_{12} & a_{22} \end{pmatrix}$$

Further recall that the *cofactor* of a matrix **A**, written A_{ij}, is found by first

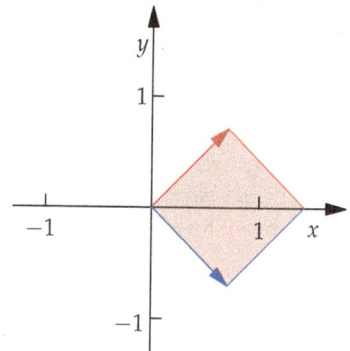

Figure 14.10: The magnitude of the determinant of the 2×2 orthogonal matrix **A** is $|\mathbf{A}| = 1$. Geometrically, the magnitude of $|\mathbf{A}|$ is the shaded area of the unit square, defined by the orthonormal column vectors forming **A**, which is unity.

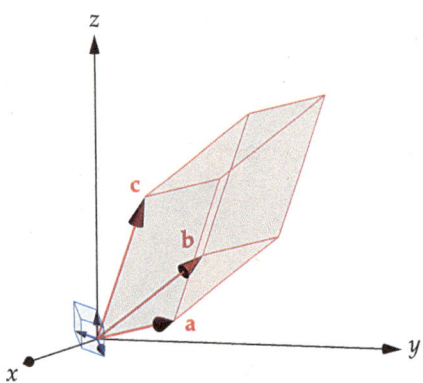

Figure 14.11: The magnitude of the determinant $|\mathbf{A}|$ is the volume of the larger parallelepiped (shaded red) while the magnitude of the determinant $|\mathbf{A}^{-1}|$ is the volume of the smaller parallelepiped (shaded blue). The two are related by $|\mathbf{A}^{-1}| = |\mathbf{A}|^{-1}$.

removing the ith row and jth column of the original matrix $\mathbf{A}$, then taking the determinant of the resulting $(n-1) \times (n-1)$ square matrix, and multiplying by $(-1)^{i+j}$. Any $n \times n$ square matrix $\mathbf{A}$ has $n \times n$ cofactors A_{ij}. Those cofactors form the elements of the cofactor matrix, defined as

$$\mathbf{A}_{cof} = \begin{pmatrix} A_{11} & A_{12} \\ A_{21} & A_{22} \end{pmatrix}$$

where each element is the corresponding cofactor of $\mathbf{A}$, A_{ij}. The inverse of a *non-singular matrix* $\mathbf{A}$, for which the determinant $|\mathbf{A}| \neq 0$, can be written in the compact form

$$\mathbf{A}^{-1} = \frac{1}{|\mathbf{A}|} \mathbf{A}_{cof}^{T}$$

where $\mathbf{A}_{cof}^{T}$ is the transpose of the cofactor matrix. Geometrically, the magnitude of the determinant of the matrix $\mathbf{A}$ and that of its matrix inverse $\mathbf{A}^{-1}$ are related by

$$\left| \mathbf{A}^{-1} \right| = |\mathbf{A}|^{-1}$$

The geometric interpretation of this properties is depicted in Figure 14.11.

Let's use this definition to determine the inverse of the matrix

$$\mathbf{A} = \begin{pmatrix} 1 & 2 \\ 3 & 4 \end{pmatrix} \tag{14.11}$$

for which the determinant is

$$|\mathbf{A}| = \begin{vmatrix} 1 & 2 \\ 3 & 4 \end{vmatrix} = 4 - 6 = -2$$

The matrix cofactors are $A_{11} = (-1)^{1+1}\, 4 = 4$, $A_{21} = (-1)^{2+1}\, 2 = -2$, $A_{12} = (-1)^{1+2}\, 3 = -3$, and $A_{22} = (-1)^{2+2}\, 1 = 1$. As such, the cofactor matrix is

$$\mathbf{A}_{cof} = \begin{pmatrix} A_{11} & A_{12} \\ A_{21} & A_{22} \end{pmatrix} = \begin{pmatrix} 4 & -3 \\ -2 & 1 \end{pmatrix}$$

Taking the transpose of the cofactor matrix and dividing by the determinant leads to the final result

$$\mathbf{A}^{-1} = \frac{1}{|\mathbf{A}|}\mathbf{A}^T_{cof} = \frac{1}{-2}\begin{pmatrix} 4 & -2 \\ -3 & 1 \end{pmatrix} = \begin{pmatrix} -2 & 1 \\ \frac{3}{2} & -\frac{1}{2} \end{pmatrix} \qquad (14.12)$$

We can validate our solution by determining the product of the matrix $\mathbf{A}$ and its matrix inverse $\mathbf{A}^{-1}$ as

$$\mathbf{AA}^{-1} = \begin{pmatrix} 1 & 2 \\ 3 & 4 \end{pmatrix}\begin{pmatrix} -2 & 1 \\ \frac{3}{2} & -\frac{1}{2} \end{pmatrix} = \begin{pmatrix} -2+3 & 1-1 \\ -6+6 & 3-2 \end{pmatrix} = \begin{pmatrix} 1 & 0 \\ 0 & 1 \end{pmatrix} = \mathbf{I}$$

or

$$\mathbf{A}^{-1}\mathbf{A} = \begin{pmatrix} -2 & 1 \\ \frac{3}{2} & -\frac{1}{2} \end{pmatrix}\begin{pmatrix} 1 & 2 \\ 3 & 4 \end{pmatrix} = \begin{pmatrix} -2+3 & -4+4 \\ \frac{3}{2}-\frac{3}{2} & 3-2 \end{pmatrix} = \begin{pmatrix} 1 & 0 \\ 0 & 1 \end{pmatrix} = \mathbf{I}$$

Given that $\mathbf{A}^{-1}\mathbf{A} = \mathbf{AA}^{-1} = \mathbf{I}$, it follows that the commutator

$$\left[\mathbf{A}^{-1},\mathbf{A}\right] = \mathbf{A}^{-1}\mathbf{A} - \mathbf{AA}^{-1} = \mathbf{I} - \mathbf{I} = \mathbf{O}$$

so that a matrix $\mathbf{A}$ and its matrix inverse $\mathbf{A}^{-1}$ always commute.

The same procedure can be followed to determine the matrix inverse of a 3×3 matrix. Consider the matrix

$$\mathbf{B} = \begin{pmatrix} 3 & 0 & -1 \\ 2 & 2 & 1 \\ 0 & -1 & 2 \end{pmatrix} \qquad (14.13)$$

We first compute the matrix determinant

$$|\mathbf{B}| = 3(4+1) - 0(4-0) - 1(-2-0) = 15+0+2 = 17$$

to demonstrate that the matrix is not singular. The corresponding cofactor matrix is determined by evaluating the nine cofactor matrix elements. The resulting cofactor matrix is[17]

$$\mathbf{B}_{cof} = \begin{pmatrix} 5 & -4 & -2 \\ 1 & 6 & 3 \\ 2 & -5 & 6 \end{pmatrix}$$

Multiplying the transpose of the cofactor matrix by the inverse of the matrix determinant, we find the matrix inverse

$$\mathbf{B}^{-1} = \frac{1}{|\mathbf{B}|}\mathbf{B}^T_{cof} = \frac{1}{17}\begin{pmatrix} 5 & 1 & 2 \\ -4 & 6 & -5 \\ -2 & 3 & 6 \end{pmatrix} \qquad (14.14)$$

This approach to finding the matrix inverse is generally applicable to any $n \times n$ non-singular square matrix. We will find that determining the matrix inverse has great utility in solving algebraic equations involving matrices that commonly arise in the physical sciences.

[17] For example, the cofactor B_{11} of the matrix $\mathbf{B}$ is

$$B_{11} = (-1)^{1+1}\begin{vmatrix} 2 & 1 \\ -1 & 2 \end{vmatrix} = 1(4+1) = 5$$

A similar procedure is followed to evaluate the other eight elements of the cofactor matrix $\mathbf{B}_{cof}$.

14.2.6 *Multiplication of matrices by vectors*

By the rules of matrix algebra the multiplication of a matrix of order $m \times k$ by a matrix of order $k \times n$ results in a matrix of order $m \times n$. For example, consider the multiplication of a 2×3 matrix $\mathbf{A}$ and a 3×2 matrix $\mathbf{B}$:

$$
\begin{aligned}
\mathbf{AB} &= \begin{pmatrix} a_{11} & a_{12} & a_{13} \\ a_{21} & a_{22} & a_{23} \end{pmatrix} \begin{pmatrix} b_{11} & b_{12} \\ b_{21} & b_{22} \\ b_{31} & b_{32} \end{pmatrix} \\
&= \begin{pmatrix} a_{11}b_{11} + a_{12}b_{21} + a_{13}b_{31} & a_{11}b_{12} + a_{12}b_{22} + a_{13}b_{32} \\ a_{21}b_{11} + a_{22}b_{21} + a_{23}b_{31} & a_{21}b_{12} + a_{22}b_{22} + a_{23}b_{32} \end{pmatrix}
\end{aligned}
$$

This results in a product matrix of order 2×2.

A special case of multiplication of matrices of differing order involves the multiplication of a square matrix by a row or column vector, such as

$$
\begin{aligned}
\mathbf{CD} &= \begin{pmatrix} c_{11} & c_{12} & c_{13} \\ c_{21} & c_{22} & c_{23} \\ c_{31} & c_{32} & c_{33} \end{pmatrix} \begin{pmatrix} d_{11} \\ d_{21} \\ d_{31} \end{pmatrix} \\
&= \begin{pmatrix} c_{11}d_{11} + c_{12}d_{21} + c_{13}d_{31} \\ c_{21}d_{11} + c_{22}d_{21} + c_{23}d_{31} \\ c_{31}d_{11} + c_{32}d_{21} + c_{33}d_{31} \end{pmatrix} = \begin{pmatrix} e_{11} \\ e_{21} \\ e_{31} \end{pmatrix} = \mathbf{E}
\end{aligned}
$$

where a 3×3 matrix $\mathbf{C}$ is multiplied by a 3×1 column vector $\mathbf{D}$ to yield a new 3×1 column vector $\mathbf{E}$. Similarly, multiplying a 1×2 row vector $\mathbf{F}$ by a 2×2 square matrix $\mathbf{G}$ results in a 1×2 row vector $\mathbf{H}$:

$$
\begin{aligned}
\mathbf{FG} &= \begin{pmatrix} f_{11} & f_{12} \end{pmatrix} \begin{pmatrix} g_{11} & g_{12} \\ g_{21} & g_{22} \end{pmatrix} \\
&= \begin{pmatrix} f_{11}g_{11} + f_{12}g_{21} & f_{11}g_{12} + f_{12}g_{22} \end{pmatrix} \\
&= \begin{pmatrix} h_{11} & h_{12} \end{pmatrix} = \mathbf{H}
\end{aligned}
$$

These examples demonstrate how multiplication of a $m \times k$ matrix by a $k \times n$ results in a matrix of order $m \times n$.

Let's practice multiplying matrices by vectors while exploring the properties of rotation matrices in two and three dimensions. Recall the 2×2 rotation matrix

$$
\mathbf{R}(\theta) = \begin{pmatrix} \cos\theta & -\sin\theta \\ \sin\theta & \cos\theta \end{pmatrix}
$$

Now consider $\mathbf{R}(\theta)$ multiplied by a vector $\mathbf{x}_1$ in two-dimensional cartesian space as

$$
\mathbf{R}(\theta)\mathbf{x}_1 = \mathbf{x}_2
$$

The rotation matrix acts as a *matrix operator* by transforming vector $\mathbf{x}_1$ into a new vector $\mathbf{x}_2$ by way of a counter-clockwise rotation in space through an angle θ. For example, suppose $\mathbf{x}_1$ is aligned along the x-axis and $\theta = \frac{\pi}{2}$.

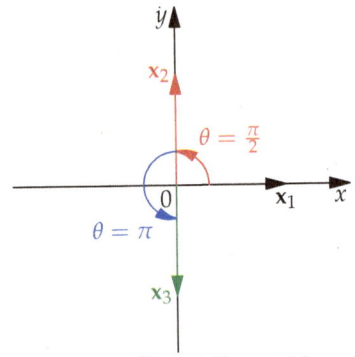

Figure 14.12: The rotation matrix $\mathbf{R}(\frac{\pi}{2})$ transforms a vector $\mathbf{x}_1$ into a new vector $\mathbf{x}_2 = \mathbf{R}(\frac{\pi}{2})\mathbf{x}_1$ by rotating $\mathbf{x}_1$ counter-clockwise through an angle $\frac{\pi}{2}$ in the xy-plane. The vector $\mathbf{x}_2$ is transformed as $\mathbf{x}_3 = \mathbf{R}(\pi)\mathbf{x}_2$.

Accordingly, we find

$$\mathbf{R}(\theta)\mathbf{x}_1 = \begin{pmatrix} 0 & -1 \\ 1 & 0 \end{pmatrix} \begin{pmatrix} 1 \\ 0 \end{pmatrix} = \begin{pmatrix} 0 \\ 1 \end{pmatrix} = \mathbf{x}_2$$

The rotation matrix acts to rotate vector $\mathbf{x}_1$, aligned along the positive x-axis, counter-clockwise through an angle $\theta = \frac{\pi}{2}$ to form $\mathbf{x}_2$, aligned along the positive y-axis (see Figure 14.12).

Now suppose $\mathbf{R}(\theta)$ acts on $\mathbf{x}_2$, aligned along the positive y-axis, with $\theta = \pi$. The result is

$$\mathbf{R}(\theta)\mathbf{x}_2 = \begin{pmatrix} -1 & 0 \\ 0 & -1 \end{pmatrix} \begin{pmatrix} 0 \\ 1 \end{pmatrix} = \begin{pmatrix} 0 \\ -1 \end{pmatrix} = \mathbf{x}_3$$

where the rotation matrix acts to rotate $\mathbf{x}_2$, aligned with the positive y-axis, counter-clockwise through an angle $\theta = \pi$ to form $\mathbf{x}_3$, aligned with the negative y-axis.

Now consider the extension of the rotation matrix in two dimensions to three dimensions. In two dimensions, we rotate about a single axis. In three dimensions, three matrices are required for rotation about the x, y, or z axes. Consider the matrix for counter-clockwise rotation about the positive x-axis:[18]

$$\mathbf{R}_x(\theta) = \begin{pmatrix} 1 & 0 & 0 \\ 0 & \cos\theta & -\sin\theta \\ 0 & \sin\theta & \cos\theta \end{pmatrix}$$

Note that the terms in red mark the components of the two-dimensional rotation matrix that is embedded in the rotation matrix in three dimensions. The unit element acts to maintain the current value of x while the red terms act to rotate the vector in the yz-plane.

Suppose $\mathbf{x}_1$ is aligned along the positive z-axis and $\theta = \pi$. Accordingly, we find

$$\mathbf{R}_x(\theta)\mathbf{x}_1 = \begin{pmatrix} 1 & 0 & 0 \\ 0 & -1 & 0 \\ 0 & 0 & -1 \end{pmatrix} \begin{pmatrix} 0 \\ 0 \\ 1 \end{pmatrix} = \begin{pmatrix} 0 \\ 0 \\ -1 \end{pmatrix} = \mathbf{x}_2$$

where the rotation matrix acts to rotate $\mathbf{x}_1$, aligned with the positive z-axis, through an angle $\theta = \pi$ to form $\mathbf{x}_2$, aligned with the negative z-axis.

14.2.7 Applications involving complex matrices

Until now we have largely considered matrices for which the elements are real. However, as we found in the case of *Pauli matrices* and *quaternions*, we can also form matrices where the elements are imaginary or complex. In fact, matrices with complex elements are common in the physical sciences, and complex matrices have many useful properties.

For a complex number $z = x + iy$, the *absolute value* is

$$|z| = \sqrt{z^*z} = \sqrt{(x - iy)(x + iy)} = \sqrt{x^2 + y^2}$$

where $z^* = x - iy$ is the *complex conjugate* of z. Consider the same relation

[18] The rotation matrix for counter-clockwise rotation about the positive y-axis is written

$$\mathbf{R}_y(\theta) = \begin{pmatrix} \cos\theta & 0 & \sin\theta \\ 0 & 1 & 0 \\ -\sin\theta & 0 & \cos\theta \end{pmatrix}$$

and for counter-clockwise rotation about the positive z-axis

$$\mathbf{R}_z(\theta) = \begin{pmatrix} \cos\theta & -\sin\theta & 0 \\ \sin\theta & \cos\theta & 0 \\ 0 & 0 & 1 \end{pmatrix}$$

You may also encounter rotation matrices defined in terms of *clockwise* rotation. Those matrices have the same form where $-\sin\theta$ is substituted for $\sin\theta$.

where we write $z = re^{i\theta}$ and $z^* = re^{-i\theta}$. We find

$$|z|^2 = z^*z = x^2 + y^2 = r^2 = re^{-i\theta}re^{i\theta}$$

where $z = re^{i\theta}$ and $z^* = re^{-i\theta}$. We think of $|z|$ as the magnitude of the complex number z.

Recall that for a vector $\mathbf{a}$, the absolute value $|\mathbf{a}|$ is defined in terms of the dot product

$$|\mathbf{a}|^2 = \mathbf{a} \cdot \mathbf{a}$$

where $|\mathbf{a}|$ is the magnitude. If we have a complex column vector $\mathbf{c}$, we can define the absolute value of $\mathbf{c}$ by combining the concepts of the complex conjugate and the vector magnitude as

$$|\mathbf{c}|^2 = \mathbf{c}^{*T}\mathbf{c} = \begin{pmatrix} a^* & b^* \end{pmatrix} \begin{pmatrix} a \\ b \end{pmatrix} = a^*a + b^*b$$

where the result is a real scalar.

Consider the example of the complex vector

$$\mathbf{c} = \begin{pmatrix} c_{11} \\ c_{21} \\ c_{31} \end{pmatrix} = \begin{pmatrix} 2i \\ 4 \\ 2+i \end{pmatrix}$$

for which the absolute value $|\mathbf{c}|$ is defined in terms of

$$\begin{aligned} |\mathbf{c}|^2 = \mathbf{c}^{*T}\mathbf{c} &= \begin{pmatrix} -2i & 4 & 2-i \end{pmatrix} \begin{pmatrix} 2i \\ 4 \\ 2+i \end{pmatrix} \\ &= (-2i)(2i) + (4)(4) + (2-i)(2+i) \\ &= 4 + 16 + (4+1) = 25 \end{aligned}$$

As such the absolute value and magnitude $|\mathbf{c}| = \sqrt{|\mathbf{c}|^2} = \sqrt{25} = 5$.

For matrices composed of real elements, a *symmetric matrix* is equal to its matrix transpose:[19]

$$\mathbf{A} = \mathbf{A}^T$$

[19] A symmetric matrix satisfies the equality $[\mathbf{A}]_{ij} = [\mathbf{A}]_{ji}$.

For a complex matrix, there is a corresponding property relating a matrix $\mathbf{C}$ and its complex conjugate $\mathbf{C}^*$. When a matrix $\mathbf{C}$ is equal to the transpose of its complex conjugate,[20] such that

$$\mathbf{C} = (\mathbf{C}^*)^T$$

[20] In compact notation this is written $[\mathbf{C}]_{ij} = [\mathbf{C}^*]_{ji}$.

we say that $\mathbf{C}$ is a *hermitian matrix*.[21] The matrix $(\mathbf{C}^*)^T$ is called the *hermitian conjugate* of $\mathbf{C}$, just as z^* is referred to as the *complex conjugate* of z.

[21] Note that all real, symmetric matrices are hermitian matrices.

Hermitian matrices are regularly encountered in the physical sciences and play a special role in quantum theory. As such, a compact notation has been developed to express hermitian matrices:

$$\mathbf{C}^\dagger = (\mathbf{C}^*)^T$$

As such, we can write

$$\mathbf{C}^{\dagger}\mathbf{C} = (\mathbf{C}^{*})^{T}\mathbf{C}$$

When a matrix $\mathbf{D}$ is both hermitian and orthogonal, it is called a *unitary matrix*. In that case, the matrix has the property that

$$\mathbf{D}^{\dagger}\mathbf{D} = \mathbf{D}\mathbf{D}^{\dagger} = \mathbf{I}$$

The unitary matrix is composed of columns having the properties of orthonormal vectors. Namely, they are normalized and mutually orthogonal.

Consider the complex matrix

$$\mathbf{E} = \frac{1}{5}\begin{pmatrix} -1+2i & -4-2i \\ 2-4i & -2-i \end{pmatrix}$$

We can demonstrate that $\mathbf{E}$ is a unitary matrix by proving that the set of column vectors composing the matrix has the properties of a set of orthonormal vectors, namely individual normalization

$$\frac{1}{5^2}\begin{pmatrix} -1-2i & 2+4i \end{pmatrix}\begin{pmatrix} -1+2i \\ 2-4i \end{pmatrix} = \frac{1}{25}[(1+4)+(4+16)] = 1$$

$$\frac{1}{5^2}\begin{pmatrix} -4+2i & -2+i \end{pmatrix}\begin{pmatrix} -4-2i \\ -2-i \end{pmatrix} = \frac{1}{25}[(16+4)+(4+1)] = 1$$

and mutual orthogonality

$$\frac{1}{5^2}\begin{pmatrix} -1-2i & 2+4i \end{pmatrix}\begin{pmatrix} -4-2i \\ -2-i \end{pmatrix} = \frac{1}{25}[(4-4)+(-4+4)] = 0$$

$$\frac{1}{5^2}\begin{pmatrix} -4+2i & -2+i \end{pmatrix}\begin{pmatrix} -1+2i \\ 2-4i \end{pmatrix} = \frac{1}{25}[(4-4)+(-4+4)] = 0$$

Furthermore, as expected for a unitary matrix, we find that the product of the matrix $\mathbf{E}$ and its hermitian conjugate $(\mathbf{E}^{*})^{T}$ results in the identity matrix $\mathbf{I}$, regardless of the order of multiplication:

$$(\mathbf{E}^{*})^{T}\mathbf{E} = \frac{1}{5^2}\begin{pmatrix} -1+2i & -4-2i \\ 2-4i & -2-i \end{pmatrix}\begin{pmatrix} -1-2i & 2+4i \\ -4+2i & -2+i \end{pmatrix}$$

$$= \frac{1}{25}\begin{pmatrix} (1+4)+(16+4) & (-2-8)+(8+2) \\ (-2-8)+(8+2) & (4+16)+(4+1) \end{pmatrix} = \begin{pmatrix} 1 & 0 \\ 0 & 1 \end{pmatrix}$$

and

$$\mathbf{E}(\mathbf{E}^{*})^{T} = \frac{1}{5^2}\begin{pmatrix} -1-2i & 2+4i \\ -4+2i & -2+i \end{pmatrix}\begin{pmatrix} -1+2i & -4-2i \\ 2-4i & -2-i \end{pmatrix}$$

$$= \frac{1}{25}\begin{pmatrix} (1+4)+(4+16) & (4-4)+(-4+4) \\ (4-4)+(-4+4) & (16+4)+(4+1) \end{pmatrix} = \begin{pmatrix} 1 & 0 \\ 0 & 1 \end{pmatrix}$$

This example demonstrates how the properties of real orthogonal matrices can be generalized to the case of complex matrices.

14.3 Solving coupled linear equations using Cramer's rule

OUR STUDY OF LINEAR ALGEBRA BEGAN with the solution of two coupled linear equations and led to the definition of the determinant. We return to the topic to apply our methods of matrix algebra to the general solution of systems of linear equations. We will develop a general method for solving for an arbitrary number of independent linear equations with the same number of unknown variables.

14.3.1 Solving coupled linear equations using determinants

Systems of linear equations are commonly encountered in the physical sciences. In the process of solving coupled linear equations for a set of unknown variables in terms of some constant coefficients, algebraic terms equivalent to a determinant naturally appear. Let's see why.

Consider Equation 14.2, consisting of two equations with two unknown variables x and y:

$$a_{11}x + a_{12}y = d_1$$
$$a_{21}x + a_{22}y = d_2$$

Now consider the determinant:

$$\mathcal{D} = \begin{vmatrix} a_{11} & a_{12} \\ a_{21} & a_{22} \end{vmatrix} = a_{11}a_{22} - a_{12}a_{21}$$

We know that if we multiply a column of the matrix by a constant x, the determinant of the new matrix will be

$$\begin{vmatrix} a_{11}x & a_{12} \\ a_{21}x & a_{22} \end{vmatrix} = a_{11}a_{22}x - a_{12}a_{21}x = \mathcal{D}x$$

Furthermore, we know that if we add one column of the matrix to another, the determinant is unchanged, as

$$\begin{vmatrix} a_{11}x + a_{12}y & a_{12} \\ a_{21}x + a_{22}y & a_{22} \end{vmatrix} = (a_{11}x + a_{12}y)\,a_{22} - a_{12}\left(a_{21}x + a_{22}y\right)$$

$$= a_{11}a_{22}x + a_{12}a_{22}y - a_{12}a_{21}x - a_{12}a_{22}y$$

$$= a_{11}a_{22}x - a_{12}a_{21}x = \mathcal{D}x$$

Noting that $a_{11}x + a_{12}y = d_1$ and $a_{21}x + a_{22}y = d_2$, we can reform the determinant as

$$\mathcal{D}x = \begin{vmatrix} a_{11}x + a_{12}y & a_{12} \\ a_{21}x + a_{22}y & a_{22} \end{vmatrix} = \begin{vmatrix} d_1 & a_{12} \\ d_2 & a_{22} \end{vmatrix}$$

which can be solved for x as

$$x = \frac{1}{\mathcal{D}} \begin{vmatrix} d_1 & a_{12} \\ d_2 & a_{22} \end{vmatrix} = \frac{d_1 a_{22} - d_2 a_{12}}{a_{11}a_{22} - a_{21}a_{12}}$$

Similar manipulations lead to a solution for y, as

$$y = \frac{1}{\mathcal{D}} \begin{vmatrix} a_{11} & d_1 \\ a_{21} & d_2 \end{vmatrix} = \frac{d_2 a_{11} - d_1 a_{21}}{a_{11} a_{22} - a_{12} a_{21}}$$

This solution agrees with our prior results in Equations 14.3 and 14.4. Our general solution expressed through determinants is known as *Cramer's rule*.[22]

[22] Named for Swiss mathematician *Gabriel Cramer* (1704-1752).

14.3.2 *Generalization of Cramer's rule to arbitrary dimension*

Using our knowledge of matrix algebra and the properties of determinants, our derivation of Cramer's rule can be generalized to n coupled equations in n variables. We start by generalizing the expression for our coupled equations in Equation 14.2 to an arbitrary number of variables:

$$\mathbf{Ax} = \mathbf{d} \tag{14.15}$$

In this expression, $\mathbf{x}$ is a vector of n unknown variables $x_1, x_2, \ldots, x_n$, written as

$$\mathbf{x} = \begin{pmatrix} x_1 \\ x_2 \\ \vdots \\ x_n \end{pmatrix}$$

and $\mathbf{A}$ is a $n \times n$ matrix of coefficients, written as

$$\mathbf{A} = \begin{pmatrix} a_{11} & a_{12} & \cdots & a_{1n} \\ a_{21} & a_{22} & \cdots & a_{2n} \\ \vdots & \vdots & \ddots & \vdots \\ a_{n1} & a_{n2} & \cdots & a_{nn} \end{pmatrix} \tag{14.16}$$

Finally, the vector $\mathbf{d}$ represents the n constants in the linear equations:

$$\mathbf{d} = \begin{pmatrix} d_1 \\ d_2 \\ \vdots \\ d_n \end{pmatrix}$$

Through $\mathbf{x}$, $\mathbf{A}$, and $\mathbf{d}$, the n linear equations can be represented as $\mathbf{Ax} = \mathbf{d}$.

The general solution for the n unknown variables can be written as

$$x_k = \frac{1}{|\mathbf{A}|} |\mathbf{A}_k| \tag{14.17}$$

where $\mathbf{A}_k$ is the matrix $\mathbf{A}$ in which the kth column is replaced by the vector $\mathbf{d}$:

$$\mathbf{A}_k = \begin{vmatrix} a_{11} & a_{12} & \cdots & a_{1(k-1)} & d_1 & a_{1(k+1)} & \cdots & a_{1n} \\ a_{21} & a_{22} & \cdots & a_{2(k-1)} & d_2 & a_{2(k+1)} & \cdots & a_{2n} \\ \vdots & \vdots & \ddots & \vdots & \vdots & \vdots & \ddots & \vdots \\ a_{n1} & a_{n2} & \cdots & a_{n(k-1)} & d_n & a_{n(k+1)} & \cdots & a_{nn} \end{vmatrix} \tag{14.18}$$

In principle, Cramer's rule allows us to solve an arbitrarily large system of linear equations defining the same number of unknown variables. However, note that for a system of n equations with n unknowns, Cramer's rule requires the evaluation of $n + 1$ determinants. As such, the application of Cramer's rule can be costly for systems of equations where n is large. In those cases, more efficient approaches can be considered. A method known as *Gaussian elimination* achieves the same result with the effort of computing a single determinant. That method is explored in the complements.

14.3.3 *Applications of Cramer's rule*

Let's explore an application of Cramer's rule by solving the following set of coupled linear equations:

$$3x - y = 6$$
$$x + 2y = 5 \tag{14.19}$$

We can rewrite these equations in matrix form as

$$\mathbf{Ax} = \begin{pmatrix} 3 & -1 \\ 1 & 2 \end{pmatrix} \begin{pmatrix} x \\ y \end{pmatrix} = \begin{pmatrix} 6 \\ 5 \end{pmatrix} = \mathbf{d}$$

Using Cramer's rule, the solution for x and y can be written as

$$x = \frac{1}{|\mathbf{A}|} |\mathbf{A}_1| = \frac{\begin{vmatrix} 6 & -1 \\ 5 & 2 \end{vmatrix}}{\begin{vmatrix} 3 & -1 \\ 1 & 2 \end{vmatrix}} = \frac{12 + 5}{6 + 1} = \frac{17}{7}$$

and

$$y = \frac{1}{|\mathbf{A}|} |\mathbf{A}_2| = \frac{\begin{vmatrix} 3 & 6 \\ 1 & 5 \end{vmatrix}}{\begin{vmatrix} 3 & -1 \\ 1 & 2 \end{vmatrix}} = \frac{15 - 6}{6 + 1} = \frac{9}{7}$$

These solutions can be verified by inserting the values of x and y into in Equation 14.19 to show

$$3x - y = 3\left(\frac{17}{7}\right) - \frac{9}{7} = 6 \qquad x + 2y = \frac{17}{7} + 2\left(\frac{9}{7}\right) = 5$$

Now consider the set of equations

$$x + 2y + 3z = -5$$
$$3x + y - 3z = 4$$
$$-3x + 4y + 7z = -7 \tag{14.20}$$

Note that each line in Equation 14.20 defines a plane in three-dimensional cartesian space. Geometrically, we seek the point in space (x, y, z) representing the intersection of the three planes defined by the three independent equations of three variables. The point of intersection is the solution to our set of coupled equations (see Figure 14.13).

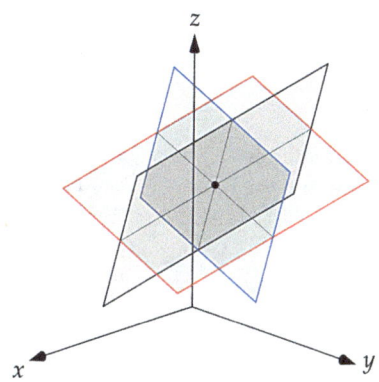

Figure 14.13: In three dimensions, points representing the solution to a linear equation form a plane. The point of intersection of the three planes (black dot) represents the common solution (x, y, z) to the three linear equations.

Applying the general form of Cramer's rule provides expressions for the values of the variables x, y, and z, which define the point of intersection of the three planes:

$$x = \frac{1}{|\mathbf{A}|}\,|\mathbf{A}_1| = \frac{\begin{vmatrix} -5 & 2 & 3 \\ 4 & 1 & -3 \\ -7 & 4 & 7 \end{vmatrix}}{\begin{vmatrix} 1 & 2 & 3 \\ 3 & 1 & -3 \\ -3 & 4 & 7 \end{vmatrix}} = \frac{-40}{40} = -1$$

$$y = \frac{1}{|\mathbf{A}|}\,|\mathbf{A}_2| = \frac{\begin{vmatrix} 1 & -5 & 3 \\ 3 & 4 & -3 \\ -3 & -7 & 7 \end{vmatrix}}{\begin{vmatrix} 1 & 2 & 3 \\ 3 & 1 & -3 \\ -3 & 4 & 7 \end{vmatrix}} = \frac{40}{40} = 1$$

$$z = \frac{1}{|\mathbf{A}|}\,|\mathbf{A}_3| = \frac{\begin{vmatrix} 1 & 2 & -5 \\ 3 & 1 & 4 \\ -3 & 4 & -7 \end{vmatrix}}{\begin{vmatrix} 1 & 2 & 3 \\ 3 & 1 & -3 \\ -3 & 4 & 7 \end{vmatrix}} = \frac{-80}{40} = -2$$

This result can be verified by substituting $x = -1$, $y = 1$, and $z = -2$ into Equation 14.20:

$$(-1) + 2(1) + 3(-2) = -5$$
$$3(-1) + (1) - 3(-2) = 4$$
$$-3(-1) + 4(1) + 7(-2) = -7 \tag{14.21}$$

These examples demonstrate the general applicability of Cramer's rule.

A$_{14}$ Applications of determinants in Hückel theory

Matrix algebra is used in the study of *Hückel theory* used to estimate the electronic energy of conjugated organic molecules.[23] In that context, we find matrix equations such as

$$\begin{vmatrix} \alpha - E & \beta - SE \\ \beta - SE & \alpha - E \end{vmatrix} = 0$$

where E is a variable, and α, β, and S are constants. The determinant is

$$\begin{vmatrix} \alpha - E & \beta - SE \\ \beta - SE & \alpha - E \end{vmatrix} = (\alpha - E)^2 - (\beta - SE)^2 = 0$$

For the case $S = 0$, this expression reduces to

$$(\alpha - E)^2 = \beta^2$$

The roots of this equation can be solved for E using the *quadratic formula*, resulting in

$$E_{\pm} = \alpha \pm \beta$$

In Hückel theory we find that both α and β are negative numbers. Therefore, the lower root is $E_+ = \alpha + \beta$ and the higher root is $E_- = \alpha - \beta$

Let's consider another example involving a 4×4 determinant:

$$\begin{vmatrix} \alpha - E & \beta & 0 & 0 \\ \beta & \alpha - E & \beta & 0 \\ 0 & \beta & \alpha - E & \beta \\ 0 & 0 & \beta & \alpha - E \end{vmatrix} = 0$$

Expanding the 4×4 determinant, we find

$$(\alpha - E) \begin{vmatrix} \alpha - E & \beta & 0 \\ \beta & \alpha - E & \beta \\ 0 & \beta & \alpha - E \end{vmatrix} - \beta \begin{vmatrix} \beta & \beta & 0 \\ 0 & \alpha - E & \beta \\ 0 & \beta & \alpha - E \end{vmatrix} = 0$$

Expanding the 3×3 determinants, we find

$$(\alpha - E)^2 \begin{vmatrix} \alpha - E & \beta \\ \beta & \alpha - E \end{vmatrix} - (\alpha - E)\beta \begin{vmatrix} \beta & \beta \\ 0 & \alpha - E \end{vmatrix}$$

$$- \beta^2 \begin{vmatrix} \alpha - E & \beta \\ \beta & \alpha - E \end{vmatrix} + \beta^2 \begin{vmatrix} 0 & \beta \\ 0 & \alpha - E \end{vmatrix} = 0$$

This can further be reduced to the polynomial equation in E

$$(\alpha - E)^4 - (\alpha - E)^2 \beta^2 - (\alpha - E)^2 \beta^2 - \beta^2 (\alpha - E)^2 + \beta^4 = 0$$

where

$$(\alpha - E)^4 - 3(\alpha - E)^2 \beta^2 + \beta^4 = 0$$

[23] Named for German physicist and physical chemist *Erich Armand Arthur Joseph Hückel* (1896-1990) who made fundamental contributions to the molecular orbital theory of unsaturated organic molecules and our understanding of ionic solutions.

Simplifying the equation by making the substitution

$$x = \left(\frac{\alpha - E}{\beta}\right)^2$$

results in

$$x^2 - 3x + 1 = 0$$

This expression yields the roots

$$x_\pm = \frac{3 \pm \sqrt{9 - 4}}{2} = \begin{cases} 2.62 \\ 0.38 \end{cases}$$

which leads to the final result

$$E = \begin{cases} \alpha \pm 1.62\beta \\ \alpha \pm 0.62\beta \end{cases}$$

since $\sqrt{2.62} = 1.62$ and $\sqrt{0.38} = 0.62$. This matrix algebra can be used to study the energetics and spectroscopy of conjugated organic molecules.

Additional examples of computing determinants derived from Hückel theory are explored in the exercises at the end of this chapter. Further exploration of Hückel theory as an eigenvalue problem can be found in the exercises at the end of Chapter 15.

B$_{14}$ Solving coupled linear equations using Gaussian elimination

We explored Cramer's rule as a reliable means to solve n coupled linear equations with n unknown variables. However, it was noted that Cramer's rule requires the calculation of $n + 1$ determinants. In this complement, we explore the alternative method of *Gaussian elimination*, also known as *row reduction*, which requires an effort equivalent to the calculation of a single determinant.

Let's return to the example of Equation 14.19 having two equations with two unknown variables that we solved using Cramer's rule:

$$3x - y = 6$$
$$x + 2y = 5$$

We can write these equations in matrix form as

$$\begin{pmatrix} 3 & -1 \\ 1 & 2 \end{pmatrix} \begin{pmatrix} x \\ y \end{pmatrix} = \begin{pmatrix} 6 \\ 5 \end{pmatrix}$$

Let's form a 2×3 *augmented matrix* from the coefficients of the linear equations:

$$\left[\begin{array}{cc|c} 3 & -1 & 6 \\ 1 & 2 & 5 \end{array} \right]$$

The method of Gaussian elimination consists of performing elementary row operations on the augmented matrix in order to form an upper triangular matrix. Those operations may include multiplying rows by a constant, adding or subtracting rows, or swapping rows of the matrix. From the upper triangular matrix, we can determine y. From y, we can determine x. Let's see how this works.

We begin by multiplying the first row by $-\frac{1}{3}$ and adding it to the second row

$$\begin{bmatrix} 3 & -1 & 6 \\ 1-\frac{1}{3}(3) & 2-\frac{1}{3}(-1) & 5-\frac{1}{3}6 \end{bmatrix}$$

which results in

$$\begin{bmatrix} 3 & -1 & 6 \\ 0 & \frac{7}{3} & 3 \end{bmatrix}$$

Returning to the matrix form of our linear equations, we now find

$$\begin{pmatrix} 3 & -1 \\ 0 & \frac{7}{3} \end{pmatrix} \begin{pmatrix} x \\ y \end{pmatrix} = \begin{pmatrix} 6 \\ 3 \end{pmatrix}$$

From the second row, we see that

$$\frac{7}{3}y = 3$$

or $y = \frac{9}{7}$. With a knowledge of y, we can determine x by substituting y into the first equation as

$$3x - y = 3x - \frac{9}{7} = 6$$

so that $3x = \frac{51}{7}$ and $x = \frac{17}{7}$. This is the same solution we found using Cramer's rule. So you see how Gaussian elimination works.

As a second example, let's return to Equation 14.20 which is a set of three equations with three unknown variables:

$$x + 2y + 3z = -5$$
$$3x + y - 3z = 4$$
$$-3x + 4y + 7z = -7$$

We solved these equations for x, y, and z using Cramer's by calculating four 3×3 determinants. To solve these equations using Gaussian elimination, we write the equations in matrix form as

$$\begin{pmatrix} 1 & 2 & 3 \\ 3 & 1 & -3 \\ -3 & 4 & 7 \end{pmatrix} \begin{pmatrix} x \\ y \\ z \end{pmatrix} = \begin{pmatrix} -5 \\ 4 \\ -7 \end{pmatrix}$$

and form a 3×4 augmented matrix from the coefficients of the linear equations:

$$\begin{bmatrix} 1 & 2 & 3 & -5 \\ 3 & 1 & -3 & 4 \\ -3 & 4 & 7 & -7 \end{bmatrix}$$

We now perform elementary row operations on the augmented matrix, which may include multiplying rows by a constant, adding or subtracting rows, or swapping rows of the matrix. The goal is to create an upper triangular matrix where all elements of the matrix below the diagonal are zero.

We first subtract three times the first row from the second row as

$$\left[\begin{array}{ccc|c} 1 & 2 & 3 & -5 \\ 3-3(1) & 1-3(2) & -3-3(3) & 4-3(-5) \\ -3 & 4 & 7 & -7 \end{array}\right]$$

which results in

$$\left[\begin{array}{ccc|c} 1 & 2 & 3 & -5 \\ 0 & -5 & -12 & 19 \\ -3 & 4 & 7 & -7 \end{array}\right]$$

We then add three times the first row to the third row as

$$\left[\begin{array}{ccc|c} 1 & 2 & 3 & -5 \\ 0 & -5 & -12 & 19 \\ -3+3(1) & 4+3(2) & 7+3(3) & -7+3(-5) \end{array}\right]$$

which results in

$$\left[\begin{array}{ccc|c} 1 & 2 & 3 & -5 \\ 0 & -5 & -12 & 19 \\ 0 & 10 & 16 & -22 \end{array}\right]$$

Finally, we add two times the second row to the third row to find

$$\left[\begin{array}{ccc|c} 1 & 2 & 3 & -5 \\ 0 & -5 & -12 & 19 \\ 0+2(0) & 10+2(-5) & 16+2(-12) & -22+2(19) \end{array}\right] = \left[\begin{array}{ccc|c} 1 & 2 & 3 & -5 \\ 0 & -5 & -12 & 19 \\ 0 & 0 & -8 & 16 \end{array}\right]$$

Returning to the matrix form of our linear equations, we now find

$$\begin{pmatrix} 1 & 2 & 3 \\ 0 & -5 & -12 \\ 0 & 0 & -8 \end{pmatrix} \begin{pmatrix} x \\ y \\ z \end{pmatrix} = \begin{pmatrix} -5 \\ 19 \\ 16 \end{pmatrix}$$

From the third row, we find that

$$-8z = 16$$

and $z = -2$. Substituting $z = -2$ into the second equation leads to

$$-5y - 12z = -5y - 12(-2) = 19$$

so that $-5y = -5$ and $y = 1$. Finally, substituting $y = 1$ and $z = -2$ into the first equation leads to

$$x + 2y + 3z = x + 2(1) + 3(-2) = -5$$

so that $x - 4 = -5$ and $x = -1$. This is the same solution we found using Cramer's rule, after computing four 3×3 determinants. This result validates the method of Gaussian elimination and demonstrates its relative efficiency.

As these examples make clear, Gaussian elimination provides a general and relatively efficient means of solving coupled linear equations. For very large matrices, which may involve millions of elements, Gaussian elimination may be replaced by iterative methods.

C_{14} Finding the matrix inverse using Gauss-Jordan elimination

We explored a reliable method for the determination of a matrix inverse, when one exists, through the composition of the cofactor matrix. In this complement, we explore the alternative method of *Gauss-Jordan elimination*, also known as *row reduction*, for the calculation of a matrix inverse.

Let's return to the example of inverting the 2×2 matrix defined in Equation 14.11:

$$\mathbf{A} = \begin{pmatrix} 1 & 2 \\ 3 & 4 \end{pmatrix}$$

We can form a 2×4 *augmented matrix* from the matrix $\mathbf{A}$ and the identity matrix:

$$\left[\begin{array}{cc|cc} 1 & 2 & 1 & 0 \\ 3 & 4 & 0 & 1 \end{array} \right]$$

To determine the matrix inverse, we perform elementary row operations on the augmented matrix in order to transform the original matrix, forming the left half of the augmented matrix, into the identity matrix. Those operations may include multiplying rows by a constant, adding or subtracting rows, or swapping rows of the matrix. In the process, the right half of the augmented matrix, originally the identity matrix, is transformed into the matrix inverse. Let's see how this works.

We begin by multiplying the first row by -3 and adding it to the second row

$$\left[\begin{array}{cc|cc} 1 & 2 & 1 & 0 \\ 3-3(1) & 4-3(2) & 0-3(1) & 1-3(0) \end{array} \right]$$

which results in

$$\left[\begin{array}{cc|cc} 1 & 2 & 1 & 0 \\ 0 & -2 & -3 & 1 \end{array} \right]$$

We then add the second row to the first row as

$$\left[\begin{array}{cc|cc} 1+0 & 2-2 & 1-3 & 0+1 \\ 0 & -2 & -3 & 1 \end{array} \right]$$

which results in

$$\left[\begin{array}{cc|cc} 1 & 0 & -2 & 1 \\ 0 & -2 & -3 & 1 \end{array} \right]$$

Finally, we multiply the second row by $-\frac{1}{2}$ to find

$$\left[\begin{array}{cc|cc} 1 & 0 & -2 & 1 \\ 0 & 1 & \frac{3}{2} & -\frac{1}{2} \end{array} \right]$$

Through our elementary row operations, we have transformed the original matrix, forming the left half of the augmented matrix, into the identify matrix. The right half of the augmented matrix is the matrix inverse of **A**:

$$\mathbf{A}^{-1} = \begin{pmatrix} -2 & 1 \\ \dfrac{3}{2} & -\dfrac{1}{2} \end{pmatrix} \tag{14.22}$$

This matrix inverse agrees with our earlier result in Equation 14.14 based on composing the cofactor matrix. So you see how this works.

As a second example, let's find the matrix inverse of the 3×3 matrix defined in Equation 14.13:

$$\mathbf{B} = \begin{pmatrix} 3 & 0 & -1 \\ 2 & 2 & 1 \\ 0 & -1 & 2 \end{pmatrix} \tag{14.23}$$

We form a 3×6 *augmented matrix* from the matrix **B** and the identity matrix:

$$\left[\begin{array}{ccc|ccc} 3 & 0 & -1 & 1 & 0 & 0 \\ 2 & 2 & 1 & 0 & 1 & 0 \\ 0 & -1 & 2 & 0 & 0 & 1 \end{array} \right]$$

To determine the matrix inverse using Gauss-Jordan elimination, we begin by multiplying the first row by $-\frac{2}{3}$ and adding it to the second row

$$\left[\begin{array}{ccc|ccc} 3 & 0 & -1 & 1 & 0 & 0 \\ 2-\dfrac{2}{3}(3) & 2-\dfrac{2}{3}(0) & 1-\dfrac{2}{3}(-1) & 0-\dfrac{2}{3}(1) & 1-\dfrac{2}{3}(0) & 0-\dfrac{2}{3}(0) \\ 0 & -1 & 2 & 0 & 0 & 1 \end{array} \right]$$

which results in

$$\left[\begin{array}{ccc|ccc} 3 & 0 & -1 & 1 & 0 & 0 \\ 0 & 2 & \dfrac{5}{3} & -\dfrac{2}{3} & 1 & 0 \\ 0 & -1 & 2 & 0 & 0 & 1 \end{array} \right]$$

We then add the second row multiplied by $\frac{1}{2}$ to the third row as

$$\left[\begin{array}{ccc|ccc} 3 & 0 & -1 & 1 & 0 & 0 \\ 0 & 2 & \dfrac{5}{3} & -\dfrac{2}{3} & 1 & 0 \\ 0+\dfrac{1}{2}(0) & -1+\dfrac{1}{2}(2) & 2+\dfrac{1}{2}(\dfrac{5}{3}) & 0+\dfrac{1}{2}(-\dfrac{2}{3}) & 0+\dfrac{1}{2}(1) & 1+\dfrac{1}{2}(0) \end{array} \right]$$

which results in

$$\left[\begin{array}{ccc|ccc} 3 & 0 & -1 & 1 & 0 & 0 \\ 0 & 2 & \dfrac{5}{3} & -\dfrac{2}{3} & 1 & 0 \\ 0 & 0 & \dfrac{17}{6} & -\dfrac{1}{3} & \dfrac{1}{2} & 1 \end{array} \right]$$

At this point, it is useful to multiply the third row by $\frac{6}{17}$ to find

$$\left[\begin{array}{ccc|ccc} 3 & 0 & -1 & 1 & 0 & 0 \\ 0 & 2 & \dfrac{5}{3} & -\dfrac{2}{3} & 1 & 0 \\ 0 & 0 & 1 & -\dfrac{2}{17} & \dfrac{3}{17} & \dfrac{6}{17} \end{array}\right]$$

Now let's add the third row multiplied by $-\frac{5}{3}$ to the second row, and also add the third row to the first row as

$$\left[\begin{array}{ccc|ccc} 3+0 & 0+0 & -1+1 & 1-\dfrac{2}{17} & 0+\dfrac{3}{17} & 0+\dfrac{6}{17} \\ 0-\dfrac{5}{3}(0) & 2-\dfrac{5}{3}(0) & \dfrac{5}{3}-\dfrac{5}{3}(1) & -\dfrac{2}{3}-\dfrac{5}{3}(-\dfrac{2}{17}) & 1-\dfrac{5}{3}(\dfrac{3}{17}) & 0-\dfrac{5}{3}(\dfrac{6}{17}) \\ 0 & 0 & 1 & -\dfrac{2}{17} & \dfrac{3}{17} & \dfrac{6}{17} \end{array}\right]$$

which results in

$$\left[\begin{array}{ccc|ccc} 3 & 0 & 0 & \dfrac{15}{17} & \dfrac{3}{17} & \dfrac{6}{17} \\ 0 & 2 & 0 & -\dfrac{8}{17} & \dfrac{12}{17} & -\dfrac{10}{17} \\ 0 & 0 & 1 & -\dfrac{2}{17} & \dfrac{3}{17} & \dfrac{6}{17} \end{array}\right]$$

Finally, we multiply the first row by $\frac{1}{3}$ and the second row by $\frac{1}{2}$ to find

$$\left[\begin{array}{ccc|ccc} 1 & 0 & 0 & \dfrac{5}{17} & \dfrac{1}{17} & \dfrac{2}{17} \\ 0 & 1 & 0 & -\dfrac{4}{17} & \dfrac{6}{17} & -\dfrac{5}{17} \\ 0 & 0 & 1 & -\dfrac{2}{17} & \dfrac{3}{17} & \dfrac{6}{17} \end{array}\right]$$

Through our elementary row operations, we have transformed the original matrix, forming the left half of the augmented matrix, into the identify matrix. The right half of the augmented matrix is the matrix inverse of $\mathbf{B}$:

$$\mathbf{B}^{-1} = \frac{1}{17}\begin{pmatrix} 5 & 1 & 2 \\ -4 & 6 & -5 \\ -2 & 3 & 6 \end{pmatrix} \tag{14.24}$$

which agrees with our earlier result in Equation 14.14 based on composing the cofactor matrix. Note that while the calculation of the matrix determinant is not a required step in the application of row reduction to find a matrix inverse, it is worth doing to know that the matrix determinant is not zero.

The examples provided in this and the preceding complement show the power of row reduction to solve sets of linear equations or determine a matrix inverse. Further applications of Gaussian elimination, for the solution of linear coupled equations, and Gauss-Jordan elimination, for the calculation of the matrix inverse, are explored in the exercises at the end of this chapter.

D_{14} End-of-chapter problems

You shouldn't do science just to improve wealth - do science for the sake of human culture and knowledge. There must be some purpose in life that is higher than just surviving.

Gerhard Herzberg

Warm-ups

14.1 Evaluate the following determinants

(a) $\begin{vmatrix} 1 & 2 & 3 \\ 3 & 0 & 1 \\ -1 & 4 & 2 \end{vmatrix}$

(b) $\begin{vmatrix} 4 & 2 & 1 \\ -1 & 6 & 3 \\ -1 & 5 & -1 \end{vmatrix}$

(c) $\begin{vmatrix} x & 1 & 0 \\ 1 & x & 1 \\ 0 & 1 & x \end{vmatrix}$

(d) $\begin{vmatrix} 4 & 3 & 1 & -1 \\ 6 & 1 & 0 & -3 \\ 1 & 5 & 2 & -2 \\ 8 & 6 & -5 & 0 \end{vmatrix}$

(e) $\begin{vmatrix} x & b & 0 & 0 \\ b & x & b & 0 \\ 0 & b & x & b \\ 0 & 0 & b & x \end{vmatrix}$

(f) $\begin{vmatrix} \sec\theta & \tan\theta \\ \tan\theta & \sec\theta \end{vmatrix}$

14.2 Solve the following determinants for x

(a) $\begin{vmatrix} x & 1 \\ 1 & x \end{vmatrix} = 0$

(b) $\begin{vmatrix} x & -2 \\ 1 & x \end{vmatrix} = 6$

(c) $\begin{vmatrix} x & 1 & 1 \\ 1 & x & 1 \\ 1 & 1 & x \end{vmatrix} = 2$

(d) $\begin{vmatrix} x & 1 & 1 & 1 \\ 1 & x & 0 & 0 \\ 1 & 0 & x & 0 \\ 1 & 0 & 0 & x \end{vmatrix} = 0$

14.3 Add the pair of matrices

$$\begin{pmatrix} 0 & 1 & 2 \\ 2 & 4 & -3 \\ 6 & 3 & 5 \end{pmatrix} + \begin{pmatrix} 6 & 3 & -7 \\ -1 & 1 & -1 \\ -5 & 2 & 7 \end{pmatrix}$$

14.4 Multiply the following pairs of matrices

(a) $\begin{pmatrix} 1 & 4 \\ 3 & 2 \end{pmatrix} \begin{pmatrix} 6 & -3 \\ -3 & 1 \end{pmatrix}$

(b) $\begin{pmatrix} 1 & 0 \\ 0 & 1 \end{pmatrix} \begin{pmatrix} 4 & -1 \\ 2 & 3 \end{pmatrix}$

(c) $\begin{pmatrix} 3 & 0 & 3 \\ 4 & -1 & -1 \\ 1 & 2 & 5 \end{pmatrix} \begin{pmatrix} 1 & 1 & 1 \\ -2 & 1 & 6 \\ 3 & 4 & 5 \end{pmatrix}$

(d) $\begin{pmatrix} 1 & 8 & 4 \\ -2 & 3 & 0 \\ 5 & -1 & -1 \end{pmatrix} \begin{pmatrix} x \\ y \\ z \end{pmatrix}$

14.5 Given the matrices

$$A = \begin{pmatrix} 1 & 1 & 4 \\ 2 & -6 & 10 \\ 4 & -1 & -1 \end{pmatrix} \qquad B = \begin{pmatrix} 6 & 1 & 0 \\ 4 & 2 & -1 \\ 8 & -4 & 3 \end{pmatrix}$$

calculate the matrix defined by the commutator $[A, B] = AB - BA$.

14.6 Find the inverse of the following matrix

$$A = \begin{pmatrix} 1 & 0 & -1 \\ 3 & 2 & 4 \\ -2 & 1 & 0 \end{pmatrix}$$

where you determine the matrix inverse A^{-1} in terms of the cofactor matrix. Validate your result by multiplying the original matrix by the matrix inverse as $A^{-1}A = AA^{-1} = I$ where I is the identity matrix.

14.7 Solve the following sets of equations using Cramer's rule

(a)
$$\begin{aligned} x + y &= 3 \\ 4x - 3y &= 5 \end{aligned}$$

(b)
$$\begin{aligned} x\sin\theta + y\cos\theta &= x' \\ -x\cos\theta + y\sin\theta &= y' \end{aligned}$$

Check your solutions by inserting them into the original equations.

Homework exercises

14.8 Consider the rotation matrix

$$R(\theta) = \begin{pmatrix} \cos\theta & -\sin\theta \\ \sin\theta & \cos\theta \end{pmatrix}$$

and the vector

$$x_1 = \begin{pmatrix} x_1 \\ y_1 \end{pmatrix}$$

Apply the rotation matrix to the following vectors x_1 to determine the vectors $x_2 = R(\theta)x_1$ for the stated value of θ.
(a) $x_1 = (1 \quad 0)^T$ for $\theta = \pi$, (b) $x_1 = (0 \quad 1)^T$ for $\theta = \pi/4$, (c) $x_1 = (1 - 1)^T$ for $\theta = \pi/2$.

14.9 Prove that for

$$R(\theta) = \begin{pmatrix} \cos\theta & -\sin\theta \\ \sin\theta & \cos\theta \end{pmatrix}$$

the matrix product $R(\theta)R(\theta) = R^2$ equals

$$R^2 = \begin{pmatrix} \cos 2\theta & -\sin 2\theta \\ \sin 2\theta & \cos 2\theta \end{pmatrix} = R(2\theta)$$

14.10 Consider the matrix

$$A = \begin{pmatrix} 1 & -1 & 0 \\ 0 & 1 & 1 \\ 2 & 2 & 0 \end{pmatrix}$$

(a) Find the matrix inverse A^{-1}.
(b) Verify your answer by demonstrating that $A^{-1}A = AA^{-1} = I$ where I is the identity matrix.

14.11 The rotation matrix in three dimensions for counter-clockwise rotation about the positive z-axes can be written as

$$\mathbf{R}(\theta) = \begin{pmatrix} \cos\theta & -\sin\theta & 0 \\ \sin\theta & \cos\theta & 0 \\ 0 & 0 & 1 \end{pmatrix}$$

(a) Show that $|\mathbf{R}| = 1$.

(b) Show that $\mathbf{R}$ is composed of column vectors that are orthonormal.

(c) Show that $\mathbf{R}^T = \mathbf{R}^{-1}$ by showing that $\mathbf{R}^T\mathbf{R} = \mathbf{I}$.

14.12 Consider the matrices

$$\mathbf{A} = \begin{pmatrix} 2 & -1 \\ 1 & 2 \end{pmatrix} \qquad \mathbf{B} = \begin{pmatrix} 3 & 1 \\ 1 & 3 \end{pmatrix}$$

and show that $|\mathbf{AB}| = |\mathbf{BA}| = |\mathbf{A}||\mathbf{B}|$.

14.13 An *orthogonal matrix* is composed of columns that form a set of orthonormal vectors that are individually normalized and mutually orthogonal. Orthogonal matrices have the property that the transpose of the matrix is also its inverse or $\mathbf{A}^T = \mathbf{A}^{-1}$.

Consider the matrix

$$\mathbf{A} = \frac{1}{9}\begin{pmatrix} 1 & 8 & -4 \\ 4 & -4 & -7 \\ 8 & 1 & 4 \end{pmatrix}$$

(a) Prove orthogonality by taking each of the three column vector pairs and showing the dot product is zero.

(b) Prove normalization by showing that each column vector has a magnitude of unity.

(c) Demonstrate that the matrix transpose $\mathbf{A}^T = \mathbf{A}^{-1}$ by showing that $\mathbf{A}^T\mathbf{A} = \mathbf{AA}^T = \mathbf{I}$.

14.14 A real *orthogonal matrix* has the property that

$$\mathbf{A}^T = \mathbf{A}^{-1}$$

For matrices with complex elements the equivalent property is

$$(\mathbf{A}^*)^T = \mathbf{A}^{-1}$$

where $\mathbf{A}^*$ is the complex conjugate of $\mathbf{A}$, in which each element of $\mathbf{A}^*$ is the complex conjugate of the corresponding element of $\mathbf{A}$. A matrix satisfying this property is called a *unitary matrix*. Show that the complex matrix

$$\mathbf{A} = \frac{1}{6}\begin{pmatrix} 2 - 4i & 4i \\ -4i & -2 - 4i \end{pmatrix}$$

is a unitary matrix by demonstrating that $(\mathbf{A}^*)^T\mathbf{A} = \mathbf{I}$.

14.15 Show that the Pauli matrices

$$\sigma_1 = \begin{pmatrix} 0 & 1 \\ 1 & 0 \end{pmatrix} \quad \sigma_2 = \begin{pmatrix} 0 & -i \\ i & 0 \end{pmatrix} \quad \sigma_3 = \begin{pmatrix} 1 & 0 \\ 0 & -1 \end{pmatrix}$$

are unitary matrices for which $(\mathbf{A}^*)^T\mathbf{A} = \mathbf{I}$.

14.16 Solve the following sets of equations using Cramer's rule

(a)
$$\begin{aligned} x &+ 2y &+ 3z &= -5 \\ -x &- 3y &+ z &= -14 \\ 2x &+ y &+ z &= 1 \end{aligned}$$

(b)
$$\begin{aligned} x &+ 2y &- z &+ t &= 2 \\ x &- 2y &+ z &- 3t &= 6 \\ 2x &+ y &+ 2z &+ t &= -4 \\ 3x &+ 3y &+ z &- 2t &= 10 \end{aligned}$$

Check your solutions by inserting them into the original equations.

14.17 Consider two points (x_1, y_1) and (x_2, y_2) in two-dimensional cartesian coordinates. Suppose we are asked to determine the parameters m and b defining the equation of a line $y(x) = mx + b$ containing the two points. We can do this using Cramer's rule.

As each point is on the line defined by $y = ax + b$ we know that

$$y_1 = mx_1 + b$$
$$y_2 = mx_2 + b$$

or that

$$-mx_1 + y_1 = b$$
$$-mx_2 + y_2 = b$$

which can be written in the matrix form $\mathbf{Ax} = \mathbf{b}$ as

$$\mathbf{Ax} = \begin{pmatrix} -mx_1 & y_1 \\ -mx_2 & y_2 \end{pmatrix} = \begin{pmatrix} x_1 & y_1 \\ x_2 & y_2 \end{pmatrix} \begin{pmatrix} -m \\ 1 \end{pmatrix} = \begin{pmatrix} b \\ b \end{pmatrix} = \mathbf{b}$$

(a) Use Cramer's rule to solve for the parameters m and b in terms of x_1, x_2, y_1, and y_2.

(b) Determine the equation for a line $y(x) = mx + b$ passing through the two points $(5, 2)$ and $(3, -4)$ by determining the values of m and b using Cramer's rule.

14.18 Consider the three points (x_1, y_1, z_1), (x_2, y_2, z_2), and (x_3, y_3, z_3) in three-dimensional cartesian coordinates. Suppose we are asked to determine the parameters a, b and c defining the equation of a plane $z(x, y) = ax + by + c$ containing the three points. We can do this using Cramer's rule.

As each point is on the plane defined by $z = ax + by + c$, we know that

$$z_1 = ax_1 + by_1 + c$$
$$z_2 = ax_2 + by_2 + c$$
$$z_3 = ax_3 + by_3 + c$$

or that

$$-ax_1 - by_1 + z_1 = c$$
$$-ax_2 - by_2 + z_2 = c$$
$$-ax_3 - by_3 + z_3 = c$$

which can be written in the matrix form $\mathbf{Ax} = \mathbf{c}$ as

$$\mathbf{Ax} = \begin{pmatrix} -ax_1 - by_1 + z_1 \\ -ax_2 - by_2 + z_2 \\ -ax_3 - by_3 + z_3 \end{pmatrix} = \begin{pmatrix} x_1 & y_1 & z_1 \\ x_2 & y_2 & z_2 \\ x_3 & y_3 & z_3 \end{pmatrix} \begin{pmatrix} -a \\ -b \\ 1 \end{pmatrix} = \begin{pmatrix} c \\ c \\ c \end{pmatrix} = \mathbf{c}$$

(a) Use Cramer's rule to solve for the parameters a, b, and c in terms of $x_1, x_2, x_3, y_1, y_2, y_3, z_1, z_2,$ and z_3.

(b) Determine the equation for a plane $z(x, y) = ax + by + c$ that contains the three points $(5, 2, 1)$, $(3, -4, 2)$, and $(-2, 1, -3)$ by determining the values of a, b and c using Cramer's rule.

14.19* Determinants commonly appear in the *Hückel theory* of molecular orbitals of π-systems. Determine which values of x satisfy each of the following equations expressed in the form of determinants.

(a) The determinant equation for cyclobutadiene

$$\begin{vmatrix} x & 1 & 0 & 1 \\ 1 & x & 1 & 0 \\ 0 & 1 & x & 1 \\ 1 & 0 & 1 & x \end{vmatrix} = 0$$

(b) The determinant equation for triethylenemethane

$$\begin{vmatrix} x & 1 & 1 & 1 \\ 1 & x & 0 & 0 \\ 1 & 0 & x & 0 \\ 1 & 0 & 0 & x \end{vmatrix} = 0$$

(c) The determinant equation for benzene

$$\begin{vmatrix} x & 1 & 0 & 0 & 0 & 1 \\ 1 & x & 1 & 0 & 0 & 0 \\ 0 & 1 & x & 1 & 0 & 0 \\ 0 & 0 & 1 & x & 1 & 0 \\ 0 & 0 & 0 & 1 & x & 1 \\ 1 & 0 & 0 & 0 & 1 & x \end{vmatrix} = 0$$

14.20* Solve the following sets of equations using Gaussian elimination

(a)
$$\begin{aligned} x + y &= 3 \\ 4x - 3y &= 5 \end{aligned}$$

(b)
$$\begin{aligned} x \sin\theta + y \cos\theta &= x' \\ -x \cos\theta + y \sin\theta &= y' \end{aligned}$$

Check your solutions by inserting them into the original equations.

14.21* Solve the following sets of equations using Gaussian elimination

(a)
$$\begin{aligned} x + 2y + 3z &= -5 \\ -x - 3y + z &= -14 \\ 2x + y + z &= 1 \end{aligned}$$

(b)
$$\begin{aligned} x + 2y - z + t &= 2 \\ x - 2y + z - 3t &= 6 \\ 2x + y + 2z + t &= -4 \\ 3x + 3y + z - 2t &= 10 \end{aligned}$$

Check your solutions by inserting them into the original equations.

14.22* Consider the following matrix

$$\mathbf{A} = \begin{pmatrix} 1 & 0 & -1 \\ 3 & 2 & 4 \\ -2 & 1 & 0 \end{pmatrix}$$

(a) Find the matrix inverse $\mathbf{A}^{-1}$ using Gauss-Jordan elimination.

(b) Verify your answer by demonstrating that $\mathbf{A}^{-1}\mathbf{A} = \mathbf{A}\mathbf{A}^{-1} = \mathbf{I}$ where $\mathbf{I}$ is the identity matrix.

14.23* Consider the following matrix

$$\mathbf{A} = \begin{pmatrix} -4 & 1 & 8 \\ -7 & 4 & -4 \\ 4 & 8 & 1 \end{pmatrix}$$

(a) Find the matrix inverse $\mathbf{A}^{-1}$ using Gauss-Jordan elimination.

(b) Verify your answer by demonstrating that $\mathbf{A}^{-1}\mathbf{A} = \mathbf{A}\mathbf{A}^{-1} = \mathbf{I}$ where $\mathbf{I}$ is the identity matrix.

(c) Do you note any special properties of the matrix inverse $\mathbf{A}^{-1}$ and its relationship to $\mathbf{A}$?

14.24* In analyzing the determinant $|\mathbf{A}|$ of the 2×2 matrix

$$\mathbf{A} = \begin{pmatrix} a & c \\ b & d \end{pmatrix}$$

we found that the magnitude of the determinant was equal to the area of a parallelogram formed by the vectors $\mathbf{a} = a\,\hat{\mathbf{x}} + b\,\hat{\mathbf{y}}$ and $\mathbf{b} = c\,\hat{\mathbf{x}} + d\,\hat{\mathbf{y}}$. Prove that the area of the parallelogram is equal to the magnitude of the determinant $|ad - bc|$.

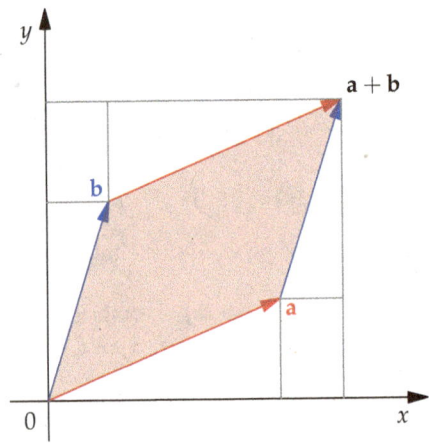

HINT: Consider making a geometric argument using the partitioning of space suggested in the figure above.

14.25* This question explores the application of the Gram-Schmidt orthogonalization procedure, introduced in a complement to Chapter 4. Suppose you are given two non-orthogonal vectors

$$\mathbf{v}_1 = \begin{pmatrix} 1 \\ -2 \end{pmatrix} \qquad \mathbf{v}_2 = \begin{pmatrix} 3 \\ 2 \end{pmatrix}$$

(a) Use the Gram-Schmidt orthogonalization procedure, and matrix algebra discussed in this chapter, to construct the orthogonal vectors $\mathbf{u}_1$ and $\mathbf{u}_2$. Assume $\mathbf{u}_1 = \mathbf{v}_1$.

(b) Determine the orthonormal vectors

$$\mathbf{e}_1 = \frac{1}{\sqrt{\mathbf{u}_1 \cdot \mathbf{u}_1}}\mathbf{u}_1 \qquad \mathbf{e}_2 = \frac{1}{\sqrt{\mathbf{u}_2 \cdot \mathbf{u}_2}}\mathbf{u}_2$$

14.26* This question explores the application of the Gram-Schmidt orthogonalization procedure, introduced in a complement to Chapter 4. Suppose you are given three non-orthogonal vectors

$$\mathbf{v}_1 = \begin{pmatrix} 1 \\ -2 \\ 1 \end{pmatrix} \qquad \mathbf{v}_2 = \begin{pmatrix} 1 \\ -1 \\ 0 \end{pmatrix} \qquad \mathbf{v}_3 = \begin{pmatrix} 0 \\ 3 \\ -2 \end{pmatrix}$$

(a) Use the Gram-Schmidt orthogonalization procedure, and the matrix algebra discussed in this chapter, to construct the orthogonal vectors $\mathbf{u}_1$, $\mathbf{u}_2$, and $\mathbf{u}_3$. Assume $\mathbf{u}_1 = \mathbf{v}_1$.

(b) Determine the orthonormal vectors

$$\mathbf{e}_1 = \frac{1}{\sqrt{\mathbf{u}_1 \cdot \mathbf{u}_1}}\mathbf{u}_1 \qquad \mathbf{e}_2 = \frac{1}{\sqrt{\mathbf{u}_2 \cdot \mathbf{u}_2}}\mathbf{u}_2 \qquad \mathbf{e}_3 = \frac{1}{\sqrt{\mathbf{u}_3 \cdot \mathbf{u}_3}}\mathbf{u}_3$$

15 EIGENVALUES AND EIGENVECTORS

15.1 Matrix eigenvalues and eigenvectors

MANY PROBLEMS IN THE PHYSICAL SCIENCES CAN BE EXPRESSED AS A SET OF LINEAR EQUATIONS. A special class of homogeneous linear equations are known as eigenvalue problems. In an eigenvalue problem, an operator acts on a function resulting in the same function (the eigenfunction) times a constant (the eigenvalue). Matrix eigenvalue problems can be solved using the special properties of determinants and matrix algebra. In this section, we develop general methods for solving matrix eigenvalue problems.

15.1.1 The characteristic equation and eigenvalues

In the physical sciences we commonly encounter problems that can be written in the form of the *homogeneous equation*[1]

$$\mathbf{A}\mathbf{x} = \lambda \mathbf{x} \qquad (15.1)$$

[1] The equation is homogeneous as every term is proportional to **x**.

The matrix **A** multiplies the vector **x**, resulting in the same vector **x** times a constant λ. In practice, the constant can be real, imaginary, or complex. Let's suppose λ is real and positive. We find that the operation **Ax** acts to scale the length of the vector **x** by a constant λ:

$$|\mathbf{A}\mathbf{x}| = |\lambda \mathbf{x}| = \lambda |\mathbf{x}|$$

However, the orientation defined by the unit vector $\hat{\mathbf{x}}$ is unchanged, since[2]

$$\frac{1}{|\mathbf{A}\mathbf{x}|}\mathbf{A}\mathbf{x} = \frac{1}{\lambda |\mathbf{x}|}\lambda \mathbf{x} = \frac{1}{|\mathbf{x}|}\mathbf{x} = \hat{\mathbf{x}}$$

These qualitative properties of Equation 15.1 and its solution are shown in Figure 15.1.

It is convenient to rewrite Equation 15.1 as

$$\mathbf{A}\mathbf{x} = \lambda \mathbf{I}\mathbf{x}$$

where **I** is the *identity matrix*, and we have used the fact that $\mathbf{x} = \mathbf{I}\mathbf{x}$. Subtracting $\lambda \mathbf{I}\mathbf{x}$ from each side results in

$$\mathbf{A}\mathbf{x} - \lambda \mathbf{I}\mathbf{x} = (\mathbf{A} - \lambda \mathbf{I})\,\mathbf{x} = \mathbf{0} \qquad (15.2)$$

where **0** is the zero vector in which all elements of the vector are zero. This expression, which is equivalent to our original Equation 15.1, can be used to determine eigenvectors from known eigenvalues.

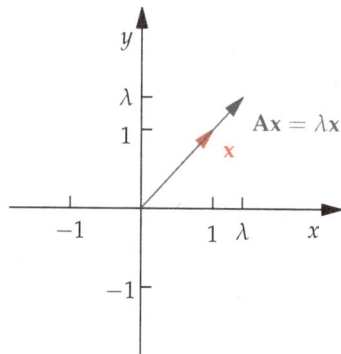

Figure 15.1: The matrix **A** operates on **x** (red arrow) to form **Ax** (gray arrow). The length of **x** is scaled by a factor of λ while the orientation is unchanged.

[2] Here the notation $|\mathbf{a}|$ represents the magnitude of the vector **a**, as distinct from the determinant $|\mathbf{A}|$ of a square matrix **A**.

Let's consider the specific example

$$\mathbf{A}\mathbf{x} = \begin{pmatrix} a_{11} & a_{12} \\ a_{21} & a_{22} \end{pmatrix} \begin{pmatrix} c_1 \\ c_2 \end{pmatrix} = \lambda \begin{pmatrix} c_1 \\ c_2 \end{pmatrix} = \lambda \mathbf{x}$$

where

$$\mathbf{x} = \begin{pmatrix} c_1 \\ c_2 \end{pmatrix}$$

The goal is to solve for $\mathbf{x}$ by identifying c_1 and c_2. Returning to our example we find

$$(\mathbf{A} - \lambda \mathbf{I}) \mathbf{x} = \begin{pmatrix} a_{11} - \lambda & a_{12} \\ a_{21} & a_{22} - \lambda \end{pmatrix} \begin{pmatrix} c_1 \\ c_2 \end{pmatrix} = \begin{pmatrix} 0 \\ 0 \end{pmatrix} = \mathbf{o}$$

A general solution for the elements of $\mathbf{x}$ is provided by *Cramer's rule* (see Chapter 14) as

$$c_k = \frac{1}{|\mathbf{A} - \lambda \mathbf{I}|} |(\mathbf{A} - \lambda \mathbf{I})_k| \tag{15.3}$$

where $(\mathbf{A} - \lambda \mathbf{I})_k$ is the matrix $(\mathbf{A} - \lambda \mathbf{I})$ in which the kth column has been replaced by the vector $\mathbf{o}$. For our specific example, we find

$$c_1 = \frac{1}{|\mathbf{A} - \lambda \mathbf{I}|} \begin{vmatrix} 0 & a_{12} \\ 0 & a_{22} - \lambda \end{vmatrix} \qquad c_2 = \frac{1}{|\mathbf{A} - \lambda \mathbf{I}|} \begin{vmatrix} a_{11} - \lambda & 0 \\ a_{21} & 0 \end{vmatrix}$$

However, since the determinants

$$|\mathbf{A} - \lambda \mathbf{I}|_1 = |\mathbf{A} - \lambda \mathbf{I}|_2 = 0$$

it appears that we have only the *trivial solution* $c_1 = c_2 = 0$.

Not so fast! It turns out that our equations for c_1 and c_2 can have a *non-trivial solution* when the denominator

$$|\mathbf{A} - \lambda \mathbf{I}| = 0$$

Viewed through Cramer's rule, the *non-trivial solution* for c_k results from dividing $|(\mathbf{A} - \lambda \mathbf{I})_k| = 0$ by $|\mathbf{A} - \lambda \mathbf{I}| = 0$ to arrive at a non-zero result for c_k.

Based on these observations, we can form a general rule. The condition for identifying a non-trivial solution to our original Equation 15.1

$$\mathbf{A}\mathbf{x} = \lambda \mathbf{x}$$

is expressed

$$|\mathbf{A} - \lambda \mathbf{I}| = 0 \tag{15.4}$$

This is known as the *characteristic equation*.[3]

As the matrix $\mathbf{A}$ defines the particular problem we wish to solve, the characteristic equation is used to identify those values of λ, known as *eigenvalues*, for which a non-trivial solution for $\mathbf{x}$ exists. As such, we refer to Equation 15.1 as an *eigenvalue problem* and the characteristic equation, Equation 15.4, is the starting point for its solution.

[3] The term *characteristic equation* is derived from the work of French mathematician and physicist *Augustin-Louis Cauchy* (1789-1857).

15.1.2 Eigenvalues and eigenvectors

Let's see how this works for the specific case $\mathbf{Ax} = \lambda\mathbf{x}$ where the matrix

$$\mathbf{A} = \begin{pmatrix} a & 1 \\ 1 & a \end{pmatrix} \tag{15.5}$$

We expect a non-trivial solution for $\mathbf{x}$ only for values of λ satisfying the characteristic equation defined by Equation 15.4. We first form the matrix

$$\mathbf{A} - \lambda\mathbf{I} = \begin{pmatrix} a & 1 \\ 1 & a \end{pmatrix} - \lambda\begin{pmatrix} 1 & 0 \\ 0 & 1 \end{pmatrix} = \begin{pmatrix} a-\lambda & 1 \\ 1 & a-\lambda \end{pmatrix}$$

Taking the determinant, we find

$$|\mathbf{A} - \lambda\mathbf{I}| = \begin{vmatrix} a-\lambda & 1 \\ 1 & a-\lambda \end{vmatrix} = (a-\lambda)^2 - 1 = 0$$

This results in a *quadratic equation* in the parameter λ

$$\lambda^2 - 2a\lambda + (a^2 - 1) = 0$$

with roots defined by the *quadratic formula*[4]

$$\lambda_\pm = a \pm \sqrt{a^2 - (a^2 - 1)} = a \pm 1$$

[4] We can also find these roots by noting that $(a - \lambda)^2 = 1$ so that $a - \lambda = \mp 1$ and $\lambda = a \pm 1$.

The two values of λ for which the characteristic equation is satisfied are $\lambda_1 = a + 1$ and $\lambda_2 = a - 1$. As such, there are two equations

$$\mathbf{Ax}_1 = \lambda_1\mathbf{x}_1 \qquad \mathbf{Ax}_2 = \lambda_2\mathbf{x}_2$$

leading to two solutions $\mathbf{x}_1$ and $\mathbf{x}_2$. The scalars λ_1 and λ_2 are the *eigenvalues* of the matrix $\mathbf{A}$, while $\mathbf{x}_1$ and $\mathbf{x}_2$ are the corresponding *eigenvectors*.[5] Let's determine the explicit form of the eigenvectors $\mathbf{x}_1$ and $\mathbf{x}_2$.

[5] In general, for an $n \times n$ matrix $\mathbf{A}$ there will be n eigenvalues λ_k and n eigenvectors $\mathbf{x}_k$.

15.1.3 Eigenvectors and their properties

Basis sets of orthonormal vectors, such as $\hat{\mathbf{x}}$, $\hat{\mathbf{y}}$, and $\hat{\mathbf{z}}$, are useful when expressing and analyzing a given vector. Similarly, it is useful to express solutions to eigenvalue problems in terms of a linear combination of eigenvectors. To determine the eigenvectors $\mathbf{x}_1$ and $\mathbf{x}_2$ we turn to Equation 15.2. For $\lambda_1 = a + 1$ and $\mathbf{x}_1$, we have

$$(\mathbf{A} - \lambda_1\mathbf{I})\,\mathbf{x}_1 = \begin{pmatrix} a-\lambda_1 & 1 \\ 1 & a-\lambda_1 \end{pmatrix}\begin{pmatrix} c_1 \\ c_2 \end{pmatrix}$$

$$= \begin{pmatrix} -1 & 1 \\ 1 & -1 \end{pmatrix}\begin{pmatrix} c_1 \\ c_2 \end{pmatrix} = \begin{pmatrix} -c_1 + c_2 \\ c_1 - c_2 \end{pmatrix} = \begin{pmatrix} 0 \\ 0 \end{pmatrix}$$

with the solution $c_1 = c_2$. To normalize our eigenvector, we impose the condition

$$\mathbf{x}_1{}^T\mathbf{x}_1 = \begin{pmatrix} c_1 & c_2 \end{pmatrix}\begin{pmatrix} c_1 \\ c_2 \end{pmatrix} = c_1^2 + c_2^2 = 1$$

Solving for $c_1 = c_2 = 1/\sqrt{2}$ leads to the final result for the eigenvector

$$\mathbf{x}_1 = \frac{1}{\sqrt{2}} \begin{pmatrix} 1 \\ 1 \end{pmatrix}$$

For $\lambda_2 = a - 1$ and $\mathbf{x}_2$, we have

$$(\mathbf{A} - \lambda_2 \mathbf{I}) \mathbf{x}_2 = \begin{pmatrix} a - \lambda_2 & 1 \\ 1 & a - \lambda_2 \end{pmatrix} \begin{pmatrix} c_1 \\ c_2 \end{pmatrix}$$

$$= \begin{pmatrix} 1 & 1 \\ 1 & 1 \end{pmatrix} \begin{pmatrix} c_1 \\ c_2 \end{pmatrix} = \begin{pmatrix} c_1 + c_2 \\ c_1 + c_2 \end{pmatrix} = \begin{pmatrix} 0 \\ 0 \end{pmatrix}$$

with the solution $c_1 = -c_2$. We want our eigenvectors to have an absolute value of unity, so we also demand that

$$\mathbf{x}_2{}^T \mathbf{x}_2 = \begin{pmatrix} c_1 & c_2 \end{pmatrix} \begin{pmatrix} c_1 \\ c_2 \end{pmatrix} = c_1^2 + c_2^2 = 1$$

Solving for $c_1 = -c_2 = 1/\sqrt{2}$ leads to the final result for the eigenvector

$$\mathbf{x}_2 = \begin{pmatrix} \dfrac{1}{\sqrt{2}} \\ -\dfrac{1}{\sqrt{2}} \end{pmatrix} = \frac{1}{\sqrt{2}} \begin{pmatrix} 1 \\ -1 \end{pmatrix}$$

In addition to being individually normalized, our eigenvectors are mutually orthogonal, as

$$\mathbf{x}_1{}^T \mathbf{x}_2 = \frac{1}{2} \begin{pmatrix} 1 & 1 \end{pmatrix} \begin{pmatrix} 1 \\ -1 \end{pmatrix} = \mathbf{x}_2{}^T \mathbf{x}_1 = \frac{1}{2} \begin{pmatrix} 1 & -1 \end{pmatrix} \begin{pmatrix} 1 \\ 1 \end{pmatrix} = 1 - 1 = 0$$

Finally, we insert $\mathbf{x}_1$ and $\mathbf{x}_2$ into the original Equation 15.5 and find

$$\mathbf{A}\mathbf{x}_1 = \begin{pmatrix} a & 1 \\ 1 & a \end{pmatrix} \begin{pmatrix} \dfrac{1}{\sqrt{2}} \\ \dfrac{1}{\sqrt{2}} \end{pmatrix} = \frac{1}{\sqrt{2}} \begin{pmatrix} a & 1 \\ 1 & a \end{pmatrix} \begin{pmatrix} 1 \\ 1 \end{pmatrix} = \frac{1}{\sqrt{2}} \begin{pmatrix} a+1 \\ 1+a \end{pmatrix}$$

$$= \frac{1}{\sqrt{2}} (a+1) \begin{pmatrix} 1 \\ 1 \end{pmatrix} = \lambda_1 \mathbf{x}_1$$

and

$$\mathbf{A}\mathbf{x}_2 = \begin{pmatrix} a & 1 \\ 1 & a \end{pmatrix} \begin{pmatrix} \dfrac{1}{\sqrt{2}} \\ -\dfrac{1}{\sqrt{2}} \end{pmatrix} = \frac{1}{\sqrt{2}} \begin{pmatrix} a & 1 \\ 1 & a \end{pmatrix} \begin{pmatrix} 1 \\ -1 \end{pmatrix} = \frac{1}{\sqrt{2}} \begin{pmatrix} a-1 \\ 1-a \end{pmatrix}$$

$$= \frac{1}{\sqrt{2}} (a-1) \begin{pmatrix} 1 \\ -1 \end{pmatrix} = \lambda_2 \mathbf{x}_2$$

These results validate that $\mathbf{x}_1$ and $\mathbf{x}_2$ are eigenvector solutions to $\mathbf{A}\mathbf{x} = \lambda \mathbf{x}$ with corresponding eigenvalues $\lambda_1 = a + 1$ and $\lambda_2 = a - 1$.

Our final result takes the form of a pair of eigenvalues, λ_1 and λ_2, and the corresponding eigenvectors, $\mathbf{x}_1$ and $\mathbf{x}_2$. Under the transformation

$$\mathbf{Ax} = \lambda\mathbf{x}$$

the orientation of each vector satisfying the transformation is preserved. This is a special property of specific eigenvectors $\mathbf{x}$ that is valid only for particular eigenvalues λ. For all other vectors $\mathbf{x}$ and for all other constants λ, we find that

$$\mathbf{Ax} \neq \lambda\mathbf{x}$$

The transformation of vectors that are and are not eigenvectors is depicted graphically in Figure 15.2. Note that eigenvectors of the transformation maintain their direction but may be scaled in magnitude. In contract, vectors that are not eigenvectors are reoriented in the plane.

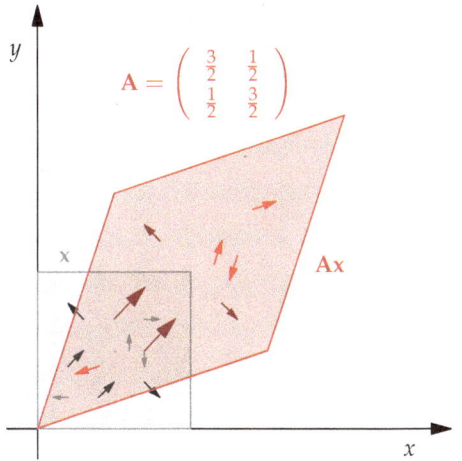

Figure 15.2: The matrix $\mathbf{A}$ transforms vectors $\mathbf{x}$ (gray space) into vectors $\mathbf{Ax}$ (red space). The dark vectors (shown in red and black) are eigenvectors of $\mathbf{A}$ satisfying $\mathbf{Ax} = \lambda\mathbf{x}$ as vector orientation is preserved while the magnitude is scaled. For the light vectors (shown in pink and gray) $\mathbf{Ax} \neq \lambda\mathbf{x}$ as the transformation alters the vector orientation.

Just as polynomial equations have characteristic roots, the eigenvalues are the roots of the characteristic equation. In the next section, we will explore the special properties and utility of a matrix composed of eigenvectors when those eigenvectors are mutually orthogonal.

15.1.4 Eigenvectors and the eigenvector matrix

Let's build a matrix $\mathbf{C}$ having columns composed of our eigenvectors $\mathbf{x}_1$ and $\mathbf{x}_2$. It takes the form

$$\mathbf{C} = \begin{pmatrix} \dfrac{1}{\sqrt{2}} & \dfrac{1}{\sqrt{2}} \\ \dfrac{1}{\sqrt{2}} & -\dfrac{1}{\sqrt{2}} \end{pmatrix} = \frac{1}{\sqrt{2}}\begin{pmatrix} 1 & 1 \\ 1 & -1 \end{pmatrix}$$

Given the fact that the eigenvectors are orthonormal, the resulting matrix $\mathbf{C}$ is an *orthogonal matrix*. The columns form a set of vectors that are individually normalized and mutually orthogonal. $\mathbf{C}$ is known as the *eigenvector matrix*.

As **C** is an orthogonal matrix, its transpose matrix is also its inverse

$$\mathbf{C}^T = \begin{pmatrix} \dfrac{1}{\sqrt{2}} & \dfrac{1}{\sqrt{2}} \\ \dfrac{1}{\sqrt{2}} & -\dfrac{1}{\sqrt{2}} \end{pmatrix} = \frac{1}{\sqrt{2}} \begin{pmatrix} 1 & 1 \\ 1 & -1 \end{pmatrix} = \mathbf{C}^{-1}$$

which we can demonstrate as

$$\mathbf{CC}^{-1} = \left[\frac{1}{\sqrt{2}} \begin{pmatrix} 1 & 1 \\ 1 & -1 \end{pmatrix} \right] \left[\frac{1}{\sqrt{2}} \begin{pmatrix} 1 & 1 \\ 1 & -1 \end{pmatrix} \right]$$

$$= \frac{1}{2} \begin{pmatrix} 1 & 1 \\ 1 & -1 \end{pmatrix} \begin{pmatrix} 1 & 1 \\ 1 & -1 \end{pmatrix}$$

$$= \frac{1}{2} \begin{pmatrix} 1+1 & 1-1 \\ 1-1 & 1+1 \end{pmatrix} = \frac{1}{2} \begin{pmatrix} 2 & 0 \\ 0 & 2 \end{pmatrix} = \begin{pmatrix} 1 & 0 \\ 0 & 1 \end{pmatrix} = \mathbf{I}$$

Similarly, $\mathbf{C}^{-1}\mathbf{C} = \mathbf{I}$. This suggests an alternative approach to the solution of our original problem $\mathbf{Ax} = \lambda\mathbf{x}$ where $\mathbf{A}$ is defined by Equation 15.5.

Suppose we multiply the coefficient matrix **A** from the right by our eigenvector matrix **C**. The result is

$$\mathbf{AC} = \frac{1}{\sqrt{2}} \begin{pmatrix} a & 1 \\ 1 & a \end{pmatrix} \begin{pmatrix} 1 & 1 \\ 1 & -1 \end{pmatrix} = \frac{1}{\sqrt{2}} \begin{pmatrix} (a+1) & (a-1) \\ (1+a) & (1-a) \end{pmatrix}$$

Now suppose we multiply the resulting matrix **AC** from the left by the inverse of our eigenvector matrix $\mathbf{C}^{-1}$. We find the remarkable result

$$\mathbf{C}^{-1}\mathbf{AC} = \frac{1}{\sqrt{2}} \begin{pmatrix} 1 & 1 \\ 1 & -1 \end{pmatrix} \left[\frac{1}{\sqrt{2}} \begin{pmatrix} (a+1) & (a-1) \\ (1+a) & (1-a) \end{pmatrix} \right]$$

$$= \frac{1}{2} \begin{pmatrix} 2(a+1) & 0 \\ 0 & 2(a-1) \end{pmatrix}$$

$$= \begin{pmatrix} a+1 & 0 \\ 0 & a-1 \end{pmatrix}$$

The matrix transformation $\mathbf{C}^{-1}\mathbf{AC}$ leads to a diagonal matrix where the elements are the eigenvalues $\lambda_1 = a+1$ and $\lambda_2 = a-1$ of the matrix **A**. We refer to the result as the *eigenvalue matrix*:

$$\mathbf{C}^{-1}\mathbf{AC} = \mathbf{\Lambda} = \begin{pmatrix} \lambda_1 & 0 \\ 0 & \lambda_2 \end{pmatrix} \tag{15.6}$$

If we know the orthonormal eigenvectors $\mathbf{x}_1$ and $\mathbf{x}_2$, we can form the eigenvector matrix **C** and determine the eigenvalues λ_1 and λ_2 by evaluating

$$\mathbf{C}^{-1}\mathbf{AC} = \mathbf{\Lambda}$$

This general approach is useful in the evaluation of quadratic polynomials of many variables known as *quadratic forms*.[6] We will explore the use of quadratic polynomials in eigenvalue problems modeling vibrational motion.

[6] We can express multidimensional quadratic functions in the form of the equation

$$\mathbf{x}^T\mathbf{Ax}$$

where **x** is a coordinate vector, $\mathbf{x}^T$ is its vector transpose, and **A** is a matrix of constant coefficients. By identifying the eigenvectors of the matrix **A**, building the eigenvector matrix **C**, and employing the identity $\mathbf{CC}^{-1} = \mathbf{C}^{-1}\mathbf{C} = \mathbf{I}$ we can write

$$\mathbf{x}^T\mathbf{Ax} = \mathbf{x}^T\mathbf{IAIx} = \mathbf{x}^T\mathbf{CC}^{-1}\mathbf{ACC}^{-1}\mathbf{x}$$

which can also be written

$$\mathbf{y}^T\mathbf{\Lambda y}$$

where $\mathbf{C}^{-1}\mathbf{AC} = \mathbf{\Lambda}$ is a diagonal matrix, $\mathbf{y} = \mathbf{C}^{-1}\mathbf{x}$ is the normal coordinate vector, and $\mathbf{y}^T = \left(\mathbf{C}^{-1}\mathbf{x}\right)^T = \mathbf{x}^T\mathbf{C}$. In this way, we can transform a set of coupled quadratic equations in **x** into a set of uncoupled quadratic equations in the normal coordinates **y**.

15.1.5 Solving eigenvalue problems of arbitrary dimension

The method we have developed for solving *eigenvalue problems* of the form

$$\mathbf{A}\mathbf{x} = \lambda\mathbf{x} \tag{15.7}$$

is generally applicable to the case of n coupled equations in n unknown variables where $\mathbf{A}$ is an $n \times n$ matrix of coefficients:

$$\mathbf{A} = \begin{pmatrix} a_{11} & a_{12} & \cdots & a_{1n} \\ a_{21} & a_{22} & \cdots & a_{2n} \\ \vdots & \vdots & \ddots & \vdots \\ a_{n1} & a_{n2} & \cdots & a_{nn} \end{pmatrix} \tag{15.8}$$

λ is the associated eigenvalue and $\mathbf{x}$ is the corresponding eigenvector:

$$\mathbf{x} = \begin{pmatrix} c_1 \\ c_2 \\ \vdots \\ c_n \end{pmatrix}$$

The eigenvalues are solutions to the *characteristic equation*:

$$|\mathbf{A} - \lambda\mathbf{I}| = \begin{vmatrix} a_{11} - \lambda & a_{12} & \cdots & a_{1n} \\ a_{21} & a_{22} - \lambda & \cdots & a_{2n} \\ \vdots & \vdots & \ddots & \vdots \\ a_{n1} & a_{n2} & \cdots & a_{nn} - \lambda \end{vmatrix} = 0 \tag{15.9}$$

which is an nth order polynomial equation in λ having n roots corresponding to a set of n eigenvalues $\{\lambda_k\}$. The eigenvalues have special properties. First, the sum of all eigenvalues equals the *trace* of $\mathbf{A}$:[7]

$$\sum_{k=1}^{n} \lambda_k = \text{Tr}\,\mathbf{A}$$

Second, the product of all eigenvalues equals the determinant of $\mathbf{A}$:

$$\prod_{k=1}^{n} \lambda_k = |\mathbf{A}|$$

Each of the n eigenvalues λ_k is then used to define a corresponding eigenvector $\mathbf{x}_k$ through Equation 15.2:

$$(\mathbf{A} - \lambda_k\mathbf{I})\,\mathbf{x}_k = \mathbf{0} \tag{15.10}$$

For a symmetric matrix $\mathbf{A}$, the set of orthonormal eigenvectors $\{\mathbf{x}_k\}$ can be used to form the columns of the orthogonal *eigenvector matrix* $\mathbf{C}$. The eigenvector matrix can be used to diagonalize the matrix $\mathbf{A}$:

$$\mathbf{C}^{-1}\mathbf{A}\mathbf{C} = \boldsymbol{\Lambda} \tag{15.11}$$

The n eigenvalues and eigenvectors form a complete solution to our problem.

[7] The sum of diagonal elements of a matrix

$$\sum_i [\mathbf{A}]_{ii} = \text{Tr}\,\mathbf{A}$$

is known as the *trace*. The trace of a matrix is identical for the matrix and its transpose $\text{Tr}\,\mathbf{A} = \text{Tr}\,\mathbf{A}^T$.

15.2 Matrix methods for coupled differential equations

SETS OF HOMOGENEOUS COUPLED LINEAR ALGEBRAIC EQUATIONS can be solved using matrix methods when formed as an eigenvalue problem. The same methods can be used to solve sets of coupled linear ordinary differential equations. In this section, we use matrix methods to solve eigenvalue problems involving coupled linear ordinary differential equations describing physical kinetics and dynamics.

15.2.1 Chemical kinetics revisited (as an eigenvalue problem)

In our explorations of *first order differential equations* in Chapter 10, we analyzed the chemical kinetic scheme

$$A \xrightarrow{k_1} B \xrightarrow{k_2} C$$

by solving the set of three coupled linear first order ordinary differential equations for the time-dependent concentrations of each species originally presented in Equation 10.11:

$$\frac{d}{dt}A(t) = -k_1 A(t)$$

$$\frac{d}{dt}B(t) = k_1 A(t) - k_2 B(t)$$

$$\frac{d}{dt}C(t) = k_2 B(t)$$

The initial conditions define the concentration of each species at time $t = 0$ as $A(0) = A_0$ and $B(0) = C(0) = 0$. With our knowledge of matrix algebra, we can rewrite this set of coupled linear differential equations as

$$\frac{d}{dt}\begin{pmatrix} A(t) \\ B(t) \\ C(t) \end{pmatrix} = \begin{pmatrix} -k_1 & 0 & 0 \\ k_1 & -k_2 & 0 \\ 0 & k_2 & 0 \end{pmatrix} \begin{pmatrix} A(t) \\ B(t) \\ C(t) \end{pmatrix}$$

which can be reformed as

$$\frac{d}{dt}\mathbf{x}(t) = \mathbf{K}\mathbf{x}(t) \tag{15.12}$$

where we have defined the *composition vector*

$$\mathbf{x}(t) = \begin{pmatrix} A(t) \\ B(t) \\ C(t) \end{pmatrix}$$

and the *rate constant matrix*

$$\mathbf{K} = \begin{pmatrix} -k_1 & 0 & 0 \\ k_1 & -k_2 & 0 \\ 0 & k_2 & 0 \end{pmatrix}$$

[8] This follows from the fact that the time derivative of $\mathbf{x}(t)$ is proportional to $\mathbf{x}(t)$ itself, a property of the exponential function.

We anticipate exponential time dependence for the concentrations.[8] As such, we substitute into Equation 15.12 an *ansatz* for $\mathbf{x}(t)$ in the form of

$$e^{\alpha t}$$

and find[9]

$$\frac{d}{dt}\mathbf{x}(t) = \alpha\mathbf{x}(t) = \mathbf{K}\mathbf{x}(t) \tag{15.13}$$

This equation has the form of an eigenvalue problem (see Equation 15.7). It can be solved using the general methods we have developed for the solution of ordinary differential equations and matrix eigenvalue problems.

Equation 15.13 can be rewritten in the form of Equation 15.10 as

$$(\mathbf{K} - \alpha\mathbf{I})\,\mathbf{x}(t) = \mathbf{0}$$

This expression will have non-trivial solutions only for cases where the determinant is zero:

$$|\mathbf{K} - \alpha\mathbf{I}| = 0$$

which is the *characteristic equation* defined in Equation 15.9.

Inserting the rate constant matrix $\mathbf{K}$ in the characteristic equation and evaluating the determinant leads to a cubic polynomial equation in α:

$$|\mathbf{K} - \alpha\mathbf{I}| = \begin{vmatrix} -k_1 - \alpha & 0 & 0 \\ k_1 & -k_2 - \alpha & 0 \\ 0 & k_2 & 0 - \alpha \end{vmatrix} = -(k_1 + \alpha)(k_2 + \alpha)\alpha = 0$$

The roots of this equation are the three eigenvalues $\alpha_1 = 0$, $\alpha_2 = -k_1$, and $\alpha_3 = -k_2$. We can now identify the eigenvectors $\mathbf{x}_k$ corresponding to each of the eigenvalues α_k using Equation 15.10 where

$$(\mathbf{K} - \alpha_k\mathbf{I})\,\mathbf{x}_k = \mathbf{0}$$

For $\alpha_1 = 0$, we have

$$(\mathbf{K} - \alpha_1)\,\mathbf{x}_1 = \begin{pmatrix} -k_1 & 0 & 0 \\ k_1 & -k_2 & 0 \\ 0 & k_2 & 0 \end{pmatrix} \begin{pmatrix} c_1 \\ c_2 \\ c_3 \end{pmatrix} = \begin{pmatrix} 0 \\ 0 \\ 0 \end{pmatrix} = \mathbf{0}$$

leading to a set of equations defining the coefficients:

$$-k_1 c_1 = 0$$
$$k_1 c_1 - k_2 c_2 = 0$$
$$k_2 c_2 = 0$$

The first equation leads to $c_1 = 0$, while the third equation leads to $c_2 = 0$. No constraint is placed on c_3. We set it to be unity. The eigenvector is

$$\mathbf{x}_1 = \begin{pmatrix} 0 \\ 0 \\ 1 \end{pmatrix}$$

This eigenvector corresponds to the solution achieved at long times, in which all reactants A have been converted to products C.

[9] Taking $\mathbf{x}(t) = \mathbf{x}_0\,e^{\alpha t}$ we find

$$\frac{d}{dt}\mathbf{x}(t) = \mathbf{x}_0\frac{d}{dt}e^{\alpha t} = \alpha\,\mathbf{x}_0\,e^{\alpha t} = \alpha\mathbf{x}(t)$$

by design.

For $\alpha_2 = -k_1$, we have

$$(\mathbf{K} - \alpha_2)\mathbf{x}_2 = \begin{pmatrix} 0 & 0 & 0 \\ k_1 & -k_2 + k_1 & 0 \\ 0 & k_2 & k_1 \end{pmatrix} \begin{pmatrix} c_1 \\ c_2 \\ c_3 \end{pmatrix} = \begin{pmatrix} 0 \\ 0 \\ 0 \end{pmatrix} = \mathbf{o}$$

leading to a set of equations defining the coefficients:

$$0 = 0$$
$$k_1 c_1 - (k_2 - k_1)c_2 = 0$$
$$k_2 c_2 + k_1 c_3 = 0$$

We arbitrarily set $c_1 = 1$ to find $c_2 = k_1/(k_2 - k_1)$ and $c_3 = -k_2/(k_2 - k_1)$, leading to the eigenvector

$$\mathbf{x}_2 = \begin{pmatrix} 1 \\ \dfrac{k_1}{k_2 - k_1} \\ \dfrac{-k_2}{k_2 - k_1} \end{pmatrix}$$

Finally, for $\alpha_3 = -k_2$, we have

$$(\mathbf{K} - \alpha_3)\mathbf{x}_3 = \begin{pmatrix} -k_1 + k_2 & 0 & 0 \\ k_1 & 0 & 0 \\ 0 & k_2 & k_2 \end{pmatrix} \begin{pmatrix} c_1 \\ c_2 \\ c_3 \end{pmatrix} = \begin{pmatrix} 0 \\ 0 \\ 0 \end{pmatrix} = \mathbf{o}$$

leading to a set of equations defining the coefficients:

$$-(k_2 - k_1)c_1 = 0$$
$$k_1 c_1 = 0$$
$$k_2 c_2 + k_2 c_3 = 0$$

The first and second equations lead to $c_1 = 0$. We arbitrarily set $c_2 = -k_1/(k_2 - k_1)$ to find $c_3 = k_1/(k_2 - k_1)$, leading to the eigenvector

$$\mathbf{x}_3 = \begin{pmatrix} 0 \\ \dfrac{-k_1}{k_2 - k_1} \\ \dfrac{k_1}{k_2 - k_1} \end{pmatrix}$$

Having defined the three eigenvalues and corresponding eigenvectors, we can write the solution for the time dependence of the composition vector:

$$\mathbf{x}(t) = \begin{pmatrix} A(t) \\ B(t) \\ C(t) \end{pmatrix} = d_1 \mathbf{x}_1 e^{\alpha_1 t} + d_2 \mathbf{x}_2 e^{\alpha_2 t} + d_3 \mathbf{x}_3 e^{\alpha_3 t}$$

$$= d_1 \begin{pmatrix} 0 \\ 0 \\ 1 \end{pmatrix} + d_2 \begin{pmatrix} 1 \\ \dfrac{k_1}{k_2 - k_1} \\ \dfrac{-k_2}{k_2 - k_1} \end{pmatrix} e^{-k_1 t} + d_3 \begin{pmatrix} 0 \\ \dfrac{-k_1}{k_2 - k_1} \\ \dfrac{k_1}{k_2 - k_1} \end{pmatrix} e^{-k_2 t}$$

The coefficients d_1, d_2, and d_3 must be determined by the initial conditions. Returning to the initial conditions $A(0) = A_0$ and $B(0) = C(0) = 0$ at $t = 0$, we find

$$\mathbf{x}(0) = d_1 \begin{pmatrix} 0 \\ 0 \\ 1 \end{pmatrix} + d_2 \begin{pmatrix} 1 \\ \dfrac{k_1}{k_2 - k_1} \\ \dfrac{-k_2}{k_2 - k_1} \end{pmatrix} + d_3 \begin{pmatrix} 0 \\ \dfrac{-k_1}{k_2 - k_1} \\ \dfrac{k_1}{k_2 - k_1} \end{pmatrix} = \begin{pmatrix} A_0 \\ 0 \\ 0 \end{pmatrix}$$

leading to a set of equations defining the coefficients:

$$d_2 = A_0$$

$$d_2 \frac{k_1}{k_2 - k_1} - d_3 \frac{k_1}{k_2 - k_1} = 0$$

$$d_1 - d_2 \frac{k_2}{k_2 - k_1} + d_3 \frac{k_1}{k_2 - k_1} = 0$$

The first equation sets $d_2 = A_0$, the second equation sets $d_3 = d_2 = A_0$, and the third equation defines

$$d_1 = A_0 \frac{k_2}{k_2 - k_1} - A_0 \frac{k_1}{k_2 - k_1} = A_0$$

This leads to our result for the time dependence of the reaction composition:

$$\mathbf{x}(t) = \begin{pmatrix} A(t) \\ B(t) \\ C(t) \end{pmatrix} = A_0 \begin{pmatrix} 0 \\ 0 \\ 1 \end{pmatrix} + A_0 \begin{pmatrix} 1 \\ \dfrac{k_1}{k_2 - k_1} \\ \dfrac{-k_2}{k_2 - k_1} \end{pmatrix} e^{-k_1 t} + A_0 \begin{pmatrix} 0 \\ \dfrac{-k_1}{k_2 - k_1} \\ \dfrac{k_1}{k_2 - k_1} \end{pmatrix} e^{-k_2 t}$$

The individual time-dependent concentrations are

$$A(t) = A_0 e^{-k_1 t}$$

$$B(t) = A_0 \frac{k_1}{k_2 - k_1} \left(e^{-k_1 t} - e^{-k_2 t} \right)$$

$$C(t) = A_0 + A_0 \frac{k_1}{k_2 - k_1} \left(e^{-k_1 t} - \frac{k_2}{k_1} e^{-k_2 t} \right)$$

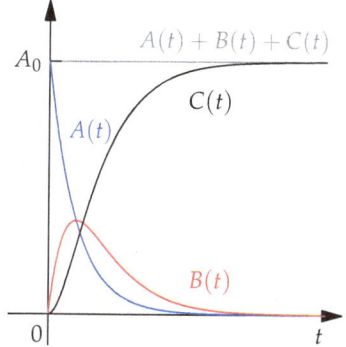

Figure 15.3: The general solution for $A(t)$ (blue), $B(t)$ (red), and $C(t)$ (black) as a function of time.

The result is shown in Figure 15.3. Note the monotonic decrease in reactants A, the monotonic increase in products C, and the increase and decrease of the intermediates B with increasing time.

In the limit that $t = 0$, the solution respects the initial conditions $A(0) = A_0$ and $B(0) = C(0) = 0$. Over time, the solution also respects the condition that

$$A(t) + B(t) + C(t) = A_0$$

In the long time limit, we find $A(\infty) = B(\infty) = 0$ and $C(\infty) = A_0$. This result, derived using matrix algebra as an eigenvalue problem, is identical to Equations 10.12, 10.14, and 10.16, derived in Chapter 10 by successively solving the series of first order ordinary differential equations.

15.2.2 *Oscillating masses on springs revisited (as an eigenvalue problem)*

In our explorations of *second order differential equations* in Chapter 11, we analyzed the dynamics of a mass on a spring governed by Newton's equation of motion

$$m\frac{d^2}{dt^2}x(t) = -\kappa\left(x(t) - x_0\right)$$

where $x(t)$ is the displacement of the oscillator from its mechanically stable position at x_0, and κ is the force constant (see Figure 15.4). We solved this linear second order ordinary differential equation by first inserting the *ansatz* for the time dependence of the solution $e^{\alpha t}$. That led to the auxiliary equation with roots $\alpha_\pm = \pm i\sqrt{\dfrac{\kappa}{m}}$. Using these roots, we found the final solution

$$x(t) = x_0 + (x(0) - x_0)\cos\left(\sqrt{\frac{\kappa}{m}}t\right)$$

for an oscillator initially in position $x(0)$ with velocity $\dfrac{d}{dt}x(0) = 0$.

Now consider the case of two masses connected to solid supports and to each other by springs (see Figure 15.5). The potential energy is

$$V(x_1, x_2) = \frac{1}{2}\kappa x_1^2 + \frac{1}{2}\kappa\left(x_2 - x_1\right)^2 + \frac{1}{2}\kappa x_2^2$$

where x_1 and x_2 are the displacements of the oscillators from the points of mechanical equilibrium at which the potential energy $V(0,0) = 0$ is a minimum. The equations of motion for the displacements of the two masses are

$$m\frac{d^2}{dt^2}x_1(t) = -\frac{dV}{dx_1} = -\kappa x_1 + \kappa\left(x_2 - x_1\right) = -2\kappa x_1 + \kappa x_2$$

$$m\frac{d^2}{dt^2}x_2(t) = -\frac{dV}{dx_2} = -\kappa\left(x_2 - x_1\right) - \kappa x_2 = \kappa x_1 - 2\kappa x_2$$

We can rewrite these equations in compact form as

$$\frac{d^2}{dt^2}\mathbf{x}(t) = -\mathbf{K}\mathbf{x}(t) \qquad (15.14)$$

where we have defined the *displacement vector* as

$$\mathbf{x}(t) = \begin{pmatrix} x_1(t) \\ x_2(t) \end{pmatrix}$$

and the *force constant matrix* as

$$\mathbf{K} = \begin{pmatrix} \dfrac{2\kappa}{m} & -\dfrac{\kappa}{m} \\ -\dfrac{\kappa}{m} & \dfrac{2\kappa}{m} \end{pmatrix}$$

Equation 15.14 has the form of an eigenvalue problem defined by Equation 15.7 and can be treated using the general methods we have developed for solving ordinary differential equations and matrix eigenvalue problems.

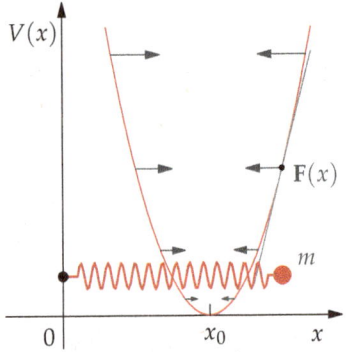

Figure 15.4: The potential energy function $V(x) = \frac{1}{2}\kappa(x - x_0)$ for a mass on a spring with corresponding force $F(x) = -\kappa(x - x_0)$.

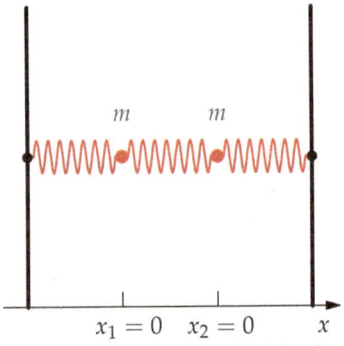

Figure 15.5: Two masses (red dots) are connected by springs to each other and fixed points (black dots) on supporting walls (thick lines). The positions are measured as displacements from the points of mechanical equilibrium defined by $x_1 = 0$ and $x_2 = 0$.

As we anticipate sinusoidal time dependence for the mass displacements, we substitute into Equation 15.14 an *ansatz* for $\mathbf{x}(t)$ of the form[10]

$$e^{i\omega t}$$

and find[11]

$$\frac{d^2}{dt^2}\mathbf{x}(t) = -\omega^2 \mathbf{x}(t) = -\mathbf{K}\mathbf{x}(t) \tag{15.15}$$

which can be rewritten as

$$\left(\mathbf{K} - \omega^2 \mathbf{I}\right)\mathbf{x}(t) = 0$$

This expression will have non-trivial solutions only for cases where the determinant is

$$\left|\mathbf{K} - \omega^2 \mathbf{I}\right| = 0$$

which is the *characteristic equation* defined in Equation 15.9.

Inserting the force constant matrix $\mathbf{K}$ in the characteristic equation and evaluating the determinant leads to a quartic polynomial equation in ω:

$$\left|\mathbf{K} - \omega^2 \mathbf{I}\right| = \begin{pmatrix} \dfrac{2\kappa}{m} - \omega^2 & -\dfrac{\kappa}{m} \\ -\dfrac{\kappa}{m} & \dfrac{2\kappa}{m} - \omega^2 \end{pmatrix} = \left(\frac{2\kappa}{m} - \omega^2\right)^2 - \frac{\kappa^2}{m^2} = 0$$

The roots of this quartic equation in ω are two pairs of eigenvalues $\pm\omega_1 = \pm\sqrt{\frac{\kappa}{m}}$ and $\pm\omega_2 = \pm\sqrt{\frac{3\kappa}{m}}$. We can now identify the eigenvectors corresponding to each of the eigenvalue pairs ω_k^2 by solving the eigenvector equation defined by Equation 15.10, where

$$\left(\mathbf{K} - \omega_k^2 \mathbf{I}\right)\mathbf{x}_k = \mathbf{0}$$

and the solutions are the eigenvectors $\mathbf{x}_k$.

For $\omega_1^2 = \dfrac{k}{m}$, we have

$$\left(\mathbf{K} - \omega_1^2\right)\mathbf{x}_1 = \begin{pmatrix} \dfrac{\kappa}{m} & -\dfrac{\kappa}{m} \\ -\dfrac{\kappa}{m} & \dfrac{\kappa}{m} \end{pmatrix}\begin{pmatrix} c_1 \\ c_2 \end{pmatrix} = \begin{pmatrix} 0 \\ 0 \end{pmatrix}$$

leading to a set of equations defining the coefficients:

$$\frac{\kappa}{m}c_1 - \frac{\kappa}{m}c_2 = 0$$
$$-\frac{\kappa}{m}c_1 + \frac{\kappa}{m}c_2 = 0$$

Both equations lead to $c_1 = c_2$ and the eigenvector

$$\mathbf{x}_1 = \begin{pmatrix} 1 \\ 1 \end{pmatrix}$$

The displacement of the masses is symmetric as shown in Figure 15.6.

[10] The second derivative with respect to time of $\mathbf{x}(t)$ is proportional to $-\mathbf{x}(t)$, a property of sinusoidal functions.

[11] Taking $\mathbf{x}(t) = \mathbf{x}_0\, e^{i\omega t}$ we find

$$\frac{d^2}{dt^2}\mathbf{x}(t) = \mathbf{x}_0 \frac{d^2}{dt^2}e^{i\omega t} = -\omega^2\, \mathbf{x}_0\, e^{i\omega t}$$
$$= -\omega^2 \mathbf{x}(t)$$

by design.

For $\omega_2^2 = \dfrac{3k}{m}$, we have

$$\left(\mathbf{K} - \omega_2^2\right) \mathbf{x}_2 = \begin{pmatrix} -\dfrac{\kappa}{m} & -\dfrac{\kappa}{m} \\ -\dfrac{\kappa}{m} & -\dfrac{\kappa}{m} \end{pmatrix} \begin{pmatrix} c_1 \\ c_2 \end{pmatrix} = \begin{pmatrix} 0 \\ 0 \end{pmatrix}$$

leading to a set of equations defining the coefficients:

$$-\frac{\kappa}{m}c_1 - \frac{\kappa}{m}c_2 = 0$$
$$-\frac{\kappa}{m}c_1 - \frac{\kappa}{m}c_2 = 0$$

Both equations lead to $c_1 = -c_2$ and the eigenvector

$$\mathbf{x}_2 = \begin{pmatrix} 1 \\ -1 \end{pmatrix}$$

The displacement of the masses is asymmetric as shown in Figure 15.7.

We have identified the two eigenvalues, $-\omega_1^2$ and $-\omega_2^2$, and the corresponding eigenvectors, $\mathbf{x}_1$ and $\mathbf{x}_2$. Now let's build the general solution for the displacement vector $\mathbf{x}(t)$. For the differential equation $\ddot{x}(t) = -\omega^2 x(t)$ having eigenvalue $-\omega^2$, we expect the oscillatory solution $x(t) = c_1 \cos(\omega t) + c_2 \sin(\omega t)$. As such, we can write the general solution for the time-dependence of the displacement vector $\mathbf{x}(t)$ as

$$\mathbf{x}(t) = \mathbf{x}_1 \left(d_1 \cos(\omega_1 t) + d_2 \sin(\omega_1 t)\right) + \mathbf{x}_2 \left(d_3 \cos(\omega_2 t) + d_4 \sin(\omega_2 t)\right) \quad (15.16)$$

where the coefficients d_1, d_2, d_3, and d_4 must be determined by the initial conditions.

The initial conditions for the displacements and velocities of the masses are

$$x_1(0), \ \frac{d}{dt}x_1(0) = 0 \qquad x_2(0), \ \frac{d}{dt}x_2(0) = 0$$

where the masses are initially at rest. Applying the initial conditions for the displacements, we find

$$\mathbf{x}(0) = d_1 \begin{pmatrix} 1 \\ 1 \end{pmatrix} + d_3 \begin{pmatrix} 1 \\ -1 \end{pmatrix} = \begin{pmatrix} x_1(0) \\ x_2(0) \end{pmatrix}$$

leading to a set of equations defining the coefficients d_1 and d_3 in terms of the initial displacements:

$$x_1(0) = d_1 + d_3$$
$$x_2(0) = d_1 - d_3$$

The resulting coefficients are

$$d_1 = \frac{1}{2}\left(x_1(0) + x_2(0)\right) \qquad d_3 = \frac{1}{2}\left(x_1(0) - x_2(0)\right)$$

with magnitudes defined by the initial displacements.

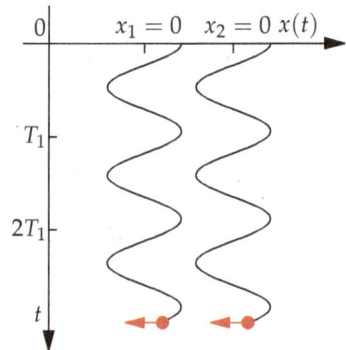

Figure 15.6: The trajectory of a symmetric displacement of the two masses leading to *in-phase* sinusoidal oscillation with angular frequency $\omega_1 = \sqrt{k/m}$ and period $T_1 = 2\pi\sqrt{m/k}$.

Applying the initial conditions for the velocities of the displacements, we find

$$\frac{d}{dt}\mathbf{x}(0) = d_2\omega_1 \begin{pmatrix} 1 \\ 1 \end{pmatrix} + d_4\omega_2 \begin{pmatrix} 1 \\ -1 \end{pmatrix} = \begin{pmatrix} 0 \\ 0 \end{pmatrix}$$

and $d_2 = d_4 = 0$. Inserting our results in Equation 15.16, we find that our final result for the general solution to Equation 15.14 is

$$\mathbf{x}(t) = \begin{pmatrix} x_1(t) \\ x_2(t) \end{pmatrix} = d_1 \begin{pmatrix} 1 \\ 1 \end{pmatrix} \cos(\omega_1 t) + d_2 \begin{pmatrix} 1 \\ -1 \end{pmatrix} \cos(\omega_2 t) \qquad (15.17)$$

which can also be written as

$$\mathbf{x}(t) = \frac{1}{2}(x_1(0)+x_2(0)) \begin{pmatrix} 1 \\ 1 \end{pmatrix} \cos\left(\sqrt{\frac{k}{m}}t\right) + \frac{1}{2}(x_1(0)-x_2(0)) \begin{pmatrix} 1 \\ -1 \end{pmatrix} \cos\left(\sqrt{\frac{3k}{m}}t\right)$$

Note that our general solution has two frequencies of motion defined by ω_1 and $\omega_2 = \sqrt{3}\omega_1$. In general, the motion of the two masses will be defined by oscillatory motion on both time scales.

To gain insight into our solution, let's analyze the time dependence for two initial displacements. Suppose we choose the initial condition $x_1(0) = x_2(0) = \Delta x(0)$. Our general solution reduces to

$$\mathbf{x}(t) = \begin{pmatrix} x_1(t) \\ x_2(t) \end{pmatrix} = \Delta x(0) \begin{pmatrix} 1 \\ 1 \end{pmatrix} \cos\left(\sqrt{\frac{k}{m}}t\right)$$

where the displacement of each mass oscillates with a period

$$T_1 = 2\pi\sqrt{\frac{m}{\kappa}}$$

during which both masses oscillate in phase as shown in Figure 15.6.

Now suppose we choose the initial condition $x_1(0) = -x_2(0) = \Delta x(0)$. Our general solution reduces to

$$\mathbf{x}(t) = \begin{pmatrix} x_1(t) \\ x_2(t) \end{pmatrix} = \Delta x(0) \begin{pmatrix} 1 \\ -1 \end{pmatrix} \cos\left(\sqrt{\frac{3k}{m}}t\right)$$

where the displacement of each mass oscillates with a period

$$T_2 = 2\pi\sqrt{\frac{m}{3\kappa}} = \frac{1}{\sqrt{3}}T_1$$

during which both masses oscillate *out-of-phase* as shown in Figure 15.7.

These oscillations represent *normal modes* of vibration. Each mode has a unique frequency, ω_1 or ω_2, and corresponding eigenvector, $\mathbf{x}_1$ or $\mathbf{x}_2$. The eigenvectors are mutually orthogonal, and can be normalized as

$$\mathbf{x}_1 = \frac{1}{\sqrt{2}} \begin{pmatrix} 1 \\ 1 \end{pmatrix} \qquad \mathbf{x}_2 = \frac{1}{\sqrt{2}} \begin{pmatrix} 1 \\ -1 \end{pmatrix}$$

These modes have the characteristics of the vibrations of atoms in solids.

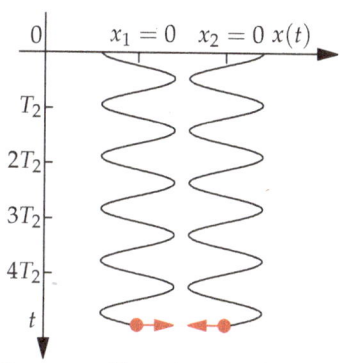

Figure 15.7: The trajectory of an asymmetric displacement of the two masses leading to *out-of-phase* sinusoidal oscillation with angular frequency $\omega_2 = \sqrt{3\kappa/m}$ and period $T_2 = T_1/\sqrt{3}$.

The lower frequency in-phase oscillation has the characteristic of a lower frequency *acoustic mode*. Atoms move together in-phase (see Figure 15.6). The higher frequency oscillation has the characteristic of a higher frequency *optical mode*. Atoms move out-of-phase (see Figure 15.7).

Note that the same time evolution of the displacements can be viewed on the two-dimensional plane where the abscissa is the displacement x_1 and the ordinate the displacement x_2, as shown in Figure 15.8. The diagonal red line represents the dynamics of the in-phase oscillation depicted in Figure 15.6 and the diagonal blue line represents the dynamics of the out-of-phase oscillation depicted in Figure 15.7.

Now consider the initial displacement $x_1(0) = x(0)$ and $x_2(0) = 0$ shown in Figure 15.9. The oscillators are coupled, and the initial displacement of x_1 away from the point of mechanical equilibrium, $x_1 = 0$ and $x_2 = 0$, leads to the displacement of x_2. As the trajectory defined by the point $(x_1(t), x_2(t))$ evolves in time, it explores an area defined by the set of displacements consistent with the constraint that the total energy of the system is constant in time. The bounding lines known as *caustics* form a square region in the xy-plane.

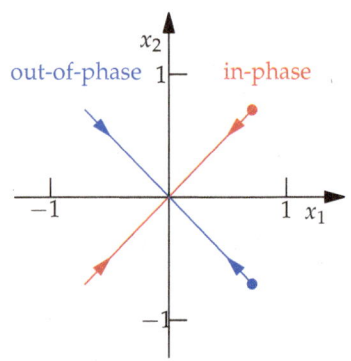

Figure 15.8: The displacements, x_1 and x_2, of two masses connected to solid supports and each other through harmonic springs (see Figure 15.5), showing in-phase motion (red line) and out-of-phase motion (blue line).

Figure 15.9: The trajectory of the time evolution of the displacements x_1 and x_2 of the two masses starting from $x_1(0) = 1$ and $x_2(0) = 0$ (black dot).

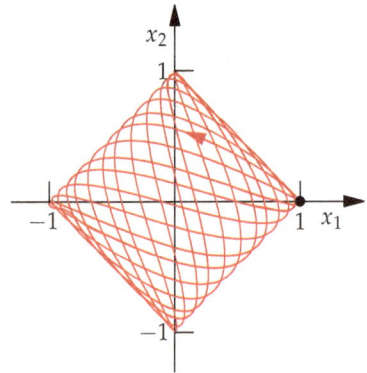

Let's return to Equation 15.14 defining the motion of our two masses:

$$\frac{d^2}{dt^2}\mathbf{x}(t) = -\mathbf{K}\mathbf{x}(t) \tag{15.18}$$

Suppose we form an *orthogonal matrix* from our two orthonormal eigenvectors $\mathbf{x}_1$ and $\mathbf{x}_2$ as

$$\mathbf{C} = \begin{pmatrix} \dfrac{1}{\sqrt{2}} & \dfrac{1}{\sqrt{2}} \\ \dfrac{1}{\sqrt{2}} & -\dfrac{1}{\sqrt{2}} \end{pmatrix} = \frac{1}{\sqrt{2}} \begin{pmatrix} 1 & 1 \\ 1 & -1 \end{pmatrix}$$

The inverse of our orthogonal matrix is its transpose $\mathbf{C}^{-1} = \mathbf{C}^T$. Now recall that $\mathbf{C}$ and $\mathbf{C}^{-1}$ can be used to diagonalize the force constant matrix $\mathbf{K}$ using Equation 15.6

$$\mathbf{C}^{-1}\mathbf{K}\mathbf{C} = \mathbf{\Lambda} \tag{15.19}$$

where

$$\mathbf{C}^{-1}\mathbf{KC} = \frac{1}{2}\begin{pmatrix} 1 & 1 \\ 1 & -1 \end{pmatrix}\begin{pmatrix} \dfrac{2\kappa}{m} & -\dfrac{\kappa}{m} \\ -\dfrac{\kappa}{m} & \dfrac{2\kappa}{m} \end{pmatrix}\begin{pmatrix} 1 & 1 \\ 1 & -1 \end{pmatrix}$$

$$= \frac{1}{2}\begin{pmatrix} 1 & 1 \\ 1 & -1 \end{pmatrix}\begin{pmatrix} \dfrac{\kappa}{m} & \dfrac{3\kappa}{m} \\ \dfrac{\kappa}{m} & -\dfrac{3\kappa}{m} \end{pmatrix}$$

$$= \frac{1}{2}\begin{pmatrix} \dfrac{2\kappa}{m} & 0 \\ 0 & \dfrac{6\kappa}{m} \end{pmatrix} = \mathbf{\Lambda}$$

This solution defines the diagonal matrix

$$\mathbf{\Lambda} = \begin{pmatrix} \dfrac{\kappa}{m} & 0 \\ 0 & \dfrac{3\kappa}{m} \end{pmatrix} = \begin{pmatrix} \omega_1^2 & 0 \\ 0 & \omega_2^2 \end{pmatrix}$$

which has diagonal elements equal to the eigenvalues found through the solution of the secular equation. We see that one way of finding the eigenvalues is to diagonalize the force constant matrix $\mathbf{K}$.

Now let's use the eigenvector matrix $\mathbf{C}$ to transform the equation of motion in Equation 15.14. We first multiply from the left by $\mathbf{C}^{-1}$ as

$$\frac{d^2}{dt^2}\mathbf{C}^{-1}\mathbf{x}(t) = -\mathbf{C}^{-1}\mathbf{K}\mathbf{x}(t)$$

We then insert the identity matrix $\mathbf{I} = \mathbf{CC}^{-1}$ between $\mathbf{K}$ and $\mathbf{x}(t)$ as

$$\frac{d}{dt}\mathbf{C}^{-1}\mathbf{x}(t) = -\mathbf{C}^{-1}\mathbf{KI}\mathbf{x}(t) = -\mathbf{C}^{-1}\mathbf{KCC}^{-1}\mathbf{x}(t)$$

We have shown that $\mathbf{C}^{-1}\mathbf{KC} = \mathbf{\Lambda}$, so it follows that

$$\frac{d}{dt}\mathbf{C}^{-1}\mathbf{x}(t) = -\mathbf{\Lambda}\mathbf{C}^{-1}\mathbf{x}(t)$$

Finally, let's define a new displacement vector

$$\mathbf{y}(t) = \begin{pmatrix} y_1 \\ y_2 \end{pmatrix} = \mathbf{C}^{-1}\mathbf{x}(t) = \frac{1}{\sqrt{2}}\begin{pmatrix} 1 & 1 \\ 1 & -1 \end{pmatrix}\begin{pmatrix} x_1 \\ x_2 \end{pmatrix} = \frac{1}{\sqrt{2}}\begin{pmatrix} x_1 + x_2 \\ x_1 - x_2 \end{pmatrix}$$

Following this transformation, the equations of motion for our two masses take on the simple form

$$\frac{d^2}{dt^2}\mathbf{y}(t) = \frac{d^2}{dt^2}\begin{pmatrix} y_1 \\ y_2 \end{pmatrix} = -\mathbf{\Lambda}\mathbf{y}(t) \tag{15.20}$$

where

$$\frac{d^2}{dt^2}\begin{pmatrix} y_1 \\ y_2 \end{pmatrix} = -\begin{pmatrix} \dfrac{\kappa}{m} & 0 \\ 0 & \dfrac{3\kappa}{m} \end{pmatrix}\begin{pmatrix} y_1 \\ y_2 \end{pmatrix}$$

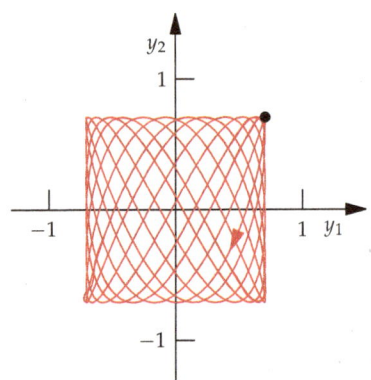

Figure 15.10: The trajectory shown in Figure 15.7 is plotted in normal mode coordinates y_1 and y_2 for the trajectory starting from $x_1(0) = 1$ and $x_2(0) = 0$ or equivalently $y_1(0) = y_2(0) = 1/\sqrt{2}$ (black dot).

where the components are known as the *normal modes* of oscillation.

We began with two coupled second order ordinary differential equations describing the displacements $x_1(t)$ and $x_2(t)$ known as *local modes*. Following the transformation defined by the orthogonal eigenvector matrix $\mathbf{C}$, we arrive at two uncoupled second order ordinary differential equations for the motion of the normal modes:

$$y_1(t) = \frac{1}{\sqrt{2}}\left(x_1(t) + x_2(t)\right) \qquad y_2(t) = \frac{1}{\sqrt{2}}\left(x_1(t) - x_2(t)\right)$$

The trajectory shown in Figure 15.9 as a function of local modes x_1 and x_2 is represented as a function of normal modes y_1 and y_2 in Figure 15.10. Transformation of the local mode displacement vector $\mathbf{x}(t)$ to the normal mode displacement vector $\mathbf{y}(t)$ leads to a rotation such that the bounding caustics form a square with sides parallel to the normal mode axes.

Finally, let's return to the symmetric and asymmetric motions displayed in Figures 15.6, 15.7, and 15.8. The same time evolution of the displacements can be viewed on the two-dimensional plane, where the abscissa is the displacement y_1 and the ordinate the displacement y_2, as shown in Figure 15.11. The horizontal red line represents the dynamics of the in-phase oscillation depicted in Figure 15.6 and the vertical blue line represents the dynamics of the out-of-phase oscillation depicted in Figure 15.7. When compared with Figure 15.8, we see that the transformation to normal mode coordinates converts coupled motion of two local mode coordinates to uncoupled motion of two one-dimensional normal mode coordinates.

We can extend this approach to the general case of N masses coupled to one another in three dimensions. In that case, the positions of the N masses are described by a vector $\mathbf{x}$ with $3N$ elements. When the coupling is quadratic, we express the equations of motion in terms of a $3N \times 3N$ force constant matrix $\mathbf{K}$. Solution of the secular equation

$$\left|\mathbf{K} - \omega^2\mathbf{I}\right| = 0$$

results in $3N$ eigenvalues and corresponding eigenvectors, forming the frequencies and coordinates of $3N$ independent normal modes of motion.

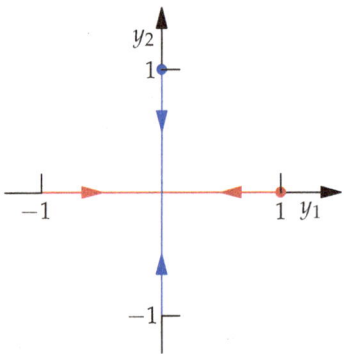

Figure 15.11: The displacements shown in Figure 15.8 plotted in normal mode coordinates y_1 and y_2 for initial conditions $x_1(0) = x_2(0) = 1/\sqrt{2}$ or equivalently $y_1(0) = 1$ and $y_2(0) = 0$ (red dot) and $x_1(0) = -x_2(0) = 1/\sqrt{2}$ or equivalently $y_1(0) = 0$ and $y_2(0) = 1$ (blue dot).

15.3 Scalar operators and eigenfunctions

EIGENVALUE PROBLEMS ARE NOT RESTRICTED TO THE SOLUTION OF MATRIX EQUATIONS DEFINED BY REAL COEFFICIENTS. More generally, eigenvalue problems may involve scalar operators rather than matrices and scalar functions rather than vectors. Moreover, eigenvalue problems are broadly defined for both real and complex matrices and functions. In this section, we explore the diverse applicability of the eigenvalue problem.

15.3.1 Eigenvalues and eigenfunctions of scalar operators

In Chapter 5, we explored the use of *linear operators*, including *scalar operators*, such as $\hat{d}_x = d/dx$, and *vector operators*, such as the gradient ∇. The notion of a linear operator can be extended to the matrix eigenvalue problem

$$\mathbf{Ax} = \lambda \mathbf{x} \qquad (15.21)$$

where the matrix $\mathbf{A}$ can be thought of as a *matrix operator*.[12]

Conversely, we can imagine forming an eigenvalue problem when no matrices are involved. For example, consider

$$\hat{d}_x \varphi(x) = \frac{d}{dx} \varphi(x) = a \varphi(x) \qquad (15.22)$$

where the scalar operator $\hat{d}_x$ acts on the scalar function $\varphi(x)$, resulting in the same function multiplied by a constant a. We consider $\varphi(x)$ to be an *eigenfunction* and a to be the corresponding *eigenvalue*.[13]

Let's look at a few examples. Consider the operator $\hat{p}_x = -i\hbar \hat{d}_x$, where $\hbar$ is a constant. We can write the eigenvalue problem as

$$\hat{p}_x \varphi(x) = -i\hbar \frac{d}{dx} \varphi(x) = \hbar k \varphi(x) \qquad (15.23)$$

where k is another constant and the goal is to identify the eigenfunction $\varphi(x)$. Using our knowledge of the solution of first order ordinary differential equations, we propose a solution of the form

$$\varphi(x) = \exp(ikx) \qquad (15.24)$$

which has $\hbar k$ as its corresponding eigenvalue.

Now consider the eigenvalue problem

$$\hat{M}\Phi(\varphi) = \frac{d^2}{dx^2}\Phi(\varphi) = -m^2\Phi(\varphi) \qquad (15.25)$$

where we seek the eigenfunction of the operator $\hat{M}$. Using our knowledge of the solution of second order ordinary differential equations, we propose a solution of the form

$$\Phi(\varphi) = \exp(im\varphi) \qquad (15.26)$$

Inserting this function into our equation, we demonstrate that it is an eigenfunction with the associated eigenvalue $-m^2$.

[12] This equation can be written using operator notation as

$$\hat{\mathbf{A}}\mathbf{x} = \lambda \mathbf{x}$$

where $\hat{\mathbf{A}}$ is a matrix operator.

[13] Examples include the Schrödinger equation in quantum theory

$$\frac{d^2\psi(x)}{dx^2} = -\frac{2mE}{\hbar^2}\psi(x)$$

Newton's equation for a mass on a spring

$$\frac{d^2x(t)}{dt^2} = -\omega^2 x(t)$$

and rate equations in chemical kinetics

$$\frac{dc(t)}{dt} = -kc(t)$$

where an operator acting on a function results in the function times a constant.

As a final example, consider the eigenvalue problem arising in the quantum theory of a *particle in a box*, written as

$$\hat{H}\psi(x) = -\frac{\hbar^2}{2m}\frac{d^2}{dx^2}\psi(x) = E\psi(x) \tag{15.27}$$

The operator $\hat{H}$ is the energy operator that acts on the wave function $\psi(x)$, resulting in the function times the constant E, which is the energy of the particle. We propose the sinusoidal function

$$\psi(x) = \sqrt{\frac{2}{L}}\sin\left(\frac{\pi}{L}x\right) \tag{15.28}$$

which we can demonstrate to be an eigenfunction of the operator $\hat{H}$, as

$$\hat{H}\psi(x) = -\frac{\hbar^2}{2m}\frac{d^2}{dx^2}\sqrt{\frac{2}{L}}\sin\left(\frac{\pi}{L}x\right) = -\frac{\hbar^2}{2m}\left[-\left(\frac{\pi}{L}\right)^2\sqrt{\frac{2}{L}}\sin\left(\frac{\pi}{L}x\right)\right]$$

$$= \frac{\hbar^2}{2m}\left(\frac{\pi}{L}\right)^2\psi(x) = E\psi(x)$$

This allows us to identify the corresponding eigenvalue

$$E = \frac{\hbar^2}{2m}\left(\frac{\pi}{L}\right)^2 = \frac{h^2}{8mL^2}$$

15.3.2 *Special properties and forms of hermitian operators*

In Chapter 14, we explored the properties of complex matrices. In particular, we defined a *hermitian matrix*[14] as any matrix $\mathbf{C}$ that is equal to the transpose of its complex conjugate:[15]

$$\mathbf{C} = (\mathbf{C}^*)^T$$

it is known as a *hermitian matrix*. The matrix $(\mathbf{C}^*)^T$ is called the *hermitian conjugate* of $\mathbf{C}$, and can also be expressed as $\mathbf{C}^\dagger$.

Consider the eigenvalue problem

$$\hat{\sigma}_x\psi_{x+} = \lambda\psi_{x+}$$

with the matrix operator

$$\hat{\sigma}_x = \begin{pmatrix} 0 & 1 \\ 1 & 0 \end{pmatrix}$$

The operator σ_x is hermitian, since

$$(\sigma_x^*)^T = \begin{pmatrix} 0 & 1 \\ 1 & 0 \end{pmatrix} = \hat{\sigma}_x$$

The normalized eigenvector is defined

$$\psi_{x+} = \frac{1}{\sqrt{2}}\begin{pmatrix} 1 \\ 1 \end{pmatrix}$$

[14] Note that all real, symmetric matrices are hermitian matrices.

[15] In compact notation this is written $[C]_{ij} = [C^*]_{ji}$.

Evaluating the expression

$$\hat{\sigma}_x \boldsymbol{\psi}_{x+} = \frac{1}{\sqrt{2}} \begin{pmatrix} 0 & 1 \\ 1 & 0 \end{pmatrix} \begin{pmatrix} 1 \\ 1 \end{pmatrix} = \frac{1}{\sqrt{2}} \begin{pmatrix} 1 \\ 1 \end{pmatrix} = \boldsymbol{\psi}_{x+}$$

demonstrates that the eigenvalue is the real number 1. In fact, for any eigenvalue problem in which the operator is hermitian, the eigenvalues are real.

The property that the eigenvalues of hermitian operators are real plays an important role in quantum science. In quantum mechanics, we often encounter eigenvalue problems such as

$$\hat{\mathbf{A}}\mathbf{x} = \lambda\mathbf{x} \tag{15.29}$$

in which $\hat{\mathbf{A}}$ is an *observable operator*, meaning that the associated eigenvalue λ is a physical property that can be experimentally measured. As such, the eigenvalue should be a real number. Therefore, we expect that operators used to calculate properties such as the momentum, kinetic energy, position, potential energy, or total energy are hermitian operators with real eigenvalues.

With this property of hermitian operators in mind, let's review a few prior examples. In the previous section we considered the eigenvalue problem

$$\hat{p}_x \varphi(x) = -i\hbar \frac{d}{dx} \varphi(x) = \hbar k \varphi(x) \tag{15.30}$$

where the imaginary operator $\hat{p}_x$ is the *linear momentum operator* in quantum theory. The complex eigenfunction

$$\varphi(x) = \exp(ikx) \tag{15.31}$$

is associated with the real eigenvalue $\hbar k$.

The real *kinetic energy operator* in quantum theory is hermitian

$$\hat{T}\psi(x) = -\frac{\hbar^2}{2m}\frac{d^2}{dx^2}\psi(x) = E_K\psi(x) \tag{15.32}$$

and the real eigenfunction

$$\psi(x) = \sqrt{\frac{2}{L}}\sin\left(\frac{\pi}{L}x\right) \tag{15.33}$$

has the real eigenvalue $E_K = \frac{\hbar^2}{8mL^2}$. The real *potential energy operator* is also hermitian

$$\hat{V}\psi(x) = V(x)\psi(x) \tag{15.34}$$

The operation consists of a simple multiplication by the real potential energy function.

Finally, since the operators $\hat{T}$ and $\hat{V}$ are hermitian, it follows that the total *energy operator* $\hat{H} = \hat{T} + \hat{V}$ is hermitian and the eigenvalues E of the eigenfunction ψ where

$$\hat{H}\psi = E\psi \tag{15.35}$$

are real. These examples demonstrate the property that eigenvalue problems involving hermitian operators yield real eigenvalues.

A$_{15}$ End-of-chapter problems

"Mathematizing" may well be a creative activity of man, like language or music, of primary originality, whose historical decisions defy complete objective rationalization.

Herman Weyl

Warm-ups

15.1 Show that the eigenvector matrix

$$C = \begin{pmatrix} \dfrac{\sqrt{6}}{\sqrt{15}} & \dfrac{\sqrt{6}}{\sqrt{10}} \\ \dfrac{3}{\sqrt{15}} & \dfrac{-2}{\sqrt{10}} \end{pmatrix}$$

has the inverse matrix

$$C^{-1} = \begin{pmatrix} \dfrac{\sqrt{6}}{\sqrt{15}} & \dfrac{3}{\sqrt{15}} \\ \dfrac{\sqrt{6}}{\sqrt{10}} & \dfrac{-2}{\sqrt{10}} \end{pmatrix}$$

by proving that $C^{-1}C = I$.

15.2 Show that the eigenvector matrix

$$C = \begin{pmatrix} \dfrac{\sqrt{6}}{\sqrt{15}} & \dfrac{\sqrt{6}}{\sqrt{10}} \\ \dfrac{3}{\sqrt{15}} & \dfrac{-2}{\sqrt{10}} \end{pmatrix}$$

will diagonalize the matrix

$$A = \begin{pmatrix} 1 & \sqrt{6} \\ \sqrt{6} & 2 \end{pmatrix}$$

by demonstrating that $C^{-1}AC = \Lambda$, where Λ is the diagonal eigenvalue matrix.

15.3 Show that c and $\lambda\,c$, where λ is a scalar, are parallel vectors in space. HINT: Make use of the definition of the scalar dot product.

15.4 Consider the matrix

$$A = \begin{pmatrix} -1 & 2 \\ 2 & 2 \end{pmatrix}$$

(a) Determine the eigenvalues of A by finding the roots, λ_1 and λ_2, of the characteristic equation

$$|A - \lambda I| = 0$$

(b) For each eigenvalue, λ_1 and λ_2, find the corresponding normalized eigenvector, x_1 and x_2.

15.5 Consider the matrix

$$\mathbf{A} = \begin{pmatrix} 1 & 1 \\ 1 & 1 \end{pmatrix}$$

(a) Determine the eigenvalues of $\mathbf{A}$ by finding the roots, λ_1 and λ_2, of the characteristic equation

$$|\mathbf{A} - \lambda\mathbf{I}| = 0$$

(b) For each eigenvalue, λ_1 and λ_2, find the corresponding normalized eigenvector, $\mathbf{x}_1$ and $\mathbf{x}_2$.

15.6 Consider the matrix

$$\mathbf{A} = \begin{pmatrix} 1 & 0 & 1 \\ 0 & 1 & 0 \\ 1 & 0 & 1 \end{pmatrix}$$

(a) Determine the eigenvalues of $\mathbf{A}$ by finding the roots, λ_1, λ_2 and λ_3, of the characteristic equation

$$|\mathbf{A} - \lambda\mathbf{I}| = 0$$

(b) For each eigenvalue, λ_1, λ_2 and λ_3, find the corresponding normalized eigenvector, $\mathbf{x}_1$, $\mathbf{x}_2$, and $\mathbf{x}_3$.

15.7 Show that $y = \sin ax$ is not an eigenfunction of the operator d/dx, but is an eigenfunction of the operator d^2/dx^2.

15.8 Show that the function $\Phi(\varphi) = A\,e^{im\varphi}$, where i, m, h, and A are constants, is an eigenfunction of the operator

$$\hat{M}_z = \frac{h}{2\pi i}\frac{\partial}{\partial\varphi}$$

Determine the eigenvalue.

Homework exercises

15.9 Consider the matrix

$$\mathbf{A} = \begin{pmatrix} 1 & 0 & -1 \\ 0 & 1 & 0 \\ -1 & 0 & 1 \end{pmatrix}$$

(a) Determine the eigenvalues of $\mathbf{A}$ by finding the roots, λ_1, λ_2, and λ_3, of the characteristic equation

$$|\mathbf{A} - \lambda\mathbf{I}| = 0$$

(b) For each eigenvalue, λ_1, λ_2, and λ_3, find the corresponding normalized eigenvector, $\mathbf{x}_1$, $\mathbf{x}_2$, and $\mathbf{x}_3$.

15.10 Solve the simultaneous equations

$$\frac{d}{dt}\mathbf{x} = \mathbf{A}\mathbf{x} = \frac{1}{2}\begin{pmatrix} 1 & \sqrt{5} \\ \sqrt{5} & 1 \end{pmatrix}\mathbf{x}$$

for the initial condition $\mathbf{x}(0) = \begin{pmatrix} 1 & 0 \end{pmatrix}^{\mathrm{T}}$.

15.11 Solve the simultaneous equations

$$\frac{d}{dt}\mathbf{x} = \mathbf{A}\mathbf{x} = \begin{pmatrix} -1 & 1 & 0 \\ 1 & -2 & 1 \\ 0 & 1 & -1 \end{pmatrix}\mathbf{x}$$

for the initial condition $\mathbf{x}(0) = (1 \ \ 0 \ \ 0)^{\mathrm{T}}$.

15.12 Show that the matrix $\mathbf{C}$ and the corresponding inverse matrix $\mathbf{C}^{-1}$ defined as

$$\mathbf{C} = \begin{pmatrix} \frac{1}{2} & \frac{\sqrt{2}}{2} & \frac{1}{2} \\ \frac{\sqrt{2}}{2} & 0 & -\frac{\sqrt{2}}{2} \\ \frac{1}{2} & -\frac{\sqrt{2}}{2} & \frac{1}{2} \end{pmatrix} \qquad \mathbf{C}^{-1} = \begin{pmatrix} \frac{1}{2} & \frac{\sqrt{2}}{2} & \frac{1}{2} \\ \frac{\sqrt{2}}{2} & 0 & -\frac{\sqrt{2}}{2} \\ \frac{1}{2} & -\frac{\sqrt{2}}{2} & \frac{1}{2} \end{pmatrix}$$

will put the matrix

$$\mathbf{A} = \begin{pmatrix} \alpha - E & \beta & 0 \\ \beta & \alpha - E & \beta \\ 0 & \beta & \alpha - E \end{pmatrix}$$

into diagonal form and that the diagonal elements are the roots of the characteristic equation $|\mathbf{A} - \lambda\mathbf{I}| = 0$.

15.13 Consider the case of two masses moving on a line and connected by springs to solid supports and each other (see Figure 15.5). The potential energy of interaction is given by

$$V(x_1, x_2) = \frac{k}{2}x_1^2 + \frac{k}{2}(x_2 - x_1)^2 + \frac{k}{2}x_2^2$$

where x_1 and x_2 are the displacements of each mass from the minimum of energy $V(0,0) = 0$. Rewrite the potential energy in terms of the variables $y_1 = x_1 + x_2$ and $y_2 = x_1 - x_2$. Interpret your result.

15.14 Consider the equation of motion for two masses moving on a line and coupled by springs to solid supports and each other (see Figure 15.5)

$$\frac{d^2}{dt^2}\mathbf{x} = \frac{d^2}{dt^2}\begin{pmatrix} x_1 \\ x_2 \end{pmatrix} = -\frac{k}{m}\begin{pmatrix} 2 & -1 \\ -1 & 2 \end{pmatrix}\mathbf{x} = -\mathbf{K}\mathbf{x}$$

(a) Show that the corresponding orthogonal eigenvector matrix

$$\mathbf{C} = \frac{1}{\sqrt{2}}\begin{pmatrix} 1 & 1 \\ 1 & -1 \end{pmatrix}$$

can be used to diagonalize $\mathbf{K}$ as

$$\mathbf{C}^{-1}\mathbf{K}\mathbf{C} = \Lambda$$

where Λ is the eigenvalue matrix the elements of which are the eigenvalues.

(b) Prove that the equation of motion can be rewritten in diagonalized form

$$\frac{d^2}{dt^2}\mathbf{y} = \Lambda\mathbf{y}$$

where

$$\mathbf{y}(t) = \mathbf{Cx} = \begin{pmatrix} y_1(t) \\ y_2(t) \end{pmatrix}$$

(c) Interpret the character of y_1 and y_2 by relating the nature of the motion (in-phase or out-of-phase) to the time scales of motion implied by the corresponding eigenvalues.

15.15 Consider two masses moving on a line and coupled by springs to solid supports and each other (see Figure 15.5). The potential energy is given by

$$V(x_1, x_2) = 5k\,x_1^2 + 6k\,(x_2 - x_1)^2 + \frac{3k}{2}\,x_2^2$$

where x_1 and x_2 are the displacements of each mass from the minimum of energy $V(0,0)$. Take the masses to be $m_1 = m_2 = m = 1$. The corresponding equations are motion for the two masses are

$$\frac{d^2}{dt^2}\mathbf{x} = \frac{d^2}{dt^2}\begin{pmatrix} x_1 \\ x_2 \end{pmatrix} = -\frac{k}{m}\begin{pmatrix} 22 & -12 \\ -12 & 15 \end{pmatrix}\mathbf{x} = -\mathbf{Kx}$$

Note that a matrix

$$\begin{pmatrix} a+b & -b \\ -b & b+c \end{pmatrix}$$

for which $a - c$, $2b$, and $\sqrt{(a-c)^2 + 4b^2}$ form a *Pythagorean triple* will result in integer eigenvalues. In this case, $a = 10$, $b = 12$, and $c = 3$ forming the Pythagorean triple $(7, 24, 25)$ since $7^2 + 24^2 = 25^2$. As such, we expect to find integer eigenvalues.

(a) Solve the characteristic equation

$$|\mathbf{K} - \omega^2\mathbf{I}| = 0$$

to find the eigenvalues represented as the allowed squared frequencies ω_1^2 and ω_2^2.

(b) Use the eigenvector equation

$$\left(\mathbf{K} - \omega_k^2\mathbf{I}\right)\mathbf{x}_k = \mathbf{o}$$

to identify the eigenvector corresponding to each eigenvalue.

15.16 Consider the generalization of the system with two masses shown in Figure 15.5 to a system consisting of three masses moving on a line and connected by springs to solid supports and each other. The system is shown below.

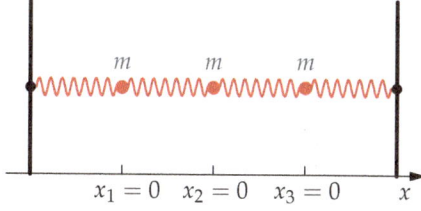

The potential energy of interaction is given by

$$V(x_1, x_2, x_3) = kx_1^2 + \frac{k}{2}(x_2 - x_1)^2 + \frac{k}{2}x_2^2 + \frac{k}{2}(x_3 - x_2)^2 + kx_3^2$$

where x_1, x_2, and x_3 are the displacements of each mass from the minimum of energy $V(0,0,0) = 0$.

(a) Derive the equations of motion

$$m_1\frac{d^2x_1}{dt^2} = -\frac{\partial V}{\partial x_1} \qquad m_2\frac{d^2x_2}{dt^2} = -\frac{\partial V}{\partial x_2} \qquad m_3\frac{d^2x_3}{dt^2} = -\frac{\partial V}{\partial x_3}$$

where $m_1 = m_2 = m_3 = m = 1$ are the three masses.

(b) Rewrite the equations of motion in the matrix format

$$\frac{d^2}{dt^2}\mathbf{x} = -\mathbf{Kx}$$

where $\mathbf{x}$ is the displacement vector

$$\mathbf{x}(t) = \begin{pmatrix} x_1(t) \\ x_2(t) \\ x_3(t) \end{pmatrix}$$

and $\mathbf{K}$ is the force constant matrix.

(c) Propose a solution of the form

$$\mathbf{x}(t) = \begin{pmatrix} c_1 \\ c_2 \\ c_3 \end{pmatrix} e^{i\omega t}$$

and insert it into the equation you derived in (b). Rearrange the terms to find the equation

$$\omega^2 \mathbf{x} = \mathbf{Kx}$$

(d) Solve the characteristic equation

$$|\mathbf{K} - \omega^2\mathbf{I}| = 0$$

to find the eigenvalues represented as the allowed squared frequencies ω_1^2, ω_2^2, and ω_3^2.

(e) Use the eigenvector equation

$$\left(\mathbf{K} - \omega_k^2\mathbf{I}\right)\mathbf{x}_k = \mathbf{o}$$

to identify the eigenvector corresponding to each eigenvalue.

(f) Interpret the motions associated with each eigenvector $\mathbf{x}_1$, $\mathbf{x}_2$, and $\mathbf{x}_3$ and its corresponding vibrational frequency ω_1, ω_2, and ω_3.

15.17 Consider the generalization of the system described in the previous problem to a system consisting of three masses moving on a line and connected by springs to each other, but not connected to a solid support. The system is shown below.

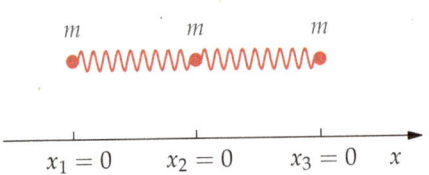

The potential energy of interaction is given by

$$V(x_1, x_2, x_3) = \frac{k}{2}(x_2 - x_1)^2 + \frac{k}{2}(x_3 - x_2)^2$$

where x_1, x_2, and x_3 are the displacements of each mass from the minimum of energy $V(0,0,0) = 0$.

(a) Derive the equations of motion

$$m_1 \frac{d^2 x_1}{dt^2} = -\frac{\partial V}{\partial x_1} \qquad m_2 \frac{d^2 x_2}{dt^2} = -\frac{\partial V}{\partial x_2} \qquad m_3 \frac{d^2 x_3}{dt^2} = -\frac{\partial V}{\partial x_3}$$

where $m_1 = m_2 = m_3 = m = 1$ are the three masses.

(b) Rewrite the equations of motion in the matrix format

$$\frac{d^2}{dt^2}\mathbf{x} = -\mathbf{Kx}$$

where $\mathbf{x}$ is the displacement vector

$$\mathbf{x}(t) = \begin{pmatrix} x_1(t) \\ x_2(t) \\ x_3(t) \end{pmatrix}$$

and $\mathbf{K}$ is the force constant matrix.

(c) Solve the characteristic equation

$$|\mathbf{K} - \omega^2 \mathbf{I}| = 0$$

to find the eigenvalues represented as the allowed squared frequencies ω_1^2, ω_2^2, and ω_3^2.

(d) Use the eigenvector equation

$$\left(\mathbf{K} - \omega_k^2 \mathbf{I}\right) \mathbf{x}_k = \mathbf{0}$$

to identify the eigenvector corresponding to each eigenvalue.

(e) Interpret the motions associated with each eigenvector $\mathbf{x}_1$, $\mathbf{x}_2$, and $\mathbf{x}_3$ and its corresponding vibrational frequency ω_1, ω_2, and ω_3.

15.18 Consider a diatomic molecule such as dioxygen physisorbed to a metallic surface. We model the system as two masses constrained to move on a line. The two masses are bonded together by a spring, and one of the masses is bonded to the solid surface by a second spring. The system is shown below.

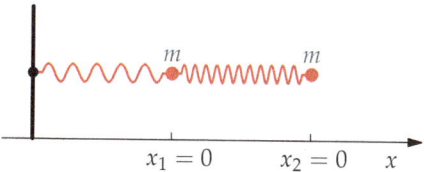

The potential energy of interaction is given by

$$V(x_1, x_2) = \frac{39}{2}kx_1^2 + \frac{380}{2}k(x_2 - x_1)^2$$

where x_1 and x_2 are the displacements of each mass relative to the minimum energy position $(0,0)$ where $V(0,0) = 0$. The masses of the two atoms are $m_1 = m_2 = m$. The equation of motion in matrix form is

$$\frac{d^2}{dt^2}\mathbf{x} = \frac{d^2}{dt^2}\begin{pmatrix} x_1 \\ x_2 \end{pmatrix} = -\frac{k}{m}\begin{pmatrix} 419 & -380 \\ -380 & 380 \end{pmatrix}\mathbf{x} = -\mathbf{Kx}$$

(a) Propose a solution of the form

$$\mathbf{x}(t) = \begin{pmatrix} c_1 \\ c_2 \end{pmatrix} e^{i\omega t}$$

and insert it into the equation of motion above. Rearrange the terms to find the equation

$$\omega^2\mathbf{x} = \mathbf{Kx}$$

(b) Solve the characteristic equation

$$|\mathbf{K} - \omega^2\mathbf{I}| = 0$$

to find the eigenvalues represented as the allowed squared frequencies ω_1^2 and ω_2^2.

(c) Use the eigenvector equation

$$\left(\mathbf{K} - \omega_k^2\mathbf{I}\right)\mathbf{x}_k = \mathbf{0}$$

to identify the eigenvector corresponding to each eigenvalue.

(d) Interpret the motions associated with each eigenvector $\mathbf{x}_1$ and $\mathbf{x}_2$ and its corresponding vibrational frequency ω_1 and ω_2.

15.19 Radioactive decay of a certain nucleus satisfies the following kinetic equations

$$\frac{dA}{dt} = -2A \qquad \frac{dB}{dt} = 2A - B \qquad \frac{dC}{dt} = B$$

(a) By defining the concentration vector

$$\mathbf{x}(t) = \begin{pmatrix} A(t) \\ B(t) \\ C(t) \end{pmatrix}$$

the rate equations can be written in the matrix form

$$\frac{d}{dt}\mathbf{x} = \mathbf{Kx}$$

where $\mathbf{K}$ is the rate coefficient matrix. What is $\mathbf{K}$?

(b) By proposing a solution of the form

$$\mathbf{x}(t) = \begin{pmatrix} c_1 \\ c_2 \\ c_3 \end{pmatrix} e^{\alpha t}$$

where a, b, c and α are constants, the rate equation can be written

$$\frac{d}{dt}\mathbf{x} = \alpha\mathbf{x} = \mathbf{Kx}$$

which puts it in the form of an eigenvalue problem

$$\mathbf{K}\mathbf{x} = \alpha\,\mathbf{x}$$

Solve the characteristic equation

$$|\mathbf{K} - \alpha\mathbf{I}| = 0$$

to find the three eigenvalues α_1, α_2, and α_3.

(c) Find the eigenvectors $\mathbf{x}_1$, $\mathbf{x}_2$, and $\mathbf{x}_3$ corresponding to the eigenvalues α_1, α_2, and α_3.

(d) The general solution capturing the time-dependent populations of the nuclear species can be written

$$\mathbf{x}(t) = \begin{pmatrix} A(t) \\ B(t) \\ C(t) \end{pmatrix} = d_1\mathbf{x}_1\,e^{\alpha_1 t} + d_2\mathbf{x}_2\,e^{\alpha_2 t} + d_3\mathbf{x}_3\,e^{\alpha_3 t}$$

where d_1, d_2, and d_3 are constant coefficients. Determine the values of the three coefficients given the initial conditions $A(0) = A_0$ and $B(0) = C(0) = 0$.

15.20* Two pendula of length l and mass m are coupled by a spring with force constant k. The equations of motion for the displacements s_1 and s_2 along the arc of motion of each pendulum are

$$m\frac{d^2}{dt^2}s_1 = -\frac{mg}{l}s_1 + k(s_2 - s_1) \qquad m\frac{d^2}{dt^2}s_2 = -\frac{mg}{l}s_2 - k(s_2 - s_1)$$

(a) First divide each equation by the mass, m. Reformulate the equations in matrix notation as

$$\frac{d^2}{dt^2}\mathbf{x} = -\mathbf{K}\mathbf{x}$$

where the displacement vector

$$\mathbf{x}(t) = \begin{pmatrix} s_1(t) \\ s_2(t) \end{pmatrix}$$

(b) By proposing a solution of the form

$$\mathbf{x}(t) = \begin{pmatrix} c_1 \\ c_2 \end{pmatrix} e^{i\omega t}$$

the equation of motion can be written

$$\frac{d^2}{dt^2}\mathbf{x} = -\omega^2\mathbf{x} = -\mathbf{K}\mathbf{x}$$

leading to the characteristic equation

$$|\mathbf{K} - \omega^2\mathbf{I}| = 0$$

Solve the characteristic equation to find the two eigenvalues ω_1 and ω_2.

(c) Find the eigenvectors $\mathbf{x_1}$ and $\mathbf{x_2}$ corresponding to the eigenvalues ω_1 and ω_2.

The general solution capturing the time-dependent displacements of the pendula can be written

$$\mathbf{x}(t) = \begin{pmatrix} s_1(t) \\ s_2(t) \end{pmatrix} = d_1\mathbf{x}_1\,e^{i\omega_1 t} + d_2\mathbf{x}_2\,e^{i\omega_2 t}$$

(d) What are the coefficients d_1 and d_2 for $s_1(0) = s_2(0) = s_0$? Describe the corresponding motion.

(e) What are the coefficients d_1 and d_2 for $s_1(0) = -s_2(0) = s_0$? Describe the corresponding motion.

15.21 Show that the functions

$$\psi_n(x) = \sqrt{\frac{2}{L}} \sin \frac{n\pi x}{L}$$

where L is a constant and $n = 1, 2, 3, \ldots$ are eigenfunctions of the operator

$$\hat{H} = -\frac{h^2}{8\pi^2 m} \frac{d^2}{dx^2}$$

when h and m are constants. What are the eigenvalues?

15.22 Show that the function $\varphi(x) = x e^{ax}$ is an eigenfunction of the operator

$$\hat{O} = \frac{d^2}{dx^2} - \frac{2a}{x}$$

where a is a constant. What is the eigenvalue?

15.23 Consider the Pauli matrices

$$\sigma_x = \begin{pmatrix} 0 & 1 \\ 1 & 0 \end{pmatrix} \qquad \sigma_y = \begin{pmatrix} 0 & -i \\ i & 0 \end{pmatrix} \qquad \sigma_z = \begin{pmatrix} 1 & 0 \\ 0 & -1 \end{pmatrix}$$

(a) Prove the matrices are hermitian by showing that $(\mathbf{A}^*)^T = \mathbf{A}$ for each matrix.

(b) Consider the eigenvalue problems

$$\sigma_x \psi_{x+} = \lambda \psi_{x+} \qquad \sigma_y \psi_{y+} = \lambda \psi_{y+} \qquad \sigma_z \psi_{z+} = \lambda \psi_{z+}$$

for the eigenvectors

$$\psi_{x+} = \frac{1}{\sqrt{2}} \begin{pmatrix} 1 \\ 1 \end{pmatrix} \qquad \psi_{y+} = \frac{1}{\sqrt{2}} \begin{pmatrix} 1 \\ i \end{pmatrix} \qquad \psi_{z+} = \begin{pmatrix} 1 \\ 0 \end{pmatrix}$$

What is the corresponding eigenvalue for each eigenvector?

(c) For the eigenvalue problems

$$\sigma_x \psi_{x-} = -1 \psi_{x-} \qquad \sigma_y \psi_{y-} = -1 \psi_{y-} \qquad \sigma_z \psi_{z-} = -1 \psi_{z-}$$

determine the normalized eigenvectors ψ_{x-}, ψ_{y-} and ψ_{z-}.

In the complements to Chapter 14, we explored the application of determinants in Hückel theory, which is used to approximate the electronic energy of conjugated organic molecules. The next two questions reconsider Hückel theory as an eigenvalue problem.

15.24* Consider the Hückel theory matrix for the ethene molecule:

$$\mathbf{H} = \begin{vmatrix} \alpha & \beta \\ \beta & \alpha \end{vmatrix}$$

where α and β are constants.

(a) Determine the eigenvalues of $\mathbf{H}$ by finding the roots, E_1 and E_2, of the characteristic equation

$$|\mathbf{H} - E\mathbf{I}| = 0$$

(b) For each eigenvalue, E_1 and E_2, find the corresponding normalized eigenvector, $\mathbf{x}_1$ and $\mathbf{x}_2$.

15.25[*] Consider the Hückel theory matrix for the cyclobutadiene molecule:

$$\mathbf{H} = \begin{vmatrix} \alpha & \beta & 0 & \beta \\ \beta & \alpha & \beta & 0 \\ 0 & \beta & \alpha & \beta \\ \beta & 0 & \beta & \alpha \end{vmatrix}$$

where α and β are constants.

(a) Determine the eigenvalues of $\mathbf{H}$ by finding the roots, E_1, E_2, E_3, and E_4, of the characteristic equation

$$|\mathbf{H} - E\mathbf{I}| = 0$$

(b) For each eigenvalue, E_1, E_2, E_3, and E_4, find the corresponding normalized eigenvector, $\mathbf{x}_1$, $\mathbf{x}_2$, $\mathbf{x}_3$, and $\mathbf{x}_4$.

15.26[*] Consider the following set of linear equations arising in Hückel theory

$$
\begin{aligned}
(\alpha - E)c_1 &+& \beta c_2 & & &=& 0 \\
\beta c_1 &+& (\alpha - E)c_2 &+& \beta c_3 &=& 0 \\
& & \beta c_2 &+& (\alpha - E)c_3 &=& 0
\end{aligned}
$$

for the variable E in terms of the constant parameters α and β.

(a) Set the determinant of the matrix of coefficiencts equal to zero and solve for E in terms of α and β. You should find three values E_1, E_2, and E_3.

(b) The column vectors formed by the coefficients are normalized as

$$\sum_{k=1}^{3} c_k^2 = 1$$

Determine the set of coefficients c_1, c_2, and c_3 that satisfy the linear equations for each value of E.

15.27[*] Consider the eigenvalue problem

$$\hat{\sigma}\psi = \lambda\psi$$

where $\hat{\sigma}$ is a hermitian operator, ψ is the eigenvector, and λ is the corresponding eigenvalue. Prove that the eigenvalue λ is real.

15.28[*] Consider the eigenvalue problems

$$\hat{\sigma}\psi_1 = \lambda_1\psi_1 \qquad \hat{\sigma}\psi_2 = \lambda_2\psi_2$$

where $\hat{\sigma}$ is a hermitian operator, ψ_1 and ψ_2 are eigenvectors, λ_1 and λ_2 are the corresponding eigenvalues, and $\lambda_1 \neq \lambda_2$. Prove that the eigenvectors ψ_1 and ψ_2 are orthogonal.

16 GEOMETRIC TRANSFORMS AND MOLECULAR SYMMETRY

16.1 Eigenvectors, geometric transforms, and symmetry

MATRICES CAN BE USED TO TRANSFORM ONE ORIENTATION OF A VECTOR INTO ANOTHER. In this section, we survey matrices describing fundamental geometric transformations in two dimensions, including scaling, inversion, rotation, and reflection.

16.1.1 Using eigenvalues and eigenvectors to explore symmetry

In two dimensions, geometric objects can be related to one another using transformations that scale the size of an object, or that invert, reflect, or rotate its shape. Those transformations can be defined in terms of a *geometric mapping* defined by

$$\mathbf{Tx} = \mathbf{x}'$$ (16.1)

where one point on the xy-plane, defined by the vector $\mathbf{x}$, is mapped into another, $\mathbf{x}'$, by a transformation defined by the matrix $\mathbf{T}$.

Let's explore how fundamental geometric transformations on the xy-plane can be performed using 2×2 matrices and how those transformations can be characterized in terms of matrix eigenvalues and eigenvectors.

Scaling

Consider the *scaling transformation*, defined by the matrix

$$\mathbf{S}_v = \begin{pmatrix} k_1 & 0 \\ 0 & k_2 \end{pmatrix}$$

where k_1 and k_2 are constants. This matrix transforms one point $\mathbf{x}$ into another $\mathbf{x}'$ as

$$\mathbf{S}_v\mathbf{x} = \begin{pmatrix} k_1 & 0 \\ 0 & k_2 \end{pmatrix} \begin{pmatrix} x \\ y \end{pmatrix} = \begin{pmatrix} k_1 x \\ k_2 y \end{pmatrix} = \mathbf{x}'$$

leading to scaling by a factor of k_1 in the x-direction and k_2 in the y-direction (see Figure 16.1).

Let's characterize the transformation matrix $\mathbf{S}_v$ in terms of its eigenvalues and eigenvectors. The eigenvalues are defined by the *characteristic equation*, Equation 15.9, and determined by solving the *characteristic polynomial* that is a

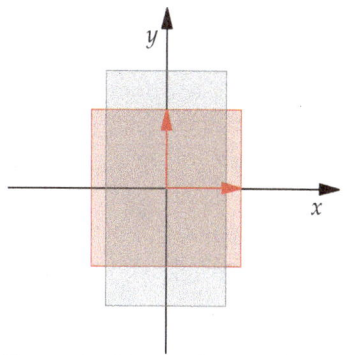

Figure 16.1: Non-uniform scaling of an area (red) into the transformed area (gray). The eigenvectors of the 2×2 scaling matrix $\mathbf{S}_v$ (red arrows) define the principal directions of the non-uniform scaling transform. The degree of scaling in the direction of a given eigenvector is defined by the associated eigenvalue.

quadratic equation in λ:

$$|\mathbf{S}_v - \lambda\mathbf{I}| = \begin{vmatrix} k_1 - \lambda & 0 \\ 0 & k_2 - \lambda \end{vmatrix} = (k_1 - \lambda)(k_2 - \lambda) = 0$$

The solutions of the characteristic polynomial are the eigenvalues $\lambda_1 = k_1$ and $\lambda_2 = k_2$. The corresponding eigenvectors are defined by Equation 15.10 as

$$(\mathbf{S}_v - \lambda_1\mathbf{I})\,\mathbf{x}_1 = \begin{pmatrix} 0 & 0 \\ 0 & k_2 - k_1 \end{pmatrix}\begin{pmatrix} c_1 \\ c_2 \end{pmatrix} = \begin{pmatrix} 0 \\ 0 \end{pmatrix}$$

and

$$(\mathbf{S}_v - \lambda_2\mathbf{I})\,\mathbf{x}_2 = \begin{pmatrix} k_1 - k_2 & 0 \\ 0 & 0 \end{pmatrix}\begin{pmatrix} c_1 \\ c_2 \end{pmatrix} = \begin{pmatrix} 0 \\ 0 \end{pmatrix}$$

This leads to the eigenvectors

$$\mathbf{x}_1 = \begin{pmatrix} 1 \\ 0 \end{pmatrix} \qquad \mathbf{x}_2 = \begin{pmatrix} 0 \\ 1 \end{pmatrix}$$

We see that the orthonormal eigenvectors define the principal scaling directions, and the eigenvalues define the degree of scaling in each direction.

Shear scaling

Consider the *shear transformation*, defined by the matrix

$$\mathbf{S} = \begin{pmatrix} 1 & k \\ 0 & 1 \end{pmatrix}$$

This matrix transforms one point $\mathbf{x}$ into another $\mathbf{x}'$ as

$$\mathbf{S}\mathbf{x} = \begin{pmatrix} 1 & k \\ 0 & 1 \end{pmatrix}\begin{pmatrix} x \\ y \end{pmatrix} = \begin{pmatrix} x + ky \\ y \end{pmatrix} = \mathbf{x}'$$

representing the effect of a shear in the x-direction (see Figure 16.2). The characteristic equation is

$$|\mathbf{S} - \lambda\mathbf{I}| = \begin{vmatrix} 1 - \lambda & k \\ 0 & 1 - \lambda \end{vmatrix} = (1 - \lambda)^2 = 0$$

and the solutions of the characteristic polynomial are a single root $\lambda_1 = \lambda_2 = 1$. Since the eigenvalues are equal, there is a single eigenvector defined by the solution to

$$(\mathbf{S} - \lambda\mathbf{I})\,\mathbf{x} = \begin{pmatrix} 0 & k \\ 0 & 0 \end{pmatrix}\begin{pmatrix} c_1 \\ c_2 \end{pmatrix} = \begin{pmatrix} 0 \\ 0 \end{pmatrix}$$

This leads to the eigenvector

$$\mathbf{x} = \begin{pmatrix} 1 \\ 0 \end{pmatrix}$$

Transformation by $\mathbf{S}$ leads to shear scaling. For a given vector $\mathbf{x}$, the y-

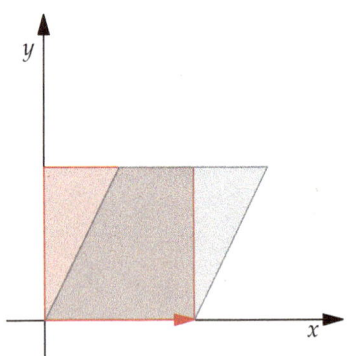

Figure 16.2: Shear scaling of an area (red) resulting in the sheared area (gray). The single eigenvector of the 2×2 shear matrix $\mathbf{S}$ is shown for $k = 1/2$ (red arrow).

component is unchanged, while the x-component is increased by an additive factor ky (see Figure 16.2).

Uniform scaling

Now consider the *uniform scaling transformation* matrix

$$\mathbf{S}_v = \begin{pmatrix} k & 0 \\ 0 & k \end{pmatrix} = k\mathbf{I}$$

where $\mathbf{I}$ is the identity matrix. This matrix transforms one point $\mathbf{x}$ into another $\mathbf{x}'$ as

$$\mathbf{S}_v\mathbf{x} = \begin{pmatrix} k & 0 \\ 0 & k \end{pmatrix} \begin{pmatrix} x \\ y \end{pmatrix} = \begin{pmatrix} kx \\ ky \end{pmatrix} = \mathbf{x}'$$

This is a special case of the general *scaling transformation* in which the diagonal elements are identical. The characteristic equation is defined by

$$|\mathbf{S}_v - \lambda\mathbf{I}| = \begin{vmatrix} k-\lambda & 0 \\ 0 & k-\lambda \end{vmatrix} = (k-\lambda)^2 = 0$$

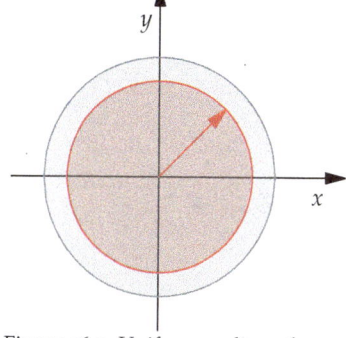

Figure 16.3: Uniform scaling of an area (red) resulting in the larger transformed area (gray). Any vector in the xy-plane is an eigenvector of the 2×2 scaling matrix $\mathbf{S}_v$ defining uniform scaling.

and the solutions of the characteristic polynomial are the eigenvalues $\lambda_1 = \lambda_2 = k$. The corresponding eigenvectors are defined by

$$(\mathbf{S}_v - \lambda_1\mathbf{I})\,\mathbf{x}_1 = \begin{pmatrix} 0 & 0 \\ 0 & 0 \end{pmatrix} \begin{pmatrix} c_1 \\ c_2 \end{pmatrix} = \begin{pmatrix} 0 \\ 0 \end{pmatrix}$$

and an identical equation for $\mathbf{x}_2$. Any values of c_1 and c_2 satisfy the equation. As such, any vectors in the xy-plane are eigenvectors of the uniform scaling transformation, which acts to scale all elements of any vector by a factor of k (see Figure 16.3).

Note that for $k = 1$, the uniform scaling transformation becomes the *identity transformation*

$$\mathbf{S}_v = \mathbf{E} = \begin{pmatrix} 1 & 0 \\ 0 & 1 \end{pmatrix}$$

Normally we represent the identity matrix as $\mathbf{I}$. In the context of geometric transforms, $\mathbf{E}$ is most commonly used rather than $\mathbf{I}$ to represent the identity transformation.[1]

[1] The E is taken from the German *einheit*, meaning unity.

Rotation

Let's evaluate the *rotation transformation*, defined by

$$\mathbf{R}(\theta) = \begin{pmatrix} \cos\theta & -\sin\theta \\ \sin\theta & \cos\theta \end{pmatrix}$$

which we recognize to be the orthogonal 2×2 rotation matrix defining counterclockwise rotation through an angle θ in the xy-plane.[2] The characteristic

[2] Eigenvectors were originally discovered by Italian mathematician and astronomer *Joseph-Louis Lagrange* (1736-1813) in the context of studying rigid body rotational motion.

equation is

$$|\mathbf{R}(\theta) - \lambda\mathbf{I}| = \begin{vmatrix} \cos\theta - \lambda & -\sin\theta \\ \sin\theta & \cos\theta - \lambda \end{vmatrix} = (\cos\theta - \lambda)^2 + \sin^2\theta = 0$$

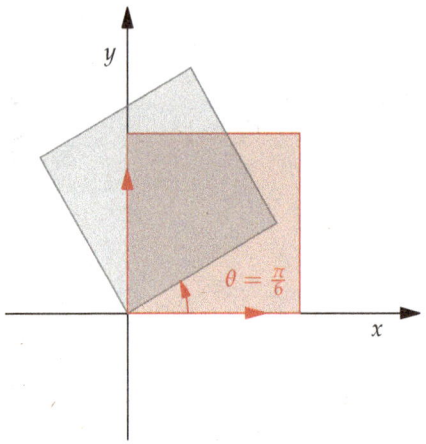

Figure 16.4: Transformation of an initial area (red) to the final area (gray) on the xy-plane defined by the 2×2 rotation matrix for counter-clockwise rotation through an angle $\theta = \pi/6$. The real part of the complex eigenvectors are shown (red arrows).

Expanding the square leads to

$$\lambda^2 - 2\lambda\cos\theta + \cos^2\theta + \sin^2\theta = \lambda^2 - 2\lambda\cos\theta + 1 = 0$$

and the roots

$$\lambda = \cos\theta \pm \sqrt{\cos^2\theta - 1} = \cos\theta \pm \sqrt{-\sin^2\theta}$$
$$= \cos\theta \pm i\sin\theta = e^{\pm i\theta}$$

This result is the complex eigenvalues $\lambda_1 = e^{-i\theta}$ and $\lambda_2 = e^{i\theta}$. For $\lambda_1 = e^{-i\theta}$, the eigenvector is defined by

$$(\mathbf{R}(\theta) - \lambda_1\mathbf{I})\,\mathbf{x}_1 = \begin{pmatrix} i\sin\theta & -\sin\theta \\ \sin\theta & i\sin\theta \end{pmatrix} \begin{pmatrix} c_1 \\ c_2 \end{pmatrix} = \begin{pmatrix} 0 \\ 0 \end{pmatrix}$$

with the solution $c_1 = 1$ and $c_2 = i$. For $\lambda_2 = e^{i\theta}$, the eigenvector is defined by

$$(\mathbf{R}(\theta) - \lambda_2\mathbf{I})\,\mathbf{x}_2 = \begin{pmatrix} -i\sin\theta & -\sin\theta \\ \sin\theta & -i\sin\theta \end{pmatrix} \begin{pmatrix} c_1 \\ c_2 \end{pmatrix} = \begin{pmatrix} 0 \\ 0 \end{pmatrix}$$

with the solution $c_1 = i$ and $c_2 = 1$. The normalized complex eigenvectors can be written as

$$\mathbf{x}_1 = \frac{1}{\sqrt{2}}\begin{pmatrix} 1 \\ i \end{pmatrix} \qquad \mathbf{x}_2 = \frac{1}{\sqrt{2}}\begin{pmatrix} i \\ 1 \end{pmatrix}$$

independent of the value of θ.

Transformation by $\mathbf{R}(\theta)$ leads to counter-clockwise rotation through an angle θ in the xy-plane that preserves the length of the vector (see Figure 16.4). Note that the eigenvectors of the rotation matrix $\mathbf{x}_1$ and $\mathbf{x}_2$ are normalized and orthogonal.[3]

[3] Note that the eigenvectors are orthonormal, as $(\mathbf{x}_1{}^*)^T\mathbf{x}_1 = (\mathbf{x}_2{}^*)^T\mathbf{x}_2 = 1$ and $(\mathbf{x}_1{}^*)^T\mathbf{x}_2 = (\mathbf{x}_2{}^*)^T\mathbf{x}_1 = 0$.

Reflection

Let's evaluate the *reflection transformation*, defined by

$$\sigma(\theta) = \begin{pmatrix} \cos 2\theta & \sin 2\theta \\ \sin 2\theta & -\cos 2\theta \end{pmatrix}$$

where $\sigma(\theta)$ is the *reflection matrix*, which performs a reflection across a line passing through the origin and making an angle θ with the x-axis (see Figure 16.5). The characteristic equation is

$$|\sigma(\theta) - \lambda \mathbf{I}| = \begin{vmatrix} \cos 2\theta - \lambda & \sin 2\theta \\ \sin 2\theta & -\cos 2\theta - \lambda \end{vmatrix}$$
$$= -(\cos 2\theta - \lambda)(\cos 2\theta + \lambda) - \sin^2 2\theta = 0$$

Expanding the square leads to

$$\lambda^2 - \cos^2 2\theta - \sin^2 2\theta = \lambda^2 - 1 = (\lambda + 1)(\lambda - 1) = 0$$

with roots defining the eigenvalues $\lambda_1 = 1$ and $\lambda_2 = -1$.

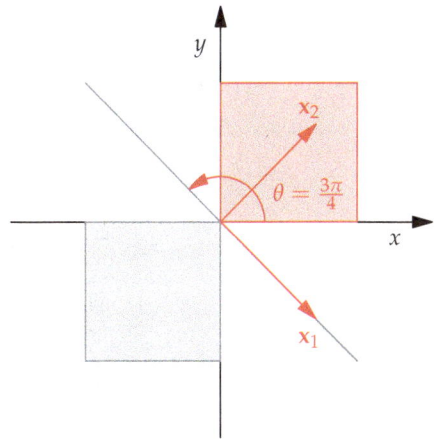

Figure 16.5: Transformation of an initial area (red) to the final area (gray) on the xy-plane defined by the 2×2 reflection matrix $\sigma(\theta)$ for an angle $\theta = 3\pi/4$ leading to reflection across the diagonal defined by $x = -y$ (gray line). In general, the eigenvectors of the reflection transformation matrix (red arrows) are dependent on θ.

Consider the specific case of $\theta = 3\pi/4$. The eigenvector corresponding to $\lambda_1 = 1$ is defined by

$$\left(\sigma\left(\frac{3\pi}{4}\right) - \lambda_1 \mathbf{I}\right)\mathbf{x}_1 = \begin{pmatrix} -1 & -1 \\ -1 & -1 \end{pmatrix}\begin{pmatrix} c_1 \\ c_2 \end{pmatrix} = \begin{pmatrix} 0 \\ 0 \end{pmatrix}$$

with the solution $c_1 = -c_2 = 1$. For λ_2, the eigenvector is defined by

$$\left(\sigma\left(\frac{3\pi}{4}\right) - \lambda_2 \mathbf{I}\right)\mathbf{x}_2 = \begin{pmatrix} 1 & -1 \\ -1 & 1 \end{pmatrix}\begin{pmatrix} c_1 \\ c_2 \end{pmatrix} = \begin{pmatrix} 0 \\ 0 \end{pmatrix}$$

with the solution $c_1 = c_2 = 1$. The normalized eigenvectors are

$$\mathbf{x}_1 = \frac{1}{\sqrt{2}}\begin{pmatrix} 1 \\ -1 \end{pmatrix} \qquad \mathbf{x}_2 = \frac{1}{\sqrt{2}}\begin{pmatrix} 1 \\ 1 \end{pmatrix}$$

This transformation leads to a reflection across the diagonal defined by $x = -y$ that preserves the length of the vector (see Figure 16.5).

How can we interpret the eigenvalues and eigenvectors of $\sigma(\theta)$? Note that x_1 defines the line of reflection, while x_2 is perpendicular to the line of reflection. Any vector parallel to the line of reflection is unchanged, as $\sigma(\theta)x_1 = x_1$, while any vector perpendicular to the line of reflection is reversed in orientation, as $\sigma(\theta)x_2 = -x_2$.

Inversion

Let's evaluate the *inversion transformation*, defined by

$$\mathbf{i} = \begin{pmatrix} -1 & 0 \\ 0 & -1 \end{pmatrix}$$

where $\mathbf{i}$ is the *inversion matrix*, resulting in inversion through the origin. This matrix transforms one point $\mathbf{x}$ into another $\mathbf{x}'$ as

$$\mathbf{ix} = \begin{pmatrix} -1 & 0 \\ 0 & -1 \end{pmatrix} \begin{pmatrix} x \\ y \end{pmatrix} = \begin{pmatrix} -x \\ -y \end{pmatrix} = \mathbf{x}'$$

inverting the x and y components through the origin. The characteristic equation is

$$|\mathbf{i} - \lambda\mathbf{I}| = \begin{vmatrix} -1 - \lambda & 0 \\ 0 & -1 - \lambda \end{vmatrix}$$

$$= (1 + \lambda)^2 = 0$$

with roots defining the eigenvalues $\lambda_1 = \lambda_2 = -1$. The corresponding eigenvectors are defined as

$$(\mathbf{i} - \lambda_1\mathbf{I})\,\mathbf{x}_1 = \begin{pmatrix} 0 & 0 \\ 0 & 0 \end{pmatrix} \begin{pmatrix} c_1 \\ c_2 \end{pmatrix} = \begin{pmatrix} 0 \\ 0 \end{pmatrix}$$

with an identical equation for $\mathbf{x}_2$. Both equations are satisfied by any values of c_1 and c_2. As such, any vectors in the xy-plane are eigenvectors of the transformation.

The inversion transformation is described graphically in Figure 16.6. The transformation leads to an inversion of any vector in the xy-plane, preserving the vector magnitude. An object is said to possess inversion symmetry when the transformation $\mathbf{x} \rightarrow -\mathbf{x}$ leaves the object unchanged.

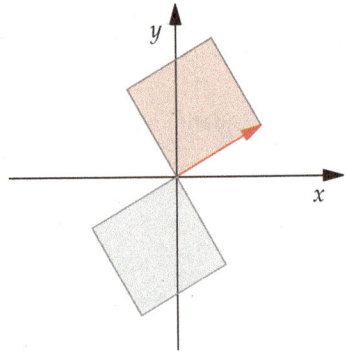

Figure 16.6: Transformation of an initial area (red) to final area (gray) on the xy-plane defined by the 2×2 inversion matrix. As in the case of uniform scaling, any vector is an eigenvector of the inversion transformation.

16.1.2 Geometric transformations in three dimensions

The geometric transformations that characterize scaling, rotation, reflection, and inversion have been defined in two dimensions. In this section, we extend our definitions of symmetry operations from two dimensions to three dimensions.

Consider the *identity transformation* in three dimensions. The identity transformation is a special case of the uniform scaling transformation that leaves the coordinate vector unchanged, as $\mathbf{Ex} = \mathbf{x}$. As such, the identity transformation defined by the *identity matrix*

$$\mathbf{E} = \begin{pmatrix} 1 & 0 & 0 \\ 0 & 1 & 0 \\ 0 & 0 & 1 \end{pmatrix}$$

Now consider the *rotation transformation* in three dimensions. The matrix

$$\mathbf{R}_z(\theta) = \begin{pmatrix} \cos\theta & -\sin\theta & 0 \\ \sin\theta & \cos\theta & 0 \\ 0 & 0 & 1 \end{pmatrix}$$

leads to counter-clockwise rotation of a vector through an angle θ about the z-axis. The x and y coordinates are transformed by counter-clockwise rotation in the xy-plane, while the z-coordinate is left unchanged. Similar matrices lead to counter-clockwise rotation of a vector about the x-axis or y-axis:

$$\mathbf{R}_x(\theta) = \begin{pmatrix} 1 & 0 & 0 \\ 0 & \cos\theta & -\sin\theta \\ 0 & \sin\theta & \cos\theta \end{pmatrix} \qquad \mathbf{R}_y(\theta) = \begin{pmatrix} \cos\theta & 0 & \sin\theta \\ 0 & 1 & 0 \\ -\sin\theta & 0 & \cos\theta \end{pmatrix}$$

Now consider the *reflection transformation* in three dimensions. The matrix

$$\sigma_{xy} = \begin{pmatrix} 1 & 0 & 0 \\ 0 & 1 & 0 \\ 0 & 0 & -1 \end{pmatrix}$$

leads to reflection of a vector across the xy-plane. Similar reflection matrices lead to reflection of a vector across the yz-plane or xz-plane:

$$\sigma_{yz} = \begin{pmatrix} -1 & 0 & 0 \\ 0 & 1 & 0 \\ 0 & 0 & 1 \end{pmatrix} \qquad \sigma_{xz} = \begin{pmatrix} 1 & 0 & 0 \\ 0 & -1 & 0 \\ 0 & 0 & 1 \end{pmatrix}$$

In addition, compound *improper rotations* can be defined by successive rotation and reflection transformations.

Finally, consider the *inversion transformation* in three dimensions, defined by the *inversion matrix*

$$\mathbf{i} = \begin{pmatrix} -1 & 0 & 0 \\ 0 & -1 & 0 \\ 0 & 0 & -1 \end{pmatrix}$$

which results in inversion through the origin. The inversion transformation reverses the x, y, and z coordinates, as $\mathbf{i}\mathbf{x} = -\mathbf{x}$.

For molecular geometries, the principal symmetries of interest are characterized by the identity, rotation, reflection, and inversion transformations. By defining a molecular conformation in terms of vectors representing the position of each atom in three-dimensional cartesian coordinates, we can use the transformation matrices defined above to rotate, reflect, or invert the atomic positions and identify the underlying molecular symmetries.

16.2 Matrix transformations and molecular symmetry

SYMMETRY PLAYS A CRITICAL ROLE IN DEFINING THE PROPERTIES OF MOLECULAR SYSTEMS. Just as matrices can be used to transform one orientation of a vector into another, matrices can be used to transform one orientation of a molecule into another. In this section, we explore how vectors can be used to describe molecular orientations in three dimensions and how matrix transforms can be used to define and assess molecular symmetry.

16.2.1 Using matrices to define symmetry in molecules

The basic structure of a molecule can be defined in terms of geometries such as trigonal planar, tetrahedral, trigonal bipyramidal, and octahedral. Those geometries can be further analyzed in terms of symmetry transformations such as inversion, rotation, and reflection. The arrangement of atoms in a crystal, such as the carbon atoms forming a diamond lattice (see Figure 16.7), can also be described in terms of basic symmetries. In addition, symmetry plays a critical role in the interpretation of molecular spectra. We will use our knowledge of transformation matrices to explore and define molecular symmetry.

16.2.2 The symmetries of H_2O (water)

The point group of a molecule is defined by the set of symmetry transformations that result in a final orientation of the molecule that is indistinguishable from its initial orientation. The term *point group* is derived from the fact that all planes of reflection symmetry, axes of rotational symmetry, and planes and axes of improper rotational symmetry intersect at a single point.

For example, consider the water molecule (see Figure 16.8). We describe its conformation by using vectors to indicate the positions of atoms in relation to each other. We must first choose which point to make the origin. As the oxygen atom is most central to the molecule, we place the oxygen atom at the origin:

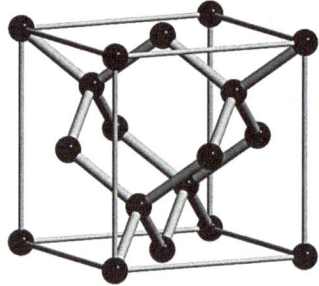

Figure 16.7: The network of carbon atoms (black) in the unit cell of a diamond crystal lattice.

$$\mathbf{x}_O = \begin{pmatrix} 0 \\ 0 \\ 0 \end{pmatrix}$$

We can then assign vectors to describe the position of each hydrogen atom relative to the central oxygen atom:[4]

$$\mathbf{x}_{H_1} = \begin{pmatrix} c_1 \\ 0 \\ c_2 \end{pmatrix} \qquad \mathbf{x}_{H_2} = \begin{pmatrix} -c_1 \\ 0 \\ c_2 \end{pmatrix}$$

In this example, we placed the hydrogen atoms in the xz-plane so that $y = 0$. This is an arbitrary choice. Placing the hydrogen atoms in the yz-plane so that $x = 0$, or in any other plane, will lead to the same results regarding the molecular symmetries of the water molecule.

[4] The O−H bond lengths are $r_{OH} = |\mathbf{x}_{H_1}| = |\mathbf{x}_{H_2}| = \sqrt{c_1^2 + c_2^2} = 96\text{pm}$ and the H−O−H angle is $\theta_{HOH} = \cos^{-1}\left(\mathbf{x}_{H_1} \cdot \mathbf{x}_{H_2} / |\mathbf{x}_{H_1}| |\mathbf{x}_{H_2}|\right) = 104.5°$.

Figure 16.8: The water molecule H_2O with the central oxygen atom (red) at the origin and hydrogen atoms (white) located in the xz-plane (shaded red).

Let's first explore the *identity symmetry* transformation defined by the *identify matrix*

$$E = \begin{pmatrix} 1 & 0 & 0 \\ 0 & 1 & 0 \\ 0 & 0 & 1 \end{pmatrix} \tag{16.2}$$

The identity transformation applied to the position of each atom leads to

$$E\, x_O = x_O \quad E\, x_{H_1} = x_{H_1} \quad E\, x_{H_2} = x_{H_2}$$

The final orientation of the water molecule is identical to and indistinguishable from the initial orientation. As such, we say that the water molecule possesses E identity symmetry.[5] In fact, every molecule we consider will possess identity symmetry.

[5] The transformation matrix is written E while the corresponding symmetry is written E.

We can also identify *rotational symmetry* of the water molecule. Suppose we apply a two-fold counter-clockwise rotation about the positive z-axis (the black rod in Figure 16.9) through an angle $\theta = \pi$. This transformation can be defined by the 3×3 rotation matrix

$$C_2 = \begin{pmatrix} -1 & 0 & 0 \\ 0 & -1 & 0 \\ 0 & 0 & 1 \end{pmatrix} \tag{16.3}$$

We apply this rotational transformation to each atom's position vector using the following matrix multiplications for oxygen

$$C_2\, x_O = \begin{pmatrix} -1 & 0 & 0 \\ 0 & -1 & 0 \\ 0 & 0 & 1 \end{pmatrix} \begin{pmatrix} 0 \\ 0 \\ 0 \end{pmatrix} = \begin{pmatrix} 0 \\ 0 \\ 0 \end{pmatrix} = x_O$$

and for the two hydrogens

$$C_2\, x_{H_1} = \begin{pmatrix} -1 & 0 & 0 \\ 0 & -1 & 0 \\ 0 & 0 & 1 \end{pmatrix} \begin{pmatrix} c_1 \\ 0 \\ c_2 \end{pmatrix} = \begin{pmatrix} -c_1 \\ 0 \\ c_2 \end{pmatrix} = x_{H_2}$$

$$C_2\, x_{H_2} = \begin{pmatrix} -1 & 0 & 0 \\ 0 & -1 & 0 \\ 0 & 0 & 1 \end{pmatrix} \begin{pmatrix} -c_1 \\ 0 \\ c_2 \end{pmatrix} = \begin{pmatrix} c_1 \\ 0 \\ c_2 \end{pmatrix} = x_{H_1}$$

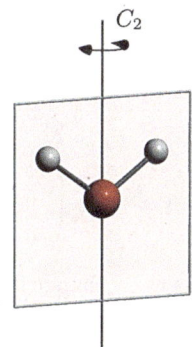

Figure 16.9: The water molecule H_2O with the two-fold C_2 rotational axis shown as a black rod.

We find rotation through an angle $\theta = \pi$ leaves the position of the oxygen atom unchanged, while exchanging the positions of the two hydrogen atoms. Since we consider the hydrogen atoms to be indistinguishable, the transformed orientation of the water molecule is indistinguishable from the initial orientation. As a result, we say that the water molecule possesses C_2 rotational symmetry.

We can also consider *reflection symmetry*. Reflection transformations are defined by the matrices σ_{xz}, for reflection across the xz-plane, and σ_{yz}, for reflection across the yz-plane:

$$\sigma_{xz} = \begin{pmatrix} 1 & 0 & 0 \\ 0 & -1 & 0 \\ 0 & 0 & 1 \end{pmatrix} \qquad \sigma_{yz} = \begin{pmatrix} -1 & 0 & 0 \\ 0 & 1 & 0 \\ 0 & 0 & 1 \end{pmatrix} \qquad (16.4)$$

Let's see how this works by applying the reflection transformations defined by σ_{xz} and σ_{yz} to the vectors defining the positions of the water molecule.

As the atoms of the water molecule are in the xz-plane (the red surface in Figure 16.10) we expect that reflection across the xz-plane will leave the orientation of the water molecule unchanged. Applying σ_{xz} to the position vector of the oxygen atom as

$$\sigma_{xz}\, \mathbf{x}_O = \begin{pmatrix} 1 & 0 & 0 \\ 0 & -1 & 0 \\ 0 & 0 & 1 \end{pmatrix} \begin{pmatrix} 0 \\ 0 \\ 0 \end{pmatrix} = \begin{pmatrix} 0 \\ 0 \\ 0 \end{pmatrix} = \mathbf{x}_O$$

results in no change in the position of the oxygen atom. Similarly, we can apply σ_{xz} to the position vectors of the hydrogen atoms as

$$\sigma_{xz}\, \mathbf{x}_{H_1} = \begin{pmatrix} 1 & 0 & 0 \\ 0 & -1 & 0 \\ 0 & 0 & 1 \end{pmatrix} \begin{pmatrix} c_1 \\ 0 \\ c_2 \end{pmatrix} = \begin{pmatrix} c_1 \\ 0 \\ c_2 \end{pmatrix} = \mathbf{x}_{H_1}$$

and

$$\sigma_{xz}\mathbf{x}_{H_2} = \begin{pmatrix} 1 & 0 & 0 \\ 0 & -1 & 0 \\ 0 & 0 & 1 \end{pmatrix} \begin{pmatrix} -c_1 \\ 0 \\ c_2 \end{pmatrix} = \begin{pmatrix} -c_1 \\ 0 \\ c_2 \end{pmatrix} = \mathbf{x}_{H_2}$$

This leads to no change in the positions of the hydrogen atoms. As the reflection transformation σ_{xz} leads to no change in the orientation of the water molecule, we say that the water molecule possesses σ_{xz} reflection symmetry.

Now let's explore reflection across the yz-plane (the gray surface in Figure 16.10), defined by σ_{yz}. Applying σ_{yz} to vectors defining the positions of atoms of the water molecule leads to

$$\sigma_{yz}\, \mathbf{x}_O = \begin{pmatrix} -1 & 0 & 0 \\ 0 & 1 & 0 \\ 0 & 0 & 1 \end{pmatrix} \begin{pmatrix} 0 \\ 0 \\ 0 \end{pmatrix} = \begin{pmatrix} 0 \\ 0 \\ 0 \end{pmatrix} = \mathbf{x}_O$$

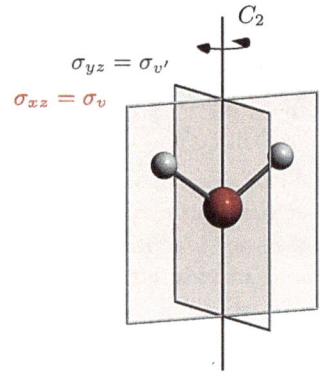

Figure 16.10: The water molecule H_2O where the xz-plane (shaded red) and yz-plane (shaded gray) are planes of reflection symmetry σ_{xz} and σ_{yz} also called σ_v and σ'_v, respectively.

and similarly

$$\sigma_{yz}\, \mathbf{x}_{H_1} = \begin{pmatrix} -1 & 0 & 0 \\ 0 & 1 & 0 \\ 0 & 0 & 1 \end{pmatrix} \begin{pmatrix} c_1 \\ 0 \\ c_2 \end{pmatrix} = \begin{pmatrix} -c_1 \\ 0 \\ c_2 \end{pmatrix} = \mathbf{x}_{H_2}$$

and

$$\sigma_{yz}\, \mathbf{x}_{H_2} = \begin{pmatrix} -1 & 0 & 0 \\ 0 & 1 & 0 \\ 0 & 0 & 1 \end{pmatrix} \begin{pmatrix} -c_1 \\ 0 \\ c_2 \end{pmatrix} = \begin{pmatrix} c_1 \\ 0 \\ c_2 \end{pmatrix} = \mathbf{x}_{H_1}$$

The transformation by σ_{yz} leads to exchange of the positions of the two hydrogen atoms. However, as the hydrogen atoms are indistinguishable, there is no *distinguishable* change in the orientation of the water molecule. We say that the water molecule possesses σ_{yz} reflection symmetry.

Finally, let's see what happens when evaluating a symmetry that the water molecule does not possess. Consider *inversion symmetry* that is possessed by molecules that are unchanged by inverting the positions of all atoms through a central origin. Without any calculations, we can imagine the outcome for the water molecule as shown in Figure 16.11.

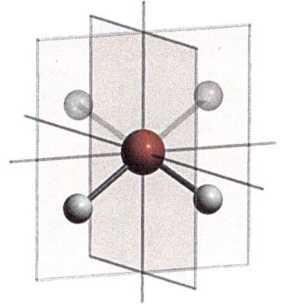

Figure 16.11: The water molecule H_2O with the central oxygen atom (red) at the origin and hydrogen atoms (white) subjected to the inversion transformation. The translucent water molecule represents the initial orientation. The opaque water molecule represents the final distinguishable orientation.

The resulting orientation of the water molecule is distinguishably different from the initial orientation. We say that the water molecule does not possess inversion symmetry. Let's see how this works by applying the inversion transformation matrix to the positions of the water atoms.

The inversion transformation is defined by the inversion matrix

$$\mathbf{i} = \begin{pmatrix} -1 & 0 & 0 \\ 0 & -1 & 0 \\ 0 & 0 & -1 \end{pmatrix} \tag{16.5}$$

We apply this transformation to the water molecule by first transforming the position of the oxygen as

$$\mathbf{i}\, \mathbf{x}_O = \begin{pmatrix} -1 & 0 & 0 \\ 0 & -1 & 0 \\ 0 & 0 & -1 \end{pmatrix} \begin{pmatrix} 0 \\ 0 \\ 0 \end{pmatrix} = \begin{pmatrix} 0 \\ 0 \\ 0 \end{pmatrix} = \mathbf{x}_O$$

followed by the two hydrogens

$$\mathbf{i}\,\mathbf{x}_{H_1} = \begin{pmatrix} -1 & 0 & 0 \\ 0 & -1 & 0 \\ 0 & 0 & -1 \end{pmatrix} \begin{pmatrix} c_1 \\ 0 \\ c_2 \end{pmatrix} = \begin{pmatrix} -c_1 \\ 0 \\ -c_2 \end{pmatrix}$$

$$\mathbf{i}\,\mathbf{x}_{H_2} = \begin{pmatrix} -1 & 0 & 0 \\ 0 & -1 & 0 \\ 0 & 0 & -1 \end{pmatrix} \begin{pmatrix} -c_1 \\ 0 \\ c_2 \end{pmatrix} = \begin{pmatrix} c_1 \\ 0 \\ -c_2 \end{pmatrix}$$

Overall, we find that the structure of the water molecule possesses identity symmetry E, two-fold rotational symmetry C_2, and reflection symmetries σ_{xz} and σ_{yz}. These symmetries characterize the *point group* C_{2v}. The name C_{2v} is derived from the fact that reflection across the xz-plane containing the atoms is also known as σ_v, where v is for *vertical* (see Figure 16.10).[6]

16.2.3 *The symmetries of* BH_3 *(borane)*

Let's consider the case of the BH_3 molecule known as borane (see Figure 16.12).

Figure 16.12: The borane molecule BH_3 with the central boron atom (green) at the origin and hydrogen atoms (white) located in the xy-plane.

We describe the orientation of the borane molecule by defining the positions of each atom in terms of a vector. We place the boron atom at the origin

$$\mathbf{x}_B = \begin{pmatrix} 0 \\ 0 \\ 0 \end{pmatrix}$$

and assign a vector to describe the position of each hydrogen atom relative to the central boron atom:

$$\mathbf{x}_{H_1} = \begin{pmatrix} c_1 \\ 0 \\ 0 \end{pmatrix} \qquad \mathbf{x}_{H_2} = \begin{pmatrix} -\dfrac{1}{2}c_1 \\ \dfrac{\sqrt{3}}{2}c_1 \\ 0 \end{pmatrix} \qquad \mathbf{x}_{H_3} = \begin{pmatrix} -\dfrac{1}{2}c_1 \\ -\dfrac{\sqrt{3}}{2}c_1 \\ 0 \end{pmatrix}$$

We have placed the hydrogen atoms in the xy-plane so that $z = 0$.[7]

Based on our experience in evaluating the symmetries of the water molecule, we expect to find three-fold *rotational symmetry* of the BH_3 molecule,

[7] The $B-H$ bond lengths are $r_{BH} = |\mathbf{x}_{H_1}| = |\mathbf{x}_{H_2}| = |\mathbf{x}_{H_3}| = c_1 = 119\text{pm}$ and the H-B-H angles are $\theta_{HBH} = \cos^{-1}\left(\mathbf{x}_{H_1} \cdot \mathbf{x}_{H_2} / |\mathbf{x}_{H_1}| |\mathbf{x}_{H_2}|\right) = \cos\left(-1/2\right) = 2\pi/3$.

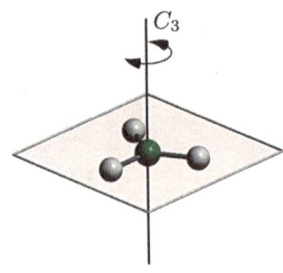

Figure 16.13: The borane molecule BH_3 showing the C_3 rotational axis (black rod).

defined by counter-clockwise rotation about the positive z-axis (the black rod in Figure 16.13) through an angle $\theta = \frac{2\pi}{3}$, and represented by the matrix

$$\mathbf{C_3} = \frac{1}{2} \begin{pmatrix} -1 & -\sqrt{3} & 0 \\ \sqrt{3} & -1 & 0 \\ 0 & 0 & 2 \end{pmatrix} \tag{16.6}$$

We can demonstrate this symmetry by applying $\mathbf{C_3}$ to vectors describing the position of each atom in the BH_3 molecule.[8] First, we apply the $\mathbf{C_3}$ rotation to the boron atom as

$$\mathbf{C_3}\,\mathbf{x_B} = \frac{1}{2} \begin{pmatrix} -1 & -\sqrt{3} & 0 \\ \sqrt{3} & -1 & 0 \\ 0 & 0 & 2 \end{pmatrix} \begin{pmatrix} 0 \\ 0 \\ 0 \end{pmatrix} = \begin{pmatrix} 0 \\ 0 \\ 0 \end{pmatrix} = \mathbf{x_B}$$

We then apply the same rotation transformation to the three hydrogen atoms

$$\mathbf{C_3}\,\mathbf{x_{H_1}} = \frac{1}{2} \begin{pmatrix} -1 & -\sqrt{3} & 0 \\ \sqrt{3} & -1 & 0 \\ 0 & 0 & 2 \end{pmatrix} \begin{pmatrix} c_1 \\ 0 \\ 0 \end{pmatrix} = \begin{pmatrix} -\frac{1}{2}c_1 \\ \frac{\sqrt{3}}{2}c_1 \\ 0 \end{pmatrix} = \mathbf{x_{H_2}}$$

$$\mathbf{C_3}\,\mathbf{x_{H_2}} = \frac{1}{2} \begin{pmatrix} -1 & -\sqrt{3} & 0 \\ \sqrt{3} & -1 & 0 \\ 0 & 0 & 2 \end{pmatrix} \begin{pmatrix} -\frac{1}{2}c_1 \\ \frac{\sqrt{3}}{2}c_1 \\ 0 \end{pmatrix} = \begin{pmatrix} -\frac{1}{2}c_1 \\ -\frac{\sqrt{3}}{2}c_1 \\ 0 \end{pmatrix} = \mathbf{x_{H_3}}$$

$$\mathbf{C_3}\,\mathbf{x_{H_3}} = \frac{1}{2} \begin{pmatrix} -1 & -\sqrt{3} & 0 \\ \sqrt{3} & -1 & 0 \\ 0 & 0 & 2 \end{pmatrix} \begin{pmatrix} -\frac{1}{2}c_1 \\ -\frac{\sqrt{3}}{2}c_1 \\ 0 \end{pmatrix} = \begin{pmatrix} c_1 \\ 0 \\ 0 \end{pmatrix} = \mathbf{x_{H_1}}$$

These results demonstrate that transformation of the borane molecule by rotation through an angle $\theta = \frac{2\pi}{3}$ leaves the position of the boron atom unchanged while the positions of the hydrogen atoms are exchanged. Since we consider the hydrogen atoms to be indistinguishable, the transformed orientation of the borane molecule is indistinguishable from the initial orientation. As a result, we say that the BH_3 molecule possesses C_3 rotational symmetry (see Figure 16.14).

The BH_3 molecule possesses additional symmetries, including identity symmetry and reflection symmetry for reflection across the xy-plane (see red surface in Figure 16.14), which is defined by

$$\sigma_{xy} = \begin{pmatrix} 1 & 0 & 0 \\ 0 & 1 & 0 \\ 0 & 0 & -1 \end{pmatrix} \tag{16.7}$$

and known as σ_h symmetry.[9] The borane molecule is also symmetric to two-

[8] For the general case of n-fold rotation

$$\mathbf{C_n} = \begin{pmatrix} \cos\left(\frac{2\pi}{n}\right) & -\sin\left(\frac{2\pi}{n}\right) & 0 \\ \sin\left(\frac{2\pi}{n}\right) & \cos\left(\frac{2\pi}{n}\right) & 0 \\ 0 & 0 & 1 \end{pmatrix}$$

defines counter-clockwise rotation about the positive z-axis through an angle $\theta = 2\pi/n$. You may also encounter rotation matrices defined in terms of *clockwise* rotation. Those matrices have the same form where $-\sin\theta$ is substituted for $\sin\theta$.

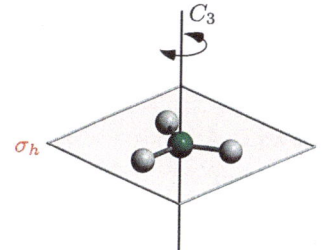

Figure 16.14: The borane molecule BH_3 showing the C_3 rotational axis and xy-plane of reflection symmetry $\sigma_{xy} = \sigma_h$.

[9] Reflection across the xy-plane containing the atoms of BH_3 is also known as σ_h symmetry, where h is for *horizontal*.

fold rotation about any of the three axes defined by the three B—H bonds (see red rods in Figure 16.15), known as C_2 symmetry, and reflection across any of the three planes defined by the z-axis and one of the three B—H bond axes (see gray surfaces in Figure 16.15), known as σ_v symmetry.

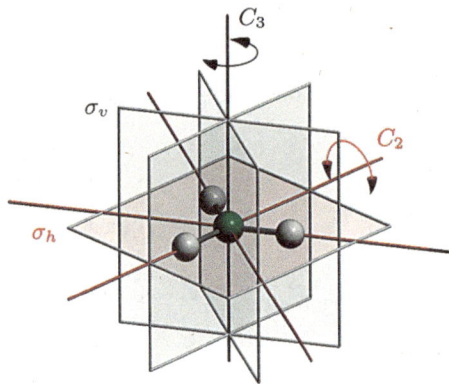

Figure 16.15: The borane molecule BH_3 with the central boron atom (green) at the origin and hydrogen atoms (white) located in the xy-plane. The two-fold (red rods) and three-fold (black rod) axes of rotational symmetry are shown along with planes of vertical (gray) and horizontal (red) reflection symmetry. The three-fold rotational symmetry C_3 and reflection symmetry σ_{xy} are combined to form the improper rotational symmetry S_3.

Finally, there is one additional symmetry of the BH_3 molecule, known as an *improper rotation*. The improper rotation is a *compound transformation* that combines a rotation about an axis and a reflection across a plane perpendicular to that axis. This particular improper rotation consists of counter-clockwise rotation by $\mathbf{C}_3$, through an angle $\frac{2\pi}{3}$ about the positive z-axis, followed by reflection by σ_{xy}, across the xy-plane perpendicular to the z-axis (see red surface in Figure 16.15). We can represent this compound transformation with the matrix product[10]

$$\mathbf{S}_3 = \sigma_{xy}\, \mathbf{C}_3 = \frac{1}{2}\begin{pmatrix} 1 & 0 & 0 \\ 0 & 1 & 0 \\ 0 & 0 & -1 \end{pmatrix}\begin{pmatrix} -1 & -\sqrt{3} & 0 \\ \sqrt{3} & -1 & 0 \\ 0 & 0 & 2 \end{pmatrix}$$

$$= \frac{1}{2}\begin{pmatrix} -1 & -\sqrt{3} & 0 \\ \sqrt{3} & -1 & 0 \\ 0 & 0 & -2 \end{pmatrix}$$

[10] As the matrices σ_{xy} and C_3 commute, the resulting compound transformation is independent of the order of operations. Rotation followed by reflection is the same as reflection followed by rotation and $\mathbf{S}_3 = \sigma_{xy}\, \mathbf{C}_3 = \mathbf{C}_3\, \sigma_{xy}$. However, in general, matrices do not commute and the order of operations can matter.

The orientation of the borane molecule is not distinguishably changed by the improper rotation defined by $\mathbf{S}_3$. As such, we say that borane has S_3 improper rotational symmetry.

In general, the $\mathbf{S}_n$ improper rotation involves rotating the molecule by $\mathbf{C}_n$ through $\frac{1}{n}2\pi$ radians, followed by reflection through the plane perpendicular to the C_n rotational axis. In this case, rotation by $\mathbf{C}_3$, through an angle $\frac{2\pi}{3}$ radians about the z-axis, was followed by reflection by σ_{xy}, across the xy-plane perpendicular to the z-axis.

Overall, we find that the structure of the BH_3 molecule possesses identity symmetry E, rotational symmetries C_2 and C_3, reflection symmetries σ_h and σ_v, and improper rotational symmetry S_3. These symmetries characterize the *point group* D_{3h}. In the next section, we develop a systematic approach to the identification of the point group of any molecule.

16.3 Point groups and the symmetry decision tree

POINT GROUPS REPRESENT A SET OF SYMMETRIES that are shared by all members of a given group, which may include identity, inversion, reflection, rotation, and improper rotation symmetries. The point group of a given molecule in a specific conformation can be determined through a series of yes or no decisions determining whether the molecule does or does not possess a specific symmetry. Those decisions can be organized in a decision tree that can be used to uniquely identify the symmetry of the molecule. In this section, the symmetry decision tree is introduced and explored through a variety of examples.

16.3.1 The symmetry decision tree

The point group of a molecule can be determined by answering a series of questions evaluating the presence or absence of fundamental symmetries, including identity, inversion, reflection, rotation, and improper rotation symmetries. Those questions are organized in the *symmetry decision tree* shown in Figure 16.16.

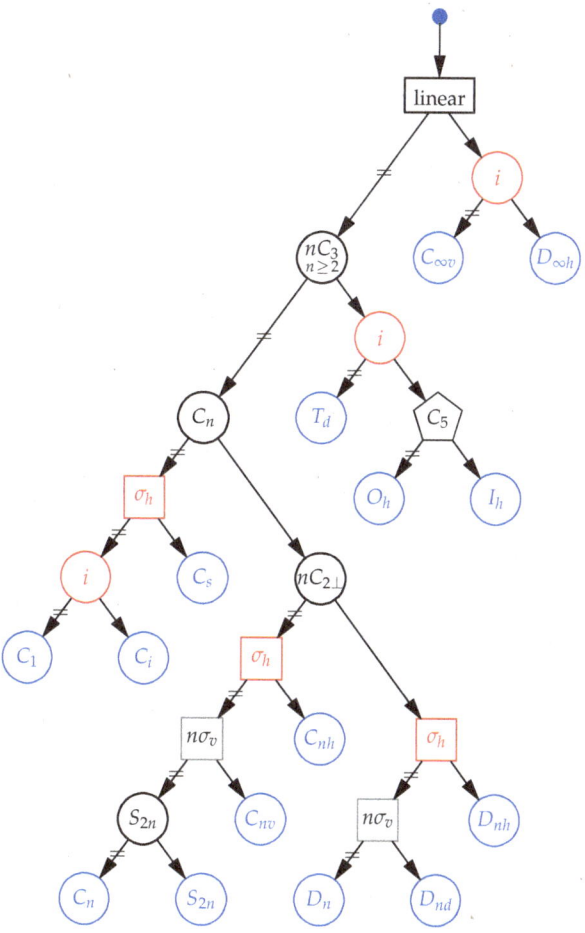

Figure 16.16: The point group defining the symmetries of a molecule can be uniquely determined using the symmetry decision tree. Positive decisions are noted by arrows branching right, while negative decisions are marked by double-hatched arrows branching left. The most critical decisions are noted by thick circles. The 15 resulting point groups are shown in blue.

Starting at the top of the decision tree, each branch point represents a decision related to whether the molecule possesses a particular symmetry (in which case we move to the right) or does not (in which case we move to the left). At each successive branch point another question is answered, until we reach a terminal point at a branch tip identifying the symmetry point group for that molecule.

The decision tree begins with the separation of linear (branch right) from non-linear (branch left) molecules. In the subset of linear molecules possessing C_∞ rotational symmetry, there are those lacking or possessing inversion symmetry i leading to point groups $C_{\infty v}$ or $D_{\infty h}$.

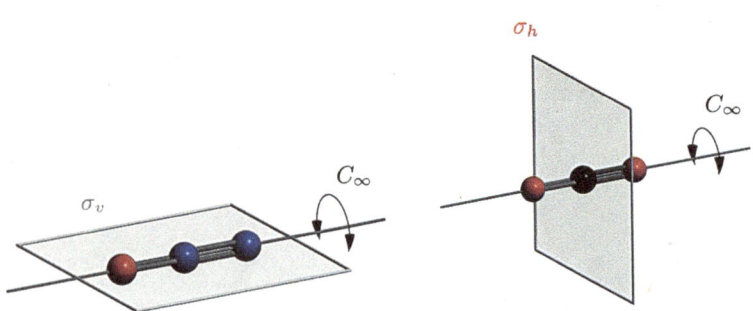

Figure 16.17: The nitrous oxide molecule, N_2O (left), and carbon dioxide molecule, CO_2 (right), belonging to point groups $C_{\infty v}$ and $D_{\infty h}$, respectively.

Consider the linear molecules shown in Figure 16.17. The nitrous oxide molecule, N_2O, has identity symmetry E, C_∞ rotational symmetry, and σ_v reflection symmetry. From the perspective of the symmetry decision tree (see Figure 16.18), N_2O is linear (branch to the right) but lacks inversion symmetry (branch to the left), leading to point group $C_{\infty v}$.

The carbon dioxide molecule, CO_2, has identity symmetry E, inversion symmetry i, and C_∞ rotational symmetry. It is also invariant to S_∞ improper rotation defined by C_∞ rotation about the bond axis and σ_h reflection. From the perspective of the symmetry decision tree, CO_2 is linear (branch right) and possesses inversion symmetry (branch right), leading to point group $D_{\infty h}$.

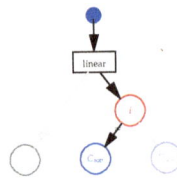

Figure 16.18: The path of decisions identifying the point group $C_{\infty v}$ for the nitrous oxide molecule, N_2O.

From the set of non-linear molecules, a special subset of molecules possessing $n \geq 2$ equivalent axes of n-fold rotational symmetry (C_n) are selected. Those molecules form the point groups T_d, O_h, and I_h characterizing the three underlying symmetries of the five *platonic solids*, the tetrahedron (T_d), cube (O_h), octahedron (O_h), dodecahedron (I_h), and icosahedron (I_h). These five solids are the only regular convex polyhedra formed by identical faces, of equal size, shape, and angles, and having an equal number of faces that meet at each vertex. This regularity leads to exceptionally high symmetry, with multiple rotational axes meeting at the center of the solid.

Consider the sulfur hexafluoride molecule, SF_6, shown in Figure 16.19. The molecule is octahedral and highly symmetric, having inversion symmetry i, multiple rotational symmetries C_4, C_3, and C_2, and multiple reflection symmetries σ_h and σ_v. In fact, there are four C_3 rotational axes (one for every two opposing triangular faces of the octahedron), three C_4 rotational axes

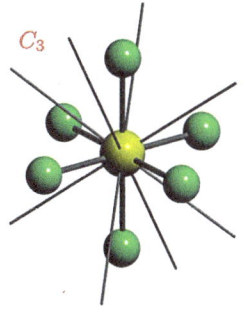
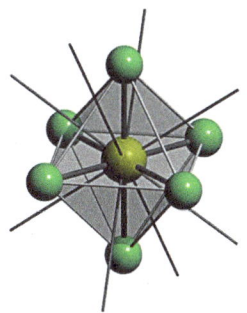

Figure 16.19: The octahedral sulfurhexafluoride molecule, SF_6, with the central sulfur atom (yellow) at the origin and six fluorine atoms (green), possesses four axes of C_3 rotational symmetry (left). Each C_3 axis is normal to two trigonal faces of the octahedron (right).

(one for every two opposing vertices of the octahedron), three planes of σ_h reflection symmetry (one perpendicular to each of the C_4 axes), and six planes of σ_v reflection symmetry (two parallel to each of the C_4 axes).

From the perspective of the symmetry decision tree, SF_6 is not linear (left), possesses multiple C_3 axes (right, see Figure 16.19), possesses inversion symmetry (right), and lacks C_5 symmetry (left), leading to point group O_h. This set of decisions is summarized in Figure 16.20.

The non-linear molecules lacking a symmetry of the *platonic solids* are separated into groups possessing or lacking a single principal axis of n-fold rotational symmetry, C_n. Molecules lacking C_n symmetry are divided into point groups characterizing molecules having reflection symmetry, C_s, molecules lacking reflection symmetry but possessing inversion symmetry, C_i, and molecules possessing neither, C_1. The point group C_1 characterizes the least symmetric of all molecules, those possessing only identity symmetry, E.

Let's evaluate the symmetry of the fluorochlorobromomethane molecule, CHFClBr, shown in Figure 16.21, using the decision tree.

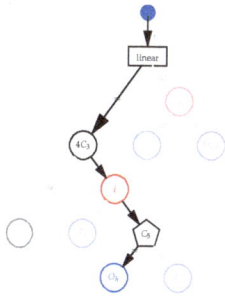

Figure 16.20: The path of decisions identifying the O_h point group for the sulfur hexafluoride molecule, SF_6.

Figure 16.21: The fluorochlorobromomethane molecule, CHFClBr. Its symmetry is characterized by the point group C_1.

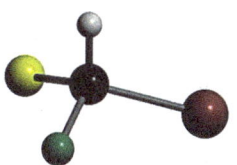

In CHFClBr, the central carbon atom (black) is surrounded by hydrogen (white), fluorine (green), chlorine (yellow), and bromine (brown) atoms. The molecule is not linear (left), it does not have multiple C_3 axes (left), it does not have a principal axis of rotational symmetry C_n (left), it does not have horizontal reflection symmetry σ_h (left), nor does it have inversion symmetry i (left), making it a member of point group C_1 (see Figure 16.22).

The non-linear molecules possessing a single principal axis of n-fold rotational symmetry are divided into groups of molecules possessing or

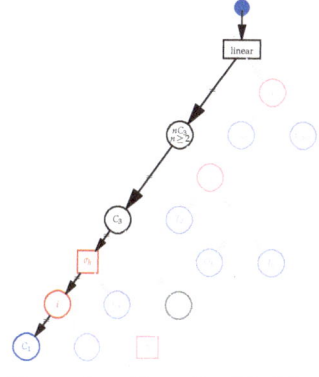

Figure 16.22: The path of decisions for the fluorochlorobromomethane molecule, CHFClBr, characterized by the point group C_1.

lacking n axes of C_2 rotational symmetry perpendicular to the principal axis. Molecules possessing this special symmetry are further divided into a group possessing a plane of σ_h reflection symmetry perpendicular to the principal axis, D_{nh}, a group lacking a horizontal plane of reflection symmetry but possessing n vertical planes of σ_v reflection symmetry, D_{nv}, and a group having no reflection symmetry, D_n.

Consider the allene molecule, C_3H_4, possessing a principal axis of C_2 rotational symmetry (see Figure 16.23).

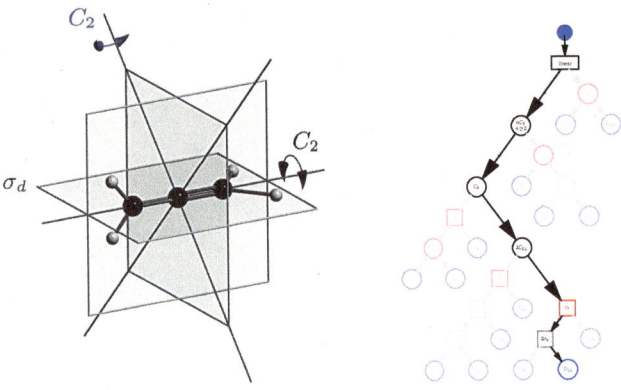

Figure 16.23: The allene molecule, C_3H_4, formed by three carbon atoms (black) and four hydrogen atoms (white), with the path of decisions leading to the identification of the D_{2d} point group.

According to the symmetry decision tree, the allene molecule is not linear (left), does not have multiple C_3 rotational axes (left), has a principal axis of C_2 symmetry (right), has two axes of C_2 rotational symmetry perpendicular to the principal axis (right), does not have a horizontal plane of σ_h reflection symmetry (left), and does have two planes of σ_v symmetry (right), making it a member of the D_{2d} point group (see Figure 16.23).

The remaining non-linear molecules are divided into a group possessing a plane of σ_h reflection symmetry perpendicular to the principal axis, C_{nh}, and a group lacking a horizontal plane of reflection symmetry but possessing n vertical planes of reflection symmetry, C_{nv}. Finally, the subset of molecules possessing a single axis of n-fold rotational symmetry but lacking any reflection symmetry are divided into one group of molecules possessing improper rotational symmetry, S_{2n}, and another group lacking it, C_n.

Consider the boric acid molecule, $B(OH)_3$, possessing a principal axis of C_3 rotational symmetry (see Figure 16.24). According to the symmetry decision tree, the boric acid molecule is not linear (left), does not have multiple C_3 rotational axes (left), has a principal axis of C_3 rotational symmetry (right), does not have axes of C_2 rotational symmetry perpendicular to the principal axis (left), and does have a horizontal plane of σ_h reflection symmetry (right), making it a member of the C_{3h} point group.

By answering at most seven questions, related to fundamental inversion, reflection, rotation, or improper rotation symmetries of a given molecule, the point group of any molecule may be uniquely defined. Let's practice the use of the decision tree by identifying the point groups of a number of common organic molecules.

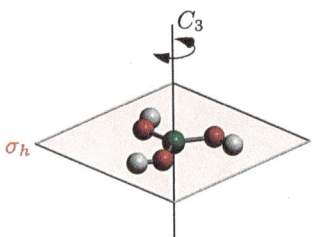

Figure 16.24: The boric acid molecule $B(OH)_3$ with the central boron atom (green) at the origin and three OH groups (red oxygen and white hydrogen) located in the xy-plane.

16.3.2 The symmetries of CH_4 (methane) and substituted methanes

Let's identify the point group of the tetrahedral methane molecule, CH_4, using the symmetry decision tree (see Figure 16.25). The molecule is nonlinear (left), has four axes of C_3 symmetry (right), and does not have inversion symmetry (left), leading to the tetrahedral point group, T_d.

Overall, the methane molecule has four axes of C_3 symmetry (right), three axes of C_2 symmetry, six planes of σ_d reflection symmetry, and three axes of S_4 improper rotational symmetry. Now let's consider how the symmetry is impacted by the gradual substitution of H atoms with Cl atoms.

Substitution of a single Cl atom for a H atom results in chloromethane, CH_3Cl (see Figure 16.26). The chloromethane molecule is nonlinear (left), has a single principal axis of C_3 rotational symmetry (left then right), has no axes of C_2 rotational symmetry perpendicular to the principal axis (left), does not have σ_h reflection symmetry (left), and does have three planes of σ_v reflection symmetry (right), leading to the C_{3v} point group. The substitution of a H atom for a Cl atom broke the tetrahedral symmetry of the methane molecule, leading to the C_{3v} point group, which has fewer symmetry operations than the T_d point group.

Figure 16.25: The methane molecule, CH_4, characterized by point group T_d.

Figure 16.26: A series of chloromethane molecules from CH_3Cl to CCl_4. The substitution of H (white) with Cl (yellow) initially diminishes and finally restores the overall level of symmetry.

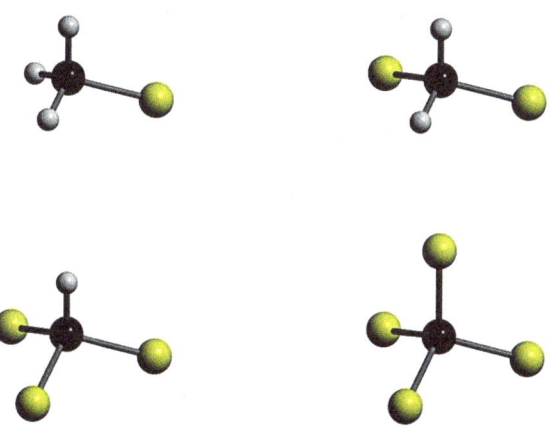

Substituting a second Cl atom for a H atom results in dichloromethane, CH_2Cl_2 (see Figure 16.26). The CH_2Cl_2 molecule is nonlinear (left), has no axis of C_3 rotational symmetry (left), has a single principal axis of C_2 rotational symmetry (right), has no axes of C_2 rotational symmetry perpendicular to the principal axis (left), does not have σ_h reflection symmetry (left), and does have two planes of σ_v reflection symmetry (right), leading to the C_{2v} point group. The substitution of a second H atom for a Cl atom lowered the symmetry from the C_{3v} point group to the C_{2v} point group.

Substituting a third Cl atom for a H atom results in trichloromethane, $CHCl_3$, also known as *chloroform* (see Figure 16.26). Note that chloroform shares the symmetries of chloromethane, making it a member of the C_{3v} point group. We observe that the third substitution of a H atom for a Cl atom

enhances the overall symmetry. That trend continues upon substitution of
the sole remaining H atom for a Cl atom, forming the tetrachloromethane
molecule CCl_4, also known as *carbon tetrachloride* (see Figure 16.26). Like
methane, the tetrahedral carbon tetrachloride molecule is highly symmetric
and characterized by the point group T_d.

16.3.3 *The symmetries of C_2H_6 (ethane) and substituted ethanes*

Consider the ethane molecule, C_2H_6, in the staggered conformation (see
Figure 16.27). Let's identify the symmetry point group using the symmetry
decision tree.

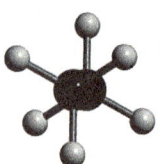

Figure 16.27: The ethane molecule,
C_2H_6, formed from carbon (black)
and hydrogen (white) atoms,
in the staggered conformation
characterized by D_{3d} symmetry.

This molecular conformation is nonlinear (left), does have a single principal axis of C_3 rotational symmetry (left then right), does have three axes of
C_2 rotational symmetry perpendicular to the principal axis (right), does not
have σ_h reflection symmetry (left), and does have three planes of σ_v reflection
symmetry. In addition, the molecule possesses inversion symmetry, i. We
conclude that ethane in the staggered conformation is characterized by the
D_{3d} point group.

Now consider the impact of substituting four H atoms with two F and Br
atoms to form $CH_2F_2Br_2$ as shown in Figure 16.28.

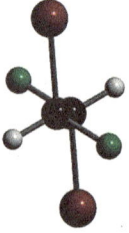
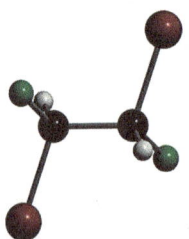

Figure 16.28: The substituted ethane
molecule, $C_2H_2F_2Br_2$, in the staggered conformation, formed from
carbon (black), hydrogen (white),
fluorine (green), and bromine
(brown) atoms, and characterized by
C_i symmetry.

These substitutions break the C_3 and C_2 rotational symmetries of ethane
in the staggered conformation. In addition, these substitutions eliminate
all planes of σ_v reflection symmetry. Applying the symmetry decision tree,
the molecule is nonlinear (left), has no axis of C_n rotational symmetry (left
then left), does not have a plane of σ_h reflection symmetry (left), but does
possess inversion symmetry i (right). As such, the $CH_2F_2Br_2$ molecule in
the staggered conformation shown in Figure 16.28 is characterized by the C_i
point group.

16.3.4 The symmetries of C_6H_{12} (cyclohexane)

It is possible to extend our analysis of molecular symmetry to more complex organic molecules. Let's consider the example of the cyclohexane molecule, C_6H_{12}, in the chair conformation (see Figure 16.29).

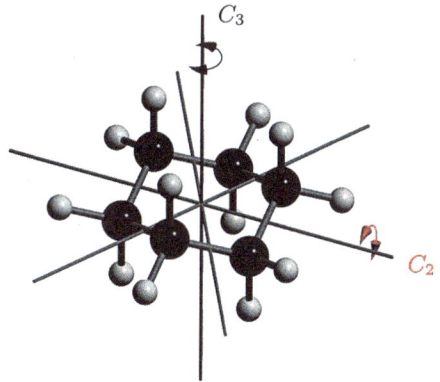

Figure 16.29: The cyclohexane molecule C_6H_{12} in the chair conformation, formed from six carbon (black) and twelve hydrogen (white) atoms, showing the one C_3 (black) and three C_2 (red) rotational axes.

We describe the conformation of the molecule in terms of vectors defining the positions of the six carbon and twelve hydrogen atoms. In the cases of H_2O and BH_3, we placed one atom at the origin. In the case of cyclohexane, the z-axis passes through the center of the six-membered carbon ring. Six C−H bonds are oriented parallel to the z-axis, three in the positive z-direction, and three in the negative z-direction. Finally, we place the first and fourth carbon atoms in the xz-plane.

We observe that the structure of the molecule possesses identity symmetry, E, as well as inversion symmetry, i. Looking down the z-axis we observe that the structure is symmetric to three-fold rotation defined by $\mathbf{C}_3$ about the z-axis (see Figure 16.29). In addition, we identify three axes of two-fold C_2 rotational symmetry. Each axis bisects the center of two opposing C−C bonds, connecting the bond formed by the $(i, i+1)$ carbons and the bond formed by the $(i+3, i+4)$ carbons.

Changing reference frames (see Figure 16.30), we identify other symmetries of the cyclohexane molecule in its chair conformation. For example, the structure lacks σ_h reflection symmetry, as it is altered by reflection across the horizontal plane. However, the structure is invariant to reflection across three vertical planes parallel to the z-axis and passing through opposing $(i, i+3)$ carbon atoms. These planes are known as dihedral planes, and the reflection symmetry operation is σ_d.

The structure of cyclohexane in the chair conformation is also unchanged by the $\mathbf{S}_6$ improper rotational transformation (see Figure 16.31). This compound transformation is defined by rotation through an angle $\pi/3$ about the z-axis, as found in a $\mathbf{C}_6$ rotational transformation, followed by reflection across the xy-plane defined by σ_h.

Note that while the chair conformation of cyclohexane does not posses C_6 rotational symmetry, it does possess S_6 improper rotational symmetry

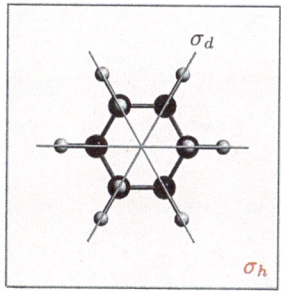

Figure 16.30: The cyclohexane molecule, C_6H_{12}, viewed from the positive z-axis, showing three σ_d reflection planes (gray) .

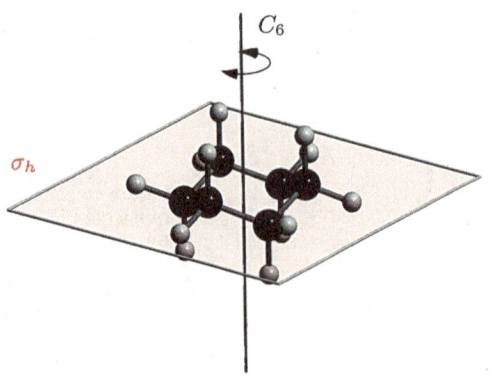

Figure 16.31: The S_6 improper rotation is a compound transformation combining C_6 rotation (black axis) with σ_h reflection in the horizontal plane (red).

combining C_6 rotation with σ_h reflection. The structure of cyclohexane in the boat conformation does not possess improper rotational symmetry. These observations can be applied to the symmetry decision tree to determine the point group of cyclohexane in the chair conformation. The path through the decision tree is shown in Figure 16.32.

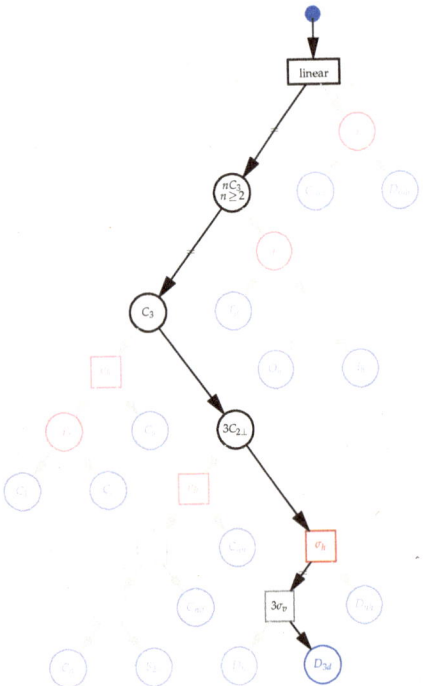

Figure 16.32: The path of decisions for the cyclohexane molecule in the chair conformation leading to the identification of the D_{3d} point group.

Starting at the top, the molecule is not linear (left), does not have two or more axes of C_3 symmetry (left), does have a principal axis of C_3 symmetry (right), does have three axes of C_2 symmetry perpendicular to the principal axis (right), does not have horizontal σ_h reflection symmetry (left), and does have three planes of vertical σ_v reflection symmetry. These symmetries characterize the D_{3d} point group.

Overall, we find that the structure of the C_6H_{12} cyclohexane molecule in the chair conformation possesses identity symmetry E, inversion symmetry i, rotational symmetries C_2 and C_3, reflection symmetry σ_d, and improper rotational symmetry S_6.

16.3.5 The symmetries of the Posner molecule $Ca_9(PO_4)_6$

Consider the Posner molecule, $Ca_9(PO_4)_6$ (see Figure 16.33). The central calcium ion, Ca^{2+} (orange), is surrounded by six phosphate groups, PO_4^{3-}, each composed of one phosphorus (yellow) and four oxygen (red) atoms. Each of the six phosphate groups coordinates a single calcium ion. Two additional calcium ions flank the central calcium, making the molecule charge neutral. Let's identify the symmetry point group of the Posner molecule using the symmetry decision tree.

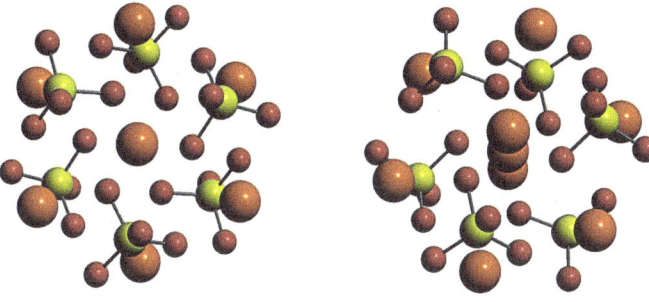

Figure 16.33: The Posner molecule, $Ca_9(PO_4)_6$, from two perspectives. The central calcium ion (orange) is surrounded by six phosphate groups, each composed of one phosphorus (yellow) and four oxygen (red) atoms, coordinating a single calcium ion. Two additional calcium ions flank the central calcium, making the molecule's total charge neutral. The Posner molecule is characterized by the point group S_6.

The Posner molecule is not linear (left), does not have two or more axes of C_3 rotational symmetry (left), does have a principal axis of C_3 rotational symmetry (right), does not have three axes of C_2 rotational symmetry perpendicular to the principal axis (left), does not have horizontal σ_h reflection symmetry (left), does not have three planes of vertical σ_v reflection symmetry (left), and does have S_6 improper rotational symmetry. These symmetries characterize the point group S_6.

The examples above demonstrate a reliable method for determining the point group of any simple or complex molecule using the symmetry decision tree. Identification of the point group provides a foundation of knowledge for the study of many fundamental properties of molecules and solids. For example, molecular symmetry can be used to define a molecule's properties, including whether a molecule has a dipole moment or whether a spectroscopic transition is allowed. In addition, the identification of molecular symmetry is essential in the application of Hückel theory, ligand field theory, and the Woodward-Hoffmann rules.

Repeated practice and problem solving will allow for mastery of the mathematical methods presented in this text. In turn, those mathematical methods will provide a foundation for the study of the structure, dynamics, thermodynamics, and kinetics of molecular systems. I wish you the very best in your explorations of the beauty and wonder of molecular science.

A_{16} End-of-chapter problems

> The miracle of the appropriateness of the language of mathematics for the formulation of the laws of physics is a wonderful gift which we neither understand nor deserve.
>
> Eugene Paul Wigner

Warm-ups

16.1 Consider the *scaling transformation* defined by

$$S_v = \begin{pmatrix} 5 & 0 \\ 0 & 2 \end{pmatrix}$$

(a) For the vector $x = (1,2)^T$ determine $x' = S_v x$.

(b) Solve the characteristic equation $|S_v - \lambda I| = 0$ to determine the eigenvalues λ_1 and λ_2.

(c) Determine the eigenvectors x_1 and x_2 satisfying the equations $(S_v - \lambda_1 I) x_1 = o$ and $(S_v - \lambda_2 I) x_2 = o$.

16.2 Consider the *shear transformation* defined by

$$S = \begin{pmatrix} 1 & 2 \\ 0 & 1 \end{pmatrix}$$

(a) For the vector $x = (1,2)^T$ determine $x' = S x$.

(b) Solve the characteristic equation $|S - \lambda I| = 0$ to determine the eigenvalues λ_1 and λ_2.

(c) Determine the eigenvectors x_1 and x_2 satisfying the equations $(S - \lambda_1 I) x_1 = o$ and $(S - \lambda_2 I) x_2 = o$.

16.3 Consider the *rotation transformation* defined by

$$R(\theta) = \begin{pmatrix} \cos\theta & -\sin\theta \\ \sin\theta & \cos\theta \end{pmatrix}$$

(a) Take $\theta = \dfrac{\pi}{4}$. For the vector $x = (\dfrac{1}{\sqrt{2}}, \dfrac{1}{\sqrt{2}})^T$ determine $x' = R\left(\dfrac{\pi}{4}\right) x$.

(b) Solve the characteristic equation $\left| R\left(\dfrac{\pi}{4}\right) - \lambda I \right| = 0$ to determine the eigenvalues λ_1 and λ_2.

(c) Determine the eigenvectors x_1 and x_2 satisfying the equations $\left(R\left(\dfrac{\pi}{4}\right) - \lambda_1 I\right) x_1 = o$ and $\left(R\left(\dfrac{\pi}{4}\right) - \lambda_2 I\right) x_2 = o$.

16.4 Consider the *reflection transformation* defined by

$$\sigma(\theta) = \begin{pmatrix} \cos 2\theta & \sin 2\theta \\ \sin 2\theta & -\cos 2\theta \end{pmatrix}$$

(a) Take $\theta = \dfrac{\pi}{2}$. For the vector $x = (\dfrac{1}{\sqrt{2}}, \dfrac{1}{\sqrt{2}})^T$ determine $x' = \sigma\left(\dfrac{\pi}{2}\right) x$.

(b) Solve the characteristic equation $\left| \sigma\left(\frac{\pi}{2}\right) - \lambda\mathbf{I} \right| = 0$ to determine the eigenvalues λ_1 and λ_2.

(c) Determine the eigenvectors $\mathbf{x}_1$ and $\mathbf{x}_2$ satisfying the equations $\left(\sigma\left(\frac{\pi}{2}\right) - \lambda_1\mathbf{I} \right) \mathbf{x}_1 = \mathbf{o}$ and $\left(\sigma\left(\frac{\pi}{2}\right) - \lambda_2\mathbf{I} \right) \mathbf{x}_2 = \mathbf{o}$.

16.5 Consider the *inversion transformation* defined by

$$\mathbf{i} = \begin{pmatrix} -1 & 0 \\ 0 & -1 \end{pmatrix}$$

(a) For the vector $\mathbf{x} = (1,2)^T$ determine $\mathbf{x}' = \mathbf{i}\,\mathbf{x}$.

(b) Solve the characteristic equation $|\mathbf{i} - \lambda\mathbf{I}| = 0$ to determine the eigenvalues λ_1 and λ_2.

(c) Determine the eigenvectors $\mathbf{x}_1$ and $\mathbf{x}_2$ satisfying the equations $(\mathbf{i} - \lambda_1\mathbf{I}) \mathbf{x}_1 = \mathbf{o}$ and $(\mathbf{i} - \lambda_2\mathbf{I}) \mathbf{x}_2 = \mathbf{o}$.

16.6 Consider the ammonia molecule, NH_3, shown below. The nitrogen atom is placed at the origin and one hydrogen atom is placed in the xz-plane so that

$$\mathbf{x}_N = \begin{pmatrix} 0 \\ 0 \\ 0 \end{pmatrix} \quad \mathbf{x}_{H1} = \begin{pmatrix} c_1 \\ 0 \\ c_2 \end{pmatrix} \quad \mathbf{x}_{H2} = \begin{pmatrix} -\frac{1}{2}c_1 \\ \frac{\sqrt{3}}{2}c_1 \\ c_2 \end{pmatrix} \quad \mathbf{x}_{H3} = \begin{pmatrix} -\frac{1}{2}c_1 \\ -\frac{\sqrt{3}}{2}c_1 \\ c_2 \end{pmatrix}$$

where we have taken the N-H bond length to be $\sqrt{c_1^2 + c_2^2}$.

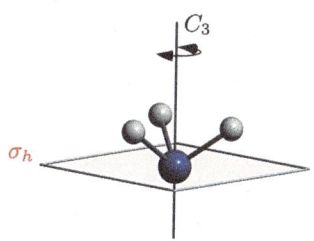

(a) Show that a three-fold rotation about the z-axis passing through the N atom transforms the molecule into an orientation indistinguishable from the initial orientation. Do this by applying

$$C_3 = \frac{1}{2}\begin{pmatrix} -1 & -\sqrt{3} & 0 \\ \sqrt{3} & -1 & 0 \\ 0 & 0 & 2 \end{pmatrix}$$

to each atom's position to find $C_3\,\mathbf{x}_N = \mathbf{x}'_N$, $C_3\,\mathbf{x}_{H1} = \mathbf{x}'_{H1}$, $C_3\,\mathbf{x}_{H2} = \mathbf{x}'_{H2}$, and $C_3\,\mathbf{x}_{H3} = \mathbf{x}'_{H3}$ where $\mathbf{x}'_N = \mathbf{x}_N$, $\mathbf{x}'_{H1} = \mathbf{x}_{H2}$, $\mathbf{x}'_{H2} = \mathbf{x}_{H3}$, and $\mathbf{x}'_{H3} = \mathbf{x}_{H1}$.

(b) Consider reflection of the ammonia molecule across the xy-plane containing the central N atom. This results in the inversion of the ammonia molecule. Do this by applying

$$\sigma_h = \begin{pmatrix} 1 & 0 & 0 \\ 0 & 1 & 0 \\ 0 & 0 & -1 \end{pmatrix}$$

to each atom's position $\sigma_h\, \mathbf{x}_N = \mathbf{x}'_N$, $\sigma_h\, \mathbf{x}_{H1} = \mathbf{x}'_{H1}$, $\sigma_h\, \mathbf{x}_{H2} = \mathbf{x}'_{H2}$, and $\sigma_h\, \mathbf{x}_{H3} = \mathbf{x}'_{H3}$. Show that this transformation leaves the molecule in an orientation that is distinguishable from the initial orientation, thereby demonstrating that ammonia lacks σ_h reflection symmetry.

(c) List all symmetry transformations that result in a final orientation of the NH_3 molecule that is indistinguishable from its initial orientation. Determine the point group for NH_3.

16.7 Consider the ethane molecule, C_2H_6, in the staggered conformation (see Figure 16.27) with symmetries characterized by the D_{3d} point group. Then consider the related molecules

(a)   1-chloroethane, C_2H_5Cl

(b)   *gauche*-1, 2-dichloroethane, $C_2H_4Cl_2$

(c)   *trans*-1, 2-dichloroethane, $C_2H_4Cl_2$

(d)   1,1-dichloroethane, $C_2H_4Cl_2$

For each molecule, use the symmetry decision tree to determine the point group defining the symmetry of the molecule in its staggered conformation.

Homework exercises

16.8 Consider the xenon tetrafluoride molecule, XeF_4, shown below. The xenon atom is placed at the origin with the fluorine atoms at the corners of a square in the xy-plane.

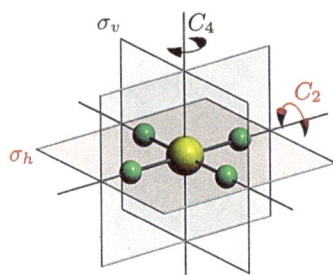

One fluorine atom is placed on the positive x-axis so that

$$\mathbf{x}_{Xe} = \begin{pmatrix} 0 \\ 0 \\ 0 \end{pmatrix} \qquad \mathbf{x}_{F1} = \begin{pmatrix} c_1 \\ 0 \\ 0 \end{pmatrix}$$

where we have taken the Xe-F bond length to be c_1.

(a) Construct the remaining vectors $\mathbf{x}_{F2}, \mathbf{x}_{F3}$, and $\mathbf{x}_{F4}$ defining the positions of the fluorine atoms.

(b) Prove that the xenon tetrafluoride molecule exhibits C_4 rotational symmetry by showing that a four-fold $\mathbf{C}_4$ rotation about the z-axis passing through the Xe atom transforms the molecule into an orientation that is indistinguishable from the initial orientation.

(c) List all symmetry transformations that result in a final orientation of the XeF_4 molecule that is indistinguishable from its initial orientation. Determine the point group for XeF_4.

16.9 Consider the ethane molecule, C_2H_6, in the eclipsed conformation shown below with symmetries characterizing the D_{3h} point group.

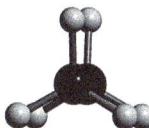 

Now consider the related molecules

(a) 1-chloroethane, C_2H_5Cl

(b) *cis*-1,2-dichloroethane, $C_2H_4Cl_2$

(c) *gauche*-1,2-dichloroethane, $C_2H_4Cl_2$

(d) 1,1-dichloroethane, $C_2H_4Cl_2$

For each molecular conformation, use the symmetry decision tree to determine the point group.

16.10 We considered the structure of the cyclohexane molecule, C_6H_{12}, in the chair conformation shown below. It was found it to possess identity symmetry E, inversion symmetry i, rotational symmetries C_2 and C_3, reflection symmetry σ_d, and improper rotational symmetry S_6. The symmetry decision tree was used to identify the point group D_{3d}.

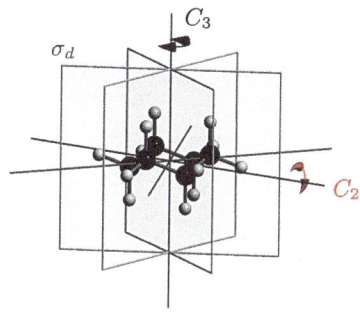

Use the symmetry decision tree to identify the molecular symmetries and point group of the cyclohexane molecule in the boat conformation.

16.11 Consider the ethene molecule, C_2H_4, shown below with symmetries characterizing the D_{2h} point group.

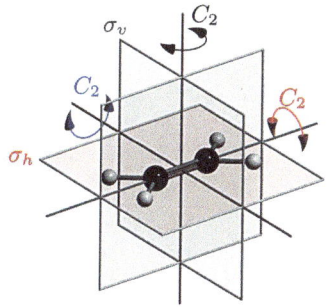

The figure depicts three axes of two-fold rotation and three planes of reflection symmetry.

Now consider the related molecules

(a)   1-chloroethene C_2H_3Cl
(c)   *trans*-1,2-dichloroethene $C_2H_2Cl_2$

(b)   *cis*-1,2-dichloroethene $C_2H_2Cl_2$
(d)   1,1-dichloroethene $C_2H_2Cl_2$

For each molecular conformation, use the symmetry decision tree to determine the point group.

16.12   The cyclopropane molecule, C_3H_6, is shown below. It has symmetry characterized by the D_{3h} point group.

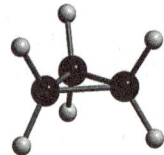

Now consider the related molecules

(a)   1-fluorocyclopropane, C_3H_5F
(c)   *cis*-1,2-difluorocyclopropane, $C_3H_4F_2$

(b)   1,1-difluorocyclopropane, $C_3H_4F_2$
(d)   *trans*-1,2-difluorocyclopropane, $C_3H_4F_2$

For each molecular conformation, use the symmetry decision tree to determine the point group.

16.13   Consider the sulfurhexafluoride molecule, SF_6, with symmetries characterizing the O_h point group.

(a)

(b)

(c)

(d)

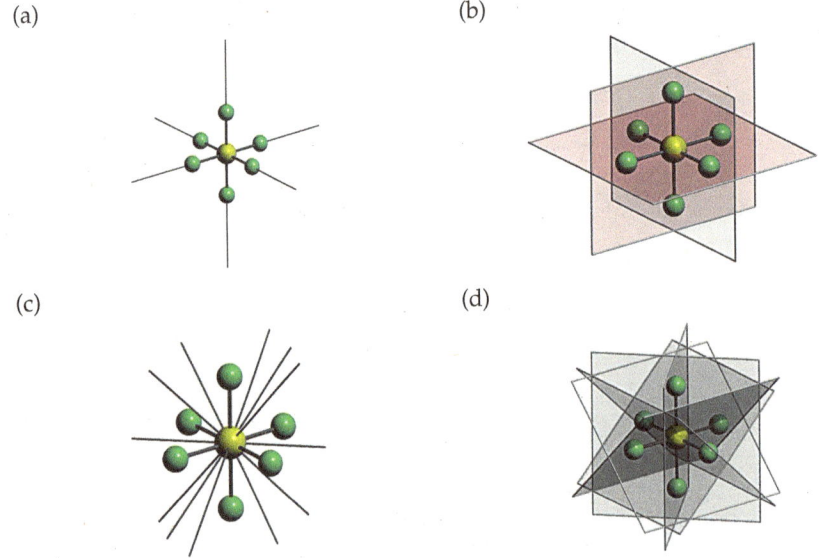

In each of the figures above, rods or planes are shown to imply a particular rotational or reflection symmetry. In each case, identify the corresponding symmetry operation.

16.14 Consider the cyclooctatetrene molecule, C_8H_8, shown below from two perspectives.

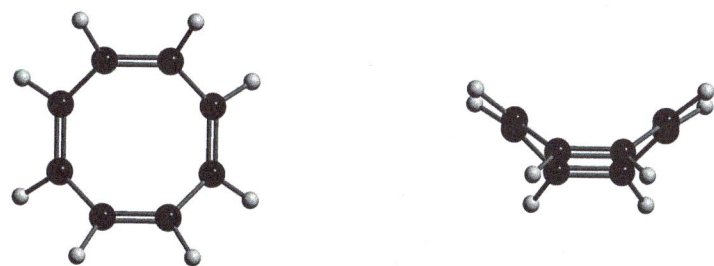

Use the symmetry decision tree to determine the point group of C_8H_8. Consider the various symmetry operations, including inversion, reflection, rotation, and improper rotation. Identify all symmetry operations that leave the C_8H_8 structure unchanged.

16.15* Consider the cyclooctasulfur molecule, S_8, shown below.

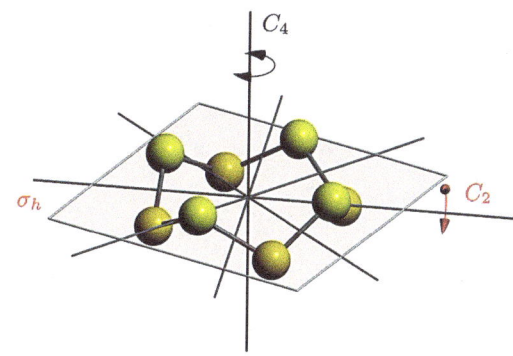

(a) An improper rotation is a combination of a rotation about an axis and a reflection in a plane perpendicular to that axis. For example, we found that borane has S_3 improper rotational symmetry defined by the transformation $S_3 = \sigma_{xy} C_3 = C_3 \sigma_{xy}$. Define the improper rotational symmetry operation S_8 in terms of elementary operations of rotation and reflection.

(b) Determine the number of dihedral reflection planes σ_d that contain the principal C_4 rotation axis and bisect two horizontal C_2 rotation axes.

(c) List all symmetry transformations that result in a final orientation of the S_8 molecule that is indistinguishable from its initial orientation. Determine the point group for S_8.

Supplements

S_1 Notes on notation

COMMON CHARACTERS AND MATHEMATICAL SYMBOLS used throughout the text are collected here.

Greek characters (lower case and unique capitals)

α	Alpha	ι	Iota			ρ	Rho
β	Beta	κ	Kappa			$\sigma\ \Sigma$	Sigma
$\gamma\ \Gamma$	Gamma	$\lambda\ \Lambda$	Lambda			τ	Tau
$\delta\ \Delta$	Delta	μ	Mu			$\upsilon\ Y$	Upsilon
ε	Epsilon	ν	Nu			$\varphi\ \Phi$	Phi
ζ	Zeta	$\xi\ \Xi$	Xi			χ	Chi
η	Eta	o	Omicron (=o, not used)			$\psi\ \Psi$	Psi
$\theta\ \Theta$	Theta	$\pi\ \Pi$	Pi			$\omega\ \Omega$	Omega

Mathematical symbols (basic)

$=$	Equal	$>$	Greater than	$<$	Less than
$\neq$	Not equal	$\geq$	Greater than or equal	$\leq$	Less than or equal
$\approx$	Approximately equal	$\gg$	Much greater than	$\ll$	Much less than
$\propto$	Proportional	$+, -$	Plus, Minus	$\rightarrow$	Approaches
$\equiv$	Equivalent	$\pm$	Plus-Minus	$\in$	In the set
$\forall$	For all	$\mp$	Minus-Plus	$\notin$	Not in the set

Mathematical symbols (linear algebra)

a	Scalar	$\mathbf{a} \cdot \mathbf{b}$	Dot product	$\lvert a \rvert$	Absolute value
$\mathbf{a}$	Vector	$\mathbf{a} \times \mathbf{b}$	Cross product	$\lvert \mathbf{a} \rvert$	Norm
$\mathbf{A}$	Matrix	$\mathbf{AB}$	Matrix product	$\lvert \mathbf{A} \rvert$	Determinant
$\mathbf{A}^T$	Matrix transpose	$\mathbf{A}^{-1}$	Matrix inverse	$\mathbf{A}^\dagger$	Hermitian conjugate

Mathematical symbols (number theory)

z	Complex number	$\mathrm{Re}(z)$	Real part	$\mathrm{Im}(z)$	Imaginary part
$\mathbb{Z}$	Integers	$\mathbb{R}$	Real numbers	$\mathbb{C}$	Complex numbers

Mathematical symbols (exponentials and logarithms)

e^x	Exponential	x^n	nth power of x	$\sqrt[n]{x}$	nth root of x
$\ln(x)$	Natural logarithm	$\log_2(x)$	Binary logarithm	$\log(x)$	Base 10 logarithm

Mathematical symbols (sums, products, and combinatorial factors)

$\displaystyle\sum_{n=1}^{N} a_n$	Definite sum	$\displaystyle\prod_{n=1}^{N} a_n$	Definite product
$n! = n \times n-1 \times \ldots \times 1$	Factorial	$n!! = n \times n-2 \times \ldots \times 1$	Double factorial (n odd)
$\Gamma(n) = (n-1)!$	Gamma function	$n!! = n \times n-2 \times \ldots \times 2$	Double factorial (n even)
$\displaystyle\binom{N}{n} = \frac{N!}{n!(N-n)!}$	Binomial coefficient	$\displaystyle\frac{N!}{n_1!n_2!\ldots n_m!}$	Multinomial coefficient

Mathematical symbols (calculus)

$\displaystyle\lim_{x\to a} f(x)$	Limit as $x \to a$	$\displaystyle\lim_{x\to a^\pm} f(x)$	Limit as $x \to a$ from $+$ or $-$
$\displaystyle\frac{df}{dx} \equiv f' \equiv \hat{D}_x f$	Derivative	$\displaystyle\frac{\partial f}{\partial x} \equiv f_x \equiv \hat{d}_x f$	Partial derivative
$\displaystyle\frac{d^2 f}{dx^2} \equiv f'' \equiv \hat{D}_x^2 f$	Second derivative	$\displaystyle\frac{\partial^2 f}{\partial y \partial x} \equiv \frac{\partial}{\partial y}\frac{\partial f}{\partial x} \equiv f_{xy}$	Second partial derivative
$\displaystyle\frac{dy}{dt} \equiv \dot{y}$	Time derivative	$\displaystyle\frac{d^2 y}{dt^2} \equiv \ddot{y}$	Second time derivative
$\displaystyle\int f(x)\,dx$	Integral	$\displaystyle\iint f(x,y)\,dx dy$	Double integral
$x \in (a,b)$	$a < x < b$	$x \in (a,b]$	$a < x \le b$
$x \in [a,b)$	$a \le x < b$	$x \in [a,b]$	$a \le x \le b$

Mathematical symbols (vector calculus)

∇V	Gradient of V	$\nabla \cdot \mathbf{f}$	Divergence of $\mathbf{f}$
$\nabla \times \mathbf{f}$	Curl of $\mathbf{f}$	$\nabla^2 V$	Laplacian of V

Mathematical constants

$e = 2.718281828\ldots$	Euler's number	$\gamma = 0.5772156649\ldots$	Euler-Mascheroni constant
$\pi = 3.141592654\ldots$	Pi	$\varphi = 1.6180339887\ldots$	Golden ratio constant

S₂ Formulas from geometry

GEOMETRIC RELATIONS ARE ESSENTIAL IN MODELING PHYSICAL SYSTEMS and problem solving in the physical sciences. A brief compilation of commonly used geometric identities is provided for convenience. The following notation is used to define these relations, where r is the radius, h is the height (or altitude), b (or a) is the base length, A is the area, C is the circumference, V is the volume, S is surface area (curved surface), and B is the base area. Volumes of polyhedra are defined in terms of polyhedron edge vectors $\mathbf{a}$, $\mathbf{b}$, and $\mathbf{c}$.

Two dimensions

$$\begin{aligned}
\text{Triangle} \quad & A = \frac{1}{2}bh \\
\text{Circle} \quad & A = \pi r^2,\ C = 2\pi r \\
\text{Ellipse} \quad & A = \pi ab,\ C \approx \pi\sqrt{2(a^2 + b^2)} \\
\text{Parallelogram} \quad & A = bh \\
\text{Trapezoid} \quad & A = \frac{1}{2}(a + b)h
\end{aligned}$$

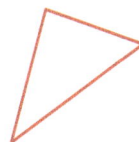  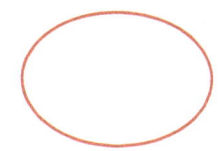  

Three dimensions (Platonic solids)

$$\begin{aligned}
\text{Tetrahedron (regular)} \quad & V = \frac{1}{12}\sqrt{2}a^3,\ S = \sqrt{3}a^2 \\
\text{Cube} \quad & V = a^3,\ S = 6a^2 \\
\text{Octahedron (regular)} \quad & V = \frac{1}{3}\sqrt{2}a^3,\ S = 2\sqrt{3}a^2 \\
\text{Dodecahedron (regular)} \quad & V = \frac{1}{4}\left(15 + 7\sqrt{5}\right)a^3,\ S = 3\sqrt{25 + 10\sqrt{5}}a^2 \\
\text{Icosahedron (regular)} \quad & V = \frac{5}{12}\left(3 + \sqrt{5}\right)a^3,\ S = 5\sqrt{3}a^2
\end{aligned}$$

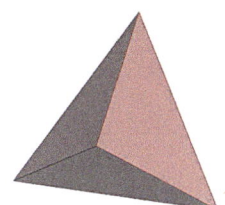 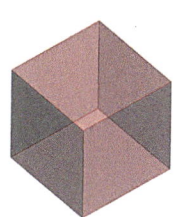 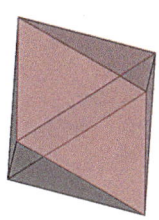 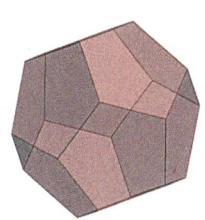 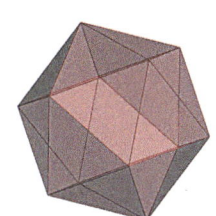

Three dimensions (other solids)

Sphere $\qquad V = \dfrac{4}{3}\pi r^3,\ S = 4\pi r^2$

Right Circular Cylinder $\qquad V = \pi r^2 h,\ S = 2\pi r h$

Right Circular Cone $\qquad V = \dfrac{1}{3}\pi r^2 h,\ S = \pi r \sqrt{r^2 + h^2}$

Prism (rectangular) $\qquad V = abh,\ S = 2ab + 2bh + 2ah$

Pyramid (square) $\qquad V = \dfrac{1}{3}Bh,\ S = 2b\sqrt{(b/2)^2 + h^2} + b^2$

Tetrahedron (general) $\qquad V = \dfrac{1}{3!}|(\mathbf{a} \times \mathbf{b}) \cdot \mathbf{c}|$

Parallelepiped $\qquad V = |(\mathbf{a} \times \mathbf{b}) \cdot \mathbf{c}|,\ S = 2ab + 2bh + 2ah$

   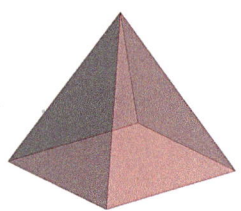 

S₃ Formulas from trigonometry

TRIGONOMETRIC RELATIONS ARE COMMONLY USED in solving problems in the physical sciences. Familiarity with the *unit circle* and a knowledge of the most commonly encountered values of trigonometric functons is essential to efficient calculation. A brief summary of useful trigonometric identities is provided for convenience. The following notation uses standard abbreviations opp ≡ opposite, adj ≡ adjacent, and hyp ≡ hypotenuse, as well as sin ≡ sine, cos ≡ cosine, tan ≡ tangent, cot ≡ cotangent, sec ≡ secant, and csc ≡ cosecant.

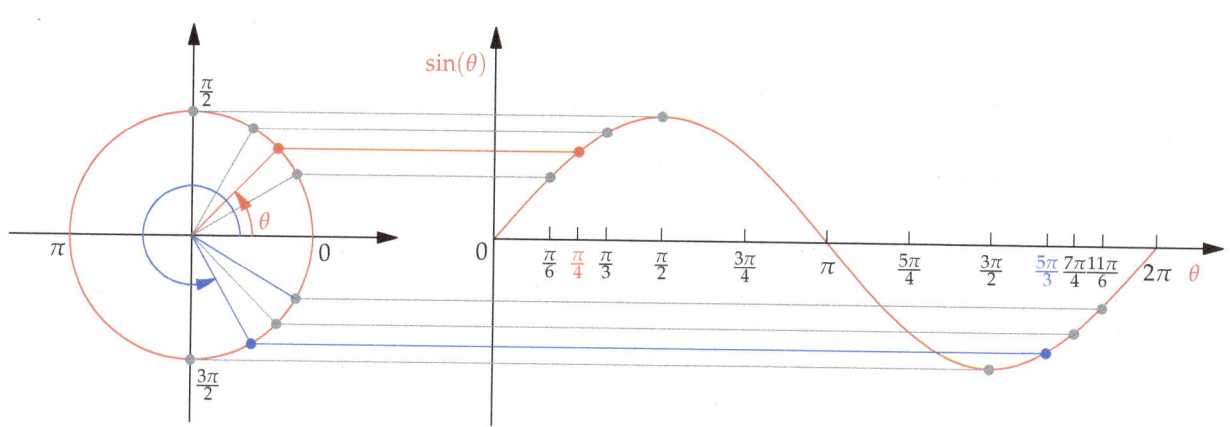

θ (rad)	0	$\frac{\pi}{6}$	$\frac{\pi}{4}$	$\frac{\pi}{3}$	$\frac{\pi}{2}$	$\frac{2\pi}{3}$	$\frac{3\pi}{4}$	$\frac{5\pi}{6}$	π	$\frac{7\pi}{6}$	$\frac{5\pi}{4}$	$\frac{3\pi}{2}$	$\frac{5\pi}{3}$	$\frac{7\pi}{4}$	$\frac{11\pi}{6}$	2π
θ (deg)	0	30°	45°	60°	90°	120°	135°	150°	180°	210°	225°	270°	300°	315°	330°	360°
$\sin(\theta)$	0	$\frac{1}{2}$	$\frac{1}{\sqrt{2}}$	$\frac{\sqrt{3}}{2}$	1	$\frac{\sqrt{3}}{2}$	$\frac{1}{\sqrt{2}}$	$\frac{1}{2}$	0	$-\frac{1}{2}$	$-\frac{1}{\sqrt{2}}$	-1	$-\frac{\sqrt{3}}{2}$	$-\frac{1}{\sqrt{2}}$	$-\frac{1}{2}$	0
$\cos(\theta)$	1	$\frac{\sqrt{3}}{2}$	$\frac{1}{\sqrt{2}}$	$\frac{1}{2}$	0	$-\frac{1}{2}$	$-\frac{1}{\sqrt{2}}$	$-\frac{\sqrt{3}}{2}$	-1	$-\frac{\sqrt{3}}{2}$	$-\frac{1}{\sqrt{2}}$	0	$\frac{1}{2}$	$\frac{1}{\sqrt{2}}$	$\frac{\sqrt{3}}{2}$	1
$\tan(\theta)$	0	$\frac{1}{\sqrt{3}}$	1	$\sqrt{3}$		$-\sqrt{3}$	-1	$-\frac{1}{\sqrt{3}}$	0	$\frac{1}{\sqrt{3}}$	1		$-\sqrt{3}$	-1	$-\frac{1}{\sqrt{3}}$	0

Function definitions

$$\sin\theta = \frac{\text{opp}}{\text{hyp}} \qquad \csc\theta = \frac{\text{hyp}}{\text{opp}} \qquad \tan\theta = \frac{\text{opp}}{\text{adj}}$$

$$\cos\theta = \frac{\text{adj}}{\text{hyp}} \qquad \sec\theta = \frac{\text{hyp}}{\text{adj}} \qquad \cot\theta = \frac{\text{adj}}{\text{opp}}$$

Function relations

$$\tan x = \frac{\sin x}{\cos x} \qquad \cot x = \frac{\cos x}{\sin x}$$

$$\csc x = \frac{1}{\sin x} \qquad \sec x = \frac{1}{\cos x} \qquad \cot x = \frac{1}{\tan x}$$

Negative angle formulas

$$\sin(-x) = -\sin x \qquad \cos(-x) = \cos x \qquad \tan(-x) = -\tan x$$

$$\csc(-x) = -\csc x \qquad \sec(-x) = \sec x \qquad \cot(-x) = -\cot x$$

Addition formulas

$$\sin(x \pm y) = \sin x \cos y \pm \cos x \sin y$$

$$\cos(x \pm y) = \cos x \cos y \mp \sin x \sin y$$

$$\tan(x \pm y) = \frac{\tan x \pm \tan y}{1 \mp \tan x \tan y}$$

Double angle formulas

$$\sin 2x = 2 \sin x \cos x$$

$$\cos 2x = \cos^2 x - \sin^2 x = 1 - 2 \sin^2 x = 2 \cos^2 x - 1$$

$$\tan 2x = \frac{2 \tan x}{1 - \tan^2 x} = \frac{2 \sin x \cos x}{\cos^2 x - \sin^2 x}$$

Half-angle formulas

$$\sin^2 \frac{x}{2} = \frac{1 - \cos x}{2} \qquad\qquad \cos^2 \frac{x}{2} = \frac{1 + \cos x}{2}$$

$$\tan \frac{x}{2} = \frac{1 - \cos x}{\sin x} = \frac{\sin x}{1 + \cos x}$$

Product formulas

$$\sin x \cos y = \frac{1}{2} \left(\sin(x + y) + \sin(x - y) \right)$$

$$\cos x \sin y = \frac{1}{2} \left(\sin(x + y) - \sin(x - y) \right)$$

$$\cos x \cos y = \frac{1}{2} \left(\cos(x + y) + \cos(x - y) \right)$$

$$\sin x \sin y = \frac{1}{2} \left(\cos(x - y) - \cos(x + y) \right)$$

Factoring formulas

$$\sin x \pm \sin y = 2 \cos \frac{x \mp y}{2} \sin \frac{x \pm y}{2}$$

$$\cos x + \cos y = 2 \cos \frac{x + y}{2} \cos \frac{x - y}{2}$$

$$\cos x - \cos y = 2 \sin \frac{x + y}{2} \sin \frac{x - y}{2}$$

S4 Table of power series

POWER SERIES REPRESENTATIONS OF FUNCTIONS are useful in approximation as well as the exact solution of differential and integral equations. This table summarizes power series that most commonly appear in the physical sciences.

1. $\dfrac{1}{1-x} = 1 + x + x^2 + x^3 + \ldots = \displaystyle\sum_{n=0}^{\infty} x^n$

2. $e^x = \exp(x) = 1 + x + \dfrac{1}{2!}x^2 + \dfrac{1}{3!}x^3 + \ldots = \displaystyle\sum_{n=0}^{\infty} \dfrac{1}{n!}x^n$

3. $e^{-x} = \exp(-x) = 1 - x + \dfrac{1}{2!}x^2 - \dfrac{1}{3!}x^3 + - \ldots = \displaystyle\sum_{n=0}^{\infty} (-1)^n \dfrac{1}{n!}x^n$

4. $e^{-x^2} = \exp(-x^2) = 1 - x^2 + \dfrac{1}{2!}x^4 - \dfrac{1}{3!}x^6 + - \ldots = \displaystyle\sum_{n=0}^{\infty} (-1)^n \dfrac{1}{n!}x^{2n}$

5. $\sin(x) = x - \dfrac{1}{3!}x^3 + \dfrac{1}{5!}x^5 - \dfrac{1}{7!}x^7 + - \ldots = \displaystyle\sum_{n=0}^{\infty} (-1)^n \dfrac{1}{(2n+1)!}x^{2n+1}$

6. $\cos(x) = 1 - \dfrac{1}{2!}x^2 + \dfrac{1}{4!}x^4 - \dfrac{1}{6!}x^6 + - \ldots = \displaystyle\sum_{n=0}^{\infty} (-1)^n \dfrac{1}{(2n)!}x^{2n}$

7. $\sinh(x) = x + \dfrac{1}{3!}x^3 + \dfrac{1}{5!}x^5 + \dfrac{1}{7!}x^7 + \ldots = \displaystyle\sum_{n=0}^{\infty} \dfrac{1}{(2n+1)!}x^{2n+1}$

8. $\cosh(x) = 1 + \dfrac{1}{2!}x^2 + \dfrac{1}{4!}x^4 + \dfrac{1}{6!}x^6 + \ldots = \displaystyle\sum_{n=0}^{\infty} \dfrac{1}{(2n)!}x^{2n}$

9. $\ln(1+x) = x - \dfrac{1}{2}x^2 + \dfrac{1}{3}x^3 - \dfrac{1}{4}x^4 + - \ldots = - \displaystyle\sum_{n=1}^{\infty} (-1)^n \dfrac{1}{n}x^n$

10. $\tan^{-1}(x) = x - \dfrac{1}{3}x^3 + \dfrac{1}{5}x^5 - \dfrac{1}{7}x^7 + - \ldots = \displaystyle\sum_{n=0}^{\infty} (-1)^n \dfrac{1}{2n+1}x^{2n+1}$

S$_5$ Table of definite integrals

CERTAIN INTEGRALS ARE PERVASIVE throughout the physical sciences. A brief table of the most frequently encountered definite integrals is provided for convenience. The integral identities listed below are derived in the text. Note the constant $a > 0$.

1. $\displaystyle\int_0^\infty x^n e^{-ax}\, dx = \frac{n!}{a^{n+1}}$

2. $\displaystyle\int_{-\infty}^\infty e^{-ax^2}\, dx = 2\int_0^\infty e^{-ax^2}\, dx = \left(\frac{\pi}{a}\right)^{1/2}$

3. $\displaystyle\int_0^\infty x^2 e^{-ax^2}\, dx = \frac{1}{4a}\left(\frac{\pi}{a}\right)^{1/2}$

4. $\displaystyle\int_0^\infty x^{2n} e^{-ax^2}\, dx = \frac{1\cdot 3\cdot 5\cdots(2n-1)}{2^{n+1}a^n}\left(\frac{\pi}{a}\right)^{1/2}$

5. $\displaystyle\int_0^\infty x e^{-ax^2}\, dx = \frac{1}{2a}$

6. $\displaystyle\int_0^\infty x^3 e^{-ax^2}\, dx = \frac{1}{2a^2}$

7. $\displaystyle\int_0^\infty x^{2n+1} e^{-ax^2}\, dx = \frac{n!}{2}\left(\frac{1}{a^{n+1}}\right)$

8. $\displaystyle\int_{-\infty}^\infty x^{2n} e^{-ax^2}\, dx = \frac{1\cdot 3\cdot 5\cdots(2n-1)}{2^n a^n}\left(\frac{\pi}{a}\right)^{1/2}$

9. $\displaystyle\int_{-\infty}^\infty x^{2n+1} e^{-ax^2}\, dx = 0$

10. $\displaystyle\int_{-\infty}^\infty e^{-ax^2+bx}\, dx = \left(\frac{\pi}{a}\right)^{1/2} e^{b^2/4a}$

11. $\displaystyle\int_{-\infty}^\infty e^{-a(x-x_1)^2} e^{-b(x-x_2)^2}\, dx = \left(\frac{\pi}{a+b}\right)^{1/2}\exp\left[-\frac{ab}{a+b}(x_1-x_2)^2\right]$

12. $\displaystyle\int_{-\infty}^\infty \frac{1}{x}\sin(ax)\, dx = \pi$

13. $\displaystyle\int_{-\infty}^\infty \frac{a}{a^2-x^2}\, dx = \pi$

14. $\displaystyle\int_0^\infty x^k e^{-\lambda x^m}\, dx = \frac{1}{m}\lambda^{-\frac{k+1}{m}}\Gamma\left(\frac{k+1}{n}\right)$

S₆ Table of indefinite integrals

INTEGRALS INVOLVING A VARIETY OF ALGEBRAIC, TRIGONOMETRIC, AND EXPONENTIAL FUNCTIONS appear throughout the physical sciences. A brief table of the most frequently encountered indefinite integrals is provided for convenience.

Basic forms and most common functions (of x)

1. $\int x^p \, dx = \dfrac{1}{p+1} x^{p+1} + C, \quad$ if $p \neq -1$

2. $\int \dfrac{1}{x} \, dx = \ln |x| + C$

3. $\int (ax+b)^p \, dx = \dfrac{(ax+b)^{p+1}}{a(p+1)} + C, \quad$ if $p \neq -1$

4. $\int \dfrac{1}{ax+b} \, dx = \dfrac{1}{a} \ln |ax+b| + C$

5. $\int \dfrac{1}{x^2+a^2} \, dx = \dfrac{1}{a} \arctan \dfrac{x}{a} + C$

6. $\int \dfrac{1}{x^2-a^2} \, dx = \dfrac{1}{2a} \ln \left| \dfrac{x-a}{x+a} \right| + C = -\dfrac{1}{a} \coth^{-1} \dfrac{x}{a} + C$

7. $\int \ln |ax+b| \, dx = \dfrac{ax+b}{a} \ln |ax+b| - x + C$

8. $\int e^{ax} \, dx = \dfrac{1}{a} e^{ax} + C$

9. $\int e^{ax} \sin bx \, dx = \dfrac{1}{a^2+b^2} \, e^{ax} (a \sin bx - b \cos bx) + C$

10. $\int e^{ax} \cos bx \, dx = \dfrac{1}{a^2+b^2} \, e^{ax} (a \cos bx + b \sin bx) + C$

11. $\int a^{bx} \, dx = \dfrac{a^{bx}}{b \ln a} + C, \quad$ if $a > 0, a \neq 1$

12. $\int x a^{bx} \, dx = \dfrac{x a^{bx}}{b \ln a} - \dfrac{a^{bx}}{b^2 (\ln a)^2} + C, \quad$ if $a > 0, a \neq 1$

13. $\int \sin(ax+b) \, dx = -\dfrac{1}{a} \cos(ax+b) + C$

14. $\int \sin^2(ax+b) \, dx = \dfrac{x}{2} - \dfrac{1}{2a} \cos(ax+b) \sin(ax+b) + C = \dfrac{x}{2} - \dfrac{1}{4a} \sin 2(ax+b) + C$

15. $\int \cos(ax+b) \, dx = \dfrac{1}{a} \sin(ax+b) + C$

16. $\int \cos^2(ax+b) \, dx = \dfrac{x}{2} + \dfrac{1}{2a} \cos(ax+b) \sin(ax+b) + C = \dfrac{x}{2} + \dfrac{1}{4a} \sin 2(ax+b) + C$

17. $\int \sin(ax+b) \cos(ax+b) \, dx = \dfrac{1}{2a} \sin^2(ax+b) + C$

Algebraic functions (of x and ax + b)

18. $\displaystyle\int \frac{x}{ax+b}\,dx = \frac{1}{a^2}\left[ax - b\ln|ax+b|\right] + C$

19. $\displaystyle\int \frac{x}{(ax+b)^2}\,dx = \frac{1}{a^2}\left(\ln|ax+b| + \frac{b}{ax+b}\right) + C$

20. $\displaystyle\int \frac{x}{(ax+b)^3}\,dx = \frac{1}{a^2}\left(\frac{b}{2(ax+b)^2} - \frac{1}{ax+b}\right) + C$

21. $\displaystyle\int x(ax+b)^p\,dx = \frac{1}{a^2(p+2)}(ax+b)^{p+2} - \frac{b}{a^2(p+1)}(ax+b)^{p+1} + C, \quad \text{if } p \neq -1,\, p \neq -2$

22. $\displaystyle\int \frac{x^2}{ax+b}\,dx = \frac{1}{a^3}\left(\frac{1}{2}(ax+b)^2 - 2b(ax+b) + b^2\ln|ax+b|\right) + C$

23. $\displaystyle\int \frac{x^2}{(ax+b)^2}\,dx = \frac{1}{a^3}\left(ax+b - 2b\ln|ax+b| - \frac{b^2}{ax+b}\right) + C$

24. $\displaystyle\int \sqrt{ax+b}\,dx = \frac{2}{3a}\sqrt{(ax+b)^3} + C$

25. $\displaystyle\int \frac{1}{\sqrt{ax+b}}\,dx = \frac{2\sqrt{ax+b}}{a} + C$

26. $\displaystyle\int x\sqrt{ax+b}\,dx = \frac{2(3ax - 2b)\sqrt{(ax+b)^3}}{15a^2} + C$

27. $\displaystyle\int \frac{1}{x}\sqrt{ax+b}\,dx = 2\sqrt{ax+b} + b\int \frac{dx}{x\sqrt{ax+b}}$

28. $\displaystyle\int \frac{1}{x\sqrt{ax+b}}\,dx = \frac{1}{\sqrt{b}}\ln\left|\frac{\sqrt{ax+b} - \sqrt{b}}{\sqrt{ax+b} + \sqrt{b}}\right| + C = -\frac{1}{\sqrt{b}}\tanh^{-1}\sqrt{\frac{ax+b}{b}} + C, \quad \text{if } b > 0$

29. $\displaystyle\int \frac{1}{x\sqrt{ax-b}}\,dx = \frac{2}{\sqrt{b}}\arctan\sqrt{\frac{ax-b}{b}} + C, \quad \text{if } b > 0$

30. $\displaystyle\int x^2\sqrt{ax+b}\,dx = \frac{2}{105\,a^3}\sqrt{(ax+b)^3}\left(15a^2x^2 - 12abx + 8b^2\right) + C$

31. $\displaystyle\int \frac{x^2}{\sqrt{ax+b}}\,dx = \frac{2}{a(2n+1)}\left[x^n\sqrt{ax+b} - nb\int \frac{x^{n-1}dx}{\sqrt{ax+b}}\right]$

Algebraic functions (of x and ax² + b)

32. $\displaystyle\int \frac{1}{ax^2+b}\,dx = \frac{1}{\sqrt{ab}}\arctan\frac{x\sqrt{ab}}{b} + C, \quad \text{if } ab > 0$

33. $\displaystyle\int \frac{1}{ax^2+b}\,dx = \frac{1}{2\sqrt{-ab}}\ln\left|\frac{b + x\sqrt{-ab}}{b - x\sqrt{-ab}}\right| + C = \frac{1}{\sqrt{-ab}}\tanh^{-1}\frac{x\sqrt{-ab}}{b} + C, \quad \text{if } ab < 0$

34. $\displaystyle\int \frac{1}{ax^2-b}\,dx = \frac{1}{2\sqrt{ab}}\ln\left|\frac{x\sqrt{a} - \sqrt{b}}{x\sqrt{a} + \sqrt{b}}\right| + C, \quad \text{if } a > 0,\, b > 0$

35. $\displaystyle\int x(ax^2+b)^p\,dx = \frac{1}{2a}\frac{(ax^2+b)^{p+1}}{p+1} + C, \quad \text{if } p \neq -1$

36. $\displaystyle\int \frac{x}{ax^2+b}\,dx = \frac{1}{2a}\ln|ax^2+b| + C$

37. $\displaystyle\int \frac{x}{(ax^2+b)^{p+1}}\,dx = \frac{-1}{2pa(ax^2+b)^p} + C, \quad \text{if } p > 0$

38. $\displaystyle\int \frac{1}{x(ax^n + b)}\, dx = \frac{1}{bn} \ln \left| \frac{x^n}{ax^n + b} \right| + C$

39. $\displaystyle\int \sqrt{ax^2 + b}\, dx = \frac{x}{2} \sqrt{ax^2 + b} + \frac{b}{2\sqrt{a}} \ln \left| x\sqrt{a} + \sqrt{ax^2 + b} \right| + C, \quad \text{if } a > 0$

40. $\displaystyle\int \sqrt{ax^2 + b}\, dx = \frac{x}{2} \sqrt{ax^2 + b} + \frac{b}{2\sqrt{-a}} \arcsin \left(x\sqrt{\frac{-a}{b}} \right) + C, \quad \text{if } a < 0$

Algebraic functions (of x and $ax^2 + bx + c$) $X \equiv ax^2 + b + c, Y \equiv 4ac - b^2$

41. $\displaystyle\int \frac{1}{ax^2 + bx + c}\, dx = \frac{2}{\sqrt{Y}} \arctan \frac{2ax + b}{\sqrt{Y}} + C, \quad \text{if } Y > 0$

42. $\displaystyle\int \frac{1}{ax^2 + bx + c}\, dx = \frac{2}{\sqrt{-Y}} \ln \left| \frac{2ax + b - \sqrt{-Y}}{2ax + b + \sqrt{-Y}} \right| + C = \frac{-2}{\sqrt{-Y}} \tanh^{-1} \frac{2ax + b}{\sqrt{-Y}} + C, \quad \text{if } Y < 0$

43. $\displaystyle\int \frac{1}{\sqrt{ax^2 + bx + c}}\, dx = \frac{1}{\sqrt{a}} \ln \left| X + x\sqrt{a} + \frac{b}{2\sqrt{a}} \right| + C, \quad \text{if } a > 0$

44. $\displaystyle\int \frac{1}{\sqrt{ax^2 + bx + c}}\, dx = \frac{1}{\sqrt{-a}} \arcsin \frac{-2ax - b}{\sqrt{-Y}} + C, \quad \text{if } a < 0$

45. $\displaystyle\int \frac{x}{ax^2 + bx + c}\, dx = \frac{1}{2a} \left[\ln |X| - b \int \frac{dx}{X} \right]$

Algebraic functions (of x and $x^2 \pm a^2$)

46. $\displaystyle\int \sqrt{x^2 \pm a^2}\, dx = \frac{1}{2} \left[x\sqrt{x^2 \pm a^2} \pm a^2 \ln \left| x + \sqrt{x^2 \pm a^2} \right| \right] + C$

47. $\displaystyle\int \frac{1}{\sqrt{x^2 \pm a^2}}\, dx = \ln \left| x + \sqrt{x^2 \pm a^2} \right| + C$

48. $\displaystyle\int x\sqrt{x^2 \pm a^2}\, dx = \frac{1}{3} \sqrt{(x^2 \pm a^2)^3} + C$

49. $\displaystyle\int \frac{1}{x} \sqrt{x^2 + a^2}\, dx = \sqrt{x^2 + a^2} - a \ln \left| \frac{a + \sqrt{x^2 + a^2}}{x} \right| + C$

50. $\displaystyle\int \frac{1}{x} \sqrt{x^2 - a^2}\, dx = \sqrt{x^2 - a^2} - a \, \mathrm{arcsec}\frac{x}{a} + C$

51. $\displaystyle\int \frac{x}{\sqrt{x^2 \pm a^2}}\, dx = \sqrt{x^2 \pm a^2} + C$

52. $\displaystyle\int \frac{1}{x\sqrt{x^2 + a^2}}\, dx = -\frac{1}{a} \ln \left| \frac{\sqrt{x^2 + a^2} + a}{x} \right| + C = -\frac{1}{a} \sinh^{-1} \frac{a}{x} + C$

53. $\displaystyle\int \frac{1}{x\sqrt{x^2 - a^2}}\, dx = \frac{1}{a} \mathrm{arcsec}\frac{x}{a} + C = \frac{1}{a} \arccos \frac{a}{x} + C$

54. $\displaystyle\int \sqrt{(x^2 \pm a^2)^3}\, dx = \frac{1}{8} \left[2x\sqrt{(x^2 \pm a^2)^3} \pm 3a^2 x \sqrt{x^2 \pm a^2} + 3a^4 \ln \left| x - \sqrt{x^2 \pm a^2} \right| \right] + C$

55. $\displaystyle\int \frac{1}{\sqrt{(x^2 \pm a^2)^3}}\, dx = \frac{\pm x}{a^2 \sqrt{x^2 \pm a^2}} + C$

56. $\displaystyle\int x\sqrt{(x^2 \pm a^2)^3}\, dx = \frac{1}{5} \sqrt{(x^2 \pm a^2)^5} + C$

57. $\displaystyle\int \frac{x}{(x^2 \pm a^2)^3}\, dx = \frac{-1}{\sqrt{x^2 \pm a^2}} + C$

58. $\displaystyle\int \frac{1}{a^2 - x^2}\, dx = \frac{1}{2a} \ln\left|\frac{a+x}{a-x}\right| + C = \frac{1}{a} \tanh^{-1}\frac{x}{a} + C$

59. $\displaystyle\int \frac{1}{\sqrt{a^2 - x^2}}\, dx = \arcsin\frac{x}{a} + C = -\arccos\frac{x}{a} + C$

Algebraic functions (of x and $a^2 - x^2$)

60. $\displaystyle\int \sqrt{a^2 - x^2}\, dx = \frac{1}{2}\left[x\sqrt{a^2 - x^2} + a^2 \arcsin\frac{x}{a}\right] + C$

61. $\displaystyle\int x\sqrt{a^2 - x^2}\, dx = -\frac{1}{3}\sqrt{(a^2 - x^2)^3} + C$

62. $\displaystyle\int \frac{1}{x}\sqrt{a^2 - x^2}\, dx = \sqrt{a^2 - x^2} - a\ln\left|\frac{a + \sqrt{a^2 - x^2}}{x}\right| + C$

63. $\displaystyle\int \frac{x}{\sqrt{a^2 - x^2}}\, dx = -\sqrt{a^2 - x^2} + C$

64. $\displaystyle\int \frac{1}{x\sqrt{a^2 - x^2}}\, dx = -\frac{1}{a}\ln\left|\frac{a + \sqrt{a^2 - x^2}}{x}\right| + C = -\frac{1}{a}\cosh^{-1}\frac{a}{x} + C$

65. $\displaystyle\int \sqrt{(a^2 - x^2)^3}\, dx = \frac{1}{8}\left[2x\sqrt{(a^2 - x^2)^3} + 3a^2 x\sqrt{a^2 - x^2} + 3a^4 \arcsin\frac{x}{a}\right] + C$

66. $\displaystyle\int \frac{1}{\sqrt{(a^2 - x^2)^3}}\, dx = \frac{x}{a^2\sqrt{a^2 - x^2}} + C$

67. $\displaystyle\int x\sqrt{(a^2 - x^2)^3}\, dx = -\frac{1}{5}\sqrt{(a^2 - x^2)^5} + C$

68. $\displaystyle\int \frac{x}{\sqrt{(a^2 - x^2)^3}}\, dx = \frac{1}{\sqrt{a^2 - x^2}} + C$

Other algebraic functions

69. $\displaystyle\int \sqrt{2ax - x^2}\, dx = \frac{1}{2}\left[(x - a)\sqrt{2ax - x^2} + a^2 \arcsin\frac{x - a}{a}\right] + C$

70. $\displaystyle\int \frac{1}{\sqrt{2ax - x^2}}\, dx = \arccos\frac{a - x}{a} + C$

71. $\displaystyle\int \sqrt{\frac{1 + x}{1 - x}}\, dx = \arcsin x - \sqrt{1 - x^2} + C$

Logarithmic functions

72. $\displaystyle\int \log_a |x|\, dx = x\log_a |x| - \frac{x}{\ln a} + C, \quad \text{if } a \neq 1, a > 0$

73. $\displaystyle\int (\ln |x|)^2\, dx = x(\ln |x|)^2 - 2x\ln |x| + 2x + C$

74. $\displaystyle\int x\ln |x|\, dx = \frac{x^2}{2}\ln |x| - \frac{x^2}{4} + C$

75. $\displaystyle\int \frac{1}{x\ln |x|}\, dx = \ln |\ln |x|| + C$

76. $\displaystyle\int x^p \ln |x|\, dx = x^{p+1}\left[\frac{\ln |x|}{p + 1} - \frac{1}{(p + 1)^2}\right] + C, \quad \text{if } p \neq -1$

77. $\displaystyle\int \frac{(\ln |x|)^p}{x}\, dx = \frac{1}{p + 1}(\ln |x|)^{p+1} + C, \quad \text{if } p \neq -1$

78. $\int \sin(\ln|x|)\, dx = \dfrac{x}{2}\left[\sin(\ln|x|) - \cos(\ln|x|)\right] + C$

79. $\int \cos(\ln|x|)\, dx = \dfrac{x}{2}\left[\sin(\ln|x|) + \cos(\ln|x|)\right] + C$

Exponential functions

80. $\int xe^{ax}\, dx = \dfrac{1}{a^2}e^{ax}(ax - 1) + C$

81. $\int x^m e^{ax}\, dx = \dfrac{1}{a}x^m e^{ax} - \dfrac{m}{a}\int x^{m-1}e^{ax}\, dx, \quad m \geq 2$

82. $\int \dfrac{1}{x}e^{ax}\, dx = \ln|x| + ax + \dfrac{1}{2(2!)}(ax)^2 + \dfrac{1}{3(3!)}(ax)^3 + \cdots + C$

83. $\int \dfrac{1}{1 + e^x}\, dx = x - \ln|1 + e^x| + C$

84. $\int \dfrac{1}{ae^{px} + b}\, dx = \dfrac{x}{b} - \dfrac{1}{bp}\ln|ae^{px} + b| + C, \quad \text{if } b \neq 0,\, p \neq 0$

85. $\int \dfrac{1}{ae^{px} + be^{-px}}\, dx = \dfrac{1}{p\sqrt{ab}}\arctan\left(e^{px}\sqrt{\dfrac{a}{b}}\right) + C, \quad \text{if } ab > 0$

86. $\int e^{ax}\ln|bx|\, dx = \dfrac{1}{a}e^{ax}\ln|bx| - \dfrac{1}{a}\int \dfrac{e^{ax}}{x}\, dx$

Trigonometric functions

87. $\int \sin^3(ax + b)\, dx = -\dfrac{1}{3a}\cos(ax + b)\left[\sin^2(ax + b) + 2\right] + C$

88. $\int \sin^4(ax + b)\, dx = \dfrac{3x}{8} - \dfrac{3}{16a}\sin 2(ax + b) - \dfrac{1}{4a}\sin^3(ax + b)\cos(ax + b) + C$

89. $\int x\sin(ax + b)\, dx = \dfrac{1}{a^2}\sin(ax + b) - \dfrac{x}{a}\cos(ax + b) + C$

90. $\int x\sin^2(ax + b)\, dx = \dfrac{x^2}{4} - \dfrac{1}{4a}x\sin 2(ax + b) - \dfrac{1}{8a^2}\cos 2(ax + b) + C$

91. $\int x^2\sin^2(ax + b)\, dx = \dfrac{1}{6}x^3 - \left(\dfrac{1}{4a}x^2 - \dfrac{1}{8a^3}\right)\sin 2(ax + b) - \dfrac{1}{4a^2}x\cos 2(ax + b) + C$

92. $\int \dfrac{1}{x}\sin ax\, dx = ax - \dfrac{1}{3(3!)}(ax)^3 + \dfrac{1}{5(5!)}(ax)^5 - + \cdots + C$

93. $\int \sin ax\sin bx\, dx = \dfrac{1}{2(a - b)}\sin(a - b)x - \dfrac{1}{2(a + b)}\sin(a + b)x + C, \quad \text{if } a^2 \neq b^2$

94. $\int \cos^3(ax + b)\, dx = \dfrac{1}{a}\sin(ax + b) - \dfrac{1}{3a}\sin^3(ax + b) + C$

95. $\int \cos^4(ax + b)\, dx = \dfrac{3x}{8} + \dfrac{1}{16a}3\sin 2(ax + b) + \dfrac{1}{4a}\cos^3(ax + b)\sin(ax + b) + C$

96. $\int x\cos(ax + b)\, dx = \dfrac{1}{a^2}\cos(ax + b) + \dfrac{x}{a}\sin(ax + b) + C$

97. $\int x\cos^2(ax + b)\, dx = \dfrac{x^2}{4} + \dfrac{1}{4a}x\sin 2(ax + b) + \dfrac{1}{8a^2}\cos 2(ax + b) + C$

98. $\int x^2\cos^2(ax + b)\, dx = \dfrac{x^3}{6} + \left(\dfrac{x^2}{4a} - \dfrac{1}{8a^3}\right)\sin 2(ax + b) + \dfrac{1}{4a^2}x\cos 2(ax + b) + C$

99. $\displaystyle\int \frac{1}{x}\cos ax\,dx = \ln|ax| - \frac{1}{2(2!)}(ax)^2 + \frac{1}{4(4!)}(ax)^4 - +\cdots + C$

100. $\displaystyle\int \frac{1}{1+\cos(ax+b)}\,dx = \frac{1}{a}\tan\left(\frac{ax+b}{2}\right) + C$

101. $\displaystyle\int \frac{1}{1-\cos(ax+b)}\,dx = -\frac{1}{a}\cot\left(\frac{ax+b}{2}\right) + C$

102. $\displaystyle\int \cos ax \cos bx\,dx = \frac{1}{2(a-b)}\sin(a-b)x + \frac{1}{2(a+b)}\sin(a+b)x + C, \quad \text{if } a^2 \neq b^2$

103. $\displaystyle\int \sin ax \cos bx\,dx = -\frac{1}{2(a-b)}\cos(a-b)x - \frac{1}{2(a+b)}\cos(a+b)x + C, \quad \text{if } a^2 \neq b^2$

104. $\displaystyle\int \sin^p(ax+b)\cos(ax+b)\,dx = \frac{1}{a(p+1)}\sin^{p+1}(ax+b) + C, \quad \text{if } p \neq -1$

105. $\displaystyle\int \sin(ax+b)\cos^p(ax+b)\,dx = -\frac{1}{a(p+1)}\cos^{p+1}(ax+b) + C, \quad \text{if } p \neq -1$

106. $\displaystyle\int \sin^2(ax+b)\cos^2(ax+b)\,dx = -\frac{1}{32a}\sin 4(ax+b) + \frac{x}{8} + C$

107. $\displaystyle\int \tan(ax+b)\,dx = -\frac{1}{a}\ln|\cos(ax+b)| + C$

108. $\displaystyle\int \tan^2(ax+b)\,dx = \frac{1}{a}\tan(ax+b) - x + C$

109. $\displaystyle\int \tan^3(ax+b)\,dx = \frac{1}{2a}\left[\tan^2(ax+b) + 2\ln|\cos(ax+b)|\right] + C$

110. $\displaystyle\int \cot(ax+b)\,dx = \frac{1}{a}\ln|\sin(ax+b)| + C$

111. $\displaystyle\int \cot^3(ax+b)\,dx = -\frac{1}{2a}\left[\cot^2(ax+b) + 2\ln|\sin(ax+b)|\right] + C$

112. $\displaystyle\int \sec(ax+b)\,dx = \frac{1}{a}\ln\left|\tan\left(\frac{ax+b}{2}+\frac{\pi}{4}\right)\right| + C$

113. $\displaystyle\int \sec^2(ax+b)\,dx = \frac{1}{a}\tan(ax+b) + C$

114. $\displaystyle\int \cot^2(ax+b)\,dx = -\frac{1}{a}\cot(ax+b) - x + C$

115. $\displaystyle\int \sec^3(ax+b)\,dx = \frac{1}{2a}\left[\sec(ax+b)\tan(ax+b) + \ln\left|\tan\left(\frac{ax+b}{2}+\frac{\pi}{4}\right)\right|\right] + C$

116. $\displaystyle\int \sec x \tan x\,dx = \sec x + C$

117. $\displaystyle\int \csc(ax+b)\,dx = \frac{1}{a}\ln\left|\tan\left(\frac{ax+b}{2}\right)\right| + C$

118. $\displaystyle\int \csc^2(ax+b)\,dx = -\frac{1}{a}\cot(ax+b) + C$

119. $\displaystyle\int \csc x \cot x\,dx = -\csc x + C$

120. $\displaystyle\int \csc^3(ax+b)\,dx = \frac{1}{2a}\left[-\csc(ax+b)\cot(ax+b) + \ln\left|\tan\left(\frac{ax+b}{2}\right)\right|\right] + C$

121. $\displaystyle\int \arcsin\frac{x}{a}\,dx = x\arcsin\frac{x}{a} + \sqrt{a^2-x^2} + C$

122. $\int x \arcsin ax \, dx = \dfrac{1}{4a^2} \left[(2a^2x^2 - 1) \arcsin ax + ax\sqrt{1 - a^2x^2} \right] + C$

123. $\int \dfrac{1}{x^2} \arcsin ax \, dx = a \ln \left| \dfrac{1 - \sqrt{1 - a^2x^2}}{ax} \right| - \dfrac{\arcsin ax}{x} + C$

124. $\int \arccos \dfrac{x}{a} \, dx = x \arccos \dfrac{x}{a} - \sqrt{a^2 - x^2} + C$

125. $\int \arctan \dfrac{x}{a} \, dx = x \arctan \dfrac{x}{a} - \dfrac{a}{2} \ln(a^2 + x^2) + C$

126. $\int \operatorname{arccot} \dfrac{x}{a} \, dx = x \operatorname{arccot} \dfrac{x}{a} + \dfrac{a}{2} \ln(a^2 + x^2) + C$

127. $\int \operatorname{arcsec} \dfrac{x}{a} \, dx = x \operatorname{arcsec} \dfrac{x}{a} - a \ln \left| x + \sqrt{x^2 - a^2} \right| + C$

128. $\int \operatorname{arccsc} \dfrac{x}{a} \, dx = x \operatorname{arccsc} \dfrac{x}{a} + a \ln \left| x + \sqrt{x^2 - a^2} \right| + C$

Hyperbolic functions

129. $\int \sinh ax \, dx = \dfrac{1}{a} \cosh ax + C$

130. $\int \sinh^2 ax \, dx = \dfrac{1}{4a} \sinh 2ax - \dfrac{1}{2}x + C$

131. $\int \cosh ax \, dx = \dfrac{1}{a} \sinh ax + C$

132. $\int \cosh^2 ax \, dx = \dfrac{1}{4a} \sinh 2ax + \dfrac{1}{2}x + C$

133. $\int \tanh ax \, dx = \dfrac{1}{a} \ln | \cosh ax | + C$

134. $\int \tanh^2 ax \, dx = x - \dfrac{1}{a} \tanh ax + C$

135. $\int \coth ax \, dx = \dfrac{1}{a} \ln | \sinh ax | + C$

136. $\int \coth^2 ax \, dx = x - \dfrac{1}{a} \coth ax + C$

137. $\int \operatorname{sech} ax \, dx = \dfrac{1}{a} \arctan(\sinh ax) + C$

138. $\int \operatorname{sech}^2 ax \, dx = \dfrac{1}{a} \tanh ax + C$

139. $\int \operatorname{csch} ax \, dx = -\dfrac{1}{a} \ln | \coth ax + \operatorname{csch} ax | + C = \dfrac{1}{a} \ln \left| \tanh \dfrac{ax}{2} \right| + C$

140. $\int \operatorname{csch}^2 ax \, dx = -\dfrac{1}{a} \coth ax + C$

141. $\int \operatorname{sech} ax \tanh ax \, dx = -\dfrac{1}{a} \operatorname{sech} ax + C$

142. $\int \operatorname{csch} ax \coth ax \, dx = -\dfrac{1}{a} \operatorname{csch} ax + C$

S7 Error function table

THE ERROR FUNCTION IS A COMMON DEFINITE INTEGRAL that frequently appears in the physical sciences

$$\text{erf}(x) = \frac{2}{\sqrt{\pi}} \int_0^x e^{-t^2}\, dt$$

A tabulation of values of the error function is provided below.

x	Hundredths digit of x									
	0	1	2	3	4	5	6	7	8	9
0.0	0.00000	0.01128	0.02256	0.03384	0.04511	0.05637	0.06762	0.07886	0.09008	0.10128
0.1	0.11246	0.12362	0.13476	0.14587	0.15695	0.16800	0.17901	0.18999	0.20094	0.21184
0.2	0.22270	0.23352	0.24430	0.25502	0.26570	0.27633	0.28690	0.29742	0.30788	0.31828
0.3	0.32863	0.33891	0.34913	0.35928	0.36936	0.37938	0.38933	0.39921	0.40901	0.41874
0.4	0.42839	0.43797	0.44747	0.45689	0.46623	0.47548	0.48466	0.49375	0.50275	0.51167
0.5	0.52050	0.52924	0.53790	0.54646	0.55494	0.56332	0.57162	0.57982	0.58792	0.59594
0.6	0.60386	0.61168	0.61941	0.62705	0.63459	0.64203	0.64938	0.65663	0.66378	0.67084
0.7	0.67780	0.68467	0.69143	0.69810	0.70468	0.71116	0.71754	0.72382	0.73001	0.73610
0.8	0.74210	0.74800	0.75381	0.75952	0.76514	0.77067	0.77610	0.78144	0.78669	0.79184
0.9	0.79691	0.80188	0.80677	0.81156	0.81627	0.82089	0.82542	0.82987	0.83423	0.83851
1.0	0.84270	0.84681	0.85084	0.85478	0.85865	0.86244	0.86614	0.86977	0.87333	0.87680
1.1	0.88021	0.88353	0.88679	0.88997	0.89308	0.89612	0.89910	0.90200	0.90484	0.90761
1.2	0.91031	0.91296	0.91553	0.91805	0.92051	0.92290	0.92524	0.92751	0.92973	0.93190
1.3	0.93401	0.93606	0.93807	0.94002	0.94191	0.94376	0.94556	0.94731	0.94902	0.95067
1.4	0.95229	0.95385	0.95538	0.95686	0.95830	0.95970	0.96105	0.96237	0.96365	0.96490
1.5	0.96611	0.96728	0.96841	0.96952	0.97059	0.97162	0.97263	0.97360	0.97455	0.97546
1.6	0.97635	0.97721	0.97804	0.97884	0.97962	0.98038	0.98110	0.98181	0.98249	0.98315
1.7	0.98379	0.98441	0.98500	0.98558	0.98613	0.98667	0.98719	0.98769	0.98817	0.98864
1.8	0.98909	0.98952	0.98994	0.99035	0.99074	0.99111	0.99147	0.99182	0.99216	0.99248
1.9	0.99279	0.99309	0.99338	0.99366	0.99392	0.99418	0.99443	0.99466	0.99489	0.99511
2.0	0.99532	0.99552	0.99572	0.99591	0.99609	0.99626	0.99642	0.99658	0.99673	0.99688
2.1	0.99702	0.99715	0.99728	0.99741	0.99753	0.99764	0.99775	0.99785	0.99795	0.99805
2.2	0.99814	0.99822	0.99831	0.99839	0.99846	0.99854	0.99861	0.99867	0.99874	0.99880
2.3	0.99886	0.99891	0.99897	0.99902	0.99906	0.99911	0.99915	0.99920	0.99924	0.99928
2.4	0.99931	0.99935	0.99938	0.99941	0.99944	0.99947	0.99950	0.99952	0.99955	0.99957
2.5	0.99959	0.99961	0.99963	0.99965	0.99967	0.99969	0.99971	0.99972	0.99974	0.99975
2.6	0.99976	0.99978	0.99979	0.99980	0.99981	0.99982	0.99983	0.99984	0.99985	0.99986
2.7	0.99987	0.99987	0.99988	0.99989	0.99989	0.99990	0.99991	0.99991	0.99992	0.99992
2.8	0.99992	0.99993	0.99993	0.99994	0.99994	0.99994	0.99995	0.99995	0.99995	0.99996
2.9	0.99996	0.99996	0.99996	0.99997	0.99997	0.99997	0.99997	0.99997	0.99997	0.99998
3.0	0.99998	0.99998	0.99998	0.99998	0.99998	0.99998	0.99998	0.99999	0.99999	0.99999
3.1	0.99999	0.99999	0.99999	0.99999	0.99999	0.99999	0.99999	0.99999	0.99999	0.99999
3.2	0.99999	0.99999	0.99999	1.00000	1.00000	1.00000	1.00000	1.00000	1.00000	1.00000

S8 Complementary error function table

THE COMPLEMENTARY ERROR FUNCTION IS A COMMON DEFINITE INTEGRAL that frequently appears in the physical sciences

$$\mathrm{erfc}(x) = \frac{2}{\sqrt{\pi}} \int_x^{\infty} e^{-t^2}\, dt$$

A tabulation of values of the error function is provided below.

	Hundredths digit of x									
x	0	1	2	3	4	5	6	7	8	9
0.0	1.00000	0.98872	0.97744	0.96616	0.95489	0.94363	0.93238	0.92114	0.90992	0.89872
0.1	0.88754	0.87638	0.86524	0.85413	0.84305	0.83200	0.82099	0.81001	0.79906	0.78816
0.2	0.77730	0.76648	0.75570	0.74498	0.73430	0.72367	0.71310	0.70258	0.69212	0.68172
0.3	0.67137	0.66109	0.65087	0.64072	0.63064	0.62062	0.61067	0.60079	0.59099	0.58126
0.4	0.57161	0.56203	0.55253	0.54311	0.53377	0.52452	0.51534	0.50625	0.49725	0.48833
0.5	0.47950	0.47076	0.46210	0.45354	0.44506	0.43668	0.42838	0.42018	0.41208	0.40406
0.6	0.39614	0.38832	0.38059	0.37295	0.36541	0.35797	0.35062	0.34337	0.33622	0.32916
0.7	0.32220	0.31533	0.30857	0.30190	0.29532	0.28884	0.28246	0.27618	0.26999	0.26390
0.8	0.25790	0.25200	0.24619	0.24048	0.23486	0.22933	0.22390	0.21856	0.21331	0.20816
0.9	0.20309	0.19812	0.19323	0.18844	0.18373	0.17911	0.17458	0.17013	0.16577	0.16149
1.0	0.15730	0.15319	0.14916	0.14522	0.14135	0.13756	0.13386	0.13023	0.12667	0.12320
1.1	0.11979	0.11647	0.11321	0.11003	0.10692	0.10388	0.10090	0.09800	0.09516	0.09239
1.2	0.08969	0.08704	0.08447	0.08195	0.07949	0.07710	0.07476	0.07249	0.07027	0.06810
1.3	0.06599	0.06394	0.06193	0.05998	0.05809	0.05624	0.05444	0.05269	0.05098	0.04933
1.4	0.04771	0.04615	0.04462	0.04314	0.04170	0.04030	0.03895	0.03763	0.03635	0.03510
1.5	0.03389	0.03272	0.03159	0.03048	0.02941	0.02838	0.02737	0.02640	0.02545	0.02454
1.6	0.02365	0.02279	0.02196	0.02116	0.02038	0.01962	0.01890	0.01819	0.01751	0.01685
1.7	0.01621	0.01559	0.01500	0.01442	0.01387	0.01333	0.01281	0.01231	0.01183	0.01136
1.8	0.01091	0.01048	0.01006	0.00965	0.00926	0.00889	0.00853	0.00818	0.00784	0.00752
1.9	0.00721	0.00691	0.00662	0.00634	0.00608	0.00582	0.00557	0.00534	0.00511	0.00489
2.0	0.00468	0.00448	0.00428	0.00409	0.00391	0.00374	0.00358	0.00342	0.00327	0.00312
2.1	0.00298	0.00285	0.00272	0.00259	0.00247	0.00236	0.00225	0.00215	0.00205	0.00195
2.2	0.00186	0.00178	0.00169	0.00161	0.00154	0.00146	0.00139	0.00133	0.00126	0.00120
2.3	0.00114	0.00109	0.00103	0.00098	0.00094	0.00089	0.00085	0.00080	0.00076	0.00072
2.4	0.00069	0.00065	0.00062	0.00059	0.00056	0.00053	0.00050	0.00048	0.00045	0.00043
2.5	0.00041	0.00039	0.00037	0.00035	0.00033	0.00031	0.00029	0.00028	0.00026	0.00025
2.6	0.00024	0.00022	0.00021	0.00020	0.00019	0.00018	0.00017	0.00016	0.00015	0.00014
2.7	0.00013	0.00013	0.00012	0.00011	0.00011	0.00010	0.00009	0.00009	0.00008	0.00008
2.8	0.00008	0.00007	0.00007	0.00006	0.00006	0.00006	0.00005	0.00005	0.00005	0.00004
2.9	0.00004	0.00004	0.00004	0.00003	0.00003	0.00003	0.00003	0.00003	0.00003	0.00002
3.0	0.00002	0.00002	0.00002	0.00002	0.00002	0.00002	0.00002	0.00001	0.00001	0.00001
3.1	0.00001	0.00001	0.00001	0.00001	0.00001	0.00001	0.00001	0.00001	0.00001	0.00001
3.2	0.00001	0.00001	0.00001	0.00000	0.00000	0.00000	0.00000	0.00000	0.00000	0.00000

S9 Table of Fourier transform pairs

FOURIER TRANSFORM PAIRS are useful in the harmonic analysis of functions and the solution of differential equations. This table summarizes Fourier transforms that most commonly appear in the physical sciences where the transform pairs are defined

$$F(\omega) = \frac{1}{\sqrt{2\pi}} \int_{-\infty}^{\infty} f(t)\, e^{-i\omega t}\, dt \qquad f(t) = \frac{1}{\sqrt{2\pi}} \int_{-\infty}^{\infty} F(\omega)\, e^{i\omega t}\, d\omega$$

	$f(t)$	$F(\omega)$
1.	$e^{-t^2/2}$	$e^{-\omega^2/2}$
2.	$e^{-at^2} \quad a > 0$	$\dfrac{1}{\sqrt{2a}}\, e^{-\omega^2/4a}$
3.	$e^{-a\lvert t\rvert} \quad a > 0$	$\sqrt{\dfrac{2}{\pi}}\left(\dfrac{a}{a^2 + \omega^2}\right)$
4.	$\left(\dfrac{a}{a^2 + t^2}\right) \quad \mathrm{Re}(a) > 0$	$\sqrt{\dfrac{\pi}{2}}\, e^{-a\lvert\omega\rvert}$
5.	$\theta(t+a) - \theta(t-a) = \begin{cases} 1 & -a < t < a \\ 0 & \text{otherwise} \end{cases}$	$\sqrt{\dfrac{2}{\pi}}\,\dfrac{\sin(a\omega)}{\omega}$
6.	$\dfrac{\sin(at)}{t} \quad a \in \mathfrak{R}$	$\sqrt{\dfrac{\pi}{2}}\,[\theta(\omega+a) - \theta(\omega-a)] = \begin{cases} \sqrt{\dfrac{\pi}{2}} & -a < \omega < a \\ 0 & \text{otherwise} \end{cases}$
7.	1	$\sqrt{2\pi}\,\delta(\omega)$
8.	$\delta(t+a)$	$\dfrac{1}{\sqrt{2\pi}}\, e^{ia\omega}$
9.	$e^{i\omega_0 t}$	$\sqrt{2\pi}\,\delta(\omega - \omega_0)$
10.	$\cos\omega_0 t \quad \omega_0 \in \mathfrak{R}$	$\sqrt{\dfrac{\pi}{2}}\,[\delta(\omega + \omega_0) + \delta(\omega - \omega_0)]$
11.	$\sin\omega_0 t \quad \omega_0 \in \mathfrak{R}$	$\sqrt{\dfrac{\pi}{2}}\, i\,[\delta(\omega + \omega_0) - \delta(\omega - \omega_0)]$
12.	$e^{-a\lvert t\rvert} \cos(\omega_0 t) \quad a > 0,\ \omega_0 \in \mathfrak{R}$	$\sqrt{\dfrac{2}{\pi}}\left[\left(\dfrac{a/2}{a^2 + (\omega - \omega_0)^2}\right) + \left(\dfrac{a/2}{a^2 + (\omega - \omega_0)^2}\right)\right]$
13.	$\dfrac{d}{dt} f(t) \quad \text{differentiable } f(t)$	$i\omega F(\omega)$
14.	$f(t)\, e^{-i\omega_0 t}$	$F(\omega + \omega_0)$
15.	$f(t + t_0)$	$F(\omega)\, e^{i\omega t_0}$
16.	$f(st) \quad s \in \mathfrak{R}$	$\dfrac{1}{\lvert s\rvert} F\left(\dfrac{\omega}{s}\right)$

17. $\displaystyle\int_{-\infty}^{\infty} f(s)g(t-s)\,ds = (f*g)(t)$ $\qquad\qquad$ $\sqrt{2\pi}\,F(\omega)G(\omega)$

18. $\qquad f(t)g(t)$ $\qquad\qquad$ $\displaystyle\frac{1}{\sqrt{2\pi}}\int_{-\infty}^{\infty} F(\alpha)G(\omega-\alpha)\,d\alpha = \frac{1}{\sqrt{2\pi}}(F*G)(\omega)$

S₁₀ Answers to end-of-chapter problems

COMPLETE ANSWERS to all end-of-chapter problems are presented here. In all cases, answers are provided. In select cases, complete solutions are presented.

CHAPTER 1

1.1   (a) $(x, y) = (-3.29, -3.29)$, (b) $(-2.57, 3.06)$, (c) $(3.00, 0)$

1.2   (a) $(r, \theta) = (2, 90°)$, (b) $(\sqrt{8}, 225°)$, (c) $(3, 0)$

1.3   (a) $(r, \theta, \varphi) = (2, 0, 0)$, (b) $(\sqrt{5}, 63.4°, 180°)$, (c) $(\sqrt{14}, 143.3°, 296.6°)$

1.4 $dA = r\,dr d\theta$, $dV = r\,dr d\theta dz$

1.5   (a) $x = 2/3$, (b) $x = 3, -4$, (c) $x = \left(n + \frac{1}{2}\right)\pi$ where $n \in \mathbb{Z}$ is an integer.

1.6   Functions plotted in plane polar coordinates where $\theta \in [0, 2\pi]$ with $r \geq 0$.

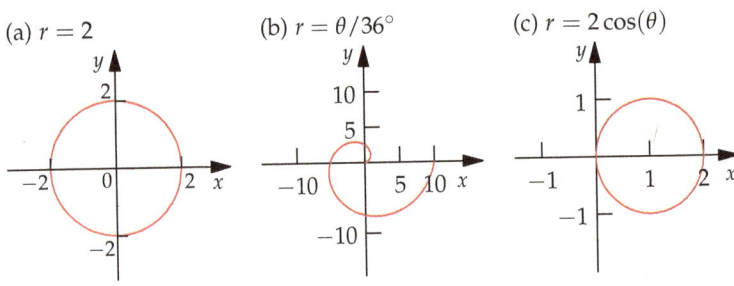

(a) $r = 2$ (b) $r = \theta/36°$ (c) $r = 2\cos(\theta)$

1.7   Functions plotted in cartesian coordinates over a range demonstrating the function's behavior.

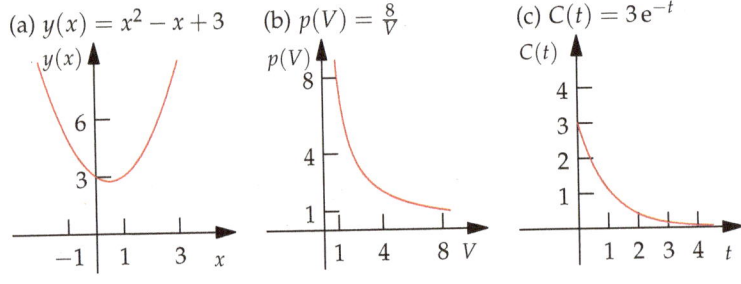

(a) $y(x) = x^2 - x + 3$ (b) $p(V) = \frac{8}{V}$ (c) $C(t) = 3e^{-t}$

1.8   Functions drawn by hand over a specified range.

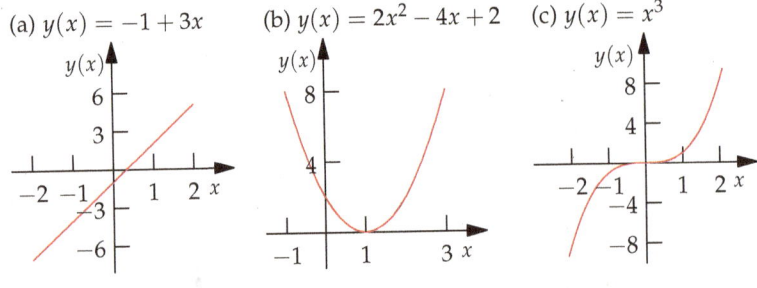

(a) $y(x) = -1 + 3x$ (b) $y(x) = 2x^2 - 4x + 2$ (c) $y(x) = x^3$

(d) $y(x) = 3\,e^{-x/2}$

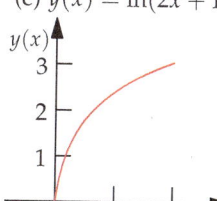

(e) $y(x) = \ln(2x+1)$

(f) $y(x) = 2\,e^{-1/x}$

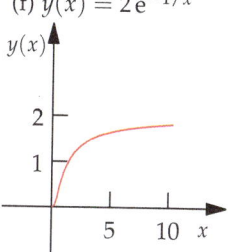

(g) $y(x) = \sin(x)$

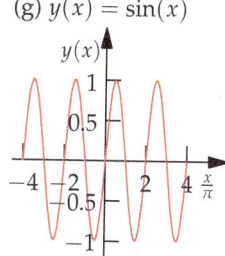

(h) $y(x) = e^{-x}\cos(2\pi x)$

(i) $y(x) = e^{-(x-1)^2}$

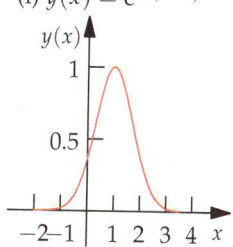

1.9 Functions plotted in cartesian coordinates over a range demonstrating the function's behavior.

(a) $V(r) = -\frac{2}{r}$

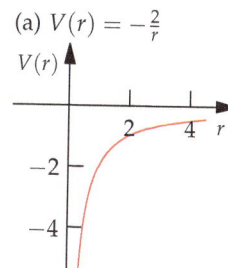

(b) $E(v) = \frac{1}{2}v^2$

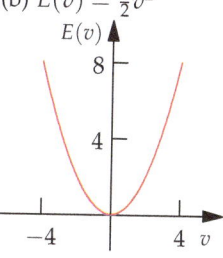

(c) $F(r) = \frac{1}{r^2}$

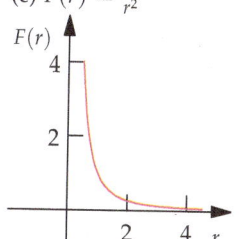

(d) $k_r(T) = e^{-\frac{1}{T}}$

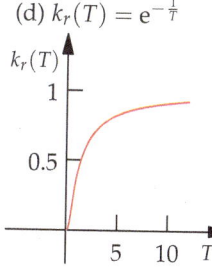

(e) $A(t) = e^{-t}$

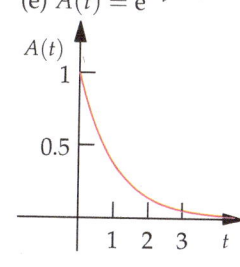

(f) $A(t) = \frac{1}{1+t}$

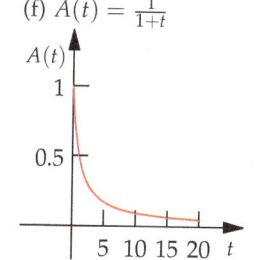

1.10 Functions plotted in cartesian coordinates over a range demonstrating the function's behavior.

(a) $V(r)$ versus $1/r$

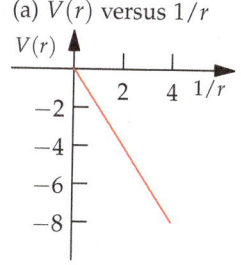

(b) $E(v)$ versus v^2

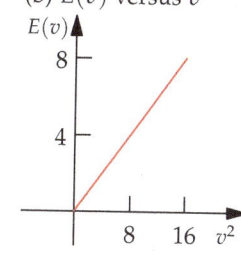

(c) $F(r)$ versus $1/r^2$

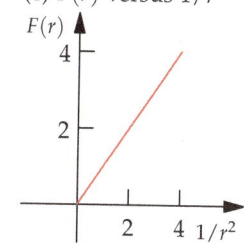

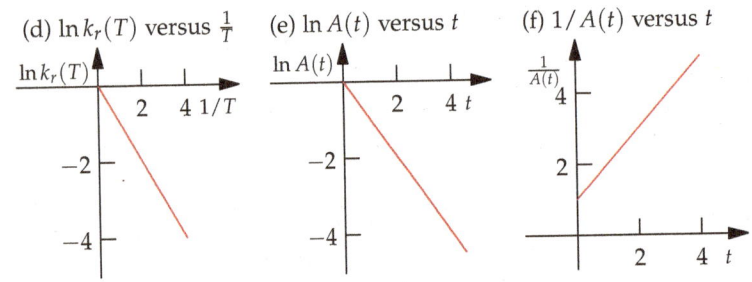

(d) $\ln k_r(T)$ versus $\frac{1}{T}$ (e) $\ln A(t)$ versus t (f) $1/A(t)$ versus t

1.11   The function $f(x) = |x| - x$ shown below is continuous everywhere and differentiable everywhere but $x = 0$.

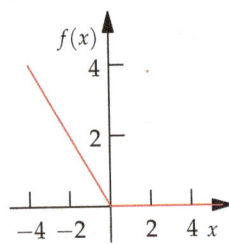

1.12 $\sinh(-x) = -\sinh(x), \cosh(x) = \cosh(-x)$

1.13   (a) odd, (b) neither, (c) even, (d) neither.

1.14   The function $f(x) = 1/(1 - e^{-\frac{1}{x}})$ shown below in red is discontinuous at $x = 0$.

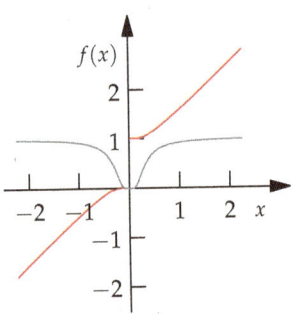

1.15   The function $y(x) = H(x) - H(x-1)$ shown below in red is also known as a "top hat" function.

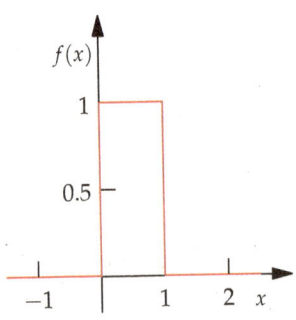

CHAPTER 2

2.1 (a) $(r,\theta) = (1,0)$, (b) $(4, \frac{\pi}{2})$, (c) $(\sqrt{2}, \frac{7\pi}{4})$, (d) $(5, 306.9°)$, (e) $(\sqrt{13}, 123.7°)$, (f) $(\sqrt{18}, \frac{5\pi}{4})$

2.2 $e^{-i\theta} = \cos(-\theta) + i\sin(-\theta) = \cos\theta - i\sin\theta$

2.3 $e^{i\theta} + e^{-i\theta} = (\cos\theta + i\sin\theta) + (\cos\theta - i\sin\theta) = 2\cos\theta$, $e^{i\theta} - e^{-i\theta} = (\cos\theta + i\sin\theta) - (\cos\theta - i\sin\theta) = 2i\sin\theta$

2.4 $\frac{1}{x+iy} = \frac{x}{x^2+y^2} - \frac{iy}{x^2+y^2}$

2.5 $|(x-1)+iy|^2 = (x-1)^2 + y^2 = 3^2$, which is the equation for a circle of radius 3 centered at $(1,0)$ shown below.

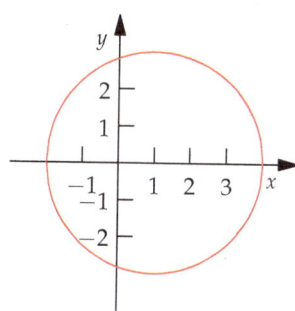

2.6 (a) 7.000, (b) 9.069, (c) 0.89, (d) −0.0607

2.7 (a) $[H^+] = 1M$, (b) $6.98 \times 10^{-3} M$, (c) 0.583M, (d) $1.33 \times 10^{-8} M$

2.8 (a) $zz^* = 1$, $z^2 = 1$, (b) $zz^* = 2$, $z^2 = 2i$, (c) $zz^* = 4$, $z^2 = -4$, (d) $zz^* = 8$, $z^2 = -8i$, (e) $zz^* = 16$, $z^2 = 16$, (f) $zz^* = 32$, $z^2 = -32i$

2.9 (a) $z = 1 + i = \sqrt{2}\, e^{i\frac{\pi}{4}}$, (b) $z = -1 - i = \sqrt{2}\, e^{i\frac{5\pi}{4}}$

2.10 $\left(e^{i\theta}\right)^n = e^{in\theta} = \cos(n\theta) + i\sin(n\theta)$

2.11 (a) $\cos(ix) = \cosh(x)$, (b) $\sinh(ix) = i\sin(x)$, (c) $\sin(ix) = i\sinh(x)$, (d) $\cosh(ix) = \cos(x)$

2.12 $(i)^i = e^{-\frac{\pi}{2}}$

2.13 (a) $\cos\alpha\cos\beta = \frac{1}{2}(e^{i\alpha} + e^{-i\alpha})\frac{1}{2}(e^{i\beta} + e^{-i\beta}) = \frac{1}{2}\cos(\alpha+\beta) + \frac{1}{2}\cos(\alpha-\beta)$, (b) $\sin\alpha\sin\beta = -\frac{1}{2}\cos(\alpha+\beta) + \frac{1}{2}\cos(\alpha-\beta)$

2.14 $w = -2.10 \times 10^3$ J

2.15 11,490 years

2.16 $\Delta S = 8.45$ J/K

2.17 (a) $\ln 1 = 0$, (b) $\ln(1+i) = 0.34 + i0.78$, (c) $\ln(2i) = 0.69 + i1.57$, (d) $\ln(-2+2i) = 1.04 + i2.36$, (e) $\ln(-4) = 1.39 + i3.14$, (f) $\ln(-4-4i) = 1.73 - i2.36$

2.18 By Landau's own estimate, Einstein was 31.6 times better than Landau.

CHAPTER 3

3.1 (a) $3x^2 + 10x - 2$, (b) $\frac{\Delta H}{RT^2}$, (c) $-\frac{M}{V^2}$, (d) $10\sec^2(2\theta)$, (e) $\frac{AB}{RT^2}e^{-\frac{B}{RT}}$, (f) $2x\left[\tan(2x) + x\sec^2(2x)\right]$,

(g) $-xe^{-x}[x\sin(x) + (x-2)\cos(x)]$, (h) $-\frac{x}{\sqrt{x^2+a^2}}e^{-\sqrt{x^2+a^2}}$, (i) $-12\frac{A}{r^{13}} + 6\frac{B}{r^7}$, (j) $5x^4\sqrt{1-e^{2x}} - \frac{x^5}{\sqrt{1-e^{2x}}}e^{2x}$,

(k) $\frac{1}{x^3}(2e^x - xe^x - 2)$, (l) $-\cos(x)e^{-\sin(x)}$, (m) $\frac{1}{x^2}[x\cos(x) - \sin(x)]$, (n) $-\ln n - 1$, (o) $\frac{1}{t}e^{-3t}(1 - 3t\ln(t))$,

(p) $-\frac{A}{p^2} + \ln p + 1$, (q) $\frac{E^2}{A}\left(2z - \frac{27}{8}\right)$, (r) $-2A\frac{n\pi}{L}\sin\left(\frac{n\pi x}{L}\right)$

3.2 (a) $\left.\frac{3}{x}\right|_{x=3} = 1$, (b) $\left.3x^2\right|_{x=3} = 27$, (c) $\left.3x^2 - 10x + 2\right|_{x=-1} = 15$, (d) $\left.\frac{x}{\sqrt{x^2-6}}\right|_{x=4} = \sqrt{\frac{8}{5}}$, (e) $\left.-5\sin(\theta)\right|_{\theta=\pi} = 0$,

(f) $\left.3\left[\cos^2(\theta) - \sin^2(\theta)\right]\right|_{\theta=\frac{\pi}{2}} = -3$

3.3 (a) $(2 - x^2)\cos(x) - 4x\sin(x)$, (b) $-2e^{-x}\cos(x)$, (c) $2\ln(x) + 3$, (d) $-\frac{e^{-x}}{(1-e^x)^2}$, (e) $-\frac{1}{(1-x^2)^{\frac{3}{2}}}$,

(f) $e^{-x^2}(4x^3 + 8x^2 - 6x - 4)$

3.4 (a) $-\frac{nRT}{V^2}$, (b) $-\frac{nRT}{(V-nb)^2} + \frac{2n^2a}{V^3}$, (c) $-\frac{pM}{RT^2}$, (d) $b + 2cT - \frac{d}{T^2}$, (e) $\frac{y}{\sqrt{x^2+y^2+z^2}}$, (f) $-r\sin(\theta)\sin(\phi)$, (g) $e^{-\frac{E_1}{kT}}\left(-\frac{1}{kT}\right)$,

(h) $e^{-\frac{E_1}{kT}}\left(\frac{E_1}{kT^2}\right) + e^{-\frac{E_2}{kT}}\left(\frac{E_2}{kT^2}\right)$

3.5 $\frac{\partial}{\partial r}\pi r^2 = 2\pi r = \frac{\partial}{\partial h}2\pi rh$

3.6 (a) $2.9 \times 10^2 \frac{m}{s}$, (b) $31.2 \times 10^{-3} \frac{J}{mol\,K}$, (c) 0.0417 K^{-1}, (d) $-8.5 \times 10^{-4} \frac{M}{min}$

3.7 $\frac{dk(T)}{dT} = \frac{E^\ddagger}{RT^2}Ae^{-\frac{E^\ddagger}{RT}}$

3.8 $\frac{\partial \rho}{\partial T} = -\frac{\rho M}{RT^2}$

3.9 $\frac{\partial Z}{\partial V} = \frac{-nb}{(V-nb)^2} + \frac{an}{V^2RT}$, $\frac{\partial Z}{\partial(1/V)} = \frac{V^2nb}{(V-nb)^2} - \frac{an}{RT}$

3.10 $\alpha = \frac{nR}{pV}$

3.11 Note $\cos(\theta) + i\sin(\theta) = e^{i\theta}$ so $\frac{df}{d\theta} = i$

3.12 $\frac{dh}{dx} = x^x(\ln x + 1)$

3.13 $\left(\frac{\partial p}{\partial T}\right)_V = \frac{R}{V-b}$, $\left(\frac{\partial p}{\partial V}\right)_T = -\frac{RT}{(V-b)^2} + \frac{2a}{V^3}$

3.14 Note $\frac{\partial V}{\partial x} = -\frac{x}{r^3}$, $\frac{\partial^2 V}{\partial x^2} = \frac{1}{r^5}(3x^2 - r^2)$ and similarly for y and z, so $\nabla^2 V = 0$

3.15 $dp(T, V) = 3.0$ J/L

3.16 $\frac{du}{dt} = e^{-3t}(1 - 3t) + e^{-2t}(2t - 2t^2) - e^{-t}$

3.17 $\frac{\partial u}{\partial s} = u(s,t)[\cos(s) + te^s]$, $\frac{\partial u}{\partial t} = u(s,t)e^s$

3.18 $\frac{\partial u}{\partial s} = u(x,y)[t\,e^s + \cos(s)]$, $\frac{\partial u}{\partial t} = u(x,y)e^s$

3.19 Note $\left(\frac{\partial T}{\partial V}\right)^{-1} = \left(\frac{p}{nR}\right)^{-1}$ so $\frac{\partial V}{\partial T} = \left(\frac{\partial T}{\partial V}\right)^{-1}$

3.20 $\frac{\partial T}{\partial V}\frac{\partial V}{\partial p}\frac{\partial p}{\partial T} = -\frac{nRT}{pV} = -1$

3.21 (a) exact, (b) exact, (c) inexact, (d) exact, (e) inexact, (f) exact

3.22 (a) $\left(\frac{\partial V}{\partial S}\right)_p = \left(\frac{\partial T}{\partial p}\right)_S$, (b) $\left(\frac{\partial S}{\partial V}\right)_T = \left(\frac{\partial p}{\partial T}\right)_V$, (c) $\left(\frac{\partial V}{\partial T}\right)_p = -\left(\frac{\partial S}{\partial p}\right)_T$

3.23 (a) $\left(\frac{\partial V}{\partial S}\right)_{p,N} = \left(\frac{\partial T}{\partial p}\right)_{S,N}$, (b) $\left(\frac{\partial S}{\partial N}\right)_{T,V} = -\left(\frac{\partial \mu}{\partial T}\right)_{V,N}$, (c) $\left(\frac{\partial \mu}{\partial T}\right)_{p,N} = -\left(\frac{\partial S}{\partial N}\right)_{T,p}$

3.24 Note $dG = dU - TdS - SdT + pdV + Vdp = Nd\mu + \mu dN$ so $-SdT + Vdp - Nd\mu = 0$

CHAPTER 4

4.1 (a) $|\mathbf{a}| = \sqrt{10}$, $\theta = 71.6°$, (b) $|\mathbf{a}| = \sqrt{8}$, $\theta = \frac{7\pi}{4} = 315°$, (c) $|\mathbf{a}| = \sqrt{13}$, $\theta = 146.3°$, (d) $|\mathbf{a}| = \sqrt{5}$, $\theta = 63.4°$, $\varphi = \pi = 180°$, (e) $|\mathbf{a}| = \sqrt{35}$, $\theta = 59.5°$, $\varphi = 258.7°$, (f) $|\mathbf{a}| = \sqrt{11}$, $\theta = 25.2°$, $\varphi = \frac{\pi}{4} = 45°$

4.2 (a) $|\mathbf{c}| = \sqrt{32}$, $\theta = 45.0°$, (b) $|\mathbf{c}| = \sqrt{17}$, $\theta = 76.0°$, (c) $|\mathbf{c}| = \sqrt{46}$, $\theta = 152.2°$, $\varphi = 198.4°$, (d) $|\mathbf{c}| = \sqrt{73}$, $\theta = 134.6°$, $\varphi = 99.5°$

4.3 (a) 8, (b) 2, (c) -36, (d) -21

4.4 (a) $|\mathbf{c}| = 5$, $\theta = \pi = 180°$, $\varphi = 0°$, (b) $|\mathbf{c}| = 2$, $\theta = \pi = 180°$, $\varphi = 0°$, (c) $|\mathbf{c}| = \sqrt{356}$, $\theta = 64.9°$, $\varphi = 339.4°$, (d) $|\mathbf{c}| = \sqrt{549}$, $\theta = 59.2°$, $\varphi = 206.6°$

4.5 (a) $\mathbf{a} \cdot \mathbf{b} = -4$, $\theta = 108.4°$, (b) $\mathbf{a} \cdot \mathbf{b} = 9$, $\theta = 15.2°$

4.6 (a) $\mathbf{a} \times \mathbf{b} = 12\hat{\mathbf{z}}$, $\theta = 71.6°$, (b) $\mathbf{a} \times \mathbf{b} = \hat{\mathbf{x}} - 2\hat{\mathbf{y}} + \hat{\mathbf{z}}$, $\theta = 15.2°$

4.7 $\mathbf{a} \cdot \mathbf{b} = a_x b_x + a_y b_y + a_z b_z = \mathbf{b} \cdot \mathbf{a}$, $\mathbf{a} \times \mathbf{b} = (a_y b_z - a_z b_y)\hat{\mathbf{x}} + (a_z b_x - a_x b_z)\hat{\mathbf{y}} + (a_x b_y - a_y b_x)\hat{\mathbf{z}} = -\mathbf{b} \times \mathbf{a}$

4.8 $\mathbf{a} \cdot \mathbf{a} = a_x^2 + a_y^2 + a_z^2 = |\mathbf{a}|^2$

4.9 $\mathbf{a} + (\mathbf{b} + \mathbf{c}) = (a_x + b_x + c_x)\hat{\mathbf{x}} + (a_y + b_y + c_y)\hat{\mathbf{y}} + (a_z + b_z + c_z)\hat{\mathbf{z}} = (\mathbf{a} + \mathbf{b}) + \mathbf{c}$

4.10 Express $\mathbf{a} \cdot (\mathbf{b} \times \mathbf{c})$ as a determinant and note that exchanging two rows changes the sign of the determinant, or expand in vector elements and cancel terms to show that $\mathbf{a} \cdot (\mathbf{b} \times \mathbf{c}) = \mathbf{b} \cdot (\mathbf{c} \times \mathbf{a}) = \mathbf{c} \cdot (\mathbf{a} \times \mathbf{b})$.

4.11 Expand $\mathbf{a} \times (\mathbf{b} \times \mathbf{c})$ in vector elements and cancel terms to arrive at $(\mathbf{a} \cdot \mathbf{c})\mathbf{b} - (\mathbf{a} \cdot \mathbf{b})\mathbf{c}$.

4.12 Expand $(\mathbf{a} \cdot \mathbf{c})(\mathbf{b} \cdot \mathbf{d}) - (\mathbf{a} \cdot \mathbf{d})(\mathbf{b} \cdot \mathbf{c})$ in vector elements and cancel terms to arrive at $(\mathbf{a} \times \mathbf{b}) \cdot (\mathbf{c} \times \mathbf{d})$.

4.13 $l_x = (yp_z - zp_y), l_y = (zp_x - xp_z), l_z = (xp_y - yp_x)$

4.14 $V = 9$

4.15 Expand $\mathbf{a} \cdot \mathbf{b}$ and cancel terms to find $\mathbf{a} \cdot \mathbf{b} = \frac{1}{4}(\mathbf{q}_1 \cdot \mathbf{q}_1 - \mathbf{q}_2 \cdot \mathbf{q}_2 + \mathbf{q}_3 \cdot \mathbf{q}_3 - \mathbf{q}_4 \cdot \mathbf{q}_4) = 0$. Similar reasoning leads to $\mathbf{a} \cdot \mathbf{a} = \mathbf{b} \cdot \mathbf{b} = 1$.

4.16 $\mathbf{F} = qE_x\hat{\mathbf{i}} + q(E_y + v_z B_x)\hat{\mathbf{j}}$

4.17 $\mathbf{l} = m(r_x v_y - r_y v_x)\hat{\mathbf{k}}$

4.18 $V = -\mu_E E_y$

4.19 $\mathbf{T} = \mu_B B_z\hat{\mathbf{j}}$

4.20   (a) $\mathbf{u}_1 = \hat{\mathbf{x}} - 2\hat{\mathbf{y}}$, $\mathbf{u}_2 = \frac{8}{5}(2\hat{\mathbf{x}} + \hat{\mathbf{y}})$, (b) $\mathbf{e}_1 = \frac{1}{\sqrt{5}}(\hat{\mathbf{x}} - 2\hat{\mathbf{y}})$, $\mathbf{e}_2 = \frac{1}{\sqrt{5}}(2\hat{\mathbf{x}} + \hat{\mathbf{y}})$

4.21   (a) $\mathbf{u}_1 = \hat{\mathbf{x}} - 2\hat{\mathbf{y}} + \hat{\mathbf{z}}$, $\mathbf{u}_2 = \frac{1}{2}(\hat{\mathbf{x}} - \hat{\mathbf{z}})$, $\mathbf{u}_3 = \frac{1}{3}(\hat{\mathbf{x}} + \hat{\mathbf{y}} + \hat{\mathbf{z}})$, (b) $\mathbf{e}_1 = \frac{1}{\sqrt{6}}(\hat{\mathbf{x}} - 2\hat{\mathbf{y}} + \hat{\mathbf{z}})$, $\mathbf{e}_2 = \frac{1}{\sqrt{2}}(\hat{\mathbf{x}} - \hat{\mathbf{z}})$, $\mathbf{e}_3 = \frac{1}{\sqrt{3}}(\hat{\mathbf{x}} + \hat{\mathbf{y}} + \hat{\mathbf{z}})$

CHAPTER 5

5.1 $2a$

5.2 $\frac{\partial^2 g}{\partial x \partial y} = \frac{\partial^2 g}{\partial y \partial x} = 2x + 1$

5.3 $\hat{p}\psi(x) = \hbar k\,\psi(x)$

5.4 $\hat{H}\psi(x) = \frac{h^2}{8mL^2}\,\psi(x)$

5.5   (a) $\hat{A}\hat{A} - \hat{A}\hat{A} = 0$, (b) $\hat{A}\hat{B} - \hat{B}\hat{A} = -(\hat{B}\hat{A} - \hat{A}\hat{B})$, (c) Expand $\hat{A}(\hat{B} + \hat{C}) - (\hat{B} + \hat{C})\hat{A}$ and reorganize terms, (d) Expand $(a + \hat{A})(b + \hat{B}) - (b + \hat{B})(a + \hat{A})$ and reorganize terms, (e) Expand $(\hat{A}\hat{B} - \hat{B}\hat{A})\hat{C} + \hat{B}(\hat{A}\hat{C} - \hat{C}\hat{A})$ and reorganize terms, (f) Expand $(\hat{A}\hat{C} - \hat{C}\hat{A})\hat{B} + \hat{A}(\hat{B}\hat{C} - \hat{C}\hat{B})$ and reorganize terms.

5.6   Evaluate $\nabla \cdot \nabla \varphi = \left(\hat{\mathbf{x}}\frac{\partial}{\partial x} + \hat{\mathbf{y}}\frac{\partial}{\partial y} + \hat{\mathbf{z}}\frac{\partial}{\partial z}\right) \cdot \left(\hat{\mathbf{x}}\frac{\partial \varphi}{\partial x} + \hat{\mathbf{y}}\frac{\partial \varphi}{\partial y} + \hat{\mathbf{z}}\frac{\partial \varphi}{\partial z}\right)$.

5.7   Evaluate $\nabla(\psi\,\varphi) = \varphi\left(\hat{\mathbf{x}}\frac{\partial \psi}{\partial x} + \hat{\mathbf{y}}\frac{\partial \psi}{\partial y} + \hat{\mathbf{z}}\frac{\partial \psi}{\partial z}\right) + \psi\left(\hat{\mathbf{x}}\frac{\partial \varphi}{\partial x} + \hat{\mathbf{y}}\frac{\partial \varphi}{\partial y} + \hat{\mathbf{z}}\frac{\partial \varphi}{\partial z}\right)$.

5.8 $\mathbf{F} = -2x\,\hat{\mathbf{x}} - 2y\,\hat{\mathbf{y}} - 2z\,\hat{\mathbf{z}}$

5.9   (a) $3x^2$, (b) $-4\cos(2x)$, (c) $a\,e^{ax}\sin(bx) + b\,e^{ax}\cos(bx)$, (d) $3x^2 y$, (e) $2y^3$, (f) $12x^3 y^2$

5.10 $[\hat{p}, \hat{x}]\varphi(x) = -i\hbar\varphi(x)$

5.11   For $V(x, y) = \ln\left(\sqrt{x^2 + y^2}\right)$ we find $\mathbf{f}(x, y) = \nabla V = \hat{\mathbf{x}}\frac{x}{x^2 + y^2} + \hat{\mathbf{y}}\frac{y}{x^2 + y^2} = \mathbf{r}\frac{1}{r^2} = \hat{\mathbf{r}}\frac{1}{r}$. It is helpful to note that $\frac{\partial}{\partial x}\ln\left(\sqrt{x^2 + y^2}\right) = \frac{x}{x^2 + y^2}$ and similarly for y. In addition, $\nabla^2 V = \nabla \cdot \mathbf{f} = \frac{\partial}{\partial x}\left(\frac{x}{x^2 + y^2}\right) + \frac{\partial}{\partial y}\left(\frac{y}{x^2 + y^2}\right) = 0$. It is helpful to note that $\frac{\partial}{\partial x}\left(\frac{x}{x^2 + y^2}\right) = \frac{1}{x^2 + y^2} - \frac{2x^2}{(x^2 + y^2)^2}$ and similarly for y.

5.12   For $V(r, \theta) = \ln(r)$ we find $\nabla V(r, \theta) = \mathbf{f}(r, \theta) = \hat{\mathbf{r}}\frac{1}{r}$. As such, $\nabla^2 V(r, \theta) = \nabla \cdot \mathbf{f}(r, \theta) = \frac{1}{r}\frac{\partial}{\partial r}(1) = 0$.

5.13   For $V(x, y, z) = \frac{1}{\sqrt{x^2 + y^2 + z^2}}$ we find $\mathbf{f}(x, y, z) = \nabla V = \hat{\mathbf{x}}\frac{-x}{(x^2 + y^2 + z^2)^{\frac{3}{2}}} + \hat{\mathbf{y}}\frac{-y}{(x^2 + y^2 + z^2)^{\frac{3}{2}}} + \hat{\mathbf{z}}\frac{-z}{(x^2 + y^2 + z^2)^{\frac{3}{2}}} = -\mathbf{r}\frac{1}{r^3} = -\hat{\mathbf{r}}\frac{1}{r^2}$. It is helpful to note that $\frac{\partial}{\partial x}(x^2 + y^2 + z^2)^{-\frac{1}{2}} = \frac{-x}{(x^2 + y^2 + z^2)^{\frac{3}{2}}}$ and similarly for y and z. In addition, $\nabla^2 V = \nabla \cdot \mathbf{f} = \frac{\partial}{\partial x}\left(\frac{-x}{(x^2 + y^2 + z^2)^{\frac{3}{2}}}\right) + \frac{\partial}{\partial y}\left(\frac{-y}{(x^2 + y^2 + z^2)^{\frac{3}{2}}}\right) + \frac{\partial}{\partial z}\left(\frac{-z}{(x^2 + y^2 + z^2)^{\frac{3}{2}}}\right) = 0$. It is helpful to note that $\frac{\partial}{\partial x}\left(-x(x^2 + y^2 + z^2)^{-\frac{3}{2}}\right) = \frac{-1}{(x^2 + y^2 + z^2)^{\frac{3}{2}}} + \frac{3x^2}{(x^2 + y^2 + z^2)^{\frac{5}{2}}}$ and similarly for y and z.

5.14   For $V(r, \theta, \varphi) = \frac{1}{r}$ we find $\nabla V(r, \theta, \varphi) = \mathbf{f}(r, \theta, \varphi) = -\hat{\mathbf{r}}\frac{1}{r^2}$. As such, $\nabla^2 V(r, \theta, \varphi) = \nabla \cdot \mathbf{f}(r, \theta, \varphi) = \frac{1}{r^2}\frac{\partial}{\partial r}(1) = 0$.

5.15 $\hat{M}_x = -i\hbar\left(y\frac{\partial}{\partial z} - z\frac{\partial}{\partial y}\right)$, $\hat{M}_y = -i\hbar\left(z\frac{\partial}{\partial x} - x\frac{\partial}{\partial z}\right)$, $\hat{M}_z = -i\hbar\left(x\frac{\partial}{\partial y} - y\frac{\partial}{\partial x}\right)$.

5.16   Start from $\mathbf{E} = -\nabla V = -\left(\hat{\mathbf{i}}\frac{\partial}{\partial x} + \hat{\mathbf{j}}\frac{\partial}{\partial y} + \hat{\mathbf{k}}\frac{\partial}{\partial z}\right)\left((\mu_x x + \mu_y y + \mu_z z)\frac{1}{r^3}\right)$. Note that $\frac{\partial}{\partial x}\left(\frac{1}{r^3}\right) = -\frac{3x}{r^5}$ and similarly for y and z. Combine to show that $\mathbf{E} = \frac{3}{r^5}(\boldsymbol{\mu} \cdot \mathbf{r})\mathbf{r} - \frac{\boldsymbol{\mu}}{r^3}$.

CHAPTER 6

6.1 (a) $x^* = \frac{5}{6}$, $y(x^*) = -\frac{1}{12}$, (b) $v^* = \sqrt{\frac{2kT}{m}}$, $p(v^*) = \frac{1}{\pi e}\sqrt{\frac{m}{2\pi kT}}$, (c) $r^* = 2^{\frac{1}{6}}\sigma$, $V(r^*) = -\epsilon$, (d) $E^* = \frac{kT}{2}$, $p(E^*) = \sqrt{\frac{2}{e\pi}}\frac{1}{kT}$

6.2 $x^* = \frac{1}{2}a$

6.3 $\left.\frac{dp}{dr}\right|_{r=r^*} = 0$ at $r^* = a_0$

6.4 $x^* = y^* = z^* = V_0^{\frac{1}{3}}$, $A(x^*, y^*, z^*) = 6V_0^{\frac{2}{3}}$

6.5 $f_x = f_y = 0$ at $(x^*, y^*) = (0,0)$, $D = -4$, and $(x^*, y^*) = (0,0)$ is a saddle point as shown as a black dot on the countour plot.

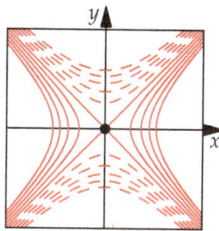

6.6 $\left.\frac{dV}{dr}\right|_{r=r^*} = 0$ at $r^* = r_0$, $\left.\frac{d^2V}{dr^2}\right|_{r=r^*} = 2AB^2 > 0$, and $r^* = r_0$ is a minimum.

6.7 (x^*, y^*) at $(1,1), (-1,1), (-1,-1)$ and $(1,-1)$ with two maxima $f(1,1) = f(-1,-1) = 1$ and two minima $f(-1,1) = f(1,-1) = -1$.

6.8 (x^*, y^*, z^*) at $\left(\frac{1}{\sqrt{2}}, \frac{1}{\sqrt{2}}, \sqrt{2}\right)$ and $\left(-\frac{1}{\sqrt{2}}, -\frac{1}{\sqrt{2}}, -\sqrt{2}\right)$ with a maximum $f\left(\frac{1}{\sqrt{2}}, \frac{1}{\sqrt{2}}, \sqrt{2}\right) = 3\sqrt{2}$ and a minimum $f\left(-\frac{1}{\sqrt{2}}, -\frac{1}{\sqrt{2}}, -\sqrt{2}\right) = -3\sqrt{2}$.

6.9 $h^* = \left(\frac{6V_0}{\pi}\right)^{\frac{1}{3}}$, $r^* = \left(\frac{3V_0}{\sqrt{2}\pi}\right)^{\frac{1}{3}}$, $A(r^*, h^*) = \sqrt{3}\pi\left(\frac{3V_0}{\sqrt{2}\pi}\right)$

6.10 $f(1,1) = 1$ maximum, $f(1,-1) = -1$ minimum, $f(-1,1) = -1$ minimum, $f(-1,-1) = 1$ maximum

6.11 (a) $\ln W = N\ln N - \sum_{k=1}^{M} n_k\ln n_k$, (b) $n_1^* = n_2^* = \ldots = n_M^* = \frac{1}{N}$

6.12 Evaluating $\left.\frac{d\rho(x)}{dx}\right|_{x=x^*} = 0$ we find $x^* = 5\left(1 - e^{-x^*}\right)$. We solve for x^* iteratively, guessing x^*, substituting

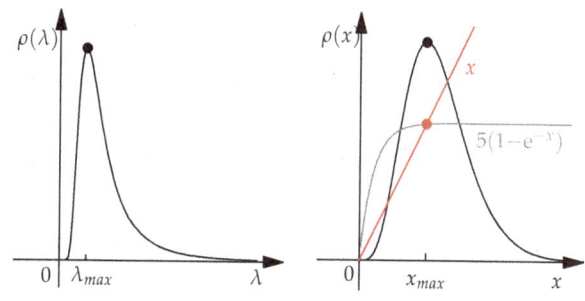

our guess into the left-hand side of $5\left(1 - e^{-x^*}\right) = x^*$ to find a new value of x^*, and repeating to convergence. We find $x_{\max} = x^* \approx 4.9651$ so that $\lambda_{\max} \approx 0.20140\,\beta hc$ (see figure above).

6.13 $r^* = \frac{e^2}{4\pi\epsilon_0 m_e(v^*)^2}$, $v^* = \frac{nh}{2\pi m_e r^*}$ (a) $r^* = \left(\frac{4\pi\epsilon_0\hbar^2}{m_e e^2}\right)n^2 = a_0\,n^2$, (b) $E(r^*, v^*) = -\left(\frac{m_e e^4}{32\pi^2\epsilon_0^2\hbar^2}\right)\frac{1}{n^2}$

CHAPTER 7

7.1 (a) $\frac{1}{2}mv^2 + C$, (b) $-\frac{1}{2x^2} + C$, (c) $-\frac{1}{3}\cos(3x) + C$, (d) $9x^4 + 40x^3 + 50x^2 + C$, (e) $-kTe^{-\frac{\epsilon}{kT}} + C$, (f) $\frac{1}{2\pi\nu}\sin(2\pi\nu t) + C$, (g) $RT\ln(p) + C$, (h) $\frac{1}{6}\kappa x^3 + C$, (i) $-\frac{q^2}{4\pi\epsilon_0 r} + C$

7.2 (a) $\frac{1}{4}\sin(2x) + \frac{1}{16}\sin(8x) + C$, (b) $-\frac{1}{\alpha^5}e^{-\alpha x}\left[\alpha^4 x^4 + 4\alpha^3 x^3 + 12\alpha^2 x^2 + 24\alpha x + 24\right] + C$, (c) $\frac{x}{2} - \left(\frac{L}{4n\pi}\right)\sin\left(\frac{2n\pi}{L}x\right) + C$, (d) $\frac{1}{n-1}(\alpha - x)^{1-n} + C$, (e) $-\sigma^2 e^{-\frac{y^2}{2\sigma^2}} + C$, (f) $\frac{-1}{\gamma^2+1}e^{-\gamma x}\left[\gamma\sin(x) + \cos(x)\right] + C$, (g) $\ln\left(\frac{4-x}{3-x}\right) + C$, (h) $-\frac{1}{4}\cos^4\varphi + C$, (i) $\frac{1}{4}\sin(2x)\left(2x^3 - 3x\right) + \frac{3}{8}\cos(2x)\left(2x^2 - 1\right) + C$.

7.3 (a) 0, (b) $RT\ln\left(\frac{p_2}{p_1}\right)$, (c) $nRT\ln\left(\frac{V_2-nb}{V_1-nb}\right) + n^2 a\left(\frac{1}{V_2} - \frac{1}{V_1}\right)$, (d) $\frac{L^3}{6}\left(1 - \frac{3}{2\pi^2 n^2}\right)$, (e) $\frac{\Delta H}{R}\left(\frac{1}{T_1} - \frac{1}{T_2}\right)$, (f) $a(T_2 - T_1) + \frac{b}{2}(T_2^2 - T_1^2) + \frac{c}{3}(T_2^3 - T_1^3) + d\ln\left(\frac{T_2}{T_1}\right)$, (g) $\left(\frac{a_0}{2}\right)^2$, (h) $2\left(\frac{kT}{m}\right)^2$, (i) $\frac{1}{a}$

7.4 (a) $\frac{1}{6}x^3 y^2 + Cy + C'$, (b) $\frac{1}{3}x^3 y + \frac{1}{3}y^3 x + Cy + C'$, (c) $\frac{1}{2}y^2\left[x\ln x - x\right] + Cx + C'$, (d) $\frac{1}{4}e^{2x}(2x^2 - 2x + 1)(y\ln y - y)z + Cyz + C'z + C''$, (e) 2, (f) $\frac{4\pi V^3}{3}$, (g) π, (h) $\left(\frac{2\pi kT}{m}\right)^{\frac{3}{2}}$

7.5 $I = \frac{8}{5}$

7.6 $E_n = \frac{n^2\hbar^2\pi^2}{2mL^2} = \frac{n^2 h^2}{8mL^2}$.

7.7 $\frac{4\pi a^4}{15}$

7.8 $\frac{\pi}{3}$

7.9 $\frac{a^2 b^4}{16}$

7.10 (a) The square well potential energy as a function of r over the range $0 < r < 2\lambda\sigma$ taking $\lambda = 2$ is shown below (red) alongside the Lennard-Jones potential (blue).

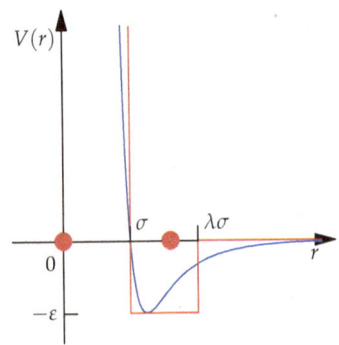

(b) $B(T) = -2\pi\int_0^\infty \left(e^{-\beta V(r)} - 1\right) r^2 dr$ as integrating over θ and φ results in a factor of 4π, (c) $B(T) = 2\pi\int_0^\sigma r^2 dr - 2\pi\int_\sigma^{\lambda\sigma}\left(e^{\beta\epsilon} - 1\right) r^2 dr = \frac{2\pi\sigma^3}{3}\left[1 - (\lambda^3 - 1)(e^{\beta\epsilon} - 1)\right]$, (d) $B_2(T) = \frac{2\pi\sigma^3}{3}$

7.11 $I_n = \int_0^\infty r^n e^{-\beta r} dr = \left(-\frac{d}{d\beta}\right)^n \int_0^\infty e^{-\beta r} dr = \left(-\frac{d}{d\beta}\right)^n \frac{1}{\beta} = \frac{n!}{\beta^{n+1}}$ so that $-\frac{d}{d\beta}\frac{1}{\beta} = \frac{1}{\beta^2}$, $\left(-\frac{d}{d\beta}\right)^2 \frac{1}{\beta} = \frac{2}{\beta^3}$, $\left(-\frac{d}{d\beta}\right)^3 \frac{1}{\beta} = \frac{2\cdot3}{\beta^4}$,

confirming the general result.

7.12 $I_{2n} = \int_0^\infty x^{2n} e^{-\alpha x^2} dx = \left(-\frac{d}{d\alpha}\right)^n \int_0^\infty e^{-\alpha x^2} dx = \left(-\frac{d}{d\alpha}\right)^n \frac{1}{2}\sqrt{\frac{\pi}{\alpha}} = \frac{(2n-1)!!}{2^{n+1}\alpha^{(2n+1)/2}}\sqrt{\pi}$, so that $-\frac{d}{d\alpha}\alpha^{-\frac{1}{2}} = \frac{1}{2}\alpha^{-\frac{3}{2}}$,

$\left(-\frac{d}{d\alpha}\right)^2 \alpha^{-\frac{1}{2}} = \frac{1\cdot3}{2^2}\alpha^{-\frac{5}{2}}$, $\left(-\frac{d}{d\alpha}\right)^3 \alpha^{-\frac{1}{2}} = \frac{1\cdot3\cdot5}{2^3}\alpha^{-\frac{7}{2}}$, confirming the general result.

7.13 $I_{2n+1} = \int_0^\infty x^{2n+1} e^{-\alpha x^2} dx = \left(-\frac{d}{d\alpha}\right)^n \int_0^\infty x e^{-\alpha x^2} dx = \left(-\frac{d}{d\alpha}\right)^n \frac{1}{2\alpha} = \frac{n!}{2\alpha^{n+1}}$, so that $-\frac{d}{d\alpha}\frac{1}{2\alpha} = \frac{1}{2\alpha^2}$, $\left(-\frac{d}{d\alpha}\right)^2 \frac{1}{2\alpha} = \frac{2}{2\alpha^3}$,

$\left(-\frac{d}{d\alpha}\right)^3 \frac{1}{2\alpha} = \frac{2\cdot3}{2\alpha^4}$, confirming the general result.

7.14 (a) $\cos(\pi) = -1$, (b) $\frac{1}{e}$, (c) $2\pi a$, (d) $4L$, (e) $16L^2$, (f) 0

7.15 (a) $|\mathbf{J}| = r^2 \sin\theta$, so that $dx\,dy\,dz = r^2 \sin\theta\,dr\,d\theta\,d\varphi$, (b) $|\mathbf{J}| = r$, so that $dx\,dy\,dz = r\,dr\,d\theta\,dz$

Chapter 8

8.1 (a) $u_n = 2n - 1$, divergent, (b) $u_n = \frac{1}{n!}$, convergent, e, (c) $u_n = (-1)^n \frac{1}{n}$, convergent, $\ln(2)$, (d) $u_n = \frac{3}{2n}$, harmonic series, ∞, (e) $u_n = 4\frac{1}{3^n}$, geometric series, 6, (f) $u_n = \frac{n}{3n+1}$, divergent

8.2 (a) $u_n = \frac{1}{2^n}, r = \frac{1}{2}$ convergent, (b) $u_n = \frac{3^n}{n}, r = 3$ divergent , (c) $u_n = \frac{n}{n+1}, r = 1$ inconclusive, (d) $u_n = \frac{2^n}{n^2}, r = 2$ divergent, (e) $u_n = \frac{n^2}{2^n}, r = \frac{1}{2}$ convergent, (f) $u_n = \frac{3^n}{n!}, r = 0$ convergent

8.3 (a) $1 + x + x^2 + \ldots = \frac{1}{1-x}, |a_n| = 1, |x| < 1$, (b) $1 - x^2 + \frac{1}{2!}x^4 - + \ldots = e^{-x^2}, |a_n| = \frac{1}{n!}, |x| < \infty$, (c) $1 + x + \frac{1}{2!}x^2 + \ldots = e^x, |a_n| = \frac{1}{n!}, |x| < \infty$, (d) $x - \frac{1}{2!}x^2 + - \ldots = 1 - e^x, |a_n| = \frac{1}{n!}, |x| < \infty$, (e) $1 - 2x + - \ldots = \frac{1}{(1+x)^2}, |a_n| = n+1, |x| < 1$, (f) $x - \frac{1}{3}x^3 + - \ldots = \tan^{-1}(x), |a_n| = \frac{1}{2n-1}, |x| < 1$

8.4 (a) $\frac{1}{1+x} = 1 - x + x^2 - x^3 + - \ldots$, (b) $\frac{1}{(1+x)^2} = 1 - 2x + 3x^2 - 4x^3 + - \ldots$, (c) $(1+x)^{\frac{1}{2}} = 1 + \frac{x}{2} - \frac{x^2}{8} + \frac{x^3}{16} - \frac{5}{128}x^4 + - \ldots$, (d) $\ln(1-x) = -x - \frac{1}{2}x^2 - \frac{1}{3}x^3 - \ldots$, (e) $e^{-x^2} = 1 - x^2 + \frac{1}{2!}x^4 - \frac{1}{3!}x^6 + - \ldots$, (f) $a^x = 1 + \ln(a)x + \frac{1}{2!}\ln^2(a)x^2 + \frac{1}{3!}\ln^3(a)x^3 + \ldots$, (g) $\cos x = 1 - \frac{1}{2!}x^2 + \frac{1}{4!}x^4 - \frac{1}{6!}x^6 + - \ldots$, (h) $(1+x)^3 = 1 + 3x + 3x^2 + x^3$

8.5 $\ln(1 - X_B) \approx -X_B - \mathcal{O}(X_B^2)$

8.6 $\sin\theta \approx \theta + \mathcal{O}(\theta^3)$

8.7 (a) $e = \sum_{n=0}^\infty \frac{1}{n!}$, (b) $\ln(2) = -\sum_{n=1}^\infty (-1)^n \frac{1}{n}$, (c) $\tan^{-1}(1) = \frac{\pi}{4} = \sum_{n=0}^\infty (-1)^n \frac{1}{2n+1}$

8.8 (a) $\sum_{n=1}^\infty nx^n$ with $a_n = n$ is convergent for $|x| < \lim_{n\to\infty}\left|\frac{n}{n+1}\right| = 1$, (b) $\frac{d}{dx}\left(\frac{1}{1-x}\right) = \frac{1}{(1-x)^2} = \frac{d}{dx}\sum_{n=0}^\infty x^n = \sum_{n=0}^\infty nx^{n-1} = 1 + 2x + 3x^2 + \ldots$ and multiplying by x we find $x\frac{1}{(1-x)^2} = x\sum_{n=1}^\infty nx^{n-1} = \sum_{n=1}^\infty nx^n$.

8.9 Starting from $\frac{x}{(1-x)^2} = \sum_{n=1}^\infty nx^n$, note that $\frac{d}{dx}\left[\frac{x}{(1-x)^2}\right] = \frac{1+x}{(1-x)^3} = \frac{d}{dx}\sum_{n=1}^\infty nx^n = \sum_{n=1}^\infty n^2 x^{n-1}$, so that $x\frac{1+x}{(1-x)^3} = x\sum_{n=1}^\infty n^2 x^{n-1} = \sum_{n=1}^\infty n^2 x^n$.

8.10   We can expand $\sinh x$ as a Maclaurin power series as $\sinh(x) = x + \frac{1}{3!}x^3 + \frac{1}{5!}x^5 + \ldots = \sum_{n=0}^{\infty} \frac{1}{(2n+1)!} x^{2n+1}$.

8.11 $\sin(x) = \sin a + \cos(a)(x-a) - \frac{1}{2!}\sin(a)(x-a)^2 - \frac{1}{3!}\cos(a)(x-a)^3 + \frac{1}{4!}\sin(a)(x-a)^4 + \ldots$

8.12   (a) $I(x) = x - \frac{x^3}{3} + \frac{x^4}{10} - \frac{x^7}{42}$, (b) $I\left(\frac{1}{3}\right) = \dfrac{147604}{459270} = \dfrac{73802}{229635} = 0.32139$, which we compare to Supplement S_7,

where $\mathrm{erf}(0.33) = 0.35928$, leading to $I\left(\frac{1}{3}\right) = \frac{\sqrt{\pi}}{2}0.35928 = 0.31840$, good to only two significant figures,

(c) $I\left(\frac{1}{2}\right) = \dfrac{4133}{8960} = 0.46127$, which we compare to the exact value 0.46128 as well as Supplement S_7, where

$\mathrm{erf}(0.50) = 0.52050$, leading to $I\left(\frac{1}{2}\right) = \frac{\sqrt{\pi}}{2}0.52050 = 0.46128$, good to five significant figures.

8.13   (a) $\ln(x) = (x-1) - \frac{1}{2}(x-1)^2 + \frac{1}{3}(x-1)^3 - + \ldots = -\sum_{n=1}^{\infty}(-1)^n \frac{1}{n}(x-1)^n$, (b) Substituting $1+x$ for x leads

to $\ln(1+x) = -\sum_{n=1}^{\infty}(-1)^n \frac{1}{n}x^n$. The Maclaurin power series expansion leads to the same result, (c) intervals of

convergence $|x| < 1$ or $|x-1| < 1$ for the series in (a) and (b), respectively.

8.14   Starting from $\rho(\nu, T) = \dfrac{8\pi h}{c^3}\dfrac{\nu^3}{e^{\beta h\nu}-1}$ and noting that $e^{\beta h\nu} \approx 1 + \beta h\nu$ we find $\rho(\nu, T) \approx \dfrac{8\pi h}{c^3}\dfrac{\nu^3}{1+\beta h\nu - 1} =$

$\dfrac{8\pi k_B T}{c^3}\nu^2$.

8.15   (a) 1, (b) $\kappa = 8$, (c) $\kappa = 4$

8.16   Starting from $C_V(T) = 3R\left(\dfrac{\Theta_E}{T}\right)^2 \dfrac{e^{-\Theta_E/T}}{(1-e^{-\Theta_E/T})^2}$ and noting that $e^{-\Theta_E/T} \approx 1 - \Theta_E/T$ we find $C_V(T) \approx$

$3R\left(\dfrac{\Theta_E}{T}\right)^2 \dfrac{e^{-\Theta_E/T}}{(1-(1-\Theta_E/T))^2} = 3R\,e^{-\Theta_E/T} \approx 3R$.

8.17   Expanding $(1+2\kappa a)^{\frac{1}{2}} = 1 + \kappa a - \frac{1}{2}(\kappa a)^2 + \frac{1}{2}(\kappa a)^3 + \ldots$ leads to $U(\kappa, a) \approx \dfrac{\kappa^3}{8\pi\beta} + \mathcal{O}(\kappa a)$.

8.18   (a) $q = \sum_{n=0}^{\infty} e^{-\beta(n+\frac{1}{2})h\nu} = e^{-\frac{1}{2}\beta h\nu}\sum_{n=0}^{\infty} e^{-n\beta h\nu} = e^{-\frac{1}{2}\beta h\nu}\dfrac{1}{1-e^{-\beta h\nu}}$, (b) $\langle\epsilon\rangle = \frac{1}{q}\,h\nu\,e^{-\frac{1}{2}\beta h\nu}\sum_{n=0}^{\infty}\left(n+\frac{1}{2}\right)e^{-n\beta h\nu}$,

evaluating the two series and combining results we find $\langle\epsilon\rangle = h\nu\left(\frac{1}{2} + \dfrac{e^{-\beta h\nu}}{1-e^{-\beta h\nu}}\right)$.

8.19   (a) After n iterations, $A_n = A_0\left[1 + \frac{1}{3}\left(1 + \frac{4}{9} + \frac{4^2}{9^2} + \ldots + \frac{4^n}{9^n}\right)\right] = A_0\left[1 + \frac{1}{3}\sum_{k=0}^{\infty}\left(\frac{4}{9}\right)^k\right]$, (b) $A_\infty = \frac{8}{5}A_0$,

(c) $P_n = N_n S_n = 3 \cdot 4^n \frac{1}{3^n}S_0$, (d) $P_\infty = \lim_{n\to\infty} 3\left(\frac{4}{3}\right)^n S_0 = \infty$.

8.20   (a) After n iterations, $A_n = 1 - \frac{1}{9}\left(1 + \frac{8}{9} + \frac{8^2}{9^2} + \ldots + \frac{8^{n-1}}{9^{n-1}}\right) = 1 - \frac{1}{9}\sum_{k=0}^{n-1}\left(\frac{8}{9}\right)^k = \left(\frac{8}{9}\right)^n$ and $H_n = 1 - A_n =$

$\frac{1}{9}\sum_{k=0}^{n-1}\left(\frac{8}{9}\right)^k = 1 - \left(\frac{8}{9}\right)^n$, (b) $H_\infty = \frac{1}{9}\left(\dfrac{1}{1-\frac{8}{9}}\right) = 1$ and $A_\infty = 0$.

8.21   (a) After n iterations, $V_n = \left(\frac{20}{27}\right)^n$, (b) $V_n = N_n L_n^3 = \left(\frac{20}{27}\right)^n$ so that $N_n = 20^n$, $L_n^3 = \left(\frac{1}{3^3}\right)^n$, and $L_n = \frac{1}{3^n}$,

(c) $d \equiv -\lim_{n\to\infty} \dfrac{\ln N_n}{\ln L_n} = \dfrac{\ln 20}{\ln 3} = \log_3(20) = 2.7268\ldots$

8.22   (a) After n iterations, $A_n = A_0 \frac{3^n}{4^n}$, (b) $A_n = N_n L_n^2$ so that $N_n = 3^n$, $L_n^2 = \frac{1}{4^n}$, and $L_n = \frac{1}{2^n}$,

(c) $d \equiv -\lim_{n\to\infty} \frac{\ln N_n}{\ln L_n} = \dfrac{\ln 3}{\ln 2} = \log_2(3) = 1.5849\ldots$

CHAPTER 9

9.1 (a) $p(5,5;10) = \frac{1}{2^{10}} \times \frac{10!}{5!5!} = \frac{1}{1024} \times \frac{3,628,800}{120 \times 120} = \frac{63}{256} \approx 0.246$, (b) $p(1,4;5) = \frac{1}{2^5} \times \frac{5!}{4!1!} = \frac{1}{120} \times \frac{120}{24 \times 1} = \frac{5}{32} \approx 0.156$,

(c) $\frac{1}{2^{10}} = \frac{1}{1024} \approx 9.76 \times 10^{-4}$

9.2

N	$N!$	$\Delta = |\ln N! - (N \ln N - N)|$	$\Delta / \ln N!$
10	$3,628,800$	2.08	0.138
50	3.04×10^{64}	2.88	0.0194
100	9.33×10^{157}	3.22	0.00886

9.3 There are $W(13, 39; 52) = \dfrac{52!}{13!(52-13)!} = 635,013,559,600$ ways to chose 13 cards from a deck of 52 cards.

9.4 We find for Euler 60, for Laplace $1,280$, and for Lagrange $10,080$.

9.5 (a) $C = \frac{1}{5}$, (b) $\bar{x} = 2\frac{1}{2}$, (c) $\sigma_x^2 = \overline{x^2} - \bar{x}^2 = \frac{25}{12} = 2.08\overline{33}$

9.6 (a) $C = \lambda$, (b) $\bar{r} = \frac{1}{\lambda}$, (c) $\sigma_r^2 = \overline{r^2} - \bar{r}^2 = \frac{1}{\lambda^2}$

9.7 (a) $\int_0^L p(x)dx = \frac{2}{L}\left(\frac{L}{2}\right) = 1$ for $n = 1, 2, 3, \ldots$, (b) $\bar{x} = \frac{L}{2}$, (c) $\sigma_x^2 = \overline{x^2} - \bar{x}^2 = \frac{L^2}{12}\left(1 - \frac{6}{n^2\pi^2}\right)$

9.8 (a) $\int_0^\infty p(\epsilon)d\epsilon = \frac{4}{\sqrt{\pi}}\left(\frac{\sqrt{\pi}}{4}\right) = 1$ where $u = \epsilon/k_B T$, (b) $\bar{\epsilon} = \frac{3}{2}k_B T$, as for an ideal gas, (c) $\sigma_\epsilon^2 = \overline{\epsilon^2} - \bar{\epsilon}^2 = \frac{3}{2}(k_B T)^2$

9.9 (a) $\bar{n} = \sum_{n=0}^N n \frac{N!}{n!(N-n)!} p^n q^{N-n} = Np \sum_{r=0}^{N-1} \frac{(N-1)!}{r!((N-1)-r)!} p^r q^{(N-1)-r} = Np$, (b) $\overline{n^2} = \sum_{n=0}^N n^2 \frac{N!}{n!(N-n)!} p^n q^{N-n}$

which can be divided into two sums $\overline{n^2} = \sum_{n=1}^N (n-1) \frac{N!}{(n-1)!(N-n)!} p^n q^{N-n} + \sum_{n=1}^N \frac{N!}{(n-1)!(N-n)!} p^n q^{N-n}$ with

some reorganization $\overline{n^2} = N(N-1)p^2 \sum_{s=0}^{N-2} \frac{(N-2)!}{s!((N-2)-s)!} p^s q^{(N-2)-s} + Np \sum_{r=0}^{N-1} \frac{(N-1)!}{r!((N-1)-r)!} p^r q^{(N-1)-r} = $

$N(N-1)p^2 + Np$ so that $\sigma_n^2 = Npq$

9.10 $\bar{r} = \frac{3}{2}a_0$

9.11 (a) $C = \frac{8}{L_x L_y L_z} = \frac{8}{V}$, (b) $\mathbf{r} = \hat{x}\frac{L_x}{2} + \hat{y}\frac{L_y}{2} + \hat{z}\frac{L_z}{2}$

9.12 (a) $C = \left(\frac{\beta m}{2\pi}\right)^{\frac{3}{2}}$, (b) $\int_{-\infty}^\infty \int_{-\infty}^\infty \int_{-\infty}^\infty du_x du_y du_z = \int_0^{2\pi} \int_0^\pi \int_0^\infty u^2 \sin\theta \, du d\theta d\varphi$, (c) $\bar{u} = \sqrt{\frac{8}{\pi\beta m}}$, (d) $\overline{u^2} = \frac{3}{\beta m}$

9.13 $\rho_{xy} = 0$

9.14 (a) $\int_{-\infty}^\infty e^{-\frac{1}{2}x^2} dx = \int_{-\infty}^\infty e^{-\frac{1}{2}y^2} dy = \sqrt{2\pi}$ so that $p(x) = \frac{1}{\sqrt{2\pi}}e^{-\frac{1}{2}x^2}$, $p(y) = \frac{1}{\sqrt{2\pi}}e^{-\frac{1}{2}y^2}$,

(b) $I = \int_{-\infty}^\infty \int_{-\infty}^\infty p(x,y)\, dx\, dy = \frac{1}{\sqrt{2\pi}} \int_{-\infty}^\infty e^{-\frac{1}{2}x^2} dx \times \frac{1}{\sqrt{2\pi}} \int_{-\infty}^\infty e^{-\frac{1}{2}y^2} dy = 1$

9.15 $\overline{V} \pm \sigma_V = 0.011 \pm 0.002\,\text{cm}^3$ will improve with a higher precision measurement of the radius.

9.16 $M \pm \epsilon_M = 112.5 \pm 0.3\,\text{g mol}^{-1}$

9.17 (a) $\epsilon_f = \left(\frac{\partial f}{\partial x}\right)_{\bar{x},\bar{y}} \epsilon_x + \left(\frac{\partial f}{\partial x}\right)_{\bar{x},\bar{y}} \epsilon_y$, (b) $\overline{\epsilon_f^2} = \left(\frac{\partial f}{\partial x}\right)_{\bar{x},\bar{y}}^2 \overline{\epsilon_x^2} + \left(\frac{\partial f}{\partial y}\right)_{\bar{x},\bar{y}}^2 \overline{\epsilon_y^2}$, (c) $\sqrt{\overline{\epsilon_f^2}} = \sigma_f \to \sqrt{N}\sigma_f$

CHAPTER 10

10.1 (a) $y(x) = e^{-3x+C}$, (b) $y(x) = \frac{1}{3}x^2 + 1 + \frac{C}{x}$, (c) $x(t) = \dfrac{ak_1}{k_1 + k_2} + Ce^{-(k_1+k_2)t}$, (d) $f(r) = \frac{1}{4} + Ce^{-2r^2}$

10.2 (a) $y(x) = \frac{1}{3} + Ce^{-x^3}$, (b) $y'(x) = -3Cx^2e^{-x^3}$

10.3 (a) $y(x) = \frac{1}{5}x^3 + \frac{2}{3}x + \frac{C}{x^2}$, (b) $y'(x) = \frac{3}{5}x^2 + \frac{2}{3} - \frac{2C}{x^3}$

10.4 $f(r) = r$

10.5 $s(t) = \left(\frac{1}{2}t^3 + t\right)e^{3t}$

10.6 $m(t) = \frac{2}{5}(20 + t) + \dfrac{1.92 \times 10^6}{(20 + t)^4}$

10.7 (a) $p(T) = C\exp\left(-\frac{\Delta H}{RT}\right)$, (b) $p(T) = p_0\exp\left[-\frac{\Delta H}{R}\left(\frac{1}{T} - \frac{1}{T_0}\right)\right]$ where $p(T_0) = p_0$ and $T_0 = \dfrac{\Delta H}{R\ln\left(\frac{p_0}{A}\right)}$

10.8 (a) $A(t) = (A_0 + B_0)\dfrac{k_b}{k_f + k_b} + \left(A_0 - (A_0 + B_0)\dfrac{k_b}{k_f + k_b}\right)e^{-(k_f+k_b)t}$, $B(t) = (A_0 + B_0) - A(t) = (A_0 +$

$B_0)\dfrac{k_f}{k_f + k_b} - \left(A_0 - (A_0 + B_0)\dfrac{k_b}{k_f + k_b}\right)e^{-(k_f+k_b)t}$, (b) $K_{eq} = \dfrac{B(\infty)}{A(\infty)} = \dfrac{k_f}{k_b}$

10.9 The rate of change in the number of moles $n(t)$ is found to be

$$\frac{dn(t)}{dt} = c(t)\frac{dV}{dt} - \frac{n(t)}{V(t)}0.50\,\text{L/s} = -\frac{n(t)}{[50.0\,\text{L} + (1.00\,\text{L/s} - 0.50\,\text{L/s})t]}0.50\,\text{L/s}$$

so that $n(t) = 2.50 \times 10^3\,\text{mol L}(50.0\,\text{L} + 0.50\,\text{L/s}\,t)^{-1}$. It follows that $c(t) = 0.05\,\text{mol/L}$ at $t = 347$ s.

10.10 (a) inexact, (b) exact

10.11 $p(T, V) = \frac{nRT}{V} + C$, $dp(T, V) = \frac{nR}{V}dT - \frac{nRT}{V^2}dV$

10.12 $E(T, V) = nC_V T - \frac{n^2a}{V} + C$, $dE(T, V) = nC_V dT + \frac{n^2a}{V^2}dV$

10.13 $dh(x, y) = f'(x)g(y)dx + f(x)g'(y)dy$, $h(x, y) = \int \frac{df(x)}{dx}g(y)dx + \int \left[f(x)\frac{dg(y)}{dy} - \frac{\partial}{\partial y}\int \frac{df(x)}{dx}g(y)dx\right]dy = f(x)g(y)$

10.14 (a) $\rho^*\rho = kx^2 + \frac{1}{m}p^2 = 2E$, (b) $\frac{d\rho}{dt} = \sqrt{k}\frac{dx}{dt} + i\frac{1}{\sqrt{m}}\frac{dp}{dt} = \frac{\sqrt{k}}{m}p - i\frac{k}{\sqrt{m}}x = -i\omega\rho$, (c) $\rho(t) = \rho(0)e^{-i\omega t} = $
$\left(\sqrt{k}x_0 + i\frac{1}{\sqrt{m}}p_0\right)e^{-i\omega t}$, (d) $\rho(t) = \left(\sqrt{k}x_0 + i\frac{1}{\sqrt{m}}p_0\right)\cos(\omega t) - i\left(\sqrt{k}x_0 + i\frac{1}{\sqrt{m}}p_0\right)\sin(\omega t) = \sqrt{k}x(t) + i\frac{1}{\sqrt{m}}p(t)$

CHAPTER 11

11.1 (a) $y(x) = \frac{1}{2}\sinh(2x)$, (b) $y(x) = \frac{1}{\sqrt{3}}\sin\left(\sqrt{3}x\right)e^{-x}$, (c) $y(x) = \frac{1}{3}\sin(3x)$, (d) $y(x) = -\frac{1}{6}(e^{-6x} - 1)$

11.2 (a) $\alpha^2 + 2\alpha + 1 = (\alpha + 1)^2$, $\alpha = -1$, (b) $y(x) = c_1e^{-x} + c_2xe^{-x}$, $y'(x) = -c_1e^{-x} + c_2e^{-x} - c_2xe^{-x}$

11.3 $y'' + a(x)y' + by = c_1y_1'' + c_2y_2'' + a(x)c_1y_1' + a(x)c_2y_2' + bc_1y_1 + bc_2y_2 = 0$

11.4 (a) $x^1 \sum_{n=0}^{\infty} na_n x^n$, (b) $\sum_{n=0}^{\infty}(n+2)(n+1)a_{n+2} x^n$, (c) $\sum_{n=0}^{\infty}(n+1)a_{n+1} x^n$

11.5 (a) $a_n = a_0(-1)^n \frac{1}{(n!)^2}$, (b) $a_n = a_0 \frac{(n+1)}{2^n}$, (c) $a_n = a_0(-1)^n \frac{2^n}{n!}$

11.6 (a) $a_{n+2} = \frac{a_n}{(n+2)(n+1)}$, (b) $a_n = a_0 \frac{1}{n!}$, (c) $y(x) = y_0(x) + y_1(x)$ where $y_0(x) = a_0 \sum_{n=0}^{\infty} \frac{1}{(2n)!} x^{2n} = a_0 \cosh(x)$ and $y_1(x) = a_1 \sum_{n=0}^{\infty} \frac{1}{(2n+1)!} x^{2n+1} = a_1 \sinh(x)$

11.7 $c_1' = (c_1 + c_2)$, $c_2' = i(c_1 - c_2)$

11.8 (a) $x(t) = \frac{u_0}{\omega} \sin(\omega t)$, (b) $x(t) = x_0 \cos(\omega t) + \frac{u_0}{\omega} \sin(\omega t)$ where $\omega T = 2\pi$ so that $T = \frac{2\pi}{\omega} = \frac{1}{\nu}$

11.9 (a) $f(x) = \frac{L}{2\pi i}\left[\exp\left(\frac{5\pi ix}{L}\right) - \exp\left(-\frac{5\pi ix}{L}\right)\right]$, (b) $f(x) = \frac{L}{\pi}\sin\left(\frac{5\pi x}{L}\right)$

11.10 (a) $m\frac{du}{dt} + \gamma u = mg$, (b) $u(t) = \frac{mg}{\gamma}\left(1 - e^{-\frac{\gamma}{m}t}\right)$, (c) $\lim_{t\to\infty} u(t) = \frac{mg}{\gamma} = u_T$

11.11 $a_n = a_0(-1)^n \frac{2^n}{n!}$, $y(x) = a_0 e^{-2x}$

11.12 (a) $a_{n+2} = -\frac{(n-2)(n-3)}{(n+2)(n+1)} a_n$, (b) $y(x) = a_0(1 - 3x^2) + a_1\left(x - \frac{1}{3}x^3\right)$, (c) $y(x) = 1 + 2x - 3x^2 - \frac{2}{3}x^3$

11.13 Noting that $\sin\left(\tan^{-1} x\right) = \frac{x}{\sqrt{x^2+1}}$ and $\cos\left(\tan^{-1} x\right) = \frac{1}{\sqrt{x^2+1}}$ we have $A\cos(\omega t + \varphi) = A\cos(\omega t)\frac{c_1}{A} - A\sin(\omega t)\left(-\frac{c_2}{A}\right) = c_1 \cos(\omega t) + c_2 \sin(\omega t)$

11.14 (a) $\alpha^2 - \alpha - 1 = 0$ so that $\alpha_+ = \frac{1}{2}(1 + \sqrt{5}) = \varphi$ and $\alpha_- = \frac{1}{2}(1 - \sqrt{5}) = (1 - \varphi)$, (b) $y(x) = \frac{1}{\sqrt{5}}\left(e^{\varphi x} - e^{(1-\varphi)x}\right)$, (c) $a_{n+2} = \frac{(n+1)a_{n+1} + a_n}{(n+2)(n+1)}$ and $a_0 = 0$, $a_1 = 1$, $a_2 = \frac{1}{2!}$, $a_3 = \frac{2}{3!}$, $a_4 = \frac{3}{4!}$, $a_5 = \frac{5}{5!}$, $a_6 = \frac{8}{6!}$ and so on proving $a_n = \frac{1}{n!} f_n$ where f_n is the nth Fibonacci number, (d) $y(x) = \frac{1}{\sqrt{5}}\left[e^{\varphi x} - e^{(1-\varphi)x}\right] = \sum_{n=0}^{\infty} \frac{1}{n!} f_n x^n$ so $f_n = \frac{1}{\sqrt{5}}\left[\varphi^n - (1 - \varphi)^n\right]$

11.15 (a) $y(x) = \sum_{n=0}^{\infty} a_n x^{n+r}$ so that $x^2 y'' + xy' + (x^2 - c^2)y = \sum_{n=0}^{\infty}\left[(n+r)(n+r-1)a_n + (n+r)a_n - c^2 a_n\right]x^{n+r} + \sum_{n=0}^{\infty} a_n x^{n+r+2} = 0$ and for $n = 0$ we have $a_0\left[r(r-1) + r - c^2\right] = 0$ so $r^2 = c^2$ and $r = \pm c$, (b) for $r = c$ and $n = 1$ we find $a_1(2c + 1) = 0$ so that $a_1 = 0$ and $a_{2n+1} = 0$ for $n \geq 1$, (c) for $n \geq 2$ we have $a_{2n} = (-1)^n \frac{a_0}{2^{2n} n!(1+c)(2+c)...(n+c)}$ and $y_+(x) = a_0 x^c\left[1 + \sum_{n=1}^{\infty} \frac{(-1)^n}{2^{2n} n!(1+c)(2+c)...(n+c)} x^{2n}\right]$, (d) repeating for $r = -c$ leads to a second solution $y_-(x) = a_0 x^{-c}\left[1 + \sum_{n=1}^{\infty} \frac{(-1)^n}{2^{2n} n!(1-c)(2-c)...(n-c)} x^{2n}\right]$.

CHAPTER 12

12.1 (a) The boundary conditions $u(0,t) = u(L,t) = 0$ result in $X_n(x) = b_n \sin(\frac{n\pi}{L}x)$ with $n = 1, 2, 3, \ldots$ so that $T_n(t) = c_n \exp[-(\frac{n\pi}{L})^2 \kappa t]$ and $u(x,t) = \sum_{n=1}^{\infty} u_n \sin(\frac{n\pi}{L}x)\exp[-(\frac{n\pi}{L})^2 \kappa t]$ where $u_n = b_n c_n$, (b) $u_4 = 2$, $u(x,t) = 2\sin(\frac{4\pi}{L})\exp[-(\frac{4\pi}{L})^2 \kappa t]$.

12.2 The boundary conditions $u(0,t) = u(L,t) = 0$ result in $u(x,t) = \sum_n \sin\left(\frac{n\pi}{L}x\right)\left[a_n \cos\left(\omega_n t\right) + b_n \sin\left(\omega_n t\right)\right]$ where $\omega_n = \frac{n\pi}{L}v$. Applying $u(x,0) = 5\sin\left(\frac{3\pi}{L}x\right)$ and $u_t(x,0) = 0$ leads to $u(x,t) = 5\sin\left(\frac{3\pi}{L}x\right)\cos\left(\frac{3\pi}{L}vt\right)$.

12.3 (a) For $c(x,t) = \frac{1}{\sqrt{4\pi Dt}} \exp\left[-\frac{(x-x_0)^2}{4Dt}\right]$, $c_t(x,t) = \left[-\frac{1}{2t} + \frac{(x-x_0)^2}{4Dt^2}\right] c(x,t)$, $c_{xx}(x,t) = \left[-\frac{1}{2Dt} + \frac{(x-x_0)^2}{(2Dt)^2}\right] c(x,t)$, and $c_t = Dc_{xx}$, (b) $\int_{-\infty}^{\infty} c(x,t)dx = \frac{1}{\sqrt{4\pi Dt}} \sqrt{4\pi Dt} = 1$ for all t, (c) $\lim_{t\to 0} c(x,t) = \lim_{t\to 0} \frac{1}{\sqrt{4\pi Dt}} \exp\left[-\frac{(x-x_0)^2}{4Dt}\right] = \delta(x-x_0)$, (d) The normalized gaussian distribution function $c(x,t)$ centered at $x_0 = 1$ is shown below for times $t = 1/4D, 2/4D$ and $4/4D$.

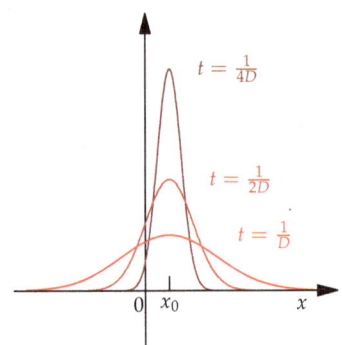

12.4 $\varphi = \sqrt{\frac{8}{L^3}} \sin\left(\frac{n_x\pi}{L}x\right) \sin\left(\frac{n_y\pi}{L}y\right) \sin\left(\frac{n_z\pi}{L}z\right)$, $\varphi_{xx} = -\left(\frac{n_x\pi}{L}\right)^2 \varphi$, $\varphi_{yy} = -\left(\frac{n_y\pi}{L}\right)^2 \varphi$, and $\varphi_{zz} = -\left(\frac{n_z\pi}{L}\right)^2 \varphi$, and $E_{n_x n_y n_z} = \frac{h^2}{8mL^2}\left(n_x^2 + n_y^2 + n_z^2\right)$.

12.5 (a) $c(0,t) = \frac{1}{\sqrt{4\pi Dt}}\left[\exp(-\frac{x_0^2}{4Dt}) - \exp(-\frac{x_0^2}{4Dt})\right] = 0$ for all time t, $c(x,0) = \lim_{t\to 0} c(x,t) = \delta(x-x_0) - \delta(x+x_0)$, $c(x,0) = \delta(x-x_0)$ for $x_0 > 0$ and $x \geq 0$, (b) $p(t) = \int_0^{\infty} c(x,t)dx = 2\sqrt{\frac{1}{\pi}} \int_0^{\frac{x_0}{\sqrt{4Dt}}} e^{-t^2} dt = \mathrm{erf}\left(\frac{x_0}{\sqrt{4Dt}}\right)$, (c) The survival probability $p(t) = \mathrm{erf}\left(\frac{x_0}{\sqrt{4Dt}}\right)$ is shown below as a function of time t for $D = 1$ for three values of x_0.

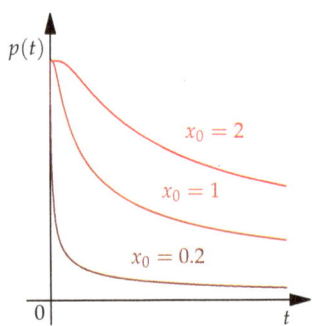

12.6 (a) $\psi = A e^{2\pi i\left(\frac{x}{\lambda} - vt\right)}$, $\psi_t = -2\pi i v\psi$, $\psi_{xx} = -\left(\frac{2\pi}{\lambda}\right)^2 \psi$, and $E = \frac{h^2}{2m\lambda^2}$, (b) $p = \frac{h}{\lambda}$.

12.7 (a) $i\hbar\varphi_t = -\frac{\hbar^2}{2m}\varphi_{xx} + V(x)\varphi = E\varphi$, $\tau = \frac{it}{\hbar}$ or $t = -i\hbar\tau$, so that $\varphi_\tau = \frac{\hbar^2}{2m}\varphi_{xx} - V(x)\varphi = -E\varphi$ for the function $\varphi(x,\tau)$, (b) $\varphi(x,\tau) = \left(\frac{m\omega}{\pi\hbar}\right)^{\frac{1}{4}} e^{-\frac{m\omega}{2\hbar}x^2} e^{-E\tau}$, $\varphi_\tau = -E\tau\varphi$, $\varphi_{xx} = \left[4x^2\left(\frac{m\omega}{2\hbar}\right)^2 - 2\left(\frac{m\omega}{2\hbar}\right)\right]\varphi$, and $E = \frac{1}{2}\hbar\omega$, (c) $D = \frac{\hbar^2}{2m}$, (d) $\psi(x,\tau) = A\exp\left[2\pi i\left(\frac{x}{\lambda} + i\hbar v\tau\right)\right] = Ae^{2\pi i\frac{x}{\lambda}}e^{-h\tau}$, $\psi_\tau = -hv\psi$, $\psi_{xx} = -\left(\frac{2\pi}{\lambda}\right)^2 \psi$, and $E = hv = \frac{h^2}{2m\lambda^2}$.

CHAPTER 13

13.1 For the periodic step function $a_0 = 1$, $a_n = 0$ for $n \geq 0$, and $b_n = \frac{2}{n\pi}$ for $n = 1, 3, 5, \ldots$ and zero otherwise. Twice the periodic step function minus unity is the square wave, so that $a_0 = 1 - 1 = 0$ and $b_n = 2 \times \frac{2}{n\pi} = \frac{4}{n\pi}$ for the square wave.

13.2 $a_n = \frac{1}{\pi} \int_{-\pi}^{\pi} f_o(x) \cos(nx)\, dx = 0$, as $\cos(nx)$ is an even function of x making the integrand an odd function of x and the integral zero. $b_n = \frac{1}{\pi} \int_{-\pi}^{\pi} f_e(x) \sin(nx)\, dx = 0$ as $\sin(nx)$ is an odd function of x making the integrand an odd function of x and the integral zero.

13.3 $f(x) = \frac{1}{2} - \frac{4}{\pi^2} \sum_{n=1,3,\ldots}^{N} \frac{1}{n^2} \cos\left(\frac{n\pi x}{L}\right)$, $a_0 = 1$, $a_n = -\frac{4}{\pi^2 n^2}$ $n = 1, 3, 5, \ldots$ and zero otherwise.

13.4 This can be shown using the Fourier transform pairs in Supplement S_9. $f(t)$ and $F(\omega)$ are shown below.

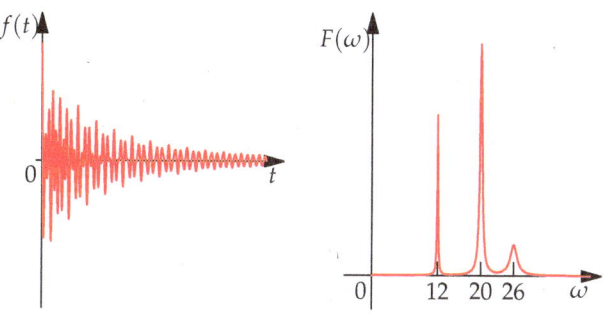

13.5 The maximum of $I(\omega)$ is located at $\omega = \omega_0$. At that point $F(\omega_0) = \frac{1}{\alpha}$.

13.6 $\frac{1}{\sqrt{2\pi}} \int_{-\infty}^{\infty} f(t + t_0) e^{-i\omega t}\, dt = \frac{1}{\sqrt{2\pi}} \int_{-\infty}^{\infty} f(s) e^{-i\omega(s - t_0)}\, ds = e^{i\omega t_0} \frac{1}{\sqrt{2\pi}} \int_{-\infty}^{\infty} f(s) e^{-i\omega s}\, ds = e^{i\omega t_0} F(\omega)$

13.7 $f\left(\frac{L}{2}\right) = \frac{1}{2} = \frac{2}{\pi} \sum_{n=1}^{\infty} \frac{(-1)^{n+1}}{n} \sin\left(\frac{n\pi}{2}\right) = \frac{2}{\pi}\left(1 - \frac{1}{3} + \frac{1}{5} - \frac{1}{7} + - \ldots\right)$ so that $\frac{\pi}{4} = 1 - \frac{1}{3} + \frac{1}{5} - \frac{1}{7} + - \ldots$

13.8 $a_0 = \frac{2\pi^2}{3}$, $a_n = (-1)^n \frac{4}{n^2}$ for $n = 1, 2, 3, \ldots$ and $b_n = 0$ for $n = 1, 2, 3, \ldots$

13.9 $F_c(\nu) = \left(\frac{a/2}{a^2 + 4\pi^2(\nu + \nu_0)^2}\right) + \left(\frac{a/2}{a^2 + 4\pi^2(\nu - \nu_0)^2}\right)$ where the figure below shows the damped cosine function (left) and its Fourier transform (right).

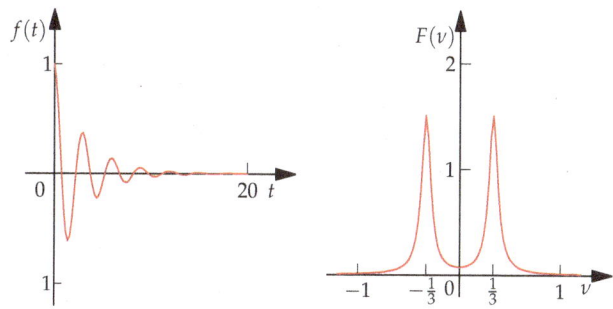

13.10 $f(x) = \frac{L}{2} - \sum_{n=1}^{\infty} \frac{L}{n\pi} \sin\left(\frac{n\pi x}{L}\right)$, $a_0 = \frac{L}{2}$, $a_n = 0$ $n > 0$, $b_n = -\frac{L}{n\pi}$ $n = 1, 2, 3, \ldots$ and are zero otherwise.

13.11 $f(t) = \sum_{n=1,3,\ldots}^{N} b_n \sin\left(\frac{n\pi t}{T}\right) = \frac{8}{\pi^2} \sum_{n=1,3,\ldots}^{N} (-1)^{\frac{n-1}{2}} \frac{1}{n^2} \sin\left(\frac{n\pi t}{T}\right)$, $b_n = (-1)^{\frac{n-1}{2}} \frac{8}{\pi^2 n^2}$ $n = 1,3,5,\ldots$

13.12 $\frac{1}{\sqrt{2\pi}} \int_{-\infty}^{\infty} f(st) e^{-i\omega t}\, dt = \frac{1}{s} \frac{1}{\sqrt{2\pi}} \int_{-\infty}^{\infty} f(\alpha) e^{-i\left(\frac{\omega}{s}\right)\alpha}\, d\alpha = \frac{1}{s} F\left(\frac{\omega}{s}\right)$

13.13 $\sqrt{2\pi}\, G(\omega) = \int_{-\infty}^{0} \left(\frac{d}{dt} e^{at}\right) e^{-i\omega t}\, dt + \int_{0}^{\infty} \left(\frac{d}{dt} e^{-at}\right) e^{-i\omega t}\, dt = a \int_{-\infty}^{0} e^{(a-i\omega)t}\, dt - a \int_{0}^{\infty} e^{-(a+i\omega)t}\, dt = 2i\omega \frac{a}{a^2+\omega^2}$,

$F(\omega) = \sqrt{\frac{2}{\pi}} \left(\frac{a}{a^2+\omega^2}\right)$, so that $G(\omega) = i\omega F(\omega)$

13.14 $f(x) = \frac{2}{\pi} + \sum_{n=1}^{\infty} \frac{4}{\pi(1-4n^2)} \cos\left(\frac{2\pi nx}{L}\right)$

13.15 $a_0 = 1$, $a_1 = \frac{1}{2}$, $a_n = 0$ for $n > 1$, $b_n = 0\,\forall\,n$, so that $f(x) = \frac{1}{2} - \frac{1}{2} \cos\left(\frac{2\pi x}{L}\right)$

13.16 $f(x) = \frac{2L^2}{3} + \frac{4L^2}{\pi^2} \sum_{n=1}^{\infty} (-1)^{n+1} \frac{1}{n^2} \cos\left(\frac{n\pi x}{L}\right)$, $a_0 = \frac{4L^2}{3}$, $a_n = (-1)^{n+1} \frac{4L^2}{n^2\pi^2}$ $n = 1,2,3,\ldots$ and zero otherwise.

$f(0) = L^2$ and $f(-L) = f(L) = 0$ where we made use of the sums $\sum_{n=1}^{\infty} \frac{1}{n^2} = \frac{\pi^2}{6}$ and $\sum_{n=1}^{\infty} (-1)^{n+1} \frac{1}{n^2} = \frac{\pi^2}{12}$.

13.17 (a) $E = \frac{a_0^2}{4} + \frac{1}{2} \sum_{n=1}^{\infty} \left(a_n^2 + b_n^2\right)$ is a result known as the *energy theorem* and more commonly *Parseval's theorem*,

(b) $\Delta E = \frac{1}{2} \sum_{n=N+1}^{\infty} a_n^2 + \frac{1}{2} \sum_{n=N+1}^{\infty} b_n^2$, (c) $E = \frac{1}{2} \sum_{n=1}^{\infty} b_n^2 = \frac{8}{\pi^2} \sum_{n=1,3,5,\ldots}^{\infty} \frac{1}{n^2} = \frac{8}{\pi^2} \frac{\pi^2}{8} = 1$ the expected result for

the integral over $f^2(t)$ for one period of the square wave function.

13.18 $g(\omega) = \sum_n a_n \delta(\omega - \omega_n)$ where the function $g(\omega)$ forms a comb with teeth occurring at discrete frequencies ω_n. As such, $g(\omega)$ is known as the *frequency comb*.

13.19

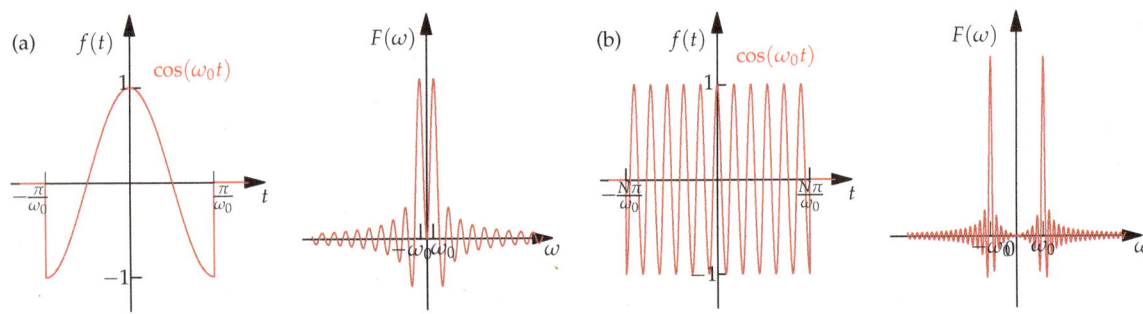

(c) $\lim_{N\to\infty} F(\omega) = \sqrt{\frac{\pi}{2}} \left(\delta(\omega + \omega_0) + \delta(\omega - \omega_0)\right)$, (d) The abrupt truncation of the sinusoidal oscillation leads to the ringing observed in the Fourier transform.

13.20 $\int_{-\infty}^{\infty} f(s) g(t-s)\, ds = \int_{-\infty}^{\infty} f(s) \left(\frac{1}{\sqrt{2\pi}} \int_{-\infty}^{\infty} G(\omega) e^{i\omega(t-s)} d\omega\right) ds = \int_{-\infty}^{\infty} G(\omega) \left(\frac{1}{\sqrt{2\pi}} \int_{-\infty}^{\infty} f(s) e^{-i\omega s} ds\right) e^{i\omega t} d\omega =$

$\int_{-\infty}^{\infty} G(\omega) F(\omega) e^{i\omega t} d\omega$ so that $\frac{1}{\sqrt{2\pi}} \int_{-\infty}^{\infty} (f * g)(t) e^{i\omega t} dt = \sqrt{2\pi}\, F(\omega) G(\omega)$.

13.21 $c_n = \frac{1}{2\pi} \int_{-\pi}^{\pi} f(x) e^{-inx} dx = \frac{1}{2\pi} \int_{-\pi}^{\pi} \left(\sum_{m=-\infty}^{\infty} c_m e^{imx}\right) e^{-inx} dx = \frac{1}{2\pi} \int_{-\pi}^{\pi} \left(\sum_{m=-\infty}^{\infty} c_m e^{i(m-n)x}\right) dx =$

$\frac{1}{2\pi} \sum_{m=-\infty}^{\infty} c_m \left(\int_{-\pi}^{\pi} e^{i(m-n)x} dx\right) = \frac{1}{2\pi} \sum_{m=-\infty}^{\infty} c_m\, 2\pi \delta_{mn} = c_n$

13.22 (a) $F(k) = i\sqrt{\frac{2}{\pi}} \left(kL\cos(kL) - \sin(kL)\right) \frac{1}{k^2 L}$ as shown below.

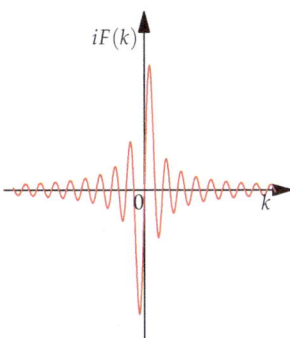

(b) $c_n = \frac{1}{2}(a_n - ib_n) = \frac{\sqrt{2\pi}}{2L} F\left(\frac{\pi n}{L}\right)$, $b_n = i\frac{\sqrt{2\pi}}{L} F\left(\frac{\pi n}{L}\right) = \frac{2}{n\pi}(-1)^{n+1}$ where the Fourier transform and Fourier coefficients are compared in the figure below.

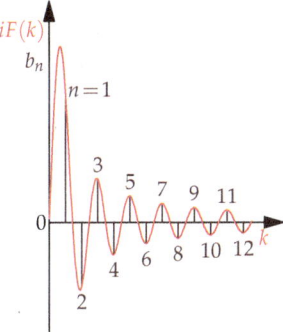

Note that the end of each bar marking the discrete value of b_n touches the continuous line formed by the scaled value of $iF(k)$ at the point $k = \frac{2\pi n}{2L} = \frac{\pi n}{L}$.

13.23 $\int_{-\infty}^{\infty} H_1(x)H_2(x)e^{-x^2}\,dx = 2\int_{-\infty}^{\infty} x(4x^2 - 4)e^{-x^2}\,dx = 8\int_{-\infty}^{\infty} x^3 e^{-x^2}\,dx - 4\int_{-\infty}^{\infty} xe^{-x^2}\,dx = 0$ where we note that $H_1(x)$ is an odd function of x, $H_2(x)$ is an even function of x, and the gaussian e^{-x^2} is an even function of x, making the product of the three functions an odd function of x, $\frac{1}{2\sqrt{\pi}}\int_{-\infty}^{\infty} [H_1(x)]^2 \, e^{-x^2}\,dx = \frac{1}{2\sqrt{\pi}}\int_{-\infty}^{\infty} 4x^2 \, e^{-x^2}\,dx = 1$, $\frac{1}{8\sqrt{\pi}}\int_{-\infty}^{\infty} [H_2(x)]^2 \, e^{-x^2}\,dx = \frac{1}{8\sqrt{\pi}}\int_{-\infty}^{\infty} (16x^4 - 16x^2 + 4) \, e^{-x^2}\,dx = 1$

13.24 $\int_0^{\infty} L_1(x)L_2(x) \, e^{-x}\,dx = \frac{1}{2}\int_0^{\infty}(-x^3 + 5x^2 - 6x + 2) \, e^{-x}\,dx = 0$, $\int_0^{\infty} [L_1(x)]^2 \, e^{-x}\,dx = \int_0^{\infty} (x^2 - 2x + 1) \, e^{-x}\,dx = 1$, $\int_0^{\infty} [L_2(x)]^2 \, e^{-x}\,dx = \frac{1}{4}\int_0^{\infty} (x^4 - 8x^3 + 20x^2 - 16x + 4) \, e^{-x}\,dx = 1$

13.25 $\int_{-1}^{1} P_1(x)P_2(x)\,dx = \frac{1}{2}\int_{-1}^{1} x(3x^2 - 1)\,dx = \frac{3}{2}\int_{-1}^{1} x^3\,dx - \frac{1}{2}\int_{-1}^{1} x\,dx = 0$, $\frac{3}{2}\int_{-1}^{1} [P_1(x)]^2\,dx = \frac{3}{2}\int_{-1}^{1} x^2\,dx = 1$, $\frac{5}{2}\int_{-1}^{1} [P_2(x)]^2\,dx = \frac{5}{8}\int_{-1}^{1}(9x^4 - 6x^2 + 1)\,dx = 1$

13.26 $\int_0^{2\pi} \int_0^\pi Y_{1,0}(\theta, \varphi) Y_{1,1}^*(\theta, \varphi) \sin\theta\, d\theta\, d\varphi = \int_0^{2\pi} \int_0^\pi \left(\frac{1}{2}\sqrt{\frac{3}{\pi}} \cos\theta \right) \left(-\frac{1}{2}\sqrt{\frac{3}{2\pi}} \sin\theta\, e^{i\varphi} \right)^* \sin\theta\, d\theta\, d\varphi =$

$-\frac{3}{4\pi\sqrt{2}} \int_0^\pi \cos\theta \sin^2\theta\, d\theta \times \int_0^{2\pi} e^{-i\varphi}\, d\varphi = 0$ as the second integral is zero. To demonstrate normaliza-

tion we evaluate $\int_0^{2\pi} \int_0^\pi Y_{1,0}(\theta, \varphi) Y_{1,0}^*(\theta, \varphi) \sin\theta\, d\theta\, d\varphi = \frac{3}{4\pi} \int_0^\pi \cos^2\theta \sin\theta\, d\theta \times \int_0^{2\pi} d\varphi = 1$ and similarly

$\int_0^{2\pi} \int_0^\pi Y_{1,1}(\theta, \varphi) Y_{1,1}^*(\theta, \varphi) \sin\theta\, d\theta\, d\varphi = \int_0^{2\pi} \int_0^\pi \left(-\frac{1}{2}\sqrt{\frac{3}{2\pi}} \sin\theta\, e^{i\varphi} \right) \left(-\frac{1}{2}\sqrt{\frac{3}{2\pi}} \sin\theta\, e^{-i\varphi} \right) \sin\theta\, d\theta\, d\varphi =$

$\frac{3}{8\pi} \int_0^\pi \sin^3\theta\, d\theta \times \int_0^{2\pi} d\varphi = 1$

13.27 We take $\varphi_0(x) = 1$ and $\phi_n(0) = 1$. $\int_0^\infty e^{-x}\phi_0(x)\phi_1(x)\, dx = 0$ leads to $c_2 = -c_3$. $\phi_1(0) = 1$ leads to $c_3 = 1$

and $\phi_1(x) = -x + 1$. $\int_0^\infty e^{-x}\phi_0(x)\phi_2(x)\, dx = 0$ leads to $2c_4 + c_5 + c_6 = 0$. $\int_0^\infty e^{-x}\phi_1(x)\phi_2(x)\, dx = 0$ leads to

$c_4 = -\frac{1}{4}c_5$. $\phi_2(0) = 1$ leads to $c_6 = 1$. Finally $c_4 = \frac{1}{2}$ and $c_5 = -2$ so that $\phi_2(x) = \frac{1}{2}(x^2 - 4x + 2)$. Overall, we

find $\phi_0(x) = 1$, $\phi_1(x) = -x + 1$, and $\phi_2(x) = \frac{1}{2}(x^2 - 4x + 2)$. These are the first three Laguerre polynomials.

13.28 We take $\varphi_0(x) = 1$ and $\int_{-\infty}^\infty e^{-x^2}\phi_n^2(x)\, dx = \sqrt{\pi}2^n n!$ which is $\sqrt{\pi}$ for $n = 0$, $2\sqrt{\pi}$ for $n = 1$, and $8\sqrt{\pi}$ for

$n = 2$. $\int_{-\infty}^\infty e^{-x^2}\phi_0(x)\phi_1(x)\, dx = 0$ leads to $c_3 = 0$. The normalization condition for $\phi_1(x)$ leads to $c_2 = 2$ so

that $\phi_1(x) = 2x$. Using $\int_{-\infty}^\infty e^{-x^2}\phi_0(x)\phi_2(x)\, dx = 0$ leads to $c_6 = -\frac{1}{2}c_4$ and $\int_{-\infty}^\infty e^{-x^2}\phi_1(x)\phi_2(x)\, dx = 0$ leads to

$c_5 = 0$. Applying the normalization condition to $\phi_2(x) = c_4(x^2 - \frac{1}{2})$ leads to $c_4 = 4$ so that $\phi_2(x) = 4x^2 - 2$.

Overall, we find $\phi_0(x) = 1$, $\phi_1(x) = 2x$, and $\phi_2(x) = 4x^2 - 2$. These are the first three Hermite polynomials.

CHAPTER 14

14.1   (a) 18, (b) -91, (c) $x^3 - 2x$, (d) 352, (e) $x^4 - 3x^2b^2 + b^4$, (f) 1

14.2   (a) $x^2 - 1 = 0, x = \pm 1$, (b) $x^2 + 2 = 6, x = \pm 2$, (c) $x^3 - 3x + 1 = 2, x = 0, \pm\sqrt{3}$, (d) $x^2(x^2 - 3), x = 0, \pm\sqrt{3}$

14.3 $\begin{pmatrix} 6 & 4 & -5 \\ 1 & 5 & -4 \\ 1 & 5 & 12 \end{pmatrix}$

14.4   (a) $\begin{pmatrix} -6 & 1 \\ 12 & -7 \end{pmatrix}$, (b) $\begin{pmatrix} 4 & -1 \\ 2 & 3 \end{pmatrix}$, (c) $\begin{pmatrix} 12 & 15 & 18 \\ 3 & -1 & -7 \\ 12 & 23 & 38 \end{pmatrix}$, (d) $\begin{pmatrix} x + 8y + 4z \\ -2x + 3y \\ 5x - y - z \end{pmatrix}$

14.5 $\begin{pmatrix} 34 & -13 & -23 \\ 64 & -43 & -1 \\ 0 & -23 & 9 \end{pmatrix}$

14.6 $|A| = -11$, $A^{-1} = \frac{1}{11} \begin{pmatrix} 4 & 1 & -2 \\ 8 & 2 & 7 \\ -7 & 1 & -2 \end{pmatrix}$, $AA^{-1} = A^{-1}A = I$

14.7   (a) $x = \frac{-14}{-7} = 2, y = \frac{-7}{-7} = 1$, (b) $x = x'\sin\theta - y'\cos\theta, y = x'\cos\theta + y'\sin\theta$

14.8   (a) $\mathbf{x}_2 = (-1, 0)^T$, (b) $\mathbf{x}_2 = (-\frac{1}{\sqrt{2}}, \frac{1}{\sqrt{2}})^T$, (c) $\mathbf{x}_2 = (1, 1)^T$

14.9 $\mathbf{R}^2(\theta) = \begin{pmatrix} \cos^2\theta - \sin^2\theta & -2\sin\theta\cos\theta \\ 2\sin\theta\cos\theta & \cos^2\theta - \sin^2\theta \end{pmatrix} = \begin{pmatrix} \cos(2\theta) & -\sin(2\theta) \\ \sin(2\theta) & \cos(2\theta) \end{pmatrix} = \mathbf{R}(2\theta)$

14.10 (a) $|\mathbf{A}| = -4$ and $\mathbf{A}^{-1} = \frac{1}{4}\begin{pmatrix} 2 & 0 & 1 \\ -2 & 0 & 1 \\ 2 & 4 & -1 \end{pmatrix}$, (b) $\mathbf{A}\mathbf{A}^{-1} = \mathbf{A}^{-1}\mathbf{A} = \mathbf{I}$

14.11 (a) $|\mathbf{R}| = \cos^2\theta + \sin^2\theta = 1$, (b) $\mathbf{x}_1 = \begin{pmatrix} \cos\theta \\ \sin\theta \\ 0 \end{pmatrix}$ $\mathbf{x}_2 = \begin{pmatrix} -\sin\theta \\ \cos\theta \\ 0 \end{pmatrix}$ $\mathbf{x}_3 = \begin{pmatrix} 0 \\ 0 \\ 1 \end{pmatrix}$,

(c) $\mathbf{R}^T\mathbf{R} = \begin{pmatrix} \cos\theta & \sin\theta & 0 \\ -\sin\theta & \cos\theta & 0 \\ 0 & 0 & 1 \end{pmatrix} \begin{pmatrix} \cos\theta & -\sin\theta & 0 \\ \sin\theta & \cos\theta & 0 \\ 0 & 0 & 1 \end{pmatrix} = \mathbf{R}\mathbf{R}^T = \mathbf{I}$

14.12 $|\mathbf{A}| = 5$, $|\mathbf{B}| = 8$, $|\mathbf{A}||\mathbf{B}| = 40$, $|\mathbf{AB}| = 40$, $|\mathbf{BA}| = 40$

14.13 $\mathbf{x}_1 = \frac{1}{9}(1,4,8)^T$, $\mathbf{x}_2 = \frac{1}{9}(8,-4,1)^T$, $\mathbf{x}_3 = \frac{1}{9}(-4,-7,4)^T$ (a) $\mathbf{x}_1 \cdot \mathbf{x}_2 = \mathbf{x}_2 \cdot \mathbf{x}_3 = \mathbf{x}_3 \cdot \mathbf{x}_1 = 0$, (b) $\mathbf{x}_1 \cdot \mathbf{x}_1 = \mathbf{x}_2 \cdot \mathbf{x}_2 = \mathbf{x}_3 \cdot \mathbf{x}_3 = 1$, (c) $\mathbf{A}\mathbf{A}^{-1} = \mathbf{I}$

14.14 $(\mathbf{A}^*)^T = \frac{1}{6}\begin{pmatrix} 2+4i & 4i \\ -4i & -2+4i \end{pmatrix}$, $(\mathbf{A}^*)^T\mathbf{A} = \mathbf{A}(\mathbf{A}^*)^T = \mathbf{I}$

14.15 (a) $(\sigma_1^*)^T\sigma_1 = \mathbf{I}$, $(\sigma_2^*)^T\sigma_2 = \mathbf{I}$, $(\sigma_3^*)^T\sigma_3 = \mathbf{I}$

14.16 (a) $x=1, y=3, z=-4$, (b) $x=1, z=-2, t=-3$

14.17 (a) $m = (y_2 - y_1)/(x_2 - x_1)$, $b = (x_1 y_2 - y_1 x_2)/(x_1 - x_2)$, (b) $y = mx + b = 3x - 13$

14.18 (a) $a = (y_3 z_2 - z_3 y_2 + y_1 z_3 - y_3 z_1 + y_2 z_1 - y_1 z_2)/(x_1 y_2 - x_1 y_3 + x_2 y_3 - x_2 y_1 + x_3 y_1 - x_3 y_2)$,
$b = (x_1 z_2 - x_1 z_3 + x_2 z_3 - x_2 z_1 + x_3 z_1 - x_3 z_2)/(x_1 y_2 - x_1 y_3 + x_2 y_3 - x_2 y_1 + x_3 y_1 - x_3 y_2)$,
$c = [x_1(y_2 z_3 - y_3 z_2) + x_2(y_3 z_1 - y_1 z_3) + x_3(y_1 z_2 - y_2 z_1)]/(x_1 y_2 - x_1 y_3 + x_2 y_3 - x_2 y_1 + x_3 y_1 - x_3 y_2)$,
(b) $z = ax + by + c = \frac{5}{8}x - \frac{3}{8}y - \frac{11}{8}$

14.19 (a) $x^4 - 4x^2 = x^2(x^2 - 4) = 0$ with roots $x = 0, \pm 2$, (b) $x^4 - 3x^2 = x^2(x^2 - 3) = 0$ with roots $x = 0, \pm\sqrt{3}$,
(c) $x^6 - 6x^4 + 9x^2 - 4 = (x+1)^2(x-1)^2(x^2-4) = 0$ with roots $x = \pm 1, \pm 2$

14.20 (a) $x=2, y=1$, (b) $x = x'\sin\theta - y'\cos\theta$, $y = x'\cos\theta + y'\sin\theta$

14.21 (a) $x=1, y=3, z=-4$, (b) $x=y=1, z=-2, t=-3$

14.22 (a) $|\mathbf{A}| = -11$, $\mathbf{A}^{-1} = \frac{1}{11}\begin{pmatrix} 4 & 1 & -2 \\ 8 & 2 & 7 \\ -7 & 1 & -2 \end{pmatrix}$, (b) $\mathbf{A}\mathbf{A}^{-1} = \mathbf{A}^{-1}\mathbf{A} = \mathbf{I}$

14.23 (a) $|\mathbf{A}| = -729$, $\mathbf{A}^{-1} = \frac{1}{81}\begin{pmatrix} -4 & -7 & 4 \\ 1 & 4 & 8 \\ 8 & -4 & 1 \end{pmatrix}$, (b) $\mathbf{A}\mathbf{A}^{-1} = \mathbf{A}^{-1}\mathbf{A} = \mathbf{I}$, (c) the matrix inverse $\mathbf{A}^{-1}$ is the

transpose of $\mathbf{A}$ as the matrix $\mathbf{A}$ is an orthogonal matrix.

14.24 The area of the rectangle containing the parallelogram is $(a+c)(b+d)$. The area within the rectangle but outside the parallelogram consists of two triangles of area $\frac{1}{2}ab$, two triangles of area $\frac{1}{2}cd$, and two rectangles of area bc. The area of the parallelogram equals the difference $(a+c)(b+d) - ab - cd - 2bc = ad - bc$ which is the value of the determinant $|\mathbf{A}| = ad - bc$.

14.25 (a) $\mathbf{u}_1 = \begin{pmatrix} 1 \\ -2 \end{pmatrix}, \mathbf{u}_2 = \frac{8}{5}\begin{pmatrix} 2 \\ 1 \end{pmatrix}$, (b) $\mathbf{e}_1 = \frac{1}{\sqrt{5}}\begin{pmatrix} 1 \\ -2 \end{pmatrix}, \mathbf{e}_2 = \frac{1}{\sqrt{5}}\begin{pmatrix} 2 \\ 1 \end{pmatrix}$

14.26 (a) $\mathbf{u}_1 = \begin{pmatrix} 1 \\ -2 \\ 1 \end{pmatrix}, \mathbf{u}_2 = \frac{1}{2}\begin{pmatrix} 1 \\ 0 \\ -1 \end{pmatrix}, \mathbf{u}_3 = \frac{1}{3}\begin{pmatrix} 1 \\ 1 \\ 1 \end{pmatrix}$, (b) $\mathbf{e}_1 = \frac{1}{\sqrt{6}}\begin{pmatrix} 1 \\ -2 \\ 1 \end{pmatrix}, \mathbf{e}_2 = \frac{1}{\sqrt{2}}\begin{pmatrix} 1 \\ 0 \\ -1 \end{pmatrix}, \mathbf{e}_3 = \frac{1}{\sqrt{3}}\begin{pmatrix} 1 \\ 1 \\ 1 \end{pmatrix}$

CHAPTER 15

15.1 $\mathbf{C}^{-1}\mathbf{C} = \mathbf{I}$

15.2 $\mathbf{C}^{-1}\mathbf{A}\mathbf{C} = \begin{pmatrix} 4 & 0 \\ 0 & -1 \end{pmatrix}$

15.3 $\mathbf{c} \cdot (\lambda\mathbf{c}) = \lambda(\mathbf{c} \cdot \mathbf{c}) = \lambda|\mathbf{c}|^2 = |\mathbf{c}||\lambda\mathbf{c}|\cos\theta$ so that $\cos\theta = 1$ and $\theta = 0$

15.4 (a) $\lambda_1 = 3, \lambda_2 = -2$, (b) $\mathbf{x}_1 = \frac{1}{\sqrt{5}}\begin{pmatrix} 1 \\ 2 \end{pmatrix}, \mathbf{x}_2 = \frac{1}{\sqrt{5}}\begin{pmatrix} 2 \\ -1 \end{pmatrix}$

15.5 $\lambda_1 = 2, \lambda_2 = 0$, (b) $\mathbf{x}_1 = \frac{1}{\sqrt{2}}\begin{pmatrix} 1 \\ 1 \end{pmatrix}, \mathbf{x}_2 = \frac{1}{\sqrt{2}}\begin{pmatrix} 1 \\ -1 \end{pmatrix}$

15.6 $\lambda_1 = 2, \lambda_2 = 1, \lambda_3 = 0$, (b) $\mathbf{x}_1 = \frac{1}{\sqrt{2}}\begin{pmatrix} 1 \\ 0 \\ 1 \end{pmatrix}, \mathbf{x}_2 = \begin{pmatrix} 0 \\ 1 \\ 0 \end{pmatrix}, \mathbf{x}_3 = \frac{1}{\sqrt{2}}\begin{pmatrix} 1 \\ 0 \\ -1 \end{pmatrix}$

15.7 $y'(x) = a\cos(ax) \neq \lambda y(x)$ while $y''(x) = -a^2\sin(ax) = -a^2 y(x) = \lambda y(x)$ where $\lambda = -a^2$

15.8 $\Phi(\varphi) = A\,e^{im\varphi}, \hat{M}_z\Phi(\varphi) = \lambda\Phi(\varphi), \lambda = \frac{mh}{2\pi}$

15.9 $\lambda_1 = 2, \lambda_2 = 1, \lambda_3 = 0$, (b) $\mathbf{x}_1 = \frac{1}{\sqrt{2}}\begin{pmatrix} 1 \\ 0 \\ -1 \end{pmatrix}, \mathbf{x}_2 = \begin{pmatrix} 0 \\ 1 \\ 0 \end{pmatrix}, \mathbf{x}_3 = \frac{1}{\sqrt{2}}\begin{pmatrix} 1 \\ 0 \\ 1 \end{pmatrix}$

15.10 $\lambda_1 = \frac{1}{2}\left(1 + \sqrt{5}\right) = \varphi, \lambda_2 = \frac{1}{2}\left(1 - \sqrt{5}\right) = 1 - \varphi$ where $\varphi = 1.6180\ldots$ is the golden ratio. The golden ratio is a special number that appears in geometry, art, architecture, and here. $\mathbf{x}_1 = \frac{1}{\sqrt{2}}\begin{pmatrix} 1 \\ 1 \end{pmatrix}, \mathbf{x}_2 = \frac{1}{\sqrt{2}}\begin{pmatrix} 1 \\ -1 \end{pmatrix}$,

$\mathbf{x}(t) = \frac{1}{2}\begin{pmatrix} 1 \\ 1 \end{pmatrix}e^{\varphi t} + \frac{1}{2}\begin{pmatrix} 1 \\ -1 \end{pmatrix}e^{(1-\varphi)t}$

15.11 $\lambda_1 = 0, \lambda_2 = -1, \lambda_3 = -3, \mathbf{x}_1 = \frac{1}{\sqrt{3}} \begin{pmatrix} 1 \\ 1 \\ 1 \end{pmatrix}, \mathbf{x}_2 = \frac{1}{\sqrt{3}} \begin{pmatrix} 1 \\ 0 \\ -1 \end{pmatrix}, \mathbf{x}_3 = \frac{1}{\sqrt{6}} \begin{pmatrix} 1 \\ -2 \\ 1 \end{pmatrix}, \mathbf{x}(t) = \frac{1}{3} \begin{pmatrix} 1 \\ 1 \\ 1 \end{pmatrix} +$

$\frac{1}{2} \begin{pmatrix} 1 \\ 0 \\ -1 \end{pmatrix} e^{-t} + \frac{1}{6} \begin{pmatrix} 1 \\ -2 \\ 1 \end{pmatrix} e^{-3t}$

15.12 $\mathbf{C}^{-1}\mathbf{A}\mathbf{C} = \begin{pmatrix} (\alpha - E) + \sqrt{2}\beta & 0 & 0 \\ 0 & (\alpha - E) & 0 \\ 0 & 0 & (\alpha - E) - \sqrt{2}\beta \end{pmatrix}, \lambda_1 = (\alpha - E) + \sqrt{2}\beta, \lambda_2 = (\alpha - E), \lambda_3 =$
$(\alpha - E) - \sqrt{2}\beta.$

15.13 $V(x_1, x_2) = k\left(x_1^2 + x_2^2 - x_1 x_2\right)$ as $V(y_1, y_2) = \frac{k}{4}\left(y_1^2 + 3y_2^2\right), y_1 = x_2 + x_1, y_2 = x_1 - x_2.$

15.14 (a) $\mathbf{C}^{-1}\mathbf{K}\mathbf{C} = \boldsymbol{\Lambda} = \frac{k}{m} \begin{pmatrix} 1 & 0 \\ 0 & 3 \end{pmatrix}$, (b) $\frac{d^2}{dt^2}\mathbf{C}^{-1}\mathbf{x} = -\mathbf{C}^{-1}\mathbf{K}\mathbf{I}\mathbf{x} = -\mathbf{C}^{-1}\mathbf{K}\mathbf{C}\mathbf{C}^{-1}\mathbf{x} = -\boldsymbol{\Lambda}\mathbf{C}^{-1}\mathbf{x}, \mathbf{y} = \mathbf{C}^{-1}\mathbf{x},$

$\frac{d^2}{dt^2}\mathbf{y} = -\boldsymbol{\Lambda}\mathbf{y}$, (c) $y_1 = x_1 + x_2$ symmetric "in-phase" stretching mode at the lower frequency, $y_2 = x_1 - x_2,$
asymmetric "out-of-phase" stretching mode at the higher frequency.

15.15 (a) $\omega_1^2 = 6k, \omega_2^2 = 31k$, (b) $\mathbf{x}_1 = \frac{1}{5} \begin{pmatrix} 3 \\ 4 \end{pmatrix}, \mathbf{x}_2 = \frac{1}{5} \begin{pmatrix} -4 \\ 3 \end{pmatrix}$

15.16 (a) $m_1\frac{d^2 x_1}{dt^2} = -3kx_1 + kx_2, m_2\frac{d^2 x_2}{dt^2} = kx_1 - 3kx_2 + kx_3, m_3\frac{d^2 x_3}{dt^2} = kx_2 - 3kx_3$, (b) $\mathbf{K} = \begin{pmatrix} 3k & -k & 0 \\ -k & 2k & -k \\ 0 & -k & 3k \end{pmatrix}$,

(c) $\frac{d^2}{dt^2}\mathbf{x} = -\omega^2 \begin{pmatrix} c_1 \\ c_2 \\ c_3 \end{pmatrix} e^{i\omega t} = -\omega^2 \mathbf{x}(t) = -\mathbf{K}\mathbf{x}$, (d) $\omega_1^2 = k, \omega_2^2 = 3k, \omega_3^2 = 4k$, (e) $\mathbf{x}_1 = \frac{1}{\sqrt{6}} \begin{pmatrix} 1 \\ 2 \\ 1 \end{pmatrix}$,

$\mathbf{x}_2 = \frac{1}{\sqrt{2}} \begin{pmatrix} -1 \\ 0 \\ 1 \end{pmatrix}, \mathbf{x}_3 = \frac{1}{\sqrt{3}} \begin{pmatrix} 1 \\ -1 \\ 1 \end{pmatrix}$, (f) $\mathbf{x}_1$ is an asymmetric stretch, $\mathbf{x}_2$ is a symmetric stretch, $\mathbf{x}_3$ is an
asymmetric stretch.

15.17 (a) $m\frac{d^2 x_1}{dt^2} = -kx_1 + kx_2, m\frac{d^2 x_2}{dt^2} = kx_1 - 2kx_2 + kx_3, m\frac{d^2 x_3}{dt^2} = kx_2 - kx_3,$

(b) $\mathbf{K} = \frac{1}{m} \begin{pmatrix} k & -k & 0 \\ -k & 2k & -k \\ 0 & -k & k \end{pmatrix}$, (c) $\omega_1^2 = 0, \omega_2^2 = k, \omega_3^2 = 3k$, (d) $\mathbf{x}_1 = \frac{1}{\sqrt{3}} \begin{pmatrix} 1 \\ 1 \\ 1 \end{pmatrix}, \mathbf{x}_2 = \frac{1}{\sqrt{2}} \begin{pmatrix} -1 \\ 0 \\ 1 \end{pmatrix}, \mathbf{x}_3 =$

$\frac{1}{\sqrt{6}} \begin{pmatrix} 1 \\ -2 \\ 1 \end{pmatrix}$, (e) $\mathbf{x}_1$ is a translational mode, $\mathbf{x}_2$ is a symmetric stretch, $\mathbf{x}_3$ is an asymmetric stretch.

15.18 (a) $\frac{d^2}{dt^2}\mathbf{x}(t) = -\omega^2\mathbf{x} = -\mathbf{Kx}$, (b) $\omega_1^2 = 19\frac{k}{m}$, $\omega_2^2 = 780\frac{k}{m}$, (c) $\mathbf{x}_1 = \frac{1}{\sqrt{761}}\begin{pmatrix} 19 \\ 20 \end{pmatrix}$, $\mathbf{x}_2 = \frac{1}{\sqrt{761}}\begin{pmatrix} -20 \\ 19 \end{pmatrix}$, (d) $\mathbf{x}_1$ is

primarily the oscillation of the center-of-mass of the diatomic molecule relative to the surface, $\mathbf{x}_2$ is primarily

the internal oscillation of the diatomic molecule.

15.19 (a) $\mathbf{K} = \begin{pmatrix} -2 & 0 & 0 \\ 2 & -1 & 0 \\ 0 & 1 & 0 \end{pmatrix}$, (b) $\alpha_1 = -2$, $\alpha_2 = -1$, $\alpha_3 = 0$, (c) $\mathbf{x}_1 = \begin{pmatrix} 1 \\ -2 \\ 1 \end{pmatrix}$, $\mathbf{x}_2 = \begin{pmatrix} 0 \\ 1 \\ -1 \end{pmatrix}$, $\mathbf{x}_3 = \begin{pmatrix} 0 \\ 0 \\ 1 \end{pmatrix}$,

(d) $\mathbf{x}(t) = A_0\left[\begin{pmatrix} 1 \\ -2 \\ 1 \end{pmatrix}e^{-2t} + 2\begin{pmatrix} 0 \\ 1 \\ -1 \end{pmatrix}e^{-t} + \begin{pmatrix} 0 \\ 0 \\ 1 \end{pmatrix}\right]$. Note that at long times $\lim_{t\to\infty}\mathbf{x}(t) = A_0\begin{pmatrix} 0 \\ 0 \\ 1 \end{pmatrix}$.

15.20 (a) $\mathbf{K} = \begin{pmatrix} \frac{g}{l} + \frac{k}{m} & -\frac{k}{m} \\ -\frac{k}{m} & \frac{g}{l} + \frac{k}{m} \end{pmatrix}$, (b) $\omega_1^2 = \frac{g}{l}$, $\omega_2^2 = \frac{g}{l} + \frac{2k}{m}$, (c) $\mathbf{x}_1 = \frac{1}{\sqrt{2}}\begin{pmatrix} 1 \\ 1 \end{pmatrix}$, $\mathbf{x}_2 = \frac{1}{\sqrt{2}}\begin{pmatrix} 1 \\ -1 \end{pmatrix}$,

(d) $\mathbf{x}(t) = s_0\begin{pmatrix} 1 \\ 1 \end{pmatrix}e^{i\omega_1 t}$, in-phase motion, (e) $\mathbf{x}(t) = s_0\begin{pmatrix} 1 \\ -1 \end{pmatrix}e^{i\omega_2 t}$, out-of-phase motion.

15.21 $\hat{H}\psi_n(x) = \lambda_n\psi_n(x)$, $\lambda_n = \frac{h^2n^2}{8mL^2}$ where $n = 1, 2, 3, \ldots$

15.22 $\hat{O}\varphi(x) = \lambda\varphi(x)$, $\lambda = a^2$.

15.23 (a) $(\sigma_x^*)^T = \begin{pmatrix} 0 & 1 \\ 1 & 0 \end{pmatrix} = \sigma_x$, $(\sigma_y^*)^T = \begin{pmatrix} 0 & -i \\ i & 0 \end{pmatrix} = \sigma_y$, $(\sigma_z^*)^T = \begin{pmatrix} 1 & 0 \\ 0 & -1 \end{pmatrix} = \sigma_z$, (b) $\sigma_x\psi_{x+} = \lambda\psi_{x+}$, $\lambda = 1$,

$\sigma_y\psi_{y+} = \lambda\psi_{y+}$, $\lambda = 1$, $\sigma_z\psi_{z+} = \lambda\psi_{z+}$, $\lambda = 1$, (c) $\psi_{x-} = \frac{1}{\sqrt{2}}\begin{pmatrix} 1 \\ -1 \end{pmatrix}$, $\psi_{y-} = \frac{1}{\sqrt{2}}\begin{pmatrix} 1 \\ -i \end{pmatrix}$, $\psi_{z-} = \begin{pmatrix} 0 \\ 1 \end{pmatrix}$

15.24 (a) $E_1 = \alpha + \beta$, $E_2 = \alpha - \beta$, (b) $\mathbf{x}_1 = \frac{1}{\sqrt{2}}\begin{pmatrix} 1 \\ 1 \end{pmatrix}$, $\mathbf{x}_2 = \frac{1}{\sqrt{2}}\begin{pmatrix} 1 \\ -1 \end{pmatrix}$

15.25 (a) $E_1 = \alpha + 2\beta$, $E_2 = E_3 = \alpha$, $E_4 = \alpha - 2\beta$, (b) $\mathbf{x}_1 = \frac{1}{\sqrt{4}}\begin{pmatrix} 1 \\ 1 \\ 1 \\ 1 \end{pmatrix}$, $\mathbf{x}_2 = \frac{1}{\sqrt{4}}\begin{pmatrix} 1 \\ 1 \\ -1 \\ -1 \end{pmatrix}$, $\mathbf{x}_3 = \frac{1}{\sqrt{4}}\begin{pmatrix} 1 \\ -1 \\ -1 \\ 1 \end{pmatrix}$, $\mathbf{x}_4 = \frac{1}{\sqrt{4}}\begin{pmatrix} 1 \\ -1 \\ 1 \\ -1 \end{pmatrix}$

15.26 (a) $E_1 = \alpha + \sqrt{2}\beta$, $E_2 = \alpha$, $E_3 = \alpha - \sqrt{2}\beta$, (b) $\mathbf{x}_1 = \frac{1}{2}\begin{pmatrix} 1 \\ \sqrt{2} \\ 1 \end{pmatrix}$, $\mathbf{x}_2 = \frac{1}{2}\begin{pmatrix} \sqrt{2} \\ 0 \\ -\sqrt{2} \end{pmatrix}$, $\mathbf{x}_3 = \frac{1}{2}\begin{pmatrix} 1 \\ -\sqrt{2} \\ 1 \end{pmatrix}$

15.27 $\hat{\sigma}\psi = \lambda\psi$, $[(\hat{\sigma}\psi)^*]^T = [(\lambda\psi)^*]^T$, $(\psi^*)^T(\hat{\sigma}^*)^T = \lambda^*(\psi^*)^T$, $(\psi^*)^T\hat{\sigma} = \lambda^*(\psi^*)^T$, $(\psi^*)^T\hat{\sigma}\psi = \lambda^*(\psi^*)^T\psi$,
$(\psi^*)^T\lambda\psi = \lambda^*(\psi^*)^T\psi$, $\lambda(\psi^*)^T\psi = \lambda^*(\psi^*)^T\psi$ so that $\lambda = \lambda^*$ is real.

15.28 Suppose $\hat{\sigma}\boldsymbol{\psi}_1 = \lambda_1\boldsymbol{\psi}_1$ and $\hat{\sigma}\boldsymbol{\psi}_2 = \lambda_2\boldsymbol{\psi}_2$, $\left(\boldsymbol{\psi}_2^*\right)^T\hat{\sigma}\boldsymbol{\psi}_1 = \lambda_1\left(\boldsymbol{\psi}_2^*\right)^T\boldsymbol{\psi}_1$, $\left(\hat{\sigma}^*\boldsymbol{\psi}_2^*\right)^T = \left(\boldsymbol{\psi}_2^*\right)^T\left(\hat{\sigma}^*\right)^T = \left(\boldsymbol{\psi}_2^*\right)^T\hat{\sigma} = \left(\lambda_2^*\boldsymbol{\psi}_2^*\right)^T = \lambda_2\left(\boldsymbol{\psi}_2^*\right)^T$, $\left(\boldsymbol{\psi}_2^*\right)^T\hat{\sigma}\boldsymbol{\psi}_1 = \lambda_2\left(\boldsymbol{\psi}_2^*\right)^T\boldsymbol{\psi}_1$, $\lambda_1\left(\boldsymbol{\psi}_2^*\right)^T\boldsymbol{\psi}_1 = \lambda_2\left(\boldsymbol{\psi}_2^*\right)^T\boldsymbol{\psi}_1$. If $\lambda_1 \neq \lambda_2$, $\left(\boldsymbol{\psi}_2^*\right)^T\boldsymbol{\psi}_1 = 0$, so that $\boldsymbol{\psi}_1$ and $\boldsymbol{\psi}_2$ are orthogonal.

CHAPTER 16

16.1 (a) $\mathbf{x}' = \begin{pmatrix} 5 \\ 4 \end{pmatrix}$, (b) $\lambda_1 = 5$, $\lambda_2 = 2$, (c) $\mathbf{x}_1 = \begin{pmatrix} 1 \\ 0 \end{pmatrix}$, $\mathbf{x}_2 = \begin{pmatrix} 0 \\ 1 \end{pmatrix}$

16.2 (a) $\mathbf{x}' = \begin{pmatrix} 5 \\ 2 \end{pmatrix}$, (b) $\lambda_1 = \lambda_2 = 1$, (c) $\mathbf{x}_1 = \begin{pmatrix} 1 \\ 0 \end{pmatrix}$

16.3 (a) $\mathbf{x}' = \begin{pmatrix} 0 \\ 1 \end{pmatrix}$, (b) $\lambda_1 = \frac{1}{\sqrt{2}}(1+i)$, $\lambda_2 = \frac{1}{\sqrt{2}}(1-i)$, (c) $\mathbf{x}_1 = \frac{1}{\sqrt{2}}\begin{pmatrix} i \\ 1 \end{pmatrix}$, $\mathbf{x}_2 = \begin{pmatrix} 1 \\ i \end{pmatrix}$

16.4 (a) $\mathbf{x}' = \frac{1}{\sqrt{2}}\begin{pmatrix} -1 \\ 1 \end{pmatrix}$, (b) $\lambda_1 = 1$, $\lambda_2 = -1$, (c) $\mathbf{x}_1 = \begin{pmatrix} 0 \\ 1 \end{pmatrix}$, $\mathbf{x}_2 = \begin{pmatrix} 1 \\ 0 \end{pmatrix}$

16.5 (a) $\mathbf{x}' = \begin{pmatrix} -1 \\ -2 \end{pmatrix}$, (b) $\lambda_1 = \lambda_2 = -1$, (c) any vector in the xy-plane

16.6 (a) $\mathbf{x}'_N = \mathbf{x}_N$, $\mathbf{x}'_{H_1} = \mathbf{x}_{H_2}$, $\mathbf{x}'_{H_2} = \mathbf{x}_{H_3}$, $\mathbf{x}'_{H_3} = \mathbf{x}_{H_1}$, (b) $\mathbf{x}'_N = \begin{pmatrix} 0 \\ 0 \\ 0 \end{pmatrix} = \mathbf{x}_N$ but $\mathbf{x}'_{H_1} = \begin{pmatrix} c_1 \\ 0 \\ -c_2 \end{pmatrix}$, $\mathbf{x}'_{H_2} = \begin{pmatrix} -\frac{1}{2}c_1 \\ \frac{\sqrt{3}}{2}c_1 \\ -c_2 \end{pmatrix}$, $\mathbf{x}'_{H_3} = \begin{pmatrix} -\frac{1}{2}c_1 \\ -\frac{\sqrt{3}}{2}c_1 \\ -c_2 \end{pmatrix}$ none of which equal $\mathbf{x}_{H_1}, \mathbf{x}_{H_2}$, or $\mathbf{x}_{H_3}$. In the figure below, the initial orientation of the ammonia molecule (translucent) is contrasted with the final orientation (opaque).

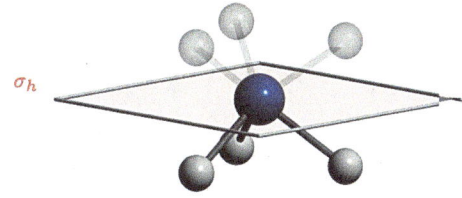

(c) E, C_2, $3\sigma_v = 3\sigma_d$ with point group C_{3v}

16.7 (a) C_s, (b) C_2, (c) C_{2h}, (d) C_s

16.8 (a) $\mathbf{x}_{Xe} = \begin{pmatrix} 0 \\ 0 \\ 0 \end{pmatrix}$, $\mathbf{x}_{F_1} = \begin{pmatrix} c_1 \\ 0 \\ 0 \end{pmatrix}$, $\mathbf{x}_{F_2} = \begin{pmatrix} 0 \\ c_1 \\ 0 \end{pmatrix}$, $\mathbf{x}_{F_3} = \begin{pmatrix} -c_1 \\ 0 \\ 0 \end{pmatrix}$, $\mathbf{x}_{F_4} = \begin{pmatrix} 0 \\ -c_1 \\ 0 \end{pmatrix}$

(b) $\mathbf{C}_4 = \mathbf{R}_z\left(\frac{\pi}{2}\right) = \begin{pmatrix} 0 & -1 & 0 \\ 1 & 0 & 0 \\ 0 & 0 & 1 \end{pmatrix}$, $\mathbf{x}'_{Xe} = \mathbf{x}_{Xe}$, $\mathbf{x}'_{F_1} = \mathbf{x}_{F_2}$, $\mathbf{x}'_{F_2} = \mathbf{x}_{F_3}$, $\mathbf{x}'_{F_3} = \mathbf{x}_{F_4}$, $\mathbf{x}'_{F_4} = \mathbf{x}_{F_1}$

(c) E, i, C_4, $4\,C_{2\perp} = 4\,C_2'$, σ_h, $4\,\sigma_v = 4\,\sigma_d$, S_4 with point group D_{4h}

16.9 (a) C_s, (b) C_{2v}, (c) C_2, (d) C_s

16.10 E, C_2, $2\,\sigma_v = 2\,\sigma_d$ with point group C_{2v}

16.11 (a) C_s, (b) C_{2v}, (c) C_{2h}, (d) C_{2v}

16.12 (a) C_s, (b) C_{2v}, (c) C_s, (d) C_2

16.13 (a) $3\,C_4$, (b) $3\,\sigma_h$, (c) $6\,C_{2\perp} = 6\,C_2'$, (d) $6\,\sigma_v = 6\,\sigma_d$

16.14 E, C_2, $2\,C_{2\perp} = 2\,C_2'$, $2\,\sigma_v = 2\,\sigma_d$, S_4 with point group D_{2d}

16.15 (a) $\mathbf{S_8} = \sigma_h\,\mathbf{C_8}$, (b) $4\,\sigma_v = 4\,\sigma_d$, (c) E, C_4, $4\,C_{2\perp} = 4\,C_2'$, $4\,\sigma_v = 4\,\sigma_d$, S_8 with point group S_8

BIBLIOGRAPHY

Milton Abramowitz and Irene A. Stegun, editors. *Handbook of Mathematical Functions with Formulas, Graphs, and Mathematical Tables [1964]*. Dover Publications, first edition, 1972. ISBN 978-0486612720.

George B. Arfken, Hans J. Weber, and Frank E. Harris. *Mathematical Methods for Physicists: A Comprehensive Guide [2012]*. Academic Press, seventh edition, 2012. ISBN 978-0123846549.

James R. Barrante. *Applied Mathematics for Physical Chemistry [1974]*. Pearson Prentice Hall, third edition, 2004. ISBN 978-0131008458.

Mary L. Boas. *Mathematical Methods in the Physical Sciences [2005]*. Wiley, third edition, 2005. ISBN 978-0471198260.

John C. Bowman and Andy Hammerlindl. Asymptote: A vector graphics language. *TUGBOAT: The Communications of the TeX Users Group*, 29:288–294, 2008.

Robert Bringhurst. *The Elements of Typographic Style*. Hartley & Marks, fourth edition, 2013. ISBN 978-0881792126.

Claude Cohen-Tannoudji. *Quantum Mechanics [1977]*. Wiley, first edition, 1977. ISBN 978-0471164333.

Ken A. Dill and Sarina Bromberg. *Molecular Driving Forces [2010]*. Garland Science, second edition, 2010. ISBN 978-0815344308.

Richard P. Feynman, Robert B. Leighton, and Matthew Sands. *The Feynman Lectures on Physics [2005]*. Addison Wesley, second edition, 2005. ISBN 978-0805390469.

Izrail S. Gradshteyn and Iosif M. Ryzhik. *Table of Integrals, Series, and Products [1943]*. Academic Press, eighth edition, 2014. ISBN 978-0123849335.

Donald E. Knuth. *The TeXbook*. Addison–Wesley, first edition, 1984. ISBN 978-0201134483.

Leslie Lamport. *LaTeX: A Document Preparation System*. Addison–Wesley, second edition, 1994. ISBN 978-0201529838.

Donald A. McQuarrie. *Quantum Chemistry [2007]*. University Science Books, second edition, 2007. ISBN 978-1891389504.

Donald A. McQuarrie. *Mathematics for Physical Chemistry [2008]*. University Science Books, first edition, 2008. ISBN 978-1891389153.

Earl W. Swokowski. *Calculus with Analytic Geometry [1979]*. Prindle, Weber & Schmidt, second edition, 1979. ISBN 978-0871502681.

David O. Tall. Visualizing differentials in two and three dimensions. *Teaching Mathematics and its Applications*, 11:1–7, 1992.

Edward R. Tufte. *Envisioning Information* [1990]. Graphics Press, 1990. ISBN 978-0961392116.

Edward R. Tufte. *Visual Explanations* [1997]. Graphics Press, 1997. ISBN 978-0961392123.

Edward R. Tufte. *The Visual Display of Quantitative Information* [1983]. Graphics Press, second edition, 2001. ISBN 978-1930824133.

Edward R. Tufte. *Beautiful Evidence* [2006]]. Graphics Press, LLC, first edition, 2006. ISBN 978-1930824164.

INDEX

COLOPHON

This book was typeset using the LaTeX document preparation system authored by Leslie Lamport and released in 1985.[1] The LaTeX output is formatted using the TeX computer typesetting system[2] designed by Donald Knuth and released in 1978. In 1986, I was a graduate student at Columbia University preparing my doctoral dissertation in LaTeX. At that time, I could call Leslie Lamport to ask why all of my tables were floating to the end of the document. Now we have *Stack Exchange*.

The text is set in the humanist Palatino font designed by Hermann Zapf and released in 1949. It includes roman, italic, text figures, and small caps. Bold fonts are avoided.[3]

The text layout was inspired by *The Feynman Lectures on Physics*, with its ample margins and margin notes, and rich variety of figures.[4] The idea of pairing chapters with complements was taken from *Quantum Mechanics* by Claude Cohen-Tannoudji.[5] The colors of text and section headings were taken from *Calculus* by Earl W. Swokowski.[6] The typography and design were informed by the work of Edward R. Tufte[7] and specifically grew from a modified version of the *Tufte Book LaTeX template*. I benefitted from the advice of Paul C. Anagnostopoulos and Chen Cheng "Cici" Qi [漆晨理] regarding font size and page dimension, which informed the final design of the text. Amy Hendrickson of *TeXnology, Inc.* served as a consultant in the development of the LaTeX source code.

Tufte has articulated design principles to be followed in the clear presentation of quantitative information, including data and mathematical functions.

> Graphical excellence is that which gives to the viewer the greatest number of ideas in the shortest time with the least ink in the smallest space.
>
> Edward R. Tufte

In the design of figures for this text, I have aspired, however imperfectly, to follow this principle.

All figures were composed using the *Asymptote* vector graphics language[8] within the LaTeX programming environment. While most figures are original compositions, others are closely derived from examples provided by the *Asymptote* resource. I have benefited from wisdom and source code shared by active *Asymptote* users through *Art of Problem Solving* and *Stack Exchange*, as well as the generous advice of John C. Bowman. Without his support, the graphics in this text would not have been possible.

[1] Leslie Lamport. *LaTeX: A Document Preparation System*. Addison–Wesley, second edition, 1994. ISBN 978-0201529838

[2] Donald E. Knuth. *The TeXbook*. Addison–Wesley, first edition, 1984. ISBN 978-0201134483

[3] Robert Bringhurst. *The Elements of Typographic Style*. Hartley & Marks, fourth edition, 2013. ISBN 978-0881792126

[4] Richard P. Feynman, Robert B. Leighton, and Matthew Sands. *The Feynman Lectures on Physics [2005]*. Addison Wesley, second edition, 2005. ISBN 978-0805390469

[5] Claude Cohen-Tannoudji. *Quantum Mechanics [1977]*. Wiley, first edition, 1977. ISBN 978-0471164333

[6] Earl W. Swokowski. *Calculus with Analytic Geometry [1979]*. Prindle, Weber & Schmidt, second edition, 1979. ISBN 978-0871502681

[7] Edward R. Tufte. *The Visual Display of Quantitative Information [1983]*. Graphics Press, second edition, 2001. ISBN 978-1930824133

[8] John C. Bowman and Andy Hammerlindl. Asymptote: A vector graphics language. *TUGBOAT: The Communications of the TeX Users Group*, 29:288–294, 2008